Springer Tracts in Advanced Robotics 113

More information about this series at http://www.springer.com/series/5208

David S. Wettergreen · Timothy D. Barfoot
Editors

Field and Service Robotics

Results of the 10th International Conference

Springer

Editors
David S. Wettergreen
The Robotics Institute
Carnegie Mellon University
Pittsburgh, PA
USA

Timothy D. Barfoot
Institute for Aerospace Studies
University of Toronto
Toronto, ON
Canada

ISSN 1610-7438 ISSN 1610-742X (electronic)
Springer Tracts in Advanced Robotics
ISBN 978-3-319-27700-4 ISBN 978-3-319-27702-8 (eBook)
DOI 10.1007/978-3-319-27702-8

Library of Congress Control Number: 2016930551

Printed on acid-free paper

This Springer imprint is published by SpringerNature
The registered company is Springer International Publishing AG Switzerland

Series Foreword

Robotics is undergoing a major transformation in scope and dimension. From a largely dominant industrial focus, robotics is rapidly expanding into human environments and is vigorously engaged in its new challenges. Interacting with, assisting, serving, and exploring with humans, the emerging robots will increasingly touch people and their lives.

Beyond its impact on physical robots, the body of knowledge robotics has produced is revealing a much wider range of applications reaching across diverse research areas and scientific disciplines, such as biomechanics, haptics, neurosciences, virtual simulation, animation, surgery, and sensor networks among others. In return, the challenges of the new emerging areas are proving an abundant source of stimulation and insights for the field of robotics. It is indeed at the intersection of disciplines that the most striking advances happen.

The *Springer Tracts in Advanced Robotics* (STAR) is devoted to bringing to the research community the latest advances in the robotics field on the basis of their significance and quality. Through a wide and timely dissemination of critical research developments in robotics, our objective with this series is to promote more exchanges and collaborations among the researchers in the community and contribute to further advancements in this rapidly growing field.

The tenth edition of *Field and Service Robotics* edited by David S. Wettergreen and Timothy D. Barfoot offers in its eight-part volume a collection of a broad range of topics ranging from fundamental concepts such as control, vision, mapping, and recognition to advanced applications such as aquatic, planetary, aerial, and underground robots. The contents of the forty-two contributions represent a cross-section of the current state of robotics research from one particular aspect: field and service applications, and how they reflect on the theoretical basis of subsequent developments. Pursuing technologies aimed at non-factory robots that operate in complex and dynamic environments, as well as at service robots that work closely with humans to help them with their lives, is the big challenge running throughout this focused collection.

Rich in topics and authoritative contributors, FSR culminates with this unique reference on the current developments and new directions in field and service robotics. A fine addition to the series!

Naples, Italy
October 2015

Bruno Siciliano
STAR Editor

Preface

Field and Service Robotics (FSR) is the leading single-track conference on applications of robotics in challenging environments. Its goal is to report and encourage the development and experimental evaluation of field and service robots, and to generate a vibrant exchange and discussion in the community. Field robots are non-factory robots, typically mobile, that operate in complex and dynamic environments: on the ground (Earth or other planets), under the ground, underwater, in the air, or in space. Service robots are those that work closely with humans to help them with their lives.

The first FSR was held in Canberra, Australia, in 1997. Since that first meeting, FSR has been held roughly every two years, cycling through Asia, the Americas, and Europe. This book presents the results of the 10th edition of Field and Service Robotics, FSR 2015, held in Toronto, Canada, from 23 to 26 June 2015. This was the first time it has been held in Canada. This year we had 63 submitted papers from which we accepted 27 for oral presentations and 15 for poster presentations.

FSR 2015 was organized by the following team:

Timothy D. Barfoot
General Chair
University of Toronto

David S. Wettergreen
Program Chair
Carnegie Mellon University

Jonathan Kelly
Local Arrangements Chair
University of Toronto

Francois Pomerleau
Website and Publicity Chair
University of Toronto

Angela Schoellig
Technical Tour Chair
University of Toronto

The FSR 2015 International Program Committee generously provided their time to carry out detailed reviews of all the papers:

Peter Corke: Queensland University of Technology, Australia
Jonathan Roberts: Queensland University of Technology, Australia
Alex Zelinsky: DSTO, Australia
Uwe Zimmer: Australian National University, Australia
Salah Sukkarieh: University of Sydney, Australia
Ben Upcroft: Queensland University of Technology, Australia
Timothy D. Barfoot: University of Toronto, Canada
Jonathan Kelly: University of Toronto, Canada
David S. Wettergreen: Carnegie Mellon University, USA
Philippe Giguere: University of Laval, Canada
Steve Waslander: University of Waterloo, Canada
Josh Marshall: Queens University, Canada
Francois Pomerleau: University of Toronto, Canada
Chris Skonieczny: Concordia University, Canada
Arto Visala: Helsinki University of Technology, Finland
Simon Lacroix: LAAS, France
Christian Laugier: INRIA, France
Cedric Pradalier: GT-Lorraine, France
Andreas Birk: Jacobs University, Germany
Keiji Nagatani: Tohoku University, Japan
Kazuya Yoshida: Tohoku University, Japan
Takashi Tsubouchi: University of Tsukuba, Japan
Genya Ishigami: Keio University, Japan
Miguel Angel Salichs: Universidad Carlos III de Madrid, Spain
Roland Siegwart: ETH Zurich, Switzerland
David P. Miller: University Oklahoma, USA
Sanjiv Singh: Carnegie Mellon University, USA
Gaurav Sukhatme: University of Southern California, USA
Alonzo Kelly: Carnegie Mellon University, USA
Chuck Thorpe: Clarkson University, USA
David Silver: Google[X], USA
Carrick Dettweiler: University of Nebraska, USA
Stewart Moorehead: John Deere Corp., USA
Steve Nuske: Carnegie Mellon University, USA
Gabe Sibley: University of Colorado, USA
Ross Knepper: Cornell University, USA
Michael Jakuba: Woods Hole, USA

In addition to the submitted papers presented at the conference, there were four excellent keynote speakers at FSR 2015 and we would like to acknowledge their excellent contributions to the conference:

- Chris Urmson, Director, Self-Driving Cars, Google[x], "Realizing Self-Driving Vehicles"
- Paul Newman, Professor, University of Oxford, "Fielding Robots with Learnt Place-Specific Excellence"
- Sanjiv Singh, Professor, Carnegie Mellon University, "As the Drone Flies: The Shortest Path from Ground to Aerial Autonomy"
- Ryan Gariepy, Chief Technology Officer, Clearpath Robotics, "The Evolution of a Robotics Company"

FSR 2015 would not have been possible without the generous support of our sponsors. In particular, Clearpath Robotics went above and beyond to provide financial and in-kind support. The University of Toronto Institute for Aerospace Studies and Faculty of Applied Science and Engineering also provided financial support.

David S. Wettergreen
Timothy D. Barfoot

Contents

Part I Aquatic

A Spatially and Temporally Scalable Approach for Long-Term Lakeshore Monitoring 3
Shane Griffith and Cédric Pradalier

Autonomous Greenhouse Gas Sampling Using Multiple Robotic Boats 17
Matthew Dunbabin

Experimental Analysis of Receding Horizon Planning Algorithms for Marine Monitoring 31
Soo-Hyun Yoo, Andrew Stuntz, Yawei Zhang, Robert Rothschild, Geoffrey A. Hollinger and Ryan N. Smith

Return to Antikythera: Multi-session SLAM Based AUV Mapping of a First Century B.C. Wreck Site 45
Stefan B. Williams, Oscar Pizarro and Brendan Foley

An Overview of MIT-Olin's Approach in the AUVSI RobotX Competition 61
Arthur Anderson, Erin Fischell, Thom Howe, Tom Miller, Arturo Parrales-Salinas, Nick Rypkema, David Barrett, Michael Benjamin, Alex Brennen, Michael DeFillipo, John J. Leonard, Liam Paull, Henrik Schmidt, Nick Wang and Alon Yaari

A Parameterized Geometric Magnetic Field Calibration Method for Vehicles with Moving Masses with Applications to Underwater Gliders 81
Brian Claus and Ralf Bachmayer

Towards Autonomous Robotic Coral Reef Health Assessment . . . 95
Travis Manderson, Jimmy Li, David Cortés Poza, Natasha Dudek, David Meger and Gregory Dudek

Part II Vision

BOR^{2}G: Building Optimal Regularised Reconstructions with GPUs (in Cubes) . . . 111
Michael Tanner, Pedro Piniés, Lina Maria Paz and Paul Newman

Online Loop-Closure Detection via Dynamic Sparse Representation . . . 125
Moein Shakeri and Hong Zhang

Large Scale Dense Visual Inertial SLAM . . . 141
Lu Ma, Juan M. Falquez, Steve McGuire and Gabe Sibley

Dense and Swift Mapping with Monocular Vision . . . 157
Pedro Piniés, Lina Maria Paz and Paul Newman

Wrong Today, Right Tomorrow: Experience-Based Classification for Robot Perception . . . 173
Jeffrey Hawke, Corina Gurău, Chi Hay Tong and Ingmar Posner

Beyond a Shadow of a Doubt: Place Recognition with Colour-Constant Images . . . 187
Kirk MacTavish, Michael Paton and Timothy D. Barfoot

Segmentation and Classification of 3D Urban Point Clouds: Comparison and Combination of Two Approaches . . . 201
A.K. Aijazi, A. Serna, B. Marcotegui, P. Checchin and L. Trassoudaine

A Stereo Vision Based Obstacle Detection System for Agricultural Applications . . . 217
Patrick Fleischmann and Karsten Berns

CoPilot: Autonomous Doorway Detection and Traversal for Electric Powered Wheelchairs . . . 233
Tom Panzarella, Dylan Schwesinger and John Spletzer

Learning a Context-Dependent Switching Strategy for Robust Visual Odometry . . . 249
Kristen Holtz, Daniel Maturana and Sebastian Scherer

Part III Planetary

System Design of a Tethered Robotic Explorer (TReX) for 3D Mapping of Steep Terrain and Harsh Environments . . . 267
Patrick McGarey, François Pomerleau and Timothy D. Barfoot

Design, Control, and Experimentation of Internally-Actuated Rovers for the Exploration of Low-Gravity Planetary Bodies 283
B. Hockman, A. Frick, I.A.D. Nesnas and M. Pavone

Considering the Effects of Gravity When Developing and Field Testing Planetary Excavator Robots 299
Krzysztof Skonieczny, Thomas Carlone, W.L. "Red" Whittaker and David S. Wettergreen

Update on the Qualification of the Hakuto Micro-rover for the Google Lunar X-Prize 313
John Walker, Nathan Britton, Kazuya Yoshida, Toshiro Shimizu, Louis-Jerome Burtz and Alperen Pala

Mobility Assessment of Wheeled Robots Operating on Soft Terrain 331
Bahareh Ghotbi, Francisco González, József Kövecses and Jorge Angeles

Taming the North: Multi-camera Parallel Tracking and Mapping in Snow-Laden Environments 345
Arun Das, Devinder Kumar, Abdelhamid El Bably and Steven L. Waslander

Four-Wheel Rover Performance Analysis at Lunar Analog Test 361
Nathan Britton, John Walker, Kazuya Yoshida, Toshiro Shimizu, Tommaso Paniccia and Kei Nakata

Energy-Aware Terrain Analysis for Mobile Robot Exploration 373
Kyohei Otsu and Takashi Kubota

Part IV Aerial

Vision and Learning for Deliberative Monocular Cluttered Flight 391
Debadeepta Dey, Kumar Shaurya Shankar, Sam Zeng, Rupesh Mehta, M. Talha Agcayazi, Christopher Eriksen, Shreyansh Daftry, Martial Hebert and J. Andrew Bagnell

Robust Autonomous Flight in Constrained and Visually Degraded Environments 411
Zheng Fang, Shichao Yang, Sezal Jain, Geetesh Dubey, Silvio Maeta, Stephan Roth, Sebastian Scherer, Yu Zhang and Stephen Nuske

Autonomous Exploration for Infrastructure Modeling with a Micro Aerial Vehicle 427
Luke Yoder and Sebastian Scherer

Long-Endurance Sensing and Mapping Using a Hand-Launchable Solar-Powered UAV 441
Philipp Oettershagen, Thomas Stastny, Thomas Mantel, Amir Melzer, Konrad Rudin, Pascal Gohl, Gabriel Agamennoni, Kostas Alexis and Roland Siegwart

Aerial Vehicle Path Planning for Monitoring Wildfire Frontiers 455
Ryan C. Skeele and Geoffrey A. Hollinger

Part V Underground

Multi-robot Mapping of Lava Tubes 471
X. Huang, J. Yang, M. Storrie-Lombardi, G. Lyzenga and C.M. Clark

Admittance Control for Robotic Loading: Underground Field Trials with an LHD 487
Andrew A. Dobson, Joshua A. Marshall and Johan Larsson

From ImageNet to Mining: Adapting Visual Object Detection with Minimal Supervision 501
Alex Bewley and Ben Upcroft

Part VI Systems

Building, Curating, and Querying Large-Scale Data Repositories for Field Robotics Applications 517
Peter Nelson, Chris Linegar and Paul Newman

Search and Retrieval of Human Casualties in Outdoor Environments with Unmanned Ground Systems—System Overview and Lessons Learned from ELROB 2014 533
Bernd Brüggemann, Dennis Wildermuth and Frank E. Schneider

Monocular Visual Teach and Repeat Aided by Local Ground Planarity 547
Lee Clement, Jonathan Kelly and Timothy D. Barfoot

In the Dead of Winter: Challenging Vision-Based Path Following in Extreme Conditions 563
Michael Paton, François Pomerleau and Timothy D. Barfoot

Non-Field-of-View Acoustic Target Estimation in Complex Indoor Environment 577
Kuya Takami, Tomonari Furukawa, Makoto Kumon and Gamini Dissanayake

Novel Assistive Device for Teaching Crawling Skills to Infants 593
Mustafa A. Ghazi, Michael D. Nash, Andrew H. Fagg, Lei Ding, Thubi H.A. Kolobe and David P. Miller

SPENCER: A Socially Aware Service Robot for Passenger Guidance and Help in Busy Airports 607
Rudolph Triebel, Kai Arras, Rachid Alami, Lucas Beyer, Stefan Breuers, Raja Chatila, Mohamed Chetouani, Daniel Cremers, Vanessa Evers, Michelangelo Fiore, Hayley Hung, Omar A. Islas Ramírez, Michiel Joosse, Harmish Khambhaita, Tomasz Kucner, Bastian Leibe, Achim J. Lilienthal, Timm Linder, Manja Lohse, Martin Magnusson, Billy Okal, Luigi Palmieri, Umer Rafi, Marieke van Rooij and Lu Zhang

Easy Estimation of Wheel Lift and Suspension Force for a Novel High-Speed Robot on Rough Terrain 623
Jayoung Kim, Bongsoo Jeon and Jihong Lee

Application of Multi-Robot Systems to Disaster-Relief Scenarios with Limited Communication 639
Jason Gregory, Jonathan Fink, Ethan Stump, Jeffrey Twigg, John Rogers, David Baran, Nicholas Fung and Stuart Young

Part I
Aquatic

A Spatially and Temporally Scalable Approach for Long-Term Lakeshore Monitoring

Shane Griffith and Cédric Pradalier

Abstract This paper provides an image processing framework to assist in the inspection and, more generally, the data association of a natural environment, which we demonstrate in a long-term lakeshore monitoring task with an autonomous surface vessel. Our domain consists of 55 surveys of a 1 km lakeshore collected over a year and a half. Our previous work introduced a framework in which images of the same scene from different surveys are aligned using visual SLAM and SIFT Flow. This paper: (1) minimizes the number of expensive image alignments between two surveys using a covering set of poses, rather than all the poses in a sequence; (2) improves alignment quality using a local search around each pose and an alignment bias derived from the 3D information from visual SLAM; and (3) provides exhaustive results of image alignment quality. Our improved framework finds significantly more precise alignments despite performing image registration over an order of magnitude fewer times. We show changes a human spotted between surveys that would have otherwise gone unnoticed. We also show cases where our approach was robust to 'extreme' variation in appearance.

1 Introduction

This paper presents an application of autonomous surface vessels (ASV) for long-term observation of lakeshore environments. A growing number of robots are being targeted for inspection tasks in natural environments, including applications in agriculture, surveillance, and environment monitoring. Yet, the variation of appearance of outdoor environments significantly limits robots in tasks requiring data association, with many papers only addressing particular aspects of the variation (e.g., illumination/shadows [4], night [19], and noise [8]). Scene structure can help provide

S. Griffith (✉)
Georgia Institute of Technology, Atlanta, USA
e-mail: sgriffith7@gatech.edu

C. Pradalier
GeorgiaTech Lorraine - UMI 2958 GT-CNRS, Atlanta, USA
e-mail: cedric.pradalier@gatech.edu

D.S. Wettergreen and T.D. Barfoot (eds.), *Field and Service Robotics*,
Springer Tracts in Advanced Robotics 113, DOI 10.1007/978-3-319-27702-8_1

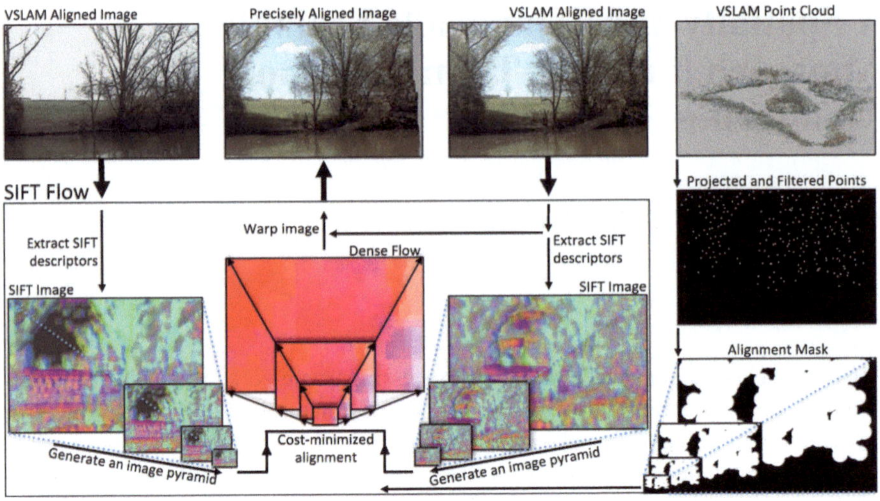

Fig. 1 The registration of two images. For each VSLAM aligned image, SIFT descriptors are computed at each pixel to form a SIFT image, which is down-sampled into an image pyramid. To avoid aligning noise due to the sky and the water, the alignment cost is biased using an image mask of the lakeshore (derived from the 3D information in the feature tracks of visual SLAM). The resulting dense flow aligns the two input images, which enables quick change detection for manual inspection tasks

robustness to natural variation of appearance [16, 18]. Few papers have, however, demonstrated robust data association across surveys of natural environments.

This paper presents a framework for achieving high resolution, pixel-level alignment between fortnightly surveys of a lakeshore. Our framework uses visual SLAM (see e.g., [1, 21, 22]) to identify images of roughly the same scene from different surveys and then it applies SIFT Flow [16] to precisely align them (see Fig. 1). Building on our previous work [6, 7], in this paper we minimize the number of expensive image alignments using a covering set of poses. To improve image alignment accuracy at a particular pose, a search for the best alignment is performed in its tight neighborhood of images. Image alignment accuracy is further improved using the 3D landmark positions from visual SLAM to bias the image registration process. Once images are precisely aligned, a human inspecting them can easily spot if something changed. At this stage, given the difficulty of automatically processing natural scenes, we are assuming the inspection task is left to a human, but we endeavor to make his/her task as efficient as possible.

To date, we have surveyed a lake a total of 55 times over 1.5 years, which represents a spatially large and a temporally long scale study using ASVs. We show our framework enabled a human to detect changes that would have otherwise gone unnoticed. We also show our approach is robust to variation in appearance of the sky, the water, changes in objects on a lakeshore, and the seasonal changes of plants. Finally, we point out failure cases, which indicate directions for future work.

2 Related Work

The field of Simultaneous Localization and Mapping (SLAM) provides a foundation for localizing a robot and mapping monitored spaces. Gaining the advantages of SLAM in natural environments requires, however, a system made to handle three particular challenges: (1) the large spatial scale; (2) the non-rigid environments (e.g., moving trees, changing water levels); and (3) the very high level of visual similarity (e.g., branches and leaves of different trees may appear to be from the same one). The variation of appearance over a long-term monitoring task further increases the difficulty of data association in surveys of an outdoor environment.

Many different techniques have been proposed to solve data association for outdoor environments. Some approaches rely on point-based features such as SIFT (e.g., [1, 9, 14]) for performing data association. Point-based feature matching is, however, often not robust to common sources of variation (see e.g., [6]). In light of this, some work has focused on directly using, or modifying, whole or parts of images. Neubert et al. [20] deals with seasonal changes by introducing a prediction step in which whole images are modified to look more like the current season. McManus et al. [18] utilize patches of images, called 'scene signatures', which are matched using classifiers and capture information about the structure of each scene. In case a particular location is stubborn to feature- and whole-image-based data association, 'multiple experiences' of the location can be accumulated until new observations are associated well [2]. This paper performs data association using SIFT Flow [16], an algorithm designed to find dense correspondences using whole images worth of point-based features. It combines the accuracy of point-based feature matching with the robustness of whole-image matching.

Traversing a lake while mapping the location of a lakeshore is an essential task of lakeshore monitoring, which some papers have already started to address. Sukhatme [10] and Subramanian et al. [23] map a lakeshore and the locations of obstacles from the visual perspectives of their ASVs. Jain et al. [12] proposed to use a drone for autonomously mapping riverine environments, which can avoid debris in the water, yet fly below dense tree cover. In case a robot repeatedly visits the same lakeshore, Hitz et al. [11] show that 3D laser scans of a shoreline can be used to delineate some types of changes. Their system distinguished the dynamic leaves from the static trunk of a willow tree in two different surveys collected in the fall and spring. This paper combines iSAM2 [13] for scalable SLAM with SIFT Flow for robust data association in a framework for long-term lakeshore monitoring.

3 Experimental Setup

We used Clearpath's Kingfisher ASV for our experiments. It is 1.35 m long and 0.98 m wide, with two pontoons, a water-tight compartment to house electronics, and an area on top for sensors and the battery. It is propelled by a water jet in each of its pontoons and turns during power differentials. It can reach a top speed of about 2 m/s, but we mostly operated it at lower speeds to maximize battery life, which is about an hour with our current payload.

Our Kingfisher is equipped with a suite of exteroceptive and interoceptive sensors. A 704×480 color pan-tilt camera captures images at 10 frames per second. Beneath it sits a single scan line laser rangefinder with a field of view of about 270°. It is pointed just above the surface of the water and provides a distance estimate for everything less than 20 m away. The watertight compartment houses a GPS, a compass, and an IMU.

The ASV was deployed on Symphony Lake in Metz, France, which is about 400 m long and 200 m wide with an 80 m-wide island in the middle. The nature of the lakeshore varied, with shrubs, trees, boulders, grass, sand, buildings, birds, and people in the immediate surroundings. People mostly kept to the walking trail and a bike path a few meters from the shore, and fishermen occasionally sat along the shore.

We used a simple set of behaviors to autonomously steer the robot around the perimeter of the lake and the island. As the boat moves at a constant velocity of about 0.4 m/s, a local planner chooses among a set of state lattice motion primitives to keep the boat 10 m away from the lakeshore on its starboard side. With this configuration, the robot is capable of performing an entire survey autonomously; however, we occasionally took control using a remote control in order to avoid fishing lines, debris, to swap batteries, etc.

We have regularly deployed the robot up to once per week since August 18, 2013. This paper analyzes data from 10 different surveys, which span seven months of variation. All 10 were chosen because they each consisted of one run around the entire lakeshore, including the island. Each survey was performed in the daytime on a weekday in sunny or cloudy weather, at various times of the day.

4 Methodology

Our framework aligns images between two surveys using a coarse-to-fine process with four main components. First visual SLAM is used to localize the trajectory of the ASV and map visual features of the shore. Second a minimum view set is identified, which covers all the sections of lakeshore with similar viewpoints in both surveys. Third, given two poses facing the same scene from two different surveys, a process of image registration using SIFT Flow is performed for the best pixel-wise alignment. In the last step images are presented to an end user in a flickering display.

A single survey represents a collection of image sequences, measurements of the camera pose, and other useful information about the robot's movement. During a survey, k, the robot acquires the tuple $\mathscr{A}^k = \{\mathscr{T}_i^k, \mathscr{I}_i^k, \hat{C}_i^k, \hat{\omega}_i^k\}_{i=1}^{|\mathscr{A}^k|}$ every 10th of a second, where $\mathscr{T}$ is the current time, $\mathscr{I}$ is the image from the pan-tilt camera, $\hat{C} \in SE(3)$ is the estimated camera pose, and $\hat{\omega}$ is the estimated angular velocity of the boat. The estimated camera pose is derived from the boat's GPS position, the boat's compass heading, and the pan and tilt positions of the camera. The IMU provides $\hat{\omega}$. Each survey is down-sampled by a factor of five to reduce data redundancy and speed up computation time.

Finding nearby images in two long surveys is possible using raw measurements of the camera pose, but because these measurements are prone to noise that could lead to trying to align images of two different scenes, we begin by using visual–inertial SLAM to improve our estimates of the camera poses.

4.1 Visual SLAM

We used generic feature tracking for visual SLAM, which is based on detecting 300 Harris corner features and then tracking them using the pyramidal Lucas–Kanade Optical Flow algorithm (from OpenCV) as the boat moves. We then apply a graph-based SLAM approach for optimizing the camera poses and the visual feature locations. A factor graph is used to represent the set of measurements of the camera poses and the landmark positions, and the different constraints between them. The GTSAM bundle adjustment framework is applied to the factor graph to reduce the error in the initial estimates of the positions [5]. See [7] for more details.

4.2 Selecting a Minimum View Set

To reduce the computational overhead of image alignment (Sect. 4.3) and to enable a manual comparison between two surveys (Sect. 4.4), we select a minimum view set from among the roughly 50,000 images of each survey. A large set of images in each survey is desirable for the optical flow step of visual SLAM and to reduce motion blur. Yet, it means the images have a lot of redundancy, which is cumbersome for a survey comparison. Ideally, a person comparing two surveys would only see a subset of these images, where each corresponds to a unique section of the shore. This section describes how we find a minimal subset of images that covers as much lakeshore as is seen in both surveys.

Another name for this is the "Set Cover Problem" (SCP) [3], which can be expressed as follows. Let $\mathscr{S}$ be the set of all the observable positions in a survey of a lakeshore. Each camera pose, i, of the survey observes a subset $\mathscr{I}_i$ of these shore points, where $\mathscr{S} = \bigcup_{i \in I} \mathscr{I}_i$. The goal is to find a set of poses J for which $\mathscr{S} = \bigcup_{j \in J} \mathscr{I}_j$ and $|J|$ is as small as possible. This Set Cover Problem is NP-Hard. It can be approximated using linear programming or a simple greedy approach, which gives sufficient performance for our application.

The set of shore points that compose $\mathscr{S}$ is identified using the optimized poses from visual SLAM. Because the robot is controlled to move at a constant distance, d, from the shore, every point $d \pm \varepsilon$ away in the camera frustum is considered part of it (where $d = 10\,\text{m}$ and $\varepsilon = 1\,\text{m}$). To get a discrete set of shore points, the shore map is rasterized into a grid, in which each non-zero cell represents the shore. An arc centered on a pose is drawn with radius d and thickness ε with an angle consistent with the camera intrinsic parameters. For each shore point in the grid, all the poses from which it was seen are identified. An example set of shore points from two different surveys and the points where they overlap are shown in Fig. 2.

The camera poses J that make up the minimal cover set must also satisfy some practical constraints. Poses with an invalid camera configuration or with a high likelihood of motion blur are rejected. Poses from two different surveys without a similar view of the lakeshore are also rejected. In this case, we estimate that two poses have a similar view if their 3D positions are similar and both have similar intersections of the camera axis with the shore at the distance, d, from the boat. The distance between the camera angles is expressed in this way to keep comparable values with the distance between the 3D positions.

The Set Cover Problem is solved with the greedy algorithm shown in Algorithm 1. The method provided the results illustrated in Fig. 3 in less than 30 s (i.e., it's tractable). Out of a survey with 50,000 images, roughly 200 are selected for the cover set of the shore, which means over an order of magnitude fewer full runs of image registration will be performed (compared to naively performing image registration for every image in a down-sampled survey). Note that the set of images does not view the entire shore; only all the shore points seen from both surveys with similar views can be covered (about 80 %).

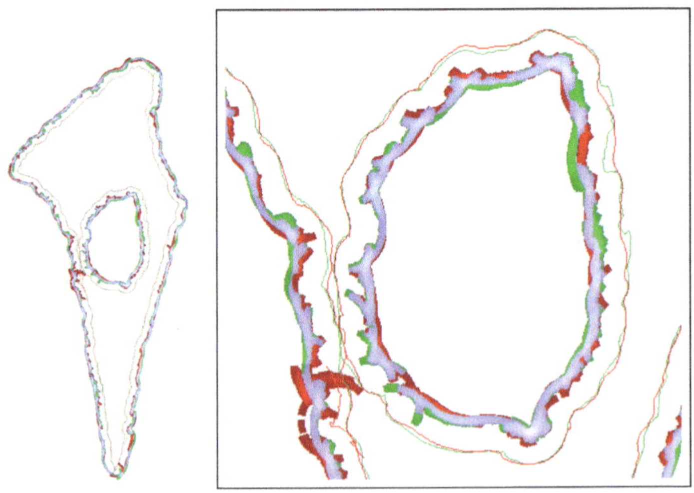

Fig. 2 The recorded trajectory of the boat and the shore points it sees for two surveys. The shore points seen from the red trajectory are displayed in *red*, those seen from the green trajectory are *green*, and those seen from both are mauve. A closeup of the island is shown on the *right*

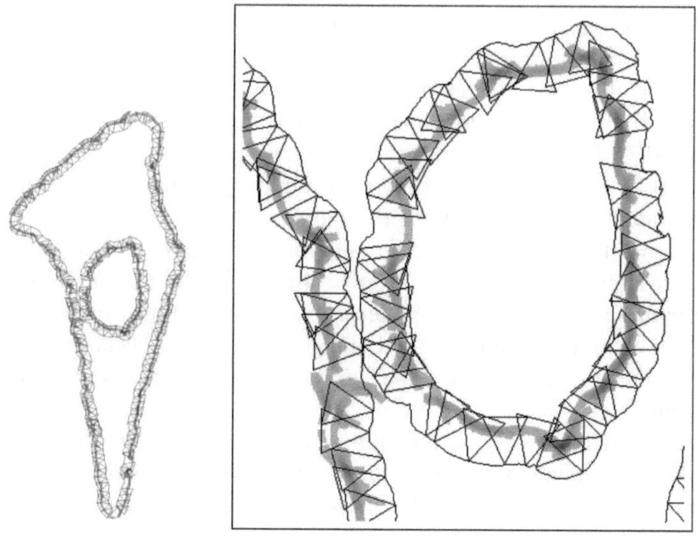

Fig. 3 The cover set for the red survey from Fig. 2, which accounts for the estimated co-visibility with the green survey. *Black triangles* indicate the visibility frustum of the selected images. A closeup of the island is shown on the *right*

```
Let L be the list of selected view points, initially empty;
while there are shore points to observe do
    Select the valid shore point P which is the least observed;
    Let V be a view point such that
            V observes P and;
            V observes the largest number of unobserved shore points;
    Remove P and all shore points observed in V from the list of points to observe;
    Append V to L;
end
return L
```

Algorithm 1: Greedy algorithm for maximizing the coverage of the shore with the minimal number of poses.

4.3 Image Registration

Given two poses viewing approximately the same scene from two different surveys, we next run image registration in a local search of several nearby images, and output the image pair with the best alignment score we find (the registration process for a single image pair is shown in Fig. 1). Image registration is performed using a modified version of the SIFT Flow scene alignment algorithm [16], which is designed for matching images with significant amounts of variation between them. SIFT Flow is named as such because a dense image of SIFT descriptors (see [17]) define the matching pattern (the data terms of an MRF) to be optimized between two images.

The algorithm is similar to optical flow in that each pixel is biased to have a similar flow to nearby pixels (a smoothness criteria), and lower degrees of flow are favored (regularization). For two images of approximately the same scene, the alignment score is minimized when the flow lines up salient structures between them.

Because SIFT Flow's cost function is designed to align the contents of a scene indiscriminately, we add a bias in favor of aligning the lakeshore rather than the sky or the water. A bias to decrease the influence of the sky and the water helps reduce the noise they add to the image alignment process. The sky and the water may compose a majority of each image to be aligned. Yet, they retain little consistent salient structure between surveys. Salient structure can appear in the water if it is reflective, which reduces the likelihood of a good alignment because the reflectivity of the water changes between surveys. The varied appearance of the sky also affects the alignment.

The bias to SIFT Flow's cost function is derived from the 3D information from visual SLAM. The location of the sky and the water in each image is approximately determined using the estimated 3D locations of tracked landmarks. Although points are mostly only tracked along the shore because most corner features occur there, some are occasionally identified in the sky and the water, which this process essentially filters out. Points with a negative elevation indicate a feature corresponds to the water. Points far away indicate a feature corresponds to the sky. The rest of the reprojected points are interpreted as part of the lakeshore. Given an image and the valid 3D landmarks from visual SLAM, an image mask is created by drawing the reprojected points on an image as a circle (an empirically determined radius, $r = 28$, gave the best performance). For each pixel in the non-zero regions, the data terms of SIFT Flow's objective function are biased (by a factor of 1.5) in favor of aligning the contents there compared to the other areas of the image.

Because belief propagation is used to find the best alignment, which can require a significant amount of computation time to converge in large graphical models, SIFT Flow uses image pyramids to speed up the process. An image pyramid progressively halves the size of the two images for several layers (four in this paper). The search for the best alignment proceeds in a search backwards down the image pyramid, with the flow from each layer bootstrapping the optimization at the next higher resolution. A search window defines the area to be considered for each pixel, and reduces in size with each successive layer.

The final output alignment is chosen after a local search around the two candidate poses to find the two images that align best. SIFT Flow seldom finds a dense correspondence between the first two coarsely aligned images we give it. The perspective difference and the optimization error between the two images is often different enough that an incorrect, high score alignment is found. A better, low score alignment is usually possible between nearby images, which have a slightly different perspective. Therefore, the local search for the best dense correspondence is performed on images at 0, ± 1.5, and ± 3.0 second offsets from the two image candidates, for a total of 25 different alignments. To speed up this search, the alignment is only performed for images at the highest level of the image pyramid.

4.4 Survey Comparison

Although we endeavor to create a system for fully autonomous lakeshore monitoring, including detecting changes autonomously, in this work change detection is left to an end user. Our user interface is designed to exploit human skill at detecting changes in flickering images of a scene. If an image pair from two different surveys is aligned at the pixel level, changes flash on and off when the images are flickered back and forth. If the precise alignment is not possible, a user can always revert to a side-by-side comparison of images. This approach enables a human to perform fast change detection (often requiring only a single flicker) for a survey comparison of a large spatial environment consisting of hundreds of images.

5 Evaluation

5.1 General Alignment Quality

We first evaluate how well our framework aligns 10 consecutive lakeshore surveys, which provides a point of reference for our system's performance. The 10 surveys are compared in consecutive order for a total of nine different comparisons. The surveys span a total time of 210 days, with seven days the shortest interval between compared surveys and 62 days the longest. For each survey, each image from its cover set and the aligned image from the following survey were flickered back-and-forth in a display. A human evaluated the alignment quality according to three criteria: (**precise**) almost the entire image is aligned well with little noise; (**coarse**) the images correspond to the same scene and some objects may be precisely aligned; and (**misaligned**) the images correspond to different scenes or it is hard to tell they come from the same scene.

The results are shown in Fig. 4. The framework in this paper significantly outperforms that of our previous work in [7], which compared surveys from June 13 to 25 and achieved 52 % precise alignments, 36 % coarse alignments, and 12 % misalignments. In all the comparisons a significant number of precise alignments are found, although some have more than others. The two cases with the fewest precise alignments involve a comparison with the July 11 survey, which had a much higher water level. The upper half of many images in these two comparisons were precisely aligned, yet because the perspective significantly changed, and the shoreline appeared very different between surveys, SIFT Flow inaccurately extended the shore downward to try to compensate for the large differences in appearance. In the other comparisons, fewer precise alignments are due to sun glare and larger intervals of time between surveys (increasing e.g., the seasonal variation of plants). For every

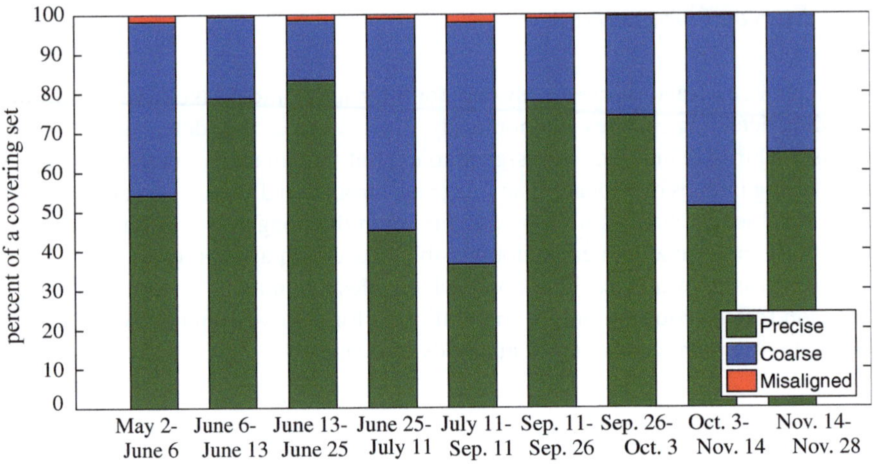

Fig. 4 Alignment quality for comparisons of 10 different surveys. All 10 were performed in 2014

case, however, the few number of misalignments indicates an end user is almost always shown images of the same scene. Thus, because the approach can find good alignments, we next demonstrate its use for change detection.

5.2 *Detected Changes*

While labeling the alignment quality of each comparison, we also saved notable changes between surveys to show our approach is useful for change detection. Six interesting examples are shown in Fig. 5. Five were found in precisely aligned images; the removed treetop was identified in coarsely aligned images. Although the difference between precise and coarse alignments is hard to spot in the figure, it is readily apparent in a flickering display. This is also true for the detection of the cut branch, which is nearly impossible to notice unless the images are precisely aligned and flickered back and forth. Except for the case with people, none of the changes were known of before this analysis. In fact, although we noticed a tree fell in the water after some heavy rain (its branches are sticking out of the water in the Sky and Water example of Fig. 6), we did not know where it came from. Because being able to spot changes depends on image alignment quality, the next section evaluates the robustness of our framework to the variation of appearance across surveys.

Fig. 5 Six notable changes a human found while comparing the 10 different lakeshore surveys

5.3 Robustness to Different Sources of Variation

Our framework can find many precise alignments in all the surveys only because it is robust to many different, combined sources of variation of appearance. Before two images are precisely aligned the appearance variation between them is often 'extreme'. Six prototypical examples of robustness to a particular source of variation are shown in Fig. 6. Perhaps the example with the most extreme amount of variation is the one labeled 'seasonal'. In addition to the foliage depletion captured in this image pair, there is also different illumination, sky, water, shadows, and a globe reflection. Maybe a precise alignment would not have been possible if there was also sun glare. However, there are many cases in which precise alignments are not found. The next section identifies common types of alignment errors.

Fig. 6 Precisely aligned image pairs for six different sources of noise, which indicates our approach can be robust to 'extreme' appearance variation

5.4 Alignment Errors

In some cases the alignment process adds significant noise to the images, which requires reverting to the unregistered image pair for performing a comparison. Six common ways the precise alignments failed are shown in Fig. 7. Image alignment does not comply with the physics of structures in each warped image, which is apparent in all the cases (and is an effect observed in other image processing work as well, e.g., texture synthesis [15]). Because each pixel is potentially warped differently than nearby pixels, the warp may be inconsistent across the image. Additionally, SIFT Flow may try to align to noise (e.g., sun glare) and changes (e.g., a high-water level water), obfuscating the scene. Notwithstanding errors, most alignments are labeled 'coarse' because they are translated versions of the same scenes.

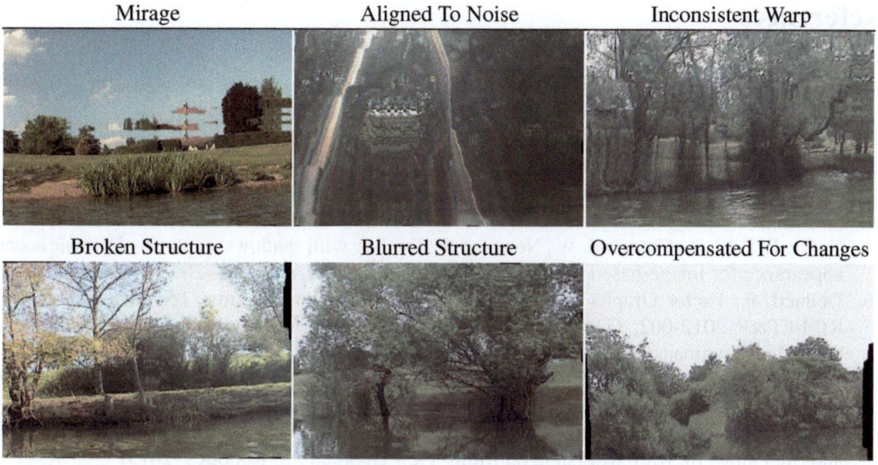

Fig. 7 Six different alignment errors made during image registration

6 Conclusion and Future Work

This paper presented a framework for spatially and temporally scalable lakeshore monitoring. Our approach is based on exploiting scene geometry, using visual SLAM and SIFT Flow, to overcome the variation in appearance of natural environments and achieve pixel-level data association. Extending prior work, this paper (1) identified a covering set of poses; (2) searched for the best alignment around each candidate pose; and (3) used the lakeshore's 3D structure to bias the image registration process. These techniques increased survey alignment accuracy with fewer expensive image alignments. This enabled an analysis of ten surveys, in which a human readily identified several changes. The number of precise alignments we obtained amidst 'extreme' appearance variation validates our approach.

In future work we plan to further improve our method's robustness to the variation in appearance between surveys. The many coarsely aligned image pairs are in reach of becoming precisely aligned. One direction is to transition from aligning mostly visual features to placing more weight on aligning the 3D structure of the lakeshore. Another direction is to remove noise (particularly sun glare) before alignment. With these extensions, precise alignments may become even more likely.

Acknowledgments Funding for this project is provided by the Lorraine Region, France.

References

1. Beall, C., Dellaert, F.: Appearance-based localization across seasons in a Metric Map. In: 6th PPNIV, Chicago, Sept 2014
2. Churchill, W., Newman, P.: Experience-based navigation for long-term localisation. IJRR **32**(14), 1645–1661 (2013)
3. Chvatal, V.: A greedy heuristic for the set-covering problem. Math. Oper. Res. **4**(3), 233–235 (1979)
4. Corke, P., Paul, R., Churchill, W., Newman, P.: Dealing with shadows: capturing intrinsic scene appearance for image-based outdoor localization. In: IROS, pp. 2085–2092. IEEE (2013)
5. Dellaert, F.: Factor Graphs and GTSAM: A Hands-on Introduction. Technical report GT-RIM-CP&R-2012-002, GT RIM, Sept 2012. https://research.cc.gatech.edu/borg/sites/edu.borg/files/downloads/gtsam.pdf
6. Griffith, A., Drews, P., Pradalier, C.: Towards autonomous lakeshore monitoring. In: ISER (2014)
7. Griffith, S., Dellaert, F., Pradalier, C.: Robot-enabled lakeshore monitoring using visual SLAM and SIFT flow. In: RSS Workshop on Multi-View Geometry in Robotics (2015)
8. Gu, J., Ramamoorthi, R., Belhumeur, P., Nayar, S.: Removing image artifacts due to dirty camera lenses and thin occluders. ACM Trans. Graph. (TOG) **28**(5), 144 (2009)
9. He, X., Zemel, R., Mnih, V.: Topological map learning from outdoor image sequences. JFR **23**(11–12), 1091–1104 (2006)
10. Heidarsson, H., Sukhatme, G.: Obstacle detection from overhead imagery using self-supervised learning for autonomous surface vehicles. In: IROS, pp. 3160–3165. IEEE (2011)
11. Hitz, G., Pomerleau, F., Colas, F., Siegwart, R.: State estimation for shore monitoring using an autonomous surface vessel. In: ISER (2014)
12. Jain, S., Nuske, S.T., Chambers, A.D., Yoder, L., Cover, H., Chamberlain, L.J., Scherer, S., Singh, S.: Autonomous river exploration. In: FSR, Dec 2013
13. Kaess, M., Johannsson, H., Roberts, R., Ila, V., Leonard, J.J., Dellaert, F.: iSAM2: incremental smoothing and mapping using the Bayes tree. IJRR, **31**(2), 216–235 (2012)
14. Košecka, J.: Detecting changes in images of street scenes. In: Computer Vision-ACCV 2012, vol. 7727 of LNCS, pp. 590–601. Springer (2013)
15. Kwatra, V., Schödl, A., Essa, I., Turk, G., Bobick, A.: Graphcut textures: image and video synthesis using graph cuts. In: ACM Transactions on Graphics (ToG), vol. 22, pp. 277–286. ACM (2003)
16. Liu, C., Yuen, J., Torralba, A.: SIFT flow: dense correspondence across scenes and its applications. PAMI **33**(5), 978–994 (2011)
17. Lowe, D.: Distinctive image features from scale-invariant keypoints. Int. J. Comput. Vis.**60**(2), 91–110 (2004)
18. McManus, C., Upcroft, B., Newman, P.: Scene signatures: localized and point-less features for localization. In: RSS, Berkeley (2014)
19. Nelson, P., Churchill, W., Posner, I., Newman, P.: From dusk till dawn: localisation at night using artificial light sources. In: ICRA (2015)
20. Neubert, P., Sünderhauf, N., Protzel, P.: Superpixel-based appearance change prediction for long-term navigation across seasons. In: RAS (2014)
21. Rogers, J., Christensen, H.: Normalized graph cuts for visual slam. In: IROS, pp. 918–923. IEEE (2009)
22. Sibley, G., Mei, C., Reid, I., Newman, P.: Vast-scale outdoor navigation using adaptive relative bundle adjustment. IJRR **29**(8), 958–980 (2010)
23. Subramanian, A., Gong, X., Riggins, J., Stilwell, D., Wyatt, C.: Shoreline mapping using an omni-directional camera for autonomous surface vehicle applications. In: OCEANS, pp 1–6. IEEE (2006)

Autonomous Greenhouse Gas Sampling Using Multiple Robotic Boats

Matthew Dunbabin

Abstract Accurately quantifying total greenhouse gas emissions (e.g. methane) from natural systems such as lakes, reservoirs and wetlands requires the spatial-temporal measurement of both diffusive and ebullitive (bubbling) emissions. Traditional, manual, measurement techniques provide only limited localised assessment of methane flux, often introducing significant errors when extrapolated to the *whole-of-system*. In this paper, we directly address these current sampling limitations and present a novel multiple robotic boat system configured to measure the spatiotemporal release of methane to atmosphere across inland waterways. The system, consisting of multiple networked Autonomous Surface Vehicles (ASVs) and capable of persistent operation, enables scientists to remotely evaluate the performance of sampling and modelling algorithms for real-world process quantification over extended periods of time. This paper provides an overview of the multi-robot sampling system including the vehicle and gas sampling unit design. Experimental results are shown demonstrating the system's ability to autonomously navigate and implement an exploratory sampling algorithm to measure methane emissions on two inland reservoirs.

1 Introduction

Quantification of greenhouse gas emissions to atmosphere is becoming an increasingly important requirement for scientists and managers to understand their total carbon footprint. Methane in particular is a powerful greenhouse gas, approximately 21 times higher global warming potential than carbon dioxide. Water storages are known emitters of methane to atmosphere [11]. The spatiotemporal variation of release is dependent on many environmental and biogeochemical parameters. Therefore, in order to accurately quantify this greenhouse gas release requires long duration and repeat monitoring of the entire water body.

M. Dunbabin (✉)
Institute for Future Environments, Queensland University of Technology,
2 George Street, Brisbane, QLD 4000, Australia
e-mail: m.dunbabin@qut.edu.au

D.S. Wettergreen and T.D. Barfoot (eds.), *Field and Service Robotics*,
Springer Tracts in Advanced Robotics 113, DOI 10.1007/978-3-319-27702-8_2

There are two primary pathways for methane to be released from water storages; (1) diffusion, and (2) ebullition (or bubbling). Diffusion is the most common pathway considered due to greater consistency across a waterway. Rates of methane ebullition represent a notoriously difficult emission pathway to quantify with highly variable spatial and temporal changes [6]. However, the importance of bubbling fluxes in terms of total emissions is increasingly recognised from a number of different globally relevant natural systems including lakes, reservoirs and wetlands. This represents a critical challenge to current manual survey efforts to quantify spatiotemporal greenhouse gas emissions and reduce the uncertainty associated with bubbling fluxes. This is where robotics can play a significant role.

In this work, a novel system for direct measurement of the combined diffusive and ebullitive methane flux and an ability to persistently monitor a wide spatial area is presented. Named the *Inference* Robotic Adaptive Sampling System, it consists of multiple (four) networked robotic boats (see Fig. 1) and provides an open architecture allowing researchers to evaluate new sampling algorithms with customisable scientific payloads on real-world processes over extended periods of time.

The contributions presented in this paper are; (1) A novel ASV system for navigating complex inland waterways, (2) a new greenhouse gas sampling system, (3) a multi-robot sampling strategy to survey a previously unseen environment, and (4) an experimental evaluation of the entire system on two inland water storages.

The remainder of this paper is as follows: Sect. 2 provides background information. Section 3 describes the *Inference* system and the gas sampling system. Section 4 describes a preliminary sampling methodology with Sect. 5 showing results from two inland water storages. Finally, Sect. 6 draws conclusions and discusses future research.

Fig. 1 The multi-robot *Inference* Robotic Adaptive Sampling System

2 Related Work

Robotic platforms capable of persistent environmental monitoring offer an efficient alternative to manual or static sensor network sampling for studying large-scale phenomena. However, in practice most applications are short-term experiments for validating existing models [3]. Recent cross-disciplinary research extensively used robots to investigate assumptions around spatiotemporal homogeneity of environmental processes such as toxic algal blooms in lakes [5] and methane production in reservoirs [6]. These studies show that combined robotic persistence and spatiotemporal sampling can provide significant new insight into environmental processes. However, there are challenges to achieving persistent robotic process monitoring, particularly in the complex environments considered here. These primarily relate to robotic platforms for persistent navigation within complex and often dynamic environments, and the ability to adaptively coordinate multiple robots to appropriately sample the process of interest.

Robotic monitoring of marine and aquatic environments has received considerable attention over the last two decades [3]. Whilst most studies have focused on underwater vehicles with restricted payloads and endurance, there is now increasing focus on Autonomous Surface Vehicles (ASVs) with greater endurance and payload carrying capacity for large-scale unsupervised environmental monitoring [12, 13, 16]. These systems are primarily designed for oceanographic surveys and are not particularly suitable for relatively unexplored inland waterways with challenging and often varying navigational requirements.

Recently, a series of ASVs have been designed and applied on inland waterways. Typically, these catamaran style vehicles are of sufficient size for carrying scientific payloads for tasks such as mapping hazards above and below the waterline [4], and water quality monitoring [1, 7]. Whilst demonstrating environmental monitoring capabilities, there is little flexibility for adding external payloads and their navigation capabilities are generally customised to the specific environment. The provision of a flexible, yet capable, robotic platform is a key consideration in this research.

Navigation around narrow inland waterways is often more challenging than for the ocean due to issues such as above, below and on-water obstacles and GPS reliability (e.g. in mountainous and forested systems). A number of sensors have been used to detect obstacles and in identifying free-space paths. Hitz et al. [7] use water depth only for detecting shallow regions, whereas Ferreira et al. [4] and Leedekerden et al. [9] use scanning laser range finders and sonar to produce high-resolution 3D maps of the above and below water environment. Cameras have also been proposed for detecting specific objects on the water [2, 4]. Scherer et al. [14] have used cameras and laser scanners (albeit on an aerial robot) to map the edges of waterways and the free-space above the water as the robot traverses them. Whilst high-resolution sensors such as lasers and sonar can provide robust navigation capabilities, for persistent monitoring their power consumption can be a particular challenge. Exploiting lower power, and cost, sensing modalities such as vision and ultrasonics to provide sufficient obstacle detection capabilities is a goal of this research.

The overall coordination of the mobile sensors (robots) is critical to accurately measure spatiotemporal environmental processes. An emerging research area for ASVs is that of mobile adaptive sampling where the ASV can alter its trajectory to improve measurement resolution in space and time (e.g. [17]). The survey paper [3] summarises advances in robotic adaptive sampling for environmental monitoring. Past research has focused primarily on the Gaussian Process-based reconstruction of stationary processes using combinations of mobile and static sensors networks [8, 17]. Whilst demonstrating the ability to capture and reconstruct various parameter distributions, these studies offer simulation only or short duration small-scale experimental validation. Larger-scale adaptive coordination of mobile sensing assets (underwater gliders) has been considered for tracking large oceanographic plumes in [10, 15]. Developing and demonstrating multi-robot adaptive sampling algorithms for the large-scale monitoring and tracking of spatiotemporal environmental processes is an over-arching goal of this research.

3 The Inference Autonomous Surface Vehicle

This section describes the current *Inference* Robotic Adaptive Sampling system and the greenhouse gas sampling payload system as applied and evaluated in this paper.

3.1 High-Level Scenario

The *Inference* Robotic Adaptive Sampling system was developed with the goal of providing a shared resource of multiple networked ASVs to allow researchers to remotely evaluate new sampling algorithms on real-world processes over extended periods of time. A typical use scenario proposed for the system is outlined below:

1. The ASVs, each carrying a scientific payload, are deployed on a water body.
2. Based on a desired sampling protocol (e.g. random, adaptive) and process modelling requirements, new sampling locations are determined. This can be achieved either from a remote centralised, or an on-board decentralised process.
3. Determine which ASV goes to each of the updated sample locations. This may involve optimising a cost function (e.g. minimising energy and/or travel time, maximising solar energy harvesting).
4. Each ASV navigates to their commanded sampling location.
5. Each ASV takes its scientific measurement and reports it back through the network.
6. Repeat steps 2–5 until a termination condition is met.

The system described in this paper is working towards achieving this goal with a preliminary experimental evaluation of this scenario using a simplified random exploration algorithm as described in Sect. 4.

Fig. 2 One of the autonomous surface vehicles from the *Inference* system. The navigation sensors, computing and batteries are located underneath the two solar panels. The scientific payload is attached to the moon-pool opening underneath the camera. Note the pan-tilt dome camera visible was not used in this study, only the smaller USB camera directly in front of it

3.2 Hardware Overview

The Autonomous Surface Vehicles used in the multi-robot *Inference* system are custom designed for persistent and cooperative operation in challenging inland waterways. The overall hull shape (see Fig. 2) has four key features; (1) A low draft allowing traversal in shallow water, (2) open sides and low curved top deck to minimise windage and the associated drift when station keeping during sampling, (3) a large top surface area angled for maximising energy harvesting from the solar panels, and (4) a moon-pool (open centre section) with standardised attachment points to mount custom sensor packages. The overall dimensional and mass specifications for the ASVs are given in Table 1.

Table 1 Physical and performance specifications of the ASVs

General specifications	
Length	1.50 m
Width	1.50 m
Height (above waterline)	0.7 m
Draft	0.15 m
Weight	33 kg (without payload)
	External payload: 4 kg
Propulsion	2 × BlueRobotics T100 brushless electric thrusters
Power	12 V 20 Ah LFP battery
	2 × 40 W solar panels
Speed	Max: 2.3 ms^{-1}
	Typical survey: 0.5–0.8 ms^{-1}

Propulsion of the ASVs is provided by two BlueRobotics T100 brushless thrusters mounted at the rear of each side of the hull. These provide the forward motion as well as steering (through differential control) of the vehicles. The system is powered by a single 20 Ah Lithium Iron Phosphate battery and two 40 W solar panels. This limited energy capacity requires advanced path-planning algorithms to coordinate the ASVs for maximising energy harvesting as well as to meet the overall sampling objectives. These algorithms are current ongoing research and not considered in this paper.

The ASVs are required to autonomously navigate inland waterways using only their on-board sensors. Each ASV has a suite of low-cost navigation sensors which include a GPS, magnetic compass with roll and pitch, and a depth sensor for measuring bathymetry. Of particular importance is the ability to detect the water's edge and potential obstacles on top of the water. The obstacle sensors used in this study are a USB camera (Microsoft LifeCam) mounted above the moon-pool, and four Maxbotix ultrasonic range sensors mounted just under the leading and trailing edges of the top deck. These sensors are used to detect the edge of the water and at-surface structure such as reeds, trees and water lilies (see Sect. 4). To minimise power consumption and cost, typical scanning laser-based or radar sensors are not currently used, although they can be added if required in future scenarios.

The ASV's thrusters are controlled via a custom designed motor and sensor interface board. This system is capable of providing waypoint control and ultrasonic and depth sensor based obstacle avoidance. To facilitate vision-based obstacle avoidance, each ASV has an Odroid C1 ARM Cortex-A5 1.5 GHz quad core CPU running the Robotic Operating System (ROS) and OpenCV.

There are two communication systems on-board the ASVs. The first is a 2.4 GHz WiFi system allowing communication to a gateway located on a floating platform on the water storage. This gateway has a wireless router and 3G modem allowing bidirectional data transfer from a centralised server located at the Queensland University of Technology. The second is a 2.4 GHz wireless embedded system (XBee IEEE 802.15.4) allowing serial communication between each vehicle as well as with existing static floating sensor nodes located on the water body.

Each ASV is capable of carrying additional custom payloads weighing up to 4 kg. The payload is mounted under the moon-pool opening via six attachment bolts. Currently available payloads include gas sampling (see Sect. 3.3), multi-beam and profiling sonars, water sampling and a winch system for water column profiling. A six pin connector is provided for use by the custom payloads. This connector provides power as well as bi-directional serial communications via a standardised protocol for triggering sampling, and reporting sample completion and possible faults.

3.3 Gas Sampling System

The goal of this study is to measure greenhouse gas emissions (efflux) from the waterway. Figure 3 shows the self-contained greenhouse Gas Sampling System (GSS) developed to autonomously measure both the diffusive and ebullitive efflux. This payload is mounted underneath the ASV via the moon-pool payload attachment points as described in Sect. 3.2.

The GSS (Fig. 3) automates the traditional manual chamber-based sampling process and consists of three primary components; (1) A frame allowing the lowering and raising of a chamber into the water, (2) a chamber fitted with a continuous methane gas (CH_4) sensor and purge valve, and (3) a physical gas sampling unit.

The process of sampling the greenhouse gas being released from the water to the atmosphere using the GSS is illustrated in Fig. 4 and consists of four steps. Firstly, the ASV navigates to the desired sampling location it goes into a *weak* station-keeping mode. This limits the control input to the motors to reduce any disturbance that may influence the CH_4 efflux at the expense of a slightly increased station bound. At this

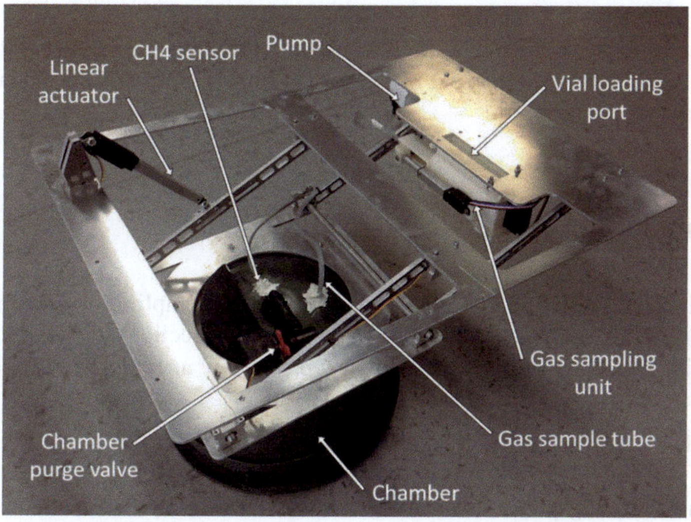

Fig. 3 The Gas Sampling System (GSS) used to measure greenhouse gas (methane) release to atmosphere from the inland water storages. The GSS is attached to the ASV as described in Sect. 3.2

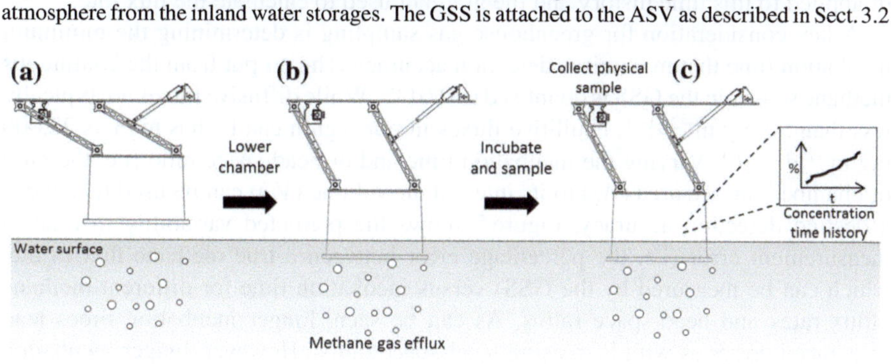

Fig. 4 The sequence of actions required to measure greenhouse gas using the GSS

point, the chamber purge valve (see Fig. 3) is opened and the chamber lowered using the linear actuator to achieve a desired air volume within the chamber (Fig. 4a–c). The second step involves closing the chamber purge valve and letting the methane concentration within the chamber increase for a predetermined *incubation* time (see Sect. 4 for a discussion on incubation time). During incubation, the methane sensor continuously measures the concentration within the chamber (Fig. 4b, c). At the end of the incubation, the third step (Fig. 4c) calculates the overall gas efflux rate from the gradient of the recorded methane concentration time history. Also a physical sample of gas from the chamber is collected for laboratory analysis using the gas sampling unit (see Fig. 3). This involves a sequence of actions that firstly purges the sample tube using the pump, then loads a pre-evacuated 12 mL vial into the sampling unit. A linear actuator on the unit drives a hypodermic needle into the vial whilst pumping gas from the chamber. Once 20 mL of gas has been pumped into the vial (over pressure sampling technique), the needle retracts and the unit discharges the vial ready for the next sample.

After sampling is completed, the final step involves opening the chamber purge valve and raising the chamber out of the water. At this point the ASV can move to the next sample location.

4 Technical Approach

This section outlines technical details relating to the sampling of greenhouse gas (methane), obstacle avoidance, and the sample site selection algorithms used for coordinating a number of the ASVs across a previously unexplored water body.

Gas Sampling Protocol

During the sampling phase, the concentration measured by the methane sensor is polled every 2 s for the entire incubation period. A linear least squares line of best fit applied to this time history and the gradient used to calculate the flux rate.

A key consideration for greenhouse gas sampling is determining the minimum incubation time that maximises detection accuracy. The output from the continuous methane sensor in the GSS is quantised to 0.01 %. While diffusive fluxes are typically less than 50 mg m^{-2} d^{-1}, ebullitive fluxes in our region can be has high as 22,000 mg m^{-2} d^{-1} [6]. Varying the incubation time and/or head-space ratio (i.e. the ratio of chamber surface area (A_c) to its internal air volume (V_c)) can be used to achieve a desired detection accuracy. Figure 5 shows the predicted variability in relative measurement error (i.e. the percentage error between a true methane flux to that which can be measured by the GSS) versus incubation time for different methane efflux rates and head-space ratios. As can be seen, longer incubation times lead to reduced errors as with increasing head-space ratios. However, longer incubation times mean less sample points can be performed per day. In this study, the primary

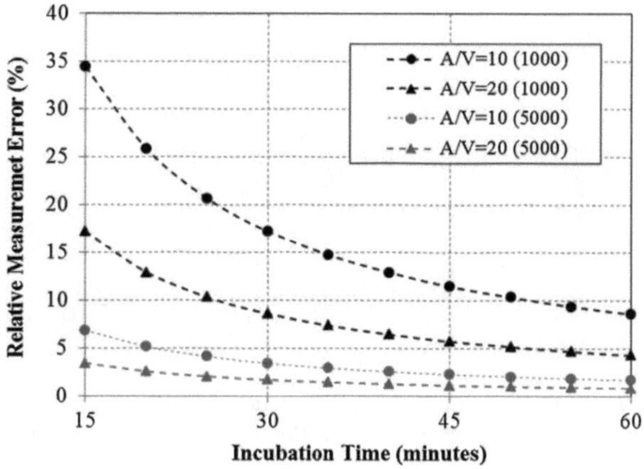

Fig. 5 The predicted percentage relative measurement error of methane flux rate with incubation time for the prototype GSS (see Sect. 3.3) with a sensor output resolution of 0.01 %. Two efflux rates are considered, 1000 and 5000 mg m^{-2} d^{-1} with head-space ratios (A_c/V_c) of 10 and 20 m^{-1}

interest is the detection of methane "hot-spots", that is where it is bubbling from the water. Therefore, incubation times of 15–20 min were chosen here to allow detection of methane rates as low as 1000 mg m^{-2} d^{-1}, albeit at lower accuracy. However, the higher the efflux rate, the more accurate the measurement.

Obstacle Avoidance

The ASVs have three sensors for obstacle avoidance; (1) ultrasonic sensors, (2) a camera, and (3) water depth sensor. The ultrasonic sensors have a maximum range of 6.5 m and are used to detect above water objects in front of the ASV such as land, reeds, trees and larger buoys. The camera, only used when moving between sample waypoints, is used to detect water lilies on the water's surface. The image stream is processed at 1 Hz. With the camera fixed to the ASV, the horizon can be approximated and only the scene below the horizon considered. Image segmentation is conducted using an empirically determined threshold on the green and blue color channels with an approximate water lily size threshold to reduce noise. Figure 6 shows an example image from an ASV and the resulting segmentation of the water lilies (shown in red).

To detect shallow, non-traversable water, the depth of water below the ASV is continuously monitored. The outputs from all obstacle sensors are parsed by the on-board controller. When a detection occurs, the ASV trajectory is modified as described in the following section.

Multi-robot Sample Site Selection

A random walk-based algorithm is proposed here for selecting locations for n ASVs to sample the environment in an attempt to identify regions with high methane gas flux. There are two key assumptions: (1) the boundary of the water body is known from sources such as GIS, and (2) the ASVs can communicate between each other

Fig. 6 Example of image segmentation from the ASV for detecting on-water obstacles such as water lilies (*Left* original image. *Right* image with detected obstacles highlighted in *red*)

and can share their list of previous and next sample locations. In this study, we do not use bathymetry but it could be used in the future to help guide the algorithm.

The selection of new sample locations is based on an online random walk and potential fields. Iterating through each robot, the basis of the algorithm is as follows:

1. All previously sampled sites for all robots are represented as 2D Gaussian potentials centred at those points with fixed amplitude and standard deviation.
2. A random position at radius r from the current position is selected. If this position is not on land, and the value from the closest Gaussian potential is less than a threshold, this becomes the next sample point for that robot. If this condition is not met, the process is iterated until a location can be found. If no location can be found after a set number of iterations, the search radius is increased by Δr and the process repeated until a site is found or some termination criteria is met.
3. To increase local intensification of sampling in methane "hot-spots", if the measured flux rate at the robot's current location exceeded some threshold, the search radius for the next sample step is set to βr where $(0 < \beta \leq 1)$ and the potential threshold trigger relaxed.

During waypoint execution each robot drives in a straight line towards the goal. If the water depth falls below a threshold (i.e., too shallow), or an obstacle is detected, the vehicle starts to move either clockwise or counter clockwise around the contour until a new straight line to the goal can be achieved. This entire process is repeated for all robots until a desired number of samples are collected or some other termination condition met.

5 Results

An experimental evaluation using two ASVs with gas sampling payloads was conducted on two water reservoirs in South East Queensland, Australia; (1) Gold Creek Dam, and (2) Little Nerang Dam. These are established study sites and selected as they exhibit regions of significant methane ebullition and provide a range of challenging operational conditions for evaluating robotic systems.

Previous studies [6] had collected georeferenced outlines of the water's edge (boundary) as well as bathymetry maps for both sites. Only the boundary was used in this study for implementing the sample site selection algorithm described in Sect. 4. Figure 7 shows the two ASVs used in this study on Gold Creek Dam.

The first experiment was conducted at Gold Creek Dam. This is a small, relatively open dam with a narrowing distal arm. The sample selection algorithm was run to collect 12 samples for each ASV, with a step radius of 100 m, and intensification factor of 0.5. The trigger was set at 1000 mg m^{-2} d^{-1} with 20 min incubations. The time to complete the sampling was approximately 5 h. Figure 8 shows the results of implementing the sample strategy for both ASVs. These results show the ASVs were capable of navigating the water storage and implementing the sample protocol. The online detections of methane exceeding the trigger threshold (markers in yellow) correspond to areas physically observed to have methane ebullition. As ebullition is essentially a point source emitter, there can be extreme variability even at short spatial and temporal scales (see [6]). Therefore, whilst ebullition can often be seen in expected regions (e.g. top image of Fig. 8) a sample within that region does not always guarantee the capture of gas bubbles sufficient to achieve high rates.

A second experiment was conducted at Little Nerang Dam. This is a longer and narrower water storage with a steep sided catchment. The sample selection was run with a total of 30 samples for each ASV, step radius of 200 m and an intensification factor of 0.5. The trigger was set at 1000 mg m^{-2} d^{-1} with 15 min incubations. The time to complete the experiment was approximately 10.5 h.

Figure 9 shows the results of implementing the sample strategy for both ASVs. These results again show the ASVs ability to implement the sample protocol and

Fig. 7 The two ASVs at the start of a sampling campaign on Gold Creek Dam, Queensland. The retracted gas sampling unit is visible underneath the ASV on the *right*

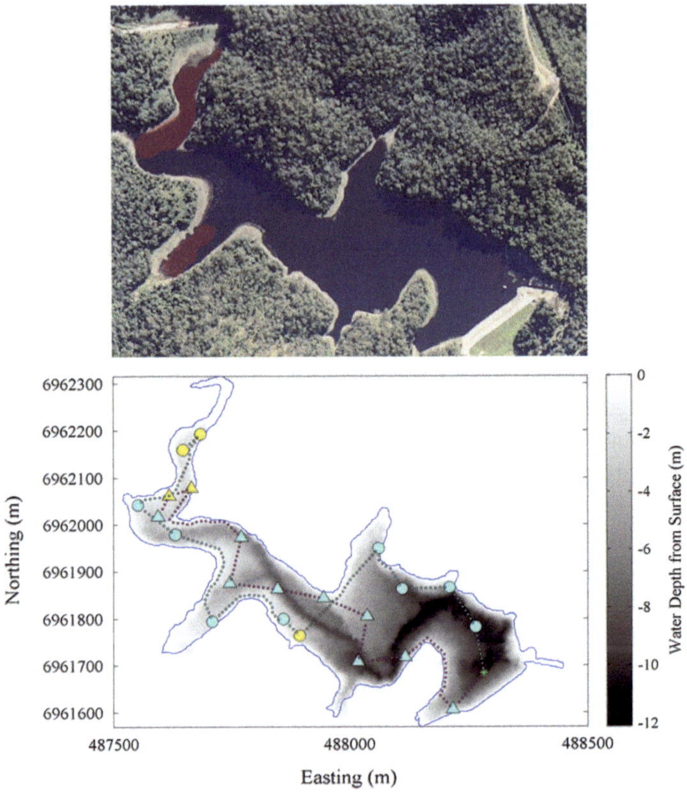

Fig. 8 Sampling locations and ebullition detections from 20 min incubations using two ASVs on Gold Creek Dam, Queensland. *Top* An aerial image of Gold Creek Dam with red overlay showing the regions of physically observed methane ebullition. *Lower* The trajectory and resulting sample locations indicated by the *circles* for ASV1 and *triangles* for ASV2. The start location for both ASVs is indicated by the *green dot*. The *circles* and *triangles* highlighted in *yellow* indicate the online chamber measurements that exceeded 1000 mg m^{-2} d^{-1}

navigate the water storage. The online detections of methane exceeding the trigger threshold (markers in yellow) are consistent with previous research at the dam [6].

Whilst these experiments demonstrated the system for real-time sampling of greenhouse gases across water bodies, the online component of gas sampling system was not optimised for detecting lower (and more common) flux rates of less than 1000 mg m^{-2} d^{-1}. Future work will look at adaptive chamber head-space control as well as higher precision sensors to improve the utility of the system for accurate quantification of the combined diffusive and ebullitive flux of greenhouse gases.

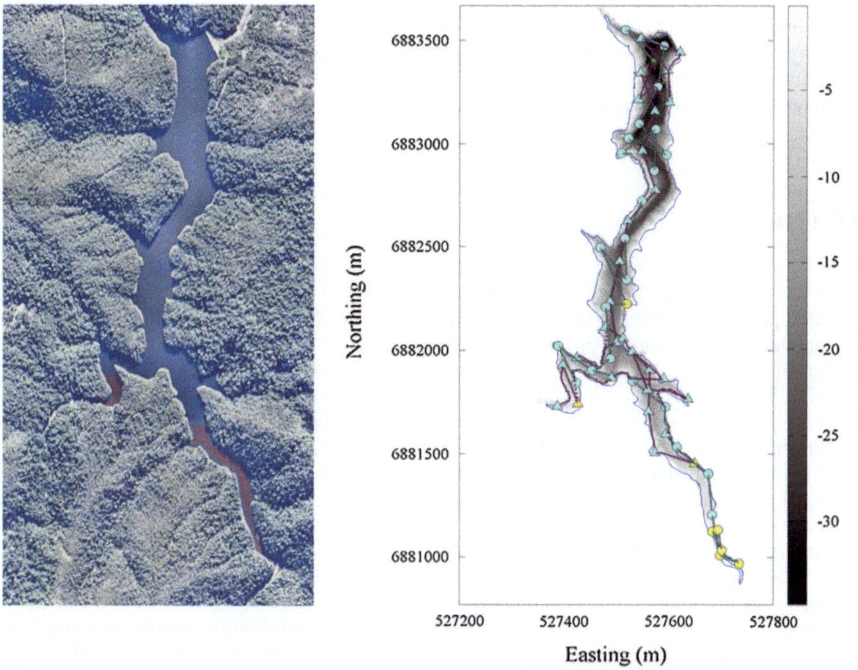

Fig. 9 Sampling locations and ebullition detections from 15 min incubations using two ASVs on Little Nerang Dam, Queensland. *Left* An aerial image of Little Nerang Dam with red overlay showing the regions of physically observed methane ebullition. *Right* The trajectory and resulting sample locations indicated by the *circles* for ASV1 and *triangles* for ASV2. The start location for both ASVs was at the dam wall located at the northern most end. The *circles* and *triangles* highlighted in *yellow* indicate the online chamber measurements that exceeded 1000 mg m^{-2} d^{-1}

6 Conclusions

This paper has presented a novel robotic sampling system for conducting large-scale, persistent monitoring on complex inland waterways. The system, named *Inference*, consists of multiple networked Autonomous Surface Vehicles (ASVs) carrying a range of scientific payloads. Experimental results demonstrate the ASV's ability to navigate complex waterways whilst executing a multi-robot online sampling protocol. Using a custom Gas Sampling System (GSS) attached to each ASV, experimental results also show the robotic system is capable of measuring and localising strong greenhouse gas release (methane) to atmosphere. Future research is focused on developing more sophisticated multi-robot adaptive sampling algorithms to achieve persistent monitoring and mapping of spatiotemporal processes whilst considering energy, speed and sampling constraints of the vehicles. Additionally, new sensors and algorithms for head-space control of the GSS are being developed to improve its lower detection limit for sampling regions with low gas flux rates.

Acknowledgments The author would like to thank Alistair Grinham and Katrin Strum from the University of Queensland for their assistance with the initial gas sampling unit and ASV prototype evaluation, and laboratory processing of gas samples. Also thanks to Riki Lamont for his assistance in the payload integration and commissioning of the ASVs.

References

1. Dunbabin, M., Grinham, A.: Experimental evaluation of an autonomous surface vehicle for water quality and greenhouse gas monitoring. In: Proceedings of International Conference on Robotics and Automation, pp. 5268–5274, May 2010
2. Dunbabin, M., Grinham, A., Udy, J.: An autonomous surface vehicle for water quality monitoring. In: Proceedings of Australasian Conference on Robotics and Automation, Dec 2009
3. Dunbabin, M., Marques, L.: Robots for environmental monitoring: significant advancements and applications. Rob. Autom. Mag. IEEE **19**(1), 24–39 (2012)
4. Ferreira, C., Almeida, A., Martins, J., Almeida, N., Dias, E., Silva, E.: Autonomous bathymetry for risk assessment with ROAZ robotic surface vehicle. In: Proceedings of Americas, Maritime Systems and Technology (2010)
5. Garneau, M.-E., Posch, T., Hitz, G., Pomerleau, F., Pradalier, C., Siegwart, R., Pernthaler, J.: Short-term displacement of planktothrix rubescens (cyanobacteria) in a pre-alpine lake observed using an autonomous sampling platform. Limnol. Oceanogr. **58**(5), 1892–1906 (2013)
6. Grinham, A., Dunbabin, M., Gale, D., Udy, J.: Quantification of ebullitive and diffusive methane release to atmosphere from a water storage. Atmos. Environ. **45**(39), 7166–7173 (2011)
7. Hitz, G., Pomerleau, F., Garneau, M.-E., Pradalier, C., Posch, T., Pernthaler, J., Siegwart, R.Y.: Autonomous inland water monitoring: design and application of a surface vessel. Rob. Autom. Mag. IEEE **19**(1), 62–72 (2012)
8. Hombal, V., Sanderson, A., Blidberg, D.R.: Multiscale adaptive sampling in environmental robotics. In: IEEE Conference on Multisensor Fusion and Integration for Intelligent Systems (MFI), pp. 80–87, Sept 2010
9. Leedekerken, J., Fallon, M., Leonard, J.: Mapping complex marine environments with autonomous surface craft. In: Experimental Robotics, Springer Tracts in Advanced Robotics, vol. 79, pp. 525–539. Springer, Berlin Heidelberg (2014)
10. Leonard, N.E., Paley, D.A., Davis, R.E., Fratantoni, D.M., Lekien, F., Zhang, F.: Coordinated control of an underwater glider fleet in an adaptive ocean sampling field experiment in monterey bay. J. Field Robot. **27**(6), 718–740 (2010)
11. Louis, V.St., Kelly, C., Duchemin, E., Rudd, J., Rosenberg, D.: Reservoir surfaces as sources of greenhouse gases to the atmosphere: a global estimate. Bioscience **50**, 766–775 (2000)
12. Manley, J., Willcox, S.: The wave glider: a persistent platform for ocean science. In: OCEANS 2010 IEEE—Sydney, pp. 1–5, May 2010
13. Rynne, P.F., von Ellenrider, K.D.: A wind and solar-powered autonomous surface vehicle for sea surface measurements. In: Proceedings of IEEE OCEANS, pp. 1–6 (2008)
14. Scherer, S., Rehder, J., Achar, S., Cover, H., Chambers, A., Nuske, S., Singh, S.: River mapping from a flying robot: state estimation, river detection, and obstacle mapping. Auton. Robots **33**(1–2), 189–214 (2012)
15. Smith, R., Das, J., Chao, Y., Caron, D., Jones, B., Sukhatme. Cooperative multi-AUV tracking of phytoplankton blooms based on ocean model predictions. In: OCEANS 2010 IEEE—Sydney, pp. 1–10 (2010)
16. Wang, J., Gu, W., Zhu, J.: Design of an autonomous surface vehicle used for marine environmental monitoring. In: Proceedings of International Conference on Advanced Computer Control (ICACC09), pp. 405–409, Jan 2008
17. Zhang, B., Sukhatme, G.S.: Adaptive sampling for estimating a scalar field using robotic boat and a sensor network. In: Proceedings of International Conference on Robotics and Automation, pp. 3673–3680, Apr 2007

Experimental Analysis of Receding Horizon Planning Algorithms for Marine Monitoring

Soo-Hyun Yoo, Andrew Stuntz, Yawei Zhang, Robert Rothschild, Geoffrey A. Hollinger and Ryan N. Smith

Abstract Autonomous surface vehicles (ASVs) are becoming more widely used in environmental monitoring applications. Due to the limited duration of these vehicles, algorithms need to be developed to save energy and maximize monitoring efficiency. This paper compares receding horizon path planning models for their effectiveness at collecting usable data in an aquatic environment. An adaptive receding horizon approach is used to plan ASV paths to collect data. A problem that often troubles conventional receding horizon algorithms is the path planner becoming trapped at local optima. Our proposed Jumping Horizon (J-Horizon) algorithm planner improves on the conventional receding horizon algorithm by jumping out of local optima. We demonstrate that the J-Horizon algorithm collects data more efficiently than commonly used lawnmower patterns, and we provide a proof-of-concept field implementation on an ASV with a temperature monitoring task in a lake.

S.-H. Yoo (✉) · Y. Zhang · G.A. Hollinger
Oregon State University, Corvallis, OR, USA
e-mail: yoos@onid.oregonstate.edu

Y. Zhang
e-mail: zhanyawe@onid.oregonstate.edu

G.A. Hollinger
e-mail: geoff.hollinger@oregonstate.edu

A. Stuntz · R. Rothschild · R.N. Smith
Fort Lewis College, Durango, CO, USA
e-mail: abstuntz@fortlewis.edu

R. Rothschild
e-mail: rarothchild@fortlewis.edu

R.N. Smith
e-mail: rnsmith@fortlewis.edu

D.S. Wettergreen and T.D. Barfoot (eds.), *Field and Service Robotics*,
Springer Tracts in Advanced Robotics 113, DOI 10.1007/978-3-319-27702-8_3

1 Introduction

Autonomous surface vehicles (ASVs) are becoming more commonly used to collect data in oceans and inland waterways using instruments such as: acoustic doppler current profilers (ADCPs); conductivity, temperature, and depth sensors (CTDs); and sidescanning sonars. These autonomous vehicles allow data collection in tight places, such as in and around glaciers or ice, as well as in close proximity to land (e.g., around river deltas) [2, 5].

Commercially-available ASVs, such as the Platypus Lutra (Fig. 1b) and Ocean-Sever Q-Boat, typically execute a simple lawnmower path to cover the area to be explored (Fig. 2). Such a path can provide high data yield, but at the expense of substantial fuel and time costs [11].

Previous work has shown that sampling in a spiral pattern is slightly more energy-efficient than doing so in a lawnmower pattern [9], but only by a margin of less than 5 %. This margin will be shown to be negligible compared to that demonstrated by J-Horizon over lawnmower, so for simplicity, the lawnmower pattern will be simulated and considered as the baseline.

(a) (b)

Fig. 1 Two commercially-available autonomous surface vehicles for aquatic sampling. **a** Q-Boat 1800P with an integrated ADCP. **b** Platypus Lutra with a dissolved oxygen and pH sensor

Fig. 2 The proprietary area search algorithm from Platypus generates a dense lawnmower pattern that is highly energy-inefficient

Here, we propose a receding horizon path-planning algorithm that, given an information or uncertainty map, generates a sampling path to maximize the information gathered or reduce the uncertainty. We compare this algorithm against the simple lawnmower path planner for a given transport budget and examine the effects of various algorithm parameters on the quality of the generated path. Furthermore, we propose a Jumping Horizon (J-Horizon) algorithm that improves on the conventional receding horizon algorithm by varying the look ahead step size if desired threshold values cannot be found within the current horizon. This allows the planner to "jump" out of local optima if higher peaks can be found elsewhere on the map. Finally, we validate our simulated results during a field trial using an ASV. An initial data set is collected to provide a base scalar field. The J-Horizon algorithm is then run over this scalar field produced, and a qualitative analysis is given. The J-Horizon planner is able to produce paths superior to the simple lawnmower pattern in simulation, and experimentally the J-Horizon path is shown to cover more area and generate a more representative scalar field.

2 Related Work

Past work has shown that a receding horizon path planner is effective at optimizing paths in "no-fly" zone environments with hard constraints [10], where the agent is prohibited from entering certain areas bounded by walls. This is a useful constraint for aerial and land vehicles that must navigate cluttered environments. However, these constraints do not apply to an ASV that must cover a large body of water such as a lake or the open ocean.

In previous work, AUVs have played a similar role as the ASV in our project. Binney et al. [1] describe an offline path planner for an uncertainty area. Hollinger and Singh [7] describe an approach for multiple agents searching for a target in a known environment. Hitz et al. [6] discuss a path planner that can choose an efficient path for measurement of fluorescent bacteria in the ocean using an ASV. To reduce computational complexity, all of these authors employ a receding horizon path planner. Besides implementation on ASVs, receding horizon algorithms are widely used in other robotics scenarios. Tisdale et al. [12] describe a receding horizon path planner for multiple unmanned aerial vehicles to search for a stationary object. For currently implemented receding horizon planners, no research exists that examines the effect of the horizon length, or the possibility of modifying this horizon based on the remaining information.

Frolov et al. [3] compares lawnmower paths to other planning algorithms using fleets of research vehicles. They come to the conclusion that lawnmower paths are only marginally worse than adaptive algorithms. They also conclude that graph-based search algorithms are actually worse than lawnmower patterns, thus cannot maximize their performance, because they are unable to adapt to prior uncertainty. Our J-Horizon algorithm adapts to the environment and removes these limitations to provide improved performance over other adaptive algorithms.

Gotovos et al. [4] propose a Level Set Estimation (LSE) algorithm that uses Gaussian Processes to estimate level sets of measured quantities and generate sampling points that reduce uncertainty around a certain threshold level. In a different context, [8] describe an incremental sampling-based motion planning algorithm. Instead of reducing the uncertainty, they try to optimize the information gathering, depreciating the information value of sampled points.

A key limitation of existing research in receding horizon planning is that none of the aforementioned works discusses the role of the parameters in the receding horizon algorithm, e.g., horizon length or adaptation based on gathered or remaining information. In addition, prior research has not focused on a single ASV performing data collection over large areas. In this paper, we address this gap in research in the aforementioned papers through the presentation of the J-Horizon algorithm. We present the application of our proposed method over different scalar fields both in simulation and through field experiments. The algorithm's performance is experimentally demonstrated to outperform existing lawnmower and traditional receding horizon methods.

3 Problem Setup

Due to the wide variety of data that is sampled, it is challenging to model the data collection in a general way. The areas of interest being surveyed by ASVs are often dynamic environments, and the data collected is often a reflection of changes in the environment. Data collection around glaciers, in river deltas, or in relatively shallow water are environments that are changing quickly. The data collected is often collected to provide a *snapshot* of the processes that are evolving in the general area, and plan for future targeted sampling. Prior surveys can provide a heuristic upon which to formulate plans for future surveys, and multiple surveys can be combined to provide a time-series evolution of the region of interest. Here, we exploit the existence of a partially known underlying field and present a method for improved sampling based on time and energy optimization while gathering data of maximal reward.

3.1 Objective Function

In this paper, the J-Horizon planner addresses the following maximization problem:

$$p^* = \underset{p \in \psi}{\arg\max}\ R(p) \quad \text{s.t.} \quad c(p) \leq B,$$

where ψ is the space of possible trajectories for the ASV, B is the initial budget (e.g., time, fuel), and R is a reward function that represents the information gathered or uncertainty reduced along the trajectory p.

Furthermore, we depreciate the value of $R(p)$ each time we sample p. That is, for intersecting partial trajectories p_a and p_b (i.e., $p_{a \cap b} \neq \emptyset$),

$$R(p_{a \cup b}) + R(p_{a \cap b}) \leq R(p_a) + R(p_b),$$

where $p_{a \cup b}$ and $p_{a \cap b}$ are the union and intersection of p_a and p_b, respectively. This makes the objective function submodular.

3.2 Experimental Setup

We first present a simulation setup that uses computer-generated scalar fields to compare the performance of J-Horizon, a conventional receding horizon, and the lawnmower planning algorithms. We then present a real-world dataset acquired from Lake Haviland outside of Durango, CO to generate a path maximizing gathered information for a given transport budget.

3.2.1 Simulation

The J-Horizon algorithm is most effective when there is a prior dataset that can be used to generate an information map. The reward function is then specified by the maximum amount of new information that could be gathered at a map location. Furthermore, the algorithm improves upon the conventional receding horizon algorithm by seeking out areas of high reward when the local map area has been exhausted of new information, resulting in its "jumping" behavior.

For our simulated testing, a MatLab script was used to randomly generate 2960 different scalar fields with varying numbers and distributions of high-reward peaks. Between 5 and 50 such peaks were randomly generated on each map with a reward value that decays as a function of distance from the peak center. One such field is visualized in Figs. 4 and 5 as contour maps.

The total reward accumulated along a path generated by J-Horizon for a given fuel budget was averaged for these scalar fields. This performance was compared with that of a lawnmower exploration pattern on the same datasets and fuel budget.

3.2.2 Hardware

A Platypus Lutra ASV was used to take physical samples from Lake Haviland. This ASV is fan-powered, maneuverable and capable of sampling data in lakes or other small bodies of water. This small ASV is an ideal platform upon which to implement our algorithm, based on the limited deployment duration and sensing capabilities for relatively large bodies of water. The ASV is capable of simultaneously sampling temperature, conductivity, pH, and Dissolved Oxygen. Additionally, it measures depth and has a side-scan sonar. The latter sensors allow for bottom mapping of the lake.

The Platypus Lutra ASV has non-holonomic constraints that limits its ability to execute some of the sharper turns produced by the J-Horizon algorithm. Thus, due to hardware limitations, it is necessary to modify the path produced by J-Horizon. These modifications allow the ASV to follow the planned path. Due to the limited locomotion of the Platypus Lutra as well as a need to simplify data collection, some assumptions have to be made:

1. The ASV is limited in its motion and has non-holonomic turning constraints.
2. That sampled scalar fields were not dramatically changing over time.
3. Distance traveled equates to using a linear and constant amount of energy.
4. Additional data sampling points at a given location correlates to better quality data.

4 Algorithm Design

We seek to maximize the reward function for a given transport budget. In reality, this budget is a combination of fuel expenditure, time, and distance, each of which are specific to the vehicle and data collection scheme in use. For simplicity, we assume these factors are linearly related and that acceleration (e.g., due to turning, data collection) has zero cost. In addition, we enhance the conventional receding horizon algorithm by increasing the look-ahead step size if none of the predicted future states satisfy a reward threshold, allowing the planner to "jump" out of low-information areas. This makes J-Horizon especially effective when the input scalar field has high local variability.

The sequence of potential future steps, as well as the final generated path, are stored in a tree wherein each node stores the state of the ASV, which consists of the cumulative reward value of the path, remaining budget, and the location of the ASV. Each look-ahead step recursively generates a number of possible future states. Of these, the best branch is chosen, and the rest are pruned. The sequence of nodes remaining after the remaining budget reaches zero is considered the optimal path. The lawnmower and J-Horizon algorithms share the same functions to calculate the information available at a map location and to depreciate the available information after sampling that location.

The following sections describe the J-Horizon implementation shown in Algorithms 1, 2, and 3 by their respective line numbers.

4.1 Algorithm 1—Main

Path planning begins with the specified transport budget B and loops over the following four steps until either the budget is expended or the planner covers the prescribed area.

6: From the current state σ, take L look-ahead steps with LOOK-AHEAD. This updates the path tree with possible future states L levels below the current node.
7: Find the location of the "best" adjacent node that will achieve the highest reward at the end of L steps through that node.
8: Prune the path tree of all descendants under the current node.
9: Add sample point nodes between current and best locations and update the current node to the latest node.

4.2 Algorithm 2—Look-ahead

Given an initial state σ and maximum recursion depth d, we recursively generate and add possible future states to the path tree. Each step is taken with a new, temporary copy of all data. During each call, it performs the following:

1: Generate set of future states S_f from σ.
3: Remove a fraction $R \sim U([0, 1))$ of the states (but not all) in S_f.
7: Recurse on each descendant node.

4.3 Algorithm 3—J-Horizon

Given a state σ and an information threshold t, probe outwards from the given location and update the map:

2: Start with a sample interval of D.
3: Calculate number of future states to generate b per some factor F.
4: While S_f is empty, perform the following:

    5: Generate b equally spaced points around a circle of radius D around σ.
    6: For each such point, if the quality of the map at that point exceeds t, then add the point to S_f.
    8–9: Increment δ by D and update b.

10: Update the information map.

Algorithm 1: MAIN Main function to run

```
loc ← (0, 0)
reward ← 0
budget ← B
σ ← {loc, reward, budget}
while σ.budget > 0 do
    look-ahead(σ, L)
    l_b ← FindBest(σ)
    Prune(σ)
    σ ← AddSamples(σ.loc, l_b)
return
```

Algorithm 2: LOOK- AHEAD Recursively look several steps ahead

```
Input: State σ, recursion depth d
Output: Number of future states from location
S_f = JHorizon(σ, T)
for i ← 1 to ⌊R · |S_f|⌋ do
    RemoveRandom(S_f)
n ← 0
foreach {σ_f ∈ S_f} do
    if R > 1 then
        n ← n + 1 + look-ahead(σ_f, d − 1)
return n
```

Algorithm 3: JHORIZON Generate frontier locations

```
Input: State σ, threshold t
Output: Set of possible future states to explore
S_f ← ∅
δ ← D
b ← F√δ
while |S_f| = 0 do
    for σ_m ∈ GenerateNew(σ, δ, b) do
        if GetInfo(σ_m) > t then
            S_f ← S_f ∪ {σ_m}
    δ ← δ + D
    b ← F√δ
Depreciate(σ)
return S_f
```

5 Results

In this section, we present the results of application of the J-Horizon algorithm on simulated data, as well as data collected during a field trial.

5.1 Simulation Results

We validate the J-Horizon planner using 2960 simulated scalar fields. We compare the quality of the paths generated by the J-Horizon planner to that of a simple lawnmower pattern using the sum of the data collected over the path given the same transport budget as the metric. For the purpose of comparison, the information gathered was arbitrarily selected to correspond to the deviation from a particular target value of the quantity of interest (Fig. 3). By identifying pertinent or interesting data, the algorithm is able to successfully maximize data collection for a given deployment region. Such quantities are representative of the uncertainty in temperature or another physical parameter that can be approximated by a scalar field.

The simulated vector field seen in Fig. 3 is representative of many types of scalar data. One of the benefits of the J-Horizon algorithm is that it can plan across any type of scalar field, e.g., temperature, humidity, pressure, salinity or dissolved oxygen. The point being that the user may specify the data being looked for and J-Horizon will attempt to maximize the data collection. Figure 3 is an example of the J-Horizon planning over a simulated scalar field.

Figures 4, 5a, and 5b show a typical lawnmower, receding horizon, and J-Horizon path, respectively, planned over a simulated scalar field. The same transport budget was used for all three paths, yet the quality of the paths were 132, 189, and 420, respectively. The J-Horizon planner outperforms lawnmower by a factor of 3.18. Lawnmower required 3.51 times the transport budget to achieve the same reward on

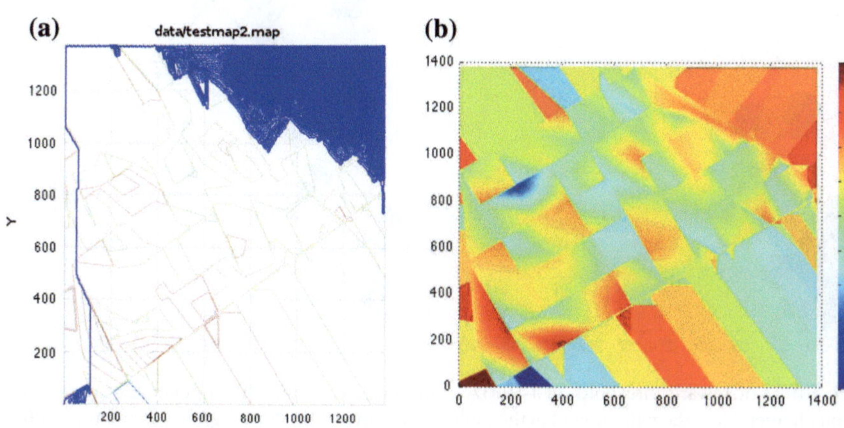

Fig. 3 J-Horizon path planned over simulated scalar field. **a** Dense path planned over reward field. **b** Field reward level map. *Red* indicates high reward

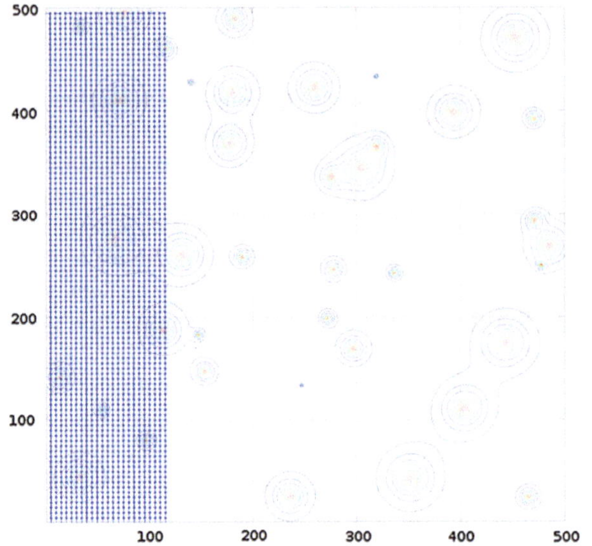

Fig. 4 Lawnmower path on generated scalar field. Reward of 132

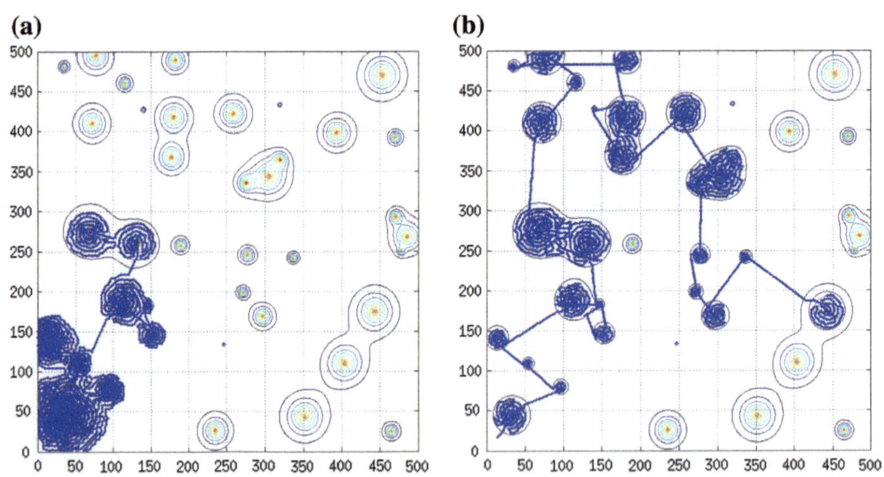

Fig. 5 Effect of reward threshold on J-Horizon jumping behavior. **a** Path generated by receding horizon planner. Note the path lingers in the high-reward area at the *lower left* for a long time before moving on to more worthwhile areas. Reward of 189. **b** Path generated by J-Horizon planner with threshold of 0.1. Once the peak at the *lower-left* has been exhausted of potential reward, the planner quickly moves on to other points of interest. Reward of 420

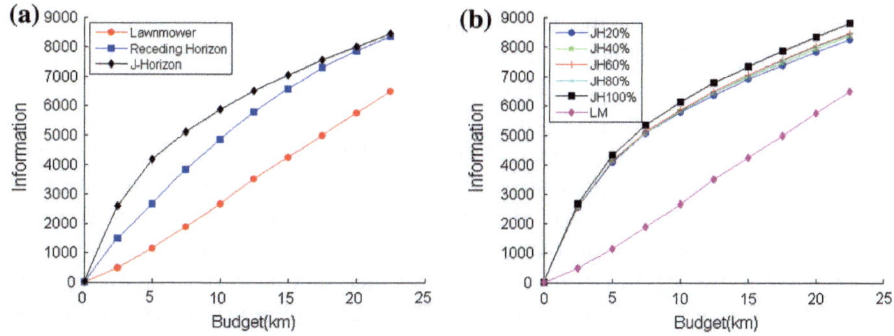

Fig. 6 Performance comparison between lawnmower, receding horizon, and J-Horizon algorithms. **a** Information gathered with increasing budget. **b** The information gathering ability of the J-Horizon planner against a lawnmower pattern as we decrease the fraction of generated future states

the same map and still fails to bring the maximum uncertainty below the threshold of 0.1.

Figure 6 demonstrates the general behavior of the J-Horizon planner, which outperforms the lawnmower planner by a factor of 4 for the first 5 Km of the 25 Km total budget and slows down as it explores the remaining, lower-reward regions of the scalar field.

Figure 6a compares the information gathering ability of the three algorithms with increasing budget. J-Horizon gathers information most rapidly. When the budget is large enough, the information gathered by the receding horizon planner is nearly equivalent to that of the J-Horizon planner.

Figure 6b shows the information gathering ability of the J-Horizon planner against a lawnmower pattern as we decrease the fraction of generated future states. For example, JH80 % indicates the algorithm generates 80 % of the usual number of future states. Even at JH20 %, J-Horizon significantly outperforms a lawnmower path. This suggests it is possible to drastically reduce computational complexity with only a minor performance penalty.

One of the most advantageous qualities of this planner is that it is not limited to any particular search space. It is capable of planning paths over anything that can be estimated by a scalar field.

5.2 *Experimental Results*

Here, we present results from field trials for the implementation of the J-Horizon planner over a scalar field of surface temperature in a small lake in Colorado.[1] Specif-

[1]The specific location of the field trials is Lake Haviland, outside of Durango, CO, located at 37°31′55″N 107°48′27″W.

ically, the goal presented to the ASV is to focus sampling at low-temperature regions, which correspond to high information reward in this case. We use the Platypus Lutra ASV, as shown in Fig. 1b, to conduct an initial survey and then use the J-Horizon planner to compute a new path with the objective to minimize the uncertainty and maximize information gain on the underlying scalar field. The initial path for representative data collection is presented in Fig. 7a, with the scalar field generated from these data and the path planned by the J-Horizon planner shown in Fig. 7b. The ASV executed the first section of the path prescribed by the J-Horizon planner, and results are shown in Fig. 8. At the terminus of this executed section, J-Horizon prescribed a *jump* to a new region for further sampling. This second section was not executed for the proof-of-concept field trial.

As seen in Fig. 7b, the J-Horizon gathers data in areas of low data yield from the initial data collection. For instance, the lower left hand corner of Fig. 7b is an area that was not covered in the initial survey and requires more data collection to accurately represent the underlying field. This is the area of focus for our execution of the J-Horizon planner path, as more than half of entire length of the planned path lies within this region. The portion of the planned path that was executed is shown in Fig. 8 by the red path.

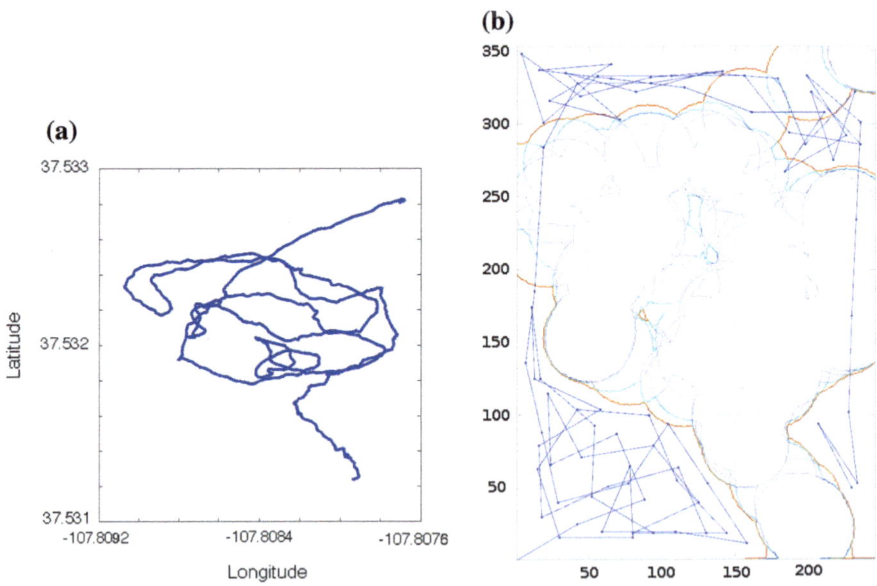

Fig. 7 Paths executed by the ASV to test and demonstrate the J-Horizon planner. **a** The ASV's initial path on Lake Haviland outside Durango, CO. **b** J-Horizon path generated on ASV path shown in (**a**)

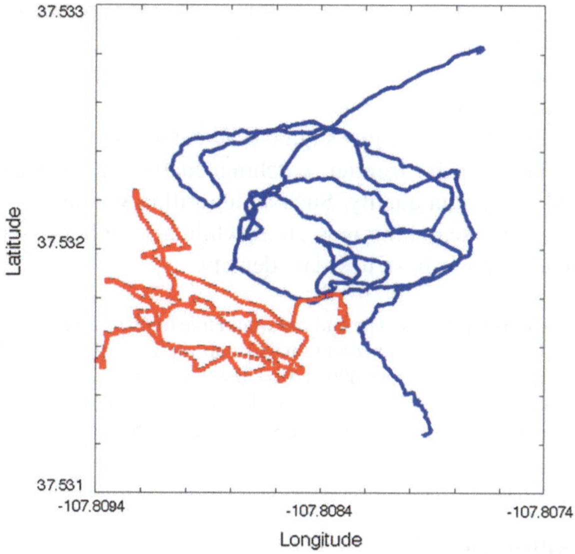

Fig. 8 The initial, data collection path (*upper, blue*) and the J-Horizon path (*lower left, red*) executed by the ASV on Lake Haviland

6 Conclusion

Improved algorithm design for autonomous vehicles operating on water has a promising future in robotics. Collecting higher quality data that can be better utilized by scientists, as well as reducing costs of the data collection, is a key goal in making autonomous monitoring a reality. In this paper, we presented a receding horizon algorithm that attempts to find an optimized path to perform costly, and sometimes difficult or dangerous, data collection in oceans or other large bodies of water. In conclusion, we have presented a novel approach to better collect data over scalar fields. Simulation results show a 14.53 % gain in reward of information collection compared to a lawnmower pattern while in simulation. The J-Horizon planner shows a 23.85 % increase in information gathered in a simple experimental trial compared to a lawnmower pattern. Both of these results show quality gains compared to a lawnmower path of equivalent length. These optimized sampling paths allow scientists to more easily collect pertinent data in the field.

7 Future Work

Extended hardware trials would verify that the simplifying assumptions made in the algorithm design are realistic. Additional performance improvements could be made by running the planning algorithm on the ASV to update the error of the scalar field and re-plan in realtime. This would serve to allow the ASV to run autonomously in highly dynamic environments for longer periods of time without having to transmit data to the shore for processing.

The obvious extension of this work is the application to Autonomous Underwater Vehicles, and sampling in three dimensions. After further testing and validation on 2-D scalar fields, we are planning to investigate problems that exist for both underwater and aerial applications.

Finally, we are investigating an extension to the J-Horizon planner that includes applications for frontier searching, enabling a robotic platform to explore areas with unknown data quality. Such an algorithm will aim to balance explore vs. exploit in missions, searching new areas while also collecting data in areas that are deemed interesting or have low data density.

Acknowledgments The authors would like to thank Jonathan Nash at Oregon State University for his insightful comments and insight into the algorithmic development. A. Stuntz was supported in part by ONR grant N00014-14-1-0490. R.N. Smith was supported in part by NSF Grant DUE-1068341 and a gift from the Fort Lewis Foundation. S. Yoo, Y. Zhang, and G. Hollinger were supported in part by ONR grant N00014-14-1-0509.

References

1. Binney, J., Krause, A., Sukhatme, G.: Informative path planning for an autonomous underwater vehicle. In: IEEE International Conference on Robotics and Automation(ICRA), pp. 4791–4796. Anchorage, Alaska (2010)
2. Curcio, J., Leonard, J., Patrikalakis, A.: Scout—a low cost autonomous surface platform for research in cooperative autonomy. Oceans, pp 725–729 (2005)
3. Frolov, S., Garau, B., Bellingham, J.: Can we do better than the grid survey: optimal synoptic surveys in presence of variable uncertainty and decorrelation scales. J. Geophys. Res. Oceans **119**(8), 5071–5090 (2014). doi:10.1002/2013JC009521
4. Gotovos, A., Casati, N., Hitz, G., Krause, A.: Active learning for level set estimation. In: International Joint Conference on Artificial Intelligence, Beijing, China (2013)
5. Grasmueck, M., Eberli, G.P., Viggiano, D.A., Correa, T., Rathwell, G., Luo, J.: Autonomous underwater vehicle (auv) mapping reveals coral mound distribution, morphology, and oceanography in deep water of the straits of florida. Geophys. Res. Lett. **33**(23) (2006)
6. Hitz, G., Gotovos, A., Pomerleau, F., Garneau, M.E., Pradalier, C., Krause, A., Siegwart, R.: Fully autonomous focused exploration for robotic environmental monitoring. In: IEEE International Conference on Robotics and Automation (ICRA), pp 2658–2664 (2014). doi:10.1109/ICRA.2014.6907240
7. Hollinger, G., Singh, S.: Proofs and experiments in scalable, near-optimal search by multiple robots. In: Robotics: Science and Systems, June 2008
8. Hollinger, G.A., Sukhatme, G.: Sampling-based motion planning for robotic information gathering. In: Robotics: Science and Systems (2013)
9. Mora, A., Ho, C., Saripalli, S.: Analysis of adaptive sampling techniques for underwater vehicles. Auton. Robots **35**(2–3), 111–122 (2013)
10. Schouwenaars, T., How, J., Feron, E.: Receding horizon path planning with implicit safety guarantees. In: Proceedings of the American Control Conference, vol 6, pp 5576–5581. IEEE (2004)
11. Stoker, C., Barch, D., Farmer, J., Flagg, M., Healy, T., Tengdin, T., Thomas, H., Schwer, K., Stakes, D.: Exploration of mono lake with an rov: a prototype experiment for the maps auv program. Autonomous Underwater Vehicle Technology AUV'96, 33–40 (1996)
12. Tisdale, J., Kim, Z., Hedrick, J.K.: Autonomous path planning and estimation using uavs. IEEE Robot. Autom. Mag. **16**(2), 35–42 (2009)

Return to Antikythera: Multi-session SLAM Based AUV Mapping of a First Century B.C. Wreck Site

Stefan B. Williams, Oscar Pizarro and Brendan Foley

Abstract This paper describes an expedition to map a first century B.C. ship wreck off the coast of the Greek island of Antikythera using an Autonomous Underwater Vehicle (AUV) equipped with a high-resolution stereo imaging system. The wreck, first discovered in 1900, has yielded a wealth of important historical artefacts from two previous interventions, including the renowned Antikythera mechanism. The deployments described in this paper aimed to map the current state of the wreck site prior to further excavation. Over the course of 10 days of operation, the AUV completed multiple dives over the main wreck site and other nearby targets of interest. This paper describes the motivation for returning to the wreck and producing a detailed map, gives an overview of the techniques used for multi-session Simultaneous Localisation and Mapping (SLAM) to stitch data from two dives into a single, composite map of the site and presents preliminary results of the mapping exercise.

1 Introduction

In September 2014 an expedition was mounted to revisit the site of a first century BC shipwreck off the coast of the Greek island of Antikythera. The project began in 2013 with multibeam mapping and diver-based search of the site of the Antikythera shipwreck in preparation for further excavation. The objective of this second phase of the project was to produce a high-resolution, 3D map of the site using the Autonomous Underwater Vehicle (AUV) Sirius operated by the University of Sydney's Australian

S.B. Williams (✉) · O. Pizarro
Australian Centre for Field Robotics (ACFR), University of Sydney,
Sydney Nsw 2006, Australia
e-mail: stefanw@acfr.usyd.edu.au

O. Pizarro
e-mail: o.pizarro@acfr.usyd.edu.au

B. Foley
Woods Hole Oceanographic Institution, Woods Hole, MA, USA
e-mail: b.foley@whoi.edu

D.S. Wettergreen and T.D. Barfoot (eds.), *Field and Service Robotics*,
Springer Tracts in Advanced Robotics 113, DOI 10.1007/978-3-319-27702-8_4

Fig. 1 The AUV Sirius conducting surveys over the wreck site on the coast of Antikythera, Greece. This frame, extracted from a video (http://tinyurl.com/q7erkcv) of the vehicle surveying the wreck, shows the footprint of the strobe on the seafloor as it travels *down* the slope *above* the wreck site (credit: Phil Short)

Centre for Field Robotics (ACFR). Figure 1 shows the vehicle at work during one of the deployments over the wreck site.

This paper outlines the AUV based mapping of the site. We describe the technical challenges that were addressed in order to facilitate this work and examine the rationale for preliminary mapping of the site, showing how robotic systems are well suited to the task of collecting data that can facilitate the documentation of the site as a historical record prior to the commencement of excavation. We also present preliminary outcomes of the surveys and examine how the resulting maps were used to facilitate subsequent diving operations.

The remainder of this paper is organised as follows. Section 2 provides background and an overview of the historical significance of the wreck site while Sect. 3 describes the tools used to deliver geo-referenced benthic imagery and associated data products. Section 4 presents results of the surveys conducted over a 10 day period during the 2014 field season and Sect. 5 presents concluding remarks and future directions for this project.

2 The Antikythera Wreck

The Antikythera Shipwreck (circa 60–80 B.C.) is one of the richest ancient wrecks ever discovered [1]. Greek sponge divers located the wreck on a rocky shelf at the base of a cliff in a depth of approximately 55 m of water on the NE coast of the island of Antikythera in 1900. They spent a year salvaging its treasures, with the help of the Hellenic Navy, in the process recovering hundred of works of art including bronze and marble statues that now fill galleries at the National Archaeological Museum in Athens. The wreck also yielded the Antikythera Mechanism, a geared device designed to calculate and display celestial information, including phases of the sun and a luni-solar calendar [2]. This mechanism has fascinated historians for the quality of the workmanship and the sophistication of the mechanism design which had to capture the retro-grade motion of the planets and sun resulting from the fact that the earth was considered to be the centre of the solar system at the time. Numerous projects have sought to reproduce the workings of the mechanism over the years [3, 4] to get a better understanding of this previously unknown mechanical system, which has been described as one of the first known examples of an analog computer.

Undersea explorer Jacques Cousteau and the Calypso crew worked at the Antikythera wreck site for several weeks in 1976, with the approval of the Ministry of Culture. Cousteau and his team recovered numerous artefacts while documenting their excavation as part of a television program following their expedition. As part of their work, the team dredged a section of the wreck to reveal more artefacts for the cameras.

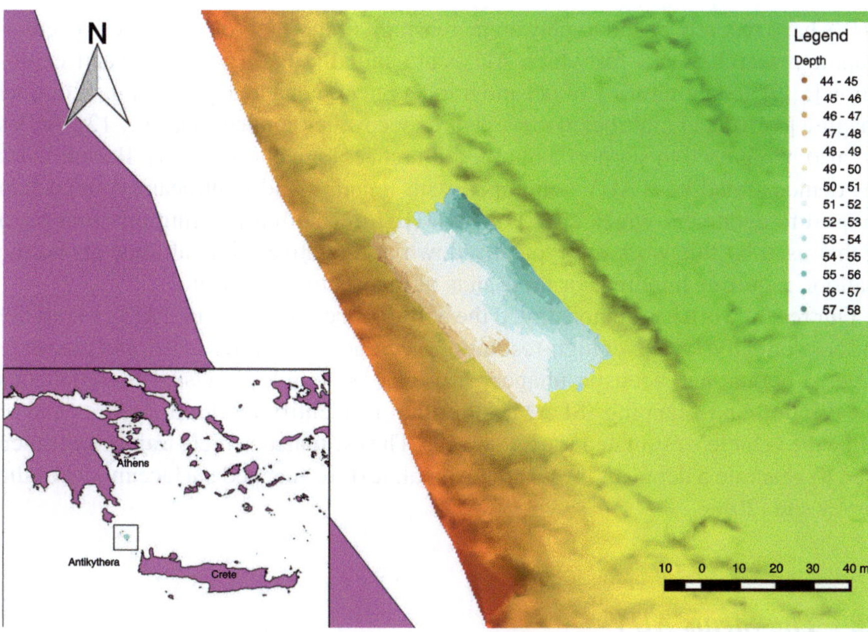

Fig. 2 The location of the AUV based visual survey of the Antikythera wreck. The small island of Antikythera is located between the larger islands of Crete to the SW and Kythera to the NW. The AUV mission covered an area of approximately 70 m × 50 m at depths of 44–58 m on a shelf adjacent to the coastline. The vehicle's estimated trajectory during two dives is shown with the location of each pose coloured by seafloor depth based on the vehicle's depth sensor and altimeter measurements. The underlying bathymetric map of the site was produced using ship-borne multibeam collected during the 2013 field season

In 2013 a team from the Hellenic Ministry of Culture and Sports, the Ephorate of Underwater Antiquities and the Woods Hole Oceanographic Institution (WHOI) returned to Antikythera to survey the island and map the site of the wreck. As part of the expedition, divers conducted visual census of the site, in the process uncovering what was thought to be a second, previously unknown wreck to the south of the main wreck site. Figure 2 shows the location of the island in the Greek archipelago and features a portion of the ship-borne multibeam map and the path of the AUV used in this work to map the site. The decision was made to seek funding to support further fieldwork in order to produce detailed maps of the wreck site and to conduct excavation operations.

3 Wreck Survey Tools and Design

3.1 Autonomous Underwater Vehicles

AUVs have recently begun to play an increasingly important role in modern oceanographic research. Tasks for which AUVs are suited range from deep water exploration [5, 6] and monitoring of oceanographic phenomena to high-resolution optical imaging [7–10] and multibeam surveying in deepwater applications [11, 12]. AUVs are also being used to support a number of archaeological operations. Recent work has demonstrated how AUVs equipped with imaging and multibeam systems can be used to document wreck sites [13–15]. High-resolution imaging missions such as that used by this work are typically flown at a relatively low altitude above the seafloor, requiring hundreds or thousands of images to cover a site.

Our recent work has demonstrated the ability of benthic imaging AUVs to rapidly and cost-effectively deliver high-resolution, accurately geo-referenced, and precisely targeted optical and acoustic imagery [16–18]. We employ a visual Simultaneous Localisation and Mapping (SLAM) algorithm to identify the loop closures and to refine the vehicle's estimated trajectory [19]. The estimated vehicle trajectory is then used to generate a detailed, three dimensional, texture-mapped surface model of the survey site [20].

3.2 The Sirius AUV

The primary requirement of this expedition was to produce a high-resolution, 3D model of the wreck site prior to the diving operations. We operate an ocean going AUV called *Sirius* capable of undertaking the required high-resolution, geo-referenced survey work [18]. This platform is a modified version of a mid-size robotic vehicle called SeaBED built at the Woods Hole Oceanographic Institution [21]. This class of AUV has been designed specifically for relatively low speed, high-resolution imaging and is passively stable in pitch and roll. The submersible is equipped with a full suite of oceanographic sensors, including a high-resolution stereo imaging system (2 x Prosilica GC1380 cameras) with synchronised LED strobes, multibeam sonar, CTD, fluorometers and a comprehensive navigation suite [18].

3.2.1 Realtime Navigation

Our vehicle is equipped with a single band GPS receiver, a Doppler velocity log (DVL), a depth sensor, a magnetic compass with integrated roll and pitch sensors and an Ultra Short Baseline (USBL) Acoustic positioning system deployed by the support vessel. The observations of velocity provided by the DVL are fused with observations of attitude and depth using an Extended Kalman Filter [22]. The USBL observations, consisting of range and bearing measurements between the vessel and the vehicle, are collected on the surface and are sent together with the ship's position and attitude to the vehicle using the USBL's acoustic modem. These observations are received by the vehicle and fused into its onboard navigation filter. The heading reference used is sensitive to the magnetic signature of the rest of the vehicle [23], which can introduce distortions of several degrees into the heading estimate. Even when soft and hard iron calibration are performed, persistent heading-dependent errors of O(1 deg) are possible. While adequate to perform linear transects or broader acoustic surveys (particularly when aided by acoustic positioning from LBL or USBL), the magnitude of these errors makes an intended dense 'mow the lawn' pattern with reciprocal, closely spaced, parallel tracklines difficult for the vehicle to complete. We have recently shown that it is possible to derive a heading-dependent correction to the magnetic compass using visual data and that this correction can enable a compass-equipped AUV to perform dense visual coverage of a seafloor patch of approximately $50\,\text{m} \times 75\,\text{m}$ with 50 parallel tracklines [24]. This has resulted in a navigation suite that is capable of meeting the requirements for full coverage survey with narrow track spacing.

3.3 Simultaneous Localisation and Mapping

In order to generate accurate models of the seafloor, it is important that the estimated vehicle trajectory is self-consistent with respect to the data being collected during each survey. We employ visual Simultaneous Localisation and Mapping (SLAM) to optimally fuse uncertain navigation estimates and visual observations [19, 25]. This allows us to further refine the estimated vehicle trajectory using the environmental data, including high-resolution imagery and multibeam sonar, collected during the survey. Stereo cameras are capable of high-resolution observations such that if the same scene is imaged from different positions, it is possible to determine the relative poses of the cameras using observations of features in the scene. These constraints are fused into the vehicle's navigation solution to further refine the vehicle's estimated trajectory.

Recently, a number of authors have considered the problem of multi-session SLAM, in which data from multiple deployments of one or more robotic platforms must be registered and fused together to produce a final map of the environment. This has included multi-session SLAM work in terrestrial [26, 27], aerial [28] and underwater [29, 30] environments. In cases where there is little change in the environment between deployments, it may be sufficient to simply re-initialise the estimated vehicle location and to match features across deployments to allow this data to be fused. In other cases in which there are more significant changes, such as one might expect from deployments in different seasons or over longer periods of time, more sophisticated methods have been proposed for robust place recognition [31].

In the case considered here, deployments were completed within a period of approximately 10 days and results are shown from two dives completed two days apart. Given the small amount of time between dives, standard feature-based visual recognition was sufficient to identify matching features between subsequent dives using techniques similar to those reported in [17].

3.4 Seafloor 3D Reconstruction and Visualization

Although SLAM recovers consistent estimates of the vehicle trajectory, the estimated vehicle poses themselves do not provide a representation of the environment suitable for human interpretation. A typical dive will yield several thousand geo-referenced overlapping stereo pairs. While useful in themselves, single images make it difficult to appreciate spatial features and patterns at larger scales. We have developed a suite of tools to combine the SLAM trajectory estimates with the stereo image pairs to generate 3D meshes and place them in a common reference frame [20]. These meshes are generated once the vehicle is recovered and take on the order of the same amount of time to compute as the length of the dive allowing dive outcomes to be examined while still at a site. The resulting composite mesh allows a user to quickly and easily interact with the data while choosing the scale and viewpoint suitable for the investigation. In contrast to more conventional photomosaicking approaches [32, 33], the full three dimensional spatial relationships within the data are preserved and users can move from a high level view of the environment down to very detailed investigation of individual images and features of interest within them. This is a useful data exploration tool for the end user to examine the survey area.

3.5 Survey Design

As outlined above, the objective of the missions reported on in this paper were to produce a full coverage, texture-mapped 3D map of the wreck site using the vehicle's high-resolution stereo imaging system. Bathymetric data from the 2013 campaign and markers surveyed in by divers provided information with which to plan the dives over the wreck site. The AUV, which is capable of hovering and turning on the spot using a pair of lateral thrusters at the rear of the vehicle, was programmed with a mission consisting of four legs across the site spaced approximately 12.5 m apart followed by a dense grid survey consisting of 51 parallel tracklines, each 70 m long, spaced by 1.0 m covering the site. The initial track lines serve as candidate across-track loop closure points while the trackline spacing of the dense grid is selected to yield sufficient overlap between adjacent legs to ensure along-track loop closures are also found.

4 Results

During the Antikythera dives presented here, the vehicle completed 2 full coverage dives over the site from which data was used to generate the final site maps. The estimated pose of each stereo pair is plotted in Fig. 2, with the symbols coloured by estimated seafloor depth based on combining the vehicle's depth sensor and altimeter measurements. The underlying multibeam map shows the complex structure of the site and the proximity of the dives to the coastline.

For this particular survey, we employed two dives completed over the main wreck site. The site is at the base of a steep cliff in approximately 50 m of water, extending out across a 60 m wide shelf which then drops down to 75 m of water depth. There are a number of large boulders in the middle of the site and the north west side of the survey area comprised a dense boulder field at the base of a cliff, presenting a challenging environment in which to conduct near-bottom survey operations. During the first deployment the vehicle was programmed to maintain an altitude of 2 m above the seafloor while travelling at a speed of 0.5 m/s and capturing stereo images at 1 Hz. With a field of view of approximately 42 × 34 degrees, this yields an image footprint of approximately 1.5 m × 1.2 m and ensures an overlap of approximately 2/3 between frames along track and 1/3 across track.

The rough terrain and large obstacles caused the vehicle's altitude controller to struggle to maintain a constant height above the seafloor throughout the dive, despite it slowing down as forward obstacles were approached. This resulted in gaps in some portions of the vehicle's trajectory as a lower altitude results in a narrower image footprint on the ground. The tuning of the altitude controller was adjusted and subsequent missions were flown at a higher altitude of 3 m to ensure full coverage of the survey site, thereby increasing the footprint of the images and facilitating obstacle avoidance over the rough terrain and large boulders in the survey site. The image framerate was also increased to 1.5 Hz to increase the along track overlap between images. This increased altitude and imaging rate increased the overlap both along and across track, resulting in significantly more loop closures as shown below.

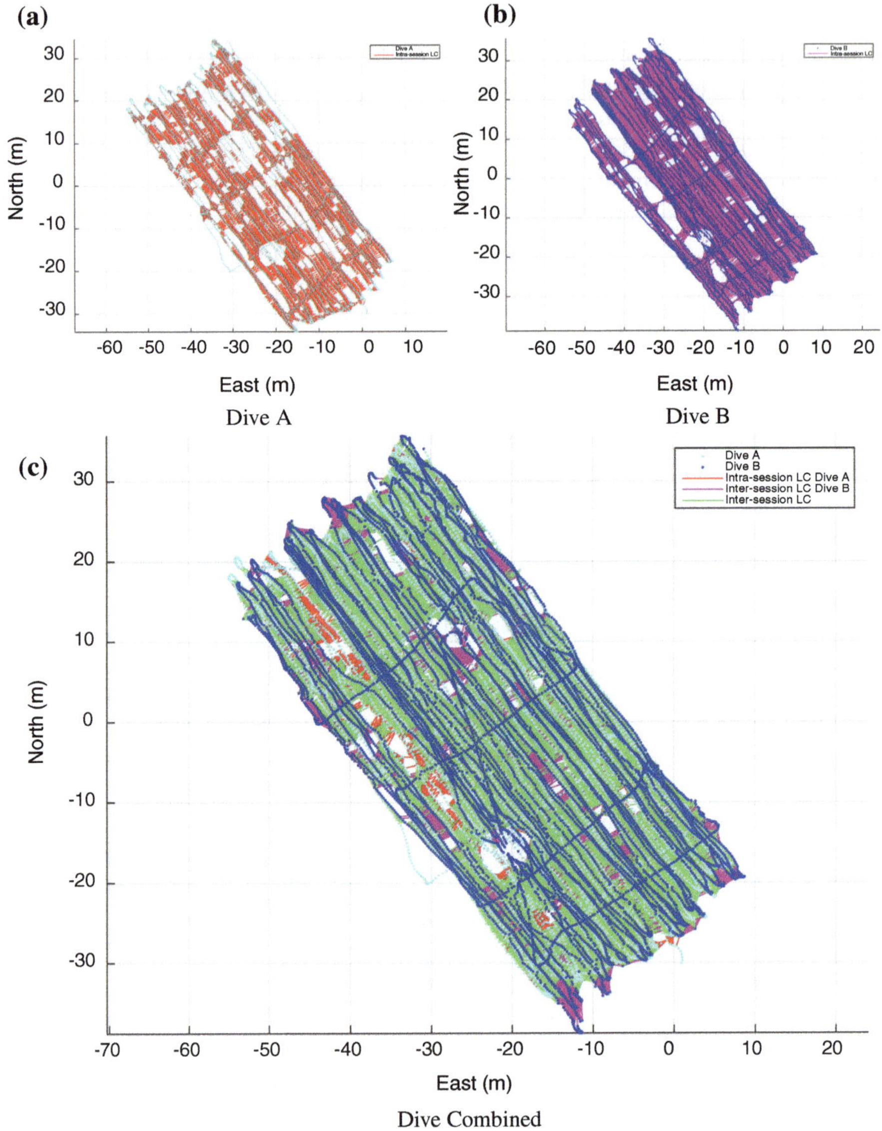

Fig. 3 Multi-session SLAM result. **a** The Dive A solution, showing estimated camera locations and intra-session loop closures **b** Dive B solution, showing estimated camera locations and intra-session loop closures **c** Multi-session SLAM solution. Loop closure links are shown in *red* and *magenta* for intra-session loop-closures and in *green* for inter-session links. Over 340,000 loop closures were identified in total to produce this model

4.1 Multi-session SLAM

In order to produce a complete map of the area, the two surveys were combined using our multi-session SLAM tool. Each dive is initialised independently using the GPS and USBL data available to the vehicle. This is sufficient to georeference the mission data to within 2–3 m between dives. However, finescale registration requires matching features between dives to co-register the dives. This step is performed automatically by matching SIFT features in a manner identical to that used for identifying loop closures from within a single dive. Figure 3 shows the result of the use of multi-session SLAM to fuse data from two dives completed over the wreck site. The figure shows the estimated vehicle trajectories for the two dives, as well as the combined estimates of the two dives. We use the terminology adopted from [27], designating loop closures from a single dive as 'intra-session' loop closures and loop closures between dives as 'inter-session' loop closures.

Table 1 presents statistics of the two dives, including the dive times, number of individual stereo pairs and loop closures identified within and between dives. As can be seen, both dives took just under two hours to complete. The second dive, completed at a higher altitude and with a higher framerate, resulted in significantly more loop closures and, as can be seen, there are a large number of inter-session loop closures that serve to co-register the dives.

4.2 Three Dimensional Surface Model

Sample reconstructions produced using data collected during the AUV surveys conducted on the Antikythera site are shown in Fig. 4. While it is possible to examine the individual images that were used to generate these 3D surface models, the spatial structure of the site is more evident in the composite mesh. Figure 5 shows examples of details from the 3D surface model, highlighting historical artefacts of interest that were visible in the model.

Table 1 Multi-session Dive statistics

Mission time Dive A	1:49
Mission time Dive B	1:57
Dive A stereo pairs	6,565
Dive B stereo pairs	10,554
Total stereo pairs	17,119
Dive A intra-session loop closures	32,098
Dive B intra-session loop closures	162,778
Inter-session loop closures	149,386
Total loop closures	344,262

Fig. 4 **a** The final texture mapped model of the site is generated by blending the imagery collected by the vehicle to produce a seamless texturemap which is draped over the 3D surface model. **b** The underlying 3D structure of the site reveals the base of the cliff to the SW and several large boulders around which artefacts, including the ship's anchor, amphorae, pottery sherds and a 2 m long bronze spear, were located

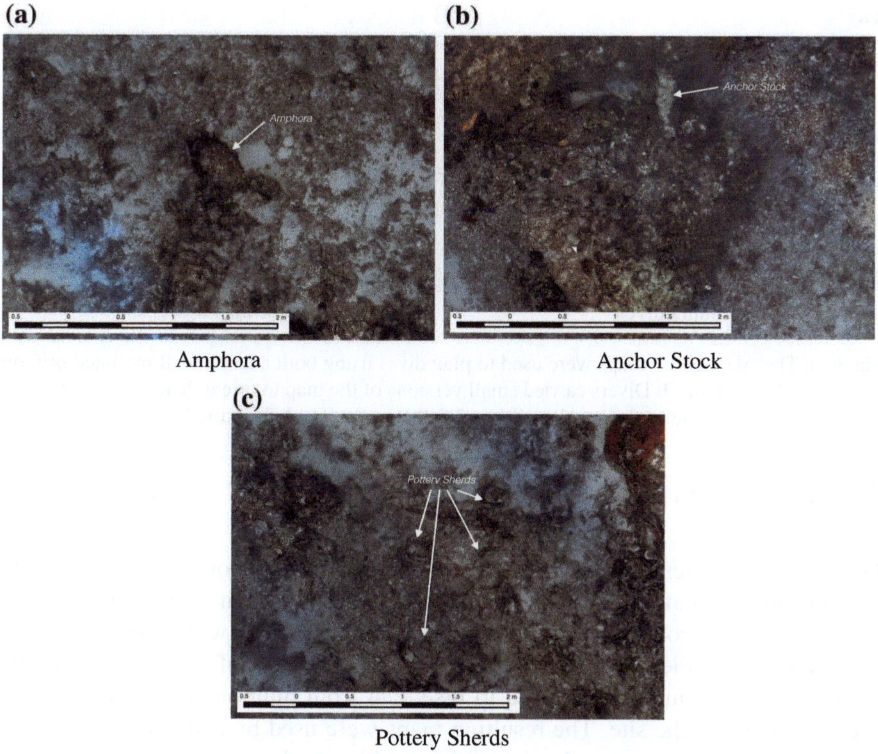

Fig. 5 Examples of the detail of the 3D texture mapped surface model including **a** an Amphora, **b** one of the ship's lead anchor stocks and **c** pottery sherds, possibly left after the Cousteau excavation in 1976

4.3 Diver Aiding

Many remaining artefacts from the wreck are thought to be buried under sediment. As part of the 2014 expedition, divers conducted surveys of the site using underwater metal detectors to identify buried metal objects. Figure 6 shows the project team (a) using a 3 m × 2 m printout of the AUV derived maps to plan dives and (b) on-site with the metal detectors and recovered artefacts. They used laminated copies of the AUV generated maps to help with in situ identification of the location of potential excavation sites. During the 2014 field season, the vessel's anchor stock, a number of small artefacts and a 2 m bronze spear were recovered.

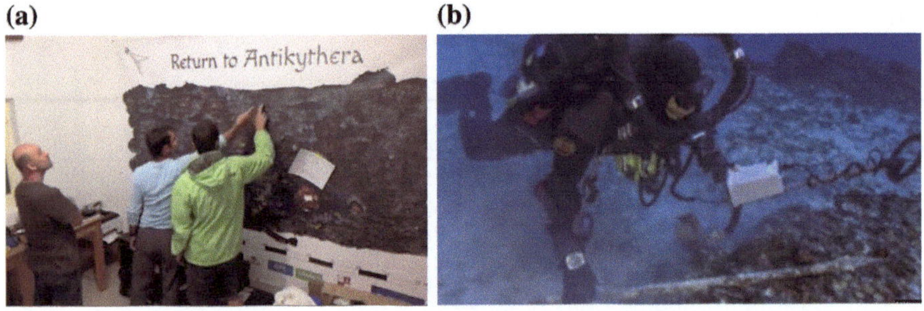

Fig. 6 **a** The AUV based maps were used to plan dives using both a large, wall-mounted printout as well as GIS systems. **b** Divers carried small versions of the map to orient themselves on the site. They conducted a visual census and used metal detectors to search for buried artefacts

5 Conclusions and Future Work

This paper has described an expedition to document the site of a first century B.C. wreck on the coast of the island of Antikythera, Greece. We conducted multiple dives using an AUV to collect tens of thousands of stereo images with which to build a detailed model of the wreck site prior to the commencement of excavation. A multi-session SLAM technique was used to fuse data from multiple dives into a single, detailed model of the site. The resulting maps were used by divers to help with in situ survey of the site and to document the resulting finds.

The ability to quickly and automatically generate detailed, texture-mapped 3D models of the site were instrumental in assessing the quality of the maps while in the field and in facilitating subsequent diving operations. Combining data from multiple dives allowed us to generate a full coverage site map. We were also able to update the vehicle's obstacle avoidance behaviour and mission parameters, including standoff altitude and imaging rate, to ensure full coverage and to avoid some of the more challenging areas of the terrain.

While this first year of AUV surveys was a success, with the production of a detailed site map and exploration of a number of other areas of interest, adverse weather limited the number of days for the archaeological dive team. However, the surveying they were able to complete revealed prospective targets both within the extent of the area surveyed by the AUV as well as immediately to the south. A potential second wreck site was also confirmed a few hundred metres to the south. Future expeditions will seek to map areas to the south of the mapped area and to conduct a more systematic metal detection survey to help identify prospective excavation targets. More extensive excavation operations are also planned for the 2015 field season.

Acknowledgments This work is supported by the Hellenic Ministry of Culture and Sports, the ARGO Civil Non-Profit Company for Scientific Research and Training, the Aikaterini Laskaridis Foundation, Hublot S.A., Swordspoint Foundation, Jane and James Orr, Jr., OTE-Cosmote, Domestic Property Committee of Kythera and Antikythera, Costa Navarino, Family of Michael Dubno and Loren Blackford, Laetitia and Richard Garriott de Cayeux. Dr. Pizarro and Prof. Williams are supported by the Australian Research Council. Thanks to EdgeTech for providing the gridded multibeam maps used to plan the AUV dives. The authors would like to thank the captain and crew of MV Glaros used to facilitate the deployment and recovery of the vehicle and the RV Poseidon who were instrumental in tracking the AUV during dives. Thanks also to Ephorate of Underwater Antiquities Director Dr. Aggeliki Simosi and EUA field director Dr. Theotokis Theodoulou, Alexandros Sotiriou for help with logistics and Phil Short of Phil Short Technical for providing video of the vehicle while conducting the survey. We also acknowledge the help of all those who have contributed to the development and operation of the AUV Sirius, including Christian Lees, Andrew Durrant, Ritesh Lal and Jeremy Randle.

References

1. de Solla Price, D.: Gears from the greeks. the antikythera mechanism: a calendar computer from ca. 80 b. c. Trans. Am. Philos. Soc. **64**(7), 1–70 (1974). http://www.jstor.org/stable/1006146
2. Freeth, T., Bitsakis, Y., Moussas, X., Seiradakis, J., Tselikas, A., Mangou, H., Zafeiropoulou, M., Hadland, R., Bate, D., Ramsey, A., Allen, M., Crawley, A., Hockley, P., Malzbender, T., Gelb, D., Ambrisco, W., Edmunds, M.: Decoding the ancient greek astronomical calculator known as the antikythera mechanism. Nature **444**, 587–591 (2006)
3. Edmunds, M.G., Morgan, P.: The antikythera mechanism: still a mystery of greek astronomy?. Astron. Geophys. **41**(6), 6.10–6.17 (2000). http://astrogeo.oxfordjournals.org/content/41/6/6.10.short
4. Wright, M.: The antikythera mechanism and the early history of the moon-phase display. Antiq. Horol. **29**(3), 319 (2006)
5. Grasmueck, M., Eberli, G.P., Viggiano, D.A., Correa, T., Rathwell, G., Luo, J.: Autonomous underwater vehicle (AUV) mapping reveals coral mound distribution, morphology, and oceanography in deep water of the Straits of Florida. Geophys. Res. Lett. **33**, 6 (2006)
6. Marthiniussen, R., Vestgard, K., Klepaker, R., Storkersen, N.: HUGIN-AUV concept and operational experiences to date. In: OCEANS '04. MTTS/IEEE TECHNO-OCEAN '04, vol. 2, pp. 846–850 (2004)
7. Singh, H., Armstrong, R., Gilbes, F., Eustice, R., Roman, C., Pizarro, O., Torres, J.: Imaging coral I: imaging coral habitats with the SeaBED AUV. Subsurf. Sens. Technol. Appl. **5**, 25–42 (2004)
8. Yoerger, D., Jakuba, M., Bradley, A., Bingham, B.: Techniques for deep sea near bottom survey using an autonomous underwater vehicle. Int. J. Robot. Res. **26**, 41–54 (2007)
9. Kunz, C., Singh, H.: Map building fusing acoustic and visual information using autonomous underwater vehicles. J. Field Robot. **30**, 763783 (2013)
10. Clarke, M.E., Tolimieri, N., Singh, H.: Using the SeaBED AUV to assess populations of groundfish in untrawlable areas. The Future Fisheries Science in North America. Fish and Fisheries Series, vol. 1, pp. 357–372. Springer, Netherlands (2009)
11. McEwen, R., Caress, D., Thomas, H., Henthorn, R., Kirkwood, W.: Performance of an autonomous underwater vehicle while mapping smooth ridge in monterey bay. In: ASLO/TOS/AGU Ocean Sciences Meeting (2006)
12. Henthorn, R., Caress, D., Thomas, H., McEwen, R., Kirkwood, W., Paull, C., Keate, R.: High-resolution multibeam and subbottom surveys of submarine canyons and gas seeps using the MBARI mapping AUV. In: Proceedings of the MTS/IEEE Oceans, pp. 1–6 (2006)

13. Foley, B. Mindell, D.: Precision survey and archaeological methodology in deep water. ENALIA The Journal of the Hellenic Institute of Marine Archaeology, 49–56 (2002)
14. Foley, B., DellaPorta, K., Sakellariou, D., Bingham, B., Camilli, R., Eustice, R., Evagelistis, D., Ferrini, V., Katsaros, M., Kourkoumelis, D., Mallios, A., Micha, P., Mindell, D., Roman, C., Singh, H., Switzer, D., Theodoulou, T.: The 2005 chios ancient shipwreck survey: new methods for underwater archaeology. Hesperia **78**(2), 269305 (2009)
15. Bingham, B., Foley, B., Singh, H., Camilli, R., Delaporta, K., Eustice, R., Mallios, A., Mindell, D., Roman, C., Sakellariou, D.: Robotic tools for deep water archaeology: surveying an ancient shipwreck with an autonomous underwater vehicle. J. Field Robot. **27**(6), 702–717 (2010)
16. Williams, S.B., Pizarro, O., Jakuba, M., Barrett, N.: AUV benthic habitat mapping in south eastern Tasmania. In: Howard, A., Iagnemma, K., Kelly, A. (eds.) Proceedings of the 7th International Conference on Field and Service Robotics, Springer Tracts in Advanced Robotics, vol. 62, pp. 275–284. Springer, Berlin (2010)
17. Williams, S.B., Pizarro, O., Jakuba, M., Mahon, I., Ling, S., Johnson, C.: Repeated AUV surveying of urchin barrens in North Eastern Tasmania. In: Proceedings IEEE International Conference on Robotics and Automation, vol. 1, pp. 293–299, 3–8 May 2010
18. Williams, S., Pizarro, O., Jakuba, M., Johnson, C., Barrett, N., Babcock, R., Kendrick, G., Steinberg, P., Heyward, A., Doherty, P., Mahon, I., Johnson-Roberson, M., Steinberg, D., Friedman, A.: Monitoring of benthic reference sites: using an autonomous underwater vehicle. IEEE Robot. Autom. Mag. **19**(1), 73–84 (2012)
19. Mahon, I., Williams, S.B., Pizarro, O., Johnson-Roberson, M.: Efficient view-based SLAM using visual loop closures. IEEE Trans. Robot. **24**, 1002–1014 (2008)
20. Johnson-Roberson, M., Pizarro, O., Williams, S.B., Mahon, I.: Generation and visualization of large-scale three-dimensional reconstructions from underwater robotic surveys. J. Field Robot. **27**(1), 21–51 (2010)
21. Singh, H., Can, A., Eustice, R., Lerner, S., McPhee, N., Pizarro, O., Roman, C.: Seabed auv offers new platform for high-resolution imaging. EOS, Trans. AGU, **85**(31), 289–295 (2004)
22. Williams, S.B., Pizarro, O., Webster, J., Beaman, R., Mahon, I., Johnson-Roberson, M., Bridge, T.: AUV-assisted surveying of drowned reefs on the shelf edge of the Great Barrier Reef, Australia. J. Field Robot. **27**(5), 675–697 (2010)
23. Caruso, M.J.: Applications of magnetic sensors for low cost compass systems. In: Position Location and Navigation Symposium, IEEE 2000, 177–184 (2000)
24. Jakuba, M.V., Pizarro, O., Williams, S.B.: High resolution, consistent navigation and 3D optical reconstructions from AUVs using magnetic compasses and pressure-based depth sensors. In: Proceedings of IEEE Oceans (2010)
25. Eustice, R., Singh, H., Leonard, J., Walter, M.: Visually mapping the RMS Titanic: Conservative covariance estimates for SLAM information filters. Int. J. Robot. Res. **25**(12), 1223–1242 (2006)
26. Latif, Y., Cadena, C., Neira, J.: Robust loop closing over time for pose graph slam. Int. J. Robot. Res. **32**(14), 1611–1626 (2013). http://ijr.sagepub.com/content/32/14/1611.abstract
27. McDonald, J., Kaess, M., Cadena, C., Neira, J., Leonard, J.: Real-time 6-dof multi-session visual SLAM over large-scale environments. Robot. Auton. Syst. **61**(10), 1144–1158 (2013). selected Papers from the 5th European Conference on Mobile Robots (ECMR 2011)
28. Forster, C., Lynen, S., Kneip, L., Scaramuzza, D.: Collaborative monocular slam with multiple micro aerial vehicles. In: IEEE/RSJ International Conference on Intelligent Robots and Systems (IROS), pp. 3962–3970, Nov 2013
29. Fallon, M., Johannsson, H., Kaess, M., Folkesson, J., McClelland, H., Englot, B., Hover, F., Leonard, J.J.: Simultaneous localization and mapping in marine environments. In: Seto, M.L. (ed.) Marine Robot Autonomy, pp. 329–372. Springer, New York (2013)
30. Ozog, P., Eustice, R.: Real-time slam with piecewise-planar surface models and sparse 3d point clouds. In: IEEE/RSJ International Conference on Intelligent Robots and Systems (IROS), pp. 1042–1049, Nov 2013
31. Pandey, G., McBride, J., Savarese, S., Eustice, R.: Toward mutual information based place recognition. In: IEEE International Conference on Robotics and Automation (ICRA), pp. 3185–3192, May 2014

32. Singh, H., Howland, J., Pizarro, O.: Advances in large-area photomosaicking underwater. IEEE J. Ocean. Eng. **29**, 872–886 (2004)
33. Ferrer, J., Elibol, A., Delaunoy, O., Gracias, N., Garcia, R.: Large-area photo-mosaics using global alignment and navigation data. In Oceans'07, pp. 1–9 (2007)

An Overview of MIT-Olin's Approach in the AUVSI RobotX Competition

Arthur Anderson, Erin Fischell, Thom Howe, Tom Miller, Arturo Parrales-Salinas, Nick Rypkema, David Barrett, Michael Benjamin, Alex Brennen, Michael DeFillipo, John J. Leonard, Liam Paull, Henrik Schmidt, Nick Wang and Alon Yaari

Abstract The inaugural RobotX competition was held in Singapore in Oct. 2014. The purpose of the competition was to challenge teams to develop new strategies for tackling unique and important problems in marine robotics. The joint team from Massachusetts Institute of Technology (MIT) and Olin College was chosen as one

A. Anderson · E. Fischell · T. Howe · T. Miller · N. Rypkema · M. Benjamin · A. Brennen · J.J. Leonard · L. Paull (✉) · H. Schmidt · N. Wang · A. Yaari
MIT, 77 Massachusetts Avenue, Cambridge, MA 02139, USA
e-mail: lpaull@mit.edu

A. Anderson
e-mail: arthura@mit.edu

E. Fischell
e-mail: emf43@mit.edumail

T. Howe
e-mail: thomhowe@mit.edu

T. Miller
e-mail: mille219@mit.edu

N. Rypkema
e-mail: rypkema@mit.edu

M. Benjamin
e-mail: mikerb@mit.edu

A. Brennen
e-mail: vab@mit.edu

J.J. Leonard
e-mail: jleonard@mit.edu

A. Parrales-Salinas
Tufts University, 419 Boston Ave, Medford, MA 02155, USA
e-mail: Arturo.Parrales_Salinas@tufts.edu

D. Barrett
Olin College, Olin Way, Needham, MA 02492, USA
e-mail: David.Barrett@olin.edu

M. DeFillipo
MIT Sea Grant, 292 Main Street, Cambridge, MA 02142, USA
e-mail: mikedef@mit.edu

D.S. Wettergreen and T.D. Barfoot (eds.), *Field and Service Robotics*,
Springer Tracts in Advanced Robotics 113, DOI 10.1007/978-3-319-27702-8_5

of 15 competing teams from five nations (USA, South Korea, Japan, Singapore and Australia). The team received the surface vehicle platform, the WAM-V (Fig. 1) in Nov. 2013 and spent a year building the propulsion, electronic, sensing, and algorithmic capabilities required to complete the five tasks that included navigation, underwater pinger localization, docking, light sequence detection, and obstacle avoidance. Ultimately the MIT/Olin team narrowly won first place in a competitive field. This paper summarizes our approach to the tasks, as well as some lessons learned in the process. As a result of the competition, we have developed a new suite of open-source tools for feature detection and tracking, realtime shape detection from imagery, bearing-only target localization, and obstacle avoidance.

1 Introduction

The inaugural RobotX competition, hosted by the association for unmanned vehicle systems international (AUVSI) Foundation, was held in Singapore in October 2014. The motivation for the competition was to increase the capabilities of marine vehicle systems to perform commercial tasks and operate in the vast and challenging ocean environment. Much larger in scope than previous competitions, such as RoboBoat and RoboSub, this was the largest autonomous surface vehicle (ASV) competition ever held. In total 15 teams competed from five different countries (USA, South Korea, Singapore, Australia, and Japan). Each of the 15 teams were provided with an identical platform, shown in Fig. 2, and were responsible for equipping it with sensors, propulsion, electrical systems, and onboard autonomy to achieve the tasks.[1]

The competition consisted of five tasks:

- Task 1: Navigate through two sets of colored buoy gates;
- Task 2: Report the location of an underwater pinger and also the color of the closest buoy to the pinger;
- Task 3: Identify the correct docking location based on a placard on the seawall and then subsequently dock;
- Task 4: Find a buoy that is emitting an LED light pattern and then report the light pattern;
- Task 5: Enter an obstacle field through a buoy gate (specified by color) and then navigate through a densely cluttered field of obstacles, and finally exit through the specified gate.

Each task had a unique scoring system and the sum of all task points was used to rank teams. After three qualification days, the top six ranked teams advanced to the finals. In the finals, the points accumulated in the last attempted run were used.

In this paper we summarize our approach to each of the five tasks. These required basic capabilities such as object detection and autonomy, as well as task-specific

[1] See http://robotx.mit.edu for more details and updates.

Fig. 1 The WAM-V [7] on the water in Singapore at the RobotX competition

Fig. 2 The "WAM-V" ASV platform used in the competition. Sensor payload includes: GPS (*green*), 3D laser scanner (*yellow*), camera (*pink*), and sonar transducer and mount (*blue*)

capabilities such as pattern recognition and acoustic target localization. The remainder of the paper is structured as follows: In Sect. 2, we detail our laser/vision based approach to object detection and tracking for navigation tasks (required for Task 1, 2, 4, and 5). In Sect. 3, we discuss our approach to specific vision-based pattern identification tasks (Task 3 and 4). In Sect. 4, we present the particle-filter based acoustic localization system (Task 2). In Sect. 5, we present an overview of our approach to autonomy and control based on behavior-based multi-objective optimization. In Sect. 6, we provide some details about the choice of hardware used. Finally we provide some of the competition results in Sect. 7 and some conclusions in Sect. 8.

2 Object Detection, Tracking, and Classification

A prerequisite for Tasks 1, 2, 4, and 5 is to be able to detect and track objects floating on the water surface. Above-water perception onboard the vehicle was achieved through a combination of 3D laser and vision. Laser-based sensing was particularly effective in this case since the water surface only produced weak returns that could be easily removed through laser intensity filtering, leaving only solid objects such as buoys.

Algorithm 1 Object Detection and Tracking

Input: Laser scan
Output: Feature List
1: Cull points outside of desired sector
2: Downsample to voxel grid
3: Euclidean clustering
4: Update the persistent cluster list
5: Transform persistent features to world frame
6: Try to assign feature color through image sub-windowing
7: Associate features and update tracked feature list

Since laser provides limited color information, object detections were fed to a vision system for classification. An overview of the approach is summarized in Algorithm 1.

2.1 Laser-Based Feature Detection

Each point in the laser scan, $p = \{r, \phi, \theta, I\}$, is a tuple consisting of a range r, an azimuth ϕ, an elevation θ, and an intensity I. One scan of laser data consists of a collection of N points $P = \{p_i\}_{i=1...N}$. The points are first culled using thresholds for minimum and maximum range and azimuth, as well as minimum intensity:

$$P_c = \{p_i | r_{min} < r_i < r_{max}, \phi_{min} < \phi_i < \phi_{max}, I_{min} < I_i\}. \qquad (1)$$

These points are then downsampled using a voxel grid and ordered into clusters, $\mathscr{C} = \{C^j\}_{j=1...J}, C^j = \{P^j, \mu^j\}$, where P^j and μ_j are the set of points and centroid of cluster j respectively. We wish to be able to detect buoys at the maximum possible range, at which point there may be only one or two returns from a buoy. In order to mitigate the impact of false returns while still being able to track small features at long distances, we use a temporal persistence filter. A persistent cluster list, $\mathscr{C}^{pcl} = \{\mathscr{C}, K\}$, is maintained, where K is the "lifetime" of the cluster. As a new set of clusters arrives at time k, they are fused with the persistent cluster list. For each new cluster, if its centroid, μ, is within ϵ of one of the centroids of the clusters in $\mathscr{C}^{pcl}$, then the associated cluster's lifetime is incremented, otherwise the cluster is added to the persistent cluster list with a lifetime of $K = 1$. A laser scan and associated camera image are shown in Fig. 3. This particular snapshot is from the obstacle avoidance task. In this case there are four persistent features.

The set of persistent clusters with a lifetime larger than K_{min} are deemed to be active objects in the world and are transformed to world coordinates and added as

Fig. 3 Object detection from point cloud. *Top* Point cloud from 3D laser with buoys identified. *Bottom* Corresponding image from camera used for buoy color detection

features, f:

$$f^j = T_l^w \mu^j \tag{2}$$

where the transformation T_l^w transforms the centroid of the cluster in the laser frame to a point in the global frame. This feature is rejected if it is outside of course boundaries.

In order to compute T_l^w we directly used the output from our GPS sensor, which provided a stable pose estimate in practice. Nevertheless, a more reliable approach would be to implement a full SLAM system, or use some other form of marine vehicle navigation [11].

2.2 *Buoy Detection*

Each reception of an object detection from the laser triggers an attempt to classify the color of the object. The feature location is back-projected into the camera frame to try and identify color [5]:

$$\left[l_u^j \; l_v^j \; 1\right]^T = \mathbf{K}\left[\frac{l_x^j}{l_z^j} \; \frac{l_y^j}{z_l^j} \; 1\right]^T, \tag{3}$$

where (l_u^j, l_v^j) and (l_x^j, l_y^j, l_z^j) are the locations of the feature in pixel and world coordinates respectively, and $\mathbf{K}$ is the camera calibration matrix. A sub-windowed image around around the landmark pixel location is created, which is then subjected to a series of thresholding operations in the hue-saturation-value (HSV) color space.

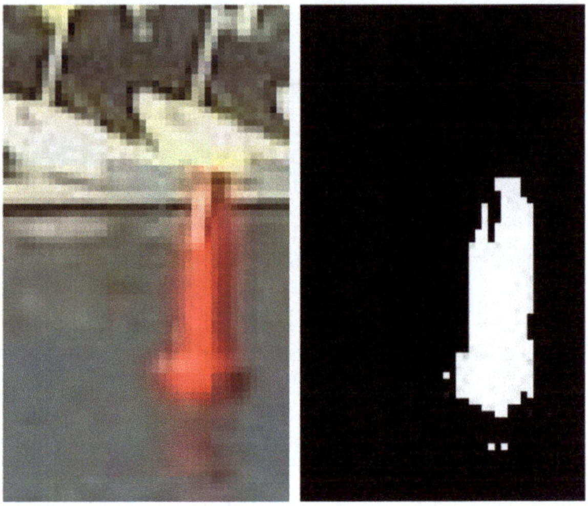

Fig. 4 *Left* Sub-windowed image from camera. *Right* Output from "red" color segmentation filter

Using the HSV colorspace is beneficial for color detection in images because it is less sensitive to lighting conditions as the majority of the color information should be contained within the hue channel and the aggressive sub-windowing was found to be critical to avoid false detections.

Figure 4 shows a sub-windowed image from one of the test trials as well as the output of the red filter showing correct color identification.

2.3 Feature Association

We use a simple nearest neighbor [1] approach to associate features. If an incoming feature is at location l^j then the feature is associated to feature i if two conditions are met:

$$\begin{aligned} l^i &= \underset{l \in L}{\arg\min}\, ||l - l^j|| \\ d_{min} &< ||l - l^j|| \end{aligned} \tag{4}$$

We refer to these associated features as "tracked features" $L^m = \{l^{1\ldots|m|}, c^{1\ldots|m|}\}$ where $|m|$ is the number of times that tracked feature L^m has been detected and the c values corresponds to the color decisions made for each detection. The final set of M distinct tracked features $\mathscr{L} = \{L^m\}_{m=1\ldots M}$ are used by the control and autonomy system (Sect. 5) to complete the specific tasks.

3 Pattern Identification

Pattern recognition was a required capability for two tasks. The first was Task 3 where a spatial pattern, either a cross, circle, or triangle, was used to identify the correct bay for docking. The second was Task 4 which required the identification of a temporal pattern. In both cases, aggressive sub-windowing in the image was performed to guide the visual search and decrease false positives while maintaining low computation.

3.1 *Placard Detection for Docking*

The key objectives for our placard detector were:

1. Robustness to degradation caused by motion, scale and perspective transformation from different viewing positions, warp and occlusion caused by wind, and variants of color from light condition, and
2. Speed and accuracy to support real-time decision-making.

We tackle this problem by using two-step pipeline. First, a *detection* phase identifies candidate regions and we subsequently process each region in a *decoding* stage to see if it matches any of the three placards.

Detection

To minimize unnecessary computation and to avoid looking for placards in nearly empty image regions, in the first stage we extract candidate regions using Extremal Regions (ERs) [9]. An ER is a region R whose outer boundary pixels ∂R have strictly higher values in the intensity channel $\mathbf{C}$ than the region R itself, i.e., $\forall p \in R, q \in \partial R : \mathbf{C}(p) < \theta < \mathbf{C}(q)$, where θ is the threshold of the ER. Let a grayscale input frame $\mathbf{I}$ be a mapping $\mathbf{I} : D \subset \mathbb{R} \rightarrow \{0, \ldots, 255\}$. $\mathcal{R}_b$ and $\mathcal{R}_w$ donate the sets of detected ERs from $\mathbf{I}$ and inverted $\mathbf{I}$, respectively. We extracted features $\mathbf{F}$ for each region in $\mathcal{R}_b$ and $\mathcal{R}_w$, and then filter according to size, aspect ratio, and number of holes. We observed that a placard is designed as a black symbol on a white board. The set of candidate regions $\mathcal{R}_c \subset \mathcal{R}_b$ is formed when a region r_b in $\mathcal{R}_b$ satisfies $r_b \cap r_w = r_b$, as well as certain conditions on relative size, location, and intensity of r_b and r_w, where $r_w \in \mathcal{R}_w$. Imposing such constraints drastically reduced false positives, and typically only the black symbols on placards are detected.

Decoding

In the decoding stage, we desired very high precision at the expense of recall since occasional missed detections are tolerable but false positives will cause significant problems. During the competition the system needed to be able to adapt quickly and there were only limited training examples of placards available. We set up one template for each of circle, triangle, and cross, and match a candidate region $r_c \in \mathcal{R}_c$ to one of the placards, when the number of good matching keypoints is higher than

Fig. 5 Placard feature detection. The three correctly detected placards are *circled* in *blue*, *red*, and *green*

a threshold. SIFT [6] and FAST [12] keypoints and SIFT descriptors are appealing choices to distinguish each placard. For example, the cross contains many SIFT keypoints, typically corners surrounded by gradients, and triangle contains FAST keypoints (typically "sharp" corners), where a a pixel p has contiguous n pixels in the circle around p brighter or darker. Figure 5 demonstrates the decoded candidate regions shown with blue, red, and green circles. The computation runs at two frames per second for an image resolution of 1280 × 720 pixels.

3.2 *Light Buoy Sequence Detection*

The light buoy color sequence consists of an LED panel mounted on top of a buoy that emits a sequence of three colors (each for half a second), followed by a two second break. Detection of color on the LED is done with a similar process as for the buoy color detection (Sect. 2.2) except that there is an added temporal component required to detect the sequence. An overview of the approach is given in Algorithm 2. An example of a sequence being detected is shown in Fig. 6.

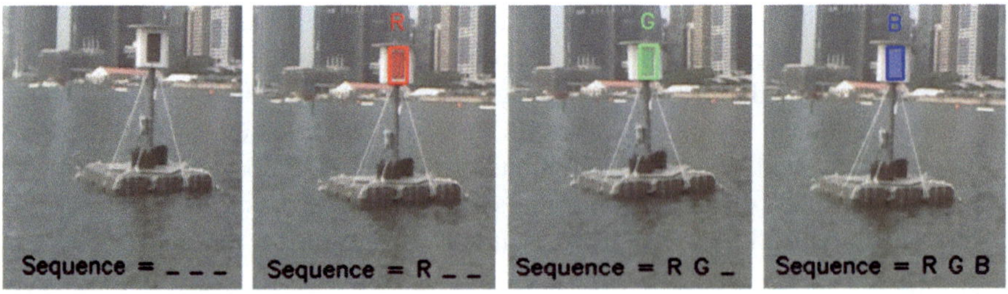

Fig. 6 Light tower sequence detection

Algorithm 2 Light Buoy Sequence Detection

Input: Video stream
Output: Light sequence Φ

1: $\Phi \leftarrow \emptyset$
2: Wait until first detection is made
3: Wait until no detection is found for 2 seconds
4: **while** $|\Phi| < 3$ **do**
5: $C \leftarrow$ color detected in image
6: **if** $C \neq$ last entry in Φ **then**
7: $\Phi \leftarrow \Phi \bigcup C$
8: **end if**
9: **if** No Detection **then**
10: Return to Step 3
11: **end if**
12: **end while**

This system requires the light buoy to be within the field of view of the camera for minimum of four seconds (the end a sequence and then one full sequence). If no detections are being made the segmentation thresholds are adapted automatically to be more admissive. Similarly, if the pause in the sequence is never being found (caused by false detections) then the thresholds are adaptively made more restrictive.

4 Acoustic Sensing

The process of localizing the pinger in Task 2 had two main components: First, relative bearing measurements are obtained from processing the signals received at the hydrophone. Second, subsequent bearing measurements are combined with a particle filter to yield a final estimate of the pinger location.

4.1 Relative Bearing Measurements

The acoustic system consisted of a 4-element hydrophone phased array, a custom amplification and filtering board (AFB), a data acquisition board (DAB), and a computer. The phased array was assembled into a 'T' shape (see Fig. 7) with uniform element spacing $d = 1.9$ cm. This formed two sub-arrays, one horizontal for use in bearing estimation and one vertical for use in elevation estimation. The signal from each hydrophone channel passed through a 10 kHz Sallen-key high-pass filter, a 2x amplifier, and then a 50 kHz Sallen-key low-pass filter on the AFB [13]. The resulting signal was converted to digital by the DAB and used to determine pinger location from four channels of hydrophone data. First, matched filtering on the first acoustic channel was used to identify if a ping of the correct frequency occurred [10]. Conventional (delay-and-sum) beamforming was applied to the array data, and the

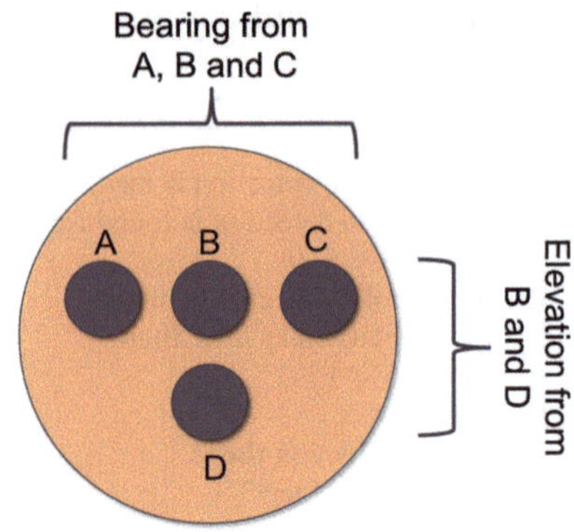

Fig. 7 "T"-shaped acoustic array

maximum value in the beampattern was used to determine bearing to the pinger [14]. Let z represent the direction along the array. The discrete array has elements at locations $\mathbf{z} = [-d, 0, d]$. The goal of beamforming is to find the angle of incidence, θ_0, of the signal from a pinger with frequency f_0. This gives a wavenumber $k_0 = 2\pi f_0/c$. The z-component of the wavenumber can be expressed in terms of 'look' direction θ:

$$k_z = k_0 \cos\theta. \tag{5}$$

This component of the wavenumber is used to calculate the delay vector $\mathbf{v}$:

$$\mathbf{v}(\theta) = e^{-j\mathbf{z}k_z}. \tag{6}$$

Delay-and-sum beamforming [14] is then applied by first multiplying the snapshot time series, $\mathbf{x} = [x_1, x_2, x_3]$, with the delay vector and then taking the Fourier transform:

$$Y = \mathscr{F}(\mathbf{x}\mathbf{v}'). \tag{7}$$

The beampattern function at look angle θ is the value of $\mathbf{Y}$ at frequency f_0, $B(\theta) = \mathbf{Y}(f_0)$. The bearing to the pinger of frequency f_0 is the look angle that results in the maximum for the beampattern:

$$\theta_0 = \arg\max_{\theta} \|B(\theta)\| \tag{8}$$

A similar process was used to determine the pinger elevation angle: conventional beamforming was applied to the vertical array (elements 1 and 3) and the beamforming angle with the maximum response identified as the elevation angle of the pinger.

4.2 Particle Filter Pinger Localization

The estimated elevation and bearing angles reported by the hydrophone system were used by a particle filter [8] to estimate the possible pinger location. An overview of the method is illustrated in Fig. 8. When the first relative bearing measurement is received, the particles are initialized uniformly along the portion of the bearing line that falls within the task boundary.

When the second and subsequent bearing measurements are received, the particles are each given a weight based on their proximity to the new bearing line based on the following equation:

$$w_i^t = w_i^{t-1} \frac{p(r_i^t|\zeta_i^t)p(\zeta_i^t|\zeta_i^{t-1})}{q(\zeta_i^t|\zeta_i^{0:t}, r_i^t)} \tag{9}$$

where ζ_i^t is the xy-positon of particle i at time t, and r_i^t is the orthogonal distance from the particle position, ζ_i^t, to the line anchored at the current vehicle position, x_t, y_t with slope corresponding to the bearing measurement, θ_0 calculated in (8). If we set the transition prior $p(\zeta_i^t|\zeta_i^{t-1})$ equal to the importance function $q(\zeta_i^t|\zeta_i^{0:t}, r_t)$ [8], and assume a normal distribution for $p(r_i^t|\zeta_i^t)$, we can simplify (9) to:

$$w_i^t = w_i^{t-1} \frac{1}{\sigma\sqrt{2\pi}} e^{-\frac{(r_i^t)^2}{2\sigma^2}} \tag{10}$$

Finally, we use sequential importance resampling to avoid particle depletion. This involves a check to determine if the effective number of particles N_{eff} has fallen

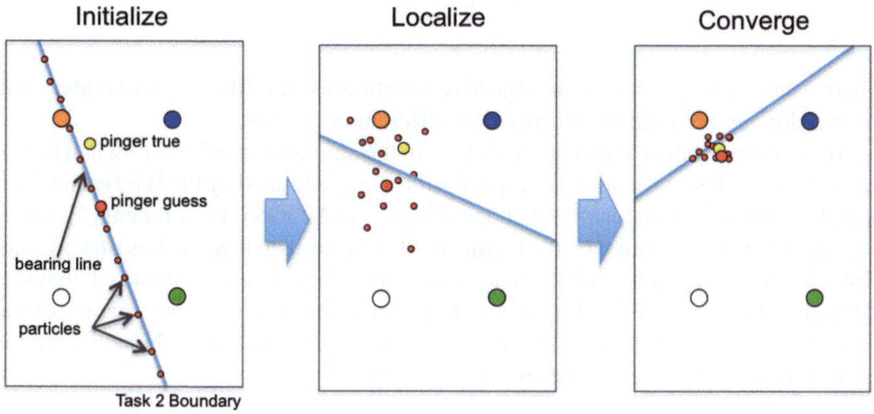

Fig. 8 Particle Filter Localization. *Left* The particles are initialized after the first bearing line is received. *Middle* As more bearings are received, the particles begin to localize on the pinger location. *Right* After more information is received the particles converge on a single point

below a threshold $N_{threshold}$. The effective number of particles is:

$$N_{eff} = \frac{1}{\sum_{i=1}^{N}(w_i^2)} \tag{11}$$

where N is the total number of particles. The best guess for the pinger location at any given time is computed is the average location of the particles.

5 Autonomy and Control

The operation of the vehicle is broadly characterized into two modes: (1) *Autonomy*, which is used for moving the boat around and avoiding obstacles, and (2) *Observation*, which is used for keeping the vehicle's sensors pointed in a specific direction.

5.1 *Autonomy*

In autonomy mode, the vehicle has to balance different objectives, such as transiting to a goal point while avoiding obstacles. This balance is achieved using multi-objective optimization with interval programming (IvP), [3, 4], where each goal is represented by a piecewise linearly-defined objective function for evaluation in conjunction with all other active objective functions. The optimization engine on-board the ASV considers and solves for the resultant maneuver (ordered course, speed) using

$$\overrightarrow{x}^* = \arg\max_{\overrightarrow{x}} \sum_{i=1}^{k}(w_i \cdot f_i(\overrightarrow{x})) \tag{12}$$

where each $f_i(x_1, \ldots, x_n)$ is an objective function for the ith of k active goal, and the weights, w_i are used to prioritize the different objectives.

An overview of the control methodology in the autonomy mode is shown in Fig. 9. The outer loop desired heading and speed values are generated by the IvP Helm which operates within the mission oriented operating suite (MOOS) environment [2]. In our case, each feature outputted by the feature tracker (described in Sect. 2) is treated as an obstacle and spawns a new obstacle avoidance behavior. These avoidance behaviors ("Avoid 1" to "Avoid N" in Fig. 9) are then used to prioritize actions that move the vehicle away from obstacles. These are weighed with a waypoint behavior that is used to steer the vehicle towards the desired goal.

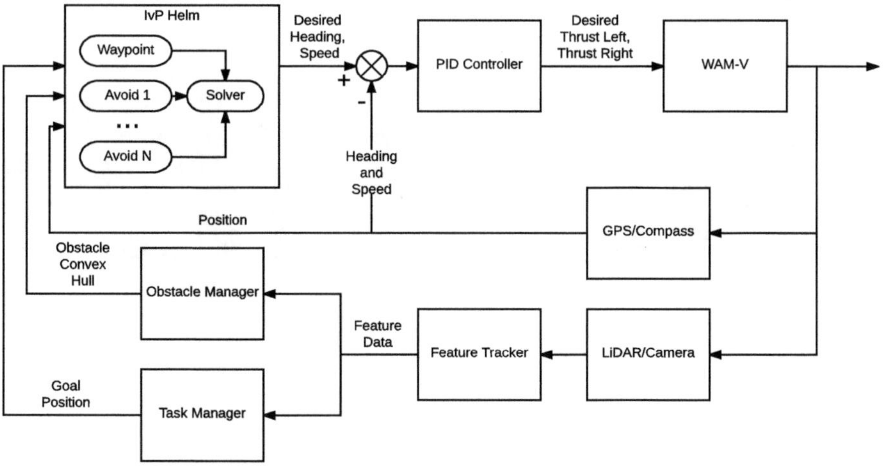

Fig. 9 Control system used in autonomy mode. Behaviors in the IvP Helm generate objective functions which are weighed at runtime to determine a best choice desired heading and speed. These values are tracked by an inner loop PID controller

5.1.1 The Obstacle Manager

The association of features is performed by the feature tracker that processes the clustered output from the laser. Due to noise in the system, as well as the fact that features (such as buoys) may be actually moving on the water surface, the reported locations of features can be variable. To be conservative, we track the history of reported feature locations and avoid all of them.

This is done in the obstacle manager by tracking all reported locations for a given feature, and then defining a convex hull for each feature as shown in Fig. 10.

The obstacle manager reports the convex hulls as polygons to the IvP Helm. The IvP Helm is configured with an obstacle avoidance behavior template that will spawn a new behavior with each new obstacle ID that is received and subsequent updates from the obstacle manager may change the shape of the polygon representing the obstacle. In Fig. 11, the vehicle is transiting through an obstacle field in a qualification run where four of the obstacles are "active" (generating objective functions) and they are shown in the figure as filled in polygons. An additional buffer is added around each obstacle but if necessary this buffer is shrunk for the vehicle to be able to fit through tight spaces. The collective objective function (Fig. 11-right) is the sum of the waypoint behavior and the four active obstacle avoidance behaviors. In the figure, colors closer to red are higher utility and closer to blue are worse. The angles on the circle denote desired headings (in the same reference frame as the picture on the right) and distance from the center of the circle denotes desired speed. The pink dot in the figure is the outputted desired heading and speed.

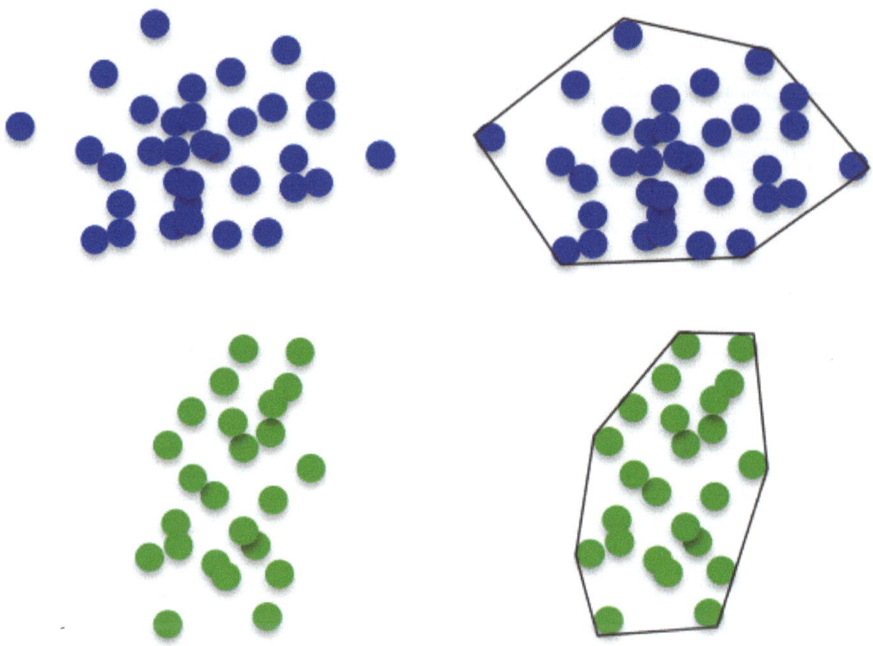

Fig. 10 Conversion of feature locations to convex hull: As new points (features) arrive, the convex hull is incrementally updated

5.2 Observation

For observations required in Tasks 2, 3, and 4, we developed a control mode that bypasses the IvP Helm and directly maintains a certain observation point within the field of view of the sensor. In this mode, the desired heading is generated by comparing the actual robot pose (observed through GPS and compass sensors) and the heading required to maintain the observation point in the field of view (Fig. 12). This value is computed in the "Pose Keeping" block (Fig. 13).

6 Hardware Setup

The platform base of the vehicle provided to the team was the WAM-V [7], which is a 13-foot, double-pontooned hull with a dynamic suspension system that supports a platform for the vehicle's sensors and electronic components above. The custom-designed power and propulsion system consisted of two Torqeedo Power 26–104 batteries, which rested at the back of the pontoons and powered two Riptide Transom 80 saltwater transom mount trolling motors. The vehicle was steered using a differential drive paradigm through a Roboteq VDC2450 motor controller. The bat-

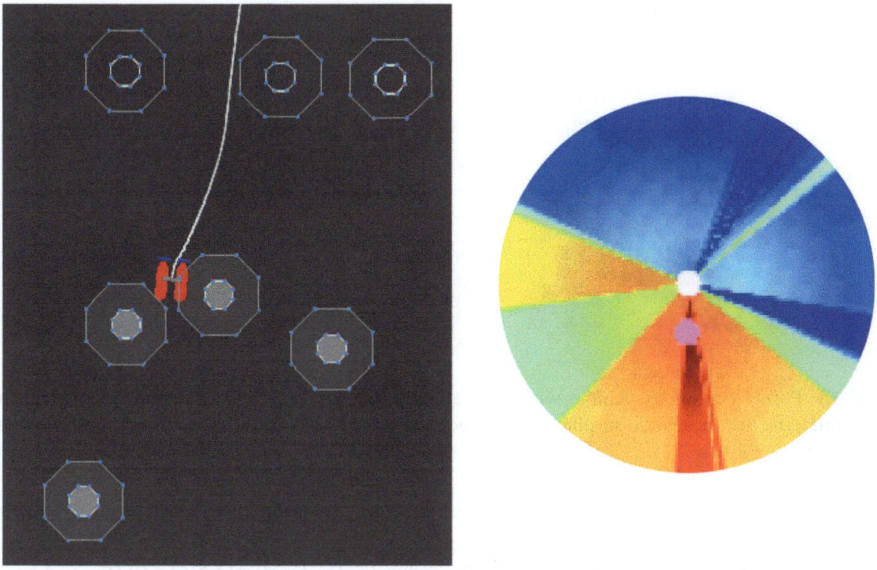

Fig. 11 *Left* The ASV navigating through a field of obstacles. The *filled polygons* are currently active and generating objecting functions. *Right* A color plot showing the sum of all objective functions where *redder colors* are higher utility and *bluer colors* are lower utility. Vehicle not to scale

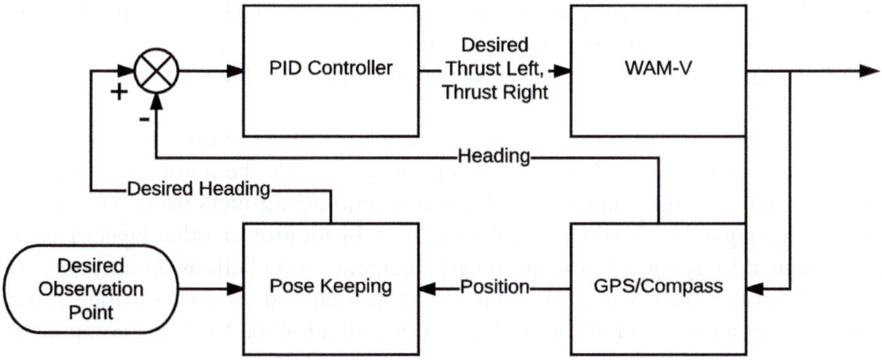

Fig. 12 Control system for maintaining observation of a fixed point. The pose keeping block is used to generate the reference heading and error is minimized through feedback PID control loop

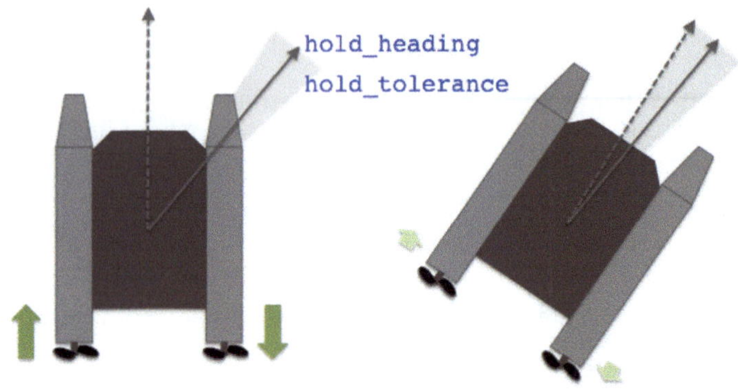

Fig. 13 Pose Keeping: A vehicle with differential thrust applies opposing thrust of equivalent magnitude to turn a vehicle in place until it achieves a desired `hold_heading`, with a given `hold_tolerance`

teries had enough capacity to last all day, and the motors provided enough thrust to be practical and proved easy for folding and stowage.

The vehicle had four computers on board: two Portwell NANO-6060's and two Intel NUC kits, which were configured to be used interchangeably. These computers provided the processing power to process the sensor data from the laser, run the autonomy system, and also communicate with a shoreside computer through a WiFi antenna. The system received location and heading data from a Vector V102 GPS system, and the acoustic data was processed on a PC-104 stack, both of which talked directly to the four main computers. This system was powered by a lead-acid battery, separate from the propulsion power system.

An emergency stop system was designed to sit between the vehicle's computers and the motor controller.

The emergency stop system can communicate directly to an operator control unit (OCU) box, which allowed a human operator to override the autonomy system at any point in time with manual control. Arduino microcontrollers using Xbee radios communicating over a 2.4 GHz signal were used. In addition, another layer of safety was designed by tying a pair of on-board emergency stop buttons directly into the motor controller. The whole emergency stop system had its own separate power source, for an added level of safety. An overview of all of the hardware components and connections is shown in Fig. 14.

7 Results and Discussion

A snapshot of the vehicle performing each task is shown in the left column of Fig. 15.[2] On the right column is a task-specific snapshot built from the data collected. For Task 1 (top), the figure shows the navigation through the buoys. We were able to reliably

[2] A video of our qualification and final has been made available (http://robotx.mit.edu/fsr_video).

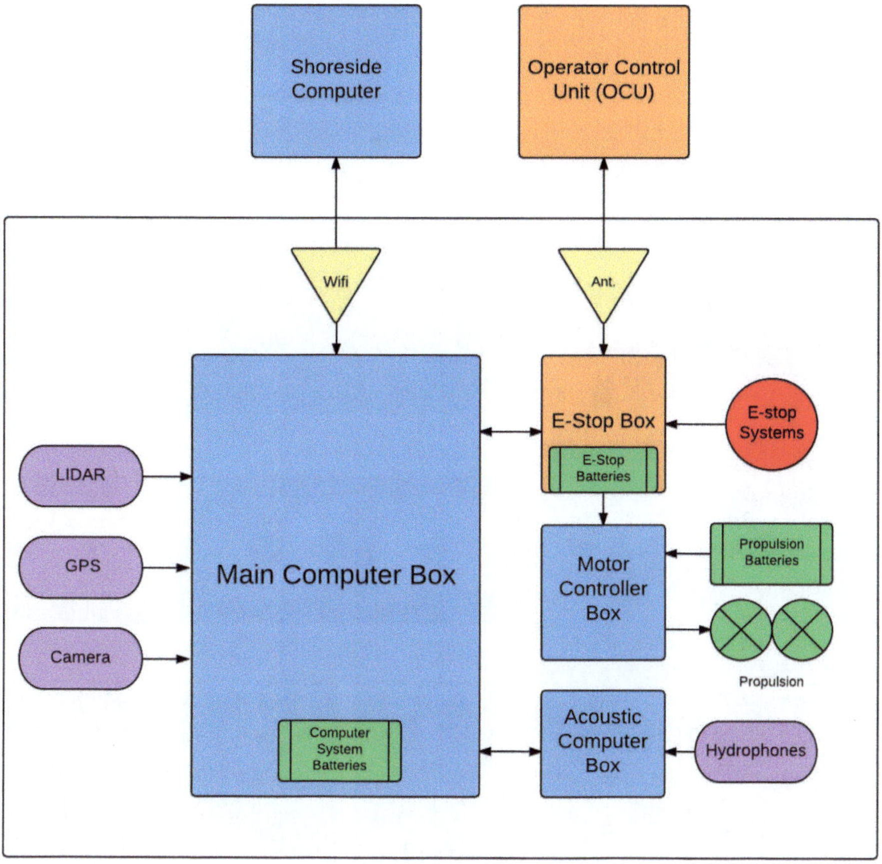

Fig. 14 RobotX hardware system layout and connections

achieve this task throughout. For Task 2 (second row) we show the output of the particle filter as well as the last bearing generated. Row three shows the docking task. Our feature detection based on the method in Sect. 3 was reliable. The fourth row shows the light buoy sequence detection. This was perhaps the most challenging task since it involved color detection in variable light conditions. Additionally, the colorful background enhanced the probability of false detection. The final task was obstacle avoidance. The feature detection and tracking system was reliable, but the overall system had some latency issues as described below.

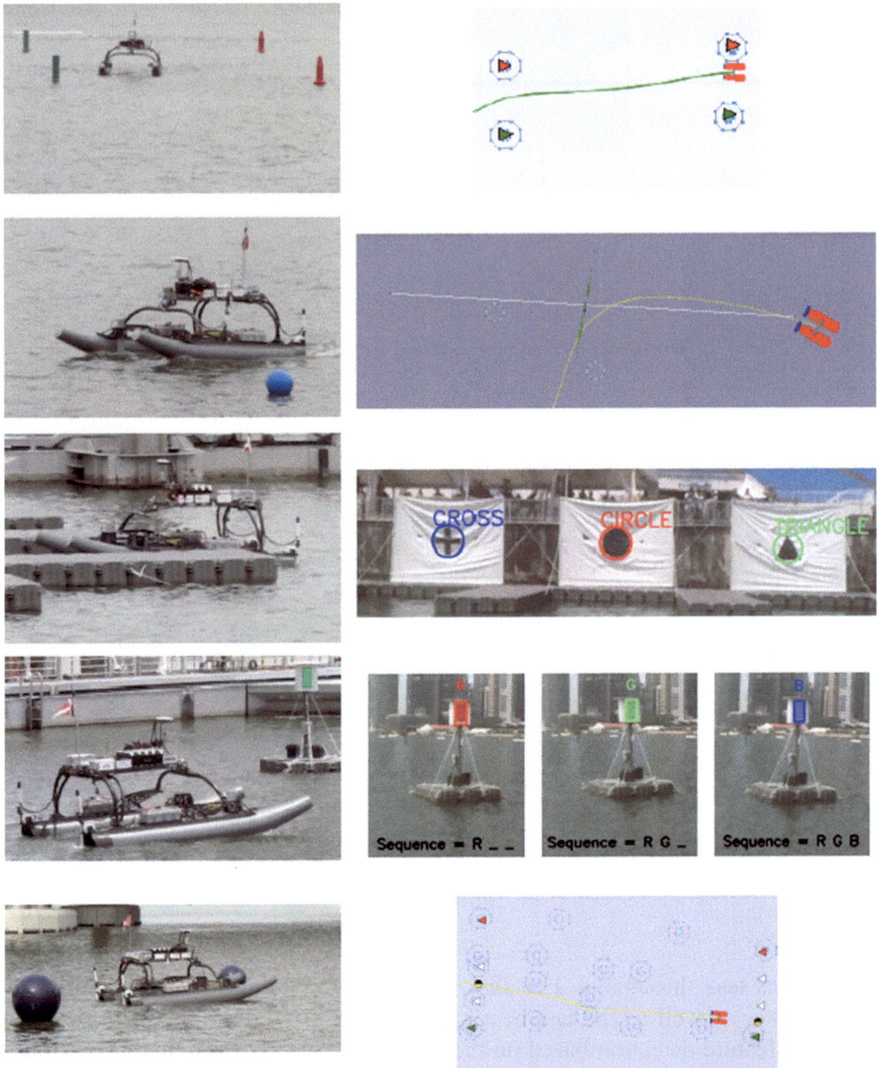

Fig. 15 *Left column* snapshots of the WAM-V robot performing each of the five tasks (Task 1 at *top* to Task 5 at *bottom*). *Right column* Row one, two and five snapshots from the pMarineViewer [2]. Row three and four show processed snapshots from the camera onboard

Table 1 RobotX final rankings

1	MIT/Olin (USA)
2	KAIST (South Korea)
3	Queensland University of Technology (Australia)
4	Embry-Riddle Aeronautical University (USA)
5	National University of Singapore (Singapore)
6	Osaka University (Japan)

7.1 What Went Wrong—Lessons Learned

We were able to successfully complete all the tasks successfully in qualifications. However, a few mishaps prevented us from completing each task on the final run. Due to time constraints, we reduced the amount of time that we would wait for the acoustic system to process data, and therefore only received two bearing measurements. This gave us partial points for identifying the color of the closest buoy but not the exact pinger location. On the docking task, we were able to correctly identify the "CIRCLE" placard which was designated at the start of the run, but our right pontoon caught the edge of the dock. This was likely due to incorrect extrinsic calibration of our camera system. At the last second before the final run, we decided to add functionality such that if the light buoy sequence was not determined before a timeout was reached, then we would move on to the final task and at least take a guess.

Unfortunately, we inputted the incorrect task number and this forced a guess to be reported when we *entered* Task 4 (the light buoy observation task), so as a result a guess was reported after docking even though post-processing of the camera data determined that we would have reported a correct sequence. This last-minute change deviated from our typically methodical approach to simulating and testing all code changes prior to deployment and really reinforced that if there is insufficient time to test a modification before deployment then it simply should not be made. We also struck a buoy in the obstacle field. It was later determined that this was due to a delay in our obstacle managing system. Although the buoy had been correctly detected, the behavior necessary to avoid it was not spawned in time to avoid collision. Despite these errors, we accumulated the highest point total in the final round. The final rankings are shown in Table 1.

8 Conclusion

This paper outlines the MIT/Olin team's approach and performance in the inaugural AUVSI RobotX competition. In the competition, each of the fifteen teams were provided with an identical marine vehicle frame and were responsible for building the propulsion, electronic, sensing, and autonomy systems required to complete a series of five tasks. Ultimately, the MIT/Olin team narrowly won first place in a very competitive field. The team's codebase and data are publicly available.

Acknowledgments The RobotX competition was supported by AUVSI and ONR. Autonomy development and autonomy test facilities used for RobotX at MIT were supported in part by Battelle Maritime Systems and the ONR, Code 311. The project was also supported by the following grants ONR N00014-13-1-0588 and N00014-14-1-0214, APS-14-12 SLIN 0001, DARPA contract No. N66001-14-C-4031, UCSD award no 43019208, and NSF Award IIS-1318392. We would also like to thank our other sponsors that contributed to the project: MIT Sea Grant, Oliver and Merideth Seikel, Chick and Sandy Corrado, CENSAM, Singapore-MIT Alliance for Research and Technology (SMART), The US Military Academy, and Portwell, Inc. In addition we would like to acknowledge the contributions of the following individuals: Jonathan Garcia-Mallen, Chrys Chyssostomidis, Devynn Diggins, Mindy Tieu, Shivali Chandra, Nikolay Lapin, Justin Poh, Paul Titchener, Alex Kessler, Madelin Perry, Jay Woo, Zoher Ghadyali, William Warner, Victoria Coleman, and Pat Marion.

References

1. Bar-Shalom, Y.: Tracking and Data Association. Academic Press Professional Inc, San Diego, CA, USA (1987)
2. Benjamin, M., Schmidt, H., Leonard, J.J.: http://www.moos-ivp.org (2014)
3. Benjamin, M.R.: Interval programming: a multi-objective optimization model for autonomous vehicle control. Ph.D. thesis, Brown University, Providence, RI (2002)
4. Benjamin, M.R., Schmidt, H., Newman, P.M., Leonard, J.J.: Nested autonomy for unmanned marine vehicles with MOOS-IvP. J. Field Robot. **27**(6), 834–875 (2010)
5. Hartley, R.I., Zisserman, A.: Multiple View Geometry in Computer Vision, 2nd edn. Cambridge University Press (2004). ISBN: 0521540518
6. Lowe, D.G.: Distinctive image features from scale-invariant keypoints. Int. J. Comput. Vision **60**(2), 91–110 (2004)
7. Marine Advanced Research: http://www.wam-v.com
8. Maskell, S., Gordon, N.: A Tutorial on Particle Filters for On-line Nonlinear/Non-Gaussian Bayesian Tracking (2001)
9. Neumann, L., Matas, J.: Real-time scene text localization and recognition. In: Proceedings of IEEE Inernational Conference Computer Vision and Pattern Recognition (CVPR) (2012)
10. Oppenheim, A., Schafer, R.: Discrete-Time Signal Processing, 3rd edn. Prentice Hall (2009)
11. Paull, L., Saeedi, S., Seto, M., Li, H.: AUV navigation and localization: a review. IEEE J. Oceanic Eng. **39**(1), 131–149 (2014)
12. Rosten, E., Drummond, T.: Machine learning for high-speed corner detection. In: European Conference on Computer Vision (ECCV), pp. 430–443. Springer (2006)
13. Sallen, R.P., Key, E.: A practical method of designing rc active filters. IRE Trans. Circuit Theory **2**(1), 74–85 (1955)
14. Trees, H.V.: Optimal Array Processing. Wiley (2002)

A Parameterized Geometric Magnetic Field Calibration Method for Vehicles with Moving Masses with Applications to Underwater Gliders

Brian Claus and Ralf Bachmayer

Abstract The accuracy of magnetic measurements performed by autonomous vehicles is often limited by the presence of moving ferrous masses. This work proposes a third order parameterized ellipsoid calibration method for magnetic measurements in the sensor frame. In this manner the ellipsoidal calibration coefficients are dependent on the locations of the moving masses. The parameterized calibration method is evaluated through field trials with an autonomous underwater glider equipped with a low power precision fluxgate sensor. These field trials were performed in the East Arm of Bonne Bay, Newfoundland in December of 2013. During these trials a series of calibration profiles with the mass shifting and ballast mechanisms at different locations were performed before and after the survey portion of the trials. The nominal ellipsoidal coefficients were extracted using the full set of measurements from a set of calibration profiles and used as the initial conditions for the third order polynomials. These polynomials were then optimized using a gradient descent solver resulting in a RMS error between the calibration measurements and the local total field of 28 and 17 nT for the first and second set of calibration runs. When the parameterized coefficients are used to correct the magnetic measurements from the survey portion of the field trials the RMS error between the survey measurements and the local total field was 124 and 69 nT when using the first and second set of coefficients.

1 Introduction

The use of underwater vehicles as a platform for oceanic research is an excellent way to collect high quality data in a challenging environment. Long range AUVs, capable of travelling thousands of kilometers before needing to be recovered are

B. Claus (✉)
Applied Ocean Physics and Engineering, Woods Hole Oceanographic Institute, Woods Hole, Massachusetts, USA
e-mail: bclaus@whoi.edu

R. Bachmayer
Faculty of Engineering, Memorial University, St. John's, Newfoundland, Canada
e-mail: bachmayer@mun.ca

D.S. Wettergreen and T.D. Barfoot (eds.), *Field and Service Robotics*,
Springer Tracts in Advanced Robotics 113, DOI 10.1007/978-3-319-27702-8_6

recently the focus of significant interest [3, 10]. Underwater gliders are a type of long range underwater vehicle, however, they require surface access for navigation, have limited speed and require vertical translation for forward movement [14]. For these vehicles minimizing energy consumption is one of the primary design and operational goals.

The use of magnetic field measurements as a heading reference for navigation in underwater vehicles has been well established [8]. In recent work earth magnetic information has also been suggested for possible use in total-field map based relative navigation techniques [7, 16]. This use of magnetic measurements for online navigational aiding is the motivation for this research. In such a system, magnetic measurements are capable of augmenting a terrain relative navigation scheme in regions of low terrain variability or when the terrain is beyond the range of the vehicle's acoustic sensors. However, an online implementation of a magnetic aided navigation system has not been realized. This lack of progress has been limited by the challenges involved in instrumenting and calibrating an underwater vehicle for accurate online magnetic measurements and the lack of suitably high resolution magnetic maps.

Scalar calibration of vector magnetometers has shown to be a robust method of calibration based on a geometric fit to an ellipsoid [2, 13, 17]. Another method relies on projecting the measurement vector onto the horizontal plane and fitting an ellipse [6, 9]. Of these methods, the second is more suited to vehicles which have limitations in the controllable degrees of freedom such as an underwater glider. However, it requires a precision attitude reference to rotate the magnetic measurements to the horizontal plane which is infeasible on an underwater glider due to their relatively large energy consumption. Additionally, long range underwater vehicles, and underwater gliders in particular, require additional effort to calibrate the magnetic field measurements. This extra effort is due to the use of an adjustable internal mass for attitude control which is typically composed of a battery pack and therefore includes hard and soft magnetic materials.

As a step towards a real time total field magnetically aided navigation system this work examines suitable methods for calibrating, instrumenting and performing magnetic measurements with an underwater glider. The variable locations of the mass shifting and ballast mechanisms on the underwater glider provide an additional challenge for calibrating the magnetic measurement system. As such, a parameterized calibration method is presented which fits polynomial functions to the calibration parameters based on the actuator locations. To this end the theory for a nominal geometric calibration and a parameterized geometric calibration method is presented and the underwater glider equipped with the magnetic instrumentation developed for this work is introduced. Lastly, the calibration procedures are demonstrated on field data gathered using the underwater glider during trials in the East Arm of Bonne Bay. The calibrated data are compared with magnetic anomaly models produced from prior aeromagnetic surveys of the region.

2 Calibration Methods

Measurements of the earth's magnetic field must be calibrated in order to remove the effects of the sensing platform. These effects can be due to instrument non-linearities as well as hard and soft magnetic effects.

2.1 Nominal Geometric Calibration

If the moving masses in the vehicle are held stationary the hard and soft magnetic effects from the vehicle as well as scaling, bias and other instrument errors may be calibrated for using geometric batch methods [2, 13, 17]. These methods assume a constant magnetic field and rely on rotations of the instrument through the calibration space such that an ellipsoid may be fit to the data.

An ideal magnetic sensor at a fixed location produces measurements with a constant magnitude resulting in the data lying on the surface of a sphere, centered on the origin with the radius equal to this magnitude. Distortions due to the sensor errors and the vehicle hard and soft magnetic effects have been shown to cause the measurements to be translated, rotated and scaled such that the sphere becomes an ellipsoid. The problem of finding this set of translation, rotation and scaling coefficients can be expressed in matrix notation as

$$[\mathbf{M}, \mathbf{S}, \mathbf{T}] = G(\mathbf{H}_r) \tag{1}$$

where $\mathbf{M}$, $\mathbf{S}$, and $\mathbf{T}$ are the rotation, scaling and translation matrices that are representative of the ellipsoidal fit $G()$ to the raw magnetic data vector $\mathbf{H}_r$. Geometrically, the translation coefficients are the distance from the center of the ellipsoid to the origin, the scaling coefficients are the magnitudes of the major and minor ellipsoid axes and the rotation coefficients are the rotations of the major and minor axes of the ellipsoid. The ellipsoid equation representing the relationship between the raw magnetic data and the corrected data is written as

$$\mathbf{H}_r = H_e^{-1}\mathbf{SMH}_c + \mathbf{T} \tag{2}$$

The raw magnetic data may then be translated, rotated and scaled accordingly by re-arranging the ellipsoid equation to

$$\mathbf{H}_c = H_e\mathbf{S}^{-1}\mathbf{M}^{-1}(\mathbf{H}_r - \mathbf{T}) \tag{3}$$

where $\mathbf{H}_c$ is the calibrated magnetic data vector in the sensor frame. This calibration procedure normalizes the magnitude of the magnetic measurements due to the product of the inverse of the scaling coefficients. To give the calibrated values units, the normalized values must be scaled by the magnitude of the local magnetic field at the calibration location H_e which often may be approximated from the International Geomagnetic Reference Field (IGRF) [5]. The IGRF does not include many of the higher frequency components and the local magnetic anomalies. If a local anomaly map is available these anomaly values may be included as in

$$H_e = ||\mathbf{H}_{IGRF}|| + H_a \tag{4}$$

where H_a is the magnitude of the magnetic anomalies at the calibration locations. The resulting values given by $\mathbf{H}_c$ are the calibrated measurements of the magnetic field for a vehicle with fixed locations of the hard and soft magnetic influences and no significant electrical currents.

2.2 *Parameterized Geometric Calibration*

For vehicles with moving hard or soft magnetic parts that have a number of steady state values a parameterized version of the geometric calibration method is proposed. In this method the nominal geometric calibration procedure from Sect. 2.1 is performed on data gathered from a number of different steady state values for each of the moving parts. The fixed calibration parameters are used as the initial conditions for an iterative gradient decent solver which optimizes a third order function with each of the moving masses as parameters. In the case of underwater gliders, the primary parameters are the moving mass mechanism used for fine control of the vehicle pitch and the ballast mechanism which is responsible for the large pitch and buoyancy changes between diving and climbing. The geometric fitting then becomes of the form

$$[\mathbf{M}, \mathbf{S}, \mathbf{T}](p_m, p_b) = G(\mathbf{H}_r(p_m, p_b)) \tag{5}$$

where each of the rotation, translation and scaling coefficients is a function of the moving mass location p_m and the ballast piston location p_b. The parameterized functions are found by fitting polynomials to the set of individual calibration coefficients found for a geometric fit to the magnetic measurements for a given moving mass and ballast location. The parameterized ellipsoid equation is similarly given as

$$\mathbf{H}_r = H_e^{-1}\mathbf{S}(p_m, p_b)\mathbf{M}(p_m, p_b)\mathbf{H}_c + \mathbf{T}(p_m, p_b) \tag{6}$$

Upon re-arranging, the raw magnetic data may be corrected by computing the translation, rotation and scaling matrices for a given moving mass and ballast location as in

$$\mathbf{H}_c = H_e\mathbf{S}(p_m, p_b)^{-1}\mathbf{M}(p_m, p_b)^{-1}(\mathbf{H}_r - \mathbf{T}(p_m, p_b)) \tag{7}$$

The polynomial functions in this case are of third order and take the form of

$$c_0 p_m^3 + c_1 p_b^3 + c_2 p_m^2 + c_3 p_b^2 + c_4 p_m^2 p_b + c_5 p_m p_b^2 \\ + c_6 p_m p_b + c_7 p_m + c_8 p_b + c_9 \tag{8}$$

resulting in a total of 90 coefficients required for a two parameter calibration problem.

3 Instrumentation

An underwater glider's energy is provided by onboard batteries which gives it an endurance of around one month when using alkaline primary cells and six months when using lithium primary cells. In a standard configuration of a vehicle equipped only with a conductivity, temperature and pressure sensor (CTD), the vehicle uses an average power of around one Watt. To not significantly impact the endurance or range of the vehicle, additional sensors should use as little power as possible. Therefore, to instrument an underwater glider with a precision magnetic sensor, the power consumption of the device must remain low to minimize the impact on the vehicle's endurance.

While progress is being made towards lower power cesium vapour magnetometers which would be well suited to integration in mobile platforms, the power consumption of presently available devices still remains on the order of Watts [12, 15]. Fluxgate sensors, on the other hand, have power requirements down to the level of 10s of milliwatts. For this reason the chosen sensor is a low power tri-axial Mag-648 fluxgate magnetometer by Bartington Instruments which consumes around 14 mW [1]. Low power fluxgates of this type are often subject to higher degrees of noise, orthogonality errors, and offset errors than higher power versions [11]. While the impact of the higher noise is mitigated through low frequency sampling requirements, the orthogonality errors and offset errors require careful calibration. Additionally, the offset error settles to a slightly different value each time the sensor is powered on requiring the sensor to remain energized once calibrated.

The fluxgate sensor is mounted in a strap-down configuration in the vehicle's payload bay. The device is powered by a set of independent batteries and is sampled using an isolated 24-bit sigma-delta analog to digital converter (ADC). This ADC uses several different internal low pass filters and modifies the filter coefficients based on the sampling rate selected. The effective resolution of the device is therefore variable with the sampling rate. The inputs to the ADC have anti-aliasing filters with a corner frequency of 0.33 Hz to mitigate high frequency noise from the electronics and other systems. The ADC uses the serial peripheral interface (SPI) to send the data to the glider payload computer where it is logged at a frequency of 0.25 Hz. The ADC used has a single digitizer and samples of each channel are taken at different times requiring the time stamp of each channel's measurement to be recorded such that the measurements may be interpolated to the same time base. The electrical current drawn by the fluxgate and its electronics is around 4.5 mA. As a result of this low energy consumption, a single set of three AA alkaline cells connected in series will power the fluxgate and its electronics for one month. The goal of not influencing the endurance of the underwater glider while staying within the size and weight requirements for the payload are therefore achieved.

4 Field Trials

Field trials using the magnetic fluxgate sensor installed on a 200 m Slocum Electric glider were performed to evaluate the efficacy of making magnetic measurements using this platform. The parameterized calibration field trials took place in December, 2013 in the East Arm of Bonne Bay, Newfoundland. In these trials the underwater glider was launched from the small aluminum boat Freezy as illustrated in Fig. 1 and after launch was controlled from the Bonne Bay Marine Station. During the deployment there were light winds and the air temperature was around −10 °C. Recovery of the vehicle was originally planned for December 12th but had to be delayed due to strong winds. The vehicle was left to loiter in the lee of the head on Norris Point until a lull in the winds on the 13th allowed the recovery of the vehicle.

After the deployment, a series of clockwise calibration spirals were performed with the vehicle commanded to set the movable battery once during each ascent or descent to achieve a certain pitch according to a look up table. In this way five different

Fig. 1 The Bonne Bay Marine Station's boat Freezy shown with the Slocum autonomous underwater glider during the parameterized trials in December 2013

Table 1 Calibration runs for the parameterized magnetic calibration trials

Run	Direction	p_b [cm^3]	Pitch [°]	p_m Trial 1 [in]	p_m Trial 2 [in]
1	Dive	−200	−14	0.272	0.226
2	Climb	200	14	−0.181	−0.139
3	Dive	−200	−18	0.380	0.274
4	Climb	200	18	−0.234	−0.191
5	Dive	−200	−22	0.428	0.375
6	Climb	200	22	−0.289	−0.246
7	Dive	−200	−26	0.491	0.400
8	Climb	200	26	−0.344	−0.300
9	Dive	−200	−30	0.527	0.472
10	Climb	200	30	−0.401	−0.348

battery locations were tested for two different ballast conditions. The ballast was also set to a single value, once for each ascent or descent. Each calibration run therefore consisted of a single spiralling descent and ascent with the ballast and battery at a fixed location and took around 30 min to complete. Another full calibration procedure was repeated prior to recovery. The calibration runs are summarized in Table 1.

The vehicle was then flown in a criss-cross pattern down into the bay and back again with a commanded pitch of plus or minus 26° and a commanded ballast of plus or minus 200 cm^3. The calibration locations along with the vehicle track-line are shown against the local residual magnetic field in Fig. 2.

To provide reference measurements, aeromagnetic data overlapping the East Arm of Bonne Bay was used from the Newfoundland and Labrador Geoscience Atlas [4]. Unfortunately, the East Arm is split in half by the boundary of two different surveys, the 2009 Corner Brook survey and the 2012 Offshore Western Newfoundland survey.

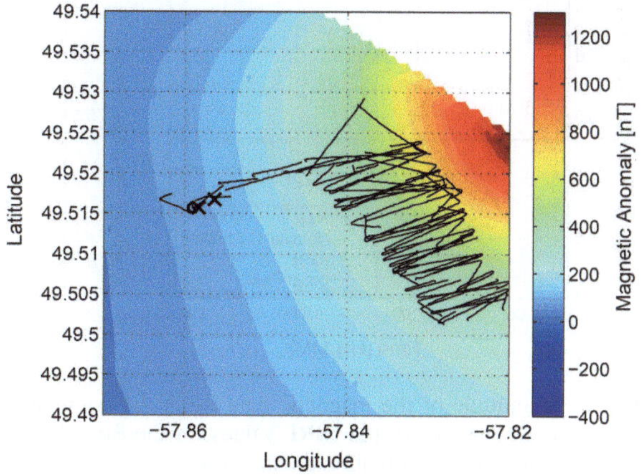

Fig. 2 Calibration locations (x's) and the Bonne Bay Trials track-line (*black line*) starting from the *circle* and proceeding to crisscross south and then north in the East Arm of Bonne Bay. The residual magnetic grid of the Bonne Bay region is shown in the background

To obtain a reference grid both residual magnetic grids were upward continued to a constant altitude of 90 m. The grids were then combined, using the average value in the regions of overlap. A mask was applied to these larger grids to limit the region to the area of the East Arm of Bonne Bay. To smooth any discontinuities, 20 passes of a 3×3 Convolution (Hanning) filter were applied to remove the high frequency content introduced by combining the grids. The resulting grid is shown in Fig. 2.

For the parameterized calibration method, an initial global fit of the nominal geometric method was performed by using the full set of raw measurements from each of the calibration runs. To constrain the ellipsoid in this initial fit it was necessary to make the x and z scaling values equal as there were no calibration measurements in the "northern hemisphere" of the calibration space. Additionally, the ellipsoid was constrained in rotation such that $\mathbf{M} = \mathbf{I}$. The global fit was then used as the initial conditions for the parameterized equations by setting the c_9 coefficients from Eq. 8 to be equal to the ellipsoid's scaling, translation and rotation coefficients. The parameterized equations were then adjusted using a gradient descent optimization scheme by minimizing the error between the local total field and measured values. In this optimization scheme the local total field was computed from the IGRF model and the magnetic anomaly value at the calibration locations. The resulting magnitude of the calibrated measurements are shown in Fig. 3.

The nominal geometric method results in a root mean square error between the total field estimate from the IGRF and aeromagnetic data and the calibrated data of 153 and 145 nT for the first and second set of calibration runs. The resulting magnitude of the calibration measurements, corrected with the parameterized coefficients are shown in Fig. 4.

The parameterized geometric method results in a root mean square error between the total field estimate from the IGRF and aeromagnetic data and the calibrated data of 29 and 17 nT for the first and second calibration trials. Each of these sets of parameters is then used to correct the magnetic data gathered during the remainder of

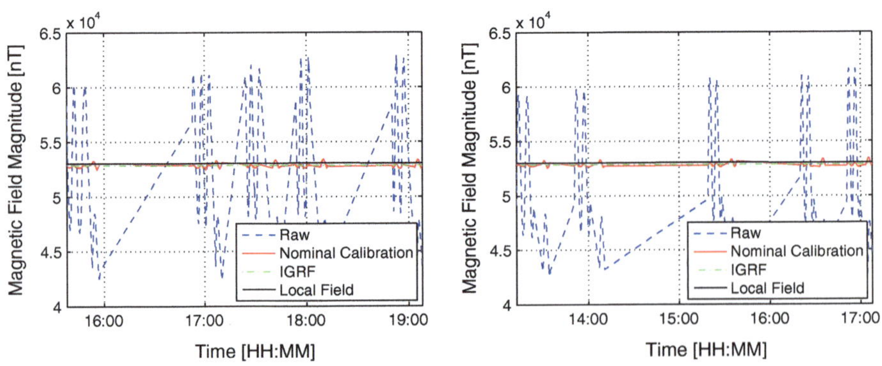

Fig. 3 Magnitude of the magnetic data using the nominal calibration method before and after correction shown against the IGRF values for the Bonne Bay field trials using the first (*left*) and second (*right*) set of calibration coefficients

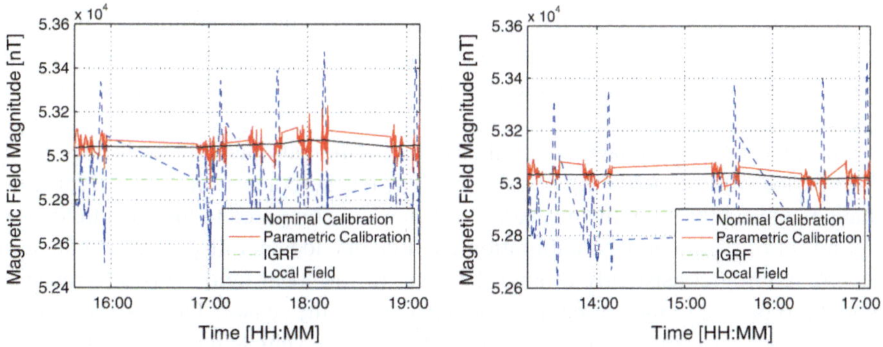

Fig. 4 Magnitude of the magnetic data using the nominal and parameterized calibration method with the data from the first (*left*) and second (*right*) set of trials shown against the IGRF and local field values for the Bonne Bay trials

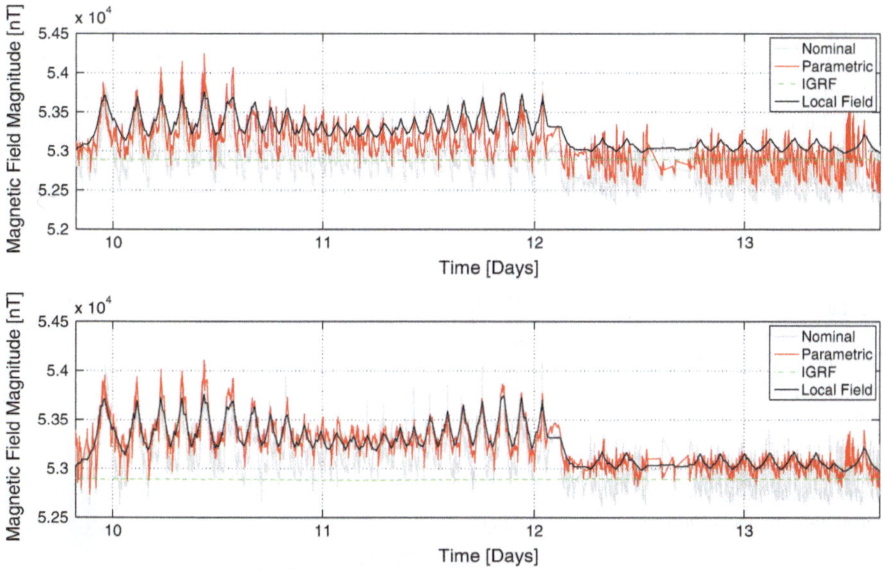

Fig. 5 Magnetic data collected during the Bonne Bay deployment in December 2013 shown against the IGRF and local field values calibrated using the first (*top*) and second (*bottom*) set of nominal and parameterized calibration coefficients

the deployment as shown in Fig. 5. In correcting this data the calibration coefficients are assumed to be constant. As such the mean of the local magnetic field at the calibration locations, H_e, is used for each set of calibration coefficients.

The calibrated magnetic measurements gathered by the glider may then be compared to the residual magnetic grids. The resulting interpolated values have a constant bias when compared to the complete set of glider magnetic measurements. Additionally, the glider data contains significantly more high frequency components than the aeromagnetic grids. These differences are attributed to the aeromagnetic data being collected at a higher altitude reducing the high frequency signatures present in the reference data as well as the significant low-pass filtering applied during the gridding operations.

The first set of parameterized calibration coefficients perform well only for a short period of time. After the first day or so of measurements, there is a significant change in bias present in the measured values when compared to the local field. The second set of parameterized calibration coefficients does not display this change in bias, remaining consistently around the level of the local field. This difference is thought to be due to the temperature dependence of the sensor. The first calibration run was performed immediately after launch while the vehicle had been at a temperature of less than −10 °C. The second calibration run was performed after the data collection before retrieval allowing the sensor adequate time to warm up to the water temperatures of around 2 °C. The measurements calibrated using the second set of parameterized coefficients were deemed more accurate for this reason and are shown next to the residual magnetic field values from the vehicle locations in Fig. 6.

The measured magnetic anomaly data calibrated using the second set of parameterized calibration coefficients is in reasonable agreement with the residual magnetic field data from the aeromagnetic surveys with RMS errors indicated in Table 2. Additionally, the parameterized geometric calibration method improves significantly upon the nominal geometric calibration method. This agreement indicates that the para-

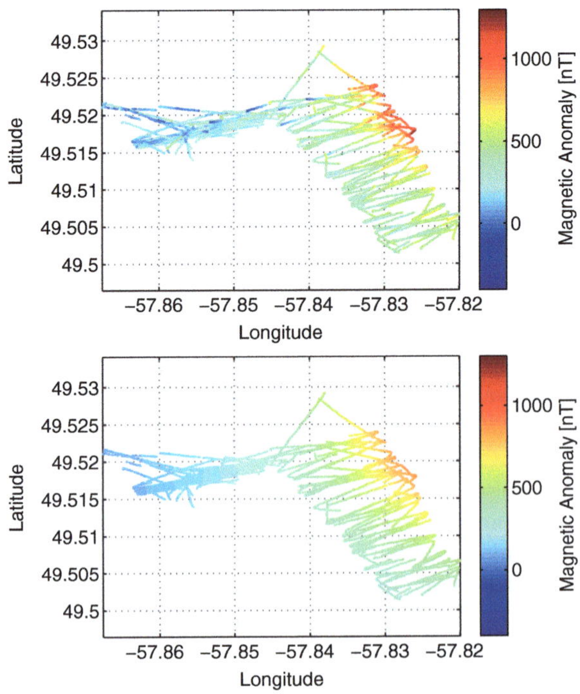

Fig. 6 Magnetic anomaly of the data collected during the Bonne Bay deployment in December 2013 calibrated using the parameterized geometric method (*top*) compared with the interpolated magnetic anomaly data from the aeromagnetic grids (*bottom*)

Table 2 The RMS errors between the magnetic anomaly map values and the calibrated measurements using the first and second set of nominal and parametric calibration coefficients during the Bonne Bay field trials

	Nominal (nT)	Parametric (nT)
Trial 1	207	124
Trial 2	136	69

meterized calibration method is effective for calibration of magnetic measurements performed from a vehicle with moving masses. The drawback of this method are the increased number of calibration runs that need to be performed over the nominal calibration method. However, while the parameterized calibration method takes longer to perform, it constrains the calibration space to a higher degree than the nominal method for the limited maneuvering space available to the underwater glider resulting in a better calibration.

5 Conclusions

Augmenting underwater relative navigation methods with total field magnetic measurements and a-priori magnetic anomaly grids has been proposed previously in several theoretical studies. Evaluating this proposition in practice is challenging due to the high levels of distortions which must be calibrated out of the magnetic measurements.

For rigid platforms with fixed components and low levels of electrical noise a geometric calibration method may be used. In this nominal geometric calibration method the raw measurements are assumed to lie on the surface of an ellipsoid. The ellipsoid's offset, radii and rotations of the major and minor axis form a set of calibration coefficients which may be used to correct the measurements in the sensor frame. For platforms with moving masses a parameterized geometric calibration method has been proposed. In this method a third order polynomial is estimated using gradient descent methods where the initial conditions are formed from the nominal geometric method parameters.

The parameterized calibration method is evaluated using an autonomous underwater glider equipped with a precision low power fluxgate magnetometer. During field trials of the system, which took place in December 2013 in the East Arm of Bonne Bay, Newfoundland, calibration runs were performed upon deployment and before recovery. For each calibration run the underwater glider performed a series of descending and ascending spirals such that the mass shifting mechanism and ballast system were each at multiple steady state locations. Between these sets of calibration runs, the underwater glider ran its mission, cris-crossing up and down the East Arm. To obtain the parameterized calibration coefficients the complete set of calibration measurements from each run was used to extract the nominal ellipsoid coefficients.

These nominal coefficients were then used as the initial conditions for the gradient descent solver which computed the third order polynomial coefficients which define each ellipsoid coefficient for the given mass shifter and ballast mechanism location.

The parameterized calibration method resulted in an RMS error between the calibration measurements and the local total field of 29 and 17 nT for the first and second set of calibration runs. During the survey portion of the field trials the first and second set of parameterized calibration coefficients resulted in a RMS error between the calibrated measurements and the local total field from the a-priori grid of 124 and 69 nT respectively.

Magnetic measurements performed in this manner are suited to the online calibration of magnetic data. This online correction is the ultimate goal of this work towards allowing the augmentation of terrain relative navigation methods with magnetic anomaly measurements.

Acknowledgments This work was supported by the Natural Sciences and Engineering Research Council (NSERC) through the NSERC Canadian Field Robotics Network (NCFRN), the Research Development Corporation, the Marine Institute and Memorial University of Newfoundland.

References

1. Bartington Instruments: Mag648 and Mag649 Low Power Three-Axis Magnetic Field Sensors, Bartington Instruments, DS2298/9 (2011). Accessed July 2011
2. Bronner, A., Munschy, M., Sauter, D., Carlut, J., Searle, R., Maineult, A.: Deep-tow 3C magnetic measurement: solutions for calibration and interpretation. Geophysics **78**(3), J15–J23 (2013)
3. Furlong, M.E., McPhail, S., Stevenson, P.: A concept design for an ultra-long-range survey class AUV. In: Proceedings of IEEE Oceans—Europe, pp. 1–6 (2007)
4. Honarvar, P., Nolan, L., Crisby-Whittle, L., Morgan, K.: The geoscience atlas. Report 13–1, Newfoundland and Labrador Department of Natural Resources (2013). Geological Survey
5. International Association of Geomagnetism and Aeronomy: Working Group V-MOD. Participating members, Finlay, C.C., Maus, S., Beggan, C.D., Bondar, T.N., Chambodut, A., Chernova, T.A., Chulliat, A., Golovkov, V.P., Hamilton, B., Hamoudi, M., Holme, R., Hulot, G., Kuang, W., Langlais, B., Lesur, V., Lowes, F.J., Lhr, H., Macmillan, S., Mandea, M., McLean, S., Manoj, C., Menvielle, M., Michaelis, I., Olsen, N., Rauberg, J., Rother, M., Sabaka, T.J., Tangborn, A., Tffner-Clausen, L., Thbault, E., Thomson, A.W.P., Wardinski, I., Wei, Z., Zvereva, T.I.: International geomagnetic reference field: the eleventh generation. Geophys. J. Int. **183**(3), 1216–1230 (2010)
6. Isezaki, N.: A new shipboard three-component magnetometer. Geophysics **51**(10), 1992–1998 (1986)
7. Kato, N., Shigetomi, T.: Underwater navigation for long-range autonomous underwater vehicles using geomagnetic and bathymetric information. Adv. Robot. **23**(7–8), 787–803 (2009)
8. Kinsey, J.C., Eustice, R.M., Whitcomb, L.L.: A survey of underwater vehicle navigation: recent advances and new challenges. In: IFAC Conference of Manoeuvering and Control of Marine Craft. Lisbon, Portugal (2006). Invited paper
9. Korenaga, J.: Comprehensive analysis of marine magnetic vector anomalies. J. Geophys. Res.: Solid Earth (1978–2012) **100**(B1), 365–378 (1995)
10. McPhail, S., Stevenson, P., Pebody, M., Furlong, M.: The NOCS long range AUV project. In: National Marine Facilities Department Seminar Series (2008)

11. Primdahl, F.: The fluxgate magnetometer. J. Phys. E: Sci. Instrum. **12**(4), 241 (1979)
12. Prouty, M., Johnson, R.: Small, low power, high performance magnetometers. In: EGM 2010 International Workshop (2010)
13. Renaudin, V., Afzal, M.H., Lachapelle, G.: Complete triaxis magnetometer calibration in the magnetic domain. J. Sensors **2010** (2010)
14. Rudnick, D., Davis, R., Eriksen, C., Fratantoni, D., Perry, M.: Underwater gliders for ocean research. Mar. Technol. Soc. J. **38**, 73–84 (2004)
15. Shah, V., Knappe, S., Schwindt, P.D.D., Kitching, J.: Sub-picotesla atomic magnetometry with a microfabricated vapour cell. Nat. Photonics **1**, 649–652 (2007)
16. Teixeira, F.C., Pascoal, A.M.: Geophysical navigation of autonomous underwater vehicles using geomagnetic information. In: 2nd IFAC Workshop Navigation, Guidance and Control of Underwater Vehicles (2008)
17. Vasconcelos, J., Elkaim, G., Silvestre, C., Oliveira, P., Cardeira, B.: Geometric approach to strapdown magnetometer calibration in sensor frame. IEEE Trans. Aerosp. Electron. Syst. **47**(2), 1293–1306 (2011)

Towards Autonomous Robotic Coral Reef Health Assessment

Travis Manderson, Jimmy Li, David Cortés Poza, Natasha Dudek, David Meger and Gregory Dudek

Abstract This paper addresses the automated analysis of coral in shallow reef environments up to 90 ft deep. During a series of robotic ocean deployments, we have collected a data set of coral and non-coral imagery from four distinct reef locations. The data has been annotated by an experienced biologist and presented as a representative challenge for visual understanding techniques. We describe baseline techniques using texture and color features combined with classifiers for two vision sub-tasks: live coral image classification and live coral semantic segmentation. The results of these methods demonstrate both the feasibility of the task as well as the remaining challenges that must be addressed through the development of more sophisticated techniques in the future.

1 Introduction

In this paper we describe a system for the automated detection and video identification of coral growths using a marine robot. Our objective is to develop a fully autonomous system that can swim over coral reefs in open water, collect video data of live coral formations, and make an estimate of coral abundance. The video is intended for examination by human specialists, but the system needs to be able to both remain resident on the reef surface and recognize coral as it is encountered to perform its mission.

Coral reefs are delicate marine environments of immense importance both ecologically and socio-economically, and yet they are under substantial threat almost everywhere they occur. One preliminary step to retaining these environments is to be able to objectively record their presence, their change over time, and their health. Such records are critical not only to any remediation effort, but also in order to present

T. Manderson (✉) · J. Li · D. Cortés Poza · D. Meger · G. Dudek
School of Computer Science, McGill University, Montréal, Canada
e-mail: travism@cim.mcgill.ca

N. Dudek
Department of Ecology and Evolutionary Biology, University of California
at Santa Cruz, Santa Cruz, USA

D.S. Wettergreen and T.D. Barfoot (eds.), *Field and Service Robotics*,
Springer Tracts in Advanced Robotics 113, DOI 10.1007/978-3-319-27702-8_7

a compelling case to law makers and law enforcement officials regarding the preservation of these ecosystems. While human divers are commonly deployed to observe reefs and measure their health, the requisite measurements need to be performed using scuba gear under conditions that present a risk to the divers involved.

In the work reported here, we use a small, portable, and high mobility underwater vehicle which is able to swim over the surface of a coral reef, hover in place, navigate in confined spaces, and collect video data from multiple cameras operating simultaneously. In our current experimental configuration the vehicle is accompanied by a human supervisor, but our approach and target scenario does not require a human operator to be present while data is being collected. This vehicle is ideally suited for reef surveillance since it can be deployed manually by a single user either from shore or in the water, does not require an associated tender (ship), can maneuver even in very shallow water, and can even land on a set of legs on sand or a reef surface with limited physical contact. Our approach to covering coral reefs requires the vehicle to be initialized over or near a reef. It can subsequently circumnavigate the reef and cover its interior using inertial navigation. In prior work we have also employed GPS data, acquired by allowing the vehicle to surface, to assist in the navigation task, but in this work navigation is accomplished while remaining underwater at the expense of global localization. This paper does not focus on coverage and navigation, but rather on the system architecture, the nature of the data we collect, and our ability to detect and recognize living coral using this vehicle.

In this paper, we propose and evaluate two critical components of the visual processing pipeline used for both the guidance and data collection for our vehicle. These operations are the classification of images that are observed as either containing live coral or not, and the subsequent segmentation of the live coral within the image. Several structured data sets used in our evaluation are described below and are available to the community.[1]

2 Background

As coral health is an issue of worldwide importance, its monitoring has been studied by many authors previously, both in the field of biology and intelligent systems. This section describes several of the most relevant contributions.

2.1 Coral Reef Biology and Reef Health

Coral reefs are majestic structures crucial to ecosystem functioning. They are home to roughly 25 % of the oceans' inhabitants, and act as a nursery, feeding ground, and shelter for thousands of marine organisms [1]. To humans, they represent

[1]Dataset hosted at: http://www.cim.mcgill.ca/mrl/data.html.

approximately US$30 billion annually in goods and services, and are the focus of many studies searching for novel biochemically active drug compounds [2]. Optimistic reports estimate that at the current rate, by 2050 some 75 % of the world's remaining reefs will be critically threatened [3]; more pessimistic estimates predict that all of Earth's coral reefs will be dead by the end of the century [4].

Some of the major driving forces behind coral decline worldwide include increasing water temperatures, ocean acidification, increase in frequency and intensity of coral diseases, and damage due to natural disasters such as hurricanes. Many anthropogenic activities are also causing direct harm to reefs, including the overfishing of essential herbivorous species of fish, increasing amounts of water pollution from terrestrial runoff, and increasing sedimentation from coastal construction [3]. Arrival of invasive species can further exacerbate the situation and lead to a dramatic decrease in reef diversity and health, such as the invasion of Indo-Pacific lionfish in the Caribbean Sea and of the crown-of-thorns seastar in Australia [5].

While little can be done on a regional scale about issues such as global warming and increasing ocean temperature, there is an increasing focus on local management and conservation of coral reefs [6]. One critical component of any successful conservation effort is being able to assess whether a particular conservation strategy results in beneficial outcomes on the system in question. In order to protect what remains of the world's coral reefs, it is essential that we design accurate and precise methods to assess the health of coral reefs without undue risk to human participants. This will not only allow us to see when conservation efforts work, but will also help determine which reefs should be conservation priorities and provide evidence to policy makers and the general public that conservation efforts are necessary to preserve the well being of coral reef ecosystems [7].

2.2 Robotic Reef Surveys

Several research groups have considered the use of autonomous underwater vehicles (AUVs) for data collection in marine environments, and even in coral reefs. Reefs are challenging environments since they are both valuable and physically delicate, and they have complex morphologies. A few vehicles have been developed that can make close approaches to the ocean floor, corals, or aquatic structures [8, 9]. This can be challenging due to several factors: (a) the propulsion systems may be unsafe to operate close to sensitive underwater environments; (b) otherwise "gentle" devices such as gliders have limited maneuverability; (c) it is difficult for humans to produce preplanned trajectories since sensor feedback underwater is often poor, communications are difficult and terrain models are rarely complete; (d) many propulsion systems are prone to disturbing bottom sediments which reduces visibility.

The problem of designing and controlling stable AUVs has been studied by several authors [10, 11] on a variety of platforms. In prior work with the Aqua class of vehicles developed in our lab, we have demonstrated a combination of small size, low weight, and high maneuverability with diverse gaits [12, 13].

Several authors have also considered using towed or autonomous surface vehicles to perform visual data collection over marine environments [14], although in the context of coral reefs such an approach is feasible only for the shallowest reef structures and depends critically on very good visibility. Deep water AUVs have been used to map the ocean floor, inspect underwater structures, and measure species diversity [15].

Australia's Integrated Marine Observation System (IMOS) is carrying out a project to deliver precisely navigated time series of seabed imagery and other variables at selected stations on Australia's continental shelf [16]. They are using UAVs to make this endeavor scalable and cost efficient.

In [17], the authors present a structure from motion framework aided by the navigation sensors for building 3D reconstructions of the ocean floor and demonstrate it on an AUV surveying over a coral reef. Their approach assumes the use of a calibrated camera and some drifting pose information (compass, depth sensor, DVL). They use the SeaBED AUV, an imaging platform designed for high resolution optical and acoustic sensing [18].

In previous work [19] we have developed a controller to allow our vehicle to autonomously move about over coral reef structures using visual feedback. In this paper we restrict our attention to the analysis of the data collected by such a system, and consider the sensing issues that arise.

2.3 Visual Coral Categorization

Our methodology has been inspired by recent successes of previous biologically relevant visual data sets. For example, the Fish Task of the recent LifeCLEF contest [20] supported progress on detecting moving fish in video and fish species identification through the release of nearly 20,000 carefully annotated images. The identification of coral using visually equipped AUVs has been studied previously [21]. While we share similar motivations to this work, we differ in deployment and algorithmic objectives. Nonetheless, the relationship is a motivation for the public release of our training and test images which could facilitate comparisons. Additionally, Girdhar et al. [22] has demonstrated a system which modifies swimming behavior on-line to follow novel visual content.

3 The MRL Coral Identification Challenge

The first contribution of this paper is a robot-collected data set of visual images from environments proximal to a number of coral reefs. This data was collected by the Aqua swimming robot during a series of field deployments in the Caribbean, where the robot's existing navigation technologies were exercised to cover each reef and its surroundings. Although our robot did not use vision to inform its navigation strategies during these trials, the images that it collected are representative of the challenge that faces a coral-seeking robot. Therefore, we have organized and annotated them to form two visual challenge tasks: live coral image classification and live coral segmentation. The remainder of this section describes the components of this effort.

3.1 Robotic Data Collection

As mentioned previously, robots require specialized hardware and capabilities in order to operate safely near coral formations. We utilized the Aqua robot [23], an amphibious hexapod that swims using the oscillations of its flippers. Aqua has been designed for use as a visual inspection device and is equipped with four cameras with a variety of properties: a forward-facing stereo pair with a narrow field-of-view (which allows recovery of depth), a front fish-eye camera (which captures a wider scene), and finally a 45° (which allows the fourth camera to capture the ocean floor directly below the robot).

In order to achieve broad coverage of the underwater environment, our robot executed a coverage pattern repeatedly over the reef. We set the parameters of this motion by hand so that the robot would pass completely over the reef as well as an equal portion of the sandy surroundings. This gives our data set a roughly equal split between the coral images we target and less desirable content, which poses an interesting classification problem for the visual processing component.

Two attitude strategies were employed, each targeted to induce ideal viewpoints for a different sub-set of Aqua's cameras. First, a *flat-swimming* maneuver controlled the robot to be aligned with gravity in both the roll and pitch rotational axes. With this attitude, the downward looking camera views the ocean bottom with an orthogonal viewpoint and the front fish-eye camera views the horizon at roughly half the image height. Second, we considered swimming with a downwards pitch of 30°. This strategy allowed the narrow-view stereo pair to view the ocean bottom slightly in front of the robot. The depths observed at this angle would allow fixed-altitude operations, which are desirable in order to prevent accidental collisions with the coral.

The robot executed five data collection runs at four distinct reef locations (one reef was visited twice). We selected reefs within the Folkstone Marine Preserve and in Heron Bay, both of which are located on the western coast of St. James, Barbados. During each run, the robot covered an area of approximately $100\,m^2$. Each reef location was an instance of the spur-and-groove coral formations that tend to present the widest range of diversity of coral species, and are thus ideal regions for collection of biologically relevant data.

Data Statistics

All of the videos are taken at 15 frames per second, with VGA resolution. The total size of visual data collected over the five collection runs is 104 gigabytes consisting of 164 min of video. Depth and IMU data are also recorded throughout.

3.2 Data Annotation

A marine biologist manually annotated the coral within a subset of the images we collected. The results of this annotation have been made available in a standardized format, and the data is being released publicly for the purposes of comparison of results and classifier training. As a variety of tasks can be considered, depending on the goals of the robot platform, we define two coral-related visual tasks and accompanying evaluation criteria. We continue by describing our annotation procedure.

Annotation for Image Classification

The first sub-task that we define is coral image classification. Given an image, the system outputs whether there is live coral in the image. To create training and testing data for this task, we extracted images at 5 s intervals from all of the videos taken by the downward-looking mirrored camera while the robot was swimming flat. Each image was then subdivided into four 320×256 quadrants to limit the diversity and facilitate ease of labeling. The biologist labeled 3704 images into one of three categories:

- **Yes**: There is live coral in the image
- **No**: There is no live coral in the image
- **Reject**: The image should be discarded because it is too difficult to tell whether there is live coral or not. This could be because the image is too blurry or the coral is too small to see clearly.

This provided us with 1087 **Yes** images, 2336 **No**, and 281 **Reject** images. Figure 1 shows some examples of **Yes** and **No** images.

Annotation for Segmentation

Secondly, we define the coral segmentation task, where the coral regions within an image must be identified, through creation of a coral mask. While some existing segmentation data sets contain pixel-wise ground truth, we lacked the resources to produce this detailed data. Instead, we have manually annotated rectangular coral regions for each of the 1087 **Yes** images from our classification data set. Examples of the selected image regions are shown in Fig. 2. Rectangular regions cause a small approximation error at region boundaries, but this task is still a reliable proxy for coral segmentation, as will be demonstrated in our results section.

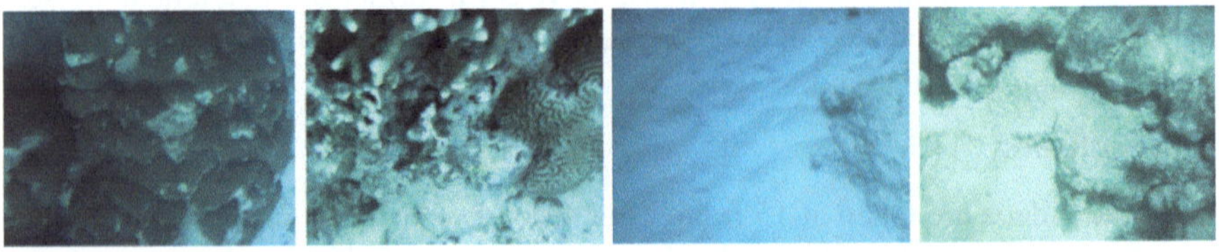

Fig. 1 Annotated images used for training a detector for the image classification task. The *left* two images are labeled as having coral and the *right* two images are labeled as not containing coral

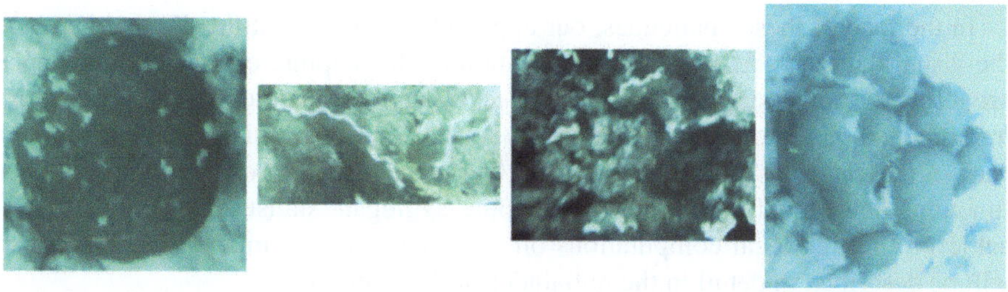

Fig. 2 Positive training images cropped to contain only coral, which is useful for training a detector for the coral segmentation task

Annotation Statistics

The final annotated data set produced by our labeler was reduced in size from the raw robot footage due to the rejection of poor quality and ambiguous images. We separated the annotated data into a training set (416 positive examples and 701 negative examples) and a test set (492 positive examples and 1544 negative examples). The training set contains images from three data collection runs at three unique reefs, and the test set contains images taken from two data collection runs at the fourth reef location. Thus, there is no overlap between the training and test sets.

We have additionally defined evaluation protocols for the use of this data, following best-practices from existing challenges such as the ILSVRC [24]. Broadly, we measure performance on each binary categorization task as prediction accuracy, normalized by the data set size. For the categorization task this represents the number of images, and for segmentation this is measured in image area. Methods cannot be optimized directly on the test data set. Rather, parameters should be refined by splitting the training set into folds and then reporting the performance after a single run on the test set. This data is being released to the public alongside this paper and we will maintain a record of the best performing techniques over time as other authors attempt the task. We now continue by describing several baseline techniques that we have developed.

4 Method

Coral identification in the ocean shares many of the typical challenges that face terrestrial vision systems, as well as several challenges unique to this task. The lighting conditions in the shallow ocean include caustics caused by the water's surface, interreflections and the absorption of low-frequency colors. This makes brightness invariance essential. The robot changes its orientation during the survey, which implies the need for orientation invariance. Small floating particles are ubiquitous in the underwater domain, causing an optical snow effect. Additionally, the appearance of the coral itself has a wide diversity and there are local variations between reef locations, so generalization must be the focus of learned methods.

In the face of these challenges, our approach to coral identification is to encode the visual data in robust feature representations that capture canonical appearance properties of coral, such as its color and texture and to learn coral classifiers from training data on top of these features. We develop two processing streams—one for each of the visual tasks described above. Our classification process employs Gabor functions and global processing to compute aggregate statistics. Segmentation is achieved through local computations on sub-regions of the image. Each approach will be described in detail in the remainder of this section.

4.1 Global Image Statistics for Coral Classification

The classification pipeline uses both global color and aggregate texture features in a classifier subsystem to learn from labeled example images and subsequently predict whether an image contains live coral. This subsystem computes two types of attributes over the entire (global) images to produce a characteristic feature vector. These vectors are then classified using a support vector machine (SVM) trained with our manually classified data. Figure 3 (top) illustrates the classification pipeline.

Our method represents texture through the use of the well-known Gabor transform. The Gabor function [25] is a sinusoid occurring within a Gaussian envelope and has inspired a class of image filters particularly suited to describing texture [26]. Our method automatically selects a sub-set of Gabor wavelets from a large family by selecting those with frequency and spatial support parameters that optimize task performance, using cross-validation on the training set.

Applying filters result in a stack of transformed images and we extract robust energy statistics from these in order to produce a vector suitable for classification.

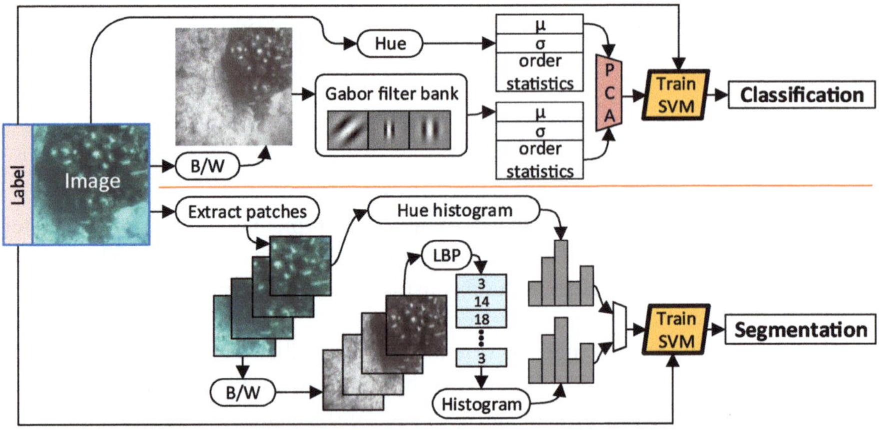

Fig. 3 Image processing pipeline. (*top*) Gabor-based classification. (*bottom*) LBP-based segmentation

The amplitude histogram of each Gabor filter provides a characterization of the image content including the presence of outlier objects. Order statistics can effectively characterize such a signal [27] and are robust to much of the noise present in our task. For this reason we characterized the energy distribution with several statistics of each Gabor filter: the mean energy, the variance of the energy distribution, and the energy at a specific set of percentiles of the cumulative distribution (5th, 20th, 80th and 95th percentiles). In order to capture color information, we additionally extracted the same robust statistics for the distribution of hue values observed in the image.

The result of both the Gabor texture filters and the color summary were concatenated into a fixed-length vector. Depending on the number of active Gabor components, this representation had between 24 and several hundred dimensions. In order to reduce computation and simplify the learning, we performed principal components analysis on these vectors to find the subspace that captures 99.99 % of the variance.

The final step in this pipeline is to predict the label of an image (live coral or not). We learn an SVM from the training images described previously and apply the resulting learned model to make coral predictions on new images.

4.2 Local Binary Pattern Based Coral Segmentation

Our coral segmentation pipeline uses LBPs [28] and color information as image descriptors, and an SVM to detect whether small patches of the images correspond to live coral or not. Unlike the Gabor filters, which are applied globally, our features and classifier are applied on small image patches, which allows fine-grained segmentation of coral regions. Figure 3 (bottom) illustrates the segmentation pipeline.

For a given pixel in the image, its LBP is computed by comparing its gray level g_c with that of a set of P samples in its neighborhood, g_p ($p = 1, 2, \ldots, P$). These samples are evenly spaced along a circle with radius R pixels, centered at g_c (see Fig. 4). For any sample that does not fall exactly in the center of a pixel, its gray value is estimated by interpolation. The LBP is computed according to

$$LBP_{P,R} = \sum_{p=0}^{P-1} \mathbb{1}_{\{g_c - g_p \geq 0\}} 2^p, \quad (1)$$

where $\mathbb{1}_{\{\cdot\}}$ is the indicator function.

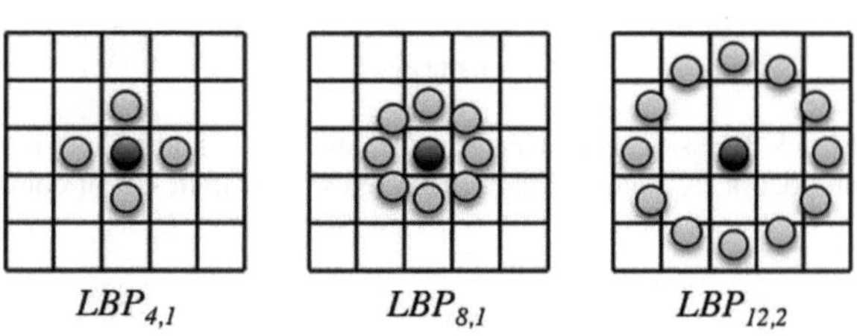

Fig. 4 Local binary pattern neighbor sets for ($P = 4, R = 1$), ($P = 8, R = 1$) and ($P = 12, R = 2$)

To achieve rotational invariance, Ojala et al. [28] proposed to label the LBPs according to their number of 0/1 transitions. LBPs with up to two transitions are called *uniform* and they are assigned a label corresponding to the number of 1's in the pattern. LBPs with more than two transitions are called *nonuniform* and they are all assigned the label $P+1$. Finally, the rotation invariant LBP image descriptor is a $P+2$ bin histogram of these labels computed across all pixels in the image. Uniform patterns are assigned to unique bins, while nonuniform patterns are all assigned to a single bin. As color is also an important feature for coral segmentation, we appended the LBP histogram with an eight bin histogram of the hue values of the pixels in the image patch.

During operation time, our learned model is used to segment an image by splitting it into patches with the same size as those used during training. Features are extracted from each patch and scored with the SVM, producing a coral segmentation mask that can be used to guide the robot during its mission.

5 Experiments and Results

5.1 Global Coral Classification

Our global classifier was tested on the data sets above using distinct testing and training sets collected over different reefs. We were able to achieve a net classifier accuracy of 89.9 % on balanced sets of images containing coral and not containing coral. This accuracy generally increased with the number of Gabor basis functions; however, since these are the primary source of computational cost, we are interested in a compromise between performance and the number of filters user. The trade-off between accuracy and the size of the filter bank is illustrated in Fig. 5. While using a bank of 24 or more filters provides maximal performance, the 80.6 % rate achieved with just 20 filters appears quite acceptable for our applications.

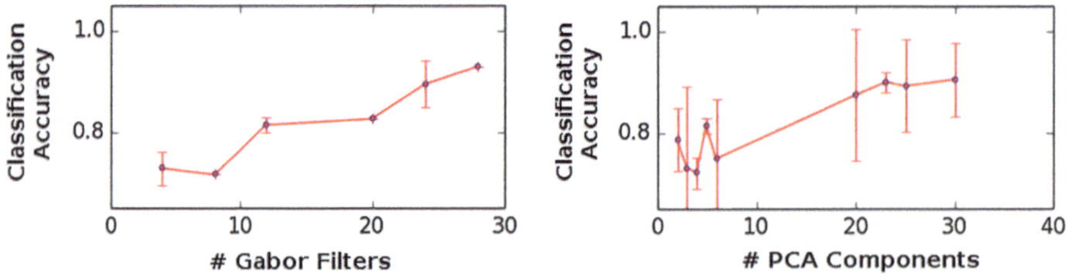

Fig. 5 Classification accuracy increases with both: (*left*) number of Gabor filters; and (*right*) number of PCA components. This reflects the trade-off of computational effort and performance

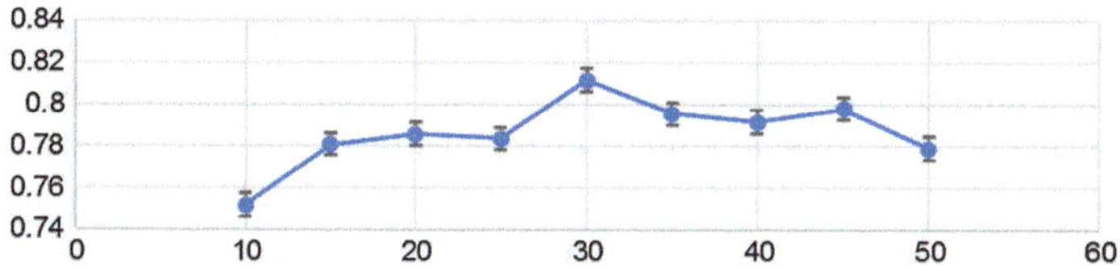

Fig. 6 Classification accuracy versus patch size (pixels)

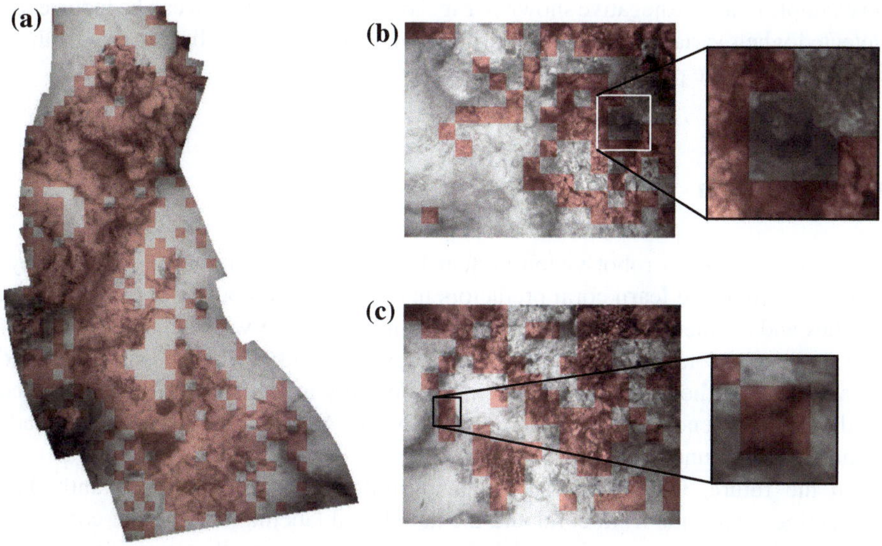

Fig. 7 Samples of live coral segmentation. **a** Test set reef segmentation. **b** Live coral segmentation. (*right*) false negative. **c** Live coral segmentation. (*left*) false positive

5.2 LBP-based Coral Segmentation

To study the effect of varying the number of points and radius (P, R) of the LBPs and the size of the patches on the segmentation, we performed a grid search on these parameters. Also, to optimize the performance of the SVM, we ran a grid search on the gamma, tolerance and regularization constant (C) parameters of the radial basis function (RBF) kernel.

The LBP parameters had very small impact on the accuracy of the classifier. We tested over the values $(P, R) = (8, 1), (16, 2), (16, 3), (24, 3), (32, 5)$ and found that the difference in accuracy between them was less than 2.1 %, regardless of the patch size. Given such a small impact, we decided to use $(P = 8, R = 1)$ for the remaining experiments.

The patch size, on the other hand, had a much larger impact on the classification accuracy, which is illustrated in Fig. 6. The maximum classification accuracy achieved was 81.16 % with the RBF kernel parameters set to $\gamma = 0.0001$, $tol = 2$ and $C = 10{,}000{,}000$. The optimal patch size was found to be 30 pixels.

In Fig. 7, we present some examples of images from the test set with an overlay (in red) showing the segmented live coral. Figure 7a is a stitched image created from several consecutive frames from the original video. We observe that the segmentation pipeline correctly finds areas of the image with live coral. We also observe areas where the classifier has problems detecting coral, such as when the texture is uniform – with an example of a false negative shown in Fig. 7b. Likewise, live coral can be incorrectly detected when variations in texture (or shadows) match that of live coral – with an example of a false positive shown in Fig. 7c.

6 Discussion

We have described a robot-vision system for performing automated coral surveys of the sea floor. We learn coral predictors that are able to robustly detect live coral patches and segment them from the background, agreeing with the assessments of an experienced coral biologist with an accuracy of 80–90 %. These results are based on a data set of thousands of labeled images of only moderate quality, confounded by the typical phenomena that confront any diver or AUV. Our data set is being made available in conjunction with this submission.

In the future, we plan to study the disambiguation of other zooxanthellae-containing organisms from coral and the automated labeling of different coral subspecies. This will require suitably labeled training data, as well as more diverse raw data sets, potentially including active illumination. Additionally, we hope to integrate coral mapping into the navigation stack of our vehicle, as we have successfully done in the past with other vision-guided navigation methods [22]. The resulting system has the potential to perform autonomous longitudinal surveys, providing biologists with an easy, quick, and accurate way of monitoring reef health. Such methods are critical for understanding how these ecosystems respond to environmental disturbances, documenting the efficacy of novel coral reef conservation and restoration efforts, and convincing policy makers to enact stringent protection measures for coral reef ecosystems.

References

1. Spalding, M.D., Ravilious, C., Green, E.P.: United Nations Environment Programme, World Conservation Monitoring Centre. World Atlas of Coral Reefs. University of California Press, Berkeley (2001)
2. Millennium Ecosystem Assessment: Ecosystems and human well-being—Synthesis report. World Resources Institute. Washington, DC (2005)
3. Burke, L.M., Reytar, K., Spalding, M., Perry, A.: Reefs at risk revisited. World Resources Institute. Washington, DC (2011)

4. Hoegh-Guldberg, O., Mumby, P.J., Hooten, A.J., Steneck, R.S., Greenfield, P., Gomez, E., Harvell, C.D., Sale, P.F., Edwards, A.J., Caldeira, K., Knowlton, N., Eakin, C.M., Iglesias-Prieto, R., Muthiga, N., Bradbury, R.H., Dubi, A., Hatziolos, M.E.: Coral reefs under rapid climate change and ocean acidification. Science **318**(5857), 1737–1742 (2007)
5. Albins, M.A., Hixon, M.A., et al.: Invasive Indo-Pacific lionfish Pterois volitans reduce recruitment of Atlantic coral-reef fishes. Mar. Ecol. Prog. Ser. **367**, 233–238 (2008)
6. Bellwood, D.R., Hughes, T.P., Folke, C., Nystrom, M.: Confronting the coral reef crisis. Nature **429**(6994), 827–833, 06 (2004)
7. Margules, C.R., Usher, M.B.: Criteria used in assessing wildlife conservation potential: a review. Biol. Conserv. **21**(2), 79–109 (1981)
8. Eskesen, J., Owens, D., Soroka, M., Morash, J., Hover, F., Chryssostomidis, C., Morash, J., Hover, F.: Design and performance of Odyssey IV: A deep ocean hover-capable AUV, MIT, Technical Report MITSG 09–08 (2009)
9. Woolsey, M., Asper, V., Diercks, A., McLetchie, K.: Enhancing NIUST's SeaBED class AUV, Mola Mola. In: Proceedings of Autonomous Underwater Vehicles, pp. 1–5 (2010)
10. Mohan, S., Thondiyath, A.: A non-linear tracking control scheme for an under-actuated autonomous underwater robotic vehicle. Int. J. Ocean Syst. Eng. **1**(3), 120–135 (2011)
11. Font, D., Tresanchez, M., Siegentahler, C., Pallej, T., Teixid, M., Pradalier, C., Palacin, J.: Design and implementation of a biomimetic turtle hydrofoil for an autonomous underwater vehicle. Sensors **11**(12), 11 168–11 187 (2011)
12. Meger, D., Shkurti, F., Cortes Poza, D., Giguere, P., Dudek, G.: 3d trajectory synthesis and control for a legged swimming robot. In: 2014 IEEE/RSJ International Conference on Intelligent Robots and Systems (IROS 2014), pp. 2257–2264. IEEE (2014)
13. Meger, D., Higuera, J.C.G., Xu, A., Dudek, G.: Learning legged swimming gaits from experience. In: International Conference on Robotics and Autonomous Systems (ICRA) (2015)
14. Williams, S., Pizarro, O., Johnson-Roberson, M., Mahon, I., Webster, J., Beaman, R., Bridge, T.: Auv-assisted surveying of relic reef sites. In: OCEANS 2008, pp. 1–7, Sept 2008
15. Maki, T., Kume, A., Ura, T., Sakamaki, T., Suzuki, H.: Autonomous detection and volume determination of tubeworm colonies from underwater robotic surveys. In: OCEANS 2010 IEEE—Sydney, pp. 1–8, May 2010
16. Williams, S., Pizarro, O., Jakuba, M.V., Johnson, C.R., Barrett, N.S., Babcock, R.C., Kendrick, G.A., Steinberg, P.D., Heyward, A.J., Doherty, P., Mahon, I., Johnson-Roberson, M., Steinberg, D., Friedman, A.: Monitoring of benthic reference sites using an autonomous underwater vehicle. IEEE Robot. Autom. Mag. **19**(1), 73–84 (2012)
17. Pizarro, O., Eustice, R.M., Singh, H.: Large area 3-D reconstructions from underwater optical surveys. IEEE J. Oceanic Eng. **34**(2), 150–169 (2009)
18. Singh, H., Armstrong, R., Gilbes, F., Eustice, R., Roman, C., Pizarro, O., Torres, J.: Imaging Coral I: imaging coral habitats with the SeaBED AUV. Subsurf. Sens. Technol. Appl. **5**(1), 25–42 (2004)
19. Giguere, P., Dudek, G., Prahacs, C., Plamondon, N., Turgeon, K.: Unsupervised learning of terrain appearance for automated coral reef exploration. In: Canadian Conference on Computer and Robot Vision, 2009. CRV '09, pp. 268–275, May 2009
20. Spampinato, C., Palazzo, S., Boom, B., Fisher, R.B.: Overview of the lifeclef 2014 fish task. In: Working Notes for CLEF 2014 Conference, pp. 616–624, Sheffield, UK (2014). http://ceur-ws.org/Vol-1180/CLEF2014wn-Life-SpampinatoEt2014.pdf. Accessed 15–18 Sept 2014
21. Johnson-Roberson, M., Kumar, S., Willams, S.: Segmentation and classification of coral for oceanographic surveys: a semi-supervised machine learning approach. In: OCEANS 2006—Asia Pacific, pp. 1–6, May 2006
22. Girdhar, Y., Whitney, D., Dudek, G.: Curiosity based exploration for learning terrain models. In: IEEE International Conference on Robotics and Automation (ICRA), 2014, pp. 578–584. IEEE (2014)
23. Dudek, G., Jenkin, M., Prahacs, C., Hogue, A., Sattar, J., Giguere, P., German, A., Liu, H., Saunderson, S., Ripsman, A., Simhon, S., Torres-Mendez, L.A., Milios, E., Zhang, P., Rekleitis, I.: A visually guided swimming robot. In: Proceedings of Intelligent Robots and Systems, Edmonton, Alberta, Canada, Aug 2005

24. Russakovsky, O., Deng, J., Su, H., Krause, J., Satheesh, S., Ma, S., Huang, Z., Karpathy, A., Khosla, A., Bernstein, M., Berg, A.C., Fei-Fei, L.: ImageNet Large Scale Visual Recognition Challenge (2014)
25. Gabor, D.: Theory of communication. J. IEEE **93**, 429–457 (1946)
26. Fogel, I., Sagi, D.: Gabor filters as texture discriminator. Biol. Cybern. **61**(2), 103–113, (1989). http://dx.doi.org/10.1007/BF00204594
27. Meer, P., Jolion, J., Rosenfeld, A.: A fast parallel algorithm for blind estimation of noise variance. IEEE Trans. Pattern Anal. Mach. Intell. **12**(2), 216–223 (1990)
28. Ojala, T., Pietikainen, M., Maenpaa, T.: Multiresolution gray-scale and rotation invariant texture classification with local binary patterns. IEEE Trans. Pattern Anal. Mach. Intell. **24**(7), 971–987 (2002)

Part II
Vision

BOR^{2}G: Building Optimal Regularised Reconstructions with GPUs (in Cubes)

Michael Tanner, Pedro Piniés, Lina Maria Paz and Paul Newman

Abstract This paper is about dense regularised mapping using a single camera as it moves through large work spaces. Our technique is, as many are, a depth-map fusion approach. However, our desire to work both at large scales and outdoors precludes the use of RGB-D cameras. Instead, we need to work with the notoriously noisy depth maps produced from small sets of sequential camera images with known inter-frame poses. This, in turn, requires the application of a regulariser over the 3D surface induced by the fusion of multiple (of order 100) depth maps. We accomplish this by building and managing a cube of voxels. The combination of issues arising from noisy depth maps and moving *through* our workspace/voxel cube, so it envelops us, rather than orbiting *around* it as is common in desktop reconstructions, forces the algorithmic contribution of our work. Namely, we propose a method to execute the optimisation and regularisation in a 3D volume which has been only partially observed and thereby avoiding inappropriate interpolation and extrapolation. We demonstrate our technique indoors and outdoors and offer empirical analysis of the precision of the reconstructions.

1 Introduction and Prior Work

Building maps and workspace acquisition are established and desired competencies in mobile robotics. Having "better maps" is loosely synonymous with better operation and workspace understanding. An important thread of work in this area is

M. Tanner (✉) · P. Piniés · L.M. Paz · P. Newman
Mobile Robotics Group, Department of Engineering Science, University of Oxford, 17 Parks Road, Oxford OX1 3PJ, UK
e-mail: mtanner@robots.ox.ac.uk; michael.tanner@new.ox.ac.uk

P. Piniés
e-mail: ppinies@robots.ox.ac.uk

L.M. Paz
e-mail: linapaz@robots.ox.ac.uk

P. Newman
e-mail: pnewman@robots.ox.ac.uk

D.S. Wettergreen and T.D. Barfoot (eds.), *Field and Service Robotics*,
Springer Tracts in Advanced Robotics 113, DOI 10.1007/978-3-319-27702-8_8

dense mapping in which, in stark contrast to the earliest sparse-point feature maps in mobile robotics, we seek to construct continuous surfaces. This is a well studied and vibrant area of research. In this paper we consider this task in the context of large scale workspace mapping—both indoors (despite depleted texture on drab walls) and outdoors (with a large range of scales) using only a mono-camera.

A precursor to many dense reconstruction techniques, including ours, are 2.5D depth maps. These can be generated using a variety of techniques: directly with RGB-D cameras, indirectly with stereo cameras, or as in our case, from a single camera undergoing known motion.

RGB-D sensor-driven work often uses Microsoft Kinect or Asus Xtion PRO devices for example [11, 17, 19, 23]. Such "RGB-D" systems provide VGA colour and depth images at around 30 Hz, but this is at the cost of range (0.8–3.5 m) and the ability to only reliably operate indoors [20], although outdoor operation is possible at night and with the same range limitation [18]. However, for the indoor environments these structured light sensors can operate in, they produce extremely accurate 3D dense reconstructions even in low-texture environments.

Stereo cameras also enable dense reconstruction but do introduce complexity and concerns around stable extrinsic calibration to the degree that they can be cost-prohibitive for low-end robotics applications [1]. An alternative approach is to leverage a sequence of mono images. In this case we do need an external method to derive, or at least seed, accurate estimates of the inter-frame motion of the camera—perhaps from an IMU-aided Visual Odometry systems or a forward kinematic model of an arm. Note that in this work, because our focus is on the reconstruction component, we assume that this is given and point the reader to [9] for an example system. With the pose estimates between sequential images as a given, the depth of each pixel can be estimated using an identical approach to that taken in creating depth maps from stereo cameras [5, 8].

Full 3D dense reconstruction has only been demonstrated in either indoor environments [14] or small-scale outdoor environments [6, 21]. Interestingly both these methods rely on a fully-observed environment in which the observer orbits the subject. In an important sense and in contrast to what we shall present, these techniques all are object-centred in situ where the camera trajectory is chosen to generate quality depth maps. In many mobile robotics applications—e.g., an autonomous vehicle limited to an on-road trajectory—the environment observations are constrained and suboptimal for these traditional dense reconstruction techniques.

RGB-D based reconstructions can rely on high quality depth maps always being available. In this case, regularisation is not required since an average of measurements in the voxel grid can provide visually appealing results. When using camera-derived depth-maps, a vital and defining point is that the depth maps are almost always noisy and ill formed in places—particularly a problem when operating in regions where there is a dearth of texture. Accordingly, regularisation techniques must be applied to reduce these effects—essentially introducing a prior over the local structure of the workspace (planar, affine, smooth, etc.) [13].

In this paper, we propose a depth map fusion approach to densely reconstruct environments using only a monocular camera as it moves *through* large work spaces.

Given a set of noisy dense depth maps from a sub set of monocular images, we formulate the 3D fusion as a regularised energy minimisation problem acting on the Truncated Signed Distance Function (TSDF) that parametrises the surface induced by the fusion of multiple depth maps. We represent our solution as the zero-crossing level of a regularised cube. Our method can execute the optimisation and regularisation in a 3D volume which has been only partially observed while avoiding inappropriate interpolation and extrapolation.

What follows is a technique that leverages many of the constructs of previous work to achieve 3D dense reconstruction with monocular cameras but with an input range from 1.0 to 75 m in regions of low texture. We do this without requiring privileged camera motion and we do it at a near-interactive rate. We begin in Sect. 2 by describing how we frame the problem in the context of an implicit 3D function, the TSDF. In Sect. 3, we formulate the solution of the depth map fusion problem as a regularised energy minimisation. Section 4 explains the theoretical insights which allow us to set new boundary conditions inside the cube. We present the main steps of algorithmic solution in Sect. 5. Quantitative and qualitative results on a synthetic data set rendering an indoor place, and real experiments on challenging indoors/outdoors are presented in Sect. 6. Finally, we draw our conclusions and future lines of research in Sect. 7.

2 Construction of the Problem Volume: The BOR^{2}G Cube

This paper is about building optimal regularised reconstructions with GPUs. Our fundamental construct is a cube of voxels, which we refer to as the BOR^{2}G Cube, into which data is assimilated.

The cube model is a discretised version of a Truncated Signed Distance Function (TSDF) $u : \Omega \rightarrow \mathbb{R}$ where $\Omega \subset \mathbb{R}^3$ represents a subset of points in 3D space and u returns the corresponding truncated distance to surfaces in the scene [4]. The TSDF is constructed in such a way that zero is the surface of an object, positive values represent empty space, and negative values correspond to the interior of objects, as shown in Fig. 1. Thus by finding the zero-crossing level-set, $u = 0$, we can arrive at a dense representation of surfaces in the workspace.

Consider first the case of operating with a single depth map D, an image in which each pixel (i, j) represents the depth $d_{i,j}$ of the closest obstacle in space along the z axis. We use the 4×4 homogeneous matrix $\mathbf{T}_{gc} \in SE(3)$ to express the depth map's camera position, c, with respect to the voxel grid's global frame, g.

For each voxel, the steps to obtain u from a single depth map D are as follows:

1. Calculate the central point $\mathbf{p}_g = [x_g, y_g, z_g]^T$ of the voxel with respect to the camera coordinate frame as $\mathbf{p}_c = \mathbf{T}_{gc}^{-1}\mathbf{p}_g$
2. Compute the pixel (i, j) in D in which the voxel is observed by projecting $\mathbf{p}_c$ into D and rounding each index to the nearest integer.

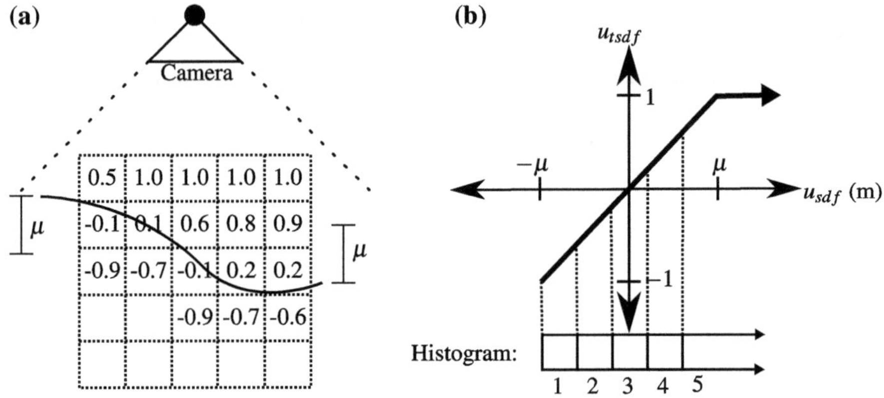

Fig. 1 A graphical depiction (**a**) of how the TSDF values represent the zero-crossing surface in a two-dimensional 'voxel' grid. In (**b**) these TSDF values are discretised into histogram bins ($n_{bins} = 5$). $u \in [-1, 1]$ which directly maps into histogram bins with indices from 1 to n_{bins}. There is no u value and no histogram bin when $u \leq -\mu$, however the n_{bins} histogram bin includes all $u > \mu$

3. If the pixel (i, j) lies within the depth image, evaluate u as the difference between $d_{i,j}$ and the z component of $\mathbf{p}_c$. If $u > 0$, the voxel is between the surface and the camera whereas $u < 0$ indicates the surface occludes the camera's view of the voxel.
4. Finally, linearly scale-and-clamp u such that any voxel for which $u > -\mu$ lies in the interval $[-1, 1]$ whereas voxels for which $u < -\mu$ are left empty. See Fig. 1.

In the next subsection we will explain how to fuse multiple depth images D_t obtained at different moments in time t.

3 Depth Map Fusion

When high-quality depth maps are available, for example depth maps obtained from a Kinect camera, data fusion can be performed by minimising, for each voxel, the following L_2 norm energy,

$$\arg\min_u \int_\Omega \sum_{t=1}^{N} ||u - f_t||_2^2 d\Omega \tag{1}$$

where N represents the number of depth maps we want to fuse, f_t is the TSDF that corresponds to depth map D_t and u is the optimised TSDF after fusing all the information available. Using a voxel grid representation for the TSDFs, the solution to this problem can be obtained by calculating the mean of all the $f_1, \ldots, f_N$ for each individual voxel. This operation can be performed in real time by sequentially integrating a new f_t when a new depth map is available [11]. The searched TSDF u does not require any additional regularisation due to the high-quality of the depth maps used in the fusion.

However, when cameras are used, the depth maps obtained are of much lower quality due, for example, to poor parallax or incorrect pixel matches. Therefore a more robust method is required. In [21] the authors propose an L_1 norm data term, which is able to cope with spurious measurements, and an additional regularisation term, based on Total Variation [16], to smooth the surfaces obtained. The energy minimised is given by,

$$\arg\min_u \int_\Omega |\nabla u|_1 + \lambda \int_\Omega \sum_{t=1}^{N} ||u - f_t||_1 d\Omega \qquad (2)$$

The first component is a *smoothness* term that penalises high-varying surfaces, while the second component, which mirrors Eq. 1, substitutes the L_2 norm with a robust L_1 energy term. The parameter $\lambda > 0$ is a weight to trade off between the regularisation and the data terms. The main drawback with this approach is that, unlike KinectFusion, we cannot just sequentially update the TSDF u when a new depth map arrives, instead, this method requires to store *all* previous history of depth values in each voxel. This greatly limits the number of depth maps that can be integrated due to memory requirements.

To overcome this limitation, since by construction the TSDFs f_t integrated are bounded to the interval $[-1, 1]$, [22] proposes to sample this interval by evenly spaced bin centres c_b (see Fig. 1) and approximate the previous data fidelity term $\sum_{t=1}^{N} |u - f_t|_1$ by $\sum_{b=1}^{n_{bins}} h_b|u - c_b|_1$ where h_b is the number of times the interval has been observed. The corresponding energy for the histogram approach is,

$$\arg\min_u \int_\Omega |\nabla u|_1 + \lambda \int_\Omega \sum_{b=1}^{n_{bins}} h_b|u - c_b|_1 d\Omega \qquad (3)$$

where the centre of the bins are calculated using,

$$c_b = \frac{2b}{n_{bins}} - 1 \qquad (4)$$

The voting process in the histogram is depicted in Fig. 1. While this voting scheme significantly reduces the memory requirements, allowing us to integrate an unlimited number of depth maps, the optimisation process carried out in [22] is not optimal. A mathematically optimal solution to this problem can be found in [10] and has been applied to histogram-based voxel grids by [6]. Before presenting this optimised solution in Sect. 5, we must introduce what we call the Ω domain.

4 Ω Domain

Since we are moving *within* the voxel grid and only observe a subset of the overall voxels, we need to develop a new technique to prevent the unobserved voxels from negatively affecting the regularisation results of the observed voxels. In order to achieve this, as illustrated in Fig. 2, we define the complete voxel grid domain as Λ and use Ω to represent the subset of voxels which have been directly observed and which will be regularised. The remaining subset, $\bar{\Omega}$, represents voxels which have never been observed. By definition, Ω and $\bar{\Omega}$ form a partition of Λ and therefore $\Lambda = \Omega \cup \bar{\Omega}$ and $\Omega \cap \bar{\Omega} = \emptyset$. All works explained in the previous section rely on a fully-observed voxel grid before regularisation and they implicitly assume that $\Lambda = \Omega$. In our mobile robotics platform, this assumption is not valid. The robot motion results in unobserved regions caused by object occlusion, field-of-view limitations, and trajectory decisions. Therefore, $\Omega \subset \Lambda$ as Fig. 2b illustrates. In this case Eq. 3 turns into,

$$\arg\min_{u} \int_{\Lambda} |\nabla u|_1 + \lambda \int_{\Omega} \sum_{b=1}^{n_{bins}} h_b |u - c_b|_1 d\Omega \tag{5}$$

Note that $\bar{\Omega}$ voxels lack the data term. As is explained in [3], this regularisation interpolates the content of voxels in $\bar{\Omega}$. Extrapolation occurs when we have unobserved voxels surrounding an observed region. To avoid this extrapolation, we use

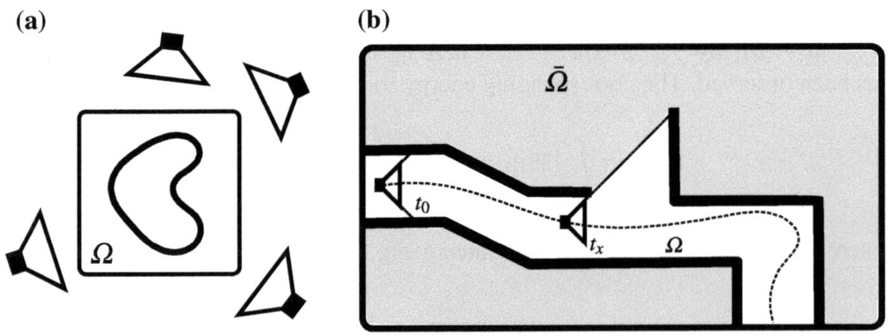

Fig. 2 Traditional voxel-grid-based reconstructions focus on object-centred applications as depicted in (**a**). In this scenario, the objects in the voxel grid are fully observed multiple times from a variety of angles. Even though the internal portion of the object has not been observed, previous regularisation techniques do not make a distinction between Ω (observed regions) and $\bar{\Omega}$ (unobserved regions). This results in spurious interpolation inside the object. However, in mobile robotics applications the world environment is traversed and observed during exploration, requiring large voxel grids (**b**) which result in significant portions never being observed. For example, at camera capture t_x, it is unknown what exists in the camera's upper field of view. Not accounting for $\bar{\Omega}$ in regularisation results in incorrect surface generation. Our technique defines Λ as the voxel grid domain while Ω is the subset we have directly observed and which will be regularised

the Ω domain boundary conditions to constrain regularisation to observed voxels, thus avoiding the indiscriminate surface creation which would occur when naively applying prior techniques.

5 Optimal Regularisation

In this section we describe the steps required to solve Eq. 3 using our Ω-domain constraint. Notice that both terms in Eq. 3 are convex but not differentiable since they depend on the L_1 norm. To solve this, we can use a Proximal Gradient method [3] which requires us to transform one of the terms into a differentiable form. We transform the Total Variation term using the Legendre-Fenchel Transform [15],

$$\min_u \int_\Omega |\nabla u|_1 d\Omega = \min_u \max_{||\mathbf{p}||_\infty \leq 1} \int_\Omega u\nabla \cdot \mathbf{p} d\Omega \tag{6}$$

where $\nabla \cdot \mathbf{p}$ is the divergence of a vector field $\mathbf{p}$ defined by $\nabla \cdot \mathbf{p} = \nabla p_x + \nabla p_y + \nabla p_z$. Applying this transformation to Eq. 3 the original energy minimisation problem turns into a saddle-point (min-max) problem that involves a new dual variable $\mathbf{p}$ and the original primal variable u,

$$\min_u \max_{||\mathbf{p}||_\infty \leq 1} \int_\Omega u\nabla \cdot \mathbf{p} + \lambda \int_\Omega \sum_{b=1}^{n_{bins}} h_b |u - c_b|_1 d\Omega \tag{7}$$

The solution to this regularisation problem was demonstrated in [6] with a Primal-Dual optimisation algorithm [3] which we briefly summarise in the following steps:

1. $\mathbf{p}$, u, and $\bar{u}$ can be initialised to $\mathbf{0}$ since the problem is convex and is guaranteed to converge regardless of the initial seed. $\bar{u}$ is a temporary variable used to reduce the number of optimisation iterations required to converge.
2. To solve the maximisation, we update the dual variable $\mathbf{p}$,

$$\begin{aligned} \mathbf{p} &= \mathbf{p} + \sigma \nabla \bar{u} \\ \mathbf{p} &= \frac{\mathbf{p}}{\max(1, ||\mathbf{p}||_2)} \end{aligned} \tag{8}$$

   where σ is the dual variable gradient-ascent step size.
3. For the minimisation problem, the primal variable u is updated by,

$$\begin{aligned} u &= u - \tau \nabla \cdot \mathbf{p} \\ W_i &= -\sum_{j=1}^{i} h_j + \sum_{j=i+1}^{n_{bins}} h_j \qquad\qquad i \in [0, n_{bins}] \\ b_i &= u + \tau \lambda W_i \\ u &= \text{median}(c_1, \ldots, c_{n_{bins}}, b_0, \ldots, b_{n_{bins}}) \end{aligned} \tag{9}$$

where τ is the gradient-descent step size, W_i is the optimal weight for histogram bin i, and b_i is the regularisation weight for histogram bin i.

4. Finally, to converge in fewer iterations, we apply a "relaxation" step,

$$\bar{u} = u + \theta(u - \bar{u}) \tag{10}$$

where θ is a parameter to adjust the relaxation step size.

Equations 8, 9, and 10 are computed for each voxel in each iteration of the optimisation loop. Since each voxel's computation is independent, we implement this as a GPU kernel which operates within the optimisation loop. The final output, u, represents the regularised TSDF distance.

As discussed in Sect. 4, applying regularisation indiscriminately within the voxel grid produces undesirable results. However, no technique to date, up to the authors' knowledge, provides a method to perform this regularisation within a voxel grid.

Without loss of generality, we describe for the x component—y and z components can be obtained by changing index i for j and k respectively—of the discrete gradient and divergence operations traditionally used to solve Eqs. 8 and 9 [2],

$$\nabla_x u_{i,j,k} = \begin{cases} u_{i+1,j,k} - u_{i,j,k} & \text{if } 1 \leq i < V_x \\ 0 & \text{if } i = V_x \end{cases} \tag{11}$$

$$\nabla_x \cdot \mathbf{p}_{i,j,k} = \begin{cases} \mathbf{p}^x_{i,j,k} - \mathbf{p}^x_{i-1,j,k} & \text{if } 1 < i < V_x \\ \mathbf{p}^x_{i,j,k} & \text{if } i = 1 \\ -\mathbf{p}^x_{i-1,j,k} & \text{if } i = V_x \end{cases} \tag{12}$$

where V_x is the number of voxels in the x dimension.

We extend the traditional gradient and divergence calculations to account for new conditions which remove the $\bar{\Omega}$ domain from regularisation. These methods can be intuitively thought of as introducing additional boundary conditions in the cube which previously only existed on the edges of the voxel grid. For an input TSDF voxel grid u, the gradient $\nabla u = [\nabla_x u, \nabla_y u, \nabla_z u]^T$ is computed by Eq. 11 with the following additional conditions,

$$\nabla_x u_{i,j,k} = \begin{cases} 0 & \text{if } u_{i,j,k} \in \bar{\Omega} \\ 0 & \text{if } u_{i+1,j,k} \in \bar{\Omega} \end{cases} \tag{13}$$

Note that the regulariser uses the gradient to diffuse information among neighbouring voxels. Our gradient definition therefore excludes $\bar{\Omega}$ voxels from regularisation.

Finally, in addition to the conditions in Eq. 12, the divergence operator must be defined such that it mirrors the modified gradient operator

$$\nabla_x \cdot \mathbf{p}_{i,j,k} = \begin{cases} 0 & \text{if } u_{i,j,k} \in \bar{\Omega} \\ \mathbf{p}^x_{i,j,k} & \text{if } u_{i-1,j,k} \in \bar{\Omega} \\ -\mathbf{p}^x_{i-1,j,k} & \text{if } u_{i+1,j,k} \in \bar{\Omega} \end{cases} \tag{14}$$

6 Results

To evaluate the performance of our technique, we performed three experiments comparing our BOR^{2}G method to a KinectFusion implementation. The dense reconstructions are executed on a NVIDIA GeForce GTX TITAN graphics card with 2,880 CUDA Cores and 6 GB of device memory.

As a proof of concept, we first carried out a qualitative analysis of our algorithm on synthetic data (Fig. 3) before performing more robust tests with real-world environments. The synthetic data set provides high-precision depth maps of indoor scenes taken at 30 Hz [7].[1,2] Our chosen scene considers both close and far objects observed from the camera with partial occlusions. The input of our 3D reconstruction pipeline is a set of truth depth maps with added Gaussian noise ($\sigma_n = 10$ cm). As can be seen in Fig. 3, where results are represented using Phong shading, there is a significant improvement in surface normals when the scene is regularised with our BOR^{2}G method compare to KinectFusion. A side-benefit of the regularised normals is that the scene can be represented with fewer vertices. We found that our BOR^{2}G scenes required 2–3 times fewer vertices than the same scene processed by KinectFusion.

To quantitatively analyse our BOR^{2}G method, we conducted two real-world experiments in large-scale environments. Again, we compare BOR^{2}G and KinectFusion fusion pipelines, but we generate our depth maps from a monocular camera using

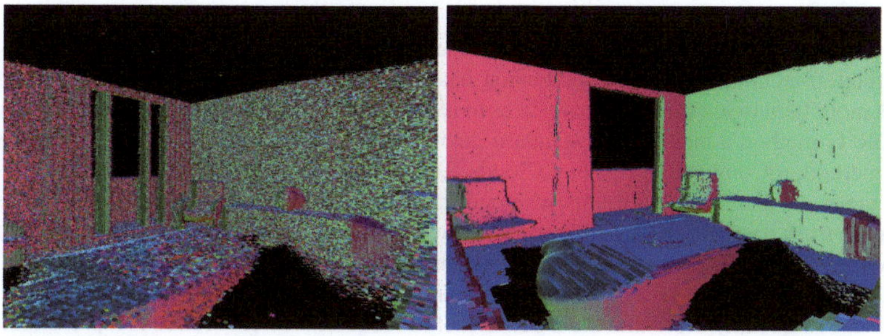

Fig. 3 Comparison of KinectFusion (*left*) and BOR^{2}G regularisation (*right*) methods for a 3D reconstruction of a synthetic [7] environment by fusing noisy depth maps. As input, we use truth depth maps with added Gaussian noise with standard deviation of $\sigma_n = 10$ cm. The Phong shading demonstrates how our regularisation produces consistent surface normals without unnecessarily adding or removing surfaces

[1] http://www.doc.ic.ac.uk/~ahanda/VaFRIC/index.html.

[2] http://www.doc.ic.ac.uk/~ahanda/HighFrameRateTracking/downloads.html.

Table 1 Timing Results of BOR^{2}G *regularisation* on an NVIDIA GeForce GTX TITAN graphics card

Experiment	Voxels	Vol. size (m)	Iterations	Reg. time (s)	Memory (MB)
Woodstock	512^3	$6 \times 25 \times 10$	100	11.09	640
Acland	512^3	$4 \times 6 \times 30$	100	11.24	640

For the configuration parameters, only the volume's dimension changed, but the number of voxels (and hence memory requirements) remained consistent between experiments

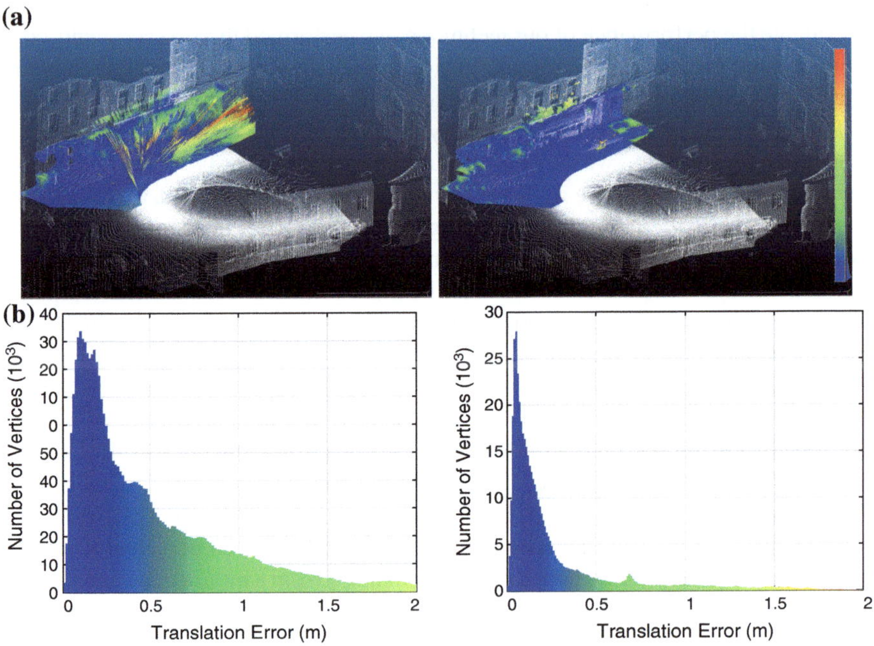

Fig. 4 Woodstock Data Set: Comparison of the KinectFusion (*left*) and BOR^{2}G (*right*) dense reconstruction techniques. The KinectFusion has a larger number spurious outlier segments and requires more than twice the number vertices to represent the structure due to its irregular surfaces. The BOR^{2}G method's median and standard deviation are approximately half that of the Kinect-Fusion method. **a** Comparison of Point Clouds. The KinectFusion implementation (*left*) produced a large range of spurious data points when compared to our BOR^{2}G method (*right*). The *white vertices* are truth data and the colour vertices correspond to the histogram bins in **b**. **b** Histograms of per-vertex-error when compared to laser-generated point clouds. The KinectFusion (*left*) has a median error of 373 mm ($\sigma = 571$ mm) while our BOR^{2}G (*right*) method has a median error of 144 mm ($\sigma = 364$ mm). Note that the BOR^{2}G method requires fewer vertices to represent the same scene

the techniques described in [12]. The first represents the 3D scene reconstruction of an urban outdoor environment in Woodstock, UK. The second is a long, textureless indoor corridor of the University of Oxford's Acland building. In both experiments,

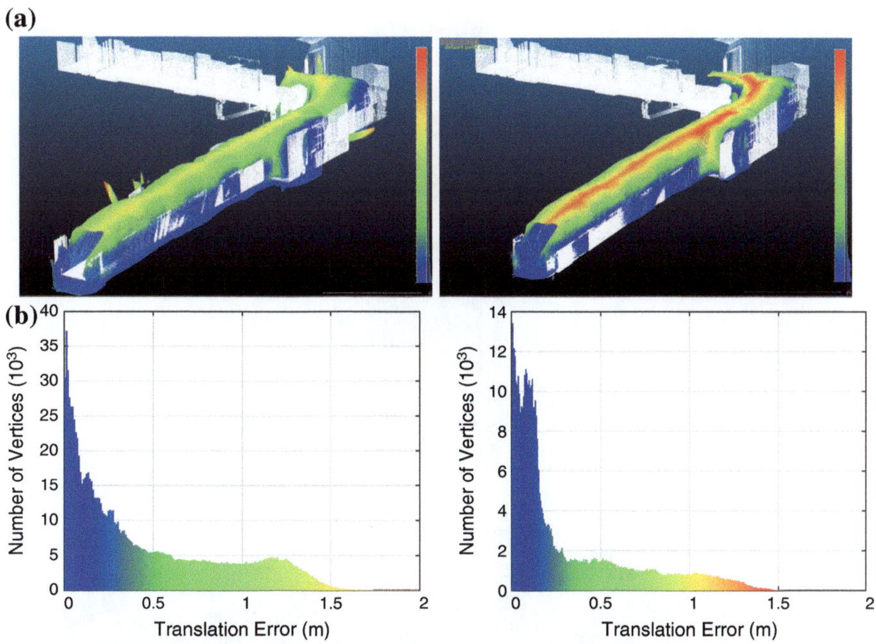

Fig. 5 Acland Data Set: Comparison of the KinectFusion (*left*) and BOR²G (*right*) dense reconstruction techniques. Note that the laser truth data was only measured depth data for the *lower-half* of the hallway. This results in the spurious errors for the *upper-half* where our depth maps produced estimates but for which there was no truth data. These errors dominate the right tail of the histograms in (**b**). As with the Woodstock data set, the BOR²G method's median and standard deviation are approximately half that of the KinectFusion method. **a** Comparison of Point Clouds. The BOR²G (*right*) method again outperformed the KinectFusion implementation (*left*). The *white vertices* are truth data and the colour vertices correspond to the histogram bins in **b**. **b** Histograms of per-vertex-error when compared to laser-generated point clouds. The KinectFusion (*left*) has a median error of 310 mm ($\sigma = 571$ mm) while our BOR²G (*right*) method had a median error of 151 mm ($\sigma = 354$ mm). Note that the BOR²G method requires fewer vertices to represent the same scene

Table 2 Error analysis comparing KinectFusion and BOR²G methods

Experiment	Median error (m)	Standard deviation (m)
Woodstock (KinectFusion)	0.3730	0.5708
Woodstock (BOR²G)	0.1441	0.3636
Acland (KinectFusion)	0.3102	0.5708
Acland (BOR²G)	0.1508	0.3537

The BOR²G error is roughly half that of KinectFusion

Fig. 6 The final 3D reconstruction of the large scale experiments using BOR^{2}G with the Acland building (*left*) and Woodstock, UK (*right*)

we used a frontal monocular camera covering a field of view of 65° × 70° and with an image resolution of 512 × 384.

For ground truth, we generated metrically consistent local 3D swathes from a 2D push-broom laser using a subset of camera-to-world pose estimates $T_{WC} \in SE(3)$ in an active time window as,

$$M_L = f(T_{WC}, T_{CL}, \mathbf{x}_L)$$

where f is a function of the total set of collected laser points $\mathbf{x}_L$ in the same time interval and T_{CL} is the extrinsic calibration between camera and laser. The resulting 3D point cloud M_L is used as ground truth for our large scale assessment.

Table 1 summarises the dimensions of the volume used for each of the experiments, the number of primal dual iterations, and the total running time required for our fusion approach. The execution time for regularisation is highly correlated to the size of the Ω space because regularisation is only performed on voxels within Ω. Figures 4 and

5 show a comparison between the ground truth and the 3D reconstructions obtained using the BOR^{2}G and the KinectFusion methods. To calculate our statistics, we perform a "point-cloud-to-model" registration of the ground truth with respect to our model estimate.[3] The key statistics comparing the methods are precisely outlined in Table 2. For both scenarios, our BOR^{2}G method was roughly two times more accurate than KinectFusion. Finally, Fig. 6 shows the obtained continuous, dense reconstructions of the indoor and outdoor environments.

7 Conclusions

In this paper we presented a new approach to reconstruct large-scale scenes in 3D with a moving monocular camera. Unlike other approaches, we do not restrict ourselves to object-centred applications or rely upon active sensors. Instead, we fuse a set of consecutive mono-generated depth maps into a voxel grid and apply our Ω-domain boundary conditions to limit our regularisation to the subset of observed voxels within the voxel grid.

Our BOR^{2}G method results in a median and standard deviation error that is roughly half that produced when using the same depth maps with the KinectFusion method.

In the future, we plan to use the Ω-domain principles to apply new boundary conditions which select portions of the voxel grid for regularisation. These subsets will be selected based on scene-segmentation heuristics. For example, we can extend the Ω domain to include enclosed "holes" which will result in the regulariser interpolating a new surface. Alternatively, we could remove a segment from Ω to prevent regularisation of a scene segment which was better estimated in the depth map (e.g., high-texture object).

References

1. Bumblebee2 FireWire stereo vision camera systems Point Grey cameras. http://www.ptgrey.com/bumblebee2-firewire-stereo-vision-camera-systems
2. Chambolle, A.: An algorithm for total variation minimization and applications. J. Math. Imaging Vision **20**(1–2), 89–97 (2004)
3. Chambolle, A., Pock, T.: A first-order primal-dual algorithm for convex problems with applications to imaging. J. Math. Imaging Vision **40**(1), 120–145 (2011)
4. Curless, B., Levoy, M.: A volumetric method for building complex models from range images. In: Proceedings of the 23rd Annual Conference on Computer Graphics and Interactive Techniques, pp. 303–312. ACM (1996)
5. Geiger, A., Roser, M., Urtasun, R.: Efficient large-scale stereo matching. In: Asian Conference on Computer Vision (ACCV) (2010)
6. Graber, G., Pock, T., Bischof, H.: Online 3D reconstruction using convex optimization. In: 1st Workshop on Live Dense Reconstruction From Moving Cameras, ICCV 2011 (2011)

[3] http://www.danielgm.net/cc.

7. Handa, A., Whelan, T., McDonald, J., Davison, A.: A benchmark for RGB-D visual odometry, 3D reconstruction and SLAM. In: IEEE International Conference on Robotics and Automation, ICRA. Hong Kong, China (2014) (to appear)
8. Hirschmuller, H.: Semi-global matching-motivation, developments and applications. http://www.hgpu.org (2011)
9. Li, M., Mourikis, A.I.: High-precision, consistent EKF-based visual–inertial odometry. Int. J. Robot. Res. **32**(6), 690–711 (2013)
10. Li, Y., Osher, S., et al.: A new median formula with applications to PDE based denoising. Commun. Math. Sci **7**(3), 741–753 (2009)
11. Newcombe, R.A., Davison, A.J., Izadi, S., Kohli, P., Hilliges, O., Shotton, J., Molyneaux, D., Hodges, S., Kim, D., Fitzgibbon, A.: KinectFusion: real-time dense surface mapping and tracking. In: 2011 10th IEEE International Symposium on Mixed and Augmented Reality (ISMAR), pp. 127–136. IEEE (2011)
12. Pinies, P., Paz, L.M., Newman, P.: Dense and swift mapping with monocular vision. In: International Conference on Field and Service Robotics (FSR). Toronto, ON, Canada (2015)
13. Pinies, P., Paz, L.M., Newman, P.: Dense mono reconstruction: living with the pain of the plain plane. In: IEEE 11th International Conference on Robotics and Automation. IEEE (2015)
14. Pradeep, V., Rhemann, C., Izadi, S., Zach, C., Bleyer, M., Bathiche, S.: MonoFusion: real-time 3D reconstruction of small scenes with a single web camera. In: 2013 IEEE International Symposium on Mixed and Augmented Reality (ISMAR), pp. 83–88 (2013)
15. Rockafellar, R.T.: Convex Analysis. Princeton University Press, Princeton, New Jersey (1970)
16. Rudin, L.I., Osher, S., Fatemi, E.: Nonlinear total variation based noise removal algorithms. In: Proceedings of the 11th Annual International Conference of the Center for Nonlinear Studies on Experimental Mathematics: Computational Issues in Nonlinear Science, pp. 259–268. Elsevier North-Holland, Inc. (1992)
17. Steinbruecker, F., Kerl, C., Sturm, J., Cremers, D.: Large-scale multi-resolution surface reconstruction from RGB-D sequences. In: IEEE International Conference on Computer Vision (ICCV). Sydney, Australia (2013)
18. Whelan, T., Kaess, M., Fallon, M.F., Johannsson, H., Leonard, J.J., McDonald, J.B.: Kintinuous: spatially extended KinectFusion. In: RSS Workshop on RGB-D: Advanced Reasoning with Depth Cameras. Sydney, Australia (2012)
19. Whelan, T., Kaess, M., Johannsson, H., Fallon, M., Leonard, J.J., McDonald, J.: Real-time large-scale dense RGB-D SLAM with volumetric fusion. Int. J. Robot. Res. 0278364914551008 (2014)
20. Xtion PRO specifications. http://www.asus.com/uk/Multimedia/Xtion_PRO/specifications/
21. Zach, C., Pock, T., Bischof, H.: A globally optimal algorithm for robust TV-L 1 range image integration. In: IEEE 11th International Conference on Computer Vision, 2007. ICCV 2007, pp. 1–8. IEEE (2007)
22. Zach, C.: Fast and high quality fusion of depth maps. In: Proceedings of the International Symposium on 3D Data Processing, Visualization and Transmission (3DPVT), 1 (2008)
23. Zeng, M., Zhao, F., Zheng, J., Liu, X.: Octree-based fusion for realtime 3D reconstruction. Graph. Models **75**(3), 126–136 (2013)

Online Loop-Closure Detection via Dynamic Sparse Representation

Moein Shakeri and Hong Zhang

Abstract Visual loop closure detection is an important problem in visual robot navigation. Successful solutions to visual loop closure detection are based on image matching between the current view and the map images. In order to obtain a solution that is scalable to large environments involving thousands or millions of images, the efficiency of a loop closure detection algorithm is critical. Recently people have proposed to apply l_1-minimization methods to visual loop closure detection in which the problem is cast as one of obtaining a sparse representation of the current view in terms of map images. The proposed solution, however, is insufficient with a time complexity worse than linear search. In this paper, we present a solution that overcomes the inefficiency by employing dynamic algorithms in l_1-minimization. Our solution exploits the sequential nature of the loop closure detection problem. As a result, our proposed algorithm is able to obtain a performance that is an order of magnitude more efficient than the existing l_1-minimization based solution. We evaluate our algorithm on publicly available visual SLAM datasets to establish its accuracy and efficiency.

1 Introduction

Autonomous mobile robots are beneficial to work in hazardous environments, or places out of range of human operators over long periods of time, such as exploration and rescue. In many environments robots have no prior knowledge about their surroundings. Therefore, it is essential for a robot to be able to map an unknown environment itself in order to perform its tasks. Simultaneous Localization And Mapping (SLAM) has been a focus of robotics research and, among the many issues of concern, is the detection of loop closures, i.e., revisits to map locations.

M. Shakeri (✉) · H. Zhang
Department of Computing Science, University of Alberta, Edmonton, Canada
e-mail: shakeri@ualberta.ca

H. Zhang
e-mail: hzhang@ualberta.ca

D.S. Wettergreen and T.D. Barfoot (eds.), *Field and Service Robotics*,
Springer Tracts in Advanced Robotics 113, DOI 10.1007/978-3-319-27702-8_9

In order to be able to handle a large environment, a loop closure detection algorithm must be efficient. The dominant approach in SLAM literature to meet this requirement is based on visual bag-of-words (BoW) that achieves efficiency through indexing. Visual BoW however often requires offline construction of a visual vocabulary, which may not be representative of the environment that a robot will visit online, and the step of keypoint detection and vector quantization can be computationally costly.

An alternative to visual BoW for loop closure detection is compact whole image descriptors that avoid the step of keypoint detection and vector quantization [13]. In this case, loop closure detection is solved as a nearest neighbor search considering the descriptor of the current view as the search key. Recently, an interesting solution based on l_1-minimization has been proposed that solves this nearest neighbor search problem through sparse reconstruction. The proposed solution, although elegant, is less efficient than linear search to find the nearest neighbor, and offers little incentive for people to adopt.

In this paper, we improves the solution based on l_1-minimization by exploiting the sequential nature of the loop closure detection problem, i.e., successive robot views look similar so that l_1-minimization does not need to be solved from scratch. We make use of recent algorithms in dynamic algorithms for l_1-minimization to achieve a solution that is an order of magnitude more efficient than the static l_1-minimization, without sacrificing accuracy. Most importantly, our solution is more efficient than linear searsh and is therefore a competitive candidate in tackling the problem of visual loop closure detection with whole-image descriptors.

The remainder of this paper is organized as follows. In Sect. 2, we discuss related works that address the loop closure detection problem. In Sect. 3, we present a brief overview of sparse representation using l_1-minimization to solve the loop closure problem. Also dynamic update of the optimization problem to avoid solving l_1-minimization for each input image is described. In Sect. 4, we explain how the proposed dynamic sparse representation can be utilized in visual robot navigation and in Sect. 5 we present the experimental results on standard datasets. Finally, we summarize our approach and offer concluding remarks in Sect. 6.

2 Related Works

Loop closure detection is a fundamental problem in SLAM and is defined as the detection of the event when a robot returns to a previously visited place. This information is necessary, since it allows the robot to reduce and bound the errors and uncertainty in the estimated pose and environment map. Loop closure detection has been extensively studied and many solutions have been proposed over the years for robot navigation. In this work we focus only on image-based methods.

One of the popular methods to address this problem is visual Bag-of-Words (BoW). The BoW approach has achieved considerable success in content-based image retrieval as well as in object recognition and image classification [5]. The solution uses an offline process in which features in training images are extracted and their descriptors are clustered. The cluster centres then serve as visual words and the collection of visual words form a visual dictionary or vocabulary [17]. Given a query image, its visual features are vector quantized through a nearest-neighbor (NN) algorithm to match with the visual words in the dictionary, and an image descriptor is built in terms of the histogram of the visual words appearing in the image. Candidate images that are similar to a query image can be retrieved efficiently using an inverted index. Because the visual dictionary is built offline, the online cost includes feature extraction, nearest neighbor search, and indexing of the query image. Although BoW has been shown to be an efficient method for producing loop closure candidates, it suffers two key weaknesses. First, an offline step is often needed to build a visual vocabulary from training images, but the training images may not represent the future views of the robot appropriately. Secondly, the step of vector quantization, which converts visual features into visual words required by indexing, can be inefficient with a linear search and may cause perceptual aliasing [14], i.e., high similarity between different locations.

Nister et al. [15] proposed "vocabulary tree" as a way of speeding up nearest neighbor search in a large database and [3] used this method for loop closure detection in visual SLAM. Vocabulary tree was introduced as a hierarchical approach to Bag of Words although a tree structure does not guarantee the exact nearest neighbor. Cummins et al. [6] proposed FAB-MAP as a probabilistic appearance based approach using BoW and showed its performance on large scale environment. Although Galvez-Lopez et al. [11] advance the method by introducing a Binary BoW (BBoW) to speed up the method, it still needs an offline process to build a dictionary.

As a competing approach to visual BoW, compact whole-image descriptors such as Gist [16] have been recently employed in performing visual loop closure detection [13]. Rather than describing an image in terms of its keypoints, a whole-image descriptor may simply use a down-sampled version of an image, its gradient information, or its response to a filter bank, to describe the image. Whole image descriptors can avoid the computational complexity of feature detection and vector quantization in BoW, but introduces the need to perform nearest neighbor search in matching the descriptor of the current view and those of the map locations. In addition, the quality of detected loop closures can be affected as the result of simple representations. These issues have been alleviated with some success with the help of the Monte Carlo technique [13].

Most recently, an interesting solution has been presented to cast the loop closure detection problem as one of sparse reconstruction [12]. The solution uses l_1-minimization algorithms and is accurate in matching the current view with the map images. In their work, the current view of the robot is matched with a small number of the all observations from the map images through convex l_1-minimization which provides a sparse solution. By using a fast convex optimization technique, they showed their method to be fast enough for a map with 8,500 images. However, since their method needs to solve l_1-minimization problem from scratch for each newly captured image, increasing the number of images as the map size grows leads to a computational complexity that can be infeasible in large scale environments.

In fact, their method has a time complexity that is worse than linear search, as we will show in the experimental result section, and this gives little incentive for one to choose this method for solving the loop closure detection problem.

To address the computational complexity issue of solving l_1-minimization from scratch, in this paper we introduce a highly efficient approach for loop closure detection by first solving a static l_1-minimization problem once and then updating the convex l_1-minimization solution dynamically to avoid solving a new optimization problem for each newly captured image. We exploit the fact that in visual SLAM the current robot view is similar to the recent previous views. We use this property of the loop closure detection problem to formulate our solution as the dynamic update of the solution to l_1-minimization in the previous step.

3 Sparse Solution for Loop-Closure Detection

Sparse Representation (SR) is a signal processing technique for reconstructing a signal by finding solutions to an underdetermined linear system and it is solved through convex optimization algorithms. SR has been extensively used for face recognition [18], denoising [9], etc. We use this framework to find the closest image in a robot map to a new observation for loop closure detection in SLAM. In this section, we first present a brief overview of sparse representation and l_1-minimization. Then we will describe the dynamic update of the convex minimization problem to approximate the solution without solving a complete new minimization.

3.1 Loop Closure Detection via Convex l_1-Minimization

Image matching is essential for loop closure detection in visual SLAM. One recent successful image matching method, especially in large datasets, is the SR method [12]. Let n be the number of images in the robot map, and m be the dimension of the descriptor of each image in the map. Further, assume y to be the current view of the robot. The map images form a matrix $A \in R^{m\times n}$, and the linear representation of y can be rewritten in terms of all map images:

$$y = Ax_0 \quad , \quad y \in R^m \tag{1}$$

where x_0 represents the contributions of the map images to the reconstruction or representation of the current view. In SR the system is underdetermined with $m < n$. Therefore, recovering x_0 constitutes a non-trivial inverse problem. A classic solution to this problem is linear least squares, which finds the minimum l_2-norm solution to this system.

$$\hat{x}_2 = argmin\|x\|_2 \quad subject\ to \quad Ax = y \tag{2}$$

Equation (2) can be easily solved, but the solution $\hat{x}_2$ is dense (i.e., all its elements are non-zero in general) as is shown in [18] and is therefore not useful to retrieve y. Due to the fact that the query image can be represented using the map images at locations similar to the current robot location—if there is loop closure—the representation is naturally sparse, i.e., all but a small number of the elements of x are 0. The sparsest solution to $y = Ax$ is obtained by the following optimization problem:

$$\hat{x}_0 = argmin\|x\|_0 \quad subject\ to \quad Ax = y \tag{3}$$

The problem of finding the sparsest solution of an under-determined system of linear equations is NP-hard and difficult even to approximate [1, 18]. The theory of sparse representation [8] shows that if the solution $\hat{x}_0$ is sparse enough, the solution of the l_0-minimization is equal to the solution of the l_1-minimization, and x_0 can be retrieved by computing the minimum l_1-norm:

$$x^* = argmin\|x\|_1 \quad subject\ to \quad Ax = y \tag{4}$$

In real applications such as image matching in visual SLAM, a true loop closing image y can only be represented by map images approximately with slightly different illuminations, translations, and rotations. In such cases $\|Ax - y\|_2 \leq \epsilon$, where $\epsilon > 0$. So, to find a sparse solution x^*, one could use conventional l_1-regularized least squares regression as follows:

$$x^* = \underset{x}{argmin}\frac{1}{2}\|Ax - y\|_2^2 + \lambda\|x\|_1 \tag{5}$$

where l_1-regularization enforces sparsity on x^*; unfortunately, the complexity of solving (5) grows polynomially with m and n.

3.1.1 Solving l_1-Minimization

In practice, for loop closure detection in visual SLAM, we have to solve (5) online and accurately. One of the fastest l_1-minimization methods is *homotopy* algorithm associated with the *basis pursuit denoising* (BPDN) [4] approach, which is applied by Latif et al. [12] and described below for the completeness of presentation.

A solution x^* to (5) should follow the condition [2, 10]:

$$\|A^T(Ax^* - y)\|_\infty \leq \lambda \tag{6}$$

In the above equation, we distinguish between the nonzero components and the zero components of x^*. We denote $\bar{x}^*$ the reduced dimensional vector built upon the nonzero components of x^*. Similarly, A_Γ denotes the associated columns in A (*Γ is a set with the indexes of nonzero elements in x^**). So, the optimality conditions for any given value of λ are as follows [10]:

$$A_{\Gamma}^{T}(Ax^{*} - y) = -\lambda z \tag{7}$$

$$\|A_{\Gamma^{c}}^{T}(Ax^{*} - y)\|_{\infty} < \lambda \tag{8}$$

where A_{Γ} is a $m \times |\Gamma|$ matrix from the columns of A indexed by Γ and vector z is signs of $\bar{x}^{*}$. $A_{\Gamma^{c}}$ denotes all columns of A that are not in A_{Γ}. From the support Γ and z, the solution x^{*} can be calculated as follows [2, 10]:

$$x^{*} = \begin{cases} (A_{\Gamma}^{T}A_{\Gamma})^{-1}(A_{\Gamma}^{T}y - \lambda z) & on\ \Gamma \\ 0, & otherwise \end{cases} \tag{9}$$

The algorithm proceeds by computing (7), (8), and (9) iteratively, until $A^{T}(Ax^{*} - y) < c$ (a small constant such as 10^{-6}) and the final x^{*} represents the solution for (5).

3.2 Dynamic Update for Homotopy

The static homotopy solution described in the previous section has a complexity that is polynomial in n and m, and can therefore be too slow for a large scale map. However, in loop closure detection, we expect the current image captured by a robot to be similar to the image that robot captured in the previous time instance. So, we can update the l_1-minimizer for the last image, described in Sect. 3.1.1, to obtain the solution to the current image without solving the optimization problem from scratch. Asif and Romberg [2] explained the problem of estimating a time varying sparse signal from a series of linear measurement vectors to update the standard BPDN homotopy dynamically. They assumed that the signal changes only slightly between measurements, so that the reconstructions will be closely related, an assumption that holds true in visual SLAM for finding the best match between the current image and the map images. This dynamic method enables us to arrive at a solution that is highly efficient and capable of handling large-scale robot environments. In the rest of this section, we apply the dynamic algorithm [2] to loop closure detection in visual SLAM.

Assume that we have solved the BPDN problem (5) for a given value of λ. Now, for a new image, we express it as a m-dimensional feature vector $\breve{y}$, and the problem we have to solve for the new image approximately is:

$$\underset{x}{argmin}\ \frac{1}{2}\|Ax - \breve{y}\|_{2}^{2} + \lambda\|x\|_{1} \tag{10}$$

with the same value of λ in (5). In classical approaches (10) is solved for each image without benefiting from the just-completed solution. Our goal is using the information from the solution of (5) to quickly compute the solution for (10). Thus, we use the homotopy formulation in [2]:

$$\underset{x}{argmin} \ \frac{1-\epsilon}{2}\|Ax-y\|_2^2 + \frac{\epsilon}{2}\|Ax-\breve{y}\|_2^2 + \lambda\|x\|_1 \tag{11}$$

where ϵ is the homotopy parameter. By increasing ϵ from 0 to 1, (11) moves from the solution of (5) to the solution of (10). By adapting the optimally conditions of (7) and (8) for (11), we have:

$$A_\Gamma^T(Ax^* - (1-\epsilon)y - \epsilon\breve{y}) = -\lambda z \tag{12}$$

$$\|A_{\Gamma^c}^T(Ax^* - (1-\epsilon)y - \epsilon\breve{y})\|_\infty < \lambda \tag{13}$$

where Γ is the support of solution x^* and z is its sign sequence on Γ. From (12) the solution x^* for (11) follows a piecewise linear path as ϵ varies. The critical point in this path occurs when an element is either added or removed from the solution x^*. Parameter ϵ increases incrementally from 0 to 1 and [2] proved that the direction of the solution x^* moves by:

$$\partial x = \begin{cases} (A_\Gamma^T A_\Gamma)^{-1} A_\Gamma^T(\breve{y} - y), & on \ \Gamma \\ 0, & otherwise \end{cases} \tag{14}$$

With the moving direction given by (14), we are able to find the step-size θ [2], which leads us to the next critical value of ϵ. Afterwards, the solutions at that point are as follows:

$$\epsilon \longleftarrow \epsilon + \theta, \qquad x^* \longleftarrow x^* + \theta\partial x \tag{15}$$

This procedure is repeated from (12) to (15) until $\epsilon = 1$ and the final x^* represents the solution for (11), which means the best matched images could be found by this dynamic updating method without solving (11) independently.

4 Implementation Details and Discussion

We have formulated loop closure detection problem with the dynamic update of BPDN homotopy algorithm in Sect. 3.2. Here, we explain how this novel formulation can be utilized in visual loop closure detection.

4.1 Initialization

To initialize our solution to loop closure detection and construct our A matrix, we use the first n images or keyframes captured by the robot where n is a small number (e.g., 20). In addition, as is customary, we exclude the last l images seen by the robot from consideration in matching the current view with the map images in order to avoid triggering false loop closure detection due to the similarity between successive images. l is another small number where $l < n$ (e.g., 15).

After constructing A, for the next query image we use standard homotopy to obtain the initial solution x^* just once, and this solution is updated via dynamic method for all the subsequent images while the robot moves and captures additional keyframes.

4.2 Selection of the Top Candidate from Solution x^*

The solution x^*, obtained by either homotopy or its dynamic update, is naturally sparse and represents candidate images from the robot map to best match with the current view. To find a unique image and potentially close a loop, we select the greatest contribution α_i from the solution $x^* = [\alpha_1, \ldots, \alpha_n]$. The index i, corresponds to the column of the matched image in the map. To improve the chance of true positive detection and reduce false alarm, we use the heuristic that if 2-norm between the matched image and the current image is less than a predefined threshold τ, a loop closure is detected ($\|A_{:,i} - \check{y}\|_2 < \tau$). τ can be chosen empirically and, as will be shown in our experiments, the precision of the detected loop closures can reach 100 %.

We should add that the proposed method accommodates the growth of the robot map, when a novel image is detected, by adding a column to the end of the matrix A, i.e. $A_k = [A_{k-1},\ f_k]$ so that the map grows incrementally similar to [12]. f_k denotes the descriptor of kth image being added to the map. The main steps of our method are summarized in Algorithm 1.

Algorithm 1 Closing Loops via the Proposed Method

Initialization:
  Preparing Matrix A (see Sect. 4.1), $\lambda = 0.5$,
For the first query image i
  Obtain x^* with Solving (5) via standard homotopy
  Closing loops (see Sect. 4.2)
  Update matrix A (see Sect. 4.2)
For all query images i to n
  Update x^* via dynamic method (see Sect. 3.2)
  Closing loops (see Sect. 4.2)
  Update matrix A (see Sect. 4.2)

4.3 Discussion

In terms of computational cost, although homotopy is one of the popular and fastest solvers for SR, the computational complexity of homotopy is still $O(dm^2 + dmn)$ to recover a d-sparse signal in d steps. Obviously, this complexity grows polynomially with m and it is expensive for large-scale datasets or maps. In contrast, in the proposed method, for each query image, the main computational cost comes from solving a $|\Gamma| \times |\Gamma|$ system of equations to compute the direction in (14). $|\Gamma|$ is equal to the number of nonzero elements of the sparse solution x^* which means its size is small enough. Therefore, the computational cost of the proposed method is significantly lower than the static homotopy method.

5 Experimental Results

In this section, we perform a set of experiments to demonstrate and validate the capability of the dynamic updating of l_1-minimization method to perform loop closure detection in visual SLAM. In particular, we evaluate the computational cost and the accuracy of the proposed method and compare it with the standard l_1-minimization in [12] and a nearest neighbor (NN) method. Since the proposed method is appropriate to detect loop closure in large scale datasets, we compare our method with FAB-MAP as well. We use three datasets: New College, City Centre, and a Google Street View dataset, with the following details.

- New College: This dataset consists of 2146 images along a 2.2 km trajectory. Each image has originally a resolution of 640×480 and is down sampled to 320×240. The dataset provides stereo images from the left and right of the robot, and we use both images from each location as query image with a combined resolution of 640×240 so that $n = 1073$.
- City Centre: This dataset consists of 2474 images and similar to New College dataset provides stereo images. Each image has a resolution of 640×480 and is down sampled to 320×240. Again, we use both images from each location as query image with a combined resolution of 640×240 so that $n = 1237$.
- Google Street View: This dataset consists of about 50,000 images captured in downtown Pittsburg by Google. This dataset has omni-directional images by four cameras at each location, and we reduce the resolution of the obtained panoramic view to 640×240. The number of locations in this dataset is around $n = 12{,}500$.

To describe an image, we use HOG [7] with $m = 576$ dimensions and a constant weighting parameter $\lambda = 0.5$. In this section, we focus on the capability of our dynamic model in comparison with the static homotopy and NN method. In all methods we use the same algorithm for closing loops. Also, to be consistent with [12] we just pick the highest α as a top candidate (the first method in Sect. 4.2).

5.1 Execution Time

In the first set of experiments, we compare the computational cost of the proposed method with the standard l_1-minimization and the NN method, when the size of the dataset is increased incrementally. We run the experiments in Matlab 2011b on a desktop computer with Core-i7 CPU of 3.40 GHz and 16 GB RAM.

Figures 1 and 2 show the execution time for finding the best candidates on "City Centre" and "New College" datasets respectively, when the images are added to the map incrementally. These figures illustrate the proposed method is faster and more stable (smaller standard deviation) than the standard L1-minimizer, by a factor of four on average. Also, our dynamic model is around two times faster than the nearest neighbor method on both datasets.

To show the capability of the proposed method on larger datasets, we compare the computational time of the dynamic method with the standard l_1-minimizer and NN method to find the best match on "Google Street View" dataset in Fig. 3. This experiment confirms that the proposed method is much faster than both the standard homotopy and NN methods when the map is large. Also, Fig. 3 demonstrates

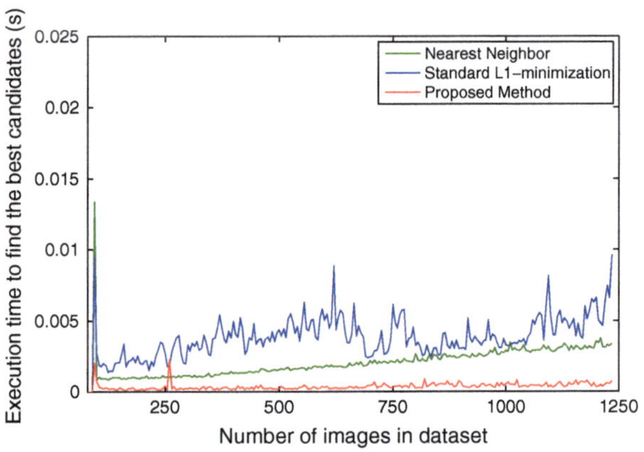

Fig. 1 Execution time comparison (in seconds) on "City Centre" dataset

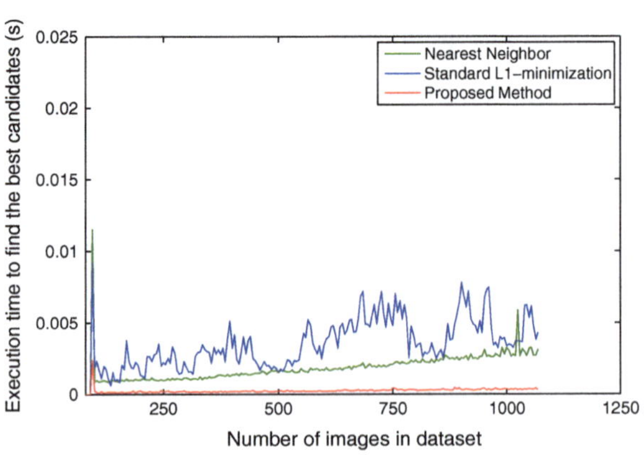

Fig. 2 Execution time comparison (in seconds) on "New College" dataset

Fig. 3 Execution time comparison (in seconds) on "Google Street View" dataset with using of 576 dimensional HOG descriptor for images

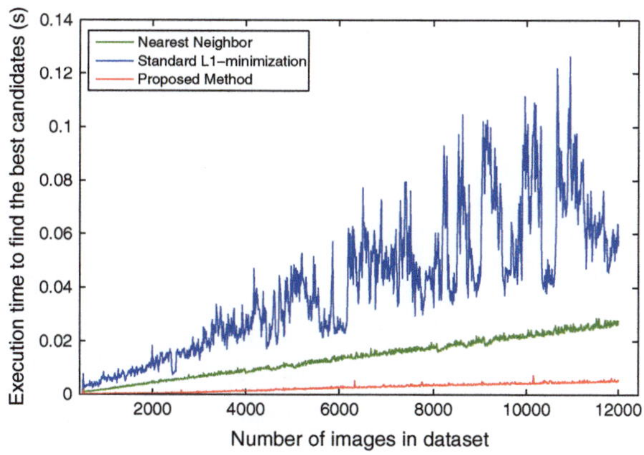

Fig. 4 Execution time comparison (in seconds) on "Google Street View" dataset with using of 81 dimensional HOG descriptor for images

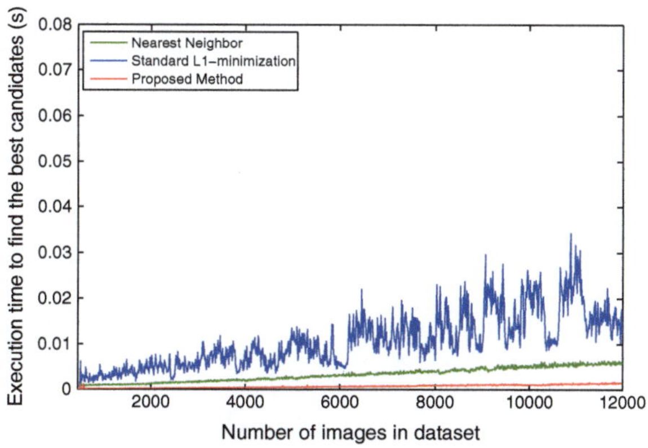

the dynamic update method has much smaller standard deviation than the standard homotopy method to obtain the solution as the map expands with additional images. Furthermore, comparison of Figs. 3 and 4 shows the scalability of our model in comparison with two other methods in terms of feature vector dimension. By increasing m as the dimension of feature vector, computation time of the two other competing methods increases at a faster rate than our model. The qualitative result on the Google Street View dataset is also shown in Fig. 5 where the blue lines represent the robot map and the red dots represent detected loop closures by the proposed method. The ground truth of loop closures (in green dots) can be found in Fig. 6.

Table 1 shows the quantitative results of Figs. 1, 2 and 3 in terms of average execution time and standard deviation on the three datasets. For small maps like City Centre or New College datasets, the proposed method is 10 times faster than the standard homotopy on average. The execution time on the Google Street View dataset in Table 1 shows the capability of our method on large maps in comparison with the standard homotopy and even NN method. The average computational time of the standard homotopy increases more than 20 times from around 3.5 to 77 ms, when the map grows 10 times from around 1200 images to 12,000 images; however, the computational time for the proposed method only increases linearly, and the standard

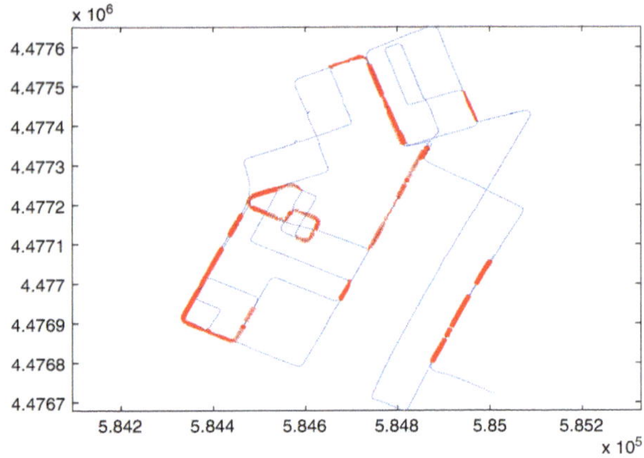

Fig. 5 Qualitative result of the proposed method on "Google Street View" dataset to find loop closures

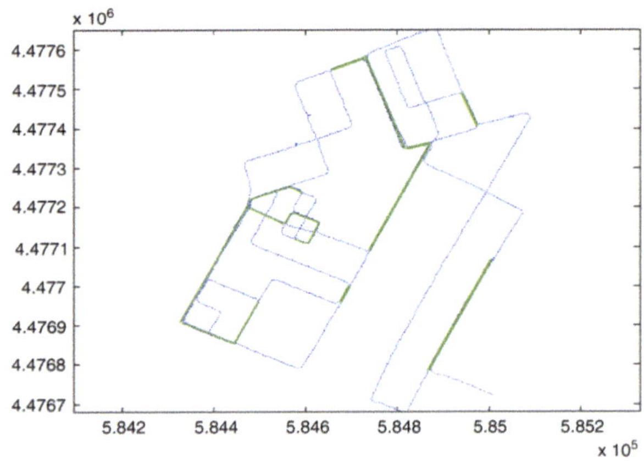

Fig. 6 Graphical ground truth for "Google Street View" dataset from Pittsburg

Table 1 Execution time statistics for one iteration of the loop detection algorithm of the proposed method, the standard homotopy, and NN method on different datasets

	Map size (K)	Nearest Neighbor		Standard homotopy		Proposed method	
		Mean (ms)	Std (ms)	Mean (ms)	Std (ms)	Mean (ms)	Std (ms)
City Centre	1.2	1.91	0.20	3.59	2.28	0.40	0.26
New College	1.1	1.74	0.23	3.32	2.52	0.39	0.22
Google Street	12.5	26.80	1.53	77.34	33.40	4.97	0.82

deviation increases approximately 3 times when the map grows 10 times. In absolute terms, with crude extrapolation, our proposed algorithm could potentially perform loop closure detection in a map with a million images in under one second.

5.2 Loop Closure Detection Accuracy

In this part, we compare the accuracy of the proposed dynamic method against the standard homotopy method for the loop closure detection. Figures 7 and 8 show the precision recall curves of the proposed method and the standard homotopy method on "City Centre" and "New College" datasets respectively. Like before, no specific verification step was used and the decision for closing loops is only based on simple thresholding of the top α_i as described in the Sect. 4.2. According to these figures, the accuracy of the proposed method using a dynamic algorithm is essentially the same as the standard l_1-minimization method. Therefore, using the proposed method, loop closure detection can be solved much faster without losing accuracy. We also compare the proposed method with FAB-MAP as a baseline method for large-scale dataset in Table 2. Although FAB-MAP is not the state-of-the-art in terms of loop detection accuracy, it has been evaluated on the same datasets as used in this work, and is therefore directly comparable. At 100 % precision, the proposed method achieves 68 and 57 % recall on "City Centre" and "New College" datasets with $\tau = 0.98$

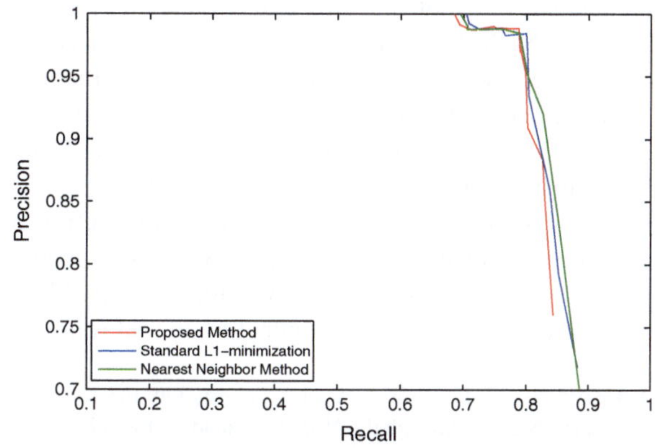

Fig. 7 Precision and recall curves of the proposed method, the standard homotopy, and NN method for loop closure detection on "City Centre" dataset

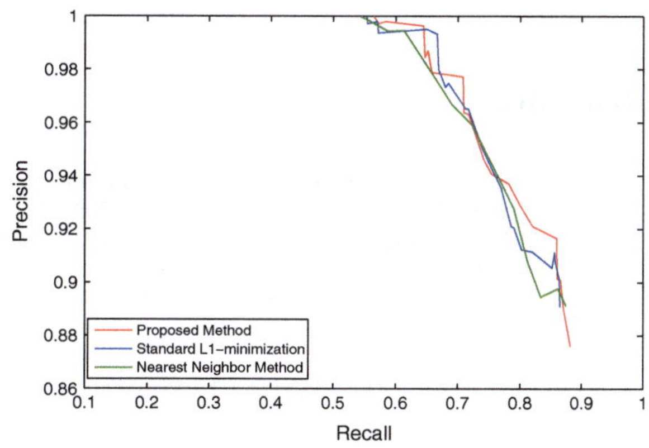

Fig. 8 Precision and recall curves of the proposed method, the standard homotopy, and NN method for loop closure detection on "New College" dataset

Table 2 Comparison of the recall between the proposed method and FAB-MAP as baseline at precision 100 %

	City Centre (%)	New College (%)
Proposed method	68	57
FAB MAP	37	48

and $\tau = 0.45$ respectively, which is higher than the recall for FAB-MAP method reported in [6]. τ is empirically chosen to allow comparison with other methods at 100 % precision.

6 Conclusion

We have presented in this paper a novel technique to detect loop closure that is highly efficient in time and competitive in detection accuracy. The proposed method formulates the loop closure detection as a sparse representation problem. Since in visual SLAM the current view of the robot is similar to the most recent previous image, we are able to update the obtained solution from one iteration of static l_1-minimization for loop closure detection using the subsequent robot view without solving the minimization problem from scratch. Using our dynamic update method, loop closure detection can be solved much faster than the static method without losing accuracy. The proposed method is therefore more scalable and able to handle larger robot maps. The reliability and efficiency of the proposed method have been validated on three different publicly available datasets. In the future, we plan to implement the proposed algorithm on real robots in an online SLAM system.

Acknowledgments This work was supported by the Natural Sciences and Engineering Research Counsil (NSERC) through the NSERC Canadian Field Robotics Network (NCFRN) and by Alberta Innovates Technology Future (AITF).

References

1. Amaldi, E., Kann, V.: On the approximability of minimizing nonzero variables or unsatisfied relations in linear systems. Theor. Comput. Sci. **209**, 237–260 (1998)
2. Asif, M.S., Romberg, J.: Dynamic updating for l1-minimization. IEEE J. Sel. Top. Signal Process. **4**(2), 421–434 (2010)
3. Callmer, J., Granstrom, K., Nieto, J., Romas, F.: Tree of words for visual loop closure detection in urban SLAM. In: Proceedings of the Australian Conference on Robotics and Automation (2008)
4. Chen, S.S., Donoho, D.L., Saunders, M.A.: Atomic decomposition by basis pursuit. SIAM J. Sci. Comput. **20**(1), 33–61 (1999)

5. Csurka, G., Dance, C.R., Fan, L., Williamowski, J., Bray, C.: Visual categorization with bags of keypoints. In: ECCV International Workshop on Statistical Learning in Computer Vision, pp. 1–22 (2004)
6. Cummins, M., Newman, P.: FAB-MAP: probabilistic localization and mapping in the space of appearance. Int. J. Robot. Res. **27**(6), 647–665 (2008)
7. Dalal, N.: Histograms of oriented gradients for human detection, In: IEEE Conference on Computer Vision and Pattern Recognition Workshops, pp. 886–893 (2005)
8. Donoho, D.: For most large underdetermined systems of linear equations the minimal l1-norm solution is also the sparsest solution. Comm. Pure Appl. Math. **59**(6), 797–829 (2006)
9. Elad, M, Figueiredo, M., Ma, Y.: On the role of sparse and redundant representations in image processing. Proc. IEEE **98**(6), 972–982 (2010)
10. Fuchs, J.: On sparse representations in arbitrary redundant bases. IEEE Trans. Inf. Theory **50**(6), 1341–1344 (2004)
11. Galvez-Lopez, D., Tardos, J.D.: Bag of binary words for fast place recognition in image sequences. IEEE Trans. Robot. **28**(5), 1188–1197 (2012)
12. Latif, Y., Huang, G., Leonard, J., Neira, J.: An online sparsity-cognizant loop-closure algorithm for visual navigation, In: Proceedings of Robotics: Science and Systems (2014)
13. Liu, Y., Zhang, H.: Visual loop closure detection with a compact image descriptor, In: IEEE/RSJ International Conference on Intelligent Robots and Systems, pp. 1051–1056 (2012)
14. Newman, P., Cole, D., Ho, K.: Outdoor SLAM using visual appearance and laser ranging. In: Proceedings of International Conference on Robotics and Automation, pp. 1180–1187 (2006)
15. Nister, D., Stewenius, H.: Scalable recognition with a vocabulary tree. In: Proceedings of IEEE Conference on Computer Vision and Pattern Recognition, vol. 2, pp. 2161–2168 (2006)
16. Oliva, A., Torralba, A.: Modelling the shape of the scene: a holistic representation of the spatial envelope. Int. J. Comput. Vision **42**(3), 145–175 (2001)
17. Sivic, J., Zisserman, A.: Video google: a text retrieval approach to object matching in videos. In: 9th IEEE ICCV, pp. 1470–1477 (2003)
18. Wright, J., Yang, A., Ganesh, A., Sastry, S., Ma, Y.: Robust face recognition via sparse representation. IEEE Trans. PAMI **31**(2), 210–227 (2009)

Large Scale Dense Visual Inertial SLAM

Lu Ma, Juan M. Falquez, Steve McGuire and Gabe Sibley

Abstract In this paper we present a novel large scale SLAM system that combines dense stereo vision with inertial tracking. The system divides space into a grid and efficiently allocates GPU memory only when there is surface information within a grid cell. A rolling grid approach allows the system to work for large scale outdoor SLAM. A dense visual inertial dense tracking pipeline incrementally localizes stereo cameras against the scene. The proposed system is tested with both a simulated data set and several real-life data in different lighting (illumination changes), motion (slow and fast), and weather (snow, sunny) conditions. Compared to structured light-RGBD systems the proposed system works indoors and outdoors and over large scales beyond single rooms or desktop scenes. Crucially, the system is able to leverage inertial measurements for robust tracking when visual measurements do not suffice. Results demonstrate effective operation with simulated and real data, and both indoors and outdoors under varying lighting conditions.

1 Introduction

Large Scale SLAM is an important research area in robotics and computer vision. Perhaps the point based approaches [1–3] are the most popular ones for large scale SLAM. Normally, such approaches use a point cloud to reconstruct the scene and cannot reconstruct connected surfaces. These approaches register the point cloud in

---

L. Ma (✉) · J.M. Falquez · S. McGuire · G. Sibley
University of Colorado - Boulder, Boulder, CO, USA
e-mail: lu.ma@colorado.edu

J.M. Falquez
e-mail: juan.falquez@colorado.edu

S. McGuire
e-mail: stephen.mcguire@colorado.edu

G. Sibley
e-mail: gabe.sibley@colorado.edu

D.S. Wettergreen and T.D. Barfoot (eds.), *Field and Service Robotics*,
Springer Tracts in Advanced Robotics 113, DOI 10.1007/978-3-319-27702-8_10

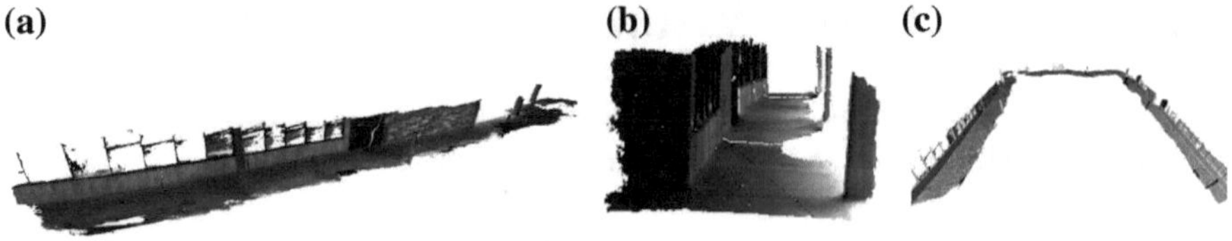

Fig. 1 An example of the reconstruction result for an outdoor scene from 7000 stereo frames (approx. 75 million vertices). **a–b** Reconstruction detail of a scene with both shadow and harsh illumination, and snow on the ground. **c** An overview of the camera path

different views and present the reconstruction result as a point cloud. However, a connected surface is important for planning and control of robots (Fig. 1).

Dense SLAM with volumetric representation have been popular in recent years [4–6]. Such techniques use a Truncated Signed Distance Function (TSDF) to represent the scene surface and incrementally refine it with the registered depth frames. Meanwhile, similar approaches have also been proposed in monocular SLAM [7, 8]. Usually, these approaches use a fixed amount of GPU memory for tracking and reconstruction; this hard constraint limits the size of the reconstructed scene and cannot be used for large scale dense SLAM.

Several systems have been proposed in order to reconstruct large scale scenes with volumetric approaches. Zeng et al. [9] and Steinbrucker et al. [10] proposed an octree based approach for indoor dense SLAM. Roth and Vona [11], Whelan et al. [12] and Finman et al. [13] used a fixed bounded volume to represent portions of the scene and incrementally reconstruct it with a rolling scheme. However, these approaches mostly focus on the indoor scene and uses RGB-D sensors, which does not perform outdoor SLAM with stereo data. Meanwhile, these approaches heavily rely on ICP for tracking which are not suitable for outdoor environments due to the quality of the depth images from the stereo sensors. Besides, a combined ICP + RGB tracking approach [11] may also fail if the scene only contain simple geometric or color information.

Here we propose a new large scale dense visual inertial SLAM system that does not rely on active depth sensing. The system uses rolling grid fusion scheme which effectively manages GPU memory and is capable of reconstructing a fully dense scene online. The system obtains depth images from stereo matching [14] and simultaneously localizes the camera based on whole image alignment and inertial data while reconstructing the scene with SDF fusion. The system automatically saves and loads data from device, host memory and hard disk, and generates a mesh (.obj, .dae, .ply formats) of the large scene (e.g. 20 millions vertices) in seconds. Given these components, a wide range of applications can be developed, especially in robotics where the proposed system is capable of providing high fidelity meshes of any outdoor environment for use in path planning and control.

Perhaps the most similar system to ours is [12, 15–17]. There are, however, key methodological differences. (1) Our approach focuses on outdoor scenes and uses stereo data while [12, 15, 17] uses an RGB-D sensor and mainly focus on indoor scenes. (2) Our system uses an dense visual inertial stereo system for tracking while other rely solely on cameras, either ICP or RGB-D approach. (3) Our approach uses a simple rolling grid SDF pipeline for reconstruction while [15, 17] used a hashing scheme, [12] used a rolling SDF scheme and [16] uses a fix grid volume scheme.

The remainder of this paper is structured as follows: Sect. 2 briefly covers preliminaries of our approach. Section 3 covers the technical details of the Rolling Grid SDF approach. Section 4 covers the dense visual inertial tracking. Section 5 offers testing methods and discusses the system performance with indoor and outdoor experiments. Section 6 addresses failure cases and limitations. Section 7 draws conclusions.

2 Overview

2.1 *Grid Based Volumetric Representation*

The proposed system uses a grid based volumetric representation, namely the *Grid SDF* $\mathscr{S}$ (see Fig. 2), to reconstruct a 3D model of the scene in the current camera view. Each cell c in the Grid SDF $\mathscr{S}$ is a small *NxNxN* dimensional TSDF (Truncated Signed Distance Function) volume and contains a pointer to GPU memory. The *Grid SDF* $\mathscr{S}$ contains (x_g, y_g, z_g) cells in the each dimension. Assuming that the resolution of each voxel is r_v, the maximum size of the scene in each dimension is the number of cells in that dimension times the size of the SDF. For example, for the x dimension we have:

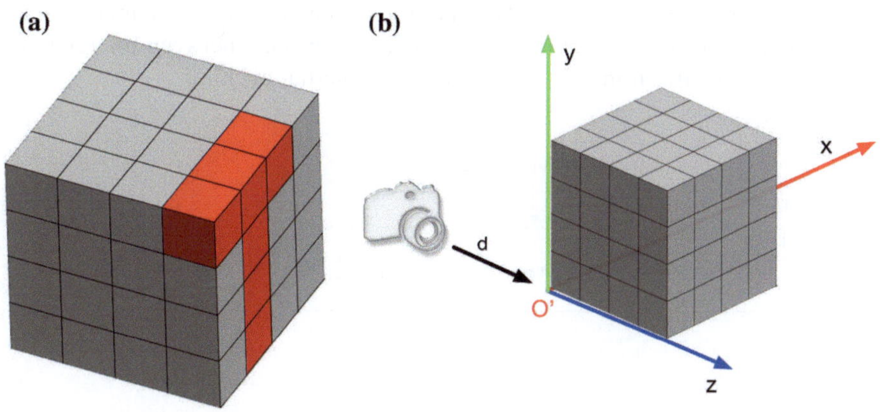

Fig. 2 **a** An example of the *Grid SDF* $\mathscr{S}$. In this example, $\mathscr{S}$ has (e.g. $(8 * 8 * 8)$) cells in the x, y, z directions. The GPU memory of a cell g in $\mathscr{S}$ is not initialized (*gray cells*) until there is actual information available corresponding to c (*red cells*). **b** An example of the pose of $\mathscr{S}$ w.r.t the camera. The z axis of the initial pose of $\mathscr{S}$ starts from the minimum distance of camera range d_{min} to r_z

$$r_x = r_v * N * x_g. \quad (1)$$

The values of x_g and y_g are usually selected depending on the horizontal and vertical field of view of the camera, and z_g is based on the maximum depth measurement desired. This can be selected dependent on the maximum expected scene depth, or ideally, thresholded by the maximum depth uncertainty desired given the rig's stereo baseline. Notice that when initializing $\mathscr{S}$, the system does not allocate any GPU memory for cell c. Meanwhile, given the camera with an initial pose T_{wc}, the system defines a *Grid SDF* $\mathscr{S}$ as in Fig. 2, where the size of $\mathscr{S}$ is (r_x, r_y, r_z).

2.2 Grid Pose Representation

The system uses P_g to represent the global pose of the whole grid, $\mathscr{S}$, with $P_g = (0, 0, 0)$ being the world pose of the initial *Grid SDF*). P_l represents the local pose of a cell c within the grid. Thus, a cell c in the current camera view can be accessed by its local index and a voxel within the cell can also be accessed by P_g and P_l.

2.3 System Structure

The following flow chart (Fig. 3) shows the structure of the proposed system.

Initialization The system first initializes a *Grid SDF* $\mathscr{S}$ without allocating any GPU memory for any cell c in $\mathscr{S}$.

Tracking Given the input stereo data, the system generates the depth images of the current frame via stereo matching and localizes the camera between the reference frame T_{wr} and the live frame T_{wl} via dense visual inertial tracking.

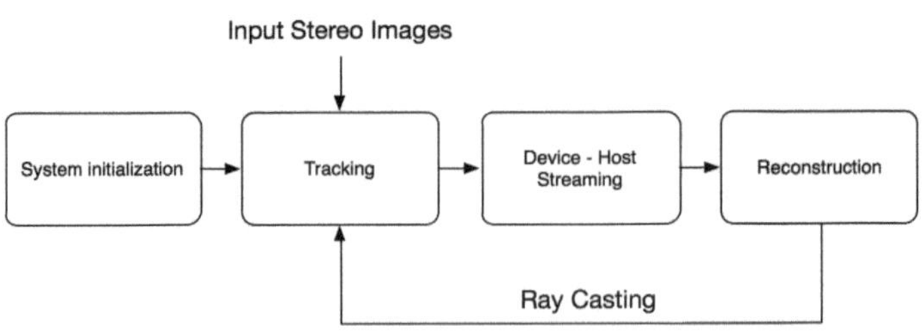

Fig. 3 Flow chart of the proposed system. After system initialization, the proposed system localizes the pose of cameras and incrementally reconstructs the scene with a rolling SDF scheme. Portions of the scene that are out of the camera view will be streamed from the GPU memory to the CPU memory (or the hard disk) directly. Such data can also be merged into a complete mesh via marching cubes

Rolling and Streaming Once the system updates the latest world pose of the camera, the system will check if rolling is needed based on the motion of the camera. If required, the system will stream the data of cells c that are out of the current camera view from the GPU memory to the CPU memory.

Reconstruction Once streaming is done, the system model can be updated via SDF fusion. Also, an updated view of the reconstructed scene is obtained via ray casting.

3 Grid Based SDF Fusion

3.1 Rolling Grid

In large scale outdoor SLAM, it is important to continuously perform mapping while at the same time reuse the GPU memory of voxels that have been taken out of the camera view. The proposed system achieves this via a rolling scheme which streams the data of cells that are currently out of the camera view into the CPU memory and reuses the GPU memory of the cells.

To address this problem, we assume the initial pose of the *Grid SDF* $\mathscr{S}$ is the origin, and $\mathscr{S}$ moves with respect to the camera motion. The global pose of $\mathscr{S}$ in the x, y, z directions will increase by 1 if the camera moves $+r_x$, $+r_y$, $+r_z$ in the corresponding direction, and -1 if in the opposite direction.

Meanwhile, under the current camera view, the system can easily access a cell c of $\mathscr{S}$ via its local pose. However, based on the motion of the camera, a cell (e.g. $c^{'}$) in $\mathscr{S}$ may have moved out of the current camera view. To reuse the GPU memory of cell $c^{'}$ for a new cell $c^{''}$ in the current camera view, the system will stream the data of $c^{'}$ from the GPU memory to the CPU memory and reuse the same allocation for the new cell. In this case, we can no longer access $c^{''}$ via its local index directly in the current $\mathscr{S}$, implying that the real index of $c^{''}$ will be different from its local index. Figure 4 shows how the system computes the real index of a cell based on its local index during rolling.

The proposed system performs rolling in a very straightforward way, as shown in Fig. 4. Assume that the initial scene the camera sees is the word *'GRID'*. If the camera moves forward in four steps, (e.g. Fig. 4a), it will see the letters *'L'*, *'L'*,*'O'*,*'R'* respectively. Here, each step (the minimum r_x/x_v, r_y/y_v, r_z/z_v) of the camera motion in a direction is considered a shift in that direction. Each time when the camera moves forward, the real index of the new cell (e.g. L in the second column of Fig. 4a) will be saved to the cell which just moved out of view, and the corresponding previous cell (letter D) will be saved to the CPU memory. Now, the local index of L in the current *Grid SDF* should be 3, but its real index is 0 instead. The following pseudocode shows how the system computes reused GPU memory by streaming cells that are out of the current camera view:

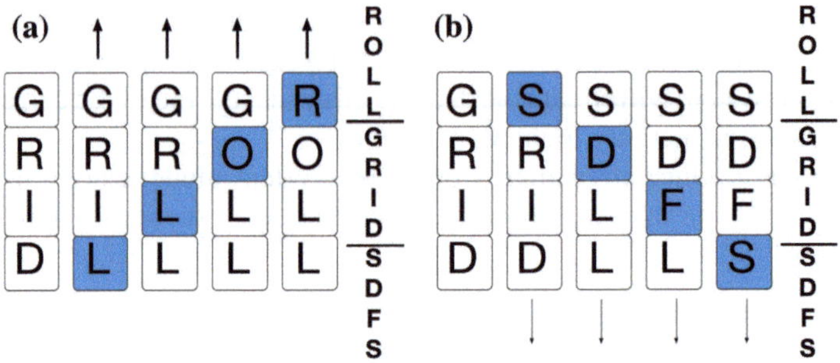

Fig. 4 An example of rolling *Grid SDF*. The camera is moving in the positive (**a**) and negative (**b**) directions; This example shows how the system reuses the GPU memory of the scene that is out of the current view. Here we assume the number of cells is 4 and the initial camera location is in the letter *D*, seeing the letter sequence *GRID* (from far to close). In each steps, the camera moves in the direction of the arrow. The *white cells* is the scene that the camera sees before, remain stationary within GPU memory. The *blue cells* store the scene that the camera sees only in the current view, while the corresponding previous GPU-located cells have been streamed to the CPU memory. For example, in (**a**), column 2, the camera moves one steps forward, sees *LGRI* (from far to close) in the current view, streams *D* from the GPU memory to the CPU memory, and then reuses the GPU memory location to store the new view *L* (*in blue*)

Meanwhile, once rolling is performed, the real index of a cell can be computed directly by algorithm 2. Notice the voxel position is the real position of the voxel (3D point) in the space in the current *Grid SDF*.

Algorithm 1 Compute the index of cells that needs to be streamed from GPU to CPU in a given direction (e.g. in the x axis)

Require: shift: s, previous shift: s_p, cell index: x, number of cells in one dimension x_g, stream flag: f
Ensure: s != 0 **and** $s < x_g$ **and** $s > -x_g$
  if $s > 0$ **then**
    if $s_p \geq 0$ **and** $x \geq s_p$ **and** $x < s_p + s$ **then**
      $f \leftarrow true$
    else if $x \geq x_g + s_p$ **and** $x < x_g + s_p + s$ **then**
      $f \leftarrow true$
    else
      $f \leftarrow false$
    end if
  else
    if $s_p < 0$ **and** $x \geq x_g + s_p + s$ **and** $x < x_g + s_p$ **then**
      $f \leftarrow true$
    else if $x \geq s_p + s$ **and** $x < s_p$ **then**
      $f \leftarrow true$
    else
      $f \leftarrow false$
    end if
  end if

Algorithm 2 Access a voxel in the *Grid SDF* by the voxel position (e.g. in the x axis)

Require: shift: s, local index: x_l, number of cells in one dimension x_v, real index: x_r
Ensure: $s < x_v$ **and** $s > -x_v$
  if $s > 0$ **then**
    if $x_l < x_v - 1 - s$ **then**
      $x_r \leftarrow x_l + s$
    else
      $x_r \leftarrow x_l - (x_v - s)$
    end if
  else
    if $x_l > -s$ **then**
      $x_r \leftarrow x_l + s$
    else
      $x_r \leftarrow x_l + x_v + s$
    end if
  end if

3.2 *SDF Fusion*

The system updates $\mathscr{S}$ by fusing every valid point from the stereo depth map I_d into $\mathscr{S}$ once T_{wc} is tracked:

$$\mathscr{S}^{'} = \mathscr{F}(\mathscr{S}, I_d, T_{wc}) \tag{2}$$

Here, $\mathscr{F}(\cdot)$ is the SDF fusion operation. T_{wc} is the world pose of the camera in the live frame (i.e. current frame). The system also generates a virtual gray image I_v^g and depth image I_v^d by ray casting $\Upsilon(\cdot)$ [4]:

$$I_v = \Upsilon(\mathscr{S}, T_{wv}), I_v = I_v^g \cup I_v^d \tag{3}$$

where T_{wv} is the pose of the virtual camera.

Notice during fusion, the system will check every valid voxel position in the *Grid SDF* and project the voxel to 2D. If there is a valid 2D pixel in the current live image with a valid depth value, the voxel will be updated (a similar operation also happens during ray casting).

3.3 *Device to Host Streaming*

The proposed system automatically streams data from device memory to the host (CPU) memory if the data present in the *Grid SDF* is out of the current camera view. Once the memory block which hold the past SDF in the CPU memory is full, the system streams data of the cells which has the furthest distance to the current camera pose from the host memory to the hard disk. See Fig. 5.

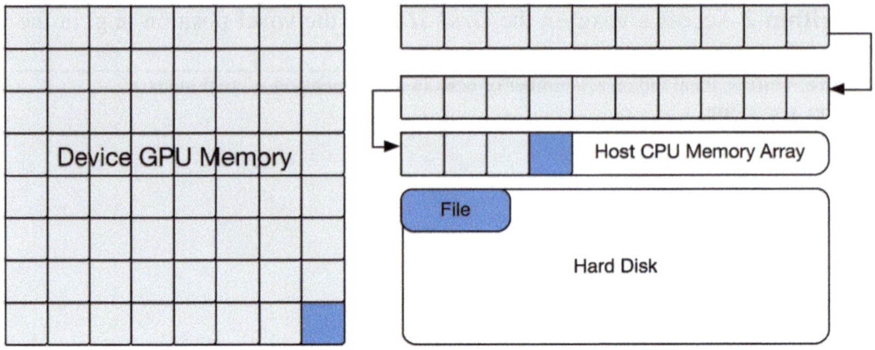

Fig. 5 Host—device streaming pipeline in the system. The *blue block* in the GPU memory will be streamed to the host CPU memory array when the data is out of the camera view and will be saved to the hard disk when the CPU memory array is full

When the camera moves to a new location, the system checks if the data in the new location previously exists in the system. If it does, the system will reuse that memory and load it back from the CPU memory or the hard disk to the GPU memory. Reloading saved data helps to complete the model of the scene from different views. Notice that each time a cell file is saved in the host memory or the hard disk, the system indexes it with a global and local index which allows fast retrieval of stored cells. Since all the SDF data is stored as individual cell files in the host memory or the hard disk, the system can merge any portion of the scene of interest into a mesh, which can be used later for any robotic application.

4 Dense Visual Inertial Tracking

Tracking is performed in a windowed dense visual inertial bundle adjuster. Visual-only frame-to-frame constraints are transformed into the IMU frame and added into the bundle adjuster as binary constraints. Inertial measurements between frames are integrated forming residuals against the estimated poses as seen in Fig. 6. Velocities and IMU biases are also estimated, and are carried through in the sliding window.

Visual tracking is performed by a Lucas-Kanade [18] style whole-image alignment algorithm via the Efficient Second Order Minimization (ESM) technique [19], and a 6-DOF camera transform is estimated by minimizing the photometric error (e_v) between a reference image and the current live image:

$$e_v = \|I_{live}\left(\varphi\left(\hat{T_{lr}}\varphi^{-1}(\mathbf{u_r}, d)\right)\right) - I_{ref}(\mathbf{u_r})\|^2. \tag{4}$$

The pixel $\mathbf{u_r}$ in the reference frame is back-projected φ^{-1} using the camera calibration parameters and the associated depth value d obtained by the stereo reconstruction

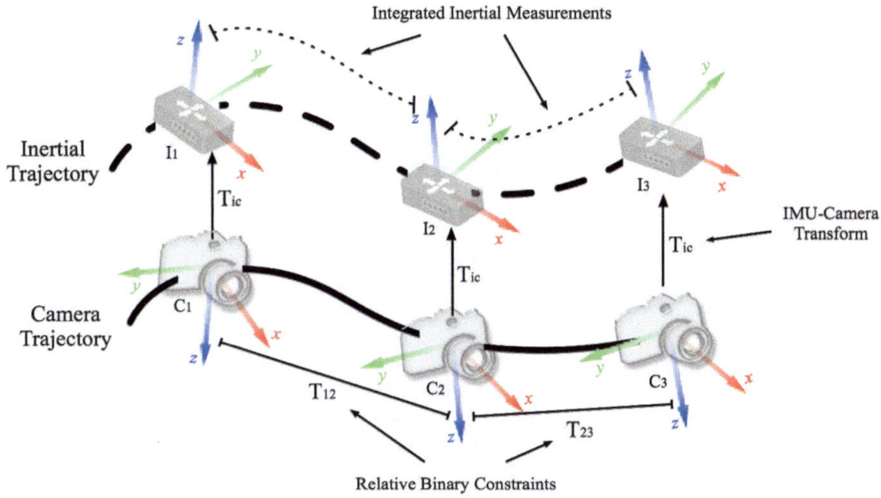

Fig. 6 Binary constraints from the visual tracker and integrated IMU poses, along with velocities and accelerometer + gyro biases are jointly optimized. The camera to IMU transform T_{ic} is calibrated offline

algorithm. The 3D point is then transferred into the live frame via the estimated transform, $\hat{T}_{lr}$, and projected φ onto the camera.

The pose covariances from the visual tracking system are then added into the bundle adjuster, which runs once a sufficient number of frames and inertial measurements are obtained. The covariance of the inertial residual between two consecutive frames is dependent on the number of measurements between images, and as such must be carried forward during the integration process (Fig. 7). Details about inertial integration and error propagation can be found in [20].

Inertial residuals between the parameters and the integrated state take the form of:

$$e_I = \left\| \begin{bmatrix} p_{wp} - \hat{p} \\ log\left(q_{wp}^{-1} \otimes \hat{q}\right) \\ v_w - \hat{v} \\ b_g - \hat{b_g} \\ b_a - \hat{b_a} \end{bmatrix} \right\|^2, \tag{5}$$

where $(p_{wp} - \hat{p})$ is the translation residual, $log\left(q_{wp}^{-1} \otimes \hat{q}\right) \in R^3$ calculates the rotation residual in $so(3)$, $(v_w - \hat{v})$ is the velocity residual, and $(b_g - \hat{b_g})$ and $(b_a - \hat{b_a})$ are the gyro and accelerometer bias residuals respectively.

A total of 15 parameters per frame are estimated during the sliding window optimization: 6 for pose parameters, 3 for velocities, 3 for accelerometer biases and 3 for gyroscope biases. Initial velocities as well as the biases are estimated and kept up to date as the sliding window shifts during execution. Given the size of the sliding win-

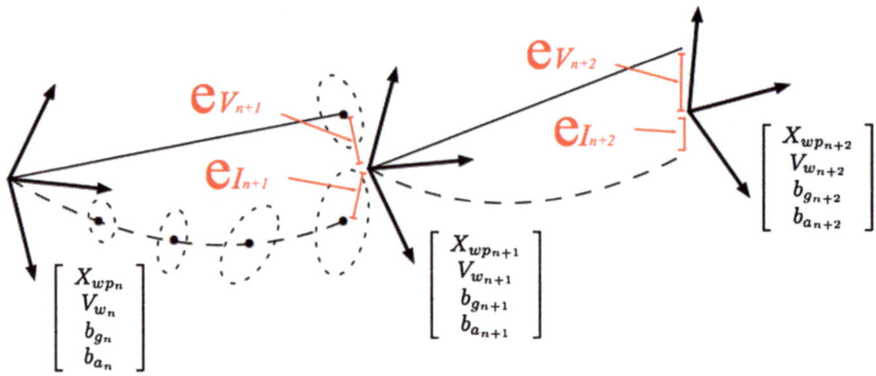

Fig. 7 Errors from the vision system (e_v) are formed by compounding the estimated relative transforms with world poses. Similarly, inertial errors (e_I) are formed by integrating inertial measurements. Uncertainties (shown as *ellipsoids*) are used to weigh in residuals for the estimation of the state parameters: world poses comprised of a translation (p) and rotation (q) vector ($X_{wp} = \left[p_{wp}\ q_{wp}\right]^T$), velocities ($V_w$), accelerometer biases (b_a) and gyroscope biases (b_g)

dow and the unambiguity of scale from the stereo vision system, no marginalization or conditioning is done on the sliding window as all parameters are observable.

The inclusion of inertial data enhances visual tracking in general, and in particular during fast camera movements and low textured areas. The addition of the IMU also speeds up visual tracking, since the typical coarse-to-fine pyramid scheme used in visual odometry is no longer required. Instead, the visual tracking is initialized with an estimated pose given by the integration of inertial measurements from the last frame up to the point where a new image is captured. In this way, only a refinement in the form of a few iterations at full image resolution is required for the final pose estimate.

5 Result and Discussions

The proposed system is tested by a hand held camera and a ClearPath Robotics Husky robot (Fig. 8) with two Ximea (MQ013MG-ON) gray scale cameras and a Microstrain 3DM-GX3-35 Inertial Measurement Unit (IMU). The camera intrinsics as well as sensor extrinsics are calibrated offline with a method similar to [21], and the rigid sensor rig is attached to the robot via a T-mount.

We implement the system using the GPU for the reconstruction pipeline and using the CPU with Intel Threaded Building Blocks for the visual inertial tracking pipeline. All the real-world datasets were captured using the stereo camera + IMU rig. The images were undistorted and scan-line rectified, and were later fed to a stereo matching technique [14] for depth map generation.

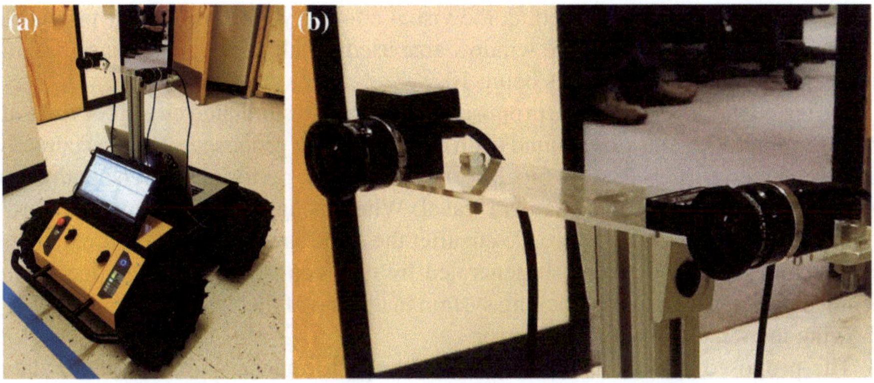

Fig. 8 An example of the system platform. **a** A clearpath Husky robot. An **b** IMU and a calibrated stereo rig or an RGB-D camera is mounted on the robot which provides stereo and inertial data during navigation

To evaluate the performance of the proposed system, we tested it with a simulated city-block dataset (15 m by 15 m in width and length, containing approximately 200 frames) with simulated IMU measurements and several real-world datasets (approx. 40–250 m in length). For the real-world datasets, we captured a variety of indoor and outdoor scenes under different lighting and weather conditions (e.g. sunny and snow). To test the robustness of the proposed visual inertial tracking system, we especially test the system in a dark office scene (Fig. 10) and in an hallway with very simple geometry (Fig. 1), where either the traditional RGB-D approach or an ICP approach would easily fail. During the experiments, we set the maximum depth of voxels that fuse into the *Grid SDF* to 15 m given the average maximum depth in all the scenes and in order to limit any potentially erroneous depth data from the stereo matching algorithm.

The dense visual inertial tracker initially performs visual odometry using a coarse to fine approach via an image pyramid. After a the minimum number of image frames is acquired, the sliding window kicks in and the image pyramid is no longer required

Fig. 9 An example of the reconstruction result for the simulated city block data. **a** The original ground truth model. **b** An overview of the reconstruction result. **c** Close view of the reconstruction result showing loop closure error

since the IMU is capable of seeding the visual odometry optimization by providing a hint of the camera's pose. The window size used for all experiments was 15, with the minimum number of frames being 10.

We tested the accuracy of the proposed system with a simulated city block dataset. When compared against the ground truth depth map, the proposed system accurately tracks and reconstructs the city block scene with online performance. The path error is approx. 8 cm after 60 m of camera travel. When using the depth from the stereo algorithm, the path error is approx. 5 cm after the same camera travel. Figure 9 shows the original mesh and the mesh generated by the proposed system. Notice in the detailed view the drift of the tracking system in the end of the reconstruction (Fig. 9c) showing the relative loop closure error.

The proposed system also shows effective performance with real-world data. While the quality of the depth images generated from stereo matching is affected significantly by different lighting, texture and weather conditions, our system is capable of successfully reconstructing all large-scale outdoor scenes with high a quality mesh.

Figure 10 shows the reconstruction result of an indoor office scene (approx. 30 m by 30 m) from 6000 stereo frames. The system has a high precision which reconstructs fine details of objects in the scene.

While visual-only tracking may easily fail in real-world scenes with simple geometry, low texture or fast motion, the proposed visual inertial tracking shows a promising tracking result. Figures 1, 11 and 12 show the system successfully tracking under several difficult frames where the inertial measurements adjust the visual tracking result.

Fig. 10 An example of the reconstruction result for an office scene (approx. 5000 stereo frames (final mesh includes approx. 6 million vertices)) from a hand held camera (*first row*) and from the Husky robot (*second row*)

Fig. 11 An example of the reconstruction result for an outdoor snow scene from approximate 5000 stereo frames (final mesh includes approx. 32 million vertices). **a** A close look at a house in the scene. **b** An overview of the scene mesh. **c** An overview of the scene texture

Fig. 12 An example of the reconstruction result for an outdoor snow yard from approximate 7000 stereo frames (final mesh includes approxi. 15 million vertices). **a** A close look of the scene. **b** An overview of the scene mesh

In general, the cell representation of the SDF volume massively saves GPU memory. When testing our simulated and real-world datasets, we set the resolution from 5–25 mm based on the dimension of the scene. In general, the proposed system requires around 650–1500 MB GPU memory to store voxels of the current camera view in a large scale scene while the regular SDF uses around 1000–3500 MB GPU memory, due to the fact that in general scenes, most of the voxels are empty.

System Run-Time. We tested our system with a single NVidia TITAN GPU, Intel i7 quad-CPU desktop, using 640 × 480 pixel resolution input images and 2.5 cm resolution of voxels. Table 1 shows our system run-time in different stages. Except for final mesh generation, the system is capable of online performance.

Table 1 System run-time

Stereo matching, 1 frame	15 ms
Tracking (CPU), 1 frame	20 ms
Reconstruction, 1 frame	32 ms
Ray casting	8 ms
Device-Host Streaming 1 cell	0.01 ms
Generate cell to mesh (e.g. 13 million vertices)	15 s

6 Failure Cases and Limitations

Although the system is robust to many real-world conditions, there are several limitations of our current work. The final reconstruction and tracking results depend heavily on the quality of the depth images which can be improved by [8]. The reconstruction can also be improved by adding loop closure by changing the local and global index of cells.

7 Conclusions

We present a large scale dense visual inertial SLAM system based on a rolling grid fusion scheme. As far as we know this is the first system to combine inertial tracking in a dense SLAM framework. The proposed system manages the space into small volume grids and only allocates GPU memory for cells if data exists. A large scale dense mapping solution is obtained via a rolling grid scheme with simple index computation while the device and the host memory automatically stream between each other in order to reuse the GPU memory. Depending on the requirements of an actual application, the system utilizes stereo cameras in both indoor and outdoor scenes. The system is tested in several outdoor and indoor scenes under different lighting (illumination changes), weather (e.g. snow, sunny), and motion conditions and shows promising results. In conclusion, the main contributions of the paper are: (1) A new large scale outdoor dense mapping system based on stereo data and (2) a new dense visual inertial dense tracking pipeline. We believe the proposed system is useful for outdoor scene reconstruction and especially for planning and control of high-speed ground vehicles.

References

1. Nüchter, A., Lingemann, K., Hertzberg, J., Surmann, H.: 6d slam3d mapping outdoor environments. J. Field Robot. **24**(8–9), 699–722 (2007)
2. Fioraio, N., Konolige, K.: Realtime visual and point cloud slam. In: Proceedings of the RGB-D Workshop on Advanced Reasoning with Depth Cameras at Robotics: Science and Systems Conference (RSS), vol. 27 (2011)
3. Strasdat, H., Davison, A.J., Montiel, J., Konolige, K.: Double window optimisation for constant time visual slam. In: IEEE International Conference on Computer Vision (ICCV), pp. 2352–2359. IEEE (2011)
4. Newcombe, R.A., Davison, A.J., Izadi, S., Kohli, P., Hilliges, O., Shotton, J., Molyneaux, D., Hodges, S., Kim, D., Fitzgibbon, A.: Kinectfusion: Real-time dense surface mapping and tracking. In: 10th IEEE International Symposium on Mixed and augmented reality (ISMAR), pp. 127–136 (2011)
5. Izadi, S., Kim, D., Hilliges, O., Molyneaux, D., Newcombe, R., Kohli, P., Shotton, J., Hodges, S., Freeman, D., Davison, A. et al.: Kinectfusion: real-time 3d reconstruction and interaction

using a moving depth camera. In: Proceedings of the 24th annual ACM symposium on User interface software and technology. ACM, pp. 559–568 (2011)
6. Keller, M., Lefloch, D., Lambers, M., Izadi, S., Weyrich, T., Kolb, A.: Real-time 3d reconstruction in dynamic scenes using point-based fusion. In: International Conference on 3D Vision-3DV 2013, pp. 1–8. IEEE (2013)
7. Newcombe, R.A., Lovegrove, S.J., Davison, A.J.: DTAM: Dense tracking and mapping in real-time. In IEEE International Conference on Computer Vision (ICCV), pp. 2320–2327 (2011)
8. Concha, A., Hussain, W., Montano, L., Civera, J.: Manhattan and piecewise-planar constraints for dense monocular mapping. In: Proceedings of Robotics: Science and Systems (RSS), (2014)
9. Zeng, M., Zhao, F., Zheng, J., Liu, X.: Octree-based fusion for realtime 3d reconstruction. Graph. Models **75**(3), 126–136 (2013)
10. Steinbrucker, F., Sturm, J., Cremers, D.: Volumetric 3d mapping in real-time on a cpu. In: IEEE International Conference on Robotics and Automation (ICRA), pp. 2021–2028. IEEE (2014)
11. Roth, H., Vona, M.: Moving volume kinectfusion. In BMVC, pp. 1–11 (2012)
12. Whelan, T., Johannsson, H., Kaess, M., Leonard, J.J., McDonald, J.: Robust tracking for real-time dense rgb-d mapping with kintinuous (2012)
13. Finman, R., Whelan, T., Kaess, M., Leonard, J.J.: Efficient incremental map segmentation in dense rgb-d maps. In: IEEE International Conference on Robotics and Automation (ICRA), pp. 5488–5494. IEEE (2014)
14. Geiger, A., Roser, M., Urtasun, R.: Efficient large-scale stereo matching. Computer Vision-ACCV 2010, pp. 25–38. Springer, Berlin (2011)
15. Nießner, M., Zollhöfer, M., Izadi, S., Stamminger, M.: Real-time 3d reconstruction at scale using voxel hashing. ACM Trans. Graph. (TOG) **32**(6), 169 (2013)
16. Sengupta, S., Greveson, E., Shahrokni, A., Torr, P.H.: Urban 3d semantic modelling using stereo vision. In: IEEE International Conference on Robotics and Automation (ICRA), pp. 580–585. IEEE (2013)
17. Prisacariu, V.A., Kähler, O., Cheng, M.M., Valentin, J., Torr, P.H., Reid, I.D., Murray, D.W.: A framework for the volumetric integration of depth images (2014). arXiv:1410.0925
18. Baker, S., Matthews, I.: Lucas-kanade 20 years on: a unifying framework. Int. J. Comput. Vis. **56**(3), 221–255 (2004)
19. Klose, S., Heise, P., Knoll, A.: Efficient compositional approaches for real-time robust direct visual odometry from RGB-D data. In: IEEE/RSJ International Conference on Intelligent Robots and Systems (IROS), November 2013
20. Keivan, N., Sibley, G.: Asynchronous adaptive conditioning for visual-inertial slam. In: International Symposium on Experimental Robotics (ISER) (2014)
21. Lovegrove, S., Patron-Perez, A., Sibley, G.: Spline fusion: a continuous-time representation for visual-inertial fusion with application to rolling shutter cameras. In Proceedings of the British machine vision conference, pp. 93.1–93.12 (2013)

Dense and Swift Mapping with Monocular Vision

Pedro Piniés, Lina Maria Paz and Paul Newman

Abstract The estimation of dense depth maps has become a fundamental module in the pipeline of many visual-based navigation and planning systems. The motivation of our work is to achieve a fast and accurate in-situ infrastructure modelling from a monocular camera mounted on an autonomous car. Our technical contribution is in the application of a Lagrangian Multipliers based formulation to minimise an energy that combines a non-convex dataterm with *adaptive pixel-wise regularisation* to yield the final local reconstruction. We advocate the use of constrained optimisation for this task—we shall show it is swift, accurate and simple to implement. Specifically we propose an *Augmented Lagrangian (AL)* method that markedly reduces the number of iterations required for convergence, more than 50 % of reduction in all cases in comparison to the state-of-the-art approach. As a result, part of this significant saving is invested in improving the accuracy of the depth map. We introduce a novel per pixel *inverse depth uncertainty estimation* that affords us to apply adaptive regularisation on the initial depth map: high informative inverse depth pixels require less regularisation, however its impact on more uncertain regions can be propagated providing significant improvement on textureless regions. To illustrate the benefits of our approach, we ran our experiments on three synthetich datasets with perfect ground truth for textureless scenes. An exhaustive analysis shows that AL can speed up the convergence up to 90 % achieving less than 4 cm of error. In addition, we demonstrate the application of the proposed approach on a challenging urban outdoor dataset exhibiting a very diverse and heterogeneous structure.

P. Piniés (✉) · L.M. Paz · P. Newman
Mobile Robotics Group, Department of Engineering Science,
University of Oxford, 17 Parks Rd., Oxford OX1 3PJ, UK
e-mail: ppinies@robots.ox.ac.uk

L.M. Paz
e-mail: linapaz@robots.ox.ac.uk

P. Newman
e-mail: pnewman@robots.ox.ac.uk

D.S. Wettergreen and T.D. Barfoot (eds.), *Field and Service Robotics*,
Springer Tracts in Advanced Robotics 113, DOI 10.1007/978-3-319-27702-8_11

1 Introduction

The creation of dense workspace models from cameras alone has long been a focus of robotics research. The mapping task is sometimes seen in a limited light as simply a precursor or at best dual for localisation. When maps were simply sparse collections of points[1] this narrow perspective was reasonable. But robots that can, through their own motion, produce *dense* reconstructions offer a new vista for autonomous and semi-autonomous plant inspection. But to do so the reconstruction process must be rapid allowing in-situ formation of the dense scene structure. This paper is about precisely that competency—creating dense depth maps *rapidly*.

Recent work has made clear the potential of variational methods in producing dense volumetric reconstructions of small workspaces under controlled lighting conditions [6, 10, 13]. In [13] the authors address the problem as a depth map estimation from a set of keyframes with corresponding camera poses obtained from a PTAM system. An energy function is optimized based on a data term that measures the photoconsistency over a set of small-baseline images, as well as total variation (TV) based regularization term. This preserves sharp depth discontinuities due to occlusion boundaries, while simultaneously enforcing smoothness of homogeneous surfaces. The problem is stated as the minimisation of an energy functional comprising both terms by using an alternation scheme with a good initial seed. A similar approach is adopted in [10]. In this case, the solution relies on a primal-dual formulation successfully applied in solving variational convex functions that arise in many image processing problems [4]. Despite the non-convex nature of the energy functional for the depth map estimation, the authors provide theoretical insights to decouple the terms leading to a two-stage optimization. Their solution is based on the application of the well known Quadratic Penalty (QP) method firstly introduced in [15] in the context of optical flow estimation with a similar energy formulation. In contrast to [13], an efficient cumulative discrete cost volume is considered to compute one of the terms allowing a robust initialization of the depth map before the optimization. While [13] avoids an exhaustive point-wise search to find a minimum solution, [10] provides strategies that accelerate the search while achieving good accuracy. A different approach was introduced in [6]. Instead of optimizing depth maps, a different energy functional over a 3D volume is formulated using a primal-dual algorithm for the minimization. The authors use an implicit truncated signed distance function (TSDF) representation to compute the globally optimal fusion using a (TV regularized) convex energy. Then the surface is extracted by finding the zero level set of the accumulated TSDF. As input, the minimization receives initial depth map estimates that are not required to be highly-accurate.

Despite these energy minimisation approaches reach soft real-time performance, their application to active tasks such as planning and obstacle avoidance is critical. For instance, in [2] the authors follow a DTAM based approach to estimate dense depth maps for live collision avoidance of a MAV. Their analysis shows that online generation of each depth map requires usually 900 primal dual iterations

[1] As in early SLAM formulations.

to converge with an estimated final error of 10 cm, requiring a significant time of 500ms for this task. More recently the works of [1, 5, 8] introduce the use of the Augmented Lagrangian (AL) in the field of video restauration and general image inverse problems. As first paper contribution, we demonstrate the efficacy of the Augmented Lagrangian method [3] for dense depth map creation using monocular cameras which at the time of writing was the first time this had been done. Our experiments show that AL method dramatically reduces the number of iterations (more than 50 %) required for the decoupling approach adopted in [10].

A second contribution lies in our consideration of how to progress from an initial guess to a final solution. In particular we need to reinforce pixels in the seed solution containing plausible depth estimates and propagate its effect over those pixels with less accurate depth estimates. We advocate that large texture-less areas of the RGB images produce noisy and often grossly misleading meaningless regions in the initial depth map that greatly impede successful optimisation. We propose a inverse depth uncertainty estimation to calculate per-pixel adaptive confidences that aid the trade-off between the data fidelity term and the regularisation term. This provides a novel approach that affords us a principled way to only seed the optimisation with pixels from regions which should yield reliable depth estimates. Furthermore, we offer an illustrative study of the effect of three different photo-consistency measures. Our motivation is to understand the degree to which each affects final solution accuracy because each determines an initial seed solution for the optimisation.

In Sect. 2 we briefly review the approach presented in [10] to build an initial depth map from monocular frames. Dealing with non-convex data terms requires careful attention, thus Sect. 3 is devoted to explain the so often used Quadratic Penalty method and the proposed Augmented Lagrangian method. How to estimate per-pixel depth uncertainties for adaptive regularisation is introduced in Sect. 4. An evaluation of the precision and convergence of the complete approach on monocular synthetic datasets with perfect Ground Truth is described in Sect. 5. Also, we demonstrate the application of the proposed approach on challenging urban outdoor dataset exhibiting a very diverse and heterogeneous structure. Finally, we draw our conlusions in Sect. 6.

2 Building an Initial Seed

As in [10], to obtain an initial depth map for our optimization algorithm, we build a cost volume $\mathbf{C}_r$ that accumulates, for a uniformly sampled set of inverse depths ξ_j, $j = 1 : d$, the photo-consistency error of overlapping images. The reason for using an inverse depth representation being that a uniform discretisation of $\boldsymbol{\xi}$ produces a uniform sampling of epipolar lines in an image.

Figure 1 shows a 2D top view of the process used to initialize each voxel of the cost volume. Given a pixel $u_i \in \mathbf{u}$ in a reference image I_r and an inverse depth ξ_j the corresponding pixel in a neighboring image $I_k \in \mathbf{I}(r)$, where $\mathbf{I}(r)$ is the set of images that overlap with I_r, is given by the warp

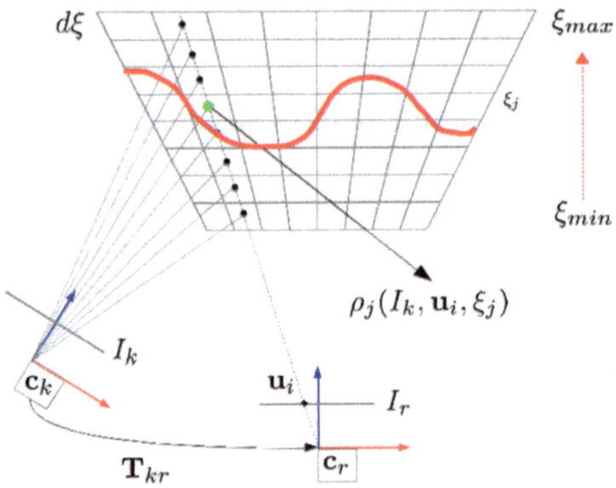

Fig. 1 This example illustrates the process of building the "data fidelity" term for our energy minimization problem. A discretised cost volume is built to accumulate the photo-consistency error: for each pixel u_i in a reference image frame I_r, we back-project the pixel along a discrete set of inverse depth distances ξ_j in the interval $[\xi_{max}\ \xi_{min}]$ obtaining the 3D pose for the center of each voxel in the cube. Then each voxel center gets projected into the current image frame (I_k, c_k) and we compare the corresponding intensities according to a predefined similarity metric ρ^*_{ij}. The results of these comparisons are stored in the corresponding cells. This process is repeated for all overlapping image frames $I_k \in \mathbf{I}(r)$ and the final average cost is calculated according to Eq. (2)

$$\mathbf{w}_k(u_i, \xi_j) = \pi(T_{kr}\pi^{-1}(u_i, \xi_j)) \tag{1}$$

where $\pi(\mathbf{x})$ describes a perspective projection of a 3D point $\mathbf{x}$, $\pi^{-1}(u_i, \xi_j)$ is the back-projection of a pixel u_i with inverse depth ξ_j and $T_{kr} \in SE(3)$ is the relative transformation between cameras corresponding to images I_k and I_r.

We have studied the effect of different photo-consistency measures in the accuracy of depth map estimates. In particular, we have tested, for different window sizes W, the Sum of Squared Differences (ρ^{SSD}), the Sum of Absolute Differences (ρ^{SAD}) and the Normalised Cross Correlation (ρ^{NCC}) which are described in Table 1.

The average photometric error $\mathbf{C}_r(u_i, \xi_j)$ for all images $I_k \in \mathbf{I}(r)$ and for each inverse depth ξ_j is given by:

$$\mathbf{C}_r(u_i, \xi_j) = \frac{1}{|\mathbf{I}(r)|} \sum_{k \in \mathbf{I}(r)} \rho^*_{ij}(I_k, u_i, \xi_j) \tag{2}$$

Table 1 Similarity metrics

Metric	Definition	Equation
Sum of square distances	ρ_{ij}^{SSD}	$\sum_{i\in W} \|I_r(u_i) - I_k(\mathbf{w}_k(u_i, \xi_j))\|_2$
Sum of absolute distances	ρ_{ij}^{SAD}	$\sum_{i\in W} \|I_r(u_i) - I_k(\mathbf{w}_k(u_i, \xi_i))\|_1$
Normalized cross correlation	ρ_{ij}^{NCC}	$\frac{\sum_{i\in W} I_r(u_i) I_k(\mathbf{w}_k(u_i,\xi_i))}{\sqrt{\sum_{i\in W} I_r^2(u_i) \sum_{i\in W} I_k^2(\mathbf{w}_k(u_i,\xi_i))}}$

where $|\mathbf{I}(r)|$ is the number of images that overlap with I_r and ρ^*_{ij} represents the chosen similarity metric.

Once the cost volume is computed, an inverse depth map $\xi_r(\mathbf{u})$ over the whole set of pixels $\mathbf{u}$ can be recovered by searching for the minimum cost for each pixel:

$$\xi_r(\mathbf{u}) = \arg\min_{\xi_j} \ \mathbf{C}_r(\mathbf{u}, \xi_j) \tag{3}$$

Since $\xi_r(\mathbf{u})$ is usually noisy, it will be used as initial seed for the optimization algorithm explained in the next section. Without loss of generality and to improve readability, we will drop the subindex r and will refer only to ξ and $\mathbf{C}$ in the remaining of the paper.

3 Dealing with Non-convex Data Terms

In this section we show how we can improve the initial crude depth map using search over a regular partitioning which replaces the so called "winner-takes-all approach" described in Eq. (3). The searched solution $\xi(\mathbf{u})^*$ minimises the energy functional:

$$\min_{\xi} E(\xi) = \int_{\Omega} w(\mathbf{u})||\nabla\xi(\mathbf{u})||_{\varepsilon} + \lambda\mathbf{C}(\mathbf{u}, \xi(\mathbf{u}))d\mathbf{u} \tag{4}$$

where $\Omega \in \mathscr{R}^2$ is the depth map domain, $w(\mathbf{u})$ is a per pixel weight based on I_r gradient that reduces the effect of regularization across image edges, $||\nabla\xi(\mathbf{u})||_{\varepsilon}$ is the Huber norm and λ is a parameter used to define the trade-off between the regulariser and the data term. After discretising the domain Ω, a depth map is redefined as the set $\boldsymbol{\xi} = [\ldots, \xi_{ij}, \ldots]$. Therefore, we can express previous equation as:

$$\min_{\boldsymbol{\xi}} E_R(\boldsymbol{\xi}) + \lambda E_D(\boldsymbol{\xi}) \tag{5}$$

where $E_R(\boldsymbol{\xi})$ is the regularization term and $E_D(\boldsymbol{\xi})$ is the data term that corresponds with the information stored in the cost volume. In order to solve Eq. (5), we will make use of the iterative Primal Dual optimization algorithm presented in [4]. This algorithm requires both the regulariser and the data term to be convex, however $E_D(\boldsymbol{\xi})$ is not a convex function. One solution to this problem is to decouple both terms and solve instead the following equivalent constrained optimization

$$\begin{array}{ll} \min_{\boldsymbol{\xi},\boldsymbol{\eta}} & E_R(\boldsymbol{\xi}) + \lambda E_D(\boldsymbol{\eta}) \\ \text{s.t.} & \boldsymbol{\xi} = \boldsymbol{\eta} \end{array} \tag{6}$$

The advantage of the decoupling approach is that it allows us to independently solve for the regulariser term using convex optimization methods and for the data term using a simple exhaustive search in the cube. Obviously, both problems are in fact coupled by the constraint. In the following subsections we will discuss the main possible solutions of the previous constraint optimization problem: The Quadratic Penalty (QP) and the Augmented Lagrangian (AL) whose numerical implementation is illustrated in Algorithm 1. The interested reader can find a more detailed discussion of these and more general techniques for constraint minimization in [3].

3.1 Quadratic Coupling Penalty

We briefly describe the algorithm proposed in [10] in order to obtain an improved $\xi(\mathbf{u})^*$ depth map solution from the initial seed in Eq. 3. This approach is based on eliminating the constraints through the use of a coupling penalty function. Popularly a

Algorithm 1 $\boldsymbol{\xi}$ = EnergyMinimisation($\boldsymbol{\eta}, \theta, \varepsilon, \boldsymbol{\alpha}$)

{Initialization of variables:}
$\tau, \sigma > 0, \gamma \in [0,\ 1], \theta \in [0,\ 1]$
{For each pixel ij}
$\xi_{ij}^0 = \eta_{ij}, \mathbf{p}_{ij}^0 = 0$
$\bar{\xi}_{ij} = \xi_{ij}^0$
while $t \leq N$ **do**
  {Update Dual}
  $\mathbf{p}_{ij}^{t+1} = \frac{\mathbf{p}_{ij}^t + \sigma w_{ij} \nabla \bar{\xi}_{ij}^t}{1+\sigma\varepsilon}$
  $\mathbf{p}_{ij}^{t+1} = \mathbf{p}_{ij}^{t+1} / \max(1, |\mathbf{p}_{ij}^{t+1}|)$
  {Update Primal}
  $\xi_{ij}^{t+1} = (\xi_{ij}^t + \tau w_{ij} \nabla \cdot \mathbf{p}_{ij}^{t+1} + \frac{\tau}{\theta^t}\eta_{ij}^t - \tau\alpha_{ij}^t)/(1 + \frac{\tau}{\theta})$
  {Relaxation}
  $\bar{\xi}_{ij}^{t+1} = \xi_{ij}^{t+1} + \gamma(\xi_{ij}^{t+1} - \xi_{ij}^t)$
  η_{ij}^{t+1} = SubpixelSearch($\boldsymbol{\xi}^{t+1}, \theta, \mathbf{C}, \lambda, \boldsymbol{\alpha}^t$)
  $\alpha_{ij}^{t+1} = \alpha_{ij}^t + \frac{1}{\theta}(\xi_{ij}^{t+1} - \eta_{ij}^{t+1})$
end while

Algorithm 2 $\boldsymbol{\eta}$ = SubpixelSearch($\boldsymbol{\xi}, \theta, \mathbf{C}, \lambda, \boldsymbol{\alpha}$)

{Accelerated search:}
$r = \sqrt{2\theta\lambda(C_{ij}^{max} - C_{ij}^{min})}$
{Exhaustive search for $\eta_{ij} \in [\xi_{ij} - r, \xi_{ij} + r]$}
$\eta_{ij}^{aux} = \arg\min_{\eta_{ij}} \frac{1}{2\theta}(\xi_{ij} - \eta_{ij})^2 + \lambda C_{ij}(\eta_{i,j}) + \alpha_{i,j}(\xi_{i,j} - \eta_{i,j})$
{Subpixel refinement:}
$\nabla E^{aux} = \lambda \nabla C_{ij}(\eta_{ij}^{aux}) + \frac{\eta_{ij}^{aux} - xi_{ij}}{\theta} - \alpha_{ij}$
$\nabla^2 E^{aux} = \lambda \nabla^2 C_{ij}(\eta_{ij}^{aux}) + \frac{1}{\theta}$
$\eta_{ij} = \eta_{ij}^{aux} - \nabla E^{aux} / \nabla^2 E^{aux}$

simple quadratic penalty function suffices. Using this approach, Eq. (6) is minimized by sequentially solving an unconstrained minimization problem of the form

$$\min_{\boldsymbol{\xi},\boldsymbol{\eta}} E_R(\boldsymbol{\xi}) + \frac{1}{2\theta}\|\boldsymbol{\xi} - \boldsymbol{\eta}\|_2^2 + \lambda E_D(\boldsymbol{\eta}) \tag{7}$$

where $E(\boldsymbol{\xi}, \boldsymbol{\eta}) \rightarrow E(\boldsymbol{\xi})$ as $\theta \rightarrow 0$. In general, the main disadvantages of this approach, reported in [3], are its slow convergence and ill-conditioning for small values of θ. Nevertheless, for the depth map estimation problem, this algorithm has shown an admirable performance in practice. Note that Lagrange multipliers play no direct role in this method. The new energy functional in Eq. (7) allows us to split the minimization into two different problems that are alternatively solved until convergence:

- First, for a fixed $\boldsymbol{\eta}$ solve:

  $$\min_{\boldsymbol{\xi}} E_R(\boldsymbol{\xi}) + \frac{1}{2\theta}\|\boldsymbol{\xi} - \boldsymbol{\eta}\|_2^2 \tag{8}$$

  which corresponds to the well known TV-ROF convex denoising problem that can be solved using a primal-dual algorithm [4]. In this case $\boldsymbol{\eta}$ represents a noisy image whereas $\boldsymbol{\xi}$ is the searched denoised result.
- Second, for a fixed $\boldsymbol{\xi}$ solve:

  $$\min_{\boldsymbol{\eta}} \frac{1}{2\theta}\|\boldsymbol{\xi} - \boldsymbol{\eta}\|_2^2 + \lambda E_D(\boldsymbol{\eta}) \tag{9}$$

  this optimization is performed by a point-wise exhaustive search followed by an accelerated subpixel refinement for each $\boldsymbol{\eta}$ in the cost volumn as it is explained in [10]. We show its general implementation of these steps in Algorithm 2. Lines 6–16 illustrate the main iterative per-pixel primal dual algorithm. Line 9: ascend gradient step to update the dual variable $\mathbf{p}$. Line 11: descend gradient step to update the primal variable ξ. The parameters τ and σ are calculated via preconditioning [11].

3.2 Lagrange Multipliers

We must now briefly mention the role of Langrange Multipliers as a precursor to our use of the "Augmented Lagrangian" in the next section. The original constrained optimization Eq. (6) can be transformed to an unconstrained minimization problem by introducing the Lagrangian function

$$E_R(\boldsymbol{\xi}) + \alpha^T(\boldsymbol{\xi} - \boldsymbol{\eta}) + \lambda E_D(\boldsymbol{\eta}) \tag{10}$$

where α is a Lagrange multiplier associated with the original constraint. In this approach the Lagrange multiplier is treated on an equal basis with the variables $\boldsymbol{\xi}$, $\boldsymbol{\eta}$, which means that in order to solve the unconstrained problem we have to iterate as well for α. Although there exist different methods to iteratively update $\boldsymbol{\xi}$, $\boldsymbol{\eta}$, α and solve the Lagrangian equation, we are going to concentrate on the Augmented Lagrangian method explained in the next subsection.

3.3 Augmented Lagrangian

The Augmented Lagrangian belongs to a class of methods called *methods of multipliers* in which the penalty regularization is combined with the Lagrange Multipliers method. The resultant objective function, called the *Augmented Lagrangian*, is sequentially minimized to obtain a solution to the original constrained problem. In our case the augmented Lagrangian is given by

$$E_R(\boldsymbol{\xi}) + \boldsymbol{\alpha}^T(\boldsymbol{\xi} - \boldsymbol{\eta}) + \frac{1}{2\theta}\|\boldsymbol{\xi} - \boldsymbol{\eta}\|_2^2 + \lambda E_D(\boldsymbol{\eta}) \tag{11}$$

The main advantages of this method over the previous ones are: First, convergence can be attained even when θ does not decrease to zero improving the stability of the algorithm. Second, there exists a simple update of the Lagrange Mulltiplier $\boldsymbol{\alpha}$ that tends to make it converge faster to its proper value than pure Lagrange Multipliers approaches [3].

As in the Quadratic Penalty section, Eq. (11) is minimized by alternatively solving the following sub-problems until convergence

- First, for a fixed $\boldsymbol{\eta}$ solve:

$$\min_{\boldsymbol{\xi}} E_R(\boldsymbol{\xi}) + \boldsymbol{\alpha}^T(\boldsymbol{\xi} - \boldsymbol{\eta}) + \frac{1}{2\theta}\|\boldsymbol{\xi} - \boldsymbol{\eta}\|_2^2 \tag{12}$$

  using a primal-dual algorithm [4] since previous optimization is convex in $\boldsymbol{\xi}$
- Second, for a fixed $\boldsymbol{\xi}$ solve:

$$\min_{\boldsymbol{\eta}} \boldsymbol{\alpha}^T(\boldsymbol{\xi} - \boldsymbol{\eta}) + \frac{1}{2\theta}\|\boldsymbol{\xi} - \boldsymbol{\eta}\|_2^2 + \lambda E_D(\boldsymbol{\eta}) \tag{13}$$

  using a point-wise exhaustive search for each $\boldsymbol{\eta}$ in the cube.
- Third, update α

$$\boldsymbol{\alpha} = \boldsymbol{\alpha} + \frac{1}{\theta}(\boldsymbol{\xi} - \boldsymbol{\eta}) \tag{14}$$

In contrast to Quadratic Penalty method, we have introduced the new variable α. Although it implies a change in the numerical implementation, the iterations required for convergence are substantially reduced as we will show in Sect. 5. In particular, it affects the update of the primal variable ξ (line 11 in algorithm 1) as well as the accelerated search and smoothing step for sup-pixel accuracy (algorithm 2, lines 4 and 6). For better readability, we have highlight in red color the differences between the QP and AL numerical implementations. Notice that the changes between both algorithms are minimal.

4 Adaptive Regularisation

In this paper we also exploit the concept of the uncertainty on the inverse depth to reinforce regularization on non-informative depth map regions. Regularization plays an important role in achieving high accurate depth-maps in small scenes. However, depending on the quality of the metric used as well as the initial depth seed, the effect of the regularisation does not necessarily provide a positive impact on the final solution. The lack of texture in some regions of the scene (blank walls, texture-less surfaces, ...) generates in fact a non-informative cost along the volume. Figure 2 bottom shows, for two different pixels u_g and u_r in the reference image, the corresponding set of cost values store in the cube along the inverse depth interval $[\xi_{max}\ \xi_{min}]$. Notice that the 1D cost functions present a low or high variability depending on whether the pixel belongs to a texture-less region u_r (flat walls, floor, roof, ...) or to

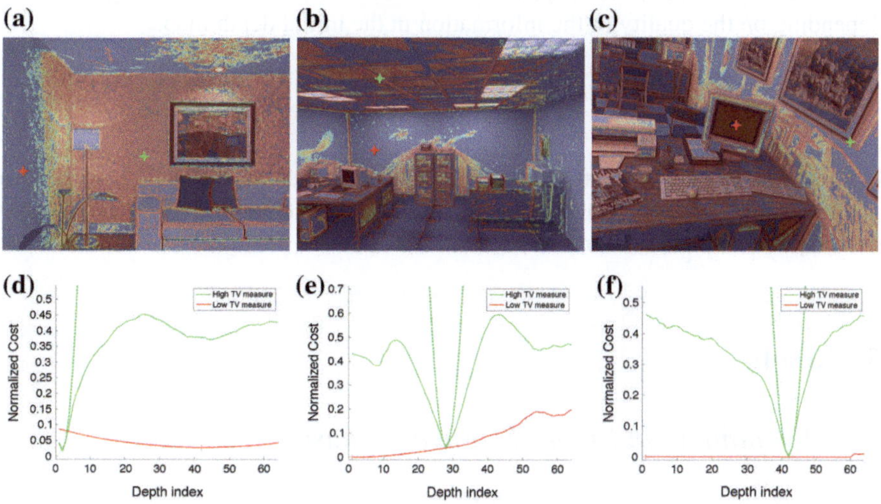

Fig. 2 Adaptive selection of λ. To weight the contribution of each pixel in the dataterm, we estimate the uncertainty of the depth represented as a Gaussian distribution on the cost along the inverse depth range, with mean centered in the depth for which the cost is minimum. First row shows the pixel-wise uncertainty overlapping the reference image for three synthetic datasets. Green crosses represent examples of highly informative pixels u_g, while red crosses determine pixels u_r with more uncertainty. Second row shows variability of the cost along 64-discrete inverse depth index values for the two examples. The fitted Gaussian is illustrated for the green case. **a** Scene 1. **b** Scene 2. **c** Scene 3. **d** $\sigma_g = 0.0684$, $\sigma_r = 1$. **e** $\sigma_g = 0.0668$, $\sigma_r = 1$. **f** $\sigma_g = 0.0760$, $\sigma_r = 1$

an informative one u_g. For each pixel $u_i \in \mathbf{u}$ in the reference image, we can estimate the inverse depth uncertainty using the following second order approximation,

$$C(u_i, \boldsymbol{\xi}) \sim C(u_i, \boldsymbol{\xi}^*) + (\boldsymbol{\xi} - \boldsymbol{\xi}^*)\nabla C(u_i, \boldsymbol{\xi})|_{\xi=\xi^*} + \frac{1}{2}\nabla^2 C(u_i, \boldsymbol{\xi})|_{\xi=\xi^*}(\boldsymbol{\xi} - \boldsymbol{\xi}^*)^2 \tag{15}$$

where $C(u_i, \boldsymbol{\xi}^*)$ represents the minimum cost along the sampled distances. Figure 2 bottom, shows the quadratic approximation of the cost function at a particular pixel. Note that the quadratic is naturally centered at the sampled depth ξ^* at which the cost is minimum. To associate uncertainties with per-pixel inverse depth estimates we look at the curvature of the correlation surface, i.e., how strong the minimum in the cost volume is at the winning inverse depth [14]. Under the assumption of small noise, photometrically calibrated images, and densely sampled inverse depth, the uncertainty is approximated by a normal distribution synthetised as follows

$$\xi(u_i) \sim \mathcal{N}(\xi^*(u_i), \Sigma_\xi) \tag{16}$$

where the variance is locally estimated by the hessian $\Sigma_\xi \propto 1/\nabla^2 C(u_i, \boldsymbol{\xi})|_{\xi=\xi^*}$ in the inverse depth point where the cost is minimum. Equation (15) allows us to calculate a per-pixel adaptive trade off $\lambda(\mathbf{u})$ between the data fidelity term and the regularisation depending on the quality of the information in the initial depth map.

$$\lambda(u_i) \propto \frac{1}{\Sigma_\xi} \tag{17}$$

Figure 2 top, shows the output image that results after the calculation of the per pixel variance for three synthetic indoor datasets. Notice that the reference image is overlapped for better interpretation.

5 Results

5.1 Evaluation on Indoor Synthetic Datasets

We have conducted our experiments on three synthetic indoor scenes that provide high precision depth maps from images taken at 30 Hz [7, 9].[2,3] Our chosen scenes consider both close and far objects from the camera and partial occlusions. We first evaluate the influence of the similarity metric used to obtain the initial solution.

---

[2] http://www.doc.ic.ac.uk/~ahanda/VaFRIC/index.html.

[3] http://www.doc.ic.ac.uk/~ahanda/HighFrameRateTracking/downloads.html.

Recall that the metrics under evaluation are the SSD, the SAD and the NCC. After executing the AL optimization algorithm for each metric, we calculate the median error of the depth-map solution with respect to the ground truth. In order to compare the accuracy of AL and QP algorithms we will calculate:

$$cost(\mathbf{u}) = median(\|\xi(\mathbf{u})_{GT} - \xi^*(\mathbf{u})\|_1) \tag{18}$$

Figure 3 shows for all scenes the median errors obtained for window sizes ranging in the interval $W = [1 \dots 15]$. This preliminary analysis shows that, for the correct

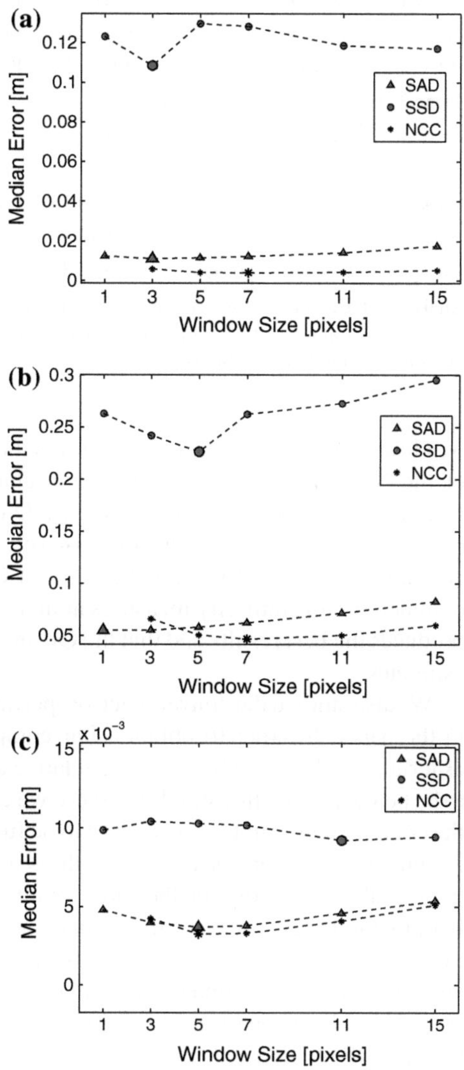

Fig. 3 Median error obtained after optimisation on three different synthetic scenes. For each similarity metric (SAD, SSD, NCC), the plots show the optimal window size to achieve the minimum error. In general, NCC yields more accurate results on all datasets (see the scale of y-axis). **a** Scene 1. **b** Scene 2. **c** Scene 3

Table 2 Convergence analysis for AL and QP

Scene 1, range = [1.655 3.445] [m]

	Median error [m]		Energy		$\|\xi - \eta\|_2$		% iter saved
Metric	AL	QP	AL	QP	AL	QP	(%)
SAD 3	0.0111	0.0107	3283.77	3233.44	0.0452	0.0500	57
SSD 3	0.1084	0.1459	1288.70	1342.70	0.0466	0.0130	74
NCC 7	0.0038	0.0032	26081.11	27804.42	0.0480	0.0497	63

Scene 2, range = [1.102 6.186] [m]

	Median error [m]		Energy		$\|\xi - \eta\|_2$		% iter saved
Metric	AL	QP	AL	QP	AL	QP	(%)
SAD 1	0.0549	0.0551	8278.24	8317.93	0.0426	0.0456	66
SSD 5	0.2264	0.2821	8141.76	8383.51	0.0406	0.0236	72
NCC 7	0.0467	0.0488	49449.81	49356.62	0.0462	0.0496	55

Scene 3, range = [0.773 5.953] [m]

	Median error [m]		Energy		$\|\xi - \eta\|_2$		% iter saved
Metric	AL	QP	AL	QP	AL	QP	(%)
SAD 5	0.0037	0.0043	11410.13	11456.60	0.0460	0.0190	84
SSD 11	0.0092	0.0089	2594.75	2577.19	0.0482	0.0098	90
NCC 5	0.0032	0.0032	75876.31	75601.43	0.0423	0.0433	67

Analysis of Errors, Energy convergence and constraint fulfill at the final solution for both the Augmented Lagrange (AL) and the Quadratic Penalty (QP) methods using different similarity measures to obtain the initial seed

window size, the NCC measure achieves the best results. This can be a consequence of the NCC invariance to illumination changes. Since the NCC is usually costly to evaluate we can also see that the SAD even with a window size of 1 performs relatively well and can be used in case of computation constraints. "Median Error" column in Table 2, shows for the AL and QP algorithms the lowest median errors obtained for all similarity measures at their optimal window size. Observe that NCC produces the best results and that the QP and AL algorithms produce similar accurate estimates.

We also studied the convergence properties of the AL and QP algorithms described in the paper. In order to obtain a fair comparison, we have applied the same stop criteria to both methods: First, the relative decrease in the energy minimization has to be below a given threshold (to assure we can not make much progress) and second, the equality constraint is considered to be fulfill if $\|\xi - \eta\| \leq 5e - 2$. Figure 4 shows the energy evolution for both algorithms using NCC with optimal window size for the initial seed. In two of the three synthetic scenes (Fig. 4 second, third column) both methods converge to similar final energy and constraint values. Notice that, in the limit, ξ and η must achieve the same values, thus the decouple energies for AL and QP should approximate very well the original energy in Eq. 4. However, the most important advantage of the AL method, which is one the contributions of this paper, over the QP method is its faster convergence requiring fewer iterations to achieve the

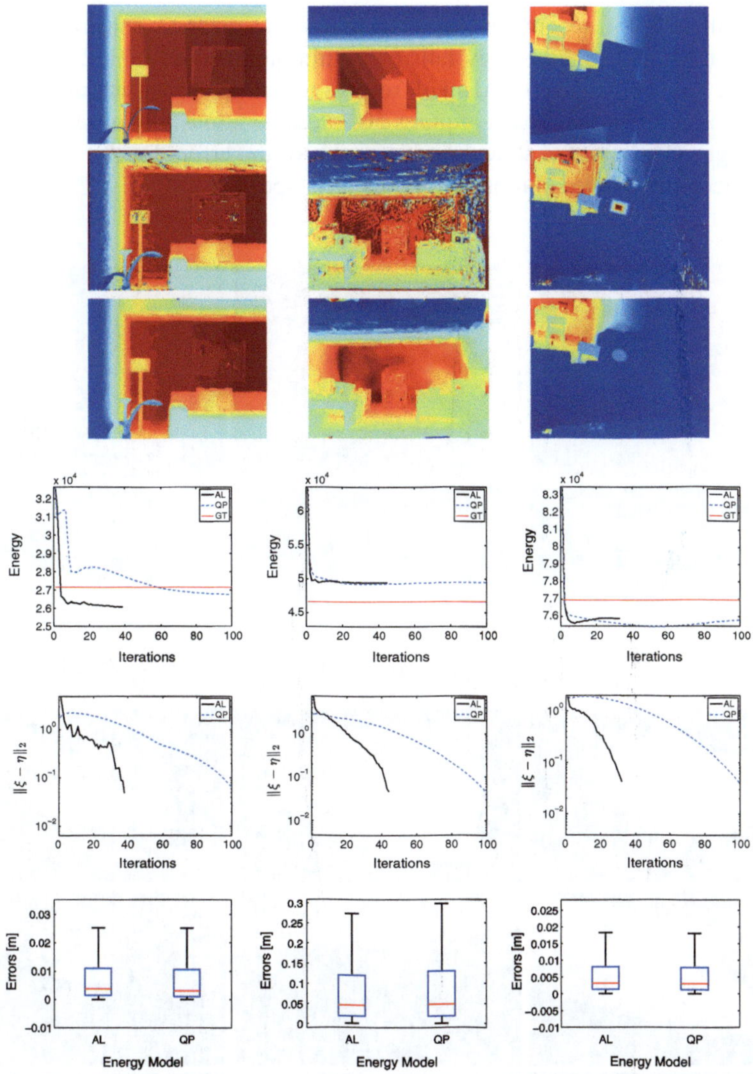

Fig. 4 Convergence and Accuracy Analysis for the proposed Augmented Lagrangian (AL) method in comparison to the common Quadratic Penalty (QP) approach. The experiments are shown for three synthetic scenes: *left*, scene 1; *middle*, scene 2; *right*, scene 3. First row, ground truth depth map. Second row, Initial seed obtained wiht a NCC-based cost volume at the optimal window size as reported in Table 2. Third row, achieved depth-map solution. Fourth row, energy evolution over all iterations. Fifth row, evolution of the constraint $\|\boldsymbol{\xi} - \boldsymbol{\eta}\|_2$ per iteration. Notice that AL (*black solid line*) outperforms QP (*blue light line*) to converge at the final solution. Energy is evaluated at the ground truth (GT) which constant value is displayed with a red line. Sixth row, boxplots of the error distributions of the per pixel inverse depth map estimates. The tops and bottoms of each box are the 25 and 75th percentiles of the samples, respectively. The distances between the *tops* and *bottoms* are the inter-quartile ranges. The line in the middle of each box is the sample median. AL and QP achieve high accurate depth-maps with similar error distributions. However, AL achieves the final solution faster than QP

same result. In Fig. 4, second row, we observe how the quadratic constraint decreases rapidly for AL and so the energy falls to its minimum value. Table 2 column nine, shows the gain percentage of AL with respect to the number of iterations required for QP. The proposed approach requires 50 % less iterations till convergence for all cases. Figure 4, sixth row, shows the histogram of the errors for AL and QP. Note that the accuracy of the solution is not traded for speed.

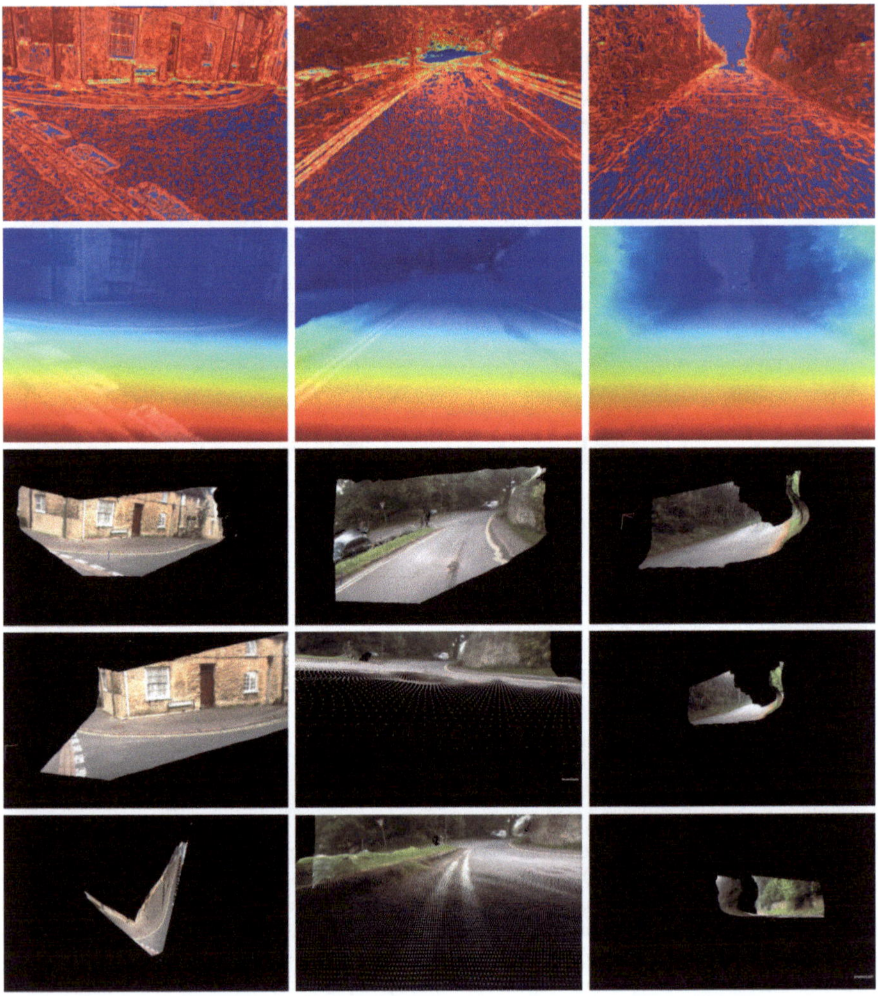

Fig. 5 3D reconstruction of outdoor scenes from monocular images. The use of Adaptive regularisation improve the appearance of the point cloud capturing the diverse shapes present in the environment. First row, pixel-wise depth uncertainty. Second row, Inverse depth map obtained after 30 primal dual iterations. Third-fifth rows, different camera views of the final 3D dense reconstruction

5.2 *Dense Reconstruction of Outdoor Scenes*

Our goal is to show that the AL method in combination with adaptive regularisation improve the appearance of the point cloud capturing the diverse shapes present in outdoor environments. Our motivation is that while a sparse map provides a compact representation for autonomous navigation, higher level robot tasks can require denser maps to improve scene understanding. We have a forwards-facing camera mounted on a car travelling forwards and sensing distant objects with a low parallax. This leads us to rely on an improved regularisation method to reinforce depth on critical parts of the scene. In our case, a suitable assumption is to expect to find many affine surfaces in the environment, like roads, pathways, building façades or vehicle surfaces.

The input to our pipeline consists of only two consecutive image frames gathered by a camera at 25 Hz. This choice enables us to estimate the depth of dynamic objects (particularly important in urban environments), which could be potentially disregarded by a long sequence integration. The sensor is mounted on a car that traverses a city environment. Figure 5, shows the reconstruction of three different scenes with heterogeneous geometry (walls, roads and vegetation). To track the camera, we employ our own scaled Visual Odometry system [12].

Figure 5 first row, shows the per pixel inverse depth uncertainty. As it is expected, road surfaces and distant regions exhibit low information. The use of the per-pixel adaptive regularisation allows us to recover most of the structure. A video showing more details of the execution of the algorithms is available at (http://youtu.be/LrNv9QCKH1s).

6 Conclusions

We have shown the efficacy of the Augmented Lagrangian method for depth map estimation using monocular cameras. As a result we can substantially reduce the number of iterations required for convergence, more than 50 % of reduction in all cases, compare to state of the art algorithms based on Quadratic Penalty methods. We have also performed an exhaustive study of different photo-consistency measures SSD, SAD and NCC and different windows sizes in order to improve the accuracy of the initial depth map used as seed in the optimization algorithm. As was expected, NCC provides the best results due to its intrinsic properties to cope with illumination changes. Finally, we introduce a novel per pixel inverse depth uncertainty estimation that affords us to apply adaptive regularisation on the initial depth map: high informative inverse depth pixels require less regularisation, however its impact on more uncertain regions can be propagated providing significant improvement on textureless regions.

References

1. Afonso, M., Dias, J., Figueiredo, M.A.T.: An augmented lagrangian approach to the constrained optimization formulation of imaging inverse problems. IEEE Trans. Image Process. **20**(3), 681–695 (2011)
2. Alvarez, H., Paz, L., Sturm, J., Cremers, D.: Collision avoidance for quadrotors with a monocular camera. In: Proceedings of The 12th International Symposium on Experimental Robotics (ISER) (2014)
3. Bertsekas, D.P.: Constrained optimization and lagrange multiplier methods. In: Computer Science and Applied Mathematics, vol. 1982, p. 1. Academic Press, Boston (1982)
4. Chambolle, A., Pock, T.: A first-order primal-dual algorithm for convex problems with applications to imaging. J. Math. Imaging Vision **40**(1), 120–145 (2011)
5. Chan, S.H., Khoshabeh, R., Gibson, K.B., Gill, P.E., Nguyen, T.Q.: An augmented lagrangian method for total variation video restoration. IEEE Trans. Image Proc. **20**(11), 3097–3111 (2011)
6. Graber, G., Pock, T., Bischof, H.: Online 3D reconstruction using Convex Optimization. In: 1st Workshop on Live Dense Reconstruction From Moving Cameras, ICCV 2011 (2011)
7. Handa, A., Newcombe, R.A., Angeli, A., Davison, A.J.: Real-time camera tracking: when is high frame-rate best? In: ECCV (2012)
8. Li, C., Yin, W., Jiang, H., Zhang, Y.: An efficient augmented lagrangian method with applications to total variation minimization. Comput. Optim. Appl. **56**(3), 507–530 (2013)
9. Nardi, L., Bodin, B., Zia, M.Z., Mawer, J., Nisbet, A., Kelly, P.H.J., Davison, A.J., Luján, M., O'Boyle, M.F.P., Riley, G., Topham, N., Furber, S.: Introducing SLAMBench, a performance and accuracy benchmarking methodology for SLAM. In: ICRA 2015 (2015). arXiv:1410.2167
10. Newcombe, R.A., Lovegrove, S.J., Davison, A.J.: In: ICCV 2011, pp. 2320–2327. IEEE Computer Society, Washington, DC, USA (2011)
11. Pock, T., Chambolle, A.: Diagonal preconditioning for first order primal-dual algorithms in convex optimization. In: IEEE International Conference on Computer Vision (ICCV), pp. 1762–1769 (2011). doi:10.1109/ICCV.2011.6126441
12. Smith, M., Baldwin, I., Churchill, W., Paul, R., Newman, P.: The new college vision and laser data set. Int. J. Rob. Res. **28**(5), 595–599 (2009)
13. Stühmer, J., Gumhold, S., Cremers, D.: Real-time dense geometry from a handheld camera. In: DAGM 2010, pp. 11–20. Darmstadt, Germany (2010)
14. Szeliski, R.: Computer Vision: Algorithms and Applications, 1st edn. Springer, New York (2010)
15. Zach, C., Pock, T., Bischof, H.: A duality based approach for realtime TV-L1 optical flow. In: Pattern Recognition (Proc. DAGM), pp. 214–223. Heidelberg, Germany (2007)

Wrong Today, Right Tomorrow: Experience-Based Classification for Robot Perception

Jeffrey Hawke, Corina Gurău, Chi Hay Tong and Ingmar Posner

Abstract This paper is about building robots that get better through use in their *particular* environment, improving their perceptual abilities. We approach this from a life long learning perspective: we want the robot's ability to detect objects in its specific operating environment to evolve and improve over time. Our idea, which we call Experience-Based Classification (EBC), builds on the well established practice of performing hard negative mining to train object detectors. Rather than cease mining for data once a detector is trained, EBC continuously seeks to learn from mistakes made while processing data observed during the robot's operation. This process is entirely *self-supervised*, facilitated by spatial heuristics and the fact that we have additional scene data at our disposal in mobile robotics. In the context of autonomous driving we demonstrate considerable object detector improvement over time using 40 Km of data gathered from different driving routes at different times of year.

1 Introduction

Object detection forms one of the cornerstones of autonomous operation in complex, dynamic environments. Whether it concerns the detection of assets for the purpose of infrastructure survey, the detection of wares and co-workers for applications in logistics, or the detection of other traffic participants in an autonomous driving context, object detectors need to provide fast, reliable performance *across* a number of

J. Hawke and C. Gurăau contributed equally to this work.

J. Hawke (✉) · C. Gurău · C.H. Tong · I. Posner
University of Oxford, Oxford OX1 3PJ, UK
e-mail: jhawke@robots.ox.ac.uk

C. Gurău
e-mail: corina@robots.ox.ac.uk

C.H. Tong
e-mail: chi@robots.ox.ac.uk

I. Posner
e-mail: ingmar@robots.ox.ac.uk

D.S. Wettergreen and T.D. Barfoot (eds.), *Field and Service Robotics*,
Springer Tracts in Advanced Robotics 113, DOI 10.1007/978-3-319-27702-8_12

workspaces. This is explicitly encouraged in the machine vision community as witnessed by, for example, the ImageNet Large Scale Visual Recognition Challenge [6].

However, while much progress is being made, error rates of state-of-the-art approaches are still prohibitive, particularly for safety critical applications (e.g. [1] for the case of pedestrian detection). This is often due to a significant amount of variation in the negative class which, in reality, is not captured in the training data. While it is relatively easy to obtain negative samples, computational limits imply that we should only include ones that have a large effect on the decision boundary. The standard method for obtaining relevant negative samples is known as hard negative mining (HNM) [13, 25], which is commonly used to bootstrap the underlying classifier used in an object detector. HNM is widely considered a mandatory part of detector training, where the classifier is first trained on the original training data and then used to perform object detection on a *labelled* dataset. False positives are identified using the ground truth labels provided and included for classifier retraining. This provides considerable improvement over the original detector, but the data used for negative mining strongly influences the resulting performance due to dataset bias [21, 28]. In robotics, where we have a limited range of operation and are not as concerned with general performance, biasing the detector's performance to our workspace is a powerful tool.

In robotics, in order to improve performance for a particular application, scene context—obtained through online sensing or contained in (semantic) map priors—is commonly leveraged as a filter (e.g. [15, 24]). Typically, this takes the form of discarding detections as spurious if certain validation criteria are not met (e.g. a car needs to be found on or near a road [22]).

In this work we also exploit scene context to validate the detections obtained. However, we advocate a radically different detector deployment model from the status quo, which leads to *self-supervised* and *environment-dependent* performance improvement over the lifetime of the detector. This reflects our desire for lifelong learning systems which excel in a robot's specific application domain instead of providing mediocre performance everywhere.

Our approach, is from one perspective, a simple and straightforward one and yet it brings remarkable and profound benefits to our problem domain: some applications of embedded perception can afford to trade generality for specificity. Robotic agents should adjust to a vanishingly small subset of all possible workspaces: the ones they operate in, or 'experience', on a daily basis.

Inspired by hard negative mining, we continue to train our detectors in a self-supervised learning by exploiting scene context from the robot's operating environment. This is achieved by continuously feeding back into the training process of the detector any false positives identified by a validation step throughout the lifetime of the system. We call this process Experience-Based Classification (EBC).

In effect, EBC automatically trains detectors for specific operating environments. While this may lead to overfitting to the background encountered, we argue that this is exactly what is required in mobile robotics where autonomous agents often traverse the same workspace over and over again. In fact, EBC relies on this behaviour and, inspired by recent work in the vision community such as [7], exploits similarities in

Fig. 1 Images from the same route at two different times of year (January and May) on which we performed pedestrian detection. While the pedestrians look similar in all images, the background class is quite different, with visible seasonal effects. This suggests the need for environment-dependent classifiers. False positives are shown in *purple*, while true positives have a *yellow* bounding box. A few iterations of EBC over the course of a few days show great improvement

geo-spatially related locations. Furthermore, the self-supervised, operational nature of this approach means that it can incorporate considerably more data in training than conventionally performing HNM on small canonical datasets. This opens up the possibility for life-long learning on robot perception, building up a collection of environment dependent object detectors over the robot's lifetime.

EBC is agnostic to the application domain, detection framework and object class considered. However, in this paper we frame the discussion in the context of pedestrian detection for autonomous driving (see Fig. 1). We utilise the fact that object detection is often performed alongside navigation and that current navigation solutions localise against a previously-acquired map [3, 14, 23]. This provides the scene context for EBC.

2 Related Work

One common approach to pedestrian detection from monocular imagery utilises a linear SVM classifier on Histogram of Oriented Gradients (HOG) features [5]. The use of a linear model permits efficient sliding window computations [9] when a sliding window detector is implemented using this classifier. More recent work in pedestrian detection has extended this to use alternative feature types such as Aggregate Channel Features (ACF) [8], or alternative classifiers such as Latent SVMs with deformable parts models [13], and decision trees with Adaboost [1]. For our sliding window detector, we elected to use the same feature type as the current state of the art pedestrian detector (Aggregate Channel Features), but with a simpler linear classifier model and a reduced number of scales.

3D scene information has been primarily used in object detection to generate Regions of Interest (ROIs). For example, a ground plane computed from stereo imagery can provide a search space for detections (e.g. [15, 24]), or enforce scale [16]. Enzweiler et al. [10] extend this idea by maintaining a height-based representation of the local environment to generate ROIs, and Ess et al. [11] jointly infer the depth, ground plane and object detections.

Instead of generating ROIs to present to our classifier, we invert the order and apply scene information after we compute detections. While both approaches provide us with a set of valid positive classifications, this ordering also allows us to obtain a set of informative negative data samples that can be used for detector improvement.

As mentioned in the introduction, the conventional approach for obtaining these hard negatives when initially training a detector is Hard Negative Mining (HNM), performed on a labelled training dataset. Initially introduced by Sung and Poggio [25] as a bootstrap method for expanding the training set, Felzenszwalb et al. [13] tailor it for structural SVMs by defining 'hard' negatives as examples that are incorrectly classified or within the margin of the SVM classifier. HNM has also been used for multiclass object detection [20], where positive samples of other classes can serve as hard negatives.

Instead of HNM, Hariharan et al. [18] suggested training Linear Discriminant Analysis (LDA) classifiers with an extremely large negative class. This was made possible for SVM classifiers by Henriques et al. [19], who used block-circulant decomposition to train with an approximation to the set of all negative samples from a series of images. In effect, training with a vast set of negatives reduces the need to specifically mine for hard negatives. While efficient, the training remains limited by computational resources, and does not escape the core requirement of labelled data.

This prior work on HNM is complementary to our work on EBC. HNM still forms a critical step when initially training a detector, and EBC builds on this to continue bootstrapping the detector to the robot's operating environment. In addition, many of these techniques to extend HNM could equally be applied to EBC. This paper is an extension of a previous workshop paper [17], which showed that EBC is comparable to HNM on the same labelled data. Here we consider the effect of place and season on life-long learning.

In all these approaches, labelled data are used to identify negative samples. While the labelling effort may be tolerated for individual datasets, real-world operation is subject to variation from seasonal, lighting and environmental changes. This has a significant impact on detection performance, but manually labelling data for all of these scenarios is impractical for life-long learning in robotics. EBC is able to meet these requirements by identifying relevant samples in a self-supervised manner.

We share some similarities with the concept of group induction [27], where self-supervised training is performed by alternating between classifying unlabelled tracks and incorporating the most confident positive classifications in retraining. Our approach differs by the fact that we use an external signal in the form of an environmental prior to provide labels for the whole scene. This allows us to focus only on hard samples and provides a means to automatically train our detectors for specific environments.

3 Framework Description

EBC augments a standard perception pipeline by introducing a scene filtering step after object detection, a memory bank of negative samples and classifier retraining. Our implementation of this system is depicted in Fig. 2. The following sections

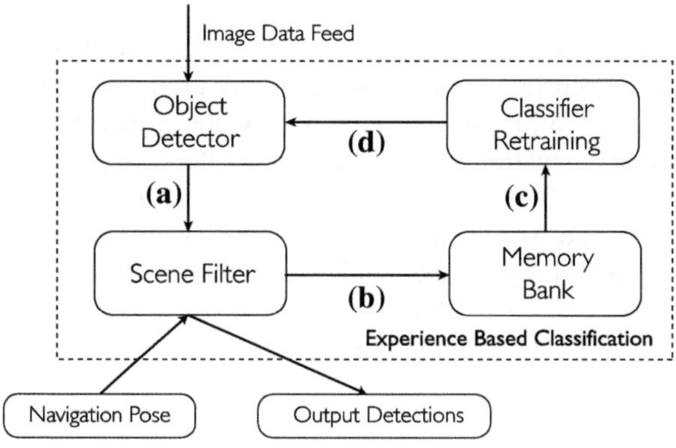

Fig. 2 The EBC architecture implemented in this paper. **a** An object detector provides detections based on the image feed. **b** A scene prior is used to filter out detections that do not touch the ground plane or have an unexpected scale. **c** Rejected samples are stored. **d** The detector is retrained at the end of an experience using additional rejected samples. The EBC detector improves through successive outings as it automatically adjusts to what it experiences

describe the function of each component in further detail and provide specific information about our implementation.

3.1 Object Detector

In general terms, an object detector processes a data stream and produces detections. EBC serves as a wrapper for this, providing additional training samples for lifelong improvement. In this work we employ a linear SVM classifier to classify whether an image patch is part of the positive class or the negative class. Given an input image, we first compute features for the entire image, and then employ a sliding window approach to obtain classification scores. Multiscale detection is performed by resizing the image and repeating the process. Finally, non-maximal suppression is used to filter out overlapping detections. The output is a set of bounding boxes which correspond to subwindows that score above a threshold, which are deemed to be positive detections. Further detail on the object detector specifics can be found in Sect. 4.

3.2 Scene Filter

The scene filter is a core component of the EBC framework. Given a set of detections, the scene filter employs local context to filter out false positives according to strong heuristics. Accepted detections are passed on to the remainder of the perception pipeline, while rejected detections are stored in the memory bank. Since the rejected samples are detections that scored highly in the previous step these are by definition hard negatives.

Given localisation information and a 3D scene prior, we first look up the local ground plane for our current location, then project the local ground plane into the image. This is used by a first filter, which rejects detections that lie off the ground plane for the current navigation frame. Our second filter then projects each remaining bounding box into the 3D scene to ensure detections are of a viable scale. The application of these heuristics is illustrated in Fig. 3. The scene filtering step should be conservative to avoid rejecting a valid detection (true positive), which may lead to semantic drift [4]. The goal of this filtering component is to reduce the number of false positives while not introducing false negatives.

3.3 Memory Bank and Retraining

The final step of the EBC cycle augments the original training set with the rejected samples and retrains the classifier model. Since these additional negatives are

Fig. 3 An illustration of the scene filter employed in this work, using localisation information and heuristics such as scale and ground correction

obtained during operation, each subsequent training cycle further adapts the classifier to the specific environment. It should be noted that data streams gathered from mobile robotic platforms tend to be spatially and temporally correlated. This can cause problems in retraining as most classifiers assume independent, identically distributed data. Subsampling may be required to avoid these issues.

4 Experimental Evaluation

4.1 Methodology

We seek to evaluate the implications of an object detector learning from the environment it experiences. We do this by taking a common baseline detector model, then use this to train separate detectors for different classes of data, comparing their performance on test datasets from these same classes. To do this, we put a baseline pedestrian detector through successive training cycles on urban driving datasets gathered from two different routes in Oxford at different times of year. We anticipate that the detector which has learned from operating data which most closely matches the test data (place and season) will perform the best, as the detector becomes fitted to the operating environment.

A single experiment consisted of the baseline detector being presented with driving data from successive days, with a detector retraining step between each dataset. This process follows the EBC system architecture diagram in Fig. 2. For a given detector model, the image data from the single specified urban driving dataset was processed to compute detections. The detections were then processed by the scene filter to validate or reject the samples according to spatial heuristics. The resulting negative samples were then sampled (taking the top 10 false positives from a random frame in every second of time), aggregated with prior rejected negative data samples, then used to retrain the detector along with the original training data. The negative data was weighted to ensure the class balance from the original training data was maintained.

Each experimental run was evaluated against a manually labelled test dataset which shared the same location and environmental conditions as one of the training categories.

4.2 Baseline Detector

Our baseline pedestrian detector used a classifier trained using LIBLINEAR [12] on the INRIA Pedestrian Dataset [5] with Aggregate Channel Features and a similar training methodology to [8]. We performed ten-fold cross validation on a training set consisting of 1237 cropped positive pedestrian samples, and 12,180 sampled negatives (10 windows per negative image). The final step in the detector training process is a bootstrapping step consisting of ten consecutive cycles of HNM using the INRIA negative images. In each HNM cycle the classifier was presented with 10,000 random negative cropped samples extracted from the negative images. Misclassified 'hard' negatives were saved and used as additional training data to retrain the classifier.

4.3 Datasets

To show the impact of environmental variation and to evaluate our self-supervised learning approach, we used twelve different urban driving training datasets gathered with a Bumblebee2 stereo camera mounted on our Wildcat vehicle (Fig. 4a) driving around Oxford. These unlabelled datasets were gathered from two different routes from successive outings at different times of year. We allocated these datasets to three categories based on the route and time of year (season), with 4 training datasets per category. These categories are referred to in this section as *North Oxford January*, *North Oxford May*, and *Central Oxford August*. A map of the routes is provided in Fig. 4b, with no overlap between the North Oxford and Central Oxford routes. For all datasets, we used only the left stereo image with a capture rate of 20 Hz.

For evaluation, we used an additional manually labelled test dataset from the *North Oxford January* category. This provided a total of 40 Km of unlabelled training data and 2 Km of labelled test data. The datasets are summarised in Table 1.

4.4 Results

We trained three different detectors from the same starting base detector, one per category. These detectors are referred to by their category name: *North Oxford January*

(a) **(b)**

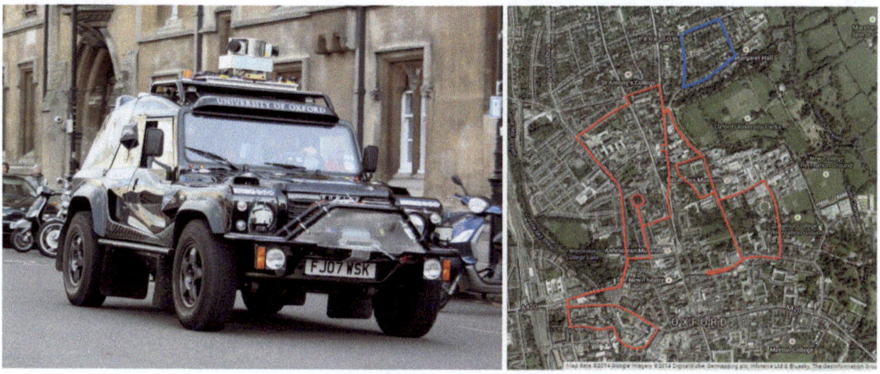

Fig. 4 The Wildcat vehicle (*left*) used to gather the image data, and a map (*right*) depicting the routes for our datasets, where we gathered images in January, May, and August. The difference in time of year provided seasonal variation, which affects the visual appearance of the scene. The two North Oxford routes, illustrated in *blue*, provided variation in season, and the Central Oxford route, depicted in *red*, provided a difference in location

Table 1 A summary of the datasets used for training and evaluation

Route		Train 1	Train 2	Train 3	Train 4	Train Total	Test
North Oxford January	Distance (km)	2.60	2.01	1.92	1.92	8.45	1.99
	Image frames	12782	9436	8172	8215	38605	9155
	Time (min)	10.7	7.86	6.81	6.84	32.2	7.63
North Oxford May	Distance (km)	1.43	1.95	1.01	1.01	5.40	–
	Image frames	6066	8676	4001	3977	22720	–
	Time (min)	5.06	7.23	3.33	3.31	18.9	–
Central Oxford	Distance (km)	6.91	6.79	6.68	6.56	26.94	–
	Image frames	36472	27720	27607	23463	115262	–
	Time (min)	30.4	23.1	23.0	19.6	96.1	–

and *North Oxford May*, and *Central Oxford August*. Each detector was evaluated against a separate test dataset also derived from the North Oxford route during January.

Firstly, the results in Fig. 5 show that we are able to improve the perceptual performance of a detector by training it on data gathered from the environment it operates in. However place is clearly important in Fig. 6. We see that both detectors trained in the same place (North Oxford) improve notably in performance, whereas the detector trained in a different place (Central Oxford) does not.

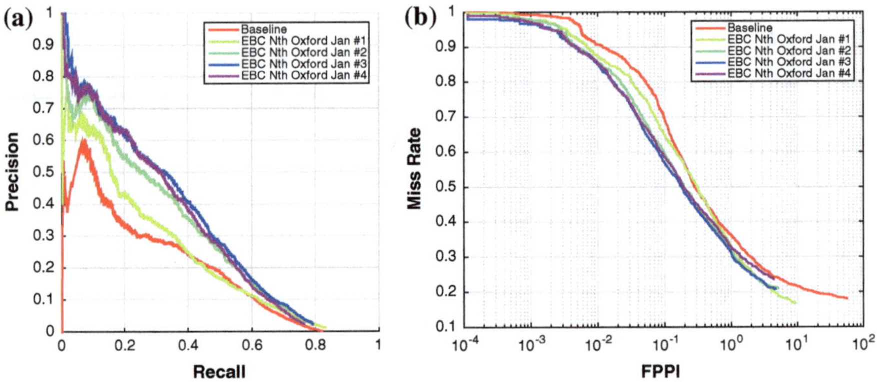

Fig. 5 The precision-recall (*left*) and miss rate-false positives per image (*right*) performance of the detector during learning, tested on the North Oxford January data. The PR curve performance increases with the first three datasets observed, moving to the *top right* corner of the graph. This then settles with a very slight performance drop on the fourth dataset. The same trend is visible in the MR-FPPI graph with the curves moving to the lower *left* corner

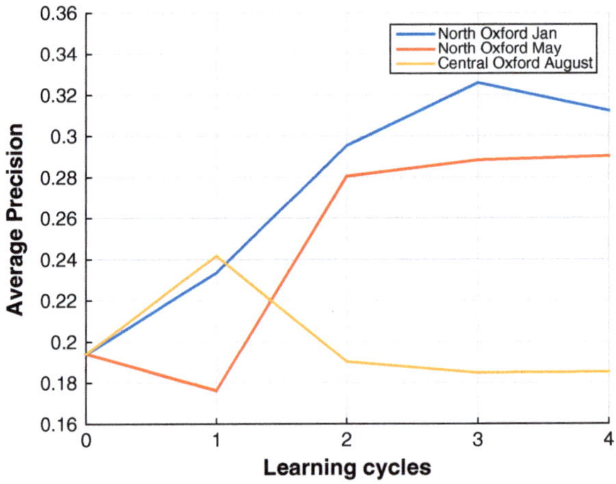

Fig. 6 The average precision (obtained by computing the area under a PR curve) for three detectors trained using EBC, evaluated on the North Oxford January test dataset. Learning from operating on the same route (North Oxford) improves performance over the baseline detector, with the detector shown data from the same season as the evaluation set performing the best (January). The detector which learned from operation on a different route and time of year (Central Oxford August) does not improve performance

Secondly, the same figure shows that there is a seasonal effect in addition to the spatial similarities. There are clear perceptual differences between the two North Oxford seasons, and a detector trained on a driving route in January has improved performance when operating in January compared to a detector which learned from the same driving route in May.

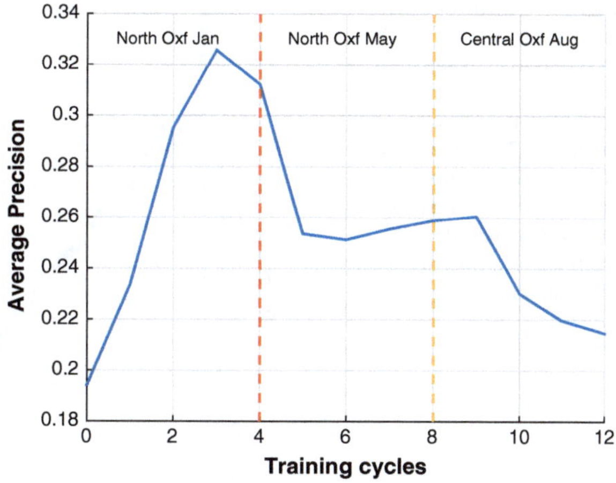

Fig. 7 The average precision (obtained by computing the area under a PR curve) for a detector trained using EBC on all datasets (both routes), evaluated on the North Oxford January test dataset. The detector shows a performance improvement with January data, but drops as it incorporates data from dissimilar environments, changing seasons to May, then changing route and season to August

These results support our argument for experience-specific classifiers. However, while it is clear that a detector trained for its operating environment is better than the general baseline detector, this raises questions around the necessary spatial and temporal resolution for these experiences in robot perception. To confirm the value of experience-specific classifiers, we also investigated the effects of simply amalgamating all the training data into one classifier, with all twelve datasets processed in temporal order (January through August). The results in Fig. 7 show that this detector is not comparable to the detector trained only on data from the same place and season as the evaluation data, with performance degrading substantially as dissimilar data is observed and learned. This result adds further weight to the argument for training place dependent classifiers, and emphasises the need for research into what defines a 'place'. Our trials considered a small set of possible places and conditions, and it is likely that experiences in robot perception will be influenced by more than simply season and route. These factors could include weather, lighting, and additional environmental changes such as traffic.

Finally, we note that we have only showed the raw detector performance in our experimental trials. Since the scene filter is already incorporated into the EBC framework, we can also validate our detections while running online if a 3D scene prior and localisation information is available. The performance increase from the scene filter on the detector's output decreases over successive training cycles, with a large initial improvement tapering off to a very small difference by the end of our trials in Fig. 8. The small difference at the end may be attributed to the fact that the ACF model is sufficiently expressive to cover what the current scene filter is able to invalidate. Further investigation is needed into the unintended slight drop in precision at higher recall when using the scene filter. Additional checks may be needed to achieve better results, potentially including computationally expensive offline checks.

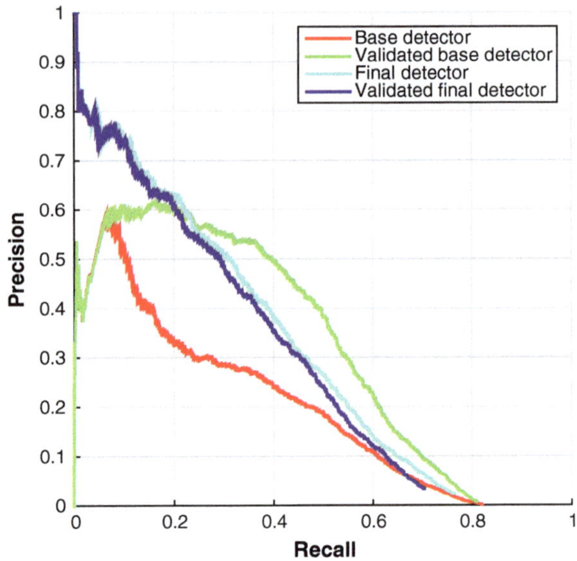

Fig. 8 Performance increase provided by the scene filter (referred to as 'validating' a classifier) when applied to both the base classifier and the final EBC detector on the North Oxford January test dataset. The invalidated data from the scene filter facilitates learning from the environment

5 Conclusions

Though general object detection remains a noble goal, applications in robotics tend to be constrained to particular operating environments. We can exploit this fact to obtain practical systems which excel in a specific application domain. This is a major step towards reliable performance for real-world safety-critical systems. In particular, we make use of scene context to validate detections, and feed the rejected samples back to retrain the detector. This augmentation to the standard perception pipeline provides self-supervised environment-dependent improvement over the lifetime of the system. We call this process Experience-Based Classification.

Using urban driving data, we demonstrate that EBC provides a means to improve a general baseline object detector beyond what conventional negative data mining on a training dataset achieves. This suggests great utility in training experience-specific classifiers, potentially leading to life-long learning in robot perception without the need for human assistance. Perceptual systems benefit from being trained to suit the local environment and their performance varies as the robot experiences different environments.

Our experimental results show that environment-specific tuning provides benefits in performance at the cost of generality, but the results raise a number of research questions, primarily around what defines a robot's perceptual experience. While we manually divided the datasets here, we require an automated method to determine when to train new classifiers based on some metric of difference between perceptual experiences. This could be achieved through localisation, with a new detector model for every small map segment. However this approach would not accommodate normal variation in weather, lighting, and seasons. We believe that there is some benefit in pursuing a data driven approach, transferring classifiers to different locations with similar observed environmental conditions. Probabilistic topic modelling [2] offers a possible mechanism for this. Finally, as we desire lifelong learning, we must address the issues of positive mining [26], further scene filter checks (including expensive offline checks), semantic drift [4], and when to 'forget' data.

Acknowledgments The authors gratefully acknowledge the support of this work by the EU project FP7-610603 (EUROPA2), EPSRC grant EP/J012017/1 and the UK Space Agency grant ST/L002981/1.

References

1. Benenson, R., Mathias, M., Timofte, R., Van Gool, L.: Pedestrian detection at 100 frames per second. In: 2012 IEEE Conference on Computer Vision and Pattern Recognition (CVPR), pp. 2903–2910 (2012)
2. Blei, D.M.: Probabilistic topic models. Commun. ACM **55**(4), 77–84 (2012)
3. Churchill, W., Newman, P.: Experience-based navigation for long-term localisation. Int. J. Robot. Res. (IJRR) **32**(14), 1645–1661 (2013)
4. Curran, J.R., Murphy, T., Scholz, B.: Minimising semantic drift with mutual exclusion bootstrapping. In: Proceedings of the 10th Conference of the Pacific Association for Computational Linguistics, pp. 172–180 (2007)
5. Dalal, N., Triggs, B.: Histograms of oriented gradients for human detection. In: Proceedings of the IEEE Computer Society Conference on Computer Vision and Pattern Recognition, vol. 1, pp. 886–893 (2005)
6. Deng, J., Dong, W., Socher, R., Li, L., Li, K., Fei-fei, L.: ImageNet: a large-scale hierarchical image database. In: Proceedings of the IEEE Conference on Computer Vision and Patter Recognition (CVPR), pp. 248–255. Miami, Florida, USA (2009)
7. Doersch, C., Singh, S., Gupta, A., Sivic, J., Efros, A.A.: What makes paris look like paris? ACM Trans. Graph. **31**(4), 101 (2012)
8. Dollár, P., Appel, R., Belongie, S., Perona, P.: Fast feature pyramids for object detection. PAMI (2014)
9. Dubout, C., Fleuret, F.: Exact acceleration of linear object detectors. In: Proceedings of the European Conference on Computer Vision (ECCV), 7574, pp. 301–311. Florence, Italy (2012)
10. Enzweiler, M., Hummel, M., Pfeiffer, D., Franke, U.: Efficient stixel-based object recognition. In: 2012 IEEE Intelligent Vehicles Symposium (IV), pp. 1066–1071 (2012)
11. Ess, A., Leibe, B., Gool, L.V.: Depth and appearance for mobile scene analysis. In: Proceedings of the International Conference on Computer Vision (ICCV) (2007)
12. Fan, R.E., Chang, K.W., Hsieh, C.J., Wang, X.R., Lin, C.J.: LIBLINEAR: a library for large linear classification. J. Mach. Learn. Res. (JMLR) **9**, 1871–1874 (2008)
13. Felzenszwalb, P., Girshick, R., McAllester, D., Ramanan, D.: Object detection with discriminatively trained part-based models. IEEE Trans. Pattern Anal. Mach. Intell. **32**(9), 1627–1645 (2010)
14. Furgale, P., Barfoot, T.D.: Visual teach and repeat for long-range rover autonomy. J. Field Rob. **27**(5), 534–560 (2010)
15. Gavrila, D.M., Munder, S.: Multi-cue pedestrian detection and tracking from a moving vehicle. Int. J. Comput. Vis. **73**(1), 41–59 (2007)
16. Gerónimo, D., Sappa, A.D., Ponsa, D., López, A.M.: 2d–3d-based on-board pedestrian detection system. Comput. Vis. Image Underst. **114**(5), 583–595 (2010)

17. Gurau, C., Hawke, J., Tong, C.H., Posner, I.: Learning on the job: Improving robot perception through experience. In: Neural Information Processing Systems (NIPS) Workshop on Autonomously Learning Robots. Montreal, Quebec, Canada (2014)
18. Hariharan, B., Malik, J., Ramanan, D.: Discriminative decorrelation for clustering and classification. European Conference on Computer Vision (2012)
19. Henriques, J., Carreira, J., Caseiro, R., Batista, J.: Beyond hard negative mining: efficient detector learning via block-circulant decomposition. In: 2013 IEEE International Conference on Computer Vision (ICCV), pp. 2760–2767 (2013)
20. Kanezaki, A., Inaba, S., Ushiku, Y., Yamashita, Y., Muraoka, H., Kuniyoshi, Y., Harada, T.: Hard negative classes for multiple object detection. In: Proceedings of the IEEE International Conference on Robotics and Automation (ICRA) (2014)
21. Khosla, A., Zhou, T., Malisiewicz, T., Efros, A.A., Torralba, A.: Undoing the damage of dataset bias. In: Computer Vision-ECCV 2012, pp. 158–171. Springer (2012)
22. Petrovskaya, A., Thrun, S.: Model based vehicle detection and tracking for autonomous urban driving. Auton. Robots **26**(2–3), 123–139 (2009)
23. Stewart, A., Newman, P.: LAPS—localisation using appearance of prior structure: 6-DOF monocular camera localisation using prior pointclouds. In: Proceedings of IEEE International Conference on Robotics and Automation (ICRA). Minnesota, USA (2012)
24. Sudowe, P., Leibe, B.: Efficient use of geometric constraints for sliding-window object detection in video. In: Proceedings of the International Conference on Computer Vision Systems (ICVS) (2011)
25. Sung, K.K., Poggio, T.: Example-based learning for view-based human face detection. IEEE Trans. Pattern Anal. Mach. Intell. **20**(1), 39–51 (1998)
26. Teichman, A., Thrun, S.: Tracking-based semi-supervised learning. Int. J. Robot. Res. (IJRR) **31**(7), 804–818 (2012)
27. Teichman, A., Thrun, S.: Group induction. In: Proceedings of the IEEE/RSJ International Conference on Intelligent Robots and Systems (IROS), pp. 2757–2763. Tokyo, Japan (2013)
28. Torralba, A., Efros, A.A.: Unbiased look at dataset bias. In: 2011 IEEE Conference on Computer Vision and Pattern Recognition (CVPR), pp. 1521–1528. IEEE (2011)

Beyond a Shadow of a Doubt: Place Recognition with Colour-Constant Images

Kirk MacTavish, Michael Paton and Timothy D. Barfoot

Abstract Colour-constant images have been shown to improve visual navigation taking place over extended periods of time. These images use a colour space that aims to be invariant to lighting conditions—a quality that makes them very attractive for place recognition, which tries to identify temporally distant image matches. Place recognition after extended periods of time is especially useful for SLAM algorithms, since it bounds growing odometry errors. We present results from the FAB-MAP 2.0 place recognition algorithm, using colour-constant images for the first time, tested with a robot driving a 1 km loop 11 times over the course of several days. Computation can be improved by grouping short sequences of images and describing them with a single descriptor. Colour-constant images are shown to improve performance without a significant impact on computation, and the grouping strategy greatly speeds up computation while improving some performance measures. These two simple additions contribute robustness and speed, without modifying FAB-MAP 2.0.

1 Introduction

Visual place recognition aims to recognize, from a stream of images, if the vehicle is revisiting a place it has previously seen. Since integrated odometry measurements drift over time, this information is especially useful if a long period of time has passed since the last visit. Over this period, lighting conditions will change, making it more difficult to recognize the matching image. To address this problem, colour-constant images transform an RGB image into a colour-space that changes less with lighting conditions than greyscale [3, 7, 10, 12, 16, 19]. This paper presents experimen-

K. MacTavish (✉) · M. Paton · T.D. Barfoot
University of Toronto Institute for Aerospace Studies, Toronto,
ON M3H 5T6, Canada
e-mail: kirk.mactavish@mail.utoronto.ca

M. Paton
e-mail: mike.paton@mail.utoronto.ca

T.D. Barfoot
e-mail: tim.barfoot@utoronto.ca

D.S. Wettergreen and T.D. Barfoot (eds.), *Field and Service Robotics*,
Springer Tracts in Advanced Robotics 113, DOI 10.1007/978-3-319-27702-8_13

tal results from a challenging multi-day dataset [16] where colour-constant images improve place recognition performance with no modification to the underlying inference algorithm, Fast Appearance-Based Mapping (FAB-MAP) 2.0 [6].

The use of colour-constant images does add a small computational overhead, since these images are used alongside the original greyscale images, increasing vocabulary size and the average number of observed features. To recover this computation effort, we use the image grouping strategy introduced by MacTavish and Barfoot [9]. This method is faster by an order of magnitude, improves some performance measures (see Sect. 4), and does not require modification or parameter tuning of the place recognition algorithm.

Similar work has been performed by Maddern and Vidas [11], who used FAB-MAP with a monochromatic and thermal camera, with a similar channel-concatenated Bag-of-Words (BoW). Collier et al. [2] address lighting change using lidar geometry and monochromatic images, running FAB-MAP separately on each sensor. MacTavish and Barfoot [9] use lidar intensity with FAB-MAP to achieve lighting invariance, requiring specialized hardware and introducing motion distortion due to a rolling shutter. Paul and Newman [17] augment visual features with spatial information using lidar. This paper focuses on improved lighting invariance without additional hardware beyond an RGB camera.

Sunderhauf et al. [20] use Sequence SLAM (SeqSLAM) [14] with monochromatic images to localize a train over 3000 km across seasons with impressive results, but do not perform full Simultaneous Localization and Mapping (SLAM) with the ability to add new places. Milford [13] shows how SeqSLAM can use very-low-resolution images to localize by leveraging sequence information. The FAB-MAP image-grouping strategy [9] used in this paper also makes use of sequence information by grouping local regions in a single descriptor.

In an effort to learn appearance change and proactively translate the image to different appearance conditions, Neubert et al. [15] introduce a super-pixel-based translation algorithm. This algorithm targets large seasonal change rather than lighting, and requires training data of the expected appearance domain. Pepperell et al. [18] blacken the sky in daytime images for better matching against those captured at night using a whole-image matching technique. Aiming to improve lighting invariance at the descriptor level, Carlevaris-Bianco and Eustice [1] train neural-net features using data from outdoor webcams. Colour-constant images improve lighting invariance without algorithm modification even at the descriptor level.

Corke et al. [3] compute image similarity scores across a small set of colour-constant images, and Maddern et al. [10] perform local metric localization; however, there has not been an evaluation of place recognition using colour-constant images. In this paper, we discuss this task and present the results of our approach on an 11 km dataset consisting of over 2000 images.

This paper presents novel results for place recognition using colour-constant images. This contribution goes beyond the simple image similarity scores that have been used in previous work to benchmark this image transform. In Sect. 2 we discuss the place recognition and image processing techniques that we have used. In Sect. 3 we discuss the field experiment, and in Sect. 4 we present and analyze the experimental results. Final conclusions and future work are discussed in Sect. 5.

2 Methodology

2.1 Place Recognition

The FAB-MAP algorithm and its extensions [4–6] have been extensively tested and widely used; in particular, FAB-MAP 2.0 has been tested on a 1000 km dataset. This paper examines the results of two input preprocessing techniques for place recognition: colour-constant images, and BoW image grouping. For place recognition itself, we use the OpenFABMAP implementation [8] of the FAB-MAP 2.0 algorithm which is summarized below.

FAB-MAP uses a BoW descriptor to describe images. To train the BoW vocabulary, Speeded Up Robust Features (SURF) descriptors are extracted from a training image dataset. These descriptors are clustered, and the BoW vocabulary is described by these cluster centers (words). An image can now be described by a BoW descriptor by quantizing its SURF features using the vocabulary, and listing which words were seen. A BoW descriptor can be represented as a binary vector of word presence, or as a list of which words were observed. To learn a factorized probability prior distribution over BoW descriptors, FAB-MAP trains a (CLT) using the BoW descriptors from the training dataset.

FAB-MAP represents a place as a vector of Bernoulli variables indicating the existence of the generator for each word in the vocabulary. The measurement model is given by the trained CLT and two user-specified parameters, and full Bayesian inference determines the posterior generator probabilities. The probability of being in a new place is determined using a Monte-Carlo approximation, sampling training images as representative new places. FAB-MAP 2.0 speeds up inference using an inverted index for each word in the vocabulary and slightly modified inference.

FAB-MAP 2.0 also uses geometric verification in the form of a 1-point Random Sample Consensus (RANSAC) test to improve precision. The results in this paper focus only on the recall task, and have not used any geometric verification, though they have used the FAB-MAP 2.0 simple motion model. Since the Visual Teach and Repeat (VT&R) algorithm [16] used to collect the dataset is already performing visual odometry, it would be straightforward to use only features that are stable over a short distance to verify geometric stability; we leave this as future work.

2.2 Colour-Constant Images

Colour-constant images were first developed in the optics community. Recent methods are based on the theory that a 1D colour space that is invariant to outdoor lighting conditions can be calculated from the channel responses of an RGB camera, given certain assumptions about the sensor and environment [7, 19]. The method presented by Ratnasingam and Collins [19] asserts that a colour-constant feature, F, can be extracted from a three-channel camera from the following:

$$F = \log(R_2) - \alpha \log(R_1) - \beta \log(R_3), \tag{1}$$

where R_i, is the approximated sensor response for channel i, and α and β are weights subject to the following constraints:

$$\frac{1}{\lambda_2} = \frac{\alpha}{\lambda_1} + \frac{\beta}{\lambda_3}, \quad \beta = (1 - \alpha), \tag{2}$$

where λ_1, λ_2, λ_3 are the peak sensor responses numbered from highest to lowest. The result of Eq. (1) is a 1D feature with much of the effect of lighting removed.

Colour-constant images have appeared in various forms [3, 10, 12, 16] in the robotics and computer vision community. The approach taken in this paper is identical to that of Paton et al. [16], which uses experimentally trained coefficients of Eq. 1 to obtain two colour-constant images: $\{F'_v, F'_r\}$ that perform well in vegetation and rocks-and-sand, respectively. Examples of these images can be seen in Fig. 1.

Fig. 1 This figure illustrates the transformation of an RGB image into a set of *greyscale* images. The *top* image is a typical *greyscale* image obtained from the *green* channel, and the *bottom* two are the colour-constant image pair $\{F'_v, F'_r\}$ [16] used in this paper to boost place recognition. By making assumptions about the sensor and environment, a weighted log-difference of the three camera channels can cancel the effect lighting has on the appearance of the scene. *Credit* [16]

These images were used to great success in an autonomous route-following algorithm presented by Paton et al. [16], which was used to collect the dataset that is used in this paper. Details on the environment and route can be found in Sect. 3.

Since FAB-MAP requires a single BoW descriptor for each observation, we can create a unified place descriptor by concatenating the BoW descriptors from each channel [11, 16]: the green channel (greyscale), F_v', and F_r'. A separate vocabulary is trained for each channel, and each is quantized into a separate BoW descriptor. These per-channel-BoW descriptors are concatenated into a stacked BoW that is used to train the CLT, and for online place recognition. We expect that there will be a strong correlation between words in each of the channels, since the channels themselves are correlated. Luckily the CLT accounts for this correlation to the extent that it is apparent in the training dataset.

2.3 *Image Grouping*

MacTavish and Barfoot [9] show that sequences of images can be grouped together and described with a single BoW descriptor. This provides two benefits: temporal smoothing, which can improve robustness if features are somewhat unstable; and a theoretical speedup of n^2 for groups of n images. The major drawback is that matches are not established at an image level. Simply adding the BoW descriptors loses sparsity as group size increases. For the CLT training to be valid, these grouped BoWs must have similar sparsity to the single-image training descriptors. We can meet this requirement by increasing the binary BoW threshold, requiring multiple observations of a word before it is considered present. A detailed description and results for this method is available by MacTavish and Barfoot [9].

3 Field Experiment

A four day field trial was conducted at the (CSA)'s Mars Emulation Terrain (MET) at Montreal, Quebec on May 12–15th, 2014, with the purpose of testing the colour-constant VT&R algorithm introduced by Paton et al. [16]. The MET, pictured in Fig. 2, is a 60×120 m manicured environment emulating the surface of Mars. It consists primarily of rock and sand, with interesting features such as outcroppings and craters. The MET is surrounded by unstructured vegetation containing trees, marshland, open fields, a small stream, and a gravel roadway.

The field trial proceeded by teaching a 1 km path, marked as a yellow line in Fig. 2, through the MET and its surrounding fields. This path was taught at approximately 11 am on the first day during sunny conditions with pronounced shadows. Over the course of the field trial, this path was autonomously traversed 26 times in varying lighting conditions. During this time the robot maintained an autonomy rate of 99.9 % of distance travelled.

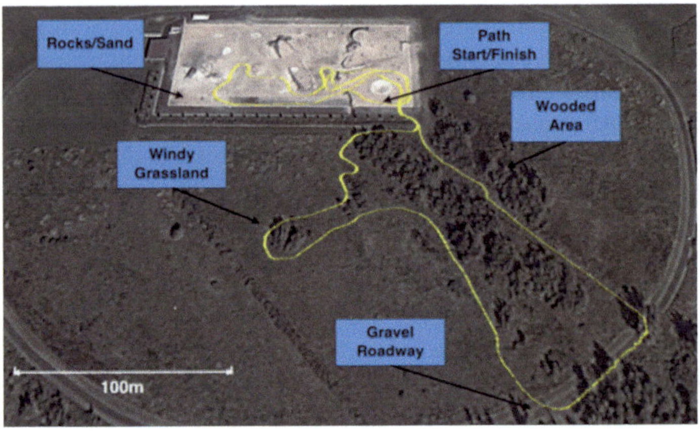

Fig. 2 Satellite imagery of the CSA MET, with the teach pass from the 2014 field trials highlighted in *yellow*, and interesting environmental features annotated. *Credit* [16]

Fig. 3 Grizzly Robotic Vehicle autonomously repeating a route during the CSA field trials, with applicable sensors highlighted. *Credit* [16]

The hardware setup used during these experiments is pictured in Fig. 3. The robot is the Clearpath Grizzly Robotic Utility vehicle. The VT&R algorithm ran on an on-board computer using a Point Grey Research Bumblebee XB3 stereo camera. GPS data was collected for the purpose of visualization only.

During the traversals of the 1 km path, the robot recorded rectified 512 × 384 stereo RGB images at 16 Hz from the Grizzly's front PGR XB3 Camera. The result is close to 1 TB of stereo data along the same path in many lighting conditions. In this paper we present results using 11 of these traversals, from dawn to dusk, selected with the intent of maximizing appearance variation. Additionally, a 247 image, 1.2 km dataset was collected in Ontario, Canada, which was used for training the place recognition algorithm.

4 Results

This section presents the place recognition results for the colour-constant and image grouping techniques. Parameter training for the FAB-MAP algorithm is covered by [5], the tuning process and results for the colour-constant images are detailed by [16], and the tuning process for image groups is explained by [9].

The CSA dataset is quite challenging for several reasons. Over the course of the experiment, the terrain was significantly modified by the vehicle, as shown in Fig. 4a, c, d. This dataset is collected by a single camera pointed forward and down, meaning a significant portion of the field of view is physically changing over the course of the experiment. As anticipated, the changing lighting conditions had a large effect—including the robot's own shadow being visible when the sun was behind (see Fig. 4b), leading to similar features being seen in different places depending on the time of day. Natural environments also tend be more challenging than urban [6], and the geometric intricacy of vegetation leads to difficult shadows as lighting changes. Finally, at times the lighting conditions were so extreme that the auto exposure was unable to produce a usable image (see Fig. 4e).

FAB-MAP is fairly sensitive to feature stability, and SURF detector thresholds had to be carefully selected for the colour-constant images, due to their far-lower dynamic range (see Fig. 1), and limited intensity information (by design). Initial results used a detector threshold that would extract a similar number of features across all image channels. This resulted in poorer performance than greyscale on its own, since many of the colour-constant features turned out to be unstable. Since colour-constant images are deliberately removing intensity information from the image to provide invariance, there is less information remaining. This leads to a noisier image, and noisier feature descriptors. The final SURF thresholds for the greyscale, F_v', and F_r' images lead to an average of 83, 9, and 22 keypoints per training image, respectively. The clustering threshold was set so that the feature-to-vocabulary-size ratio was similar for the image channels, resulting in 1017, 85 and 244 words per image type, respectively. The performance for the greyscale-and-colour-constant stack is shown in Fig. 5 as *Stack*, and for the greyscale only baseline as *Grey*. Colour-constant-only results have not been shown, as the low feature count and vocabulary size are unable support place recognition alone. For equivalent recall, the precision is strictly better using the colour-constant stack. The timing results in Table 1 show that there is a 22 % increase in computation, due to a larger vocabulary.

Fig. 4 Example images from the test dataset showing several of the challenging cases. **a** Tall grass that was flattened by the vehicle over the course of the experiment. **b** The vehicle's shadow is seen in different places depending on the time of day. **c** The same location during the first and last loop showing the terrain modification on sand. **d** The same location during the first and last loop showing the terrain modification on vegetation. **e** The same location during the first and latest-in-the-day loop showing the auto exposure struggling with low light and a still-bright sky

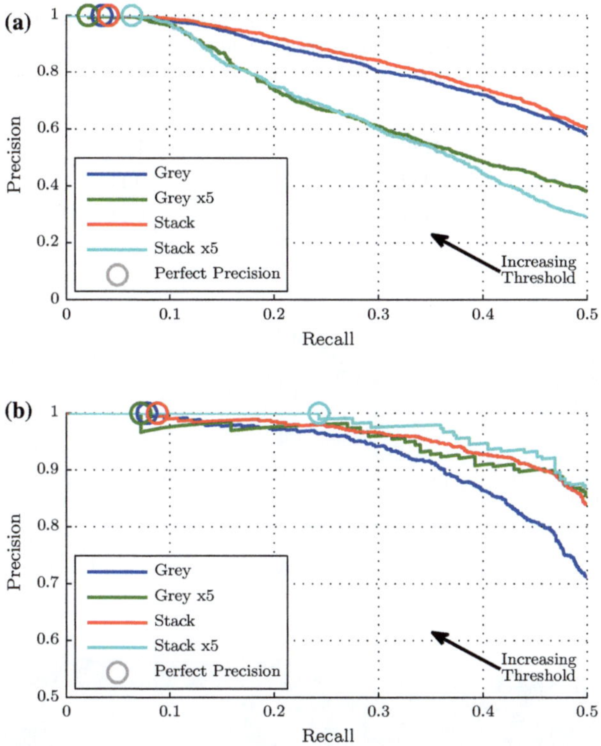

Fig. 5 Precision-Recall curves for the recall-only task (no geometric verification). *Grey* indicates only greyscale images were used, *Stack* consists of the *greyscale* as well as both colour-constant images. The *x5* indicates that sequences of 5 images were grouped and described with a single BoW descriptor. Matches are labelled as true if they are within 30 m of ground truth. **a** Precision-Recall for all matches between loops. Unfortunately, the colour-constant stack shows only modest improvement, and the image grouping fares far worse. This measure is the most common, but is not necessarily representative of the desired output. The curve below presents an alternative measure that might represent a more realistic use case. **b** Precision-Recall for *at least one* matches between loops (per query). This P-R curve represents how the system might actually be used; if every query has at least one match, the connected graph (chain of matches) will cover all of the loops even if they aren't explicit. For example, if query B matches place A and query C also matches place A, we can infer that C also matches B, without needing to explicitly label that match. Contrary to 5a, the image groupings show *improved* performance compared to their ungrouped counterparts, and the colour-constant stack is *significantly* improved over greyscale. Both techniques combined produce far better recall at 100 % precision

Table 1 Timing results show that the colour-constant stack only adds a small amount of overhead, and that the image grouping is faster by an order of magnitude

Name	# Queries	Average time (s)
Grey	2189	1.18
Stack	2189	1.44
Grey x5	437	0.11
Stack x5	437	0.15

Fig. 6 Interesting examples of successful match hypotheses with the two processing techniques. **a** A successful match at 95 % precision with the colour-constant stack that was not found using only greyscale (no image grouping). **b** A successful match at 95 % precision with the image grouping that was not found using single images (no colour-constant channels)

Figure 6a shows an example of a place that is correctly recognized by the colour-constant stack, but not by greyscale.

Image sequences were also grouped in sequences of 5 images, to illustrate the speed-up without introducing a large disparity in match specificity. MacTavish and Barfoot [9] further investigate different sized image groups. The binary BoW threshold was chosen as 2 feature occurrences per group to maintain sparsity. The mean binary BoW density for single images were 0.1357 and 0.1227; after grouping and thresholding, they were 0.1149 and 0.1639, respectively. In both cases, the speedup is approximately an order of magnitude (see Table 1). The precision-recall curves shown in Fig. 5 show that the grouping deflates the first measure, but improves the second. The first measure considers the precision-recall if the task is to identify *all*

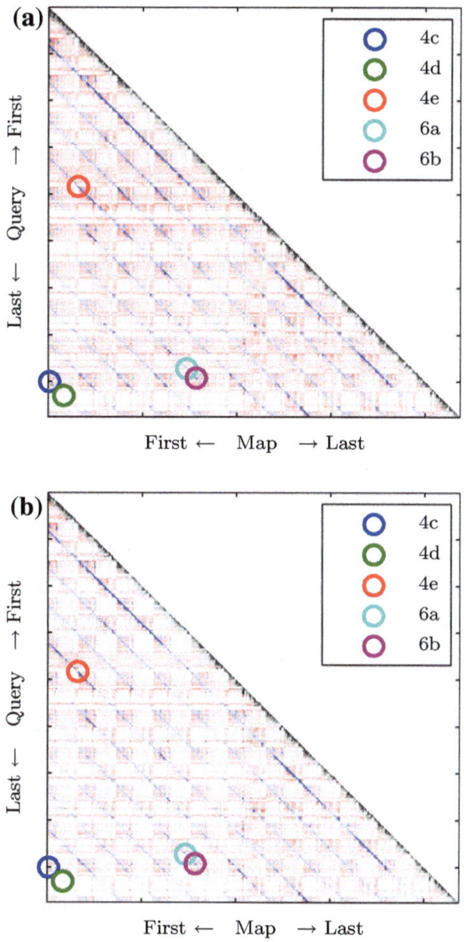

Fig. 7 Contrast-enhanced confusion matrices show the probability mass for each query (rows) over the mapped places (columns). Correct (true positive) match probability is shown in **blue**, incorrect (false positive) in **red**, and ignored matches (temporally close) in **grey**. The ground-truth for the confusion matrices is shown in Fig. 8. The circled interest points correspond to the image examples in Figs. 4 and 6. The *darker red* checkering of false positives show that the system struggled in the rocks-and-sand of the MET, which was underrepresented in the training dataset. **a** Greyscale **b** Colour-constant stack

of the possible loop closures for each query. The second measure only aims to find *at least one* of the loop closures. Due to the temporal ordering of the queries, if every query has correctly identified at least one loop closure, all possible loop closures are connected without the match being explicitly identified; e.g., B matches A and C matches B, therefore C and A must be a match.

The training for FAB-MAP must be done prior to run-time and is fairly time-consuming compared to the online algorithm. Therefore, the place recognition algorithm is trained in a geographically separate but visually similar environment. Due to geographic limitations, and since this was the first major field deployment for this robotic platform, our training dataset was restricted to 247 images over 1.2 km. It consists of a dry-run for the CSA experiment that took place in Ontario, Canada, primarily in vegetation with a very small sand portion. The confusion matrices, showing

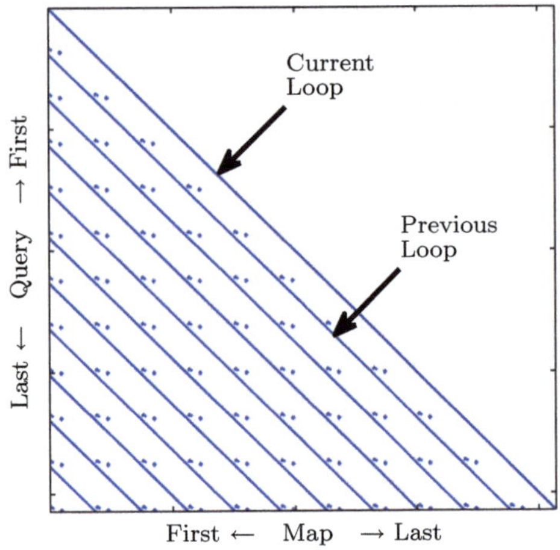

Fig. 8 Ground truth confusion matrix. Since the dataset is a repeated loop, there is diagonal banding, with the current loop on the diagonal, and previous loops on the off-diagonal bands. The *smaller dots* are regions of the MET that are re-observed toward the end of the loop

the match probabilities for each query are shown in Fig. 7. The difficult checkered square regions are the rocks-and-sand sections of the trajectory, the terrain type that was underrepresented in the training dataset.

5 Conclusion and Future Work

We can conclude that both colour-constant images and image grouping show value for place recognition in real outdoor environments. We have also shown reasonable system performance despite a very limited and not fully representative training dataset, and difficult lighting conditions that changed over the course of the day. Future work consists of improving the stability of the colour-constant image channels. By increasing the contrast of the images, the features descriptors may be less corrupted by quantization error, and the detector response may be more stable. A geometric consistency check such as the FAB-MAP 2.0 1-point RANSAC will also improve results by using more-stable features [6]. We can also verify geometric stability by only using features that have been tracked through several frames by VT&R [16].

Acknowledgments We would like to extend our deepest thanks to the Natural Sciences and Engineering Research Council (NSERC) through the NSERC Canadian Field Robotics Network (NCFRN), the Canada Foundation for Innovation, the Canadian Space Agency, and MDA Space Missions for providing us with the financial and in-kind support necessary to conduct this research.

References

1. Carlevaris-Bianco, N., Eustice, R.M.: Learning visual feature descriptors for dynamic lighting conditions. In: 2014 IEEE International Conference on Robotics and Automation (ICRA) (2014)
2. Collier, J., Se, S., Kotamraju, V., Jasiobedzki, P.: Real-time lidar-based place recognition using distinctive shape descriptors. SPIE Defense, Security, and Sensing, pp. 83870P–83870P, May 2012
3. Corke, P., Paul, R., Churchill, W., Newman, P.: Dealing with shadows: capturing intrinsic scene appearance for image-based outdoor localisation. In: IEEE International Conference on Intelligent Robots System, pp. 2085–2092 (2013)
4. Cummins, M., Newman, P.: Accelerated Appearance-Only SLAM. In: ICRA, pp. 1828–1833, May 2008
5. Cummins, M., Newman, P.: FAB-MAP: probabilistic localization and mapping in the space of appearance. Int. J. Rob. Res. **27**(6), 647–665 (2008)
6. Cummins, M., Newman, P.: Appearance-only SLAM at large scale with FAB-MAP 2.0. Int. J. Rob. Res. **30**(9), 1100–1123 (2010)
7. Finlayson, G., Hordley, S., Cheng, L., Drew, M.: On the removal of shadows from images. IEEE Trans. Pattern Anal. Mach. Intell. **28**(1), 59–68 (2006)
8. Glover, A., Maddern, W., Warren, M., Reid, S., Milford, M., Wyeth, G.: OpenFABMAP: an open source toolbox for appearance-based loop closure detection. In: 2012 IEEE International Conference on Robotics and Automation, pp. 4730–4735. IEEE, May 2012
9. MacTavish, K., Barfoot, T.D.: Towards hierarchical place recognition for long-term autonomy. In: ICRA Workshop on Visual Place Recognition in Changing Environments (2014)
10. Maddern, W., Stewart, A.D., McManus, C., Upcroft, B., Churchill, W., Newman, P.: Illumination invariant imaging: Applications in robust vision-based localisation, mapping and classification for autonomous vehicles. In: Proceedings of the Visual Place Recognition in Changing Environments Workshop, IEEE International Conference on Robotics and Automation (2014)
11. Maddern, W., Vidas, S.: Towards robust night and day place recognition using visible and thermal imaging. In: Proceedings of Robotics: Science and Systems, pp. 1–6 (2012)
12. McManus, C., Upcroft, B., Newman, P.: Scene signatures: localised and point-less features for localization. In: Proceedings of Robotics: Science and Systems, Berkely, USA (2014)
13. Milford, M.: Vision-based place recognition: how low can you go? Int. J. Rob. Res. **32**(7), 766–789 (2013)
14. Milford, M.J., Wyeth, G.F.: SeqSLAM: visual route-based navigation for sunny summer days and stormy winter nights. In: 2012 IEEE International Conference on Robotics Automation, IEEE, May 2012
15. Neubert, P., Sünderhauf, N., Protzel, P.: Superpixel-based appearance change prediction for long-term navigation across seasons. In: Robotics and Autonomous Systems (2014)
16. Paton, M., MacTavish, K., Ostafew, C.J., Barfoot, T.D.: Lighting-resistant stereo visual teach & repeat using color-constant images. In: IEEE International Conference on Robotics and Automation (ICRA) (2015)
17. Paul, R., Newman, P.: FAB-MAP 3D: topological mapping with spatial and visual appearance. In: 2010 IEEE International Conference on Robotics and Automation, pp. 2649–2656, May 2010
18. Pepperell, E., Corke, P.I., Milford, M.J.: Towards vision-based pose- and condition-invariant place recognition along routes. IEEE International Conference on Intelligent Robots and Systems (2014)
19. Ratnasingam, S., Collins, S.: Study of the photodetector characteristics of a camera for color constancy in natural scenes. J. Opt. Soc. Am. A **27**(2), 286–294 (2010)
20. Sunderhauf, N., Neubert, P., Protzel, P.: Are We There Yet? Challenging SeqSLAM on a 3000 km Journey Across All Four Seasons. In: Proceedings of Workshop Long-Term Autonomous International Conference on Robotics and Automation (2013)

Segmentation and Classification of 3D Urban Point Clouds: Comparison and Combination of Two Approaches

A.K. Aijazi, A. Serna, B. Marcotegui, P. Checchin and L. Trassoudaine

Abstract Segmentation and classification of 3D urban point clouds is a complex task, making it very difficult for any single method to overcome all the diverse challenges offered. This sometimes requires the combination of several techniques to obtain the desired results for different applications. This work presents and compares two different approaches for segmenting and classifying 3D urban point clouds. In the first approach, detection, segmentation and classification of urban objects from 3D point clouds, converted into elevation images, are performed by using mathematical morphology. First, the ground is segmented and objects are detected as discontinuities on the ground. Then, connected objects are segmented using a watershed approach. Finally, objects are classified using SVM (Support Vector Machine) with geometrical and contextual features. The second method employs a super-voxel based approach in which the 3D urban point cloud is first segmented into voxels and then converted into super-voxels. These are then clustered together using an efficient link-chain method to form objects. These segmented objects are then classified using local descriptors and geometrical features into basic object classes. Evaluated on a common dataset (real data), both these methods arethoroughly compared on three different levels:

A.K. Aijazi (✉) · P. Checchin · L. Trassoudaine
Université Blaise Pascal, Institut Pascal, BP 10448, 63000 Clermont-Ferrand, France
e-mail: kamalaijazi@gmail.com

A.K. Aijazi · P. Checchin · L. Trassoudaine
CNRS, UMR 6602, Institut Pascal, 63171 Aubière, France

P. Checchin
e-mail: paul.checchin@univ-bpclermont.fr

L. Trassoudaine
e-mail: laurent.trassoudaine@univ-bpclermont.fr

A. Serna · B. Marcotegui
MINES ParisTech, CMM – Centre de Morphologie Mathématique, 35 rue St. Honoré, 77305 Fontainebleau-Cedex, France
e-mail: andres.serna_morales@mines-paristech.fr

B. Marcotegui
e-mail: beatriz.marcotegui@mines-paristech.fr

D.S. Wettergreen and T.D. Barfoot (eds.), *Field and Service Robotics*,
Springer Tracts in Advanced Robotics 113, DOI 10.1007/978-3-319-27702-8_14

detection, segmentation and classification. After analyses, simple strategies are also presented to combine the two methods, exploiting their complementary strengths and weaknesses, to improve the overall segmentation and classification results.

1 Introduction

The segmentation and classification of 3D point clouds for the interpretation of urban scenes and detailed semantic analysis have gained major interest in recent years. This considerable attention is due to the recent advancements in 3D data acquisition technologies as well as the increasing demand for different robotics applications in the field or service industry. Presenting a fundamental problem in robotics and computer vision, different research activities pertaining to automatic interpretation of 3D urban point clouds for various field robots and autonomous vehicles operating in outdoor environments are underway such as urban accessibility analysis [23], drivable road detection [4] and point cloud classification [17].

For scene interpretation and assignment of a semantic label to each 3D point (e.g. building, ground, trees, etc.), the first step is to segment the 3D point cloud. Point cloud segmentation can support classification and further feature extraction provided that the segments are logical groups of points belonging to the same object class. Some methods, including [20, 27], employ the use of small sets of specialized features, such as local point density or height from the ground, to discriminate only few object categories in outdoor scenes, or to separate foreground from background while some segmentation methods based on surface discontinuities, such as in [15], use surface convexity in a terrain mesh as a separator between objects. Lately, segmentation has been commonly formulated as graph clustering [9, 21]. Instances of such approaches are Graph-Cuts including Normalized-Cuts and Min-Cuts. Markov Random Fields are also used to segment and label 3D point clouds [2]. Different methods, such as in [17], are introduced in order to increase their efficiency while reducing their computational time.

The next step is to extract corresponding features from the segmented 3D object. These features rely on a local 3D neighborhood which is typically chosen as a spherical neighborhood formed by a fixed number of the k closest 3D points [13], spherical neighborhood with fixed radius [12] or cylindrical neighborhood with fixed radius [7]. These features are mainly based on geometrical features (shape, size, etc.) [19], local descriptors (color, intensity, surface normals, etc.) [1] or contextual features (position with respect to neighbors, etc.) [24].

Once these features have been calculated, the next step is the classification of each 3D point. Some methods such as [1, 19] rely on pre-defined geometrical models and thresholds but classification may also be conducted via different supervised learning techniques as well, such as Support Vector Machines [22], Gaussian Mixture Models [11], Random Forests [5], AdaBoost [14] and Bayesian Discriminant Classifiers [16]. Furthermore, contextual learning approaches also utilize relationships between 3D points in a local neighborhood which is usually inferred from the training data. Such methods for classifying point cloud data have been proposed with Associative and non-Associative Markov Networks [26], Conditional Random Fields [18] and multi-stage inference procedures focusing on point cloud statistics and relational information over different scales [28], etc. In addition to the above methods, Stamos et al. [8] propose an online algorithm to classify scanned points into 6 distinct classes (ground vegetation, car, horizontal surfaces, vertical surfaces and curb regions) during data acquisition by analyzing each scan-line one-by-one relying on several efficiently computed local features.

Common problems in this detection, segmentation and classification pipeline include coping with the complexity of 3D scenes caused by the irregular sampling, a large variety of objects, occlusions caused by obstructions, density variation caused by different distances of objects from the sensors as well as the computational burden arising from large 3D point clouds and handling the various types of features. These diverse problems make it very difficult for any single method to produce the desired results. Hence, the combination of several approaches is necessary for different applications. Consequently, for effective combination, thorough evaluation and comparison is essential.

In this work, we present and compare two different approaches for segmenting and classifying 3D urban point clouds i.e. a method exploiting mathematical morphology (Sect. 2) and another based on super-voxels (Sect. 3). Evaluated on a common dataset (real data), both these methods are thoroughly compared (Sect. 4) on three different levels: detection, segmentation and classification. After analyses, simple strategies are also presented to combine the two methods, exploiting their complementary strengths and weaknesses, to improve the overall segmentation and classification results (Sect. 5).

2 Morphological Transformation Method

The method for segmenting 3D urban point cloud based on mathematical morphology is presented in [24]. It aims at developing a process to detect, segment and classify urban objects, suitable for large scale applications. In this method, the input point cloud is first mapped to a range image. This image is then interpolated in order to avoid connectivity problems and a k-flat zones algorithm is used to segment the ground (road and sidewalk). The facades and objects are extracted using morphological transformations. The method relies on facades being the highest vertical structures in the scene and objects are represented as bumps on the ground on the range image as shown in Fig. 1. Several geometrical and contextual features are computed for each object and classification is carried out using a standard SVM (Support Vector Machine). These features are summarized as follows:

- Geometrical features: object area and perimeter; bounding box area; maximum, mean, standard deviation and mode (the most frequent value) of the object height; object volume, computed as the integral of the elevation image over each object.
- Contextual features: Neighboring objects N_{neigh}, defined as the number of regions touching the object, using 8-connectivity on the elevation image. This feature is very discriminative in the case of group of trees and cars parked next to each other;

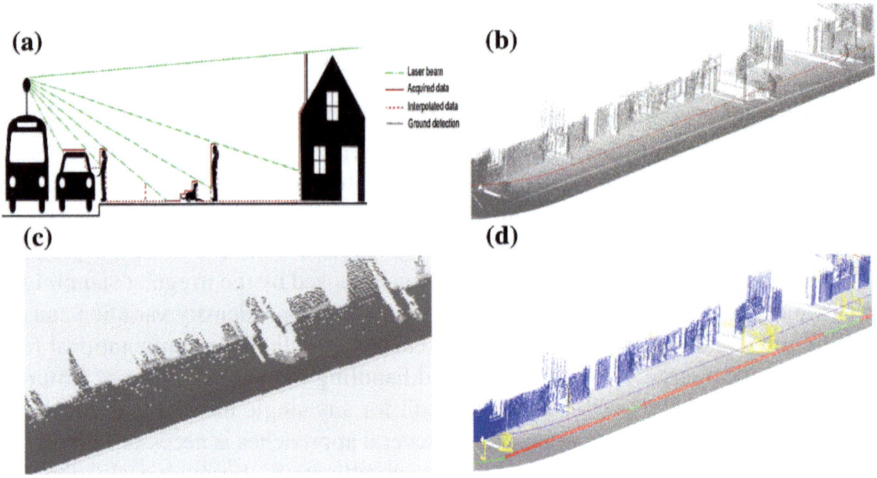

Fig. 1 **a** Segmentation of 3D point clouds based on morphological modeling. Objects are segmented out as bumps on the ground. **b** Input point cloud. **c** Range image. **d** Segmentation results

confidence index $Cind = \frac{n_{real}}{n_{real}+n_{interp}}$, where n_{real} and n_{interp} are the number of non-empty object pixels before and after elevation image interpolation, respectively. In general, occluded and far objects have a low confidence index.

Relatively fast, the method uses little a priori information, and is based on robust morphological operators and supervised classification.

3 Super-Voxel Based Segmentation and Classification Method

This method presents a super-voxel based approach in which the 3D urban point cloud is first segmented into voxels and then converted into super-voxels. These are then clustered together using an efficient link-chain method to form objects. The method as presented in [1] uses an agglomerative clustering methodology to group 3D points based on r-NN (radius Nearest Neighbor). Although the maximum voxel size is predefined, the actual voxel sizes vary according to the maximum and minimum values of the neighboring points found along each axis to ensure the profile of the structure is maintained. A voxel is then transformed into a super-voxel when properties based on its constituting points are assigned to it. These properties mainly include: geometrical center, mean R, G and B value, maximum of the variance of R, G and B values; mean intensity value; variance of intensity values; voxel size along each axis X, Y and Z and surface normals of the constituting 3D points. With the assignment of all these properties, a voxel is transformed into a super-voxel. All these properties are then used to cluster these super-voxels into objects using a link

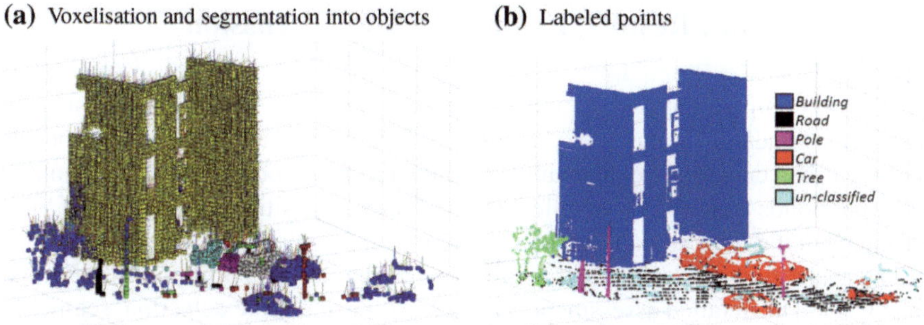

Fig. 2 **a** Super-voxel based segmentation. **b** Classified 3D points

chain method. In this method, each super voxel is considered as a link of a chain. All secondary links attached to each of these principal links are found. In the final step, all the principal links are linked together to form a continuous chain removing redundant secondary links in the process. If V_P be a principal link and V_n be the nth secondary link then each V_n is linked to V_P if and only if the following three conditions are fulfilled:

$$\left|\mathbf{V}_{P_{X,Y,Z}} - \mathbf{V}_{n_{X,Y,Z}}\right| \leq (w_D + c_D)$$

$$\left|\mathbf{V}_{P_{R,G,B}} - \mathbf{V}_{n_{R,G,B}}\right| \leq 3\sqrt{w_C}$$

$$\left|\mathbf{V}_{P_I} - \mathbf{V}_{n_I}\right| \leq 3\sqrt{w_I}$$

where, for the principal and secondary link super-voxels respectively:

- $\mathbf{V}_{P_{X,Y,Z}}$, $\mathbf{V}_{n_{X,Y,Z}}$ are the geometrical centers;
- $\mathbf{V}_{P_{R,G,B}}$, $\mathbf{V}_{n_{R,G,B}}$ are the mean R, G and B values;
- $\mathbf{V}_{P_I}$, $\mathbf{V}_{n_I}$ are the mean laser reflectance intensity values;
- w_C is the color weight equal to the maximum value of the two variances $Var(R, G, B)$, i.e. $\max(\mathbf{V}_{P_{Var(R,G,B)}}, \mathbf{V}_{n_{Var(R,G,B)}})$;
- w_I is the intensity weight equal to the maximum value of the two variances $Var(I)$.

w_D is the distance weight given as $\frac{\left(\mathbf{V}_{P_{s_{X,Y,Z}}} + \mathbf{V}_{n_{s_{X,Y,Z}}}\right)}{2}$. Here $s_{X,Y,Z}$ is the voxel size along X, Y and Z axis respectively. c_D is the inter-distance constant (along the three dimensions) added depending upon the density of points and also to overcome measurement errors, holes and occlusions, etc.

These clustered objects are then classified using local descriptors and geometrical features into 6 main classes: {Road, Building, Car, Pole, Tree, Unclassified}. These mainly include: surface normals, geometrical and barycenter, color, intensity, geometrical shape and size. The details of this method are presented in [1] while some results of this method are shown in Fig. 2. The salient features of this method are data reduction, efficiency and simplicity of approach.

4 Comparison: Results, Evaluation and Discussion

In order to compare the two approaches, we evaluated the two methods using the "Paris-Rue-Madame" dataset as presented in [25]. This database, used for benchmarking urban detection-segmentation-classification methods, consists of annotated 3D point clouds acquired by mobile terrestrial data acquisition system [10] of "Rue Madame" in the 6th Parisian district (France).

The evaluation was conducted for five common classes: {Building, Road, Pole, Tree, Car}. The detailed assessment carried out for each of the detection, segmentation and classification phase respectively are presented below.

4.1 Detection Evaluation

The detection evaluation is done to measure the capacity of the method to detect the objects present in the scene. This requires the choice of a criterion to decide if an object from the ground truth is detected or not. In order to ensure that this criterion does not bias the evaluation, the results are evaluated for a varying threshold m on the minimum object overlap as presented in [3]. In this analysis, an object *OBJ* is defined by the subset of points with the same object identifier i.e. S_{GT} and S_{AR} are the ground truth and the evaluated algorithm result subsets respectively. For any object j, S^j_{AR} is only validated as a correct detection of S^j_{GT} (a match) if the following condition is satisfied:

$$OBJ^J\,(detected) \xrightarrow{\text{iff}} \left(\frac{|S_{GT}|}{|S_{GT} \cup S_{AR}|} > m\right) \bigwedge \left(\frac{|S_{AR}|}{|S_{GT} \cup S_{AR}|} > m\right) \tag{1}$$

where $|.|$ is the cardinal (number of objects) of a set. The standard `Precision` `Pr` and `Recall` `Re` are then calculated as functions of m:

$$\texttt{Pr}(m) = \frac{\texttt{number\_of\_detected\_objects\_matched}}{\texttt{total\_number\_of\_detected\_objects}}$$

$$\texttt{Re}(m) = \frac{\texttt{number\_of\_detected\_objects\_matched}}{\texttt{total\_number\_of\_ground\_truth\_objects}}$$

These values of `Pr` and `Re` are then combined together to calculate the F-Measure as a function of m as expressed in Eq. (2).

$$F(m) = 2 \times \frac{\texttt{Pr}(m) \times \texttt{Re}(m)}{\texttt{Pr}(m) + \texttt{Re}(m)} \tag{2}$$

Figure 3 shows the values of F-Measure with the variation of m for the different object types using both methods. The value of F-Measure decreases with the increasing value of m and this decay indicates the performance quality of the detection (good performance implies slower decay). Although the super-voxel based method does

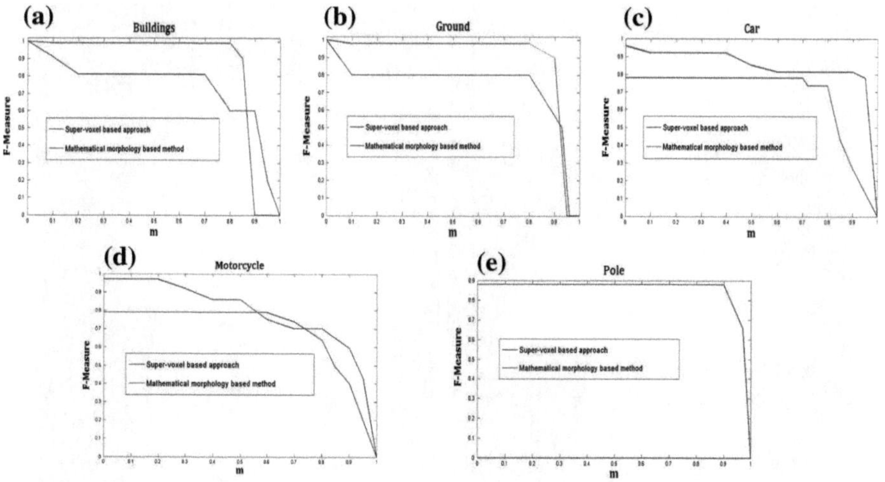

Fig. 3 Detection results for both super-voxel based and morphological transformation based methods are presented for 5 different classes in **a**–**e** respectively. **a** Buildings. **b** Ground. **c** Car. **d** Motorcycle. **e** Pole

not classify motorcycles, they were detected and classified manually to analyze their segmentation quality (discussed in the next section).

The results show that the building and ground are much better detected by the morphological transformation method while the detection quality performance for cars, poles, and other road furniture is much more superior for the super-voxel based method.

4.2 Segmentation and Classification Evaluation

The evaluation was conducted for five common classes: {Building, Road, Pole, Tree, Car} and also the motorcycle class (only segmentation results). The segmentation and classification results are presented in Fig. 4. As trees were not present in the dataset, they were not considered for analysis.

The segmentation and classification quality was evaluated point-wise i.e. the number of 3D points correctly classified as members of a particular class. The results presented in Table 1 are in the form of a confusion matrix in which rows and columns are the class labels from the ground truth and the evaluated method respectively. The matrix values are the percentage of points with the corresponding labels using the metrics defined in [1]. If T_i, $i \in \{1, \ldots, N\}$, is the total number of 3D points distributed into objects belonging to N number of different classes in the ground truth and, and let t_{j_i}, $i \in \{1, \ldots, N\}$, be the total number of 3D points classified as a particular class of type-j and distributed into objects belonging to N different classes (for example a 3D point classified as part of the building class may actually belong to a

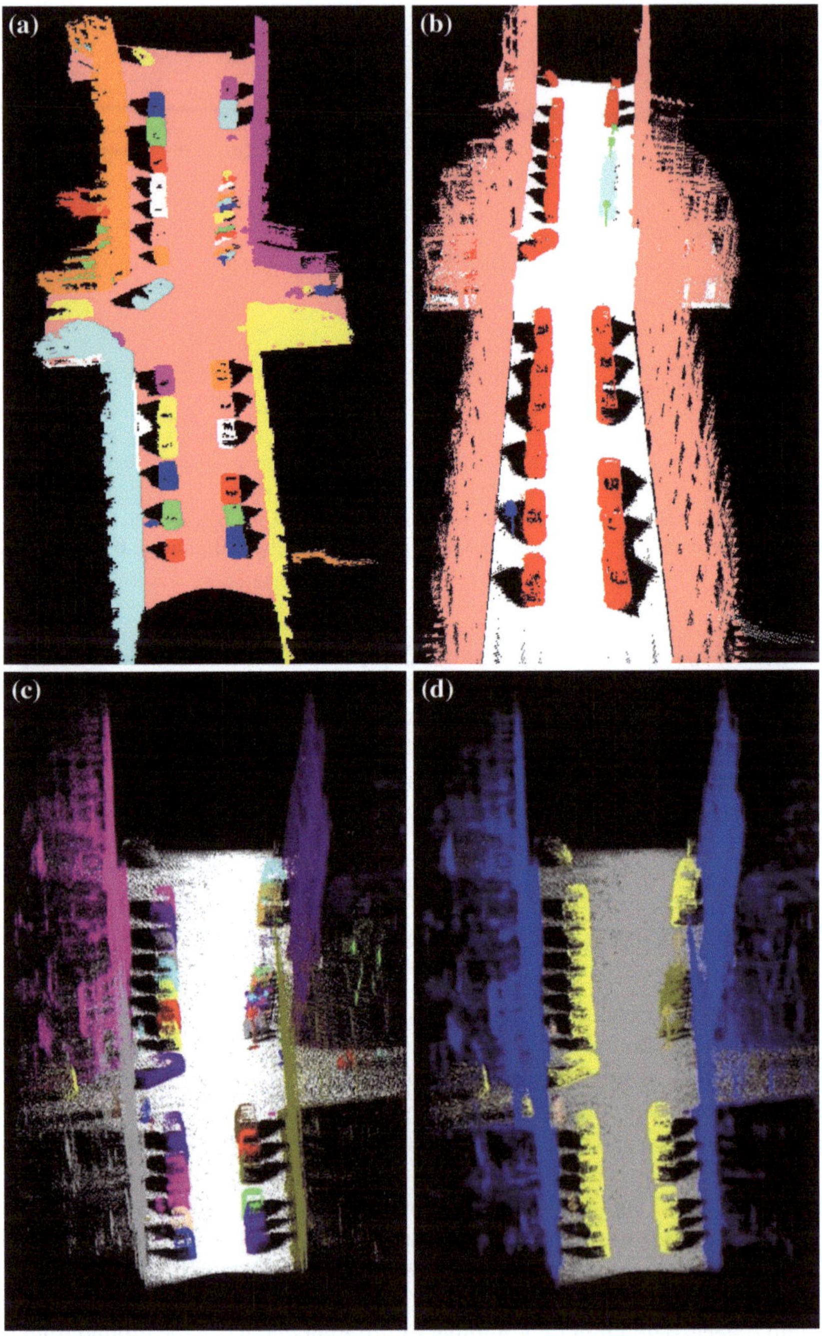

Fig. 4 **a** and **b** show the segmentation and classification results for super-voxel based method while **c** and **d** show the segmentation and classification results for morphological transformation based method respectively. In **a** and **c** every segmented object is represented by a separate color (some colors are repeated) while in **b** and **d** each class is represented by a different color

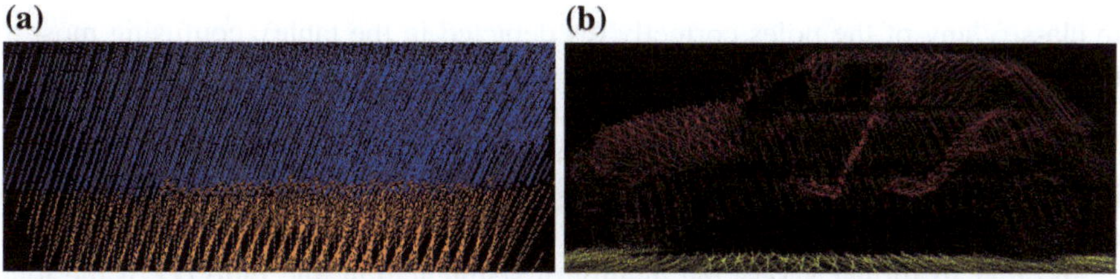

Fig. 5 **a** and **b** show the misclassification of some 3D points at boundary regions of road surface with building and car respectively for super-voxel based method

Table 1 Segmentation and classification results for both super-voxel based and mathematical morphology-based method (inside braces) are presented respectively

	Building	Road	Pole	Car	**CACC**
Building	0.914 (0.986)	0.013 (0.045)	0.000 (0.000)	0.000 (0.010)	0.950 (0.970)
Road	0.02 (0.002)	0.901 (0.940)	0.005 (0.000)	0.010 (0.002)	0.933 (0.968)
Pole	0.000 (0.000)	0.001 (0.001)	0.710 (0.000)	0.000 (0.010)	0.850 (0.495)
Car	0.000 (0.010)	0.005 (0.195)	0.000 (0.000)	0.900 (0.950)	0.950 (0.870)
Overall segmentation accuracy: **OSACC**				0.856 (0.720)	
Overall classification accuracy: **OCACC**					0.920 (0.825)

tree) then the ratio S_{jk} (j is the class type as well as the row number of the matrix and $k \in \{1, \ldots, N\}$) is given as: $S_{jk} = \frac{t_{jk}}{T_k}$.

These values of S_{jk} are calculated for each class and are used to fill up each element of the confusion matrix, row by row. The Segmentation ACCuracy (**SACC**) is represented by the diagonal of the matrix while the values of classification accuracy (**CACC**), overall segmentation accuracy (**OSACC**) and overall classification accuracy (**OCACC**) are calculated as explained in [1].

Compared to contemporary evaluation methods such as used in [17], employing a standard confusion matrix, this method is more suitable for this type of work as it provides more insight in segmentation results along with the classification results and directly gives the segmentation accuracy similar to [6]. Also as compared to standard precision and recall evaluation, the use of this metric, also accommodates for the unclassified 3D points in the results giving a more accurate result without incorporating the unclassified objects as a class in the confusion matrix.

Table 1 shows the results. It can be seen that for the super-voxel based method, some of the 3D points belonging to different object classes are found in the road class and vice versa. This was found evident at boundary regions of objects belonging to two different classes, as shown in Fig. 5, as sometimes in the voxelisation process, some of the 3D points belonging to adjacent objects are incorporated in the same voxel if they have similar color and reflectance intensity values.

Also, it was found that, for this method, one of the traffic sign post was wrongly classified as a tree resulting in a low **SACC** and **CACC** of 0.71 and 0.85 respectively. This was due to the fact that the particular sign post contained two traffic signs on the same post giving it a small tree like appearance (in 3D point cloud at least) as shown in Fig. 6. Compared to this method, the morphological transformation method failed to classify any of the poles correctly (as depicted in the table), confusing most of them with trees.

Also evident from the table, the interaction between classes is much more significant in the case of the morphological transformation method while on the other hand in the super-voxel based method the segmented objects belonging to a particular class instead of being distributed in other classes rather remain unclassified.

In order to further assess the quality of segmentation, the ratio (**f**) of the total number of objects segmented by the applied method and the total number of segmented objects in the ground truth was plotted for each of the object classes as shown in Fig. 7. A value of 1 represents overall best segmentation whereas a value greater than 1 denotes overall over-segmentation while a value less than 1 denotes overall under-segmentation. A value of 0 shows failure to detect or no detection.

The mathematical morphology based method seems to outperform the super-voxel based method in terms of segmenting building and road surface. In the super-voxel based segmentation method the road was over-segmented in 4 parts as they were found disconnected and also one of the building was over-segmented due to strong

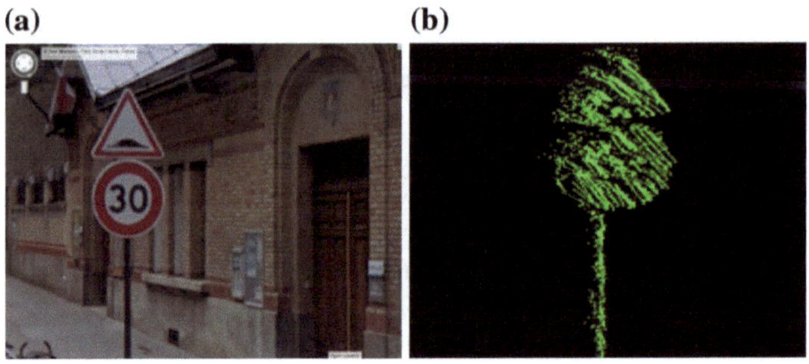

Fig. 6 **a** Google street view photo of the sign post with two traffic signs on Rue Madame. **b** Corresponding 3D points

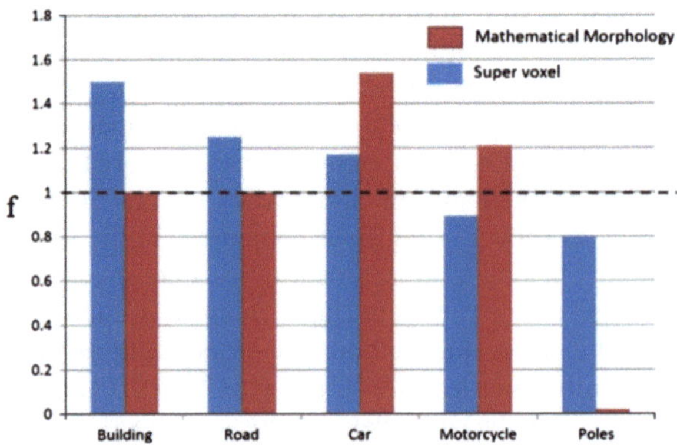

Fig. 7 Overall segmentation quality of the two methods for different object classes

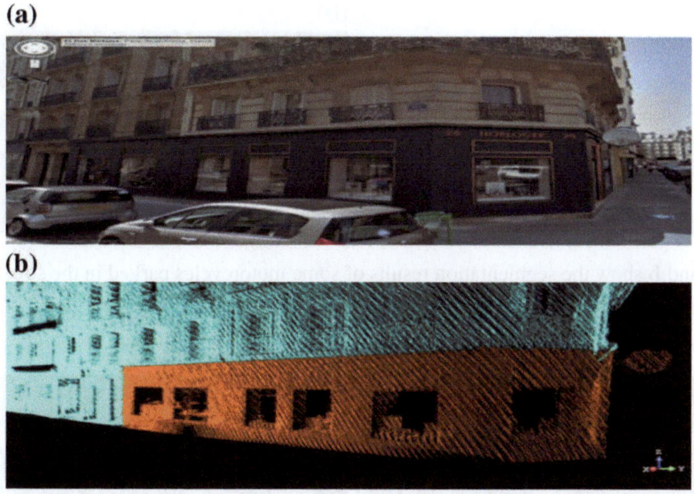

Fig. 8 **a** Google street map view photo of the building on Rue Madame with a strong variation of paint color. **b** Segmentation results of the super-voxel based method

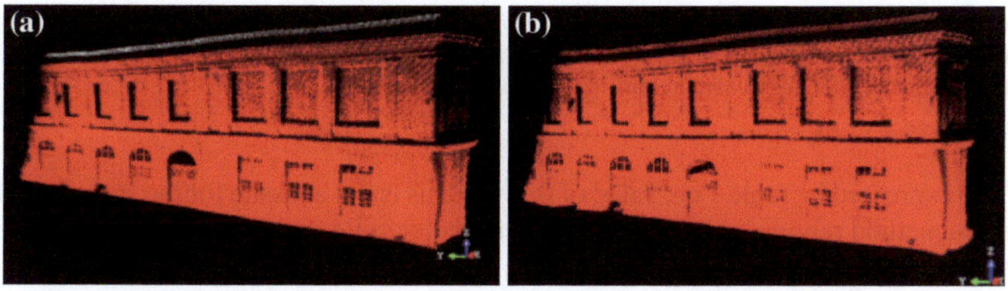

Fig. 9 **a** and **b** show the segmentation results of a particular building in Rue Madame for super-voxel based method and mathematical morphology based method respectively. In (**a**) it can be seen that part of the building that was disjoint was segmented as a separate object

variation in color and reflectance intensity values (as shown in Fig. 8) while in case of another building small part found disjoint from the main building was segmented as a separate object (shown in Fig. 9).

However, compared to the mathematical morphology based method, the super-voxel based method segments cars and other road furniture better as apart from the adjacency of the 3D points it also uses color and reflectance intensity values in the segmentation phase. Figure 10 shows the segmentation results, for both methods, of some of the motorcycles parked in the scene. For the super-voxel based method, we also find in one instance that two cars parked very close together, having similar color and reflectance intensity values, are segmented out as one single car.

The mathematical morphology based method, constrained by the generated profile, also fails to segment out 3D ground points directly under the motorcycles and car as shown in Fig. 11. These ground points are hence considered as part of the car

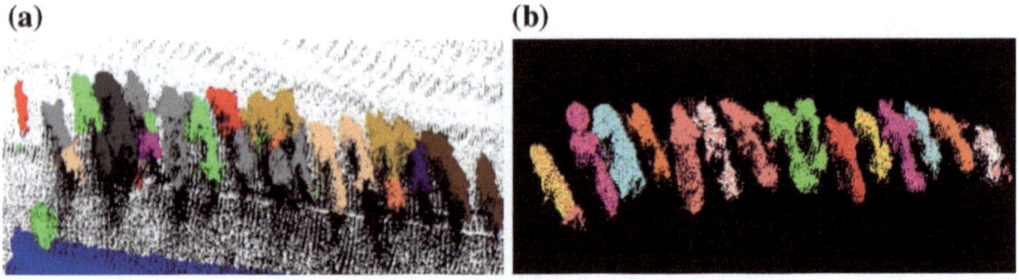

Fig. 10 **a** and **b** show the segmentation results of some motorcycles parked in the street for mathematical morphology and super-voxel based method respectively

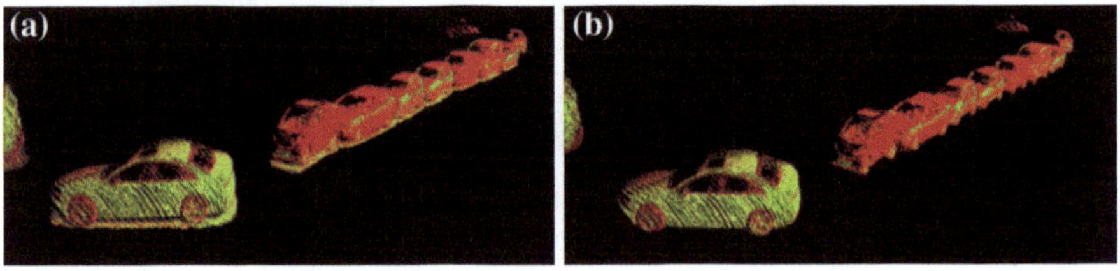

Fig. 11 **a** and **b** show the segmentation results of some cars in the street for both mathematical morphology and super-voxel based methods respectively. **a** shows some ground point directly underneath the cars, segmented as part of the cars

(also expressed in Table 1 i.e. value of 0.195). This is not an issue for the super-voxel based method relying on local descriptors i.e. color, reflectance intensity and surface normals.

5 Combining the Two Approaches

In order to exploit the strengths of the two methods and overcome their respective weaknesses, we combined the results of the two approaches. Two different types of combinations were tried which are explained below.

5.1 Direct Combination

In this combination, a simple union is applied to the segments, from the two methods, belonging to the same objects from the different object classes. A simple overlap ratio of 75 % was set (i.e. if more than 75 % of overlap between two segments,

Table 2 Segmentation and classification results for direct combination are presented

	Building	Road	Pole	Car	**CACC**
Building	0.986	0.031	0.000	0.010	0.972
Road	0.015	0.940	0.005	0.010	0.955
Pole	0.000	0.001	0.710	0.006	0.851
Car	0.001	0.110	0.000	0.950	0.912
Overall segmentation accuracy: **OSACC**				0.896	
Overall classification accuracy: **OCACC**					0.922

they are merged together as one). The improved results are presented in Table 2. We find that although the segmentation and classification results improve slightly (**OSACC = 0.896**, **OCACC = 0.922**), the overall segmentation quality decreases, due to the fact that combining of segments for each object class, in such a manner, often results in over-segmentation as shown in Fig. 12.

5.2 *Selective Combination*

In order to preserve the strengths of each method and overcome their respective weaknesses, a selective combination is proposed. Using the complimentary performances of the two approaches as discussed in Sect. 1, we combine the outputs of the two methods i.e. mathematical morphology based method for building and road surface while super-voxel based method for other classes and road furniture. The improved results are presented in Table 3. We find that not only the segmentation and classification results improve (**OSACC = 0.884**, **OSACC = 0.935**), but also the segmentation quality as shown in Fig. 12.

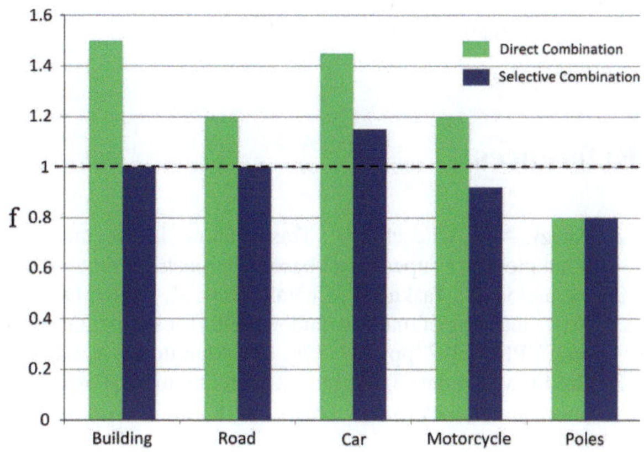

Fig. 12 Overall segmentation quality for different object classes, for the two combination methods

Table 3 Segmentation and classification results for selective combination are presented

	Building	Road	Pole	Car	**CACC**
Building	0.986	0.045	0.000	0.000	0.970
Road	0.002	0.940	0.000	0.002	0.968
Pole	0.000	0.001	0.710	0.000	0.854
Car	0.000	0.002	0.000	0.900	0.950
Overall segmentation accuracy: **OSACC**				0.884	
Overall classification accuracy: **OCACC**					0.935

6 Conclusion

In this paper, we present and compare two different approaches for segmenting and classifying of 3D urban point clouds i.e. one based on mathematical morphology while the other on super-voxels. Evaluated on a common dataset (real data), both these methods are thoroughly compared on three different levels: detection, segmentation and classification. The results show that the building and ground are much better detected by the mathematical morphology based method while the detection quality performance for cars, poles, and other road furniture is much more superior for the super-voxel based method. After analyses, simple strategies are also presented to combine the two methods, exploiting their complementary strengths and weaknesses, to improve the overall segmentation and classification results.

The same comparison methodology can be easily adapted to compare other segmentation and classification methods while the combination strategies need to be further studied and better adapted to improve upon the overall performances, for different applications.

Acknowledgments The work reported in this paper was supported by the French national research agency (ANR CONTINT iSpace & Time – ANR-10-CONT-23) and has been performed as part of Cap Digital Business Cluster TerraMobilita project.

References

1. Aijazi, A.K., Checchin, P., Trassoudaine, L.: Segmentation based classification of 3D urban point clouds: a super-voxel based approach. Remote Sens. **5**(4), 1624–1650 (2013)
2. Anguelov, D., Taskar, B., Chatalbashev, V., Koller, D., Gupta, D., Heitz, G., Ng, A.: Discriminative learning of markov random fields for segmentation of 3D scan data. In: IEEE Conference on CVPR, vol. 2, pp. 169–176. Los Alamitos, CA, USA (2005)
3. Brédif, M., Vallet, B., Serna, A., Marcotegui, B., Paparoditis, N.: Terramobilita/IQmulus urban point cloud analysis benchmark. In: IQmulus workshop in conjunction with SGP 14. Cardiff, UK (2014)
4. Byun, J., Na, K.I., Seo, B.S., Roh, M.: Drivable road detection with 3D point clouds based on the MRF for intelligent vehicle. In: Mejias, L., Corke, P., Roberts, J. (eds.) Field and Service Robotics, Springer Tracts in Advanced Robotics, vol. 105, pp. 49–60. Springer International Publishing (2015)

5. Chehata, N., Guo, L., Mallet, C.: Airborne lidar feature selection for urban classification using random forests. Int. Archiv. Photogramm. Remote Sens. Spat. Inf. Sci. **38**(3), 207–212 (2009)
6. Douillard, B., Underwood, J., Kuntz, N., Vlaskine, V., Quadros, A., Morton, P., Frenkel, A.: On the segmentation of 3D LIDAR point clouds. In: IEEE International Conference on Robotics and Automation (ICRA), p. 8. Shanghai, China (2011)
7. Filin, S., Pfeifer, N.: Segmentation of airborne laser scanning data using a slope adaptive neighborhood. ISPRS J. Photogramm. Remote Sens. **60**(2), 71–80 (2006)
8. Friedman, S., Stamos, I.: Online detection of repeated structures in point clouds of urban scenes for compression and registration. Int. J. Comput. Vis. **102**(1–3), 112–128 (2013)
9. Golovinskiy, A., Funkhouser, T.: Min-cut based segmentation of point clouds. In: IEEE Workshop on Search in 3D and Video (S3DV) at ICCV, pp. 39–46 (2009)
10. Goulette, F., Nashashibi, F., Abuhadrous, I., Ammoun, S., Laurgeau, C.: An integrated on-board laser range sensing system for on-the-way city and road modelling. In: ISPRS RFPT (2006)
11. Lalonde, J.F., Unnikrishnan, R., Vandapel, N., Hebert, M.: Scale selection for classification of point-sampled 3D surfaces. In: 5th International Conference on 3-D Digital Imaging and Modeling, pp. 285–292 (2005)
12. Lee, I., Schenk, T.: Perceptual organization of 3D surface points. In: The International Archives of the Photogrammetry, Remote Sensing and Spatial Information Sciences, vol. XXXIV, Part 3A, pp. 193–198 (2002)
13. Linsen, L., Prautzsch, H.: Global versus local triangulations. In: Roberts, J. (ed.) Procedings of Eurographics 2001, Short Presentations, pp. 257–263. Oxford, UK (2001)
14. Lodha, S., Fitzpatrick, D., Helmbold, D.: Aerial lidar data classification using adaboost. In: 6th International Conference on 3-D Digital Imaging and Modeling, 3DIM'07, pp. 435–442 (2007)
15. Moosmann, F., Pink, O., Stiller, C.: Segmentation of 3D lidar data in non-flat urban environments using a local convexity criterion. In: IEEE Intelligent Vehicles Symposium (IV), pp. 215–220 (2009)
16. Munoz, D., Bagnell, J.A.D., Vandapel, N., Hebert, M.: Contextual classification with functional max-margin Markov networks. In: IEEE Conference on CVPR, pp. 975–982 (2009)
17. Munoz, D., Vandapel, N., Hebert, M.: Onboard contextual classification of 3-D point clouds with learned high-order Markov random fields. In: IEEE International Conference on Robotics and Automation, pp. 2009–2016 (2009)
18. Niemeyer, J., Rottensteiner, F., Soergel, U.: Conditional random fields for lidar point cloud classification in complex urban areas. ISPRS Ann. Photogramm. Remote Sens. Spat. Inf. Sci. I(3), 263–268 (2012)
19. Elberink, S.O., Kemboi, B.: User-assisted object detection by segment based similarity measures in mobile laser scanner data. ISPRS - Int. Arch. Photogramm. Remote Sens. Spat. Inf. Sci. **XL-3**, 239–246 (2014)
20. Rabbani, T., van den Heuvel, F.A., Vosselmann, G.: Segmentation of point clouds using smoothness constraint. In: IEVM06 (2006)
21. Schoenberg, J., Nathan, A., Campbell, M.: Segmentation of dense range information in complex urban scenes. In: IEEE/RSJ International Conference on Intelligent Robots and Systems (IROS), pp. 2033–2038. Taipei, Taiwan (2010)
22. Secord, J., Zakhor, A.: Tree detection in urban regions using aerial lidar and image data. IEEE Geosci. Remote Sens. Lett. **4**(2), 196–200 (2007)
23. Serna, A., Marcotegui, B.: Urban accessibility diagnosis from mobile laser scanning data. ISPRS J. Photogramm. Remote Sens. **84**, 23–32 (2013)
24. Serna, A., Marcotegui, B.: Detection, segmentation and classification of 3d urban objects using mathematical morphology and supervised learning. ISPRS J. Photogramm. Remote Sens. **93**, 243–255 (2014)
25. Serna, A., Marcotegui, B., Goulette, F., Deschaud, J.E., et al.: Paris-Rue-Madame database: a 3D mobile laser scanner dataset for benchmarking urban detection, segmentation and classification methods. In: 4th International Conference on Pattern Recognition, Applications and Methods (2014)

26. Shapovalov, R., Velizhev, A., Barinova, O.: Non-associative markov networks for 3D point cloud classification. In: Photogrammetric Computer Vision and Image Analysis (PCV 2010), vol. 38, pp. 103–108 (2010)
27. Sithole, G., Vosselman, G.: Automatic structure detection in a point-cloud of an urban landscape. In: 2nd GRSS/ISPRS Joint Workshop on Remote Sensing and Data Fusion over Urban Areas, pp. 67–71 (2003)
28. Xiong, X., Munoz, D., Bagnell, J.A.D., Hebert, M.: 3-D scene analysis via sequenced predictions over points and regions. In: IEEE International Conference on Robotics and Automation (2011)

A Stereo Vision Based Obstacle Detection System for Agricultural Applications

Patrick Fleischmann and Karsten Berns

Abstract In this paper, an obstacle detection system for field applications is presented which relies on the output of a stereo vision camera. In a first step, it splits the point cloud into cells which are analyzed in parallel. Here, features like density and distribution of the points and the normal of a fitted plane are taken into account. Finally, a neighborhood analysis clusters the obstacles and identifies additional ones based on the terrain slope. Furthermore, additional properties can be easily derived from the grid structure like a terrain traversability estimation or a dominant ground plane. The experimental validation has been done on a modified tractor on the field, with a test vehicle on the campus and within the forest.

1 Introduction

According to [10], the agricultural guidance research exploring the capabilities of image sensors started in the mid-1980s in North America. With the full availability of the NAVSTAR Global Positioning System (GPS) one decade later, researchers also started to explore this new technology including its application for the agricultural sector. This research on GPS-based guidance solutions led to successful commercial products which are nowadays offered by almost all big manufacturers of agricultural products or can be bought from component suppliers. The success of this technology can be probably explained by its universal applicability. In contrast to early camera-based and specialized solutions such as crop row guidance, the GPS guidance is not restricted to individual field work or a special machine. The already mentioned, commercial products for example, offer functions such as creating a linear trajectory

P. Fleischmann (✉) · K. Berns
Robotics Research Lab, Department of Computer Science, University of Kaiserslautern, 67663 Kaiserslautern, Germany
e-mail: fleischmann@cs.uni-kl.de

K. Berns
e-mail: berns@cs.uni-kl.de

D.S. Wettergreen and T.D. Barfoot (eds.), *Field and Service Robotics*,
Springer Tracts in Advanced Robotics 113, DOI 10.1007/978-3-319-27702-8_15

defined by a waypoint and a direction. Furthermore, a complete track can be recorded and by specification of the implement's width, the system can calculate parallels to cover the whole field.

Here, a systemic disadvantage of GNSS-based (Global Navigation Satellite System) guidance systems is visible, which alone is not solvable with the GNSS technology: the calculated trajectories are not necessarily free of obstacles, which can lead to serious accidents. Accidents are caused by fatigue or inattention of the driver who has to monitor the Advanced Driver Assistance System (ADAS), where the two main reasons can be identified. On one hand, the use of an automated steering system can increase the monotony of work and thus cause fatigue—especially with large acreage. On the other hand, agricultural manufacturers are constantly increasing the working width of their machines and implements for economic reasons. For modern sprayers of 40 m width, it is difficult for the driver to estimate if the boom of the implement can be safely moved past an obstacle, especially at higher speeds.

While GNSS-based products are already very successful on the agricultural market, solutions using cameras or time-of-flight sensors are still a niche product for very specialized tasks and still in the focus of research. In the research domain stereo vision based obstacle detection for off-road and on-road is a large area, a recent survey [2] summarizes the contributions of the last decade. A very popular method is presented in [7] where obstacles are detected by analyzing the so called compatibility of the 3D points. To speed up the process, the evaluation is performed in the Disparity Space Image (DSI) where the truncated cones that have to be examined to get the compatibility turn into triangles. The well cited method has been extended and refined several times, e.g. in [13] or in [4], where the DSI was splitted into different stripes with different resolutions to allow for parallelization and to reduce the number of comparisons.

Another group of approaches can be identified which rely on a 2D grid or use a Digital Elevation Map (DEM). One recent example [5] uses the grid representation to fit B-spline surfaces into the reduced data to estimate the traversability of the ground and presence of obstacles. For road application [9] demonstrates an approach where a DEM and a density measurement of the points within a cell are used to separate the road surface from obstacles.

The QUAD-AV project [12] addresses the obstacle detection problem for agricultural vehicles by the investigation of different sensor technologies like stereo vision, thermography, ladar and a microwave radar. Along with this project, several interesting publications were made, e.g. a self-learning framework which uses geometric 3D data and color information of a trinocular camera [11] to classify the ground. Both classifiers are updated during runtime to adapt the approach to changing environments. Reference [8] describes the same framework for a multi-baseline camera but only relies on the geometric classifier. Additionally, a so called Unevenness Point Descriptor has been proposed [1] by the same research group which uses the normal vector distribution of small surfaces which are fitted using PCA.

In this paper, an obstacle detection system for the field is presented which can prevent collisions with obstacles while using a guidance system. For different reasons which are explained in detail in Sect. 2 a colored stereo camera was chosen to approach this task. As a first step after the data acquisition and pre-processing, the 3D points are sampled into a 2D grid. Initially, each grid cell is analyzed independently of its neighbors, which enables a strong parallelization of the method. Afterwards, the relations to the neighborhood are examined, the obstacles are grouped and a terrain abstraction is generated.

2 System and Scenario Description

As described in the introduction the motivation to start the research on a field obstacle detection system was driven by reported accidents with automated guidance systems. Possible and probable obstacles in this scenario can be divided into 3 categories: natural, artificial or man-made and dynamic obstacles. The first class includes any kind of vegetation which is not traversable like bushes or trees and additionally impassable terrain like ground with high slope or negative obstacles like ditches and trenches. For the field scenario, the second category includes any kind of poles (transmission, power), buildings, bridges and fences. The most difficult class contains dynamic obstacles like persons, other agricultural equipment and animals.

The system described in this paper uses a Bumblebee2 stereo vision camera by Point Grey. It has a focal length of 2.5 mm which results in a wide horizontal field of view of 97°. Furthermore, the stereo setup has a fixed 12 cm baseline and includes two Sony ICX204 1/3″ color CCD sensors providing a maximum usable resolution of 1024×768 pixels at 20 FPS. The decision to use a stereo camera instead of more precise sensor like a 3D laser scanner was influenced by the following properties. Firstly, the stereo vision system provides a very dense point cloud together with additional color information. Additional advantages like the low price of camera systems in mass production, the low energy consumption and the light weight makes the technology interesting for commercial applications. Furthermore, it could be pointed out during the tests, that the dust influence is lower than for a laser sensor which makes the device interesting for agricultural purposes.

The camera system has been mounted at a height of 2.8 m above the ground in front of the driver's cabin of a modified John Deere 6R series tractor. It was tilted downwards by about 10.5° to reduce the amount of sunlight falling into the camera. For better understanding, the mounting position together with the used Cartesian coordinate systems is shown in Fig. 1.

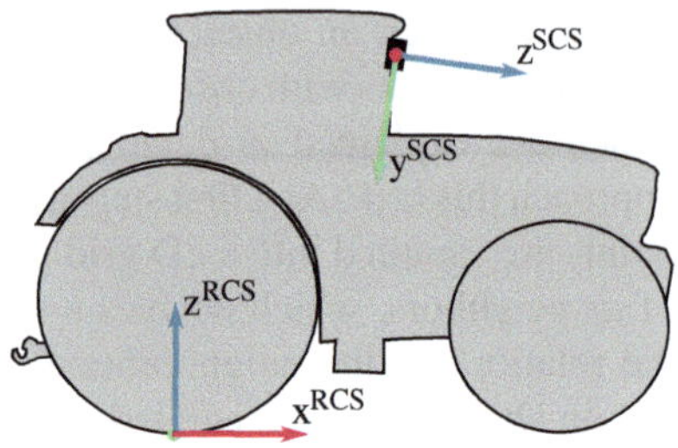

Fig. 1 Overview on the used coordinate systems: *SCS*: sensor, *RCS*: robot/tractor and the position of the camera which was mounted in front of the cabin below the roof

3 Implementation and Algorithms

3.1 Data Acquisition and Pre-processing

To grab the images from the camera, the libdc1394 library is employed. For the undistortion and rectification step, the calibration offered by Point Grey is used. Therefore, functions of the Triclops SDK[1] together with the calibration parameters stored on the sensor were used to generate lookup-tables for each camera in an offline process which map the pixels of the original image to the target. By applying these pre-computed tables, undistortion, rectification, cropping and scaling to a desired resolution can be done in one step. In addition, this enables the usage of the calibration together with other libraries, e.g. OpenCV [3] remap functionality which is applied in this case. The rectified images are then processed by a block matching algorithm which uses the sum of absolute differences as a metric to compute the disparity map. Neither the matching algorithm nor the metric are known to produce the best possible results. But its simplicity and its efficiency makes the algorithm suitable for embedded or GPU implementations. Knowing the disparity map, 3D points according to the camera reference frame (x^{SCS}, y^{SCS}, z^{SCS}) (see Fig. 1) can be calculated. In this step, the properties of the stereo vision system like the principal point, the focal length measured in pixels and the baseline is needed. All parameters are stored on the camera and can be scaled to the selected resolution. As a last step of the point cloud generation, all points are projected into the robot coordinate system (x^{RCS}, y^{RCS}, z^{RCS}). This system is originated on the ground below the kinematic center, with the x^{RCS}-axis pointing in the driving direction, z^{RCS} directed into the sky. The transformation requires the knowledge of the camera position in relation to this coordinate system. Please note that neither a statistical filtering nor a density reduction, e.g. a voxel grid filter, has been applied as it is often done in other approaches after this step.

[1]http://www.ptgrey.com/triclops.

3.2 Grid Generation and Pre-processing

The point cloud $P = \{p_1, \ldots, p_n\}$ given in the robot coordinate system is splitted into a 2D grid lying in the horizontal (x^{RCS}, y^{RCS}) plane. Each cell C_j has a parametrized dimension of $w \times h$. Due to the characteristics of the matching algorithm which produces a more dense cloud in the y-direction than in the x-direction, the width w and the height h could be set to different values. Additionally, the extend of the grid is limited in two directions $[0, x_{max}] \times [-\frac{y_{max}}{2}, \frac{y_{max}}{2}]$ as the output of a stereo vision system is only useful in a certain range, which depends on the baseline, the resulation and the focal length. The target cell index (c_x, c_y) of a point $p_i \in P,\ p_i = (p_{i_x}, p_{i_y}, p_{i_z})$ can be calculated as:

$$c_x = \left\lfloor \frac{p_{i_x}}{w} \right\rfloor \tag{1}$$

$$c_y = \left\lfloor \frac{p_{i_y}}{h} + \frac{y_{max}}{2h} \right\rfloor \tag{2}$$

To avoid errors in addressing the cells, the maximum grid dimensions should be defined as $x_{max} = a \cdot w$ and $y_{max} = 2 \cdot b \cdot h$ where $a, b \in \mathbb{N}^*$. After this step, each cell C_j contains a subset P_j of the original point cloud P. Due to the defined boundaries of the grid, the following relation applies

$$P = \left(\bigcup_{j=1}^{\frac{x_{max}}{w} \cdot \frac{y_{max}}{h}} P_j \right) \cup Q \tag{3}$$

where Q contains all points which do not belong to the grid and are not further analyzed. In a first parallelized step, the points $p_i^{(j)} \in P_j$ of each cell are sorted ascending according their $p_{i_z}^{(j)}$-coordinate to prepare the further steps which results into $P_j = \{p_i^{(j)} \mid i = 1, \ldots, n^{(j)};\ p_{i_z}^{(j)} \leq p_{(i+1)_z}^{(j)}\}$.

This is followed by a sequential extraction of the z-coordinate of the lowest point $p_1^{(j)}$ of each non-empty cell. Combined with the cell's 2D center $(m_x^{(j)}, m_y^{(j)})$, this set of lowest points is used to define an initial dominant ground plane by applying a least-squares fitting algorithm (see Sect. 3.3 and Eq. 17 for the details). Afterwards, the shortest distance between the plane and the points $(m_x^{(j)}, m_y^{(j)}, p_{1_z}^{(j)})$ is tested. If the distance is larger than a threshold t_g or in the case of empty cells, the z-value is extracted from the fitted plane. Furthermore, all these height values—either originated from $p_1^{(j)}$ or determined using the plane—are saved in a matrix whose number of rows and columns is equal to the grid. This matrix is then smoothed using a Gaussian blurring (kernel size 5×5) to reduce the influences of cells which do not provide ground points as they are containing large obstacles. Thereafter, the determined height values are stored in the corresponding grid cells as a ground guess value g_j. These height values are used to be able to separate overhangs even if no ground points are available, for instance in "obstacle shadows".

3.3 Cell Evaluation

One advantage of the presented approach is—as already mentioned—the ability to parallelize the following steps, as each subset P_j is firstly evaluated individually without incorporating the neighborhood. As a first step, the number of points assigned to a cell C_j is calculated, as it has to be above a defined threshold $|P_j| \geq \rho$ to get meaningful results. If this density of points is too low, the cell is marked as `non-evaluable`. Based on the vehicle's properties shown in Table 1, the following derived quantities can be calculated:

$$d = \sqrt{(c_x w + \frac{w}{2})^2 + (c_y h - \frac{y_{max}}{2} + \frac{h}{2})^2} \tag{4}$$

$$\alpha = \max(v_\beta, d \cdot v_\alpha) \tag{5}$$

$$z_{max} = \max(v_g, d \cdot \tan(\alpha)) \tag{6}$$

Afterwards, a decision tree is used to test if a cell contains an obstacle. If one of the rules (7), (9) or (14)–(16) applies, the label `obstacle` is assigned to the cell and the evaluation is terminated. In the other case, the next test is executed. The first rule (Eq. 7) checks if the lowest measured sample p_{1_z} (from this point on, the superscript (j) is omitted to improve the readability) is above the position which could be reached with the given maximum slope v_α. Similarly, the highest measured point p_{n_z} has to be higher than the lowest reachable position.

$$p_{1_z} > z_{max} \vee p_{n_z} < -z_{max} \tag{7}$$

In forestry scenarios it often happens that overhanging parts are detected. In combination with missing ground points, Eq. 7 would lead to many false classifications. Thus, the distance between p_{1_z} and the ground guess g_j is evaluated and the cell label is fixed to `non-evaluable` if $p_{1_z} - g_j > v_h$.

For cells which include ground as well as overhanging objects, the space between these clusters has to be examined to see if the robot can safely pass this cell. To handle this situation, the range of the z-coordinates is tested:

$$(p_{n_z} - p_{1_z}) > v_h \tag{8}$$

Table 1 Vehicle properties used to evaluate a grid cell

Symbol	Property
v_h	Height of the vehicle
v_α	Maximum slope which the vehicle can handle between two cells
v_β	Maximum desired attitude (roll and pitch)
v_g	Ground clearance of the vehicle

In the case that the range is larger than v_h, a k-means clustering algorithm is applied to the point cloud subset P_j to see if the points can be separated into ground and overhang. The number of clusters k is set to 2. Additionally, it is ensured that the center of first cluster $P_g = \{p_i \mid i = 1, \ldots, k;\ p_{i_z} \leq p_{(i+1)_z}\}$ has a lower z-value than the center of the second cluster $P_o = \{p_i \mid i = (k+1), \ldots, n;\ p_{i_z} \leq p_{(i+1)_z}\}$ which is expected to contain the points of the overhang. If both clusters P_g and P_o fulfill a density criterion, the distance between the highest point of the ground cluster P_g and the lowest of the overhang cluster is evaluated:

$$p_{(k+1)_z} - p_{k_z} < v_h \tag{9}$$

Here, the ground guess value g_j calculated during the pre-processing is used instead of p_{k_z}, if the density of P_g is too low. Furthermore, the cell label is set to `non-evaluable` if both densities are below a threshold or g_j has been applied to Eq. 9 which was evaluated to false. If the space is insufficient (Eq. 9 is true), the cell is rated as an `obstacle` in all other cases.

At this point, the remaining point cloud is either still the original one (P_j) or the overhangs have been successfully separated and only the portion P_g has to be further analyzed. To improve readability, the next steps are just explained for P_j, nevertheless the same tests will be executed on P_g if the overhang separation was conducted.

To get rid of outliers and matching errors of the stereo correspondence module, a smoothing filter as well as statistical outlier filter is applied to a copy of P_j to not lose the original measurements.

$$p_{i_z} := \frac{p_{i_z} + \mu_z}{2}, \quad \mu_z = \frac{1}{n}\sum_{i=1}^{n} p_{i_z} \tag{10}$$

The output of the smoothing filter shown in Eq. 10 is used as the input for the statistical filter shown in (13). Therefore, the indexes $q1$ of the first quartile Q_1 (also known as 25th percentile) and $q3$ of the third quartile Q_3 (75th percentile) are calculated. Using these indexes, the following boundaries are defined, where $(p_{q3_z} - p_{q1_z})$ is known as the interquartile range which contains 50 % of the data.

$$f_{min} = p_{q1_z} - 1.5 \cdot (p_{q3_z} - p_{q1_z}) \tag{11}$$

$$f_{max} = p_{q3_z} + 1.5 \cdot (p_{q3_z} - p_{q1_z}) \tag{12}$$

$$P_f = \{p_i \mid i = 0, \ldots, n;\ l \leq i \leq m;\ f_{min} \leq p_{l_z};\ p_{m_z} \leq f_{max};\ p_{i_z} \leq p_{(i+1)_z}\} \tag{13}$$

The resulting filtered point cloud P_f is tested according the following criteria. If one of the conditions apply, the `obstacle`-label is assigned to the cell.

$$|P_f| < \rho \tag{14}$$

$$p_{m_z} - p_{l_z} > \upsilon_g \tag{15}$$

$$p_{m_z} > z_{max} \vee p_{l_z} < -z_{max} \tag{16}$$

Here, the first rule is again a density check, the second rule tests if the cell range is acceptable and the third rule if the cell contains points which are above or below a reachable height.

The last and maybe strongest criterion evaluates properties of a plane fitted to the point cloud P_f. Therefore, a least squares fitting algorithm which minimizes the distance between the plane and the z-components of the points is used to find a plane defined as $z = ax + by + c$. For determination, the error $E(a, b, c) = \sum_{i=l}^{m} [(ap_{i_x} + bp_{i_y} + c) - p_{i_z}]^2$ needs to be minimized. According to [6], the following equation system (17) solves the problem, as the function $E(a, b, c)$ has its vertex when the gradient is zero.

$$\begin{pmatrix} \sum_{i=l}^{m} p_{i_x}^2 & \sum_{i=l}^{m} p_{i_x} p_{i_y} & \sum_{i=l}^{m} p_{i_x} \\ \sum_{i=l}^{m} p_{i_x} p_{i_y} & \sum_{i=l}^{m} p_{i_y}^2 & \sum_{i=l}^{m} p_{i_y} \\ \sum_{i=l}^{m} p_{i_x} & \sum_{i=l}^{m} p_{i_y} & \sum_{i=l}^{m} 1 \end{pmatrix} \begin{pmatrix} a \\ b \\ c \end{pmatrix} = \begin{pmatrix} \sum_{i=l}^{m} p_{i_x} p_{i_z} \\ \sum_{i=l}^{m} p_{i_y} p_{i_z} \\ \sum_{i=l}^{m} p_{i_z} \end{pmatrix} \tag{17}$$

After the plane is determined, the slope γ is calculated as the enclosed angle between the normal of the plane and the z-axis: If the slope is above the maximum slope α (see Eq. 5) or above the desired attitude υ_β the cell is interpreted as an obstacle. This decision can be overwritten and the cell is marked as `non-evaluable`, if the range $p_{m_z} - g_j$ is below the ground clearance value υ_g.

3.4 Neighborhood Evaluation

At this stage, only the cells have been evaluated without taking their neighborhood into account. This could lead to some misclassifications and has to be corrected in the following steps. As the methods are working on the grid structure, the neighboring cells need to be known for each cell C_j. Figure 2 shows the naming convention which is used to describe the evaluation. The full neighborhood contains 8 cells $N_{(8)}(C_j) = \{N_i(C_j)|\ i = 1, \ldots, 8\}$ while a reduced neighborhood $N_{(4)}(C_j)$—shown in red—only contains the neighboring cells with even indexes. Some extra attention is required at the borders of the grid as these cells do not have the full number of neighbors.

First, a function iterates over the whole grid and does the following analysis for each grid cell C_j which has not been labeled as `obstacle` or `non-evaluable` as described in Sect. 3.3. For each neighbor $N_i(C_j) \in N_{(8)}(C_j)$ which has been marked as potentially drivable in the cell analysis as well as for the center C_j, the mean height above the horizontal plane $\mu_z(C_j)$ is calculated based on the distance of the cell's center $(c_x w + \frac{w}{2}) + (c_y h - \frac{y_{max}}{2} + \frac{h}{2})$ to the fitted plane. Afterwards, the slope between the center and each adjoining cell N_i is determined as follows

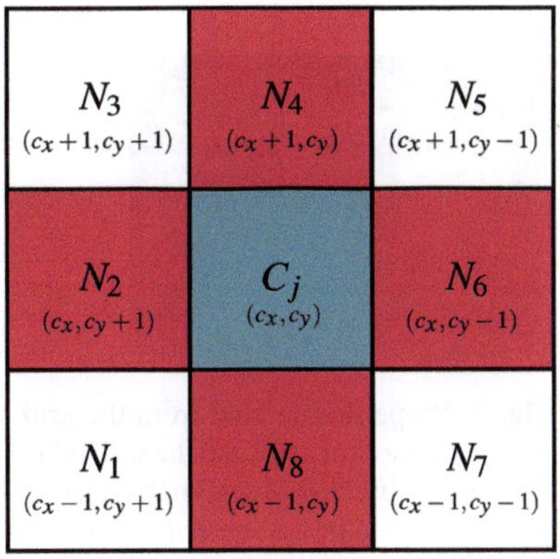

Fig. 2 Naming convention and traversing scheme of the grid cell neighborhood

$$\gamma_i(C_j, N_i(C_j)) = \text{atan2}\left[|\mu_z(C_j) - \mu_z(N_i(C_j))|, \text{dist}(C_j, N_i(C_j))\right] \tag{18}$$

where $\text{dist}(C_j, N_i(C_j))$ returns the spatial distance between two cell centers in the 2D x-y-plane. In addition, a counter is incremented for each slope measurement $\gamma_i(C_j, N_i(C_j))$ which is above the threshold v_α. If this counter is smaller than 4 after the evaluation, the cell is labeled as `obstacle` otherwise the cell is labeled as `drivable`. Some special cases have to be handled at the borders of the grid, in areas where no data points are available or if the label `non-evaluable` was assigned.

Finally, a post-processing step is executed to remove scattered `drivable` cells which are surrounded by obstacles. Therefore, a `drivable` cell close to the origin of the *RCS* is determined and used as a seed S. Furthermore, the cell is added to a list of non-isolated cells and its $N_{(4)}(S)$ neighbors are identified. For each of the neighbors $N_i(S) \in N_{(4)}(S)$ the assigned label is inspected. If it is not on the list of non-isolated cells and has been marked as `drivable` or `non-evaluable` it is used as a new seed and the method is recursively called.

3.5 Derived Properties

Based on the cell and the neighborhood evaluation different properties and views can be derived. For the presented application, the segmented obstacle view is the most important information. To generate this information, all cells tagged as `obstacle` are collected and added to an obstacle list. As long as this list contains elements, the following steps are repeated. The first element of the list generates a new obstacle cluster and is added to an auxiliary stack. Until the stack is empty, the $N_{(4)}(C_j)$ neighborhood of the top of the stack is analyzed and if it contains cells which are also on the obstacle list, they are added to the cluster as well as to an auxiliary stack and removed from the obstacle list. The process generates a collection of clusters which

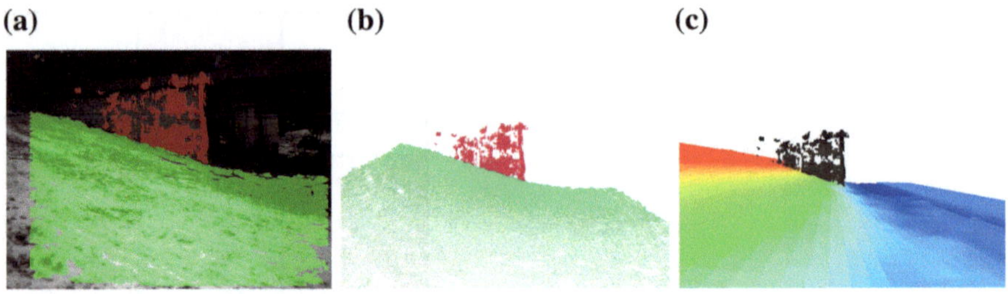

Fig. 3 Properties derived from the grid based evaluation: **a** Image of the *left camera* showing a winter scenario of a hill and the wall of a bridge. Additionally, the classification results are overlayed (*green* drivable, *red* obstacle). **b** Clustered point cloud: points belonging to the ground are shown in *green*, obstacle points are *red*, Terrain classification: the triangulated surface is color coded based on height above the x-y-plane

are enriched with some attributes like the maximum and minimum sample height within the cluster and the total number of 3D samples of the cluster. Furthermore, a polygon is calculated which describes the outer hull of the obstacle.

For the purpose of classification, the 3D points and the RGB-data of all obstacle clusters can be combined and accessed. This is possible since each cell of the grid still contains the original piece of the point cloud which was assigned to the cell. Besides the obstacle clusters, the evaluation results can also be used to divide the original point cloud into 3 separate clouds. The first one contains all points which belong to the traversable ground. The second one represents the measurements labeled as obstacles and the last one the overhanging objects which are higher than the vehicle. An example of a partitioned point cloud is shown in Fig. 3b. Here, the green points are showing all 3D points which belong to the ground. The red points are classified as an obstacle. In Fig. 3a the point cloud is shown as an overlay on the left image of a grayscale stereo camera.

For some applications, like an inverse perspective mapping which can be used to extract waysides or road markings, a dominant ground plane is a valuable information. Using the presented grid based structure, such a plane can be easily extracted. In a first step, the 2D cell center points of all `drivable` cells are collected. Afterwards, the medium cell heights of the same cells are determined using the approach depicted in Sect. 3.4 incorporating the distance to the planes fitted to the cells. Finally, the generated 3D points are used as an input for a least-squares plane fitting as described at the end of Sect. 3.3 or by applying a RANSAC plane fitting.

In addition to a binary obstacle or not-obstacle view, also the shape of the terrain is interesting as some areas might be traversable but with an increased effort or unwanted side effects like reduced wheel grip. Here, an abstract terrain model can be helpful for path planning or implement guidance. Using the grid representation, this shape of the ground can again be extracted using the small planes fitted to each cell. The algorithm to create a reduced version of a Digital Elevation Map (DEM) starts at a cell close to the *RCS*'s origin which has not been labeled as `obstacle` or `non-evaluable` (it has a reasonable height information). This starting cell is added to an auxiliary stack. In this case, a stack is required as the used grid size blasts the maximum recursion depth. While the stack is not empty, the following steps are repeated: 1. The top element is removed. 2. The $N_{(8)}(C_j)$-neighbors are calculated. 3. If the cell can provide a height value, the value is added to the DEM together with the 2D coordinates of the cell center. If it has been labeled as `obstacle` or `non-evaluable` the height is averaged using the neighboring cells. If they cannot provide valuable data, the last height value is used. 4. All neighbors which are not yet represented in the DEM are added to the stack. The final elevation map is then generated by triangulating the determined points.

Figure 3c shows the result of the traversability analysis. The hilly ground on the left side of the image has been classified as drivable ground as shown in green on Fig. 3a based of the capabilities of the vehicle. Nevertheless, the height map in Fig. 3c shows that the slope is quite high and if it is not required to go there, this area should be avoided.

4 Experiments and Results

To test and evaluate the presented obstacle detection approach, different scenarios have been recorded on the field, the campus of the University of Kaiserslautern, and the forest connected to it. For the field scenarios a John Deere 6R tractor was equipped with the stereo camera system described in Sect. 2, a differential GPS-system, an inertial measurement unit and other time-of-flight sensors to evaluate the data quality of the stereo camera. The first collection of datasets was recorded on grassland (see Fig. 4a) and on fields with different kinds and sizes of grass and weeds in summer 2014. With varying speeds from $1-15$ km/h different obstacle types and open field scenarios have been captured during different daytimes. In a second test, data has been recorded while using a stubble cultivator on a harvested grain field. This field contained several obstacles like a forest on one side, some houses at the opposite border, 2 power poles within the field and a ditch to a street nearby. Both datasets with 96,372 stereo image pairs in total were used to evaluate the obstacle detection system offline before deploying it to a real machine.

Both Figs. 4a and 4b show some typical classification results which were created in the offline analysis. In all of these scenarios a cell size of 0.5 m $\times$ 0.5 m was used to get rid of some small weeds sticking out of the ground. The grid dimensions where limited to 16 m $\times$ 16 m, as the point cloud density was too low for higher distances and the noise dramatically increased in farther regions. The 3D points belonging to the ground are summarized by the fitted planes which are shown in different colors depending on their distance to the horizontal plane. Cells which are classified as obstacles are highlighted with a red transparent box. Additionally, the individual 3D points are shown along with their color. For better understanding, the right part shows the left image of the stereo camera. The obstacle points have been back-projected to the image coordinates and are overlayed as a red mask for visualization. The hole in the center of the grid arises from the engine hood which was removed from the

(a)

(b)

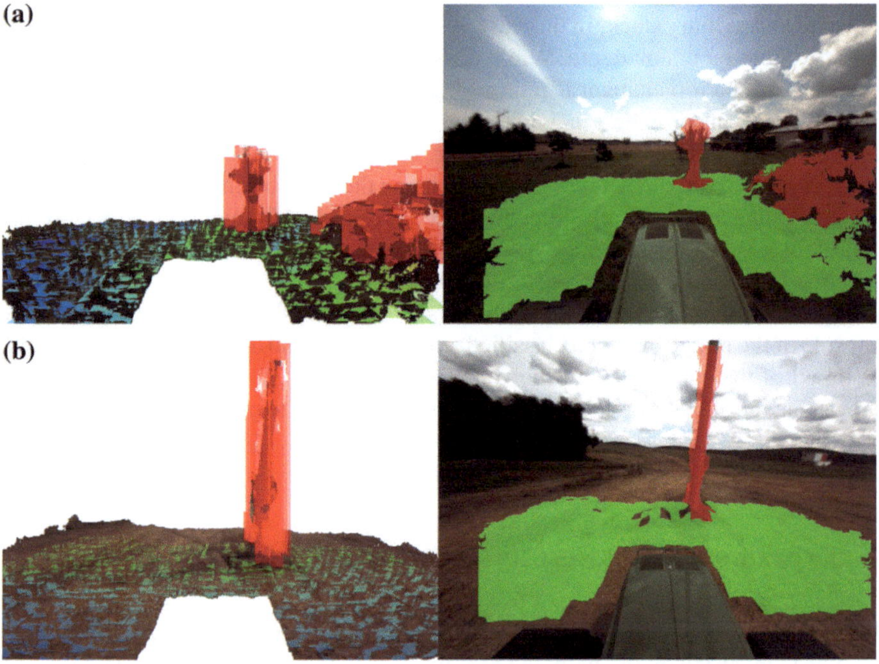

Fig. 4 Two typical obstacle situations captured on the field. The *left images* depict the results of the grid based evaluation showing the fitted planes—the color scale depends on the height above the x-y-plane—and the identified obstacles as red boxes together with the colored 3D points. *Right* resulting classification projected back to the *SCS* and visualized as an overlay on the left camera image. *Green* pixels show the drivable ground, obstacles are shown with *red* pixels. **a** Trees and a mound on grassland. **b** A power pole on a harvested grain field

point cloud before handing the data to the obstacle detection module. Figures 4a and 4b depict scenarios where the tractor was manually driven. The first figure shows the system's response to an apple tree and a small mound, the second visualization demonstrates the detection results of a large power pole which is blocking the path calculated by the GPS guidance system.

To quantitatively evaluate the detection performance, 100 randomly selected stereo pairs have been extracted from the recorded dataset described above. The ground-truth (obstacle or drivable ground) was manually set for each grid cell for all items of the selection. Using the parameters $\upsilon_h = 3.2$ m, $\upsilon_\alpha = 10^\circ$ and $\upsilon_g = 0.5$ m suitable for the tractor, it resulted in an average precision of 81.76 %, a recall of 93.16 % and an accuracy of 99.41 %. The determined false positive rate is 0.45 %. It should be mentioned, that the example images contained much more drivable cells (53,941) than obstacles (1169) as the data was collected on real fields. Furthermore, it could be seen that most of the false positive detections were caused by weeds sticking out of the ground or by cells connected to real obstacles which appear larger in the stereo cloud. Additionally, most of the false positives which were identified by the slope estimation γ were positioned at the border of the camera's field of view were the grid cells are only partially filled with 3D points.

To demonstrate the applicability of the approach, the obstacle detection was integrated into a guidance system and implemented on a modified tractor with electronically controllable steering and velocity. Here, the output of the detection was passed to a map where the results of multiple frames were combined to increase the robustness and to get rid of single misclassified cells. If an obstacle intersected the space requirements along the calculated GPS-path, the tractor was stopped by decreasing the speed to zero. The prototype was used to demonstrate an emergency stop in front of a person, a tree and a small pole while driving tracks on the field.

In addition, the methodology has also been tested at the campus and inside the Palatinate Forest as the number of obstacles and overhangs is higher than in the field scenarios and the detection has to be more precise. Two examples are given in Fig. 5a and 5b. For both sites the grid cell size was reduced to 0.25 m $\times$ 0.25 m for a better segmentation of obstacles. Additionally, the ground clearance was reduced to 0.2 m to fit to the capabilities of the testing vehicle, a John Deere Gator XUV 855D. It can been seen in both examples that the drivable ground is correctly classified

(a)

(b)

Fig. 5 Obstacle detection applied to a campus and a forest scenario using a cell size of 0.25 m $\times$ 0.25 m. For the tests, the camera was mounted on a John Deere Gator XUV 855D at a height of 2 m above the ground. For the forest scenario, a grayscale Bumblebee XB3 camera was used instead of the Bumblebee2 model employed for the other experiments. Both visualizations use the same color coding as described in the previous figure. **a** Campus scenario. **b** Forest scenario

by the system as well as the obstacles within the detection range of the camera. For the forest capture, a Bumblebee XB3 grayscale camera was used instead of the Bumblebee2. Due to the larger baseline of 24 cm also the x dimension of the grid could be increased to 35 m. For both examples the same visualization scheme as described for the field scenarios applies.

5 Summary and Future Work

The obstacle detection system presented in this paper was successfully used to detect severe obstacles on the field, the campus and inside the forest. Splitting the detection into a grid cell and a neighborhood based part allows for parallelization of the detection process which makes the approach real-time capable. Furthermore, the results are more robust than a point-wise analysis as small outliers have a reduced influence on the evaluation.

The collected data showed that further research is needed to distinguish between soft weeds sticking out of the ground and dangerous solid objects which is challenging based on the geometric data extracted by the stereo system. Thus, the system is currently extended to extract the image patches representing the obstacles found by the geometric evaluation. Afterwards, the obstacle is analyzed in the image domain to further classify the obstruction and neglect it in case of weeds.

References

1. Bellone, M., Reina, G., Giannoccaro, N.I., Spedicato, L.: 3d traversability awareness for rough terrain mobile robots. Sens. Rev. **34**(2), 220–232 (2014)
2. Bernini, N., Bertozzi, M., Castangia, L., Patander, M., Sabbatelli, M.: Real-time obstacle detection using stereo vision for autonomous ground vehicles: a survey. In: IEEE 17th International Conference on ITSC, 2014, pp. 873–878. IEEE (2014)
3. Bradski, G., Kaehler, A.: Learning OpenCV: Computer Vision with the OpenCV Library. O'Reilly Media Inc. (2008)
4. Broggi, A., Buzzoni, M., Felisa, M., Zani, P.: Stereo obstacle detection in challenging environments: the viac experience. In: IEEE/RSJ International Conference on IROS, 2011, pp. 1599–1604. IEEE (2011)
5. Broggi, A., Cardarelli, E., Cattani, S., Sabbatelli, M.: Terrain mapping for off-road autonomous ground vehicles using rational b-spline surfaces and stereo vision. In: Intelligent Vehicles Symposium (IV), 2013 IEEE, pp. 648–653. IEEE (2013)
6. Eberly, D.: Least Squares Fitting of Data. Magic Software, Chapel Hill (2015)
7. Manduchi, R., Castano, A., Talukder, A., Matthies, L.: Obstacle detection and terrain classification for autonomous off-road navigation. Auton. Robots **18**(1), 81–102 (2005)
8. Milella, A., Reina, G.: 3d reconstruction and classification of natural environments by an autonomous vehicle using multi-baseline stereo. Intell. Serv. Rob. **7**(2), 79–92 (2014)
9. Oniga, F., Nedevschi, S.: Processing dense stereo data using elevation maps: road surface, traffic isle, and obstacle detection. IEEE Trans. Veh. Technol. **59**(3), 1172–1182 (2010)
10. Reid, J.F., Zhang, Q., Noguchi, N., Dickson, M.: Agricultural automatic guidance research in north america. Comput. Electron. Agric. **25**(1), 155–167 (2000)

11. Reina, G., Milella, A.: Towards autonomous agriculture: Automatic ground detection using trinocular stereovision. Sensors **12**(9), 12405–12423 (2012)
12. Rouveure, R., Nielsen, M., Petersen, A., Reina, G., Foglia, M., Worst, R., Seyed-Sadri, S., Blas, M.R., Faure, P., Milella, A., et al.: The quad-av project: multi-sensory approach for obstacle detection in agricultural autonomous robotics. In: International Conference of Agricultural Engineering CIGR-Ageng, pp. 8–12 (2012)
13. Santana, P., Guedes, M., Correia, L., Barata, J.: A saliency-based solution for robust off-road obstacle detection. In: IEEE International Conference on ICRA, 2010, pp. 3096–3101. IEEE (2010)

CoPilot: Autonomous Doorway Detection and Traversal for Electric Powered Wheelchairs

Tom Panzarella, Dylan Schwesinger and John Spletzer

Abstract In this paper we introduce CoPilot, an active driving aid that enables semi-autonomous, cooperative navigation of an electric powered wheelchair (EPW) for automated doorway detection and traversal. The system has been cleanly integrated into a commercially available EPW, and demonstrated with both joystick and head array interfaces. Leveraging the latest in 3D perception systems, we developed both feature and histogram-based approaches to the doorway detection problem. When coupled with a sample-based planner, success rates for automated doorway traversal approaching 100 % were achieved.

1 Introduction

The U.S. Department of Health and Human Services reports that the number of people over the age of 65 will increase from 40.4 million people in 2010 to over 70 million by 2030 [16]. This rapid growth in the U.S. elder population will also increase the number of people with age-related symptoms that hamper their mobility. Such common symptoms include visual impairments, dementia, and Alzheimer's disease [14]. Providing electric-powered wheelchairs (EPWs) to seniors (and others) is a significant step in helping them live at home and maintain independent mobility. However, it is not without its own challenges. Maintaining straight paths and avoiding obstacles is often challenging—especially for drivers using alternative controls such as sip-and-puff devices, switch driving systems, chin controls, or short-throw joysticks. Additionally, traditional joystick users with impaired hand control and

T. Panzarella
Love Park Robotics, LLC, 2025 Washington Ave, Suite 217,
Philadelphia, PA 19146, USA
e-mail: tpanzarella@loveparkrobotics.com

D. Schwesinger (✉) · J. Spletzer
Lehigh University, 27 Memorial Drive West, Bethlehem, PA 18051, USA
e-mail: dts211@lehigh.edu

J. Spletzer
e-mail: josa@lehigh.edu

D.S. Wettergreen and T.D. Barfoot (eds.), *Field and Service Robotics*,
Springer Tracts in Advanced Robotics 113, DOI 10.1007/978-3-319-27702-8_16

those who rely on "latched driving" modes (i.e., cruise control) for independence and function may require additional assistance to ensure safe and comfortable mobility. To realize the home health benefits of EPWs while also maintaining safety, active safety systems for EPWs could be deployed.

To this end, we have developed CoPilot, an active driving aid that enables semi-autonomous, cooperative navigation of an EPW. Similar to active driver-assist systems in automobiles, the driver remains in primary control of the vehicle, while in the background, CoPilot uses intelligent sensing and drive control systems that work in cooperation with the driver to aid in avoiding obstacles/collisions and fine precision driving tasks. The motivation is that as an individual begins to lose cognitive, perceptive, or motor function due to age, injury, or disease, CoPilot can augment that loss because it can interpret the user's intent by seeing into the environment. This exteroceptive sensing capability is enabled by leveraging the latest in three-dimensional (3D) imaging technology. While being developed with a suite of semi-autonomous driving behaviors in mind, the focus of this paper is automated doorway detection and traversal. This functionality was motivated by discussions with physical and occupational therapists in the wheelchair space who prioritized doorway navigation as a capability that would provide real value to EPW users. CoPilot provides near 100 % effectiveness in this application.

2 Related Work

Doorway detection using 3D sensing has been accomplished in various ways. Rusu et al. used 3D point clouds to locate doors [13]. The goal was to find doors for the purpose of opening or closing them with a robotic manipulator. When the robot was at a door location, a planar model was fit to the point cloud data. The models were validated based on geometric constraints. More recently, RGB-D data has been used for the task of parsing indoor scenes [5, 11]. The goal of which is to detect and correctly label objects in indoor environments. This is a more difficult task than looking for a single category of object, in our case doorways. These algorithms are based on learning classifiers where the feature vectors are largely inspired from computer vision techniques, such as histograms of oriented features. In our work, we also leverage computer vision approaches for some aspects of doorway detection.

Early approaches of wheelchair systems capable of doorway traversal include [8, 9, 17]. For navigation, Levine et al. [8] and Yanco [17] both utilized an array of sonar sensors and Parikh et al. [9] used a planar laser scanner. While these works yielded successful demonstrations, the limitations of the sensors were not necessarily suitable for use in cluttered environments. For example, depending on sensor placement, these approaches might be susceptible to navigating through a table because the table legs could be detected but not the table top.

The work most similar to our own is Derry and Argall [3], where the goal was to detect open doorways suitable for wheelchair traversal. Their approach involved processing point cloud data to fit planar models under the assumption that gaps in the planar model correspond to doorways if they meet certain geometric criteria. A key difference in approaches is that while their focus was in processing point clouds, our algorithms emphasize processing the depth images directly. Furthermore, their investigation was limited to doorway detection. In contrast, CoPilot provides a complete solution for automated doorway navigation.

3 Development Platform

The development platform used in this research was based on the Quantum Q6 Edge electric powered wheelchair (EPW) shown in Fig. 1. The Q6 features motors with integrated encoders for measuring wheel velocities. To access these for odometry purposes, we interfaced an on-board embedded computer with the EPW's motor controller over the CAN bus. It also enabled the regulation of the EPW's linear and angular velocities via a software-based PID.

Exteroceptive sensing was from two Primesense Carmine 1.09 sensors. The Carmine 1.09 is the shorter range version of the Primesense structured lighting sensor. It has an advertised effective range between 35–140 cm (compared to 80–350 cm for the standard range Carmine 1.08). The decision to use the short range variant was to ensure that doorways and obstacles remained visible in close proximity to the chair. However, the maximum range of 1.4 m was extremely limiting. We addressed this through an intrinsic calibration procedure which extended the effective range to approximately 3 m with little degradation in accuracy. This is discussed in detail in Sect. 4.1. Two sensors were used in order to increase the total field of view. This

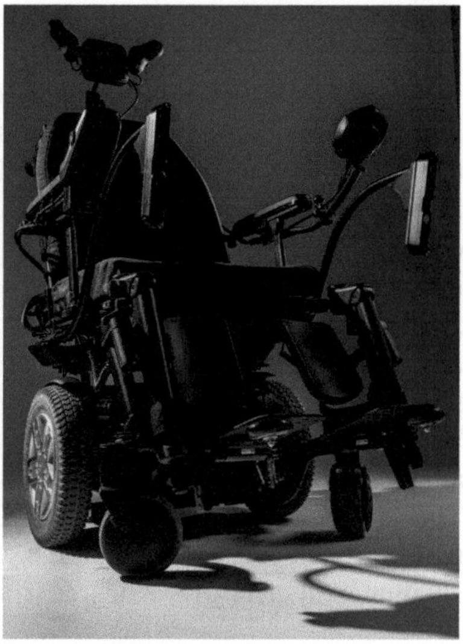

Fig. 1 CoPilot integrated into an Quantum Q6 Edge EPW

ensured better coverage of the chair footprint (to avoid collisions with obstacles), as well as facilitated doorway detection at a range of chair orientations. The mounting positions of the sensors are depicted in Fig. 1. Note that the sensors are mounted vertically rather than horizontally as this was found to be a less obstructive configuration.

The software was developed using the Robot Operating System (ROS) [10] framework and modularized based on ROS' message passing paradigm. For basic image processing and point cloud manipulation, we leveraged the OpenCV [2] and Point Cloud Library (PCL) [12] projects respectively. Processing was via a separate onboard computer with a 2.2 GHz Intel Core i7 processor and 8 GiB of RAM.

4 CoPilot Perception

4.1 Intrinsic Sensor Calibration

As alluded to in Sect. 3, the maximum advertised range of the Carmine 1.09 (1.4 m) was insufficient for effective doorway detection. While objects at depths farther than 1.4 m could still be detected, the triangulation based nature of structured light sensors induces a nonlinear noise model of the form $|\delta z| \propto z^2|\delta d|$, where δz is the error in the depth observation, z is the actual depth, and δd is the error in disparity. In other words, errors grow quadratically with depth. This can be mitigated by using an appropriate error model and adjusting the depth measurements accordingly. Unfortunately, global distortion models used for traditional camera calibration are of limited use as sensors based on the Primesense appear to have irregular distortion patterns unique to each individual sensor [15]. While they propose an unsupervised procedure to intrinsic calibration in [15], we use an alternate approach that while supervised, is fast to use and significantly less complex to implement.

Starting at the minimum effective range of the sensor, the user captures a depth image of a nominally flat wall. The sensor is then moved incrementally farther from the wall, and a new image is captured out to the maximum sensor range. For example, if the minimum and maximum ranges of interest were 0.5 m and 3.0 m respectively, depth images would be captured at nominal depths of $\mathbf{z} = [0.5, 1.0, 1.5, 2.0, 2.5, 3.0]$ meters. Note that the exact spacing is not critical. However, the accuracy of the depths $\mathbf{z}$ *is* the basis for the calibration, and must be measured accurately. This can be readily accomplished using standard tools (e.g., a tape measure or laser distance measurer). It is also important that the sensor's optical axis be roughly normal to the wall surface. To ensure this, we developed an application that provides visual feedback of the alignment error between the sensor's optical axis $\mathbf{o}$ and the wall's surface normal $\mathbf{n}_w$. This is estimated by using RANSAC [4] to automatically segment the wall plane in real-time. The user then adjusts the sensor orientation until $||\mathbf{o} \times \mathbf{n}_w|| \approx 0$. In practice, an alignment error of ≤ 1 degree is adequate for calibration, and easily obtained.

Given a set of k point cloud images $\mathbf{P} = [P_1, \ldots, P_k]$ and corresponding ground truth depth measurements $\mathbf{z}$, the remainder of the calibration process is completely automated. For each $P \in \mathbf{P}$, we recover the parameters for the respective wall planes $\Pi = [\Pi_1, \ldots, \Pi_k]$ where the relative orientation is again estimated using RANSAC and the translation using the depth measurements $\mathbf{z}$. Given robust estimates of the actual wall's relative positions and orientations Π, the point clouds $\mathbf{P}$ are adjusted to ensure that each point $p_i(i, j) \in P_i$ lies on its respective plane Π_i. This is accomplished by generating a set of scaling coefficients $\mathbf{K} = K_1, \ldots, K_k$ for each point of each point cloud. We denote the corrected point cloud set as $\mathbf{P}^*$.

The scaling coefficients $\mathbf{K}$ are to this point limited to the discrete set of ranges $\mathbf{z}$ where calibration data were collected. These are generalized to continuous space by modeling the scaling coefficients as a quadratic function of scene depth, i.e.,

$$K(i, j, z) = A(i, j)z^2 + B(i, j)z + C(i, j) \tag{1}$$

where (i, j) are the pixel coordinates of the point cloud. Thus, every sensor pixel has it's own specific quadratic function $k(i, j, z)$ that is used to determine the scaling factor at a given depth z. The quadratic coefficients $[A(i, j), B(i, j), C(i, j)]$ for each pixel (i, j) are recovered as a least squares solution minimizing the residuals between $\mathbf{P}$ and $\mathbf{P}^*$. The coefficients are calculated offline, and stored in three Look Up Tables (LUTs) A, B, C corresponding to the respective quadratic coefficients.

A point cloud P of $m \times n$ points can be described through its Euclidean coordinates $X, Y, Z \in \mathbb{R}^{m \times n}$ where each matrix entry corresponds to the x, y, z coordinates of the respective point. To calculate the corrected points, the following operations are performed on the streaming point cloud:

$$\begin{aligned}
K(i, j) &= A(i, j) * Z(i, j)^2 + B(i, j) * Z(i, j) + C(i, j) \quad \forall\ (i, j)\\
X^*(i, j) &= K(i, j) * X(i, j)\\
Y^*(i, j) &= K(i, j) * Y(i, j)\\
Z^*(i, j) &= K(i, j) * Z(i, j)
\end{aligned}$$

where X^*, Y^*, Z^* denote the corrected point set. Thus, online intrinsic calibration can be performed at a cost of only several floating point operations and array look ups per point.

We have used the calibration procedure extensively over the past year, and performance has been very good. A sample calibration run is shown at Fig. 2. The left sub-figure shows a point cloud before (top in red) and after (bottom in blue) calibration. Qualitatively, we see that both the distortion and dispersion of the points were significantly reduced. This is also reflected quantitatively in the center-right sub-figures, which show the mean error and mean standard deviation of the points versus scene depth (pre-calibration and post-calibration). The reductions in both error and variance were significant, clearly demonstrating the efficacy of the approach.

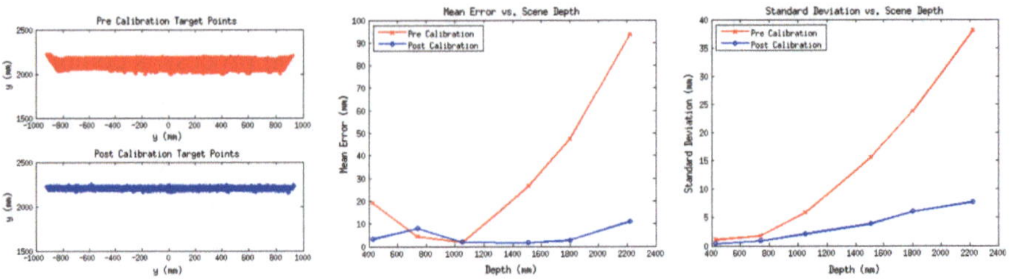

Fig. 2 *Left* Sensor points before (*top*) and after (*bottom*) intrinsic calibration. Note that both point distortions and dispersion is reduced. This is also reflected in the mean error (*center*) and standard deviation (*left*) of the points

4.2 Depth Image Warping and Fusion

Our approach to doorway segmentation relies heavily upon the observation that doorway border features are strongly vertical. We further observe that computationally, these features can be extracted most efficiently if the sensor frame is aligned vertically with the world frame, i.e., the gravity vector. An analogy would be the motivation for rectification of stereo image pairs. As a result, we warped and fused the depth image pair as a pre-processing stage.

Given two point clouds P_L, P_R associated with the left and right sensors, respectively, the first step was to warp the points to a common coordinate frame $\mathbf{F}$. We chose $\mathbf{F}$ to be centered between the actual sensor positions, and with an orientation identical to the EPW vehicle frame. Using the extrinsic calibration relating the sensor and vehicle frames, we recovered the rigid transformation between the frames and transformed the points in each point cloud

$$\hat{P}_L = {}^C R_L P_L + {}^C\mathbf{t}_L \tag{2}$$

$$\hat{P}_R = {}^C R_L P_R + {}^C\mathbf{t}_R \tag{3}$$

where $({}^C R_L, {}^C\mathbf{t}_L)$ and $({}^C R_R, {}^C\mathbf{t}_R)$ were the rigid transformations relating the left and right sensor frames to $\mathbf{F}$. Since most of our processing will be in the depth image space, we next calculated the back projection of $\hat{P}_L$, $\hat{P}_R$ to form the fused depth image I_D. In doing so, a couple of subtleties needed to be addressed. First, the back projection of points do not lie on exact pixel boundaries. As a result, we use a nearest neighbor interpolation scheme to form the depth image. Second, there was the potential that a point in both $\hat{P}_L$ and $\hat{P}_R$ would warp to the same pixel $I_D(i, j)$. In this event, the depth of the closer point was used.

The process is reflected in Fig. 3. The left-center sub-figures show the raw depth images from the left and right sensors. Note that when mounted on the EPW, the sensors were rolled approximately 90 degrees which explains the vertical orientation of the depth images. The right sub-figure shows the resulting depth image I_D after

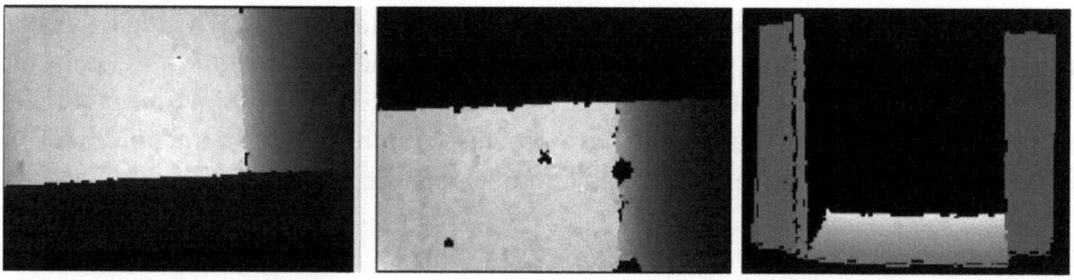

Fig. 3 *Left-Center* raw depth images of a doorway from the *left* and *right* sensors. *Right* Fused depth image

transforming and fusing the point clouds. All subsequent image and point cloud processing is done using this image as input.

4.3 Real-Time Doorway Detection

After the transformation outlined in Sect. 4.2, vertical edges in the real-world map to vertical columns in I_D. The doorway detection procedure exploits this fact to efficiently find doorway boundaries based on salient features in the depth image. We evaluated two approaches to finding doorway boundaries, a feature based approach and a histogram based approach. After a set of doorway boundaries was obtained (from either approach), they were then validated based upon geometric constraints. We now describe the process in detail.

4.3.1 Feature Based Doorway Boundary Detection

Doorways are transition features between interior and exterior space. When viewed within a depth image I_D, they appear as spatial discontinuities. This is to be expected, as there must be sufficient free space to accommodate pedestrian (or EPW) traffic across the spaces. We leveraged techniques traditionally used in 2D image processing to localize this discontinuity, and by association the doorway edges. To enhance these edges, we convolved I_D with a $[-1, 0, 1]$ kernel to generate the horizontal gradient image, and then thresholded based upon the size of the depth discontinuity to generate an edge image E_D. The next step was to identify edges of sufficient length to be classified as a doorway edge. Note that simply summing the edge pixels for each column of E_D would produce incorrect results for two reasons: (i) the edges could actually be at different depths in 3-space, possibly corresponding to multiple objects, and (ii) the resulting sum would be biased towards objects close to the sensor because they subtend more pixels.

The first problem was mitigated by calculating the median depth $\bar{z}_k$ of each column k of I_D and generating a copy of the depth image, M_D, where values in column k

are set to zero if they are not within some specified distance to $\bar{z}_k$. The idea was that true doorway edges would represent the majority of the edge length in the column, and the median value would therefore lie upon this edge. The second problem was addressed by weighting the depth measurements with the height of the unit pixel p_h subtended at the respective depth. The approach can be expressed concisely as

$$\Phi = \mathbf{1}^T (p_h \cdot E_D \odot M_D) \tag{4}$$

where $\mathbf{1}$ is a column vector of all ones, $\odot$ denotes elementwise multiplication, and Φ defines a row vector where each component corresponds to the edge height in each column. Each component in Φ was evaluated based on a minimum height requirement. The set of columns that meet the threshold were marked as potential doorway boundaries at a depth of $\bar{z}_k$.

The process is illustrated in Fig. 4. The left sub-figure shows the edge image E_D. The center image shows edge pixels overlaid on the fused RGB-D image. The right image shows edge clusters projected to the $x - y$ plane. Note that each cell represents a potential doorway boundary, so that multiple candidates can be obtained from a single doorway image. Discriminating the correct edge (e.g., the front doorway edge vs. the rear) will be discussed in Sect. 4.3.3.

We quickly determined that by themselves, doorway edges were an insufficient feature for doorway detection. For example, an inward opening door may not offer a strong edge on the hinge side as the door face can provide a smooth transition into the room. As a result, we also integrated corner features into our classifier. To do this, we first generated a 2D histogram $H(x, y)$ that bins points in 3-space to the ground plane. After applying the Harris operator to $H(x, y)$ [6], we identified the set of bins $\mathbf{C}$ in $H(x, y)$ that corresponded to corner features using an appropriate threshold. Marking a column as a potential doorway based on $\mathbf{C}$ required a small amount of effort since measurements from multiple columns could fall into the same bin. For each $C_k \in \mathbf{C}$, we found the data point $\mathbf{x}$ closest to the centroid of the bin and marked the associated column as a potential doorway boundary at a depth equal to the distance to $\mathbf{x}$.

Fig. 4 *Left* Edge image of doorway. *Center* Edge pixels identified in the scene. *Right* Top down view of door edge coordinates

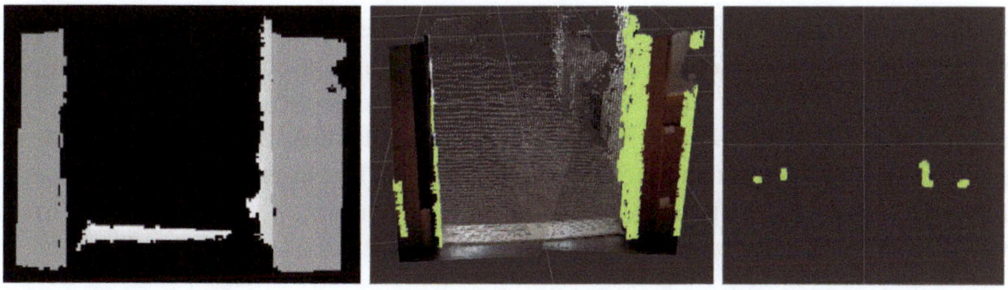

Fig. 5 *Left* Fused depth image of doorway. *Center* Corner pixels identified in the scene. *Right* Top down view of doorway corner coordinates

The corner detection process is illustrated in Fig. 5. The left sub-figure shows the fused depth image. The center image shows corner pixels overlaid on the fused RGB-D image. The right image shows valid corners projected to the $x-y$ plane.

In practice, our feature based approach was very successful at segmenting doorways. However, its computational complexity—dominated by the corner segmentation component—was of concern. This motivated our investigation into the histogram-based approach described below.

4.3.2 Occupancy Histogram Doorway Boundary Detection

Our feature-based approach attempts to directly identify the doorway boundaries, leaving only a small number of candidates as input to the validation procedure outlined in Sect. 4.3.3. However, this comes at the expense of significant up-front computation. As a result, we investigated a simpler descriptor. It is based upon the observation that the segmented edge and corner features were subsets of all columns largely occupied by a vertical object. For edge and corner features, we expend significant computational resources verifying that neighboring columns in 3-space are not occupied. But what if we simply identified each column that had a high occupancy rate as a potential doorway boundary? Undoubtedly this would lead to a much larger number of candidates for validation, but in practice the computational savings in image and point cloud processing more than makes up for this expense.

In effect, the depth image was reduced to a 1-D occupancy histogram. To accomplish this, we simplified the approach summarized in (4) to

$$\Phi = \mathbf{1}^T (p_h \cdot M_D) \tag{5}$$

which yielded a row vector where each component was the height of the object in each column corresponding to the median value $\bar{z}_k$. In other words, where in (4) we accumulated edge lengths, in (5) we are accumulating object height. Φ is now a 1D histogram of heights per bin where each bin corresponds to a column in M_D. Thresholding each component of Φ on a minimum height requirement segments every column that corresponds to a large vertical object.

When combined with the validation procedure in Sect. 4.3.3, this approach worked surprisingly well in practice. Compared to the feature-based approach, the implementation is far simpler as neither edge nor corner detection is required. It is also more efficient computationally. With a Primesense at VGA resolution (640 × 480), the feature based approach detected doorways at 12 Hz on the computer in Sect. 3. By comparison, the histogram approach ran at frame-rate (30 Hz). In the current version of CoPilot, the occupancy histogram approach is used exclusively.

4.3.3 Doorway Validation

Given the columns marked as candidate doorway boundaries and the associated depth depth values, the role of the doorway validation procedure is to find the best estimate of the relative position and orientation of the doorway. Algorithm 1 VALIDATE-DOORS outlines the procedure of reducing the set of doorway boundaries to a set of doorway candidates **D**. Each pair of doorway boundaries must meet geometric constraints based on the width of the doorway (line 5), the orientation of the EPW to the doorway (line 5) and the amount of free space beyond the sill of the doorway (lines 10–14). Guided by the American's with Disability Act (ADA) [1] accessibility guidelines, minimum and maximum doorway widths were set to 82 cm and 162 cm, respectively. The orientation constraint was set to ±45° and the free space beyond the doorway had to be sufficient to accommodate the EPW footprint. Doorway width and orientation validation are performed by the GEOMETRIC-VALIDATION sub-procedure.

Computationally, the most expensive part of VALIDATE-DOORS is the INTERSECT sub-procedure which verifies that sufficient free space exists beyond the candidate doorway via ray-tracing. In theory, there could be $O(n^2k)$ calls, where n is the number of columns and k is the number of free space tracing operations per doorway boundary pair. In practice, this will not happen due to constraints on doorway width, sensor field-of-view, and wheelchair orientation. When benchmarked with a single Primesense at VGA resolution, VALIDATE-DOORS had a mean run time of approximately 3 ms with a standard deviation of approximately 1 ms.

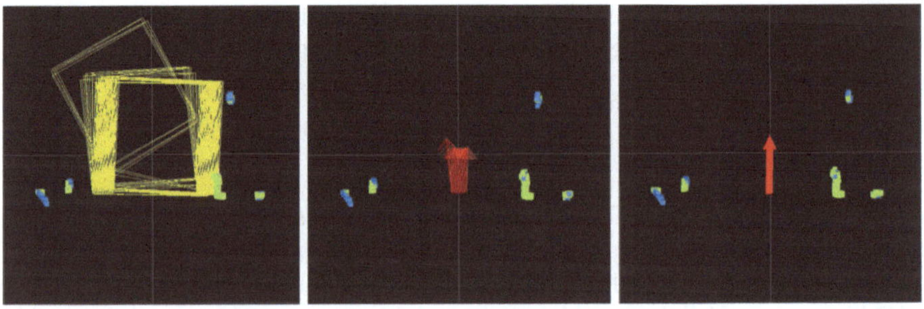

Fig. 6 *Left* Free space check for feature pairs. *Center* The set of valid doorway features. *Right* The final doorway chosen using the "nearest doorway" doorway heuristic

Algorithm 1 Door Validation

```
procedure VALIDATE-DOORS(O, B)        ▷ O: obstacle coordinates, B : boundary coordinates
    D ← ∅                                                        ▷ set of valid doorways
    for i ← 0 to n − 1 do
        for j ← i + 1 to n do
            if GEOMETRIC − VALIDATION(B[i], B[j]) then
                continue
            end if
            is_valid ← true
            for k ← i to j do                                       ▷ trace free space
                p ← INTERSECT(B[i], B[j], k)          ▷ line segment intersection point
                if ||O[k]|| < ||p|| then
                    is_valid ← false
                end if
            end for
            if is_valid = true then
                D ∪ {[x, y, θ]^T}                                   ▷ add doorway pose
            end if
        end for
    end for
end procedure
Note: The loops on B continue early when the column has no associated doorway boundary.
```

The doorway validation procedure returns the set of valid doorways **D** with the relative position of the doorway's center and its orientation. Note there is high probability that the classifier will return multiple doorway candidates. However, these will typically be variants of the actual doorway opening (e.g., front edge to rear edge, front corner to rear edge, door stop to front corner, etc.). To ensure consistent position and orientation estimates, we wish to identify only the front edges/corners of the doorway. To this end, we use a heuristic of choosing the closest doorway candidate. In practice, this has worked quite well for detecting the actual doorway.

The process is illustrated in Fig. 6. The center sub-figure shows the valid doorway candidates (red arrows), and the right sub-figure the chosen doorway. The latter well approximates the doorway position and orientation.

5 Autonomous Doorway Navigation

At the user level, the CoPilot interface is very intuitive. The user switches the EPW controller drive mode to "CoPilot" and manually drives towards the door. As soon as CoPilot detects the doorway, an icon appears on the LCD control panel. The user then pushes a single button to effect doorway traversal. Note also that the user can also steal back control from CoPilot at any time by simply touching the joystick.

At the software level, doorway navigation is decomposed into two primary subtasks: mapping the environment, and given such a map perform real-time planning and control of the EPW for safe and reliable doorway traversal.

5.1 Mapping

The local map was a 2D occupancy grid centered at the current EPW pose. We leveraged ROS for populating and clearing cells in the local map through ray tracing techniques [10]. For navigation purposes, 3D points were projected down to a 2D costmap M where the individual cells were categorized as either occupied (i.e., obstacles), free, or unknown. Each cell $M(x, y)$ was also assigned a cost $C(x, y)$ based on its proximity to obstacle cells taking into account the vehicle footprint. If the EPW were to occupy a cell (x, y), and any portion of its footprint would overlap with an obstacle cell, $C(x, y)$ would be assigned a lethal cost making it untraversable by the local planner. Otherwise, obstacles were modeled by exponential functions. The resulting costmap was then input to the local planner for trajectory planning. In our implementation, map updates were done asynchronously whenever a scan from either of the sensors was available, with an objective feedback rate of 15 Hz.

Figure 7 shows a top-down view of the costmap for the EPW staring at a doorway. The doorway edges are inflated by the potential function to define traversable regions in the costmap. Cyan colored cells correspond to regions with lethal cost, while the transition region from red to dark blue is traversable with decaying cost.

5.2 Planning

The global planner for doorway navigation is very intuitive. Given a doorway position and orientation, the it constructs an objective path down the doorway centerline with

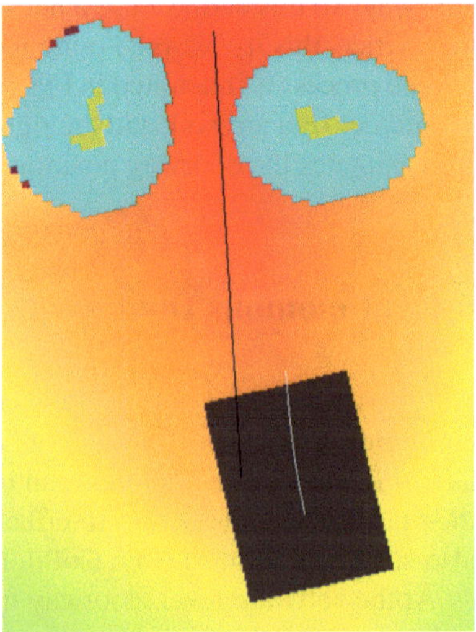

Fig. 7 Costmap $C(x, y)$ of the EPW at a doorway. The *black line* denotes the desired path

the same orientation as the doorway itself. A goal pose $G = [x_g, y_g, \theta_g]$ is then placed on this path the length of the EPW through the doorway.

For local planning, we employed a traditional sample based approach on the input space of the linear and angular control velocities (v, ω) [7]. The range of velocities sampled was $v \in [0.1, 0.4]$ m/s, and $\omega \in [-0.3, 0.3]$ rad/s. Each sampled trajectory T_i was then evaluated against a cost function

$$C(T_i, M) = C_{obst} + C_{goal} + C_{path} \quad (6)$$

C_{obst} was the maximum obstacle cost of any cell along the specified trajectory. If $C_{obst} > C_{max}$, the obstacle cost was considered fatal and the associated trajectory discarded. C_{goal} was proportional to the distance from the current EPW position to G. Similarly, C_{path} was proportional to the to the distance from the EPW position to the path derived from the doorway's centerline. The optimal trajectory $T^* = \arg\min C(T, M)$ was then selected, and the associated velocity command $(v^*, \omega^*) \in T^*$ was issued to the CoPilot motor controller.

6 Experiments

The doorway navigation behavior for CoPilot is extremely effective. ADA compliant doorways can be navigated with near 100 % reliability. The mapping capability also allows CoPilot to identify both static and dynamic obstacles in the environment, and react to these accordingly (i.e., by avoiding the obstacle or stopping when necessary). As additional anecdotal evidence of its performance, CoPilot was recently demonstrated at the headquarters of a major EPW manufacturer. The system was fully integrated into an EPW with a user-friendly interface. When placed in CoPilot driving mode, an icon would appear on the EPW's control display when a doorway was detected. The user then simply pressed a button to initiate door traversal. Although no data was collected during the demonstration, the system was tested by numerous company representatives across a large population of doors. CoPilot successfully traversed every door that the participants attempted.

To support this paper, a more formal experiment was conducted over the course of several days at various locations around the Lehigh University campus. During this time, the EPW was operated in a natural fashion with no attempt to specifically align the wheelchair into a favorable pose. A total of 100 traversals of 100 unique doorway instances were attempted. All were successful. Figure 8 depicts a sample of the doorways that were traversed. Note that CoPilot was even successful navigating through doorways where structured lighting systems might be expected to struggle, e.g., doorways with glass doors. Figure 9 shows the variety of starting EPW poses and a probability mass function of the door widths. Note also that the large majority of the doorways were at the lower range of ADA compliant doorway widths.

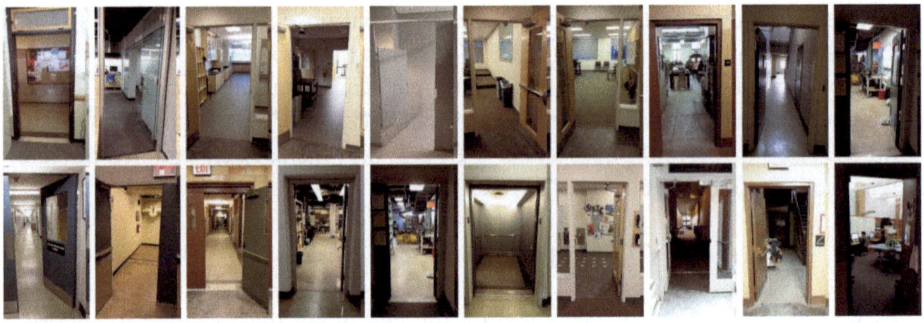

Fig. 8 Examples of the variety of doors successfully traversed

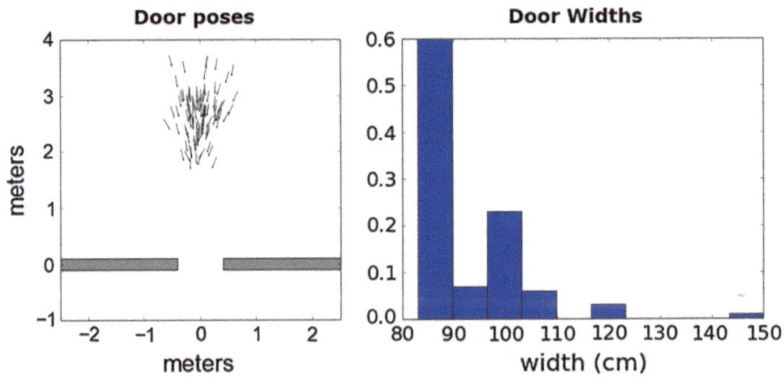

Fig. 9 (*Left*) visualization of EPW starting poses with respect to a doorway centered at the origin with an orientation of −90° and (*right*) the probability mass function of the traversed door widths

In terms of "failure modes," the doorway detection system used in CoPilot is susceptible to false positives in that clustered vertical objects meeting the geometry constraints could be interpreted as doorways. For example, two tall file cabinets with a sufficient opening in between would be segmented as a doorway. However, while some may consider this a false positive, others might consider it a feature as it generalizes CoPilot to traversing a larger range of narrow openings. We should emphasize that since migrating to the occupancy histogram approach to doorway segmentation, no false positives have been observed when attempting an actual doorway traversal.

Finally, videos demonstrating the use of CoPilot can be found at http://loveparkrobotics.com/?p=993 and http://loveparkrobotics.com/?p=997. The latter shows CoPilot integrated with a head array controller, an input device not well suited for the doorway navigation tasks. With the EPW in CoPilot mode and the doorway detected (i.e., when it puts the icon on the screen), a momentary tap of the rear switch embedded in the head-array will signal CoPilot to initiate door traversal. Just as with the Joystick mode of operation, the user can steal back control at any time by pushing the head-array switches.

7 Conclusion

In this paper we introduced CoPilot, an active driving aid that enables semi-autonomous, cooperative navigation of an EPW for automated doorway detection and traversal. The system was fully integrated into a Quantum Q6 Edge EPW using both joystick and head array controls. For doorway detection, we investigated both feature and histogram based approaches. The latter exhibited at least as good performance with significantly lower computational burden. Coupled with a sample-based planner, CoPilot demonstrated near 100 % reliability in detecting and traversing a large population of doorways when employed by a range of users. We are currently investigating the integration of additional driving aids for CoPilot, to include active braking for real-time collision avoidance and corridor following.

References

1. American's with disability act standards for accessible design. http://www.ada.gov/2010ADAstandards (2010)
2. Bradski, G.: The OpenCV library. Dr. Dobb's J. Softw. Tools (2000)
3. Derry, M., Argall, B.: Automated doorway detection for assistive shared-control wheelchairs. In: IEEE International Conference on Robotics and Automation (ICRA), pp. 1254–1259. IEEE (2013)
4. Fischler, M., Bolles, R.: Random sample consensus: a paradigm for model fitting with applications to image analysis and automated cartography. In: Communications of the ACM (1981)
5. Gupta, S., Arbelaez, P., Malik, J.: Perceptual organization and recognition of indoor scenes from rgb-d images. In: IEEE Conference on Computer Vision and Pattern Recognition (CVPR), pp. 564–571. IEEE (2013)
6. Harris, C., Stephens, M.: A combined corner and edge detector. In: Alvey Conference Vision Conference, vol. 15, p. 50. Manchester, UK (1988)
7. LaValle, S.M.: Planning Algorithms. Cambridge University Press (2006)
8. Levine, S.P., Bell, D.A., Jaros, L.A., Simpson, R.C., Koren, Y., Borenstein, J.: The navchair assistive wheelchair navigation system. IEEE Trans. Rehabil. Eng. **7**(4), 443–451 (1999)
9. Parikh, S.P., Rao, R., Jung, S.H., Kumar, V., Ostrowski, J.P., Taylor, C.J.: Human robot interaction and usability studies for a smart wheelchair. In: IEEE/RSJ International Conference on Intelligent Robots and Systems (IROS 2003). Proceedings, vol. 4, pp. 3206–3211. IEEE (2003)
10. Quigley, M., Conley, K., Gerkey, B.P., Faust, J., Foote, T., Leibs, J., Wheeler, R., Ng, A.Y.: Ros: an open-source robot operating system. In: ICRA Workshop on Open Source Software (2009)
11. Ren, X., Bo, L., Fox, D.: Rgb-(d) scene labeling: features and algorithms. In: IEEE Conference on Computer Vision and Pattern Recognition (CVPR), pp. 2759–2766. IEEE (2012)
12. Rusu, R.B., Cousins, S.: 3D is here: Point Cloud Library (PCL). In: IEEE International Conference on Robotics and Automation (ICRA). Shanghai, China (2011)
13. Rusu, R.B., Meeussen, W., Chitta, S., Beetz, M.: Laser-based perception for door and handle identification. In: International Conference on Advanced Robotics. ICAR 2009, pp. 1–8. IEEE (2009)
14. Simpson, R., LoPresti, E., Cooper, R.: How many people would benefit from a smart wheelchair? J. Rehabil. Res. Dev. **45**(1), 53–72 (2008)
15. Teichman, A., Miller, S., Thrun, S.: Unsupervised intrinsic calibration of depth sensors via SLAM. In: Robotics: Science and Systems (2013)

16. U.S. Department of Health and Human Services. Administration on Aging: A profile of older Americans: 2011 (2011)
17. Yanco, H.A.: Wheelesley: a robotic wheelchair system: indoor navigation and user interface. In: Assistive Technology and Artificial Intelligence, pp. 256–268. Springer (1998)

Learning a Context-Dependent Switching Strategy for Robust Visual Odometry

Kristen Holtz, Daniel Maturana and Sebastian Scherer

Abstract Many applications for robotic systems require the systems to traverse diverse, unstructured environments. State estimation with Visual Odometry (VO) in these applications is challenging because there is no single algorithm that performs well across all environments and situations. The unique trade-offs inherent to each algorithm mean different algorithms excel in different environments. We develop a method to increase robustness in state estimation by using an ensemble of VO algorithms. The method combines the estimates by dynamically switching to the best algorithm for the current context, according to a statistical model of VO estimate errors. The model is a Random Forest regressor that is trained to predict the accuracy of each algorithm as a function of different features extracted from the sensory input. We evaluate our method in a dataset of consisting of four unique environments and eight runs, totaling over 25 min of data. Our method reduces the mean translational relative pose error by 3.5 % and the angular error by 4.3 % compared to the single best odometry algorithm. Compared to the poorest performing odometry algorithm, our method reduces the mean translational error by 39.4 % and the angular error by 20.1 %.

1 Introduction

Autonomous aerial vehicles are often desired for performing tasks that are dangerous or impossible for humans. From urban search-and-rescue missions to remote exploration of nuclear disaster sites, these tasks often take UAVs to unknown environments that are challenging due to their diverse and dynamic nature. Among these

K. Holtz (✉) · D. Maturana · S. Scherer
Robotics Institute, Carnegie Mellon University, 5000 Forbes Ave.,
Pittsburgh, PA 15213, USA
e-mail: kholtz@andrew.cmu.edu; holtz.kristen@gmail.com

D. Maturana
e-mail: dmaturan@andrew.cmu.edu

S. Scherer
e-mail: basti@andrew.cmu.edu

D.S. Wettergreen and T.D. Barfoot (eds.), *Field and Service Robotics*,
Springer Tracts in Advanced Robotics 113, DOI 10.1007/978-3-319-27702-8_17

challenges is the likely inability of external communication, including limitations on the availability and reliability of GPS data. This requires all perception, processing, and decision-making to be made onboard. The unpredictability of the environment further contributes to the need for a more robust system that is capable of recovering from unanticipated faults. The infeasibility of considering all possible exception and errors beforehand has led to research in fault-tolerant control (FTC) and fault-tolerant perception [16].

Fault-tolerant perception can pose an especially difficult problem due to the vast diversity of environments that occur in the real world. This diversity means that for many tasks, a single method is rarely the best in all situations; instead, different methods excel in different kinds of environments. This phenomenon was shown in the task of Visual Odometry (VO) by Fang and Scherer [5], who compared the performance of different VO systems using RGB and depth data. They found that VO systems that used both kinds of information performed better, on average, than systems using only depth information. However, in dark or smoky environments, the depth-only systems would fail significantly less often than the other systems.

This motivates the main contribution of this paper, a practical and flexible framework for fault-tolerant state estimation. The main idea of our framework is to use an ensemble of algorithms, and dynamically switch between them as the vehicle moves between environments. Switching is performed by periodically selecting the algorithm expected to be the most accurate in the current context, according to a statistical model of the accuracy of each algorithm. The model is trained to predict the accuracy of the motion estimates of each algorithm as a function of various features describing different aspects of the environment and vehicle state.

We evaluate our framework in a benchmark with data from two sensors covering four challenging, real-world environments. As further contributions, this evaluation empirically shows that:

1. It is possible to improve on the performance (in terms of accuracy and robustness) of the single best algorithm in an ensemble by dynamically switching between algorithms.
2. Algorithm performance is correlated with observable characteristics of the environment and vehicle state, so it is possible to predict which algorithm will perform best in each context from the sensory input.
3. Our proposed switching strategy results in more accurate and robust estimates than any of the individual VO algorithms.

2 Related Work

Several state estimation methods can report some form of variance or confidence estimate together with their state estimate, e.g. Censi's method for ICP [4]. These variances are often used for soft fusion, e.g. in a Kalman Filter [10]. They can also be used for fusion. For example, Tomic et al. [14] use the self-reported variances of a stereo odometry algorithm and laser odometry algorithm to switch between them as the vehicle navigated between indoor and outdoor estimates. Compared to our approach, these methods have the advantage of not requiring any extra training step; however, this comes at a cost in terms of flexibility. These methods cannot take advantage of extra information our method can incorporate seamlessly. Additionally, our method does not require running an algorithm to predict its performance; this can provide significant computational benefits by turning the unused odometry algorithms "off". Finally, only some specific methods can self-report their variance, while our method is applicable to any algorithm, whether or not it has this capability.

Leishman et al. [12] propose a system that dynamically switches between different odometry methods based on context-dependent variables. The switching is based on an ad-hoc strategy based on manually selected quality thresholds for each modality. In contrast, our method uses machine learning to learn this strategy, which considerably simplifies adapting the method to different environments, sensors or algorithms.

An algorithm that has several similarities to ours is CELLO [15]. This method predicts a covariance matrix for each method as a function of past training data. The covariance matrix is predicted with a nonparametric estimator similar to nearest neighbors methods. Compared to CELLO, our method makes various choices that make it simpler and more practical. Instead of predicting a full covariance matrix, we predict error magnitudes, which are simpler to learn and sufficient in many cases. We choose a random forest-based regressor, which is faster than and more robust than nearest neighbor methods [3]. In addition, our evaluation is more exhaustive, as it has different and more challenging environments.

3 Approach

The system implementing our proposed method can be decomposed into various components:

Sensor Suite In our framework sensors serve two purposes: to serve as input for the VO estimates and to capture characteristics of the environment that will allow the model to predict which algorithm will be the most reliable in any given context.

Algorithm Ensemble Our method requires a set of base algorithms performing the same task (VO, in this paper). The algorithms should be diverse in terms of their performance characteristics across different environments.

Features In order to describe aspects of the environment that are potentially relevant for predicting accuracy, we extract various features from the sensor data and estimated vehicle state.

Error Prediction Model Using data annotated with ground truth, we train a statistical regression model to predict the accuracy of each algorithm in the ensemble from the extracted features.

Switching Planner At each time step, the switching planner selects which algorithm to run based on the predicted accuracy of each method and potentially other factors, such as computational cost.

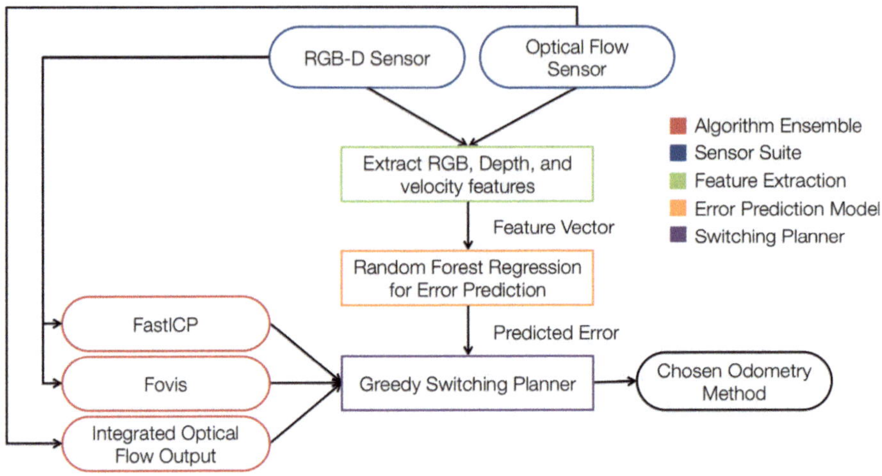

Fig. 1 Framework Outline—The adaptive architecture framework allows the robotic system to switch between different visual odometry methods. To choose whereto and when to switch between methods, we predict the error associated with each method given a feature vector extracted from current sensory information and state

We note that there is considerable flexibility in regards to the concrete implementation of each component. The concrete choices for the VO system proposed in this paper are outlined in Fig. 1. While this system worked well in our experiments, this framework easily accommodates variations for each component.

Below we describe selected components in more detail.

3.1 Sensor Suite

In the experiments for this work we use two different sensors. The first is a forward-facing RGB-D sensor, which combines a visible light camera with an active structured light system to create registered RGB and Depth (RGB-D) images.

The second is a specialized camera for optical flow [7], which faces downwards. The camera has an attached ultrasonic range sensor to estimate height relative to the ground.

Between these two sensors we have four channels of information: RGB, depth, ground optical flow and height relative to the ground. Each of these channels provide informative and complementary cues about the environment and vehicle state.

3.2 Algorithm Ensemble

A diverse yet powerful set of algorithms is crucial to maximizing the robustness and accuracy of our system as a whole. For this work we selected three representative VO methods, each using different subsets of the data:

Fovis [8] This VO algorithm uses the RGB and depth data from the RGB-D sensor. It works by detecting sparse keypoints from the RGB image and their 3D positions relative to the camera using the depth image. Then motion is estimated by robust matching of keypoints across frames using appearance information and geometric constraints.

FastICP [1] This method relies solely on depth data. It converts each depth image to a point cloud and estimates motion between frames by registering the point clouds to a local surface map using point-to-plane Iterative Closest Points (ICP).

Optical Flow [7] This method uses optical flow and height measurements to estimate motion. Unlike the other two methods, this method assumes the vehicle maintains constant height in each motion.

3.3 Feature Extraction

For each of our sensor modalities we extract multiple features designed to summarize various potentially relevant characteristics of the environment. We chose these features as they are compact, efficient to calculate and commonly used in the computer vision and point cloud processing literature.

Below we describe each feature according to the type of sensor data it describes; see Table 1 for a summary. The number corresponding to each feature will also appear in parenthesis with the description of that feature in the following text.

Table 1 Image features in the feature vector will be referred to according to the numbers in this table

RGB image	Depth and point cloud	State
1. Contrast[a]	10. Contrast[a]	18. Translational velocity
2. Harris Corners	11. Correlation[a]	19. Angular velocity
3. Shi Tomasi Corners	12. Valid depth ratio	
4. Correlation[a]	13. Energy[a]	
5. Edge Ratio	14. Homogeneity[a]	
6. Energy[a]	15. Linear-ness	
7. Entropy	16. Scattered-ness	
8. Homogeneity[a]	17. Surface-ness	
9. Mean intensity		

[a]Indicates a GLCM statistic

3.3.1 RGB Image Features

Note that as we do not expect color to be a discriminative feature in this context, we convert images to grayscale before further processing.

Luminance (9) We expect the luminance of the environment to be a useful predictor of algorithm accuracy, as methods that rely on RGB information will likely fail in dark environments. We use the Mean Intensity (9) of the grayscale image as a simple estimate of luminance.

Corners (2, 3) Keypoint-based algorithms such as Fovis rely on the presence of corner-like features in the environment; therefore, the quantity of these features may be a good predictor of the success of these algorithms. While we could use the results of Fovis' own corner detection step, this would entail running Fovis itself, and one of our goal's objectives is to reduce computation by only running algorithms we will use. Instead we simply run two corner detector algorithms from OpenCV, the Harris (2) and Shi-Tomasi (3) detectors and include the number of corners from each as in the feature vector.

Edges (5) The presence of strong intensity edges is correlated with certain kinds of environments; for example, a cluttered indoor scene will probably have a larger number of edge pixels than an empty hallway. Hence we include the number of Sobel Edge (5) pixels in each image in the feature vector.

Texture (1, 4, 6–8) The presence of salient intensity textures in an environment may aid in the extraction and tracking of keypoints, and is also a strong cue to distinguish physical environments (for example, outdoors and indoor scenes have very different textures). To succinctly describe image texture we include the entropy (7) of each image, calculated using a histogram of the 256 possible intensity values, and features extracted from the gray-level co-occurrence matrix (GLCM) [6] of the image over four different angles (0, 45, 90, and 135°). The GLCM counts how often every possible combination of gray-level pixel values occurs next to each other, and statistics of this matrix are popular texture features. The statistics we use are contrast (1), correlation (4), energy (6) and homogeneity (8).

3.3.2 Depth Features

Valid depth ratio (12) Our RGB-D sensor reports depth as a 16-bit image in which pixels are set to a special value if depth estimation is unsuccessful, deeming them invalid. If a large amount of the depth image is invalid, likely meaning it was out of range of the depth sensor, then any method using depth information may not be reliable. To quantify this we include the Valid Depth Ratio (12) of each depth image was computed to estimate the amount of information in that image.

Saliency (15–17) The three-dimensional shape of the environment may be a useful predictor of algorithm performance. For example, it might distinguish between environments that are underconstrained in depth information—particularly long, clear corridors—and environments that have several interesting depth features to track. To capture this we compute global saliency features [11], a coarse but efficient measure of shape. These features operate on point clouds, so we first project the pixels with valid depth to a 3D point cloud $\{X_i\} = \{(x_i, y_i, z_i)^\top\}_{i=1}^N$. A 3×3 covariance matrix $\sum_{i=1}^N (X_i - \bar{X})(X_i - \bar{X})^\top / N$ is computed, $\lambda_0, \lambda_1, \lambda_2$ are extracted, such that $\lambda_0 \geq \lambda_1 \geq \lambda_2$. The three saliency features of the pointcloud are the scattered-ness $f_{\text{scatter}} = \lambda_2$ (16), the linear-ness $f_{\text{linear}} = \lambda_0 - \lambda_1$ (15), and the surface-ness $f_{\text{surface}} = \lambda_1 - \lambda_2$ (17).

Depth Texture (10, 11, 13, 14) GLCM statistics, as described for the RGB texture features, were also extracted from the depth image.

3.3.3 State Features

The velocity, as reported by the visual odometry algorithms, was also included in the feature vector. Specifically, the magnitudes of both the translational (18) and angular (19) velocities of the currently active odometry method were used as features. This was added as a possible predictor for motion blur, which could affect the performance of visual odometry algorithms.

3.4 Error Prediction Model

The task of the error prediction model is to predict the trajectory errors of each algorithm at each time step given the feature vector described in the last section.

We formulate the problem as a regression problem, for which various methods may be used. Below we describe our chosen methodology.

Error Evaluation The output of our algorithm is an prediction of the trajectory errors each algorithm will make. We chose a metric based on the relative pose error (RPE) at a given time, described by [13] for VO evaluation. RPE measures the local accuracy of a trajectory, as compared to a ground truth trajectory, over a specified time interval. This measures how far the trajectory drifts in the given time interval.

$$\mathbf{E}_i := \left(\mathbf{Q}_i^{-1}\mathbf{Q}_{i+\Delta}\right)^{-1}\left(\mathbf{P}_i^{-1}\mathbf{P}_{i+\Delta}\right) \tag{1}$$

Equation (1) calculates RPE at time step i, over the time interval Δ. Here $\mathbf{Q}$ refers to the ground truth path and $\mathbf{P}$ refers to the estimated trajectory. In all experiments in this paper we use a time interval of $\Delta t = 2$ s.

As predicting structured matrices in nontrivial, instead we choose to predict two scalar quantities: the *translational error*, extracted from (1) as the Euclidean norm of the translational portion of $\mathbf{E}_i$, and the *angular error*, extracted as the absolute value of the angle of rotation from the rotation matrix portion of $\mathbf{E}_i$.

Error Prediction To predict the trajectory errors given the feature vector, we proceed as follows. First, we learn an independent regression model for each method and for each type of error (translational and angular). While joint prediction of the errors could potentially be more accurate, performing the predictions independently allows us to use virtually any off-the-shelf regression algorithm. Another advantage of using several different regressors instead of a joint regressor is that VO methods can easily be added to the algorithm ensemble without affecting regression models that have already been learned.

Regression was used instead of the potentially simpler classification. One advantage of regression over classification is that we are able to determine the magnitude of an error before switching away from a method. Regression commits less strongly to which method might be the least accurate at any time. Regression gives us more nuanced information that allows more informed decisions. For example, if two methods are performing well, classification may indicate to frequently switch between the two. With regression, we may be able to determine that the cost of switching is not worth the slight decrease in error.

For the regression model we choose Random Forests (RF) [2]. The RF algorithm is an ensemble learning method that contains many decision trees, each contributing a vote towards an answer. We chose to use RF compared to other regression algorithms because random forests are efficient, robust, and empirically among the most accurate algorithms in many problems [3]. They are also able to predict feature importance.

3.5 Switching Planner

Decision Method We obtain both translational and angular error from evaluating the RPE as above. We considered two methods of combining angular and translational error to decide which visual odometry method to use. These two methods are shown in (2) and (3). Here ε_i^τ is the translational error for method i, and ε_i^α is the angular error for method i.

$$\text{method} = \arg\min_i(\varepsilon_i^\tau \varepsilon_i^\alpha). \tag{2}$$

$$\text{method} = \arg\min_i(\beta\varepsilon_i^\tau + \varepsilon_i^\alpha). \tag{3}$$

Ultimately we decided to use the additive metric, as in Eq. 3, with $\beta = 0.5$. This showed a larger decrease in both translational and angular error than metric (2), or the metric (3) for for either $\beta = 0.25$ or $\beta = 1$.

Greedy Switching Planner We aim to improve VO estimation without greatly increasing the computational cost, as it is important for our method to run online. Therefore, it is important that we do not run multiple odometry methods in parallel, as that can be very computationally expensive. As Fovis relies on keyframes for motion estimation, and our depth-only method relies on building a map from previous point clouds, it would introduce error into the system to instantaneously switch from one method to another.

We therefore allow three image frames of overlap between the two visual odometry methods, during which both methods would be making motion estimates. We do not use the newly started visual odometry method until the overlap is completed. A more complicated switching planner may be implemented, possibly by considering the benefit of switching before committing to a switch.

4 Experiments

Datasets We test our framework in a variety of conditions that would be challenging for any individual algorithm. We looked specifically at four different environments. The *basement* datasets were taken in a dark, cluttered hallway (see Fig. 2a). This environment is particularly challenging for any algorithm relying on light-dependent RGB images or the limited optical flow information available. The *hallways* datasets were taken in an area with brightly lit, blank hallways that may be challenging for algorithms relying on depth information (see Fig. 2b). A depth cloud will be underconstrained, and therefore forward motion may be difficult to detect. The *spacious* environment included a large spacious room in which few depth features are available due to the limited range of the depth sensor (see Fig. 2c). Lastly, the *cluttered* datasets were taken in an area with many objects detectable in both depth and RGB images (see Fig. 2d).

We extract the feature vector and predict the VO error in real time, and switch between algorithms based on the predictions. For this paper, we trained RF regressions using one complete, >60 s dataset from each of the four environments and tested on the remaining four datasets, again one from each environment.

Ground Truth Because we wanted data from a variety of environments, the use of motion capture systems was infeasible. Instead, localization was performed on the datasets, matching the depth point cloud with dense 3D maps of the environment. However, due to the challenging nature of these environments, localization failed in many cases. In order to get a close estimation of ground truth, points of the path that were accurately localized were manually selected. These points were then used as landmarks for the path. The most accurate odometry method for that dataset was then smoothed using iSAM [9].

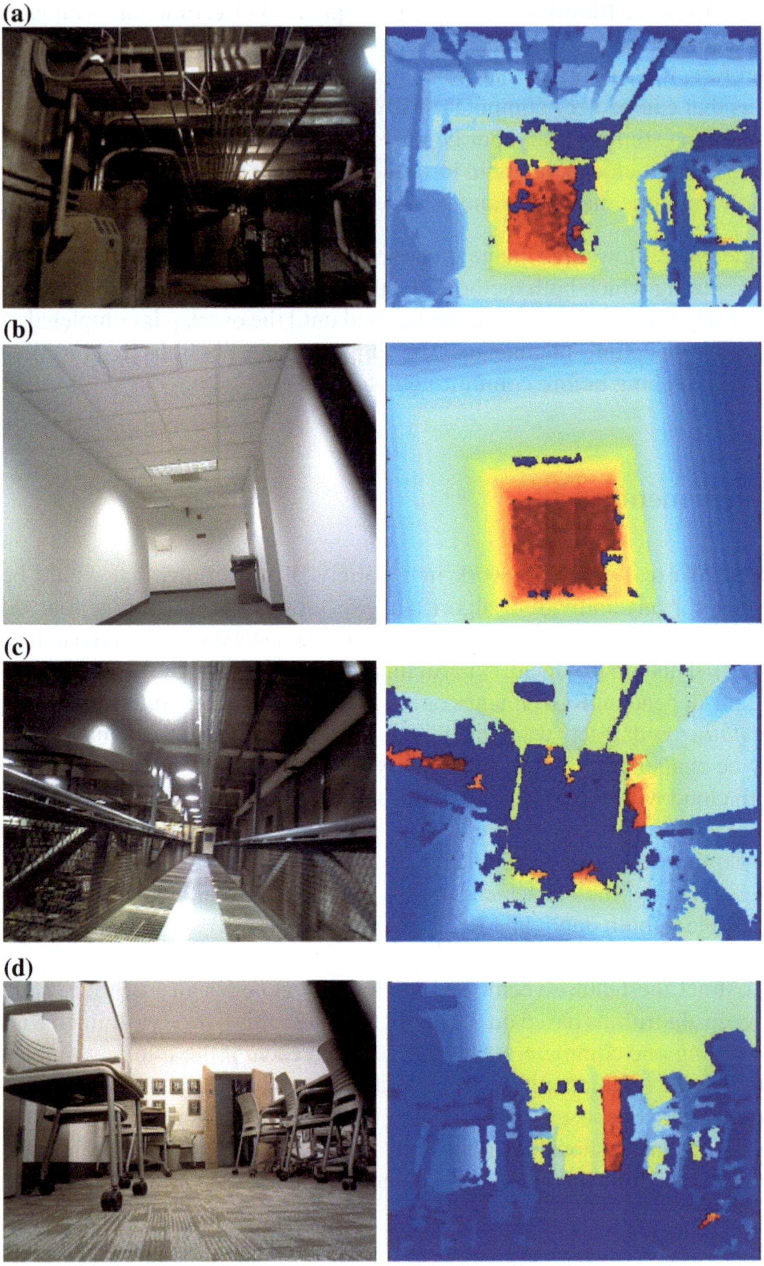

Fig. 2 Environment Examples—Four different, unique indoor environments were explored. Sample RGB and depth images are shown. **a** *Basement* Environment—RGB (*Left*) and Depth (*Right*) Images. **b** *Hallways* Environment—RGB (*Left*) and Depth (*Right*) Images. **c** *Spacious* Environment—RGB (*Left*) and Depth (*Right*) Images. **d** *Cluttered* Environment—RGB (*Left*) and Depth (*Right*) Images

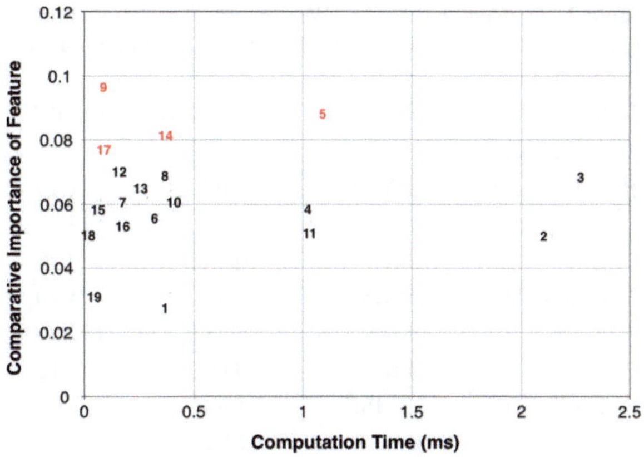

Fig. 3 Feature Importance—The importance of each feature is an estimation of how much it affects the accuracy of the random forest regression. The numbers refer to the features as numbered in Table 1. The features in *red* highlight some of the more important features

4.1 Feature Importance

The importance of each variable in the feature vector can be estimated during the training process of a random forest. We also collected information on the computation time for each feature. Computation time and variable importance were compared to determine if any features were not worthwhile to compute. In Fig. 3, the maximum importance across all six forests for each variable is plotted against the variable computation time. We see that the mean intensity of the RGB image has a higher maximum importance than the other features and relatively low computation time. It is also clear that the corner detection methods are the most time-consuming and only of moderate importance. However, the computation time remained under 3 ms for these features, which is within the limits given by the image frequency of 15 Hz. Therefore, we kept all features while moving to the next step.

4.2 Evaluation of Random Forest and Switching Performance

After training, we compared the trajectory error predicted by the random forest regression to the trajectory error previously extracted for training. We measure the effectiveness of our method by evaluating the RPE of the resultant path of switching, particularly in comparison with the lowest error odometry method.

Our results are detailed in Table 2a, b. Here the RMSE, maximum, and failure rate (FR) of the translational component of the RPE of each of the trajectories are compared. The same statistics on the angular component of the RPE is shown in Table 3a, b. Here, the ideal path is generated by making the correct selection (according to Eq. 2) at each point, using the true, extracted RPE. The switching path is generated using the architecture we have described thus far. In these Tables (2 and 3), the RMS RPE is shown as a percentage of the RMS of the ideal path's RPE.

Table 2 Translational Error—We compare translational relative pose error (RPE) of our method against raw odometry output to determine if the random forest regression can predict RPE in a useful way

Dataset	Data	Switching	FastICP	Fovis	Opt. Flow
Training data					
Basement (136 s)	RMSE	**1.09**	1.40	1.14	1.97
	Max.	**5.43**	27.38	**5.43**	18.95
	FR	**0.04**	0.16	0.05	0.44
Hallways (235 s)	RMSE	1.49	4.40	2.32	**0.93**
	Max.	25.58	172.34	90.30	**8.77**
	FR	**0.00**	0.12	**0.00**	**0.00**
Spacious (305 s)	RMSE	**1.04**	1.84	**1.04**	1.51
	Max.	**11.82**	34.67	**11.82**	18.75
	FR	**0.03**	0.30	**0.03**	0.12
Cluttered (106 s)	RMSE	**1.08**	1.63	**1.08**	2.75
	Max.	**10.95**	11.45	**10.95**	168.62
	FR	**0.01**	0.05	**0.01**	0.05
Overall	RMSE	**1.07**	1.86	1.14	1.75
	Max.	**25.58**	172.34	90.30	168.62
	FR	**0.02**	0.20	0.03	0.15
Testing data					
Basement (130 s)	RMSE	1.15	1.84	**1.13**	2.49
	Max.	24.06	24.06	**15.04**	15.46
	FR	**0.04**	0.19	**0.04**	0.59
Hallways (244 s)	RMSE	1.51	3.31	1.85	**1.10**
	Max.	15.32	34.21	15.74	**3.88**
	FR	0.02	0.10	0.03	**0.00**
Spacious (257 s)	RMSE	**1.04**	1.80	**1.04**	1.41
	Max.	9.01	39.19	9.01	**6.11**
	FR	**0.02**	0.56	**0.02**	0.27
Cluttered (93 s)	RMSE	1.14	**1.06**	1.14	1.10
	Max.	12.08	**11.93**	12.08	67.45
	FR	0.41	**0.36**	0.41	**0.36**
Overall	RMSE	**1.11**	1.76	1.13	1.55
	Max.	24.06	39.19	**15.74**	67.45
	FR	**0.07**	0.34	**0.07**	0.28

The RMS error is represented as a *fraction* of the RMS error of the *ideal* path. The Max. error is chosen as the largest % increase over the ideal path's error and is also represented as a fraction of the ideal. The failure rate (FR) is the fraction of data points for which the RPE exceeds a certain threshold. For translational error this threshold is 1 m

Table 3 Angular Error—Angular error is represented here as translational error is in Table 2

Dataset	Data	Switching	FastICP	Fovis	Opt. Flow
Training data					
Basement (136 s)	RMSE	1.53	1.51	1.77	**1.23**
	Max.	**18.35**	22.23	**18.35**	21.46
	FR	0.06	0.04	0.07	**0.01**
Hallways (235 s)	RMSE	**1.08**	1.39	1.32	**1.08**
	Max.	77.87	82.88	77.87	**10.36**
	FR	**0.00**	0.04	0.04	0.01
Spacious (305 s)	RMSE	**1.01**	1.54	**1.01**	1.62
	Max.	28.96	**20.94**	28.96	32.41
	FR	**0.01**	0.06	**0.01**	0.07
Cluttered (106 s)	RMSE	**1.09**	1.21	**1.09**	3.89
	Max.	5.89	**5.20**	5.89	60.40
	FR	**0.02**	0.03	**0.02**	0.22
Overall	RMSE	**1.13**	1.46	1.25	1.68
	Max.	77.87	82.88	77.87	**60.40**
	FR	**0.02**	0.04	0.03	0.06
Testing data					
Basement (130 s)	RMSE	1.22	**1.06**	1.30	1.13
	Max.	11.87	9.50	11.87	**9.07**
	FR	0.05	**0.01**	0.08	0.03
Hallways (244 s)	RMSE	1.15	1.23	1.19	**1.08**
	Max.	13.73	14.48	13.73	**3.26**
	FR	0.03	0.04	**0.02**	0.04
Spacious (257 s)	RMSE	1.06	1.26	**1.05**	1.33
	Max.	**11.46**	157.67	**11.46**	79.82
	FR	0.04	0.10	**0.03**	0.13
Cluttered (93 s)	RMSE	**1.01**	1.07	**1.01**	1.16
	Max.	**22.74**	29.48	**22.74**	70.82
	FR	**0.15**	0.18	**0.15**	0.18
Overall	RMSE	**1.10**	1.19	1.12	1.22
	Max.	**22.74**	157.67	**22.74**	79.82
	FR	**0.05**	0.07	**0.05**	0.09

The threshold used to calculate the failure rate (FR) was 0.5 rad

The maximum of each path is the largest percentage increase in RPE over the ideal path's RPE over all time points. The FR of each path is the fraction of data points for which RPE exceeds a given threshold. For translational error, this threshold is 1 m; for angular error it is 0.5 rad.

Our method is able to improve robustness to large faults in VO. This is demonstrated by both the maximum and failure rate metrics. The switching method almost always has the lowest rate of failure, and often avoids the largest maximums in error that correspond to large failures in the VO estimation.

Our data also shows that overall our method is able to improve accuracy by improving the RMS relative pose error. However, our method does not always outperform the best individual odometry method. Notably, our method fails to outperform optical flow odometry in translational or angular RPE for either dataset, training or testing, of the *hallways* environment. One possible explanation is that this environment was not sufficiently distinguished from others in the feature space. Future work will include analyzing the difference between these environments in the feature space and exploring potential new features that may aid distinguishing different environments.

5 Conclusions

In this paper we presented a method to robustify visual odometry by switching between algorithms based on the environment. By learning the error associated with sensory information through regression, this method aims to reduce visual odometry errors. The current results are promising in improving state estimation in various indoor environments, and particularly in avoiding large failures.

In future work, we would like to explore different methods in each component of the framework, and evaluate how they affect performance; for example, by adding more sensors and odometry algorithms, or jointly learning the features and the error prediction model.

References

1. Besl, P., McKay, N.D.: A method for registration of 3-D shapes. IEEE TPAMI **14**(2), 239–256 (1992)
2. Breiman, L.: Random forests. Mach. Learn. **45**(1), 5–32 (2001)
3. Caruana, R., Niculescu-Mizil, A.: An empirical comparison of supervised learning algorithms. In: ICML, pp. 161–168 (2006)
4. Censi, A.: An accurate closed-form estimate of ICP's covariance. In: ICRA, pp. 3167–3172 (2007)
5. Fang, Z., Scherer, S.: Experimental study of odometry estimation methods using RGB-D cameras. In: IROS, pp. 680–687 (2014)
6. Haralick, R.M., Shanmugam, K., Dinstein, I.H.: Textural features for image classification. IEEE Trans. Syst. Man Cybern. **6**, 610–621 (1973)
7. Honegger, D., Meier, L., Tanskanen, P., Pollefeys, M.: An open source and open hardware embedded metric optical flow cmos camera for indoor and outdoor applications. In: ICRA, pp. 1736–1741 (2013)
8. Huang, A.S., Bachrach, A., Henry, P., Krainin, M., Maturana, D., Fox, D., Roy, N.: Visual odometry and mapping for autonomous flight using an RGB-D camera. In: International Symposium on Robotics Research (ISRR), pp. 1–16 (2011)
9. Kaess, M., Ranganathan, A., Dellaert, F.: iSAM: incremental smoothing and mapping. IEEE Trans. Robot. **24**(6), 1365–1378 (2008)

10. Kalman, R.E.: A new approach to linear filtering and prediction problems. ASME J. Basic Eng. (1960)
11. Lalonde, J.F., Vandapel, N., Huber, D.F., Hebert, M.: Natural terrain classification using three-dimensional ladar data for ground robot mobility. J. Field Robot. **23**(10), 839–861 (2006)
12. Leishman, R.C., Koch, D.P., McLain, T.W., Beard, R.W.: Robust visual motion estimation using RGB-D cameras. In: AIAA Infotech Aerospace Conference, pp. 1–13 (2013)
13. Sturm, J., Engelhard, N., Endres, F., Burgard, W., Cremers, D.: A benchmark for the evaluation of RGB-D SLAM systems. In: IROS, pp. 573–580. IEEE (2012)
14. Tomic, T., Schmid, K., Lutz, P., Domel, A., Kassecker, M., Mair, E., Grixa, I.L., Ruess, F., Suppa, M., Burschka, D.: Toward a fully autonomous UAV: research platform for indoor and outdoor urban search and rescue. IEEE Robot. Autom. Mag. **19**(3), 46–56 (2012)
15. Vega-Brown, W., Bachrach, A., Bry, A., Kelly, J., Roy, N.: CELLO: a fast algorithm for covariance estimation. In: ICRA, pp. 3160–3167 (2013)
16. Zhang, Y., Chamseddine, A., Rabbath, C., Gordon, B., Su, C.Y., Rakheja, S., Fulford, C., Apkarian, J., Gosselin, P.: Development of advanced FDD and FTC techniques with application to an unmanned quadrotor helicopter testbed. J. Franklin Inst. **350**(9), 2396–2422 (2013)

Part III
Planetary

System Design of a Tethered Robotic Explorer (TReX) for 3D Mapping of Steep Terrain and Harsh Environments

Patrick McGarey, François Pomerleau and Timothy D. Barfoot

Abstract The use of a tether in mobile robotics provides a method to safely explore steep terrain and harsh environments considered too dangerous for humans and beyond the capability of standard ground rovers. However, there are significant challenges yet to be addressed concerning mobility while under tension, autonomous tether management, and the methods by which an environment is assessed. As an incremental step towards solving these problems, this paper outlines the design and testing of a center-pivoting tether management payload enabling a four-wheeled rover to access and map steep terrain. The chosen design permits a tether to attach and rotate passively near the rover's center-of-mass in the direction of applied tension. Prior design approaches in tethered climbing robotics are presented for comparison. Tests of our integrated payload and rover, Tethered Robotic Explorer (TReX), show full rotational freedom while under tension on steep terrain, and basic autonomy during flat-ground tether management. Extensions for steep-terrain tether management are also discussed. Lastly, a planar lidar fixed to a tether spool is used to demonstrate a 3D mapping capability during a tethered traverse. Using visual odometry to construct local point-cloud maps over short distances, a globally-aligned 3D map is reconstructed using a variant of the Iterative Closest Point (ICP) algorithm.

1 Introduction

Robotic planetary and terrestrial exploration has historically been risk-averse, favoring benign terrain in order to reduce the likelihood of mission failure [12]. Even state-of-the-art rovers deployed on Mars are not suited to access steep terrain directly, and

P. McGarey (✉) · F. Pomerleau · T.D. Barfoot
University of Toronto Institute for Aerospace Studies, Toronto, Canada
e-mail: patrick.mcgarey@robotics.utias.utoronto.ca

F. Pomerleau
e-mail: francois.pomerleau@robotics.utias.utoronto.ca

T.D. Barfoot
e-mail: tim.barfoot@utoronto.ca

D.S. Wettergreen and T.D. Barfoot (eds.), *Field and Service Robotics*,
Springer Tracts in Advanced Robotics 113, DOI 10.1007/978-3-319-27702-8_18

instead rely on remote observation [7]. As we push towards developing autonomous systems for harsh terrain, tethering (i.e., attachment of supportive electromechanical cable or climbing rope between a robot and an anchor point) will not only be necessary for safety and mobility, but also a benefit to robots requiring both assistive power and robust communication throughout challenging traverses.

Geologic exploration of steep terrain, safety inspection of walls and dams, and disaster response in resource- and communication-limited environments, are typical applications appropriate for tethered robots. Detailed point-cloud maps constructed from lidar data enable geologists to model vertical stratigraphy at high resolution [9]. Robots inspecting walls and dams allow for repeated observations of targeted areas as a means to evaluate temporal changes and reduce risk to humans [11]. Deploying robots to assist in disaster response may require that wired communication and external power sources are provided due to extended operation in resource-limited environments. The use of a tether offers a solution to power and communication requirements. However, tether management still presents a significant challenge to robots operating in dangerous conditions [8] (Fig. 1).

While prior systems have addressed these challenges with some success, it is the opinion of the authors that a lack of autonomy has attenuated continued progress in tethered mobile robotics. In order to make advancements in autonomous mobility, tether management, and environmental mapping, we have developed a new research platform, Tethered Robotic Explorer (TReX).

This paper is organized as follows. Section 2 evaluates prior tethered climbing rovers, Sect. 3 details the system design of TReX, Sect. 4 presents experimental results, Sect. 5 provides lessons learned, and Sect. 6 offers conclusions.

Fig. 1 Our Tethered Robotic Explorer (TReX) traverses the exterior of a dome structure, showing lateral motion (*green arrow*) under tension

Fig. 2 Past and present tethered climbing rovers: **a** Dante II [2], **b** TRESSA [5], **c** Axel II and **d** DuAxel [7], **e** Tetris and Moonraker [3], **f** VolcanoBot II (JPL/CalTech), and **g** vScout [13]

2 Related Work

The archetype of tethered climbing robots was Dante II (Fig. 2a), an eight-legged walking rover used to traverse the interior craters of volcanoes [15]. Dante II successfully repelled down extreme slopes and demonstrated the challenges/limitations of tethered mobility; during an ascent of a crater, Dante II was critically damaged from a fall while rotating outside the direction of applied tension.

Teamed Robots for Exploration and Science on Steep Areas (TRESSA) (Fig. 2b) was the first modular system allowing an attached flat-ground rover to access vertical terrain [14]. The off-board managed dual-tether configuration provided easy integration with different rovers and allowed some lateral motion on steep terrain. However, multiple tethers implied an increased difficulty navigating around obstacles, tether abrasion, and reduced range due to dragging (tethers were not spooled on the robot).

The most capable tethered climbing robot to date has been Axel II (Fig. 2c), a two-wheeled rover with an actuated tether caster arm [7]. Multiple Axel II rovers could be linked by a docking station to operate as either a four-wheeled rover (i.e., DuAxel), or as a redundant base station and climbing rover (Fig. 2d). Axel II's innovative configuration has been adapted by the Moonraker and Tetris robots[1] (Fig. 2e), and VolcanoBot[2] (Fig. 2f).

The vScout prototype (shown in Fig. 2g) consisted of a winch payload mounted to a Clearpath Husky A200 rover [13]. The prototype was a precursor to TReX. While the tether was not actively managed on board, vScout successfully demonstrated the maneuverability of a 50 kg commercial rover on steep slopes.

[1]Moonraker and Tetris were developed by Team Hakutu from Tohoku University.

[2]VolcanoBot II is a small rover used for mapping volcanic vents, JPL/Caltech.

Prior tethered climbing robots have shown minimal vehicle rotational range centered around the direction of applied tension on steep slopes, implying both an increased risk of entanglement with obstacles, and limitations on drivable paths. The design of TReX considered both attributes and limitations of prior designs.

3 System Design

3.1 Tethered Robotic Explorer (TReX)

In order for a tethered rover to rotate continuously under tension, a tether is connected to a freely rotating joint. When taut, the tether's tensional force is aligned with a virtual line intersecting the vehicle's center-of-mass. The on-board managed tether is wound around an actuated spool, which is mounted to a rotational joint in the center of a skid-steered Clearpath Husky A200 rover. A cut view of the TReX CAD model shown in Fig. 3 illustrates the mounting configuration of the spool on a rotating tether arm, which mechanically links to the rover using a slew bearing. The three rotating elements (rover, tether arm, and spool) are outlined in colored dashed lines. The spool, which rests on a separate turntable bearing above the tether arm, can only rotate when actuated by a motor. The motor, which is suspended in the 10 cm cavity of a rotary slip ring, is fixed to the tether arm, while its shaft is coupled to the spool. To reduce torque on the rover, a manually adjustable angled arm allows for load balancing with the vehicle's center-of-mass. The design permits the rover to rotate freely regardless of applied tension, provided there is sufficient wheel traction.

Sensor interfacing requires careful consideration of the design's rotational degrees of freedom. In order to produce three-dimensional (3D) point-clouds with a planar scanning lidar, the sensor is mounted on the rotating tether spool. This configuration allows for a single actuator (not including vehicle wheels) to be used in tether management and 3D mapping. Since the rotation of the lidar is coupled to a tether spool, 3D scanning is only possible when the vehicle is moving and the tether is actively being managed. Providing power to and receiving data from the lidar requires two Ethernet-enabled slip rings that bridge three separate rotating elements (the electronic configuration used for the lidar is adaptable to other types of Ethernet-enabled sensors if desired). A junction within the spool cylinder allows an optional electro-mechanical tether to be connected. When connected, TReX can leverage external power sources, continuous battery charging, and Power-over-Ethernet (PoE) communication. Sensors and electrical interfaces housed in the tether arm and spool are connected to the rover through the lower 30-channel slip ring. The motor's power is supplied through several high-current channels in the same 30-channel slip ring. A stereo camera mounted on the front of the vehicle is used for visual odometry, terrain imaging, and live display for tele-operation. The on-board fanless computer serves in data collection, processing, and communication with a base station.

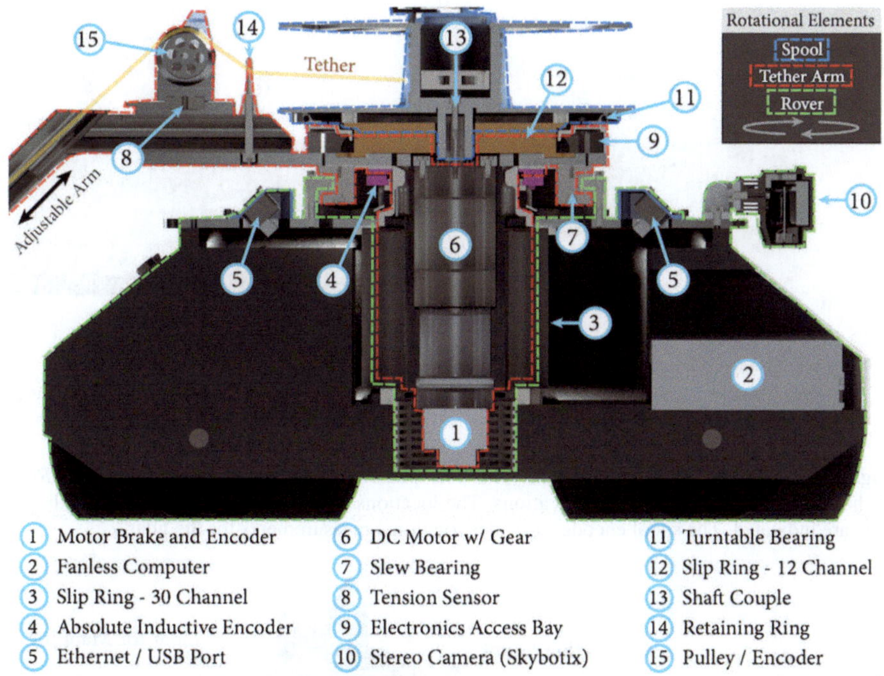

Fig. 3 Annotated CAD model (*cut view*). The three rotating elements include the rover (*green*), tether arm (*red*), and spool (*blue*). The tether spool is actuated only when motorized. Major internal components are labeled. Note that the tether arm, lidar, and rover wheels have been cropped

TReX's ability to map an environment using 3D point-clouds and to accurately measure tether orientation are illustrated in Fig. 4. In terms of 3D mapping, the *left* illustration shows the lidar scan plane, and provides specifications on the range and rotational scan spacing. The *right* image shows how tether orientation is measured. Tether yaw (i.e., bearing to current anchor point) is measured at the rotating joint between the tether arm and rover. Tether pitch is measured using the c-shaped extension mounted on the angled arm. Tether length is measured by a combined pulley-and-force-plate assembly mounted on the tether arm. The spool encoder is used to measure lidar rotation with respect to the tether arm. Given the maximum spool motor speed ($\approx$0.23 rps) and the scanning frequency of the lidar ($\approx$50 Hz), the worst-case azimuth scan spacing is 1.7° (0.029 rad) when the vehicle is under tension.

Lastly, Fig. 5 provides as-built system specifications and an image of the final build. The payload was tailored for the Clearpath Husky A200 rover due to its successful implementation in the vScout prototype.

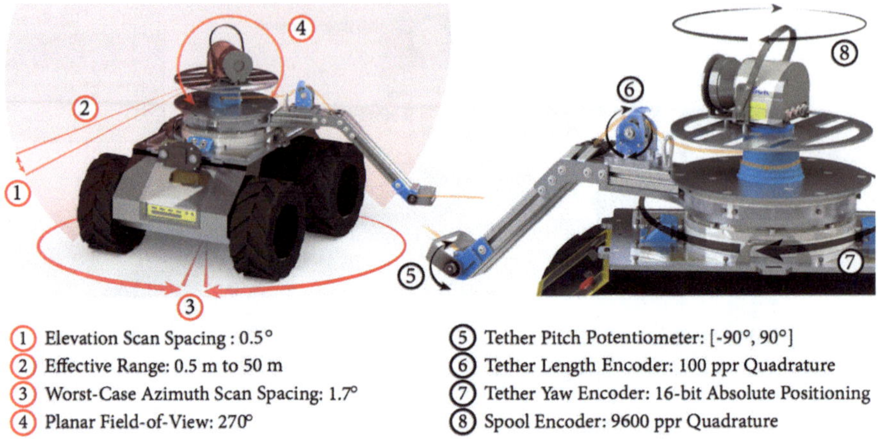

Fig. 4 *Left* 3D Mapping specifications. The lidar plane is shown by an opaque *red disk*. *Right* Tether orientation and sensor specifications. The locations of measurement for pitch, length, and yaw are indicated. The spool encoder provides rotational measurements for the lidar

Final As-Built Specifications	
Mass	92 Kg (tether included)
Dimension	1.06 x 0.67 x 0.76 m
Power	24VDC, 120VAC
Platform	Clearpath Husky A200
Actuator	MMP 170 Nm Torque
Camera	Skybotix VI-Sensor
Lidar	Sick LMS151
Computer	Logic Nuvo-3100VTC
DAQ	Lab Jack T7
Tether	50 m to 100 m

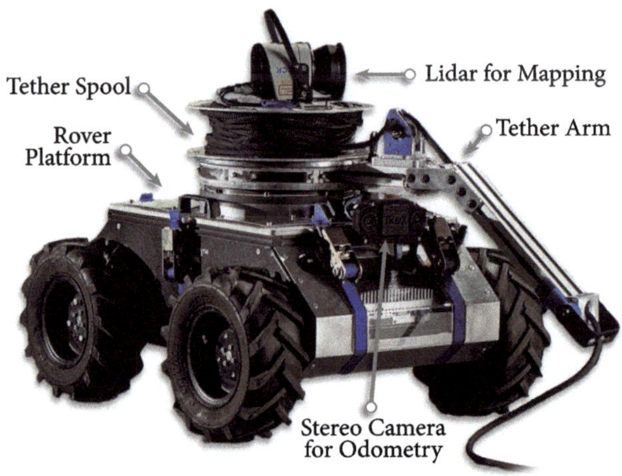

Fig. 5 *Left* Final system specifications. *Right* TReX with major systems labeled. We note that the stereo camera may be occluded by the tether arm during a traverse. However, the pivoting tether arm allows for reorientation of the rover and camera while under tension

3.2 Comparison to Prior Systems

Figure 6 presents an illustrated comparison between past tethered climbing rovers and TReX. The comparison allows for a qualitative evaluation of tethered maneuverability on steep terrain, and demonstrates the benefit of added rotational freedom while under tension.

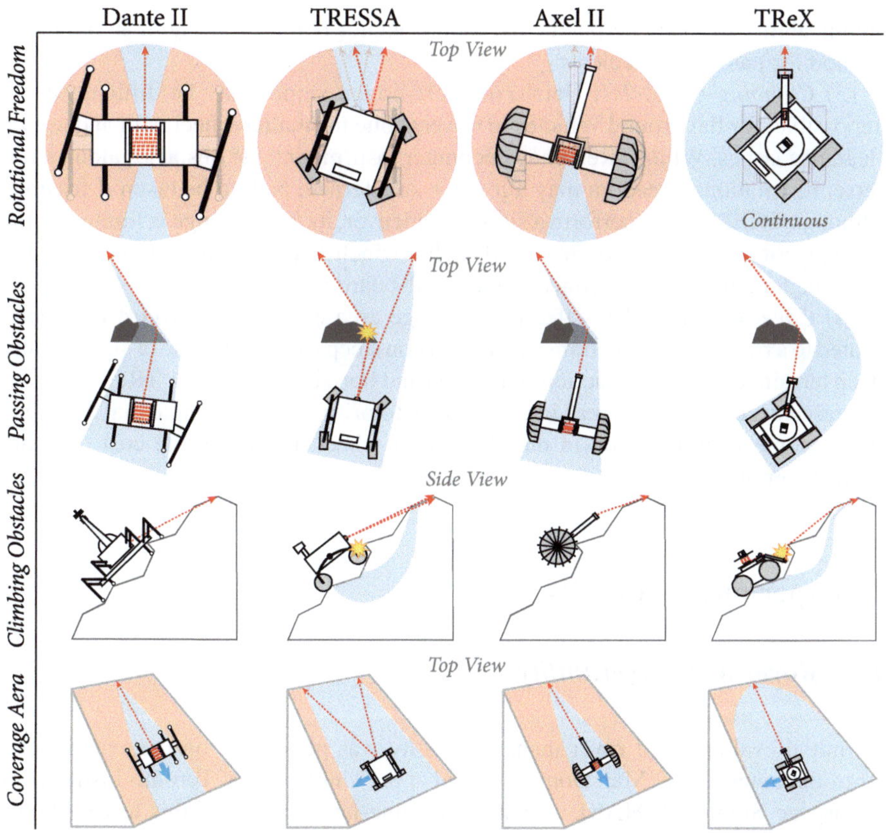

Fig. 6 Maneuverability comparison of Dante II, TRESSA, Axel II, and TReX (figure not drawn to scale). Each row represents attributes of tethered mobility: *Rotational Freedom*, *Passing Obstacles*, *Climbing Obstacles*, and *Coverage Area (single traverse)*. Each column corresponds to a different vehicle. View orientations are given by row (e.g., *top* and *side*). All vehicles with the exception of TRESSA manage tether on board. Tether is indicated by *dashed red lines*, while interactions with obstacles are shown with *yellow stars*. The *light blue* and *red colors* represent feasible and infeasible rotations/paths, respectively. *Small blue arrows* indicate vehicle heading

(1) Rotational Freedom: Prior tethered climbing robots have lacked the ability to turn significantly outside the direction of applied tension on steep terrain. TReX allows continuous rotation about a center-pivot point, which enables horizontal motion while under tension, provided there is sufficient wheel traction.

(2) Passing Obstacles: Obstacles may serve as additional anchor points for tethered rovers. Without the ability to rotate outside the direction of applied tension, obstacles serving as anchor points must be approached directly. Dante II and Axel II were designed for rough terrain and may traverse mid-sized obstacles. While TRESSA has some ability to drive laterally, its dependence on two discrete off-board winches implies higher tether abrasion due to dragging cables. TReX has the

unique ability to rotate perpendicular to any anchor point, resulting in an improved method for passing obstacles.

(3) Climbing Obstacles: Harsh, obstacle-ridden slopes may be difficult, if not impossible, for flat ground vehicles to traverse due to smaller wheel radii and terrain clearance issues. While TReX may encounter insurmountable obstacles along a traverse, its rotational freedom may allow for an alternate path to be taken if feasible. Although TRESSA can perform a similar maneuver, its lateral range is limited by the baseline configuration of top-mounted anchor winches. Furthermore, tether abrasion or entanglement become a concern due to off-board management.

(4) Coverage Area: The effective coverage area of a single traverse is directly related to a rover's ability to translate laterally on steep slopes. Outside of significantly steep terrain, causing a reduction of traction and horizontal mobility, TReX provides increased access to steep areas in comparison to prior rovers. We note that TRESSA allows some lateral motion in the absence of wheel traction at the cost of tether control complexity.

4 Experimental Results

4.1 Rover Maneuverability

An initial evaluation of rover maneuverability was performed on the exterior of a 50-m-diameter dome (MarsDome) located at the University of Toronto Institute for Aerospace Studies (UTIAS). A composite time-lapse image of this test is shown in Fig. 7. TReX was manually operated over varying slopes, demonstrating tether-assisted mobility and rotational freedom under tension. The overlaid yellow and blue arrows represent discrete sides of the lidar scan plane. The coverage area (i.e., the combined point-cloud) depends on the rotation of the spool with respect to the world.

4.2 Tether Management

Autonomy in tether management presents a significant barrier to mobility on steep slopes, especially in the presence of obstacles. To the best knowledge of the authors, no tethered climbing rover has fully implemented autonomous tether management in field experimentation. Tether management was first considered for mobile robots in the mechanical design of Dante II [6]. The developers of TRESSA and Axel II have proposed a method for tether management based on inclination, mass, and tether orientation [1, 14]. However, this has yet to be demonstrated in field testing.

Provided that tension is measured, tether management on flat ground relies on the selection of a static reference tension to maintain. On steep terrain, the influence of gravity on the rover's mass makes the selection of an appropriate reference ten-

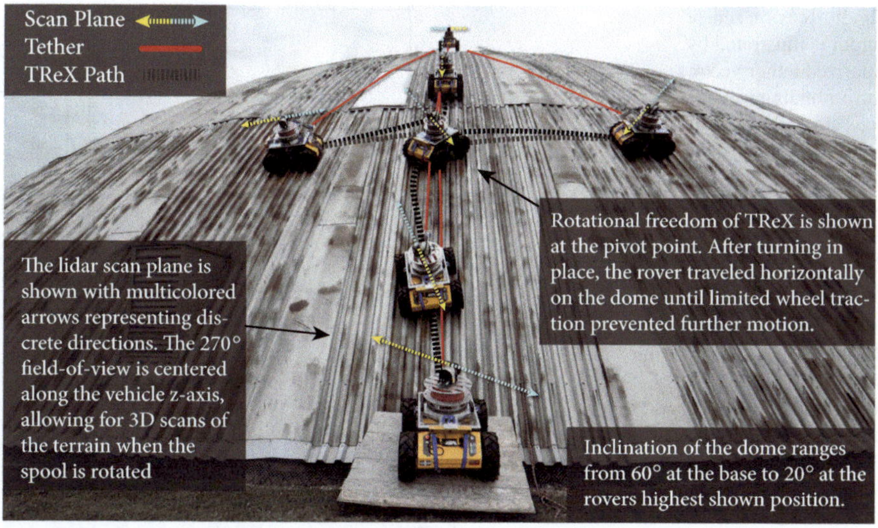

Fig. 7 TReX was manually piloted in a maneuverability test on a 50 m dome while under tension

sion nontrivial. An evaluation of tension-based tether management with the TReX platform is discussed in the following sections.

4.2.1 Tether Management on Flat Ground

Tether management on flat ground utilizes a tension-based controller, which is not reliant on feed-forward input from the rover. Figure 8 illustrates the closed-loop spool velocity controller. The overall goal is to maintain an adequate tension while preserving maneuverability.

As a basic test of the flat-ground tension-based controller, TReX was driven in the presence of a Vicon motion-capture system. The position of the rover, tether arm, and anchor point were recorded during two traverses. Figure 9 provides a colorized representation of tether tension and orientation with respect to the known rover position and anchor point. An accompanying time-lapse image of the experiment

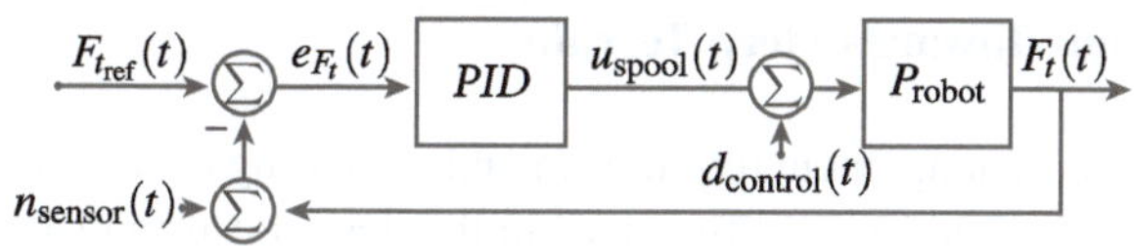

Fig. 8 Closed-loop feedback controller. The error between $F_{t_{ref}}(t)$ and $F_t(t)$ (reference and measured force) is $e_{F_t}(t)$. A gain, $K(t)$, is computed using a PID. The resulting spool control input, $u_{spool}(t)$, is the maximum spool velocity scaled by $K(t)$. The robot plant is P_{robot}. The inputs, $n_{sensor}(t)$ and $d_{control}(t)$, correspond to sensor noise and control disturbances, respectively

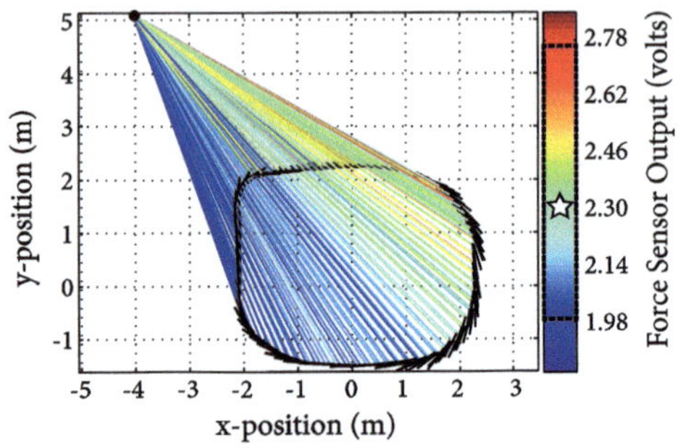

Fig. 9 Tension sensor output is illustrated by colorized tether vectors corresponding to volts. *Black lines* along the path represent vehicle headings. The *dashed box* on the color legend corresponds to the range of voltages sensed, while the *star* indicates the desired reference tension

Fig. 10 A time-lapse of the Vicon test shows TReX performing tension-based tether management. Throughout two traverses, the tether was taut with minimal sag

is shown in Fig. 10. The range of sensed volts indicated by a dashed box in the figure shows that extremes in measurement were avoided (i.e., the tether remained generally taut throughout, and at no point did it touch the ground or prevent the rover from driving its path). Volts are shown in place of kilograms force for this figure due to inaccuracies in calibration/measurement, which are discussed in the proceeding section. Further development of the controller will be necessary in order to compensate for the disparity between reeling conditions (i.e., there is currently a distinct trend in tension error when traversing to or away from the anchor point).

4.2.2 Extensions Towards Steep Terrain

Steep-terrain tether management requires an understanding of sensor performance at varying inclinations. The force sensor used in the design is provided with a factory-calibrated linear output (given in terms of volts per unit force). Unfortunately, frictional elements within the angled-arm design cause a hysteresis-influenced sensor response, where loading and unloading conditions imply different output measurements at similar inclinations. We attempted to characterize hysteresis using an angled-plane test. During the test, TReX was fixed to an anchor point and the plane angle, θ, was cycled between 0° and 90°. Three tests were performed where the tether pitch was varied with respect to the plane as shown in Fig. 11.

The result of three angled-plane tests with variations in tether pitch are shown in Fig. 12. Only test 1 shows a full cycle of the plane θ between 0° and 90°. In tests 2 and 3, the rover's rear tires left the plane due to applied torque on the tether arm at steep inclinations as noted in Fig. 12. Fortunately, the tests were still useful in determining an overall trend in sensor response due to variations in tether pitch and plane inclination.

In the first test, tether pitch was constrained parallel to the plane of inclination. While unloading, friction between the arm and tether resulted in an unchanged sensor output until stiction was overcome near 45°. During a second test, the tether pitch was set to 25°. The voltage output in test 2 generally increased in comparison to test 1, suggesting that friction had been reduced by increasing tether pitch; the

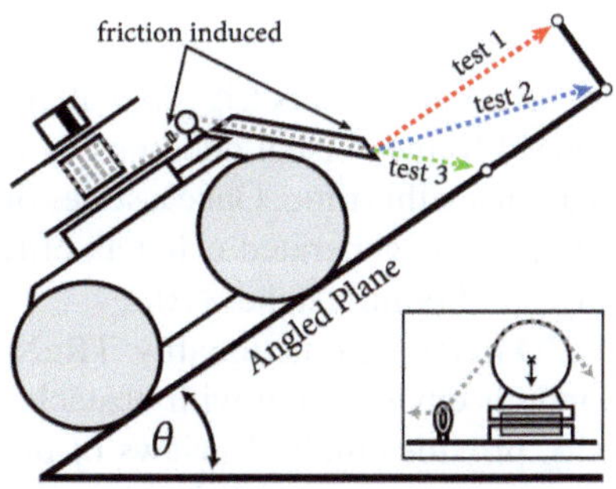

Fig. 11 Angled-plane test (illustrated). TReX was fixed statically on the plane, while variations of tether pitch and plane inclination, θ, were tested. The plane θ was cycled between 0° and 90°. The colors correspond to different tests in Fig. 12. Force sensor placement is shown by a *gray box* in the inset illustration

Fig. 12 Sensor hysteresis over three lifting cycles with varied tether pitch. The *lower* portion of the wing-shaped pattern represents loading, while the *upper* is unloading. The plot indicates that tension was lost due to friction. Test 1 displayed the most loss, suggesting that friction inducing parts should be replaced in the arm

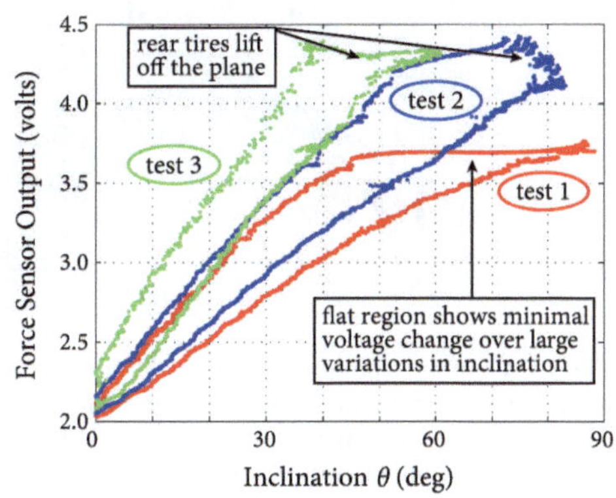

increased pitch minimized the contact surface and bending moment of the tether on the mechanical fairlead. In test 3, the fairlead was completely removed to evaluate its impact on sensor measurement. Tether pitch was constrained to 45°, matching the tilt of the angled arm. Removing the fairlead caused the entire sensor output to increase substantially from what was observed in the first two tests. However, the continued disparity between loading and unloading conditions suggested that friction was still a factor elsewhere in the tether arm design. The most likely source was the steel retaining ring located before the pulley. Unfortunately, the ring was critical to maintaining a balanced load over the pulley and force sensor, and could not be removed to test its frictional impact.

Problems in the tether arm design made a repeatable characterization of the force sensor impossible. Therefore, moving towards tension-based tether management on steep terrain will require modifications to the arm design as proposed in Sect. 5.

4.3 3D Point Cloud Mapping

The collection of a single 3D point-cloud is triggered after every 180° rotation of a lidar with respect to the world frame. Visual odometry is used to locally provide a motion estimate of the lidar during this time. Once a series of point-clouds have been recorded, a global representation is generated using an efficient variant of Iterative Closest Point (ICP) relying on libpointmatcher [10].

As an initial test of this 3D mapping functionality, TReX was anchored and manually piloted through an indoor environment with obstacles to produce a fused 3D point-cloud map. Figure 13, provides multiple views of the reconstructed environment, as well as a time-lapse image of the test. Odometry was provided by a Skybotix VI-Sensor stereo camera, which outputs a pair of calibrated images to an open-source

Fig. 13 Combined point-cloud of an indoor workshop with a time-lapse image of the test. All maps are point representations with correlated intensity. Obstacles were placed in the room to prevent TReX from performing a full scan before driving. The rover was driven around three weighted plastic bins, filling in portions of the map along the way. *Red circles* correspond to the same bin

library, Fast Odometry from VISion (FOVIS).[3] The library functions by detecting similar features in stereo images as a means to compute a velocity and output a camera pose [4].

5 Lessons Learned

With respect to the experimental results previously discussed, engineering lessons learned in the design and initial testing of TReX are summarized by category below.

Platform Maneuverability:

When rotation is constrained, linear motion is limited to the direction of tension. For Dante II, TRESSA, and Axel II, this means that vehicle motion on steep terrain is generally linear, where vehicle velocities directly correlate to tether velocities. For TReX, full range of motion implies that tether velocity is a factor of the current vehicle pose, commanded linear/angular velocities, and position of the anchor point. Excessive inclinations denote higher wheel slippage, resulting in a need for tether-assisted mobility (e.g., commanded vehicle motion is converted into tether actions). For this to occur, we must localize the robot on steep terrain and sense when and where new anchor points have been added.

Tether Management: Tension-based tether management on steep terrain was not demonstrated due to the accumulation of friction in the tether arm. The friction implied a significant variation in sensor output during repeated tests. As such, the tether arm requires modification in the form of additional pulley wheels or bearings to replace friction-inducing parts. When greater repeatability is achieved, then a characterization of the force sensor at varying inclinations should allow for a tension-based controller to be tested on moderate slopes. When the wheel traction is sufficiently reduced in steep terrain, the rover will require tether-assisted mobility. A feed-forward tether management controller, where piloted vehicle actions correspond to appropriate tether actions, will be evaluated.

3D Point Cloud Mapping: Point misalignment in the 3D map shown in Fig. 13 stems from poor visual odometry calibration as well sensor drift in the spool angle encoder. The first issue is likely related to an inaccurate measurement of the camera pose transformation to the vehicle frame. The sensor drift problem stems from measuring angular position before the gearing on the motor. This means that several hundred rotations will occur before a complete rotation is sensed. The addition of a magnetic hall effect sensor on the spool could help in reducing drift. Finally, the scan spacing of the rotating lidar is dependent on spool velocity with respect to the world frame. The impact of variances in rotating elements generate nonuniform radial densities for points, as the rotational speed of the sensor is dependent on the environment. Accordingly, an in-depth evaluation of the 3D point reconstruction pipeline is necessary.

---

[3]Package available: https://github.com/srv/fovis.

6 Conclusion

This work describes the system design and initial testing of a new tethered climbing rover. Tethered Robotic Explorer (TReX) allows for 3D mapping in steep terrain and harsh environments, and is intended to be used for cliff exploration, dam safety inspection, and disaster response. Tests of rotational freedom while under tension, tension-based tether management in varied terrain, and 3D mapping capabilities have demonstrated that the center-pivoting TReX offers improved methods for steep terrain navigation in comparison to prior tethered climbing rovers.

References

1. Abad-Manterola, P.: Axel rover tethered dynamics and motion planning on extreme planetary terrain. Ph.D. thesis, California Institute of Technology (2012)
2. Bares, J.E., Wettergreen, D.S.: Dante II: technical description, results, and lessons learned. Int. J. Robot. Res. **18**(7), 621–649 (1999)
3. Britton, N., Yoshida, K., Walker, J., Nagatani, K., Taylor, G., Dauphin, L.: Lunar micro rover design for exploration through virtual reality tele-operation. In: Field and Service Robotics, pp. 259–272. Springer (2015)
4. Huang, A.S., Bachrach, A., Henry, P., Krainin, M., Maturana, D., Fox, D., Roy, N.: Visual odometry and mapping for autonomous flight using an RGB-D camera. In: International Symposium on Robotics Research (ISRR), pp. 1–16 (2011)
5. Huntsberger, T., Stroupe, A., Aghazarian, H., Garrett, M., Younse, P., Powell, M.: Tressa: teamed robots for exploration and science on steep areas. J. Field Rob. **24**(11–12), 1015–1031 (2007)
6. Krishna, M., Bares, J., Mutschler, E.: Tethering system design for Dante II. In: Proceedings of the IEEE International Conference on Robotics and Automation 1997, vol. 2, pp. 1100–1105. IEEE (1997)
7. Matthews, J.B., Nesnas, I.A.: On the design of the axel and duaxel rovers for extreme terrain exploration. In: 2012 IEEE Aerospace Conference, pp. 1–10. IEEE (2012)
8. Nagatani, K., Kiribayashi, S., Okada, Y., Otake, K., Yoshida, K., Tadokoro, S., Nishimura, T., Yoshida, T., Koyanagi, E., Fukushima, M., et al.: Emergency response to the nuclear accident at the Fukushima Daiichi Nuclear Power Plants using mobile rescue robots. J. Field Rob. **30**(1), 44–63 (2013)
9. Osinski, G.R., Barfoot, T.D., Ghafoor, N., Izawa, M., Banerjee, N., Jasiobedzki, P., Tripp, J., Richards, R., Auclair, S., Sapers, H., et al.: Lidar and the mobile Scene Modeler (mSM) as scientific tools for planetary exploration. Planet. Space Sci. **58**(4), 691–700 (2010)
10. Pomerleau, F., Colas, F., Siegwart, R., Magnenat, S.: Comparing ICP variants on real-world data sets. Auton. Robots **34**(3), 133–148 (2013)
11. Ridao, P., Carreras, M., Ribas, D., Garcia, R.: Visual inspection of hydroelectric dams using an autonomous underwater vehicle. J. Field Robot. **27**(6), 759–778 (2010)
12. Schenker, P.S., Elfes, A., Hall, J.L., Huntsberger, T.L., Jones, J.A., Wilcox, B.H., Zimmerman, W.F.: Expanding venue and persistence of planetary mobile robotic exploration: new technology concepts for Mars and beyond. In: Photonics Technologies for Robotics, Automation, and Manufacturing, pp. 43–59. International Society for Optics and Photonics (2003)
13. Stenning, B., Bajin, L., Robson, C., Peretroukhin, V., Osinski, G.R., Barfoot, T.D.: Towards autonomous mobile robots for the exploration of steep terrain. In: Proceedings of the International Conference on Field and Service Robotics. Springer (2013)

14. Stroupe, A., Huntsberger, A., Garrett, M., Younse, P.: Robotic Mars geology with TRESSA: beyond the Mars Rovers. In: Workshop on Robotics and Challenging Environments, ICRA (2007)
15. Wettergreen, D., Thorpe, C., Whittaker, R.: Exploring Mount Erebus by walking robot. Robot. Auton. Syst. **11**(3), 171–185 (1993)

Design, Control, and Experimentation of Internally-Actuated Rovers for the Exploration of Low-Gravity Planetary Bodies

B. Hockman, A. Frick, I.A.D. Nesnas and M. Pavone

Abstract In this paper we discuss the design, control, and experimentation of internally-actuated rovers for the exploration of low-gravity (micro-g to milli-g) planetary bodies, such as asteroids, comets, or small moons. The actuation of the rover relies on spinning three *internal* flywheels, which allows all subsystems to be packaged in one sealed enclosure and enables the platform to be minimalistic, thereby reducing its cost. By controlling the flywheels' spin rates, the rover is capable of achieving large surface coverage by attitude-controlled hops, fine mobility by tumbling, and coarse instrument pointing by changing orientation relative to the ground. We discuss the dynamics of such rovers, their control, and key design features (e.g., flywheel design and orientation, geometry of external spikes, and system engineering aspects). The theoretical analysis is validated on a first-of-a-kind 6 degree-of-freedom (DoF) microgravity test bed, which consists of a 3 DoF gimbal attached to an actively controlled gantry crane.

1 Introduction

The exploration of small Solar System bodies (such as comets, asteroids, or irregular moons) has become a central objective for planetary exploration [1, 2]. In fact, recent ground- and space-based observations indicate that the exploration of small bodies would collectively address all three main science objectives prioritized by NASA's

B. Hockman (✉) · M. Pavone
(Project PI) Department of Aeronautics and Astronautics,
Stanford University, Stanford, CA, USA
e-mail: bhockman@stanford.edu

M. Pavone
e-mail: pavone@stanford.edu

A. Frick · I.A.D. Nesnas
Jet Propulsion Laboratory, California Institute of Technology, Pasadena, CA, USA
e-mail: andreas.frick@jpl.nasa.gov

I.A.D. Nesnas
e-mail: issa.a.nesnas@jpl.nasa.gov

D.S. Wettergreen and T.D. Barfoot (eds.), *Field and Service Robotics*,
Springer Tracts in Advanced Robotics 113, DOI 10.1007/978-3-319-27702-8_19

recent decadal survey: (1) the characterization of the early Solar System history, (2) the search for planetary habitats, and (3) an improved understanding about the nature of planetary processes [1]. While measurements of some chemical and physical properties can be obtained by remote sensing from space telescopes or orbiters, measurements that constrain composition (e.g., origin science) and measurements of physical properties that fill strategic knowledge gaps for human exploration require direct contact with the surface at multiple locations for extended time periods [2]. Accordingly, *controlled* mobility in low-gravity environments (micro-g to milli-g) has been identified by the National Research Council in 2012 as one of NASA's high priorities for technology development [3].

Microgravity mobility is challenging due to the virtual absence of traction. A number of approaches to mobility have been proposed in the past two decades, which can be roughly divided into four classes: mobility via thrusters, wheels, legs, and hopping. Thrusters have a number of disadvantages for mobility, including mechanical and operational complexity, limited lifetime (due to propellant limitation), and potential for surface contamination. Wheeled vehicles rely on surface normal forces to create lateral traction—a force that is orders of magnitude weaker in microgravity environments. As a result, wheeled systems are bound to extremely low speeds (1.5 mm/s per previous JPL studies [4]) and can easily lose contact with the surface when traversing rocky terrain, resulting in uncontrollable tumbling. Legged systems rely on anchoring devices at the tips, which are mechanically complex and highly dependent on (largely unknown) surface properties (the challenge of anchoring on a small body has been well illustrated by the recent Philae's landing on a comet [5, 6]). Alternatively, *hopping* systems use the low-gravity environment to their advantage. Space agencies such as NASA [4, 7], RKA [8], ESA [9], and JAXA [10] have all recognized this advantage and have designed a number of hopping rovers. However, existing platforms do not appear to allow precise traverses to designated targets in low gravity environments, as required for targeted in-situ sampling.

Statement of Contributions: In this paper we discuss our ongoing efforts toward the design of a microgravity rover aimed at controlled mobility. The platform uses internal actuation (three mutually-orthogonal flywheels) to generate reaction torques, enabling directional hopping capabilities. Specifically, by applying a controlled *internal* torque between the flywheels and the platform, one generates an angular rotation of the platform. In turn, this angular rotation gives rise to surface reaction forces at external contact points, which lead to either tumbling (i.e., pivoting around a spike tip) or hopping (when the reaction forces are large enough), as shown in Fig. 1, left. External spikes protect the platform during ground collisions and provide the primary contact interface with the surface (see Fig. 1, right). With this design, all subsystems are packaged in one sealed enclosure, which enables the platform to be minimalistic and drastically reduces its cost. Henceforth, we will refer to such a rover as *spacecraft/rover hybrid* (S/R hybrid), since it leverages flywheel actuation (typically used for spacecraft attitude control) for rover mobility.

This paper builds upon a number of previous results on microgravity internal actuation, namely [10], which first proposed the use of internal actuation (specifically a single flywheel mounted on a turntable for limited motion control), and [11, 12],

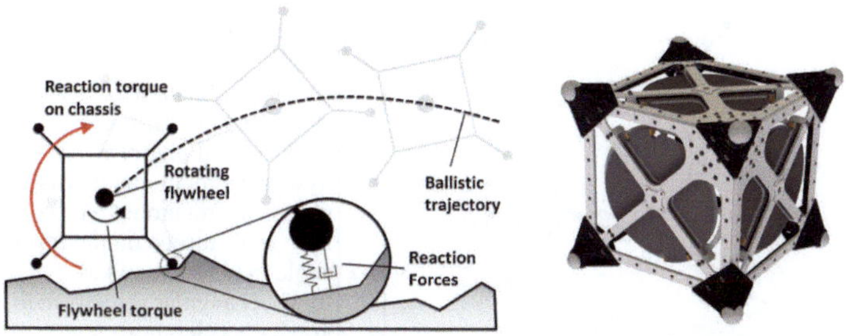

Fig. 1 *Left* By rotating internal flywheels, surface reaction forces make the rover tumble/hop. *Right* Our current prototype without avionics, covers, or solar panels. The cubical structure encloses three flywheels and is protected by external spikes on each of its corners

which consider a torque-controlled three-flywheel configuration and present experimental results on 3 degree-of-freedom (DoF) test beds. This paper is also related to [13], which considers the problem of balancing a cubic body on a corner by actuating three orthogonal flywheels.

Specifically, the contributions of this paper are threefold. First, we characterize the dynamics of the platform and develop hybrid control algorithms for precise mobility (Sect. 2). Our approach leverages a conservation of angular momentum argument, as opposed to the energy approach in [11] used to characterize hopping maneuvers. Second, in Sect. 3 we discuss the mobility platform design, with a focus on *impulsive* actuation of the flywheels to generate more efficient hopping/tumbling maneuvers as compared to [11], and present a preliminary system architecture design. Third, we validate models and control algorithms on a first-of-a-kind 6 DoF microgravity test bed in Sect. 4. The test bed consists of a 3 DoF gimbal attached to an actively controlled gantry crane, and represents, on its own, a major step toward characterizing and validating microgravity mobility (previous test beds only allowed 3DoF tests, e.g., Atwood machine [12], or only allowed tests of the first phases of motion, e.g., parabolic flights and drop towers [14]).

2 Dynamics and Control

In this section we study the dynamics and control of a S/R hybrid by considering a 2D model, i.e., the platform is modeled as a disk with equispaced rigid spikes attached to it, similar to the model commonly used in the field of passive dynamic walking [15]. At the center of mass, a motor drives a single flywheel, producing a torque on the platform (see Fig. 2). A 2D model allows us to derive useful analytical guidelines for actuation and represents a reasonable approximation for 3D configurations in which the S/R hybrid pivots about a pair of spikes.

Our analysis extends earlier studies for this class of rovers (chiefly, [11, 12]) along a number of dimensions. First, our analysis is based on a conservation of angular

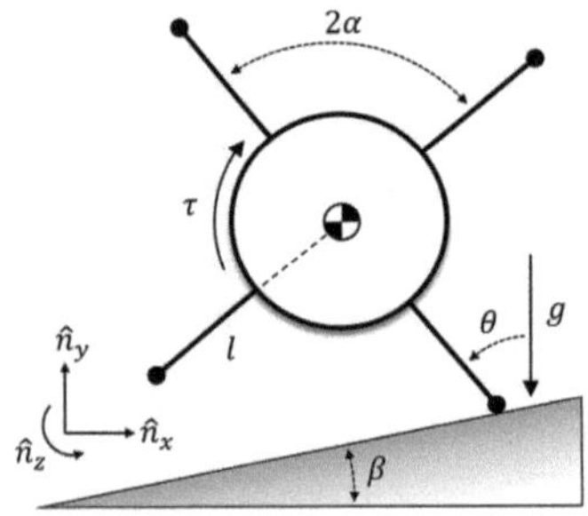

	Definition
θ	hybrid's angle
β	surface inclination
l	spikes' length from CG
m_{p}	total mass of platform
I_{f}	flywheel's rotational inertia
I_{p}	platform's rotational inertia
τ	flywheel's torque on platform
2α	angle in between spikes
ω_{f}	flywheel angular velocity in $\hat{n}_z$
g	gravity acceleration

Fig. 2 2D model: A S/R hybrid is modeled as a rigid body that pivots on an inclined surface

momentum argument, which directly accounts for energy losses. In contrast, [11, 12] mostly rely on an energy conservation approach, which, as we will show in Sect. 2.1.1, can lead to gross underestimates in required flywheel actuation. Second, we study the effect of an inclined surface. Third, and perhaps most importantly, we study in detail control strategies for the flywheel.

2.1 Dynamics of S/R Hybrids

A S/R hybrid is essentially capable of two modes of mobility: tumbling and hopping. The key assumption in our study is that the stance spike acts as a pin joint and does not slip. Under this assumption, the 2D model of a S/R hybrid is uniquely described by two states, θ and $\dot{\theta}$. See Fig. 2 for a detailed description of all parameters. A detailed study of the transition between pivoting and sliding motion of the spike tip can be found in [12] for a Coulomb friction model. One can show that modeling the spike tip as a pin joint is a reasonable approximation for coarse spike geometries where $(\theta - \beta) > \tan^{-1}(1/\mu_{\text{d}})$, where μ_{d} is the coefficient of dynamic friction. For the rubber spike tips on our current prototype, $1 < \mu_{\text{d}} < 1.5$, which, as validated via simulations in Sect. 2.2, is high enough to justify this no-slip assumption. This assumption, however, would not hold in cases where the hybrid operates on non-rigid surfaces (i.e., loose regolith), whereby the slip properties are governed by frictional interactions with granular media. This aspect is left for future research.

2.1.1 Hopping

A hopping maneuver consists of a stride phase, when the system is supported by a single stance spike, and a flight phase when the stance spike leaves the ground. We study the flywheel's torque needed to cause the platform to hop to the right at a desired speed v_{h} and angle θ_{h} (the subscript "h" denotes quantities evaluated at the hopping instant). Assume that the platform starts at rest on the inclined surface and applies a sufficient clockwise torque $\tau(t)$ that causes it to rotate about its stance spike. For the stride phase (i.e., before ground contact is lost), the equations of motion are those of an inverted pendulum and can be easily written as

$$\ddot{\theta}(t) = \frac{m_{\mathrm{p}} g l \sin(\theta(t)) - \tau(t)}{I_{\mathrm{p}} + m_{\mathrm{p}} l^2}, \tag{1}$$

as also derived in [11]. By studying the free body diagram of the system, one can readily show that in order to obtain a negative normal force (i.e., loss of ground contact) it is required that

$$|\dot{\theta}(t_{\mathrm{h}})| > \sqrt{\frac{m_{\mathrm{p}} g \cos(\beta) + \frac{\tau(t_{\mathrm{h}})}{l} \sin(\theta(t_{\mathrm{h}}) - \beta)}{m_{\mathrm{p}} l \cos(\theta(t_{\mathrm{h}}) - \beta)}}. \tag{2}$$

For a flat terrain (i.e., $\beta \to 0$) and with no input torque, $|\dot{\theta}(t_{\mathrm{h}})|_{\mathrm{min}} = \sqrt{g/\left[l \cos(\theta(t_{\mathrm{h}}))\right]}$, which corresponds to a hop distance on the order of $2l$.

Due to its simplicity, a control strategy of particular interest involves instantaneous momentum transfer from the flywheel to the platform (e.g., via impulsive braking). By equating the initial angular momentum of the flywheel $I_{\mathrm{f}}\omega_{\mathrm{f}}$ to the resulting angular momentum of the platform about the spike tip $\dot{\theta}(I_{\mathrm{p}} + m_{\mathrm{p}} l^2)$, and assuming that a hop is initiated immediately after momentum transfer (i.e., $v_{\mathrm{h}} = l\dot{\theta}(0^+)$), the resulting hop velocity, angle, and lateral distance are given by, respectively,

$$v_{\mathrm{h}} = l\omega_{\mathrm{f}} \left(\frac{I_{\mathrm{f}}}{I_{\mathrm{p}} + m_{\mathrm{p}} l^2} \right), \qquad \theta_{\mathrm{h}} = \alpha + \beta, \qquad d_{\mathrm{h}} = \frac{v_{\mathrm{h}}^2}{g} \sin(2\theta_{\mathrm{h}}). \tag{3}$$

A few interesting observations can be made from these results. First, in this regime, the hop angle is governed exclusively by the spike geometry and surface inclination. To maximize the lateral distance of the parabolic trajectory (which scales as $\sin(2\theta_{\mathrm{h}})$), a 45° hop is desired. This is one of the reasons why our current prototype is a cube (i.e., $\alpha = 45°$), see Sect. 3. Second, we define the energy transfer efficiency as

$$\eta := \frac{E(t^+)}{E(t^-)} = \frac{I_{\mathrm{f}}}{I_{\mathrm{p}} + m_{\mathrm{p}} l^2}, \tag{4}$$

where $E(t^-)$ is the energy of the system just before actuation (flywheel kinetic energy), and $E(t^+)$ is the energy just after actuation (platform kinetic energy). Interestingly, the efficiency is given by the ratio of flywheel inertia to platform inertia *about the spike tip*, which depends quadratically on the length of the spikes. Hence, there is an important trade-off between the capability of negotiating obstacles (that would require long spikes) and the actuation efficiency (that prefers short spikes). For our current prototype (augmented with dead mass as stand-in for scientific payload), $\eta \approx 0.01$. This result is critically enabled by angular momentum arguments.

2.1.2 Tumbling

The goal of a tumbling maneuver is to cause the platform to pivot to the right and land on the next consecutive spike such that its orientation is incremented by -2α and that it does not lose contact with the surface. From (1), the minimum torque required to initiate angular acceleration ($-\ddot{\theta}$) from rest on the surface is given by

$$\tau_{\min} = m_{\mathrm{p}} g l \sin(\alpha + \beta). \tag{5}$$

For typical gravity levels of interest (10–1000 μg), small motors of only a few Watts would be sufficient to exceed this torque. To characterize actuation bounds for tumbling, the actuation is regarded as an instantaneous transfer of momentum, similar to the hopping analysis in Sect. 2.1.1. Accordingly, the initial kinetic energy of the platform at $t = 0^+$ can be equated to the gravitational potential energy at the tumbling apex ($\theta = 0$). This yields an expression for the *minimum* flywheel velocity required to vault the platform over its leading spike: $\omega_{\mathrm{f,\,min}} = \sqrt{2m_{\mathrm{p}} g l (1 - \cos(\alpha + \beta)) / (\eta I_{\mathrm{f}})}$. Note that a similar result is provided in [11], but it does not directly account for energy losses or accommodate inclined surfaces. This leads to an underestimate of control input by a factor of $1/\sqrt{\eta} \approx 10$, thus illustrating the importance of an angular momentum approach.

To characterize the *maximum* flywheel velocity for tumbling, consider the hop criterion given by (2) and a zero torque input for $t \geq 0^+$. It follows that $\theta(t)$ and $|\dot{\theta}(t)|$ both decrease with time. Thus, if surface contact is lost, it will occur just after momentum transfer when $\theta(0^+) = \alpha + \beta$, and $|\dot{\theta}(0^+)| = \eta\omega_{\mathrm{f}}$. This yields the maximum flywheel velocity to perform a tumble without hopping: $\omega_{\mathrm{f,\,max}} = \sqrt{[g\cos(\beta)] / [\eta^2 l \cos(\alpha)]}$. Interestingly, there exists an inclination angle, $\beta_{\max}$, for which $\omega_{\mathrm{f,\,min}} = \omega_{\mathrm{f,\,max}}$ and tumbling is impossible. For a square geometry ($\alpha = 45°$), $\beta_{\max} \approx 30°$. Also, as expected, $\omega_{\mathrm{f,\,min}} = 0$ when $\beta = -\alpha$, which corresponds to the declination angle at which the platform freely tumbles "downhill" without actuation.

2.2 *Control of S/R Hybrids*

In this section, we study a control strategy that leverages (5) by slowly spinning up the flywheels with motor torque $\tau < \tau_{\min}$, such that the platform remains grounded. When the desired flywheel speed is achieved, a brake is applied and a hop is initiated as discussed in Sect. 2.1.1. This approach is attractive as it is simple, does not cause momentum build up in the flywheels, and generates high torques for larger hops.

With this control strategy, one can regard the initial flywheel speed ω_{f} and constant braking torque $\bar{\tau}$ as the two control variables. In bringing the flywheel to a full stop, the control variables are related by $\bar{\tau}\Delta t = I_{\mathrm{f}}\omega_{\mathrm{f}}$, where Δt is the time duration of braking. In the limit as $\Delta t \to 0$, the impulsive torque corresponds to the case of instantaneous momentum transfer discussed in Sect. 2.1, whereby Eqs. (3) and (4) can be combined to develop an expression for the flywheel speed ω_f required to cover a lateral distance d_h:

$$\omega_f(d_h) = \sqrt{\frac{d_h g}{\eta^2 l^2 \sin(2(\alpha + \beta))}}. \quad (6)$$

For a square geometry, this expression is minimized for flat terrain, but tends towards infinity as $\beta \rightarrow 45°$. This motivates the potential utility of controllable friction brakes, which can extend the duration of the stride phase and thus control the hop angle. To study the case where Δt is finite, the nonlinear differential equations of motion given by (1) must be solved numerically. However, for aggressive hops, one can assume that $\bar{\tau} \gg m_p g l \sin(\theta)$, so (1) can be well approximated by the linear second order ODE, $\ddot{\theta}(t) \approx -\bar{\tau}/(I_p + m_p l^2)$. For high enough torques, the hop criterion in (2) is not met until immediately after actuation (i.e., a hop is induced at $t_h = \Delta t = \omega_f I_f/\bar{\tau}$), so the initial hop state can be determined by integration:

$$\dot{\theta}(t_h) = \frac{-\bar{\tau}\; t_h}{I_p + ml^2} = \eta\omega_f, \qquad \theta(t_h) = \alpha - \frac{\eta I_f \omega_f^2}{2\bar{\tau}}. \quad (7)$$

Since $\theta(t_h)$ is now a function of ω_f and $\bar{\tau}$, the required torque input requires solving a nonlinear algebraic equation: $d_h = \sin\left(2\alpha - \eta I_f \omega_f^2/\bar{\tau}\right)(\eta l \omega_f)^2/g$.

To better visualize these results and validate the pivoting assumptions, numerical simulations were generated based on a full 6 DoF model, including normal spring/damper and tangential coulomb friction contact forces (as used in [12]).

The plots in Fig. 3 illustrate the hopping angle and distance relationships. Each plot represents a different flywheel speed (2000, 5000, and 10,000 rpm) and the x-axis is the braking torque $\bar{\tau}$. The kink in each curve marks the threshold of an "early hop"—the torque level $\bar{\tau}_s$ below which surface contact is lost before the flywheel is fully stopped. In other words, for a given flywheel speed, $\bar{\tau}_s$ is the minimum braking torque that should be applied to convert all of the flywheel's available kinetic energy to forward motion. This threshold (marked by a vertical line) is in very close agreement with predictions based on (2).

Figure 3b shows that for $\beta \leq 0$, travel distance increases as the torque is increased. However, the situation is different when considering inclined poses ($\beta \geq 0$), whereby high torque inputs result in high angle arching hops—an undesirable effect for distance coverage but potentially useful for getting out of pits. The peaks in these distance curves are in agreement with (6).

The duration of a single hopping maneuver can be thought of as the sum of the time to spin up the flywheels (T_{spin}), and the time of flight (T_{flight}), where

$$T_{spin} = K_S\left(\frac{\sqrt{2}\omega_f I_f}{m_p g l}\right), \quad T_{flight} = K_B\left(\frac{\sqrt{2}\eta l \omega_f}{g}\right), \quad d_{hop} = K_D\left(\frac{(\eta l \omega_f)^2}{g}\right). \quad (8)$$

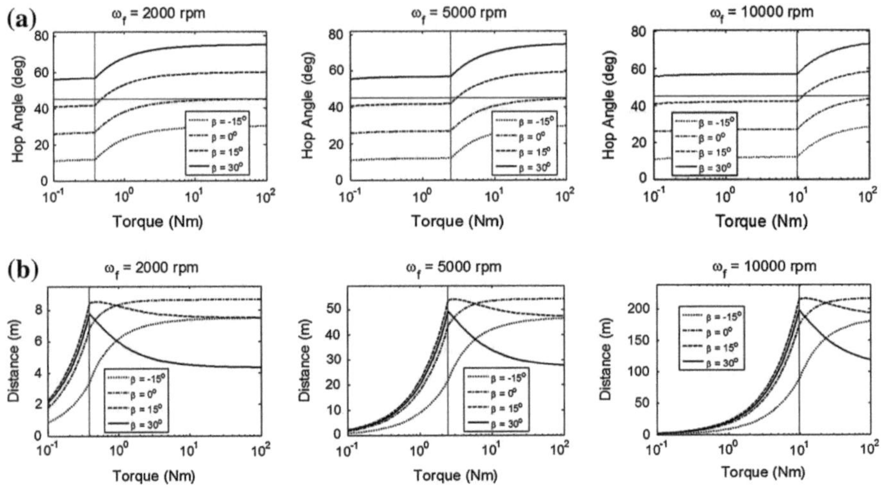

Fig. 3 Resulting hop angles and distances as functions of input torque for three initial flywheel speeds: $\omega_f = 2000, 5000$, and 10,000 rpm (the x-axis is in logarithmic scale). Each curve corresponds to a particular surface inclination β. The *vertical line* on each graph marks the minimum torque at which the flywheel can be fully stopped before a hop is initiated (see Eq. (2)). Results are based on Phobos' gravity level (0.0058 m/s^2) and parameters of our prototype (see Sect. 3.1). **a** Hopping angle (θ_h) as a function of input torque ($\bar{\tau}$). The horizontal line marks the 45° "ideal" hop angle. **b** Lateral hop distance (d_h) as a function of input torque ($\bar{\tau}$)

These equations result directly from (1) and (3), and assuming $\theta_{hop} = \alpha = 45°$. Here, K_S represents a safety factor for tipping during flywheel spin-up, K_B can be thought of as the settling time for residual bouncing as a proportional gain on the parabolic flight time, and K_D is also a proportional gain on hop distance to account for bouncing as well as for deviations in heading. Based on observations from simulations, conservative estimates are $K_B = 2$, and $K_S, K_D = 1.2$. Combining (8) and (6) yields the average expected speed:

$$\bar{V} = \frac{d_{hop}}{T_{flight} + T_{spin}} = \frac{\sqrt{2d_h g}}{2}\left(\frac{K_D \eta m_p l^2}{K_B \eta m_p l^2 + K_S I_f}\right) \approx \frac{\sqrt{2d_h g}}{2}\left(\frac{K_D}{K_B + K_S}\right). \quad (9)$$

The above approximation assumes $I_p + m_p l^2 \approx m_p l^2$, which is reasonable for our prototype ($m_p l^2 = 0.13$ and $I_p = 0.03$). Interestingly, $\bar{V}$ depends on the square root of hop distance and gravity, indicating that farther hops result in faster net motion, and motion on bodies with weaker gravity is slower. On Phobos ($g = 0.0058$ m/s^2), with the parameters of our current prototype, the parameters K_S, K_B, and K_D defined above, and for an average 10 m hop, we can expect a net speed of about 7 cm/s. However, for longer excursions, hops of 100 m are reasonable (i.e., $\omega_f = 6000$ rpm), and could increase net speed to over 20 cm/s.

3 Prototype Design

3.1 S/R Hybrid Structural Design

The prototype and CAD models for the structure and the flywheels (including the braking mechanism) are shown in Fig. 4. The three mutually orthogonal flywheels are mounted with bearing supports to adjacent internal faces of the cube to maximize their inertia (larger diameter) and allow more space for scientific payload and avionics. Each flywheel is directly driven by a small 2 W brushless DC motor (capable of $\tau_{\max} \approx 10\,\text{mNm}$) and motor controllers. Inspired by the theoretical benefits of high torque capabilities (discussed in Sect. 2.2), an impulsive braking mechanism was implemented, whereby an actuated "impact hammer" mounted to the structure collides with a protruding surface on the flywheel (earlier prototypes utilize, instead, friction brakes [11]). The spring-loaded impact hammers are jointly actuated to retract, allowing the flywheels to freely spin, and simultaneously released to snap into place for braking. An on-board microcontroller coordinates motion and collects data, and the system is powered by a 12 V DC battery. The motors have embedded hall sensors that act as velocity sensors, which can also provide torque information according to the relation $\tau = I_f \alpha_f$.

The overall structure and frame consists of a cube with an 8-in edge constructed out of lightweight laser-cut and 3-D printed parts (see Eq. (4) for motivation behind keeping m_p and I_p low), and one spike on each corner. Previous prototype iterations included more spikes [11], but it has been determined through experimentation and insights from dynamic analysis (see Sect. 2) that a cubic geometry with 8 spikes offers the best balance of protection and mobility performance. Each spike is fitted with a rubber tip to absorb impacts and increase surface friction.

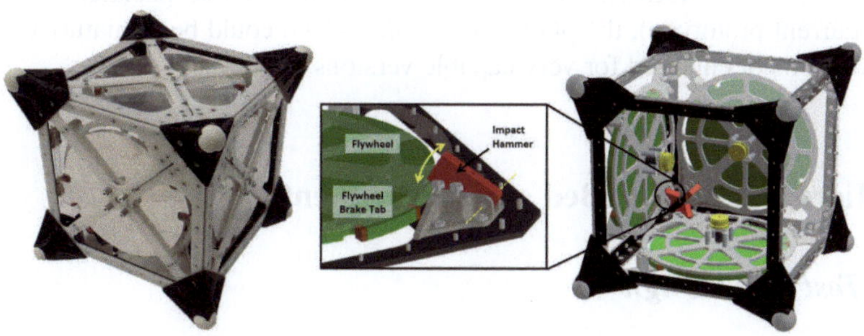

Fig. 4 Prototype and CAD models (not to scale), highlighting the impulsive braking system. The structural parameters are: $m_p = 3.75\,\text{kg}$, $l = 0.17\,\text{m}$, $I_f = 0.95\,\text{g m}^2$, $I_p = 30\,\text{g m}^2$, $\alpha = 45°$

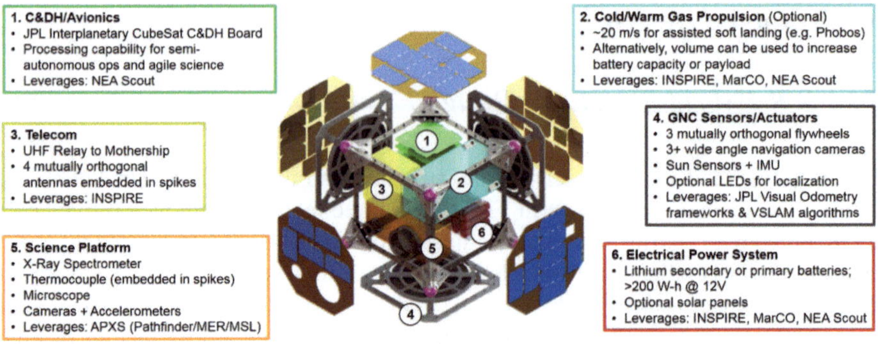

Fig. 5 Preliminary system architecture based on the current prototype design discussed in Sect. 3.1. Key subsystems include: avionics, gas propulsion system, telecommunications, sensors/actuators, power system, and scientific instruments. For the CubeSat mission acronyms, we refer the reader to [16]

3.2 *S/R Hybrid System Architecture*

Figure 5 shows a preliminary system architecture configuration where most subsystems leverage concurrent CubeSat missions under design at JPL [16] (due to space limitations, we provide here a very brief discussion). Although not required for mobility, space was allocated for a gas propulsion system to facilitate soft landing on deployment from the mothership. The deployment phase is a challenging problem but not the focus of this paper. The power system can incorporate both primary batteries (greater storage density) and secondary batteries that can be recharged by the solar panels. The rest of the space is available for avionics, telecommunication systems, sensors, and of course scientific instruments such as microscopes and an X-Ray Spectrometer (XRS). While this system is built on an 8U package size[1] (same as our current prototype), the platform is *scalable* and could be miniaturized to 1U nano-versions or enlarged for very capable versions up to 27U.

4 Microgravity Test Bed and Experiments

4.1 *Test Bed Design*

To the best of authors' knowledge, no preexisting test beds are capable of accurately emulating 6 DoF motion within a microgravity environment for an extended period of time (say, more than 20 s) and within an extended workspace (say, more than 1 m^2). ARGOS, a gravity offload system developed at NASA's Johnson Space Center, may

[1]In CubeSat's jargon, one unit, i.e., 1U, refers to the standard size $10 \times 10 \times 10$ cm volume.

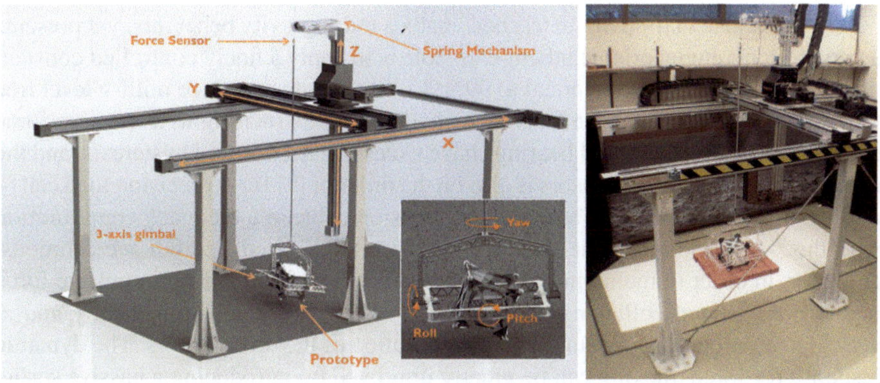

Fig. 6 *Left* 6 DoF microgravity test bed CAD rendering. The powered gantry tracks the translational motion of the platform in x, y, and z within a volume of 3 m × 1 m × 1 m respectively, while allowing for free fall in z at sub-milli-g levels. The gimbal frame allows the platform to rotate in all three axes. *Right* Image of the test bed

come the closest [17]. Used primarily for human testing in zero-g environments, ARGOS consists of an actively-controlled overhead 3-axis gantry crane that tracks the motion of the suspended subject, enabling the "free-floating" behaviors observed in space. At Stanford, we have extended this idea to create a novel 6 DoF test bed for operating rovers in *microgravity* conditions (see Fig. 6). Similar to ARGOS, this test bed is built on a powered gantry crane that permits the tracking of translational motion.

The 3-axis rotational motion is achieved by mounting the platform within a lightweight rigid gimbal frame (see Fig. 6) (Rigorously, the gimbal only enables 2.5 DoF of rotation because the roll axis is bounded to avoid ground contact with the gimbal itself.). Dead mass is fixed to the platform to geometrically center the CG such that it is precisely aligned with the three rotational axes of the gimbal for free rotation. However, this requirement can be relaxed for operation in true microgravity (i.e., on an asteroid) where the platform is no longer suspended. In this case, the control analysis in Sect. 2.2 can be modified to account for an offset CG. For example, a CG offset from the geometric centroid by 10 % of the spike radius would scale the required flywheel speed up by about 20 % on one side, and down by 20 % on the opposite side.

The gimbal-mounted platform is suspended by a (2 m) cable from an overhead attachment point on the gantry crane so that it can swing freely. By accurately measuring the relative deflection of the pendulum at 100 Hz, the x and y axes are actuated using feedback control techniques to keep the pendulum in a vertical state. In this manner, external lateral forces that act on the platform cause the whole system to accelerate as Newton's second law predicts. The sensor that performs this measurement is based on the principle of inductive sensing, whereby strategically placed inductive pick-up circuits measure the strength of the AC current-induced magnetic field emitted by the suspension cable (and thus its deflection due to $1/r$ dissipation).

The vertical actuation of the test bed enables microgravity behaviors, yet presents a very difficult engineering challenge. Its role is to apply a finely controlled constant lifting force on the platform equal to 99.9 % of its weight to induce milli-g level free fall. Applying such a precise force is a challenge in its own right, as passive force elements such as springs and bearings have excessive friction and hysteresis, and the noise floor of many force sensors is also on the order of 0.1 %. A precision load cell (4 digit resolution) is mounted along the suspension cable in a feedback configuration with the z-axis control of the gantry to produce the desired free fall accelerations. However, in order to maintain this constant offloading force during impulsive force inputs (i.e., ground collisions), the gantry must also respond immediately and at very high accelerations—a fundamental limitation of the drive motors. The dynamic response for force tracking can be greatly improved by introducing a passive spring element along the pendulum, which behaves like a series elastic actuator—a commonly used technique in robotics for high fidelity force control [18]. A low-stiffness spring/cam pulley system provides this compliance as described in [19].

Since the dynamics of the system in both the lateral and vertical axes can be simplified to an equivalent linear mass/spring/damper system *about the equilibrium*, we use standard PID control. Furthermore, because the pendulum deflections are kept very small (less than 1°), the vertical force feedback is decoupled from lateral cable deflections, allowing for three independent control loops (one for each axis).

4.2 Test Bed Validation

The test bed was validated by performing reference drop and lateral maneuvers. Specifically, drop tests with only vertical actuation exhibit a very strong parabolic fit (correlation typically above 99 %), and the noise floor on the force sensor feedback allows for effective gravity levels down to about 0.0005 g's. The lateral motion also behaves precisely as predicted without force input, remaining stationary or in a constant velocity. However, there is a small amount of drift in the signal from the lateral sensors (roughly 0.0001 g's/min), which is handled with periodic recalibration before experiments. Interestingly, the lateral signal can be intentionally biased to tilt the acceleration vector off vertical, producing an effectively inclined surface.

A more careful analysis is required to validate the test bed's response to external forces, which can be either impulsive or non-impulsive. As a first test, a constant lateral force was applied to the platform (mass m_p) mounted on the test bed with a horizontal string looped over a pulley with a known mass m_t suspended. After initial transients settle, the system tracks the expected acceleration ($a = gm_t/(m_p + m_t)$) to within 5 %. A similar test was performed in the vertical axis by simply adding small amounts of known mass to the platform, which also produces accelerations in close agreement with theoretical predictions (to within 1 %).

Characterizing the behavior under impulsive contact forces is more challenging. First of all, the elasticity of a collision depends on many factors (e.g., properties of contacting materials, speed of impact, geometry of deformation, etc.), making

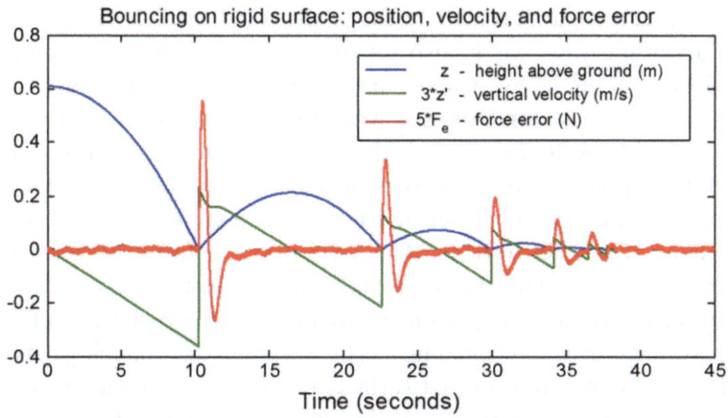

Fig. 7 Experimental results from bouncing on a rigid surface at 0.001 g's: height and velocity of test mass and error in vertical offloading force. Data is sampled at 10 kHz and the control loop runs at 100 Hz. Note that data is scaled to fit on same axes (see legend for scaling and units)

it impractical to characterize theoretically as a basis for comparison. However, as a preliminary test, a proof mass (equal to the mass of the prototype) was mounted on the test bed and dropped onto an elastic surface (basically a webbing of rubber surgical tubing acting as trampoline)—a contrived, low-stiffness interface that dissipates very little energy. In drop tests at 0.001 and 0.005 g's, the mass was released from rest a height of roughly 1 m, and it did indeed recover about 90 % of its energy after each subsequent collision (number of trials = 36, mean = 91.5 %, and standard deviation = 2.7 %).

For collisions with stiffer or even rigid surfaces, energy dissipation is much more difficult to predict. However, the deviation observed in the force signal during impact is a good indicator of fidelity. Figure 7 reports the vertical height and velocity of the proof mass during an example drop/bouncing sequence on a rigid surface, as well as the transient force errors. Although the gantry overshoots vertical position by up to a few inches after a collision, the low stiffness of the spring mechanism (≈5 N/m) results in transient force errors less than a few hundred milli-Newtons—less than 1 % of the proof mass' weight. Since the force error scales roughly linearly with impact speed, there is an upper bound at which the transient response becomes unacceptable, putting the ideal range of operation between 0.0005 and 0.005 g's.

4.3 Mobility Experiments

To further characterize the dynamics and *controllability* of the hybrid and to asses the validity of the model presented in Sect. 2.1, simple hopping experiments were performed on the test bed discussed in Sect. 4.1. As a first set of experiments, we considered a flat rigid surface and constrained the test bed motion to only two axes

(x and z) for direct comparison with the 2D analysis in Sect. 2.1. The initial platform orientation about the yaw axis is also set such that it is "facing" along the x axis, for stable pivoting about its two leading spikes. In each experiment, we executed the control approach discussed in Sect. 2.2, whereby the flywheel is slowly accelerated until a target angular velocity is reached, at which point the impulsive brakes are applied and the hopping sequence ensues unactuated. For a desired hop distance of 0.75 m in an emulated gravity level of 0.001 g's, the target flywheel velocity was calculated using (6) to be 700 rpm. The $x/y/z$ position feedback from the gantry was used in conjunction with the force and displacement signals to determine the trajectory of the hybrid in 20 experiments, four of which are plotted in Fig. 8.

An interesting observation from Fig. 8 is the variability in bouncing. Even for constrained 2D motion on a uniform flat surface, bouncing speed and angle are highly dependent on spin and orientation at the instant of impact. On the other hand, hopping angle measurements exhibit a more consistent trend and are in close agreement to the prediction of (7). The mean hop angle for the 20 experiments was 51° with a standard deviation of 4°. This is marginally higher than the 45° prediction likely due to the elastic rebound of the spike tip, which is not accounted for in the theoretical model. Specifically, instead of stopping immediately on impact as assumed in analysis, the flywheel rebounds and strikes the brake in the opposite direction, resulting in an initial hopping torque much higher than expected, shortly followed by a reverse torque. This actually causes the hybrid to counter-rotate immediately after liftoff. This explains why more energy is converted to translational motion and more distant hops than predicted. In fact, based on the 20 experiments, the mean hop distance of 1.27 m is about 70 % farther than intended—a seemingly beneficial effect. However, this presumably comes at the cost of shorter bounces due to counter-rotation. Nonetheless, correcting for hopping angle and distance discrepancies allows for *controlled* hopping with repeatable performance. We note that, although the analysis and experimental results suggest that impulsive brakes are indeed more efficient, they are

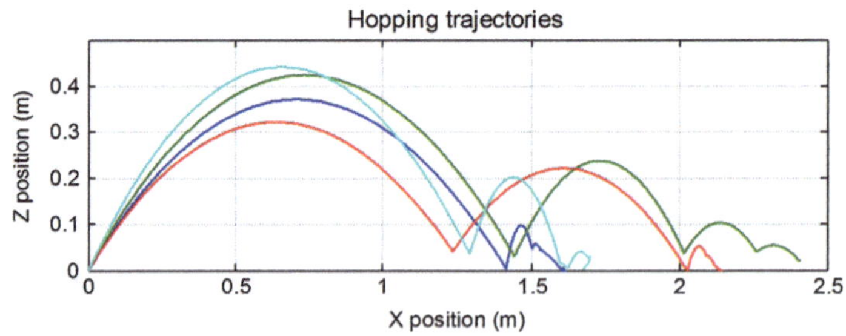

Fig. 8 Hopping trajectories of the hybrid within the microgravity test bed. The gravity level of these experiments was set to 0.001 g's, and the flywheel was commanded to 700 rpms. Position data for each experiment was shifted to start at the origin. A z position of zero corresponds to a flat stance where four spikes are in contact with the ground. Thus, bounces above zero indicate collision at a tilted orientation

also less controllable than friction brakes and induce high mechanical stresses in the structure. See http://web.stanford.edu/~pavone/movies/hop.mov for a sample video of a hopping experiment.

5 Conclusions

In this paper, we presented a planetary mobility platform that uses internal actuation to achieve controlled maneuvers for long excursions (by hopping) and short, precise traverses (by tumbling) in low-gravity environments. We have characterized the dynamics of such platforms using angular momentum arguments and developed hybrid control strategies for precise mobility. We have also presented a preliminary system architecture and prototype design, which has been used to validate control techniques in a first-of-a-kind 6 DoF microgravity test bed. Experimentation is ongoing, but the preliminary results constitute the first successful demonstration of *controlled* hopping mobility in such a high fidelity test bed.

This paper leaves numerous important extensions open for further research. First, it is important to develop more realistic contact models for interactions with loose, granular media typically found on small bodies. Second, we seek to extend the control algorithms to reliably maneuver rocky terrains and leverage them for higher level motion planning objectives. Third, from a navigation perspective, we plan to develop SLAM techniques suited for the unique and challenging environments of small bodies, and for the constantly rotating motion of the platform. Finally, future experiments will include (1) extension to all three axes, with hopping about non-symmetric orientations, (2) various surface characteristics such as inclination, rocks, sand, and fine powder, and (3) tests of the closed-loop system integrating planning, control, and navigation.

Acknowledgments The research described in this paper is supported by NASA under the Innovative Advanced Concepts program, and by the Jet Propulsion Laboratory (JPL) under the R&TD and CAP programs. The authors wish to acknowledge insightful discussions with Dr. J.C. Castillo (JPL), R. Reid (JPL), Dr. C. Zuffada (JPL), Dr. T. Cwik (JPL), Dr. J. Zmuidzinas (JPL), and B. Wilcox (JPL). Government sponsorship acknowledged.

References

1. Decadal Survey Vision and Voyages for Planetary Science in the Decade 2013–2022. Technical report, National Research Council (2011). http://solarsystem.nasa.gov/2013decadal/
2. Castillo Rogez, J.C., Pavone, M., Nesnas, I.A.D., Hoffman, J.A.: Expected science return of spatially-extended in-situ exploration at small solar system bodies. In: IEEE Aerospace Conference, pp. 1–15, Big Sky, MT, Mar 2012
3. NASA Space Technology Roadmaps and Priorities: Restoring NASA's Technological Edge and Paving the Way for a New Era in Space. Technical report, National Research Council (2012)

4. Jones, R.M.: The MUSES-CN rover and asteroid exploration mission. In: 22nd International Symposium on Space Technology and Science, pp. 2403–2410 (2000)
5. Glassmeier, K.-H., Boehnhardt, H., Koschny, D., Kührt, E., Richter, I.: The Rosetta mission: flying towards the origin of the solar system. Space Sci. Rev. **128**(1–4), 1–21 (2007)
6. Hand, E.: Philae probe makes bumpy touchdown on a comet. Science **346**(6212), 900–901 (2014)
7. Fiorini, P., Burdick, J.: The development of hopping capabilities for small robots. Auton. Robots **14**(2), 239–254 (2003)
8. Sagdeev, R.Z., Zakharov, A.V.: Brief history of the Phobos mission. Nature **341**(6243), 581–585 (1989)
9. Dietze, C., Herrmann, S., Kuß, F., Lange, C., Scharringhausen, M., Witte, L., van Zoest, T., Yano, H.: Landing and mobility concept for the small asteroid lander MASCOT on asteroid 1999 JU3. In: 61st International Astronautical Congress (2010)
10. JAXA Hayabusa mission: Technical report, JAXA (2011). http://hayabusa.jaxa.jp/e/index.html
11. Allen, R., Pavone, M., McQuin, C., Nesnas, I.A.D., Castillo Rogez, J.C., Nguyen, T.-N., Hoffman, J.A.: Internally-actuated rovers for all-access surface mobility: theory and experimentation. In: Proceedings IEEE Conference on Robotics and Automation, pp. 5481–5488, Karlsruhe, Germany, May 2013
12. Reid, R.G., Roveda, L., Nesnas, I.A.D., Pavone, M.: Contact dynamics of internally-actuated platforms for the exploration of small solar system bodies. In: i-SAIRAS, pp. 1–9, Montréal, Canada, June 2014
13. Gajamohan, M., Merz, M., Thommen, I., D'Andrea, R.: The Cubli: a cube that can jump up and balance. In: IEEE/RSJ International Conference on Intelligent Robots and Systems, pp. 3722–3727. IEEE (2012)
14. Yoshimitsu, T., Kubota, T., Nakatani, I., Adachi, T., Saito, H.: Microgravity experiment of hopping rover. In: Proceedings of IEEE Conference on Robotics and Automation, vol. 4, pp. 2692–2697 (1999)
15. McGeer, T.: Passive dynamic walking. Int. J. Robot. Res. **9**(2), 62–82 (1990)
16. JPL's Cubesats: Technical report, JAXA (2015). http://cubesat.jpl.nasa.gov/
17. Valle, P., Dungan, L., Cunningham, T., Lieberman, A., Poncia, D.: Active Response Gravity Offload System (2011)
18. Jerry, P., Krupp, B., Morse, C.: Series elastic actuators for high fidelity force control. Ind. Robot: Int. J. **29**(3), 234–241 (2002)
19. Duval, E.F.: Dual pulley constant force mechanism, 16 Mar 2010. US Patent 7,677,540

Considering the Effects of Gravity When Developing and Field Testing Planetary Excavator Robots

Krzysztof Skonieczny, Thomas Carlone, W.L. "Red" Whittaker and David S. Wettergreen

Abstract One of the challenges of field testing planetary rovers on Earth is the difference in gravity between the test and the intended operating conditions. This not only changes the weight exerted by the robot on the surface but also affects the behaviour of the granular surface itself, and unfortunatly no field test can fully address this shortcoming. This research introduces novel experimentation that for the first time subjects planetary excavator robots to gravity offload (a cable pulls up on the robot with 5/6 its weight, to simulate lunar gravity) while they dig. Excavating with gravity offload underestimates the detrimental effects of gravity on traction, but overestimates the detrimental effects on excavation resistance; though not ideal, this is a more balanced test than excavating in Earth gravity, which underestimates detrimental effects on both traction and resistance. Experiments demonstrate that continuous excavation (e.g. bucket-wheel) fares better than discrete excavation (e.g. front-loader) when subjected to gravity offload, and is better suited for planetary excavation. This key result is incorporated into the development of a novel planetary excavator prototype. Lessons learned from the prototype development also address ways to mitigate suspension lift-off for lightweight skid-steer robots, a problem encountered during mobility field testing.

K. Skonieczny (✉) · T. Carlone · W.L. "Red" Whittaker · D.S. Wettergreen
Robotics Institute, Carnegie Mellon University, 5000 Forbes Ave.,
Pittsuburgh, PA 15213, USA
e-mail: kskoniec@encs.concordia.ca

T. Carlone
e-mail: tomcarlone@gmail.com

W.L. "Red" Whittaker
e-mail: red@cmu.edu

D.S. Wettergreen
e-mail: dsw@ri.cmu.edu

D.S. Wettergreen and T.D. Barfoot (eds.), *Field and Service Robotics*,
Springer Tracts in Advanced Robotics 113, DOI 10.1007/978-3-319-27702-8_20

1 Introduction

Excavating on the Moon and Mars enables in situ resource utilization (ISRU) and extraterrestrial contruction. However, planetary excavators face unique and extreme engineering constraints relative to terrestrial counterparts. In space missions mass is always at a premium because it is the main driver behind launch costs. Lightweight planetary operation, due to low mass and reduced gravity, hinders excavation and mobility by reducing the forces a robot can effect on its environment.

This work considers lightweight excavation from the point of view of excavator configuration. It shows that continuous excavators (bucket-wheels, bucket chains, etc.) are more suitable than discrete excavators (loaders, scrapers, etc.). Figure 1 shows an example of a continuous and discrete excavator.

A wide assortment of planetary excavator prototypes have been developed in recent years, of both the continuous and discrete variety, specifically for excavation and ISRU. Muff et al. proposed a bucket-wheel excavator [15]. A Bucket-Drum Excavator, which is an adaptation of a bucket wheel [6], excavates regolith directly into a rotating drum. NASA's Regolith Advanced Surface Systems Operations Robot (RASSOR) has counter-rotating front and rear bucket drums, enabling it to balance horizontal excavation forces [13].

Examples of discrete excavator prototypes include NASA's Cratos [5], a scraper with a central bucket between its tracks. Other examples include NASA's Centaur II with front-loader bucket and Chariot with LANCE bulldozer blade [11]. The Canadian Space Agency's Juno rovers [20] can be equipped with front-end load-haul-dump scoops. The wide variability in prototypes and approaches highlights the need for a far-reaching framework to analyze, test, and classify planetary excavators.

Testing of planetary excavation has been done almost exclusively in Earth gravity with full-weight excavators. Only a single set of experiments has been published characterizing excavation with a scoop in reduced gravity [3]. A discussion of these experimental results, as well as other results pertaining to traction in reduced gravity,

Fig. 1 A robotic excavator configured for continuous (*left*) and discrete excavation (*right*)

in Sect. 2 shows why testing in Earth gravity can substantially overestimate planetary excavator performance, thus highlighting the need for a new testing methodology; a test method for gravity-offloaded excavation experiments is then presented. Section 3 predicts analytically why continuous excavators should be expected to perform better in reduced gravity than discrete excavators, and Sect. 4 uses the newly developed test methodology to provide experimental evidence supporting this result. Section 5 outlines the development of a novel prototype excavator based on the results of this research campaign, and also describes practical issues that were encountered during mobility field testing. Finally, Sect. 6 presents conclusions, lessons learned, and future work.

2 Gravity Offload Experimentation

This work presents novel experiments that for the first time subject excavators to gravity offload (a cable pulls up on the robot with 5/6 its weight, to simulate lunar gravity) while they dig. Although not fully representative of excavation on planetary surfaces (where the regolith is also subject to reduced gravity), these experiments are more representative of planetary excavation performance than testing in full Earth gravity. Testing in Earth gravity is an inadequate evaluation of planetary excavators, as it over-predicts excavator performance relative to reduced gravity. The following subsections discuss the effects of gravity on traction and excavation resistance, and explain why gravity offload testing is a more balanced approach than testing in Earth gravity. Details of the testing methodology are then described.

2.1 Effects of Reduced Gravity on Traction

A vehicle's drawbar pull is its net traction: DP = T − R (i.e. Thrust − Resistance). Note that both Thrust and Resistance depend on wheel slip. Drawbar pull at 20 % slip is a good measure of tractive performance, as pull begins to plateau around 20 % slip for many wheels (or tracks) while negative effects such as sinkage increase [21]. A non-dimensional quantity, P_{20}/W (Drawbar pull at 20 % slip, normalized by weight), has been used as a benchmark metric for lunar wheel performance from the times of Apollo [7] to today [22, 25].

The most representative test environment for planetary rovers is a reduced gravity flight, where rover and regolith are both subject to reduced g [3, 12]. Another class of tests reduces the weight of the robot, but not the regolith. NASA JPL runs mobility tests for the Curiosity rover using a full geometric scale 3/8th mass 'SCARECROW' rover [23]. SCARECROW's 3/8th mass loads the wheels with an equivalent weight to the full mass Curiosity rover in Mars gravity. Another way to achieve equivalent results is to use a full mass robot, but to 'offload gravity' by offloading a portion of the robot's weight; this is the approach used in this work.

Testing with reduced robot weight in Earth gravity does not exhibit the same mobility performance as planetary driving (or reduced-g flights), where the regolith is also subject to reduced gravity [24]. It seems to in fact over-predict traction for scenarios governed by P_{20}/W, such as pulling and slope climbing. P_{20}/W is approximately constant with changing load (i.e. changing W but keeping scale and gravity constant, as with SCARECROW or gravity offload), as has been observed experimentally [7]. This is because both thrust, T, and resistance R, are reduced under lower loads; the former due to reduced frictional shearing, the latter due to reduced sinkage. On the other hand, changing W *by* reducing gravity *reduces* P_{20}/W. Kobayashi's reduced-gravity parabolic flight experiments showed that wheel sinkage is *not* reduced when driving in low gravity [12], though thrust still is.

These results suggest that gravity offload testing underestimates detrimental effects on rover tractive performance, by maintaining constant rather than diminished P_{20}/W at conditions meant to represent lower gravity environments. However, the next subsection explains that for excavators this fact is balanced by an overestimate of the detrimental effects on excavation resistance.

2.2 Effect of Reduced Gravity of Excavation Resistance Forces

Reduced gravity increases the ratio of excavation resistance to weight in cohesive lunar regolith. Boles et al. compared excavation resistance forces measured in Earth gravity to resistance forces measured during reduced-gravity parabolic flights (for otherwise identical experiments), and showed that excavation resistance in 1/6 g could be anywhere between 1/6 and 1 of the resistance experienced in full Earth gravity ($F_{ex/E}$) [3]. This result is consistent with a theoretical analysis of excavation forces. Consider the two dominant terms of Reece's fundamental equation of earthmoving mechanics [9], based on the principles of passive earth pressure: $F_{ex} = N_{\gamma}\gamma g w d^2 + N_c c w d$ Gravitational acceleration is denoted g, γ is soil density, c is cohesion, d is cut depth, w is blade width, and the N_i are non-dimensional coefficients pertaining to different sources of resistance. The frictional part of F_{ex} is proportional to g, whereas the cohesive part is independent of g. This suggests that for a purely frictional soil F_{ex} in 1/6 g should be 1/6 of the $F_{ex/E}$, for a purely cohesive soil F_{ex} in 1/6 g should be 100 % of $F_{ex/E}$, and for typical combination soils the result should be somewhere in between. Sample data from Boles et al. shows examples of F_{ex} in 1/6 g that average 1/3 of $F_{ex/E}$.

Characterizing planetary excavators performance based on tests in Earth gravity is equivalent to assuming that excavation resistance scales down proportionally to a reduction in gravity, which Boles' experiments show is not generally, or even typically, the case. Making such an assumption would thus underestimate the detrimental effects of reduced gravity on excavation resistance.

Reducing robot weight but not regolith weight makes excavation more difficult than is to be expected in reduced gravity. Longitudinal soil-tool interactions are not directly affected by reduced robot weight, so excavation resistance force, F_{ex}, remains unchanged. Reducing weight to 1/6 thus directly increases F_{ex}/W sixfold. For planetary excavation, this corresponds to the worst possible case of purely cohesive regolith. As neither lunar nor Martian regolith is purely cohesive, excavation resistance on these planetary surfaces in not expected to scale quite so poorly.

Excavating with gravity offload thus underestimates the detrimental effects of gravity on traction, but overestimates the detrimental effects on excavation resistance. This is a more balanced and conservative test than excavating in full Earth gravity, which underestimates detrimental effects on both traction and resistance.

2.3 Experimental Setup

Gravity offloaded excavation experiments were set up at NASA Glenn Research Center's (GRC) Simulated Lunar OPErations (SLOPE) lab. The facility contains a large soil bin with GRC-1 [16] lunar simulant. This research developed an experimental apparatus for achieving gravity offload in the SLOPE lab. The main aspects of the apparatus are shown in Fig. 2. A cable pulls up on the robot, tensioned by weights acting through a 2:1 lever arm. The weights and lever assembly hang from a hoist that is pulled along a passive rail by a separate winch-driven cable. All tests are conducted in a straight line below the hoist rail. The winch speed is controlled so that the hoist is pulled along at the same speed as the robot is driving, keeping the cable vertical. For tests where excavator speed remains constant, winch speed is set

Fig. 2 Gravity offload testing with bucket-wheel (*left*) and front-loader bucket (*right*) on the Scarab robot. A cable pulls up on the robot, tensioned by weights acting through a 2:1 lever arm. The offload assembly hangs from a hoist that is pulled along a rail by a separate winch-driven cable

open loop. For tests where the excavator enters into high slip, winch speed has to be manually reduced to match the robot's decreasing speed.

Continuous bucket-wheel and discrete bucket excavation was performed using the Scarab robot (for a detailed description of the robot, see [2, 22]. With Scarab's shell removed, excavation tools were mounted to the robot's structural chassis. For continuous excavation, a bucket-wheel was mounted with its axis of rotation aligned with Scarab's driving direction. The bucket wheel is 80 cm diameter with 12 buckets, and each bucket has a width of 15 cm. The bucket used for discrete excavation is 66 cm wide, and was mounted behind Scarab's front wheels at a cutting angle of 15° down from horizontal. Figure 1 shows Scarab configured both as a continuous and as a discrete excavator.

Scarab has a mass of 312 kg (weight of 3060 N in Earth gravity) in the configuration used for these experiments. The connection point for the gravity offload cable was adjusted to preserve the robot's weight distribution (54 % on the rear wheels). This was confirmed by weighing Scarab on 4 scales (one under each wheel) before and after being connected to the gravity offload apparatus. The offloading cable was equipped with a 2-axis inclinometer and a single-axis load cell to measure cable angle and tension, respectively.

Continuous and discrete excavation experiments were conducted at equivalent nominal production rates of approximately 0.5 kg/s, and at equal speeds of 2.7 cm/s. To account for the differing geometry of the excavation tools, the rectangular discrete bucket cut at a depth of 2 cm, and the circular bucket-wheel cut at a central depth of 5 cm. Depth was set using Scarab's active suspension, which raises and lowers the central chassis. Regolith picked up by the bucket-wheel was manually collected in 5-gallon buckets not connected to the robot, and weighed. The discrete bucket collected regolith directly, and after a test that regolith was transferred into 5-gallon buckets and weighed. To capture mobility data, the excavator's position was tracked using a laser total station at a data rate of 1 Hz during all experiments.

Between each test run, soil conditions were reset using a technique developed at NASA GRC. First, the GRC-1 simulant is fully loosened by plunging a shovel approximately 30 cm deep and then levering the shovel to fluff the regolith to the surface; this is repeated every 15–20 cm in overlapping rows. Next, the regolith is leveled with a sand rake (first with tines, then the flat back edge). The regolith is then compacted by dropping a 10 kg tamper from a height of approximately 15 cm; each spot of soil is tamped 3 times. Finally, the regolith is lightly leveled again for a smooth flat finish. A cone penetrometer was used to verify that the soil preparation consistently achieved bulk density between 1700 and 1740 kg/m^3.

3 Predicted Excavation Performance

Considering that gravity offloaded excavation experiments are, on balance, more representative of planetary operating conditions, there is value in investigating cases where offloaded test results may diverge from tests in full Earth gravity; one such case is the comparison of continuous and discrete excavation. Estimates of excavation performance predict that continuous and discrete excavation should both be successful in 1 g, but that a continuous excavator achieves this with a higher performance margin. These differences in performance margin become apparent at conditions offloaded to 1/6 g, where discrete excavation is predicted to fail.

Predicted excavator performance is based on a comparison of traction and excavation forces. Excavator failure is defined as a degradation of mobility (i.e. significant increase in slip and/or sinkage), which is caused by excavation resistance forces exceeding the traction forces that the robot can sustainably produce.

The achievable traction is directly comparable for continuous and discrete excavation experiments, because in both cases Scarab is equipped with the same 'spring tire' wheels. These wheels can sustainably produce a DP/W ratio of 0.25, as measured by drawbar pull–slip experiments. Achievable traction is thus approximately equal at the start of continuous and discrete experiments, when weight is approximately equal. In the course of a discrete excavation experiment, weight and thus traction increases as regolith is collected. In continuous experiments, on the other hand, traction remains approximately constant as regolith is collected into buckets not connected to the rover. Thus in 1 g, the maximum sustainable drawbar pull for continuous excavation is 765 N, while for discrete excavation it is 765 N plus 0.25 N for every 1 N of regolith collected. Similarly in offloaded 1/6 g, the maximum sustainable drawbar pull for continuous excavation is 128 N, while for discrete excavation it is 128 N plus 0.25 N for every 1 N of regolith collected (note that collected regolith is not offloaded).

Force measurements from preliminary tests show that continuous excavation forces are bounded [18], and are in the range of 6–12 N in the case of the bucket-wheel being tested. Discrete excavation forces, on the other hand, rise approximately linearly with payload collected [1, 18], at a rate of 1.2–1.5 N per 1 N or regolith collected for a similar discrete bucket [1]. This rise in force for discrete excavation is attributable to accumulation of surcharge at the cutting edge, resisting entry of further regolith into the bucket.

Comparing continuous excavation force to achievable traction predicts consistent margins of at least 98–99 % in 1 g, and at least 90–95 % in 1/6 g. For discrete excavation, on the other hand, initially high margins are predicted to decrease to zero once 600–800 N of regolith is collected in 1 g, or once 100–140 N of regolith is collected in offloaded 1/6 g. The maximum capacity of the discrete excavation bucket is approximately 450 N of GRC-1, so in 1 g it is predicted to be filled to capacity with leftover performance margin, but in 1/6 g the zero margin condition is predicted to be reached before the discrete bucket is filled.

Analyses of these preliminary force measurements also suggest that continuous excavation is somewhat more energy efficient than discrete excavation. By integrating over a 2.5 m excavation distance, and taking into account the 1.2–1.5 N increase in excavation force per 1 N of regolith collected, 0.5 kg/s production and 2.6 cm/s forward speed, discrete excavation of 45 kg in 1 g requires 700–900 J. On the other hand, accounting for lateral and longitudinal bucket-wheel forces and displacements as well as vertical lifting of excavated soil, continuous excavation of 45 kg in 1 g requires 500–600 J; in lower g continuous excavation would be even more efficient because much of the energy goes into lifting the soil against gravity. Despite the additional actuator to turn the bucket-wheel, energy is saved due to lack of energy-sapping resistive soil accumulation.

4 Experimental Results

Experimental data support the predictions made in the previous section, highlighting the importance of including gravity offloaded experiments into testing campaigns for proposed planetary excavators. Experiments show that in 1 g continuous and discrete excavation both achieve successful performance. On the other hand, in gravity offloaded 1/6 g, discrete excavation fails from degraded mobility, while continuous excavation does not.

Three or four runs were conducted at each of the test conditions, including baseline runs of driving without digging. Total station data were analyzed to calculate excavator speed during each test, as shown in Fig. 3. The excavator maintains constant forward progress in all cases except discrete excavation with gravity offload. Average speed (as well as standard deviation) for the various test cases, is summarized in Table 1.

Tests in 1 g exhibit a slightly slower speed, because the higher weight compresses the compliant ‘spring tires’ and reduces their radius. Excavation and gravity offload both introduce a small amount of additional variability in speed compared to driving without digging in 1 g. Continuous and discrete excavation in 1 g, as well as continuous excavation in gravity offloaded 1/6 g, all collected approximately 45 kg during each 2.5 m test run. Discrete excavation in gravity offloaded 1/6 g collected only 15–20 kg, in contrast.

Gravity offload was controlled with sufficient precision to avoid pulling the excavator forward or backward. During continuous excavation, cable angle was unbiased about vertical, with a mean absolute value of just 0.1 degrees; with a cable tension of 2600 N, this corresponds to 4.5 N, or less than 1 % of offloaded excavator weight. In contrast, inducing 20 % slip in the spring tires used in the experiments would require sustained horizontal forces of 25 % of offloaded excavator weight. Transient motions of the cable did not exceed 0.8 degrees from vertical for more than a fraction of a second; this corresponds to brief transients of 35 N, or 7 % of offloaded excavator weight. Cable tension varies just ± 1 % which, amplified by the offloading ratio, corresponds to 5 % variation in the offloaded excavator weight; this also translates to no more than approximately 1 % variation in the ratio of horizontal force to offloaded excavator weight. Figure 4 shows longitudinal cable angle and cable tension for a discrete excavation test, the most challenging test due to the changing speed. Variability in angle and tension were again unbiased and small.

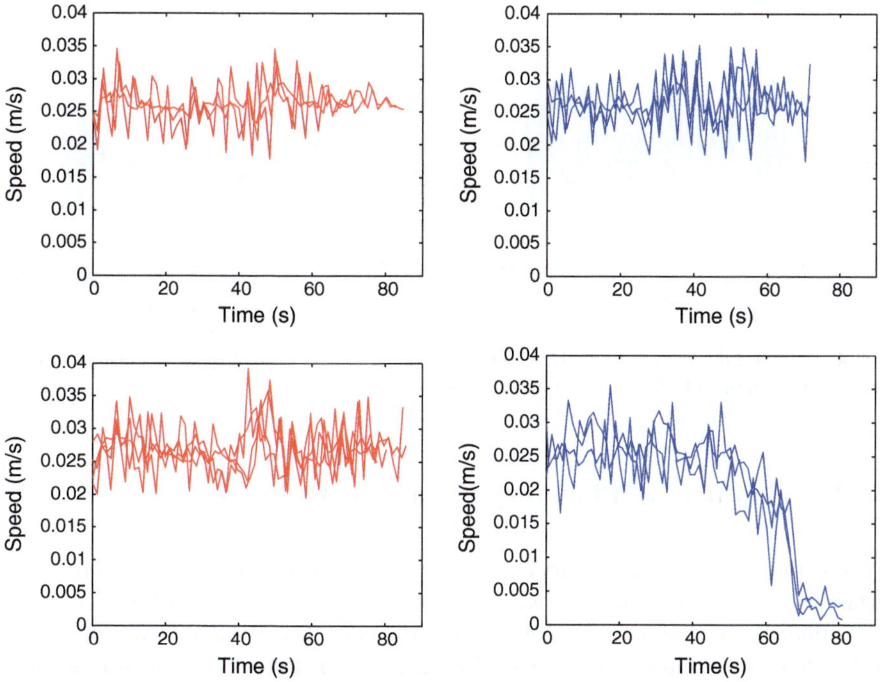

Fig. 3 Excavator forward driving speed during continuous excavation in 1 g (*top left*), discrete excavation in 1 g (*top right*), continuous excavation in gravity offloaded 1/6 g (*bottom left*), and discrete excavation in gravity offloaded 1/6 g (*bottom right*; time axes aligned at stall point). The excavator maintains constant progress in all cases except discrete excavation with gravity offload

Table 1 Discrete excavation offloaded to 1/6 g is the only test condition that does not maintain constant steady state (S/S) velocity

Excavation type	'Gravity' (g)	Average v (cm/s)	σ_v (cm/s)
Driving only	1	2.6	0.2
Continuous	1	2.6	0.3
Discrete	1	2.6	0.4
Driving only	1/6	2.7	0.3
Continuous	1/6	2.7	0.3
Discrete	1/6	No S/S	n/a

Note that σ_v represents the mean of the 3 tests' σ values, not the σ of the tests' mean v (which showed negligible variation between tests of any single set)

The gravity offload system was implemented primarily to test the hypotheses in this work and is not itself intended for extensive experimentation campaigns. Specialized gravity offload apparatus can be used to achieve even greater repeatability, and to overcome the limitations of the current system that include operating only in a straight line and lacking automatic speed adjustment.

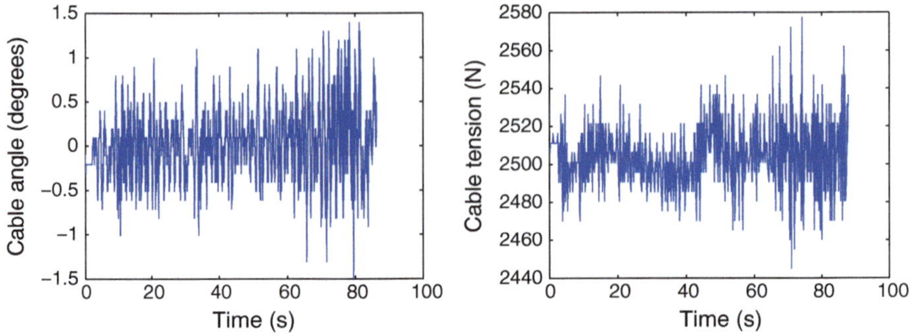

Fig. 4 Longitudinal angle (*left*) and tension (*right*) of the gravity offloading cable during a discrete excavation experiment, showing minimal variation

5 Development of Planetary Excavator Prototype

This section describes a planetary excavator prototype that incorporates the principles established by this research and addresses practical considerations of implementing a continuous excavator for planetary environments. The Polaris excavator, shown in Fig. 5, is a continuous bucket-wheel excavator. It is intended for in-situ resource utilization (ISRU), a task requiring substantial productivity. The 200 kg Polaris excavator features a nominal payload capacity of 80 kg for a payload ratio of 40 %; prior research by the authors has shown that payload ratio governs productivity [17]. To collect its payload Polaris uses continuous excavation, the benefits of which have been discussed in this paper. The entire bucket-wheel/collection bin subsystem is actuated to engage cutting with the bucket-wheel and to enable dumping at out the back of the bin at a height of 50 cm. Polaris' top driving speed is 40 cm/s.

Fig. 5 Polaris excavator featuring continuous bucket-wheel excavation and high payload ratio

5.1 Bucket-Wheel Excavator Configuration and Performance

Past planetary bucket-wheel excavator prototypes have had difficulty transferring regolith from bucket-wheel to collection bin, and as a result bucket-ladders have gained favor [10]. Bucket-ladders use chains to move buckets along easily shapeable paths, making transfer to a collection bin easy. Winners of the NASA Regolith Excavation Challenge and subsequent Lunabotics mining competitions (which require digging in lunar regolith simulant for 30 min) all employed bucket-ladders driven by exposed chains [14]. However, bucket-ladder chains are exposed directly to the soil surface and these could degrade very quickly in harsh lunar regolith and vacuum. The abrasiveness of lunar regolith rapidly degrades exposed sliding contacts or flexible materials [8, 19]. Exposed bucket-ladder chains may thus not be relevant to operation in lunar conditions.

A novel excavator configuration, with bucket-wheel mounted centrally and transverse to driving direction, achieves direct regolith transfer into a collection bin. The bucket-wheel is a single moving part, with no need for chains or conveyors. This reduces complexity and risk from regolith and dust. Once regolith has been carried to the top of the wheel in an individual bucket, it drops down out the back of the bucket and into a collection bin. This configuration offers a solution to the transfer problem for bucket-wheels identified in past literature.

The excavator prototype has demonstrated mining productivity of over 1000 kg/h. 1040 kg was produced in 58 min, with an average round trip of approximately 14 m, as demonstrated in GRC-1 at NASA Glenn's SLOPE lap. During the hour-long operation, the teleoperated excavator performed 17 dig-dump task cycles, of which approximately 1/3 of the time was spent digging. Average power draw was 470 W, with the wheels causing an average power draw of 142 W, the bucket-wheel 18 W, and lift/dump 310 W. Although this particular test was not conducted with gravity offload, the similarity in continuous excavation results in Table 1 suggests that comparable productivity may perhaps also be possible in 1/6 g. Full-scale excavation task experimentation with gravity offload is suggested for future work.

5.2 Suspension Lift-Off for Lightweight Skid Steer Rovers

Prior to integrating the excavation subsystem (consisting of bucket-wheel, dump-bed and raise/lower actuation) into Polaris, field tests were conducted to evaluate the performance of its mobility platform. These field tests revealed an undesireable phenomenon in which a wheel unintentionally lifts off the ground in a 'wheelie' fashion. Field and laboratory testing demonstrating the phenomenon, termed Suspension Lift-Off (SLO), are shown in Fig. 6. SLO occurs during skid-steering and results in reduced stability and loss of control authority; it is a problem that can be encountered with any passive differential mobility suspensions, such as rocker bogies.

Fig. 6 Field tests (*left*) led to the discovery and study of suspension lift-off (*right*)

An analytical model that relates lateral turning forces to vertical terrain-contact forces was developed, though its full details are omitted here for brevity; these details are presented in [4]. The following parameters are concluded to be root causes of SLO: a tall shoulder height to wheelbase ratio, narrow aspect ratio (i.e. ratio of lateral to longitunal wheel spacing), eccentric weight distribution, and high center of gravity. Operational factors that increase risk are high turning resistance and driving on slopes. Parameter sensitivity analysis suggests that the shoulder height to wheelbase ratio is the single most important factor.

For rovers with two shoulders, like Polaris, the effective wheelbase is the rover's actual wheelbase minus the distance between shoulders. Decreasing the effective wheelbase by separating the shoulders directly increases the shoulder height to wheelbase ratio and thus the risk of SLO. This overlooked caveat of Polaris' design was the single greatest contribution to the SLO problem encountered in field tests, particularly when the weight distribution on front and rear wheels was highly eccentric prior to integrating the excavation subsystem.

Tests compared the analytical model's predictions to experimentally measured values and found good accuracy across thirty-five long duration skid-steer trials that varied suspension geometry and weight [4]. Agreement of empirical evidence with the model suggests that SLO is predictable, and thus preventable if key design criteria are met. The mitigation is to achieve a shoulder height less than one third of the wheelbase, and a center of gravity height less than half the wheelbase. If these design criteria are met, SLO is very unlikely to occur.

The contribution of turning resistance to SLO suggests that operation in reduced gravity may exacerbate the problem. Section 2.1 discussed how sinkage does not diminish in low g for forward driving, decreasing DP/W. If sinkage also does not diminish in low g during skid-steering, this could increase the ratio of lateral resistance force to vertical contact force and lead to greater risk of SLO. Investigation of skid-steering in reduced gravity is thus suggested as a direction for future study.

6 Conclusions, Lessons Learned, and Future Work

Conclusions. The contributions of this work include the first of their kind gravity offload experiments from planetary excavators, and the conclusion that continuous excavation is more suitable for low gravity than discrete excavation. Gravity offload is an important and practical class of field or laboratory test for planetary excavator prototypes. Though not an ideal representation of low gravity operations, as the effects of gravity on regolith are not included, this is a more balanced test than excavating in full Earth gravity, which can misleadingly overpredict performance. Omitting gravity considerations from planetary excavator development misses important distinctions between classes of excavator configuration, such as the advantages of continuous excavation over discrete excavation.

The experiments presented in this work demonstrate that continuous excavation fares better than discrete excavation when subjected to low gravity. They also suggest caution in interpreting low gravity performance predictions based solely on testing in Earth gravity, where both the continuous and discrete configurations, misleadingly, operated successfully.

Lessons Learned. The key lesson learned from field testing is the need to consider suspension lift-off (SLO) for lightweight skid-steer robots. The mitigation is to achieve a shoulder height less than one third of the wheelbase, and a center of gravity height less than half the wheelbase. If the need to separate rocker arm shoulders arising in rover design, shoulder spacing should be minimized to avoid reducing the effective SLO wheelbase.

Future Work. Future research on lightweight excavation, including skid-steer testing, would benefit from testing in reduced gravity flights or drop towers. Excavation task testing would also benefit from more gravity offload testing in generalized terrain, beyond the flat straight-line tests shown here. Another important direction for future study is deep excavation in the presence of submerged rocks, which pose challenges for lightweight continuous and discrete excavators alike.

Acknowledgments This work was supported by NASA under grants NNX11CB55C and NNX07AE30G. We thank Rob Mueller for his support of this work as NASA COTR. We also thank Colin Creager, whose tireless collaboration and support at SLOPE enabled key aspects of these experiments.

References

1. Agui, J.H., Wilkinson, A.: Granular flow and dynamics of lunar simulants in excavating implements. In: ASCE Earth & Space 2010 Proceeding. Honolulu, HI (2010)
2. Bartlett, P., Wettergreen, D., Whittaker, W.: Design of the scarab rover for mobility and drilling in the lunar cold traps. In: International Symposium on Artificial Intelligence, Robotics and Automation in Space, pp. 3–6. Citeseer (2008)
3. Boles, W.W., Scott, W.D., Connolly, J.F.: Excavation forces in reduced gravity environment. J. Aerosp. Eng. **10**(2), 99–103 (1997)

4. Carlone, T.J.: Investigating suspension lift-off of skid-steer rovers with passive differential suspension. Master's thesis, Carnegie Mellon University (2013)
5. Caruso, J.J., Spina, D.C., Greer, L.C., John, W.T., Michele, C., Krasowski, M.J., Prokop, N.F.: Excavation on the moon: regolith collection for oxygen production and outpost site preparation. Technical Report 20080012503, NASA Glenn Research Center, Cleveland, Ohio 44135 (2008)
6. Clark, D.L., Patterson, R., Wurts, D.: A novel approach to planetary regolith collection: the bucket drum soil excavator. In: AIAA Space 2009 Conference and Exposition (2009)
7. Freitag, D., Green, A., Melzer, K.: Performance evaluation of wheels for lunar vehicles. Technical report, DTIC Document (1970)
8. Harrison, D.A., Ambrose, R., Bluethmann, B., Junkin, L.: Next generation rover for lunar exploration. In: 2008 IEEE Aerospace Conference, pp. 1–14. IEEE (2008)
9. Hettiaratchi, D., Witney, B., Reece, A.: The calculation of passive pressure in two-dimensional soil failure. J. Agr. Eng. Res. **11**(2), 89–107 (1966)
10. Johnson, L., Van Susante, P.: Excavation system comparison: bucket wheel vs. bucket ladder. In: Space Resources Roundtable VIII. Golden, CO (2006)
11. King, R., van Susante, P., Mueller, R.: Comparison of lance blade force measurements with analytical model results. In: Space Resources Roundtable XI/Planetary and Terrestrial Mining Sciences Symposium Proceedings (2010)
12. Kobayashi, T., Fujiwara, Y., Yamakawa, J., Yasufuku, N., Omine, K.: Mobility performance of a rigid wheel in low gravity environments. J. Terramech. **47**(4), 261–274 (2010)
13. Mueller, R.P., Smith, J.D., Cox, R.E., Schuler, J.M., Ebert, T., Nick, A.J.: Regolith advanced surface systems operations robot (rassor). In: IEEE Aerospace (2013)
14. Mueller, R.P., Van Susante, P.J.: A review of lunar regolith excavation robotic device prototypes. In: AIAA Space (2011)
15. Muff, T., Johnson, L., King, R., Duke, M.: A prototype bucket wheel excavator for the moon, mars and phobos. In: AIP Conference Proceedings, vol. 699, p. 967 (2004)
16. Oravec, H., Zeng, X., Asnani, V.: Design and characterization of grc-1: A soil for lunar terramechanics testing in earth-ambient conditions. J. Terramech. **47**(6), 361–377 (2010)
17. Skonieczny, K., Delaney, M., Wettergreen, D.S., "Red" Whittaker, W.L.: Productive lightweight robotic excavation for the moon and mars. J. Aerosp. Eng. **27**(4), 1–8 (2014)
18. Skonieczny, K., Moreland, S., Wettergreen, D., Whittaker, W.: Advantageous bucket-wheel configuration for lightweight planetary excavators. In: International Society for Terrain-Vehicle Systems International Conference (2011)
19. Stubbs, T.J., Vondrak, R.R., Farrell, W.M.: Impact of dust on lunar exploration. In: Dust in Planetary Systems, pp. 239–243. Kauai, HI (2005)
20. Theiss, R., Boucher, D., Viel, M., Roberts, D., Kutchaw, J.: Interchangeable payloads for ISRU mobility chassis. In: Space Resources Roundtable XI/Planetary and Terrestrial Mining Sciences Symposium Proceedings (2010)
21. Turnage, G., Banks, D.: Lunar surface mobility studies past and future. Technical report, DTIC Document (1989)
22. Wettergreen, D., Moreland, S., Skonieczny, K., Jonak, D., Kohanbash, D., Teza, J.: Design and field experimentation of a prototype lunar prospector. Int. J. Robot. Res. **29**(12), 1550–1564 (2010)
23. White, C.V., Frankovich, J.K., Yates, P., Wells Jr, G., Robert, L.: A capable and temporary test facility on a shoestring budget: the MSL touchdown test facility. In: 24th Aerospace Testing Seminar, 8 Apr 2008, Manhattan Beach, California. Jet Propulsion Laboratory, National Aeronautics and Space Administration, Pasadena, CA (2008)
24. Wong, J.: Predicting the performances of rigid rover wheels on extraterrestrial surfaces based on test results obtained on earth. J. Terramech. **49**(1), 49–61 (2012)
25. Wong, J., Asnani, V.: Study of the correlation between the performances of lunar vehicle wheels predicted by the nepean wheeled vehicle performance model and test data. Proc. Inst. Mech. Eng. Part D: J. Automobile Eng. **222**(11), 1939–1954 (2008)

Update on the Qualification of the Hakuto Micro-rover for the Google Lunar X-Prize

John Walker, Nathan Britton, Kazuya Yoshida, Toshiro Shimizu, Louis-Jerome Burtz and Alperen Pala

Abstract Hakuto is developing a dual rover system for the Google Lunar XPRIZE (GLXP) and exploration of a potential lava tube skylight. We designed, built and tested two rovers and a lander interface in order to prove flight-readiness. The rover architecture was iterated over several prototype phases as an academic project, and then updated for flight-readiness using space-ready Commercial Off The Shelf (COTS) parts and a program for qualifying terrestrial COTS parts as well as the overall system. We have successfully tested a robust rover architecture including controllers with performance orders of magnitude higher than currently available space-ready controllers. The test regime included component level radiation testing to 15.3 kilo-rads, integrated thermal vacuum testing to simulate the environments during the cruise phase and surface mission phases, integrated vibration testing to 10 G_{rms}, and field testing. The overall development methodology of moving from a flexible architecture composed of inexpensive parts towards a single purpose architecture composed of qualified parts was successful and all components passed testing, with only minor changes required to flight model rovers required ahead of a mid 2016 launch date.

1 Introduction

1.1 Commercial Off the Shelf Components in Space Robotics Missions

In the past several years, due to the proliferation of cubesat and micro-satellite missions, several companies have started offering off-the-shelf space-ready hardware [3]. These products offer a welcome reduction in cost but do not solve a major problem for

J. Walker (✉) · N. Britton · K. Yoshida · L.-J. Burtz · A. Pala
Tohoku University, Sendai, Miyagi, Japan
e-mail: john@astro.mech.tohouk.ac.jp

N. Britton
e-mail: nathan@astro.mech.tohoku.ac.jp

T. Shimizu
ispace Technologies, Inc., Tokyo, Japan

D.S. Wettergreen and T.D. Barfoot (eds.), *Field and Service Robotics*,
Springer Tracts in Advanced Robotics 113, DOI 10.1007/978-3-319-27702-8_21

space robotics designers: available space-ready controllers are years behind COTS microprocessors and microcontrollers in terms of performance and power consumption. For applications involving human safety or critical timing, the extra cost and difficulty of using certified space-ready hardware is justifiable.

But for some low-cost missions that require high performance, terrestrial components are increasingly being qualified and integrated. The University of Tokyo's HODOYOSHI 3 and 4 satellites have integrated readily available COTS FPGAs and microcontrollers and protected them with safeguards against Single Event Latch-up (SEL) [9]. This paper presets a lunar rover architecture that uses many COTS parts, with a focus on electrical parts and their function in and survival of various tests.

1.2 Google Lunar XPRIZE

The Google Lunar XPRIZE (GLXP) is a privately funded competition to land a rover on the surface of the Moon, travel 500 m and send HD video back to Earth. $30 million USD are available to teams who can complete these requirements, with $20 million USD for the first team to complete the requirements before December 31, 2016 [4].

In October of 2014, XPRIZE announced the Terrestrial Milestone Prize (TMP), a program for teams to be awarded for demonstrating flight-readiness to a panel of independent judges. Hakuto was selected as one of four teams to demonstrate mobility capability. Overall, five teams were selected to demonstrate achievements in mobility, imaging and lander capability. The TMP round concluded in January of 2015, and Hakuto was awarded $500 thousand USD for successfully testing its Moonraker rover with functional testing, thermal-vacuum testing, vibration testing and field testing [2].

1.3 Hakuto and Space Robotics Lab

Hakuto is the sole entrant from Japan in the GLXP competition and is developing rovers to send as payload on its landing service provider. As of 2015, it is one of 18 teams remaining in the competition. The Space Robotics Lab (SRL) is led by Professor Yoshida in the Department of Aerospace and Mechanical Engineering at Tohoku University in Sendai. It is partnered with Hakuto to design the rovers required for its mission.

1.4 Hakuto Mission and Rovers

In 2009, images from JAXA's KAGUYA (SELENE) spacecraft showed the presence of potential skylights on the surface of the moon [5]. The Lunar Reconnaissance Orbiter (LRO) has also shown several potential skylights. Hakuto's landing service provider has identified one such potential skylights as its landing target. The target is in the Lacus Mortis region at 44.95°N and 25.61°E, south of the Rimae Bürg rille. The skylight is just under 400 m in diameter, with a ramp on one side, possibly formed by a partial collapse. The minimum average slope angle is 13°, although the data from the LRO for this estimation is sparse [1].

In order to explore a skylight or cave, we developed a dual rover architecture, consisting of a one four-wheeled parent rover (code-named "Moonraker")and one two-wheeled tethered child rover (code-named "Tetris"). In this architecture, both rovers use radio communication via the third-party lander to Earth.

Moonraker will travel near the edge of a skylight with Tetris towed by a tether. The tether, up to 100 m long is wound on a motorized spool within Tetris is used to pull itself back to Moonraker after exploring steep, vertical, or any terrain that the operators wish to "scout" ahead of Moonraker (Fig. 1).

Active tethers for similar purpose have been demonstrated by the European Space Agency [6], but they are complex, requiring slip rings and multiple conductors. They would eliminate the need for solar cells or batteries, but we chose a passive tether for two types of redundancy:

Type 1 Operational redundancy: In case of failure of one rover, we can still complete the GLXP requirements.

Type 2 Lander agnosticism: Depending on the lander capabilities, one (Tetris or Moonraker) or both rovers can be integrated, maximizing the number of potential launches

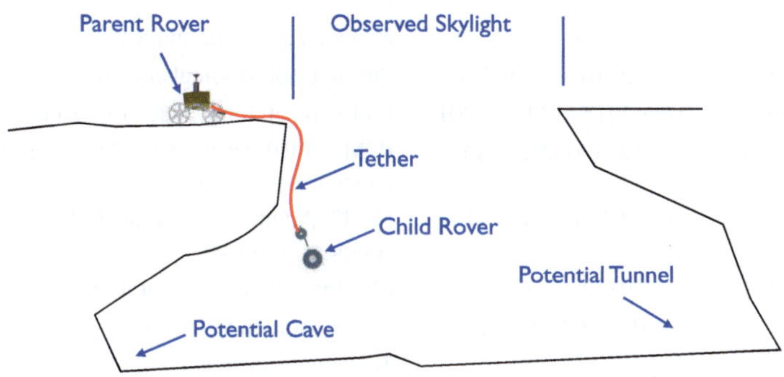

Fig. 1 Hakuto's dual rover mission architecture, with one four-wheeled parent rover and one two-wheeled tethered child rover

Because both rovers use many of the same or similar components and potentially identical controller architectures, the additional resources required for developing the dual rover system is marginal.

1.4.1 Development Phases

Hakuto has just completed the fourth development phase as described in Table 1. In this phase, within our budget and time constraints, we made the rovers as close as possible to flight configuration. There is overlap in the phases, as environmental and field testing can overlap with the design stage of a subsequent design.

Up until the end of Phase 3, Moonraker was made from an aluminum chassis with nylon body panels, and Tetris was made from an aluminum sheet metal structure. Throughout this time, small iterations to items such as wheel size, grouser length and motor power were made as a result of many field tests and lab experiments [1]. Throughout these phases, the primary goal of the rovers was academic research, with the general requirements of the GLXP used for guidance. Moonraker's development history for the GLXP project goes back to 2009. The addition of Tetris to Phase 2 created Phase 3. We plan to maintain this cycle of "major-minor" updates. Phase 4 was a major update, internally called the "Pre-Flight Model" or PFM. It was designed to the flight requirements and every component which was not a space-ready COTS component was designed or selected to qualified to flight-ready status.

A minor update to Phase 4 will also be tested. It will include the flight configuration of all electronics. In parallel, the design of Phase 6 (flight model) will be conducted, with all testing for Phase 5 completed before the Critical Design Review for Phase 6. The overall scheme is illustrated in Fig. 2.

Table 1 Description of the phases of development

Phase	Time period	Description
Phase 1	Jan 2009 to June 2010	Research and trade studies
Phase 2 (major)	June 2010 to Sept 2013	Prototype of Moonraker using COTS hardware
Phase 3 (minor)	Sept 2013 to March 2014	Prototype of Tetris added to system
Phase 4 (major)	Jan 2014 to Dec 2014	PFM: CFRP structure, COTS space-ready components and COTS terrestrial components
Phase 5 (minor)	Dec 2014 to Aug 2015	PFM2: Additional/alternate COTS candidate components added
Phase 6 (major)	Jan 2015 to Dec 2016	FM: Final flight configuration
Phase 7	April 2016 to June 2016	FM integration to lander
Launch	July 2016	Tentative launch date

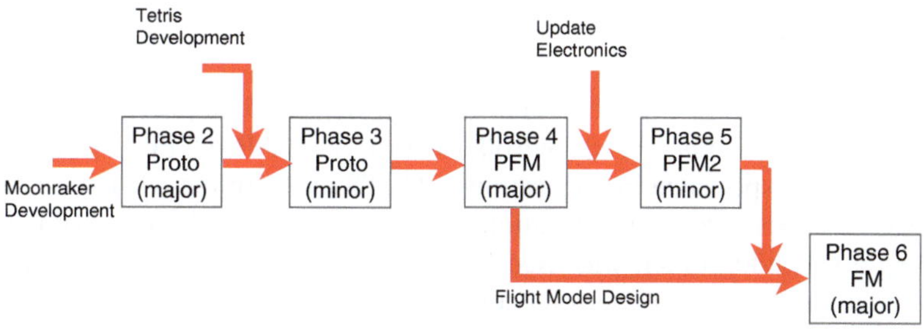

Fig. 2 Hakuto's Major-minor development strategy

2 Phase Four System Architecture

We updated the design for Phase 4 based on the design and field testing of the Phase 3 rovers. We made minimal changes to overall configuration, but performed extensive detailed design with attention to the thermal and vibration environments expected during the mission.

The criteria for component selection was: mass, power consumption, and use of components with flight heritage, especially by SRL when possible.

2.1 Rovers

The rovers we built for Phase 4 feature an aluminum substructure and Carbon Fibre Reinforced Plastic (CFRP) outer body (Fig. 3). We built these in order to meet the requirements for the Terrestrial Milestone Prize detailed in Sect. 1.2.

Fig. 3 Phase 4 Moonraker and Tetris rovers

2.1.1 Moonraker

The architecture of Phase 4 Moonraker (Fig. 4) was based on previous versions. A COTS space-ready FPGA-based controller with a "soft" ARM CPU [7] was selected due to previous experience in integration to COTS parts for the RISING-2 satellite [8]. A COTS cubesat Power Distribution Unit (PDU) and 80 Wh lithium-ion battery, including a watchdog timer, was used for the power subsystem. Solar panels were not included in Phase 4 but one solar cell was included on Moonraker to confirm its physical integration and survival of environmental testing.

The omni-directional imaging components, consisting of a COTS USB 5 mega-pixel camera, lens and parabolic mirror were retained from Phase 3. The camera points upwards to the mirror, to capture a 360° image that is manipulated by the operator to enable them to look in any direction without the complexity or lag associated with a pan-tilt mechanism [1]. We also kept a COTS laser range-finder from Phase 3 that uses a MEMS mirror to control the pan and tilt of a stationary laser to produce 3D data via a time of flight algorithm.

The main controller is not powerful enough for the real-time HD video processing required by the GLXP, so a COTS ARMv7-based controller was added to handle imaging. This is a readily-available product primarily marketed towards hobbyists, with nearly all signals from the CPU made available on two 48-pin headers making it ideal for a flexible development platform. Other COTS components were picked primarily based on flight heritage and are described in Table 2.

We made two interface boards to connect components. The "power interface" board was used to mount and connect the main controller, ethernet switch, and PDU. The "imaging interface" was used to connect the imaging controller, camera, radio. Both included minor components such as power relays, ethernet transformers, level converters and multiplexers. Many electrical connections to the interface boards were made by the pin headers factory installed on the PDU and imaging controller. We removed all connectors not designed for aerospace use, such as ethernet and USB, and replaced them with soldered "pigtail" wiring with connectors having space heritage.

2.1.2 Tetris

Tetris' planned architecture for Phase 4 was nearly identical to Moonraker's, with two wheels instead of four, no range-finder, and a tether mechanism added. The total mass of Tetris is 2629 g and the average power consumption budget is 7.3 W.

2.2 Interface to Lander

The lander interface box was made from CFRP and machined parts, with 3D printed Ultem parts in the interior to hold both rovers fixed during the launch, cruise and landing phases. Upon landing, the interface box is opened with a single Shape Mem-

Table 2 Summary of test results for Moonraker

Component	Description	Thermal	Vibration	Radiation	Field	Flight heritage	Mass (g)	Power (W)
Wheels	3D Printed Ultem	Pass	Pass	NT	Fail	No	1980	0
Structure	Machined aluminum	Pass	Marginal	NT	Pass	No	1246	0
Mechanical parts	Machined aluminum	Pass	Pass	NT	Pass	No	1793	0
Fasteners	Steel	Pass	Pass	NT	Pass	No	125	0
Body	CFRP	Pass	Marginal	NT	Pass	No	725	0
Motors	12 W brushless	Pass	Pass	NT	Pass	RISING-2	696	12
Motor controller	Custom SH2A-based	Pass	Pass	NT	Pass	RISING-2	32	1
Camera	5MP USB camera	Pass	Pass	Pass	Pass	No	10	1
Lens	COTS C-mount	NT	Pass	NT	Pass	No	105	0
Mirror	Hyperbolic mirror	NT	Pass	NT	Pass	No	78	0
Range-finder	MEMS-based laser	Pass	Pass	Pass	Pass	No	480	6
Imaging controller	ARMv7 COTS (hobbyist)	Pass	Pass	Pass	Pass	No	40	0.5
Batteries	80 Wh, 15 V COTS (cubesat)	Pass	Pass	NT	Pass	Cubesat	473	0.1
Power unit	COTS (cubesat)	Pass	Pass	NT	Pass	Cubesat	105	0.5
Solar cells	1 Triple junction cell	Pass	Pass	NT	Pass	Yes	3	0
Wiring	COTS (MIL spec, industrial)	Pass	Pass	Pass	Pass	Yes	220	0.3
Radio*	See note	Pass	Pass	Pass	Pass	No	18	1
Main controller	COTS (space-ready)	Pass	Pass	NT	Pass	No	15	1.5
Deployment switches	COTS	Pass	Pass	NT	Pass	No	20	0

(continued)

Table 2 (continued)

Component	Description	Thermal	Vibration	Radiation	Field	Flight heritage	Mass (g)	Power (W)
Debug/charge interface	COTS (MIL spec)	Pass	Pass	NT	Pass	No	30	0
Power interface board	SRL made	Pass	Pass	NT	Pass	No	35	0.1
Imaging interface board	SRL made	Pass	Pass	Pass	Pass	No	55	0.1
Mass memory	Controller on-board eMMC	Pass	Pass	Pass	Pass	No	0	0
IMU	COTS	Pass	Pass	Pass	Pass	Cubesat	10	0.1
Ethernet switch	COTS (UAV)	Pass	Pass	Pass	Pass	No	35	0
Power switches	COTS	Pass	Pass	NT	Pass	No	50	0
Charging board	SRL made	Pass	Pass	NT	Pass	No	45	0
Totals							8424	24.2

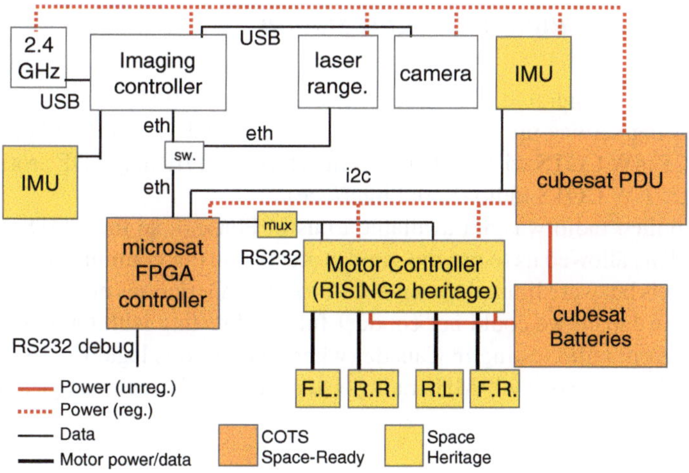

Fig. 4 Moonraker architecture for phase four of development

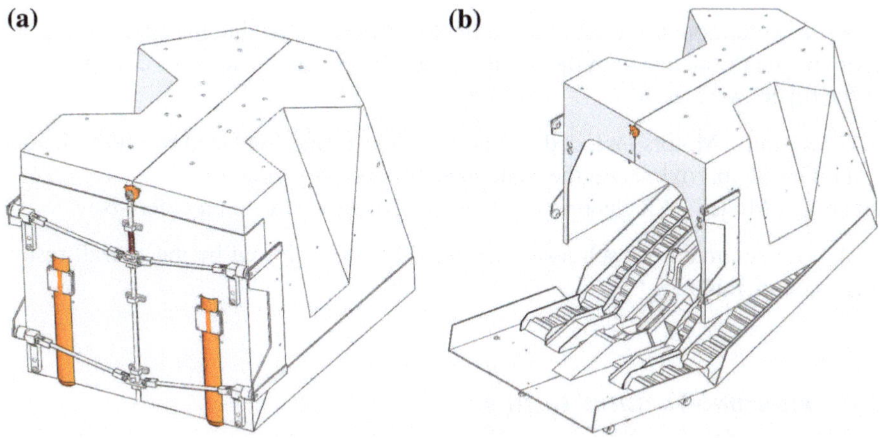

Fig. 5 View of the interface box. Moonraker and Tetris are nested inside when stowed. When deployed, the door forms a ramp for the rovers to drive down. **a** Stowed configuration. **b** Deployed configuration

ory Alloy (SMA) pin-puller actuator. The open box acts as a ramp with a slope of approximately 30° for easy egress of the rovers on to the lunar surface. Figure 5 shows the interface in the stowed configuration and deployed ramp configuration.

2.3 Communication to Ground Station

The rovers are configurable with three types of radios: a 900 MHz, 1 W radio supplied by our landing service provider (ethernet interface, TCP/IP and UDP protocols), a 2.4 GHz, 25 mW COTS wi-fi radio (USB interface, TCP/IP and UDP protocols) and a 900 MHz, 1 W COTS radio (Serial interface).

The supplied radio was not available to us in Phase 4, so the COTS wi-fi radio was used. This allowed us to use the same protocols, in our communication, as in the Flight Models but were limited in range. Due to strict restrictions on radio frequencies and power in Japan, we could not conduct full field testing with Option 3 in Japan. We did perform radio testing in Canada (where the radio is legal to use) to confirm general performance of a 900 MHz radio system at long distances and near obstacles.

3 Testing

In 2014, we thoroughly tested the Phase 4 rovers to determine the suitability of all components for inclusion in the flight model. During these tests, two configurations of the rovers were used:

MTM Model Motors included, but all other electronics replaced by representative masses of approximately the same mass and centre of gravity

Integrated Model All electronics included, except where noted otherwise

A brief summary of each test is included below, followed by the overall results presented in Table 2.

3.1 Thermal-Vacuum Testing

3.1.1 Cruise Phase Testing (MTM Model)

Thermal-Vacuum tests were performed at the Kyushu Institute of Technology established Center for Nanosatellite Testing (CeNT) in the Tobata campus of the Kyushu Institute of Technology. This is a centralized facility with test apparatuses for satellite testing, including thermal-vacuum testing (10^{-5} Pa).

The MTM model of Moonraker, Tetris and the interface box, in the stowed configuration were tested, with sensors at various internal and external points to verify thermal conductance values used in the thermal models.

We simulated the cruising phase of the mission, with the shroud temperature of the vacuum chamber at $-173\,^{\circ}$C, and the interface box wrapped in Multi-Layer Insulation (MLI). The interface box was fastened to an aluminum plate to simulate the deck of the lander. The deck was temperature controlled between 0 and $40\,^{\circ}$C.

Fig. 6 Experimental setups used for Cruise Phase and Surface Mission. **a** Cruise phase. **b** Surface Mission phase

The data from this test will be used to confirm thermal models and design heaters for the interface box in order to keep the rovers electrical components within their preferred range (with the battery having the most severe requirements of between −20 and 40 °C).

Figure 6 below shows the vacuum chamber used in the test, and the MLI-wrapped interface ready for insertion.

3.1.2 Surface Mission Phase Testing (integrated Models)

We performed integrated vacuum testing on Moonraker at Next generation Space system Technology Research Association (NESTRA) at the Kikuicho campus of Waseda University in Tokyo. This is a facility for micro-satellite integration and thermal-vacuum testing. We plan to land 12 h after sunrise (−68 °C), with deployment at 30 h after sunrise (−10 °C) with the GLXP mission complete by 75 h after sunrise (50 °C). The lens and mirror used were COTS products manufactured using the vacuum deposition coatings, so were not tested in oder to avoid contaminating the vacuum chamber through out-gassing.

Figure 6 shows the rover installed on the vacuum chamber testing baseplate. Five panel heaters were placed around the rover. During vacuum conditions, hot and cold tests, to certify operation of the rover up to 75 h after sunrise were performed. Since our current engineering model batteries do not have battery heaters installed, −20 °C was selected for the cold mode temperature. This allows us to validate our thermal model for the system without risk of damage to the batteries (minimum temperature −20 °C). 40 °C was selected for the hot mode temperature.

The data from this test will be used to confirm thermal models and design radiative cooling for the rovers during the surface mission. Although Tetris was not tested, its similar materials and design mean its thermal model can also be partially validated.

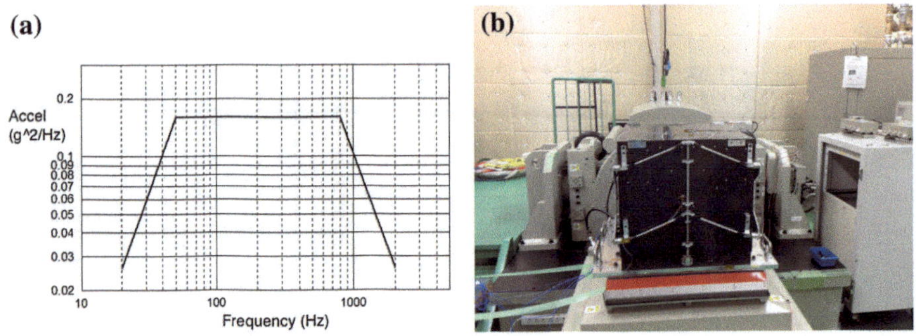

Fig. 7 Vibration testing PSD and experimental setup. **a** QT Level PSD. **b** X-Axis testing

3.2 *Vibration Testing*

We performed vibration testing to Qualification Level (QT), 14.1 G_{rms}, using motors and representative masses in place of electronics and to Acceptance Level (AT), 10.0 G_{rms}, using fully integrated rovers. These levels come from our landing service provider based on NASA standard GSFC-STD-7000A. The prescribed Power Spectral Density (PSD) is shown in Fig. 7 along with the system mounted in the X-axis configuration on a shaker table.

3.2.1 QT Level MTM Testing

Although only AT level testing was required for qualification to our landing service provider's requirements, we tested the structures only (by using the MTM models) to QT level of 14.1 G_{rms}. No damage was observed, and the overall modes of vibration were acceptable. However, five structural parts were identified with resonant frequencies near or below 40 Hz. Upon deployment the rovers could freely move down the ramp shown in Fig. 5.

3.2.2 AT Level Integrated Testing

We tested the system to AT level of 10.0 G_{rms} with all electronics disabled by holding a normally closed deployment switch open. The integrated testing to AT level also resulted in no damage. Upon deployment, the deployment switch as well as every electrical component functioned correctly, and Moonraker was commanded via a radio link and simulated ground station to leave the interface box. This test was also successful.

3.3 Component Level Radiation Test

We performed component level radiation testing at Takasaki Advanced Radiation Research Institute, Japan Atomic Energy Agency. All electrical components except those with demonstrated flight heritage were tested. Electronic subsystems were placed in front of Cobalt-60 γ source. Precise dosimeters were mounted included to provide accurate measurements of total dose. Exposure time was 4.5 h, providing a total absorbed dose of 15.3 kilo-rads $\pm 3\,\%$, about four times the expected total dose (4 kilo-rads). Testing was done with components on, and function tested continuously. All components functioned correctly and without issue except for the imaging controller. This had two reboot events, presumably caused by the effects of radiation. Correct function resumed after reboot. This result was anticipated due to the high density of transistors on the CPU, so our design relies on a watchdog timer on the PDU to reset both controllers if activity stops.

3.4 Field Testing

A field test was conducted at the Nakatajima sand dunes in Hamamatsu Japan. The sand dunes are a lunar analogue site nearly void of all vegetation except some sporadic grasses. Surface features of interest to us are long valleys of soft sand, local hills, steep cliffs, rocky as well as rock-void areas (Fig. 8).

All components functioned as expected during the field test, with no major issues. In the presence of a GLXP judge, we successfully traveled 620 m and demonstrated ability to teleoperate in realistic conditions, including a simulated time delay, and a data rate of only 100 kbps [2]. The only issue uncovered was that the grouser design can pick up rocks which become lodged in the suspension mechanism.

Fig. 8 Moonraker performance during field testing. **a** Overcoming a 15 cm high rock obstacle during field testing. **b** Climbing an approximately 30° slope on soft soil

3.5 *Radio Testing*

As described in Sect. 2.3, the third radio option could not be tested in Japan. We conducted two tests using antenna configurations and heights similar to the flight model in Vancouver, Canada to a distance of 1.5 km, and characterized the performance near obstacles up to 3 m in height so that operators can determine where to expect "dead zones" that should not be traversed [10].

3.6 *Test Results*

All of the test results are summarized in the Table 2. In this table, "NT" is used for items that weren't included in a particular test. Nearly all components passed all tests or has demonstrated flight heritage. The exceptions are shown in the first part of Table 3 with an explanation and proposed resolution.

Table 3 Phase Six changes for Moonraker

Component	Issue	Solution
Wheels	Rocks can get stuck in grouser	Modify grouser for clearance of suspension
Main controller	Some parts near 40 Hz threshold	Stiffen parts for FM design
Structure and body	Integrated structure will save mass	Remove aluminum substructure
Thermal interfaces	Integrated structure will save mass	Remove thermal paths, integrate design to structure
Deployment switches	Not radiation tested	Passive components; testing not required
Debug/charge interface	Not radiation tested	For development; not required for FM
Power interface board	Not radiation tested	Iterate design and radiation test
Power Switches	Not radiation tested	Not required for FM
Charging board	Not radiation tested	For development; not required for FM
Main interface board	New controller architecture	Change from COTS ARM-based board to custom
Wiring	New wiring standard for FM	Change connectors to MDM
Camera	Redundant architecture for FM	Change to parallel interface, add camera
Range-finder	Reduce power consumption and mass	Change from laser-based to camera-based
Debug interface	Not needed for flight configuration	Make removable debug interface
Switch interface	Not needed for flight configuration	Remove from design
Charge interface	Not needed, use solar interface	Use external connector for solar cell simulation

Fig. 9 Moonraker's internal components, with complex wiring harness

Each motor uses approximately 10 W while the rover is in motion, but in our field testing experience, the rover is stopped much of the time while operators make decisions. Therefore the average power consumption is greatly reduced, to about 12 W.

The main controller for Phase 4 was selected due to its robustness, flight heritage and SRL's experience with it. But HD imaging is a strict requirement of the GLXP competition, effectively making the architecture, as we designed it, dependent on both the imaging and main controllers functioning properly. With this result, for Phase 5 and 6 we merged the function of the two controllers and changed to a redundant computing architecture (Sect. 4).

As important as the test results was the experience of integration. The wiring shown in Fig. 9 is mostly made of a single harness with many connectors. Wiring routing, thermal paths and component placement and connector position can all be greatly improved to reduce integration time and decrease wiring mass.

4 Phase Five Architecture

We are now using the results of Phase 4 development to design and fabricate the Phase 6 rovers. Aside from the change in controller, only minor changes are specified by the test results themselves, as described in Table 2. All updates to electrical components, wiring and connectors will be tested in the Phase 5 rovers before the Critical Design Review for Phase 6.

4.1 Changes from Phase Four to Phase Five and Six

At the time the lander interface was not fixed, so the interface boards and wiring harness included options for different interconnections and protocols. This was flexible for development but now these options have been reduced so there are mass savings and opportunities for the FM. Many connectors can be removed and/or consolidated. To simplify wiring, all signal routing will take place on the interface boards. "Straight" cables with identical pin assignments on both sides are also easier to specify and purchase as items from suppliers with quality-control certifications.

HD video is a strong requirement that demands a capable controller, redundancy for both the main controller and imaging controller (and camera) is a hard requirement. Since the imaging controller is capable of the main controller functions, and passed all environental testing in Phase 4, we made a new architecture (Fig. 10) with identical controllers, each connected to a cameras. This way, redundancy is created, development time is reduced (because a heterogeneous architecture does not have to be supported).

Phase Four used a USB camera but the flight model will change to use the same imaging sensor's native parallel interface. This will eliminate the camera's on-board USB circuitry and approximately 500 mW of power consumption. We chose a 10 g, "System on Module" (SOM) board with only the components that we require. Most available SOMs include unnecessary components such as DC-DC converters and HDMI ports, or do not route all of the required interfaces to the CPU. The Phase 4 controller was approximately 40 g and included many components that use power and add failure points. Debug and charge interfaces will also be made modular so they can be removed prior to flight to save mass.

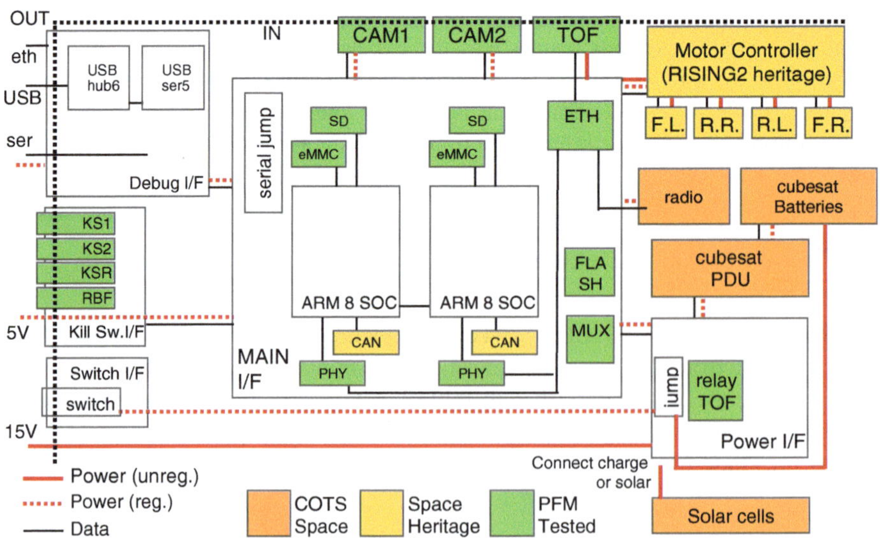

Fig. 10 Moonraker flight model architecture

The testing regimen for this phase will be similar: radiation testing, thermal-vacuum testing, vibration testing and field testing. Although architecture changes have been minimized, the change of controller described above presents a large risk, if it is not qualified before the rest of the electronics systems are designed and manufactured. This is because, due to time constraints and subsystem interdependencies, it will be difficult or impossible to change the controller. Therefore the first step of Phase 5 is fabrication of prototype boards so that component-level radiation testing can be completed ahead of detailed design.

The design target for the flight model Moonraker is a reduction of mass from 8.4 to 4.0 kg and of power from 24 to 18 W. Approximately half of the reduction in mass will be achieved by removing the aluminum substructure. The rest is achieved by small reductions in each subsystem. Reduction in power is achieved by replacing the laser rangefinder changing away from a heterogeneous controller architecture, as well as removing unnecessary interfaces (such as USB).

5 Conclusion

Through extensive radiation testing, vibration testing, thermal-vacuum testing and field testing, we have demonstrated a dual rover architecture using many space-ready and terrestrial COTS components. This architecture is capable of completing both the GLXP mission requirements and exploration of a potential lava tube skylight on the surface of the moon. We have identified five structural parts to be redesigned, and changed from a heterogeneous controller architecture using both a space-ready main controller and ARM-based imaging controller to a dual, COTS, ARM-based architecture. This has allowed us to reduce mass, number of components, power consumption and development time even while adding a redundant camera and theoretically increasing reliability of the overall system. The use of COTS components has allowed us to start from a convenient, inexpensive flexible architecture for development and arrive at purpose-built, power-efficient architecture by removing components and options for interconnections over time. The overall development strategy of alternating large overall design changes and small subsystem iterations was also effective.

References

1. Britton, N., Yoshida, K., Walker, J., Nagatani, K., Taylor, G., Dauphin, L.: Lunar micro rover for exploration through virtual reality tele-operation. In: Field and Service Robotics (2013)
2. CNET: $5.25 Million Awarded to five Google Lunar XPrize Teams. CNET Website, Feb 2015
3. cubesat.net: Suppliers [of cubesat components]. cubesat.net Website (2014)
4. Foundation, X. Overview—Google Lunar XPRIZE. GLXP Website (2015)
5. Haruyama, J., Hioki, K., Shirao, M.: Possible lunar lava tube skylight observed by selene cameras. Geophys. Res. Lett. **36**(21), 1–5 (2009)

6. Iqbal, J., Heikkila, S., Halme, A.: Tether tracking and control of rosa robotic rover. In: 10th International Conference on Control, Automation, Robotics and Vision, p. 690. IEEE (2008)
7. Mictrotec, A. OBC Lite 52X Series. AAC Mictrotec Website (2013)
8. Sakamoto, Y., Takahashi, Y., Yoshida, K., Fukuda, K., Nakano, T., Batazzo, S., Fukuhara, T., Kurihara, J.: Development of the microsatellite rising-2 by Tohoku University and Hokkaido University. In: Proceedings of the 61st International Aeronautical Congress, No. IAC-10-B4.2.12 (2010)
9. Terakura, M., Kimura, S.: High-performance low-cost image processing unit for small satellite earth observation using COTS devices. Int. J. Emerg. Trends Technol. Comput. Sci. **2**(4), 370 (2013)
10. Walker, J., Yoshida, K., Britton, N., Seo, M.: Dual rover robotic mission architecture for exploration of a potential lava tube skylight on the lunar surface. In: Proceedings of the 65th International Aeronautical Congress, No. IAC-10-B4.2.12 (2014)

Mobility Assessment of Wheeled Robots Operating on Soft Terrain

Bahareh Ghotbi, Francisco González, József Kövecses and Jorge Angeles

Abstract Optimizing the vehicle mobility is an important goal in the design and operation of wheeled robots intended to perform on soft, unstructured terrain. In the case of vehicles operating on soft soil, mobility is not only a kinematic concept, but it is related to the traction developed at the wheel-ground interface and cannot be separated from terramechanics. Poor mobility may result in the entrapment of the vehicle or limited manoeuvring capabilities. This paper discusses the effect of normal load distribution among the wheels of an exploration rover and proposes strategies to modify this distribution in a convenient way to enhance the vehicle ability to generate traction. The reconfiguration of the suspension and the introduction of actuation on previously passive joints were the strategies explored in this research. The effect of these actions on vehicle mobility was assessed with numerical simulation and sets of experiments, conducted on a six-wheeled rover prototype. Results confirmed that modifying the normal load distribution is a suitable technique to improve the vehicle behaviour in certain manoeuvres such as slope climbing.

1 Introduction

Defining robust and reliable operational strategies for wheeled robots operating on soft terrain is a challenging task. An example of this are planetary exploration rovers, one of the most demanding applications of wheeled robotics. Besides tackling the

---

B. Ghotbi (✉) · J. Kövecses · J. Angeles
McGill University, Montreal, QC H3A 0C3, Canada
e-mail: bahareh.ghotbi@mail.mcgill.ca

J. Kövecses
e-mail: jozsef.kovecses@mcgill.ca

J. Angeles
e-mail: angeles@cim.mcgill.ca

F. González
Laboratorio de Ingeniería Mecánica, University of La Coruña, Mendizábal s/n, 15403 Ferrol, Spain
e-mail: f.gonzalez@udc.es

D.S. Wettergreen and T.D. Barfoot (eds.), *Field and Service Robotics*,
Springer Tracts in Advanced Robotics 113, DOI 10.1007/978-3-319-27702-8_22

usual problems derived from operating on irregular terrain, rovers must often deal with an incomplete knowledge of the soil properties. Moreover, most missions must be accomplished in an autonomous or semi-autonomous fashion.

Optimizing the vehicle mobility is an important goal in the design and operation of wheeled robots on soft soil. In the case of wheeled robots that operate on rigid ground, mobility is a kinematic concept which can be defined based on the assumption that each wheel in the robot rolls without slipping. However, when the same robots operate on soft terrain the above mentioned assumption is generally no longer valid. Mobility can be understood in the sense of the ability to move from a certain configuration, or to move with maximum speed. This definition is close to the *trafficability* concept introduced by Apostolopoulos [1], which points to the capacity of the vehicle to overcome terrain resistance and generate traction.

Reduction of the slip at the wheel-terrain contact area has been proposed as a way to enhance the mobility of wheeled robots operating on unstructured terrain by Lamon et al. [2] and Thueer et al. [3]. In these papers, the wheel-terrain interface is modelled using the assumption of Coulomb friction while the ratio of tangential to normal forces at the wheel-ground contact is minimized with the goal of reducing the risk of developing slip. While not directly dealing with soft soil modeling, these papers highlight the need for keeping wheel slip under control in order to improve mobility.

Some publications in the literature point out that a uniform distribution of normal forces among the vehicle wheels may have a positive effect on the mobility. Grand et al. [4] state that balancing the normal loads helps the vehicle to develop a higher value of the overall traction force. Along the same lines, it is suggested by Freitas [5] that uniformly distributing the weight of the rover among the wheels is a valid strategy to achieve better mobility, when enough information about contact forces is not available. A similar conclusion was also reported in [6]: the load distribution among the wheels has to be even on flat ground to achieve the best performance. Thueer et al. [7] chose an alternative strategy to reduce the likelihood of developing wheel slip that relies on the minimization of the virtual friction coefficient $\mu^* = F_T/F_N$ where F_T is the traction and F_N the normal force at each contact. Iagnemma and Dubowsky [8] computed the normal load and the motor torque applied to each wheel as a solution of an optimization problem to enhance mobility for quasi-static motion of the rover on rough terrain.

The authors of this paper reported an experimental confirmation of the above research for a particular rover design in [9, 10]. We also introduced the normal force dispersion as performance indicator to quantify the proximity of the load distribution to ideal operation conditions [11]. This distribution can be changed by means of reconfiguration or by introducing actuation on the suspension elements. As a consequence, in some cases it is possible to obtain a more favourable load distribution that would increase the mobility for a given manoeuvre. This paper discusses the effect of two of these strategies during slope climbing and drawbar pull tests. The first one consists of relocating the vehicle centre of mass (CoM), while the second introduces redundant internal actuation between suspension elements.

2 Normal Force Dispersion as Mobility Indicator

The mobility of a wheeled robot depends on its ability to generate a required amount of drawbar pull while keeping the slip ratio low. Terramechanics theory [12] points out that the normal force F_n at each wheel of a robot affects the developed tangential force F_D and, in turn, the total drawbar pull that the vehicle provides. The terrain normal reactions have to balance the inertial and external forces applied on the rover. However, changing the normal load distribution among the wheels can result in different values of the total drawbar pull developed by the vehicle.

The effect of normal force distribution can be studied using the F_D-vs.-F_n curve. An example for a planar three-axle system in 2-D motion is shown in Fig. 1. If the three axles are moving with the same angular speed, the terrain under the vehicle is homogeneous, and the multipass effect is negligible, then the same curve can be used for all the wheels. In this case, an even normal load distribution would be the one in which $F_{n1} = F_{n2} = F_{n3} = F_n^*$. A normal load transfer between the first and second axles of the robot ($\Delta F_{n1} = -\Delta F_{n2}$) will result in $\Delta F_{D1} < 0$ and $\Delta F_{D2} > 0$ in the drawbar pull at these wheels. If the slope of the F_D-vs.-F_n curve decreases consistently with F_n, then $|\Delta F_{D2}| < |\Delta F_{D1}|$, which will yield a lower total available drawbar pull for the same slip values. In other words, in the uneven configuration the slip should become higher in order to achieve the same drawbar pull delivered by its balanced counterpart, where the normal forces are uniformly distributed among the wheels.

For the case of a wheeled robot operating on homogeneous terrain with negligible multipass effect, the F_D-vs.-F_n relation will be the same for all the wheels if they are identical and have the same slip. These assumptions can be considered close enough to reality for a broad range of operating conditions.

The *Normal Force Dispersion* (NFD) denoted by η was introduced in [11] to measure and quantify the uniformity of the normal force distribution. This performance indicator is the standard deviation of the normal forces at the wheel-terrain contact interfaces, namely,

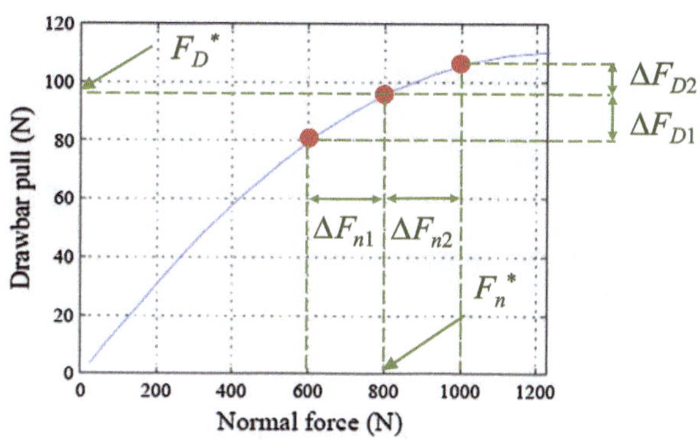

Fig. 1 Effect of non-uniform normal force distribution on the total available drawbar pull

$$\eta\left(F_{n1},\ldots,F_{np}\right)=\sqrt{\frac{1}{p}\sum_{i=1}^{p}(F_{ni}-\mu)^2} \tag{1}$$

where p is the number of wheels of the vehicle and μ is the average normal force:

$$\mu=\frac{1}{p}\sum_{i=1}^{p}F_{ni} \tag{2}$$

An even distribution of normal forces ($F_{n1}=F_{n2}=\cdots=F_{np}$) would result in $\eta=0$, which is optimum in terms of developed drawbar pull for operation on homogeneous terrain with negligible multipass effect and assuming that all the wheels of the vehicle have the same slip ratio. Quantifying the unevenness of the load distribution via NFD facilitates the comparison of different rover configurations in terms of their mobility, while it may avoid the need for a detailed knowledge of the terrain properties.

As a conclusion, it can be stated that making the normal force distribution more uniform will have a noticeable effect on the drawbar pull when the F_D-vs.-F_n curve shows an apparent sublinear relationship. This is the case of operation conditions where high slip values are expected to develop, such as in slope climbing, or in the presence of loose terrain with low values of k_ϕ.

2.1 Case Study: The RCP Rover

The normal force dispersion was used to study the mobility of the RCP, a six-wheeled rover prototype developed by the Robotics and Automation unit of MDA (MacDonald, Dettwiler and Associates Ltd.) shown in Fig. 2. The total mass of the RCP is about 125 kg. The rover main body is attached to three bogies (starboard, port, and rear), each one connected to two wheels. Every wheel can be independently steered and actuated.

Fig. 2 The RCP rover (*left*) and its multibody model (*right*)

A full-scale model of the rover was built using the generic multibody dynamics library, a multibody software tool developed by the authors [13], implemented in MATLAB. This library includes functions to evaluate the wheel-terrain interaction forces according to the terramechanics semi-empirical relations introduced in [12, 14]. Among many features of this library having access to all dynamic terms and choice of various multibody dynamics formulations and integrators make it suitable for rover analysis in complex environments. RCAST is an alternative dynamic simulation tool which has been reported to model the same rover [15].

First, the climbing manoeuvre of the RCP on a 10° slope with the terrain properties listed in Table 1 was simulated. The wheels of the rover were commanded to move with a constant angular speed $\omega = 0.4$ rad/s. In order to obtain different load distributions among the wheels of the RCP, a 22.5-kg payload was added as a movable mass element to the rover model. The simulation was repeated for different locations of the payload along the longitudinal axis of the vehicle. This resulted in variations of the position of the centre of mass (CoM) of the rover, which in turn produced different values of NFD during the climbing manoeuvre.

Figure 3 confirms that lower values of NFD resulted in less slip required to carry out the climbing, which is beneficial from the mobility and energy-consumption points of view.

Alternatively, the improvement in mobility can be quantified by the value of the maximum slope that the vehicle can negotiate. The climbing manoeuvre was simulated for a variable slope with the soil properties summarized in Table 1. In this

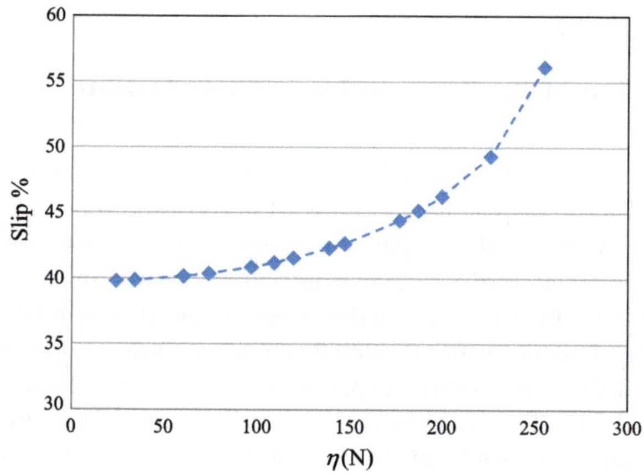

Fig. 3 Values of the slip-vs.-η index developed by the RCP while climbing a 10° slope, with a 22.5-kg payload

Table 1 Soil properties used in the simulation of the slope climbing manoeuvres

n –	c (N/m^2)	ϕ (deg)	k_c (N/m^{n+1})	k_ϕ (kN/m^{n+2})	K_d (m)
1	220	33.1	1400	2000	0.015

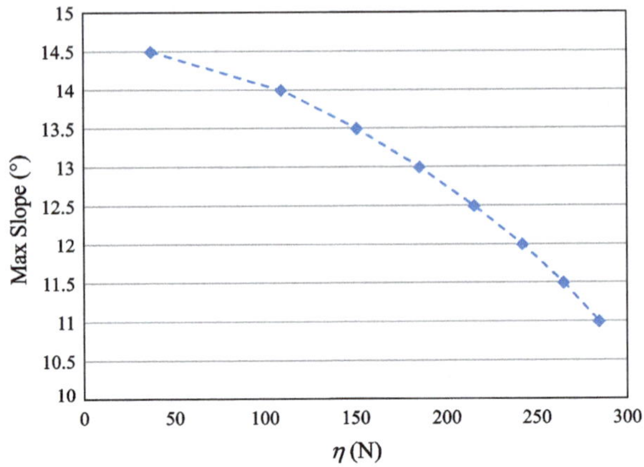

Fig. 4 Correlation between the maximum slope angle that the RCP can climb (with a 90 % slip) and the NFD

study the rover was considered unable to climb if the required slip ratio became higher than 90 %. A similar slip threshold was used in slope-climbing tests with the Dynamic Test Model of the Mars Exploration Rover [16]. The slope angle was increased until the rover was unable to complete the manoeuvre without exceeding the maximum admissible slip. Figure 4 shows that a correlation exists between the value of NFD and the maximum slope the vehicle can successfully climb. Lower values of the NFD enables the rover to climb steeper slopes which is due to the improvement of its ability to develop a greater drawbar pull for the same slip ratio.

3 Modification of the Normal Load Distribution

Two strategies to decrease the normal force dispersion were designed and tested on the real prototype of RCP: displacing the centre of mass of the vehicle and introducing actuation torques between the suspension components. In this work, the latter is refereed to *redundant internal actuation* which allows for altering the system internal forces via applying actuation forces and torques on the suspension components.

Movable mass elements were mounted on the rover chassis to obtain different sets of normal force distributions during experiments. These mass elements consisted of two weights of 22.5 kg. Two attachment positions for the weights were designated on the rover body, one at the front end of the chassis and another one on the connection to the rear bogie. Three load configurations were defined: extra mass at the front location, extra mass at the rear location, and evenly distributed extra mass. An even distribution of the normal loads, however, could not be achieved only via displacement of the CoM of the vehicle. There were limitations in terms of the placement of the movable elements and their mass. For example, the weight of these elements cannot exceed certain limits and their location must be within certain boundaries. Therefore the CoM cannot be arbitrarily displaced.

In the case of some rover designs such as the RCP, the presence of passive joints between the bogies and the chassis frame does not allow one to fully control the load distribution among all the wheels. By repositioning the CoM of the rover one can only control the load distribution between the rear wheels and the side bogies. The way in which the load of each side bogie is distributed between front and middle wheels depends on the orientation of the bogie with respect to the rover main body. Since this joint is not active, in principle the angle between the body and the bogies cannot be controlled. It is possible, however, to actuate this joint by introducing a torque between the chassis and the side bogies.

Figure 5 illustrates the effect of these two strategies on NFD. The left part of the figure represents the default configuration of RCP. In the right diagram, the CoM is displaced towards the front of the rover and redundant internal actuation is introduced between each side bogie and the main body. In this example, a 60 Nm torque in the clockwise direction is applied at each bogie joint. In these figures the lengths of the arrows that represent the reactions at the wheel-terrain interface are proportional to the magnitudes of the normal forces obtained from simulation. In the default configuration the load dispersion is $\eta = 158.1$ N and the rover is able to negotiate a maximum slope of 11°. As shown in Fig. 5, a considerable reduction of NFD was achieved with the application of the techniques described here. The normal force dispersion went down to $\eta = 24.2$ N and the rover was able to climb a 14.5° slope with the same slip as in the original configuration.

The climbing manoeuvre of the RCP with online modification of η was simulated. In this simulation the RCP climbed a 12° slope. The soil properties used in the modelling were the ones listed in Table 1 with the exception of two parameters: the parameters related to the frictional aspect of the soil were chosen as $k_\phi = 1410$ kN/m^{n+2} and $\phi = 34.1°$. The reason for this modification was to simulate a scenario

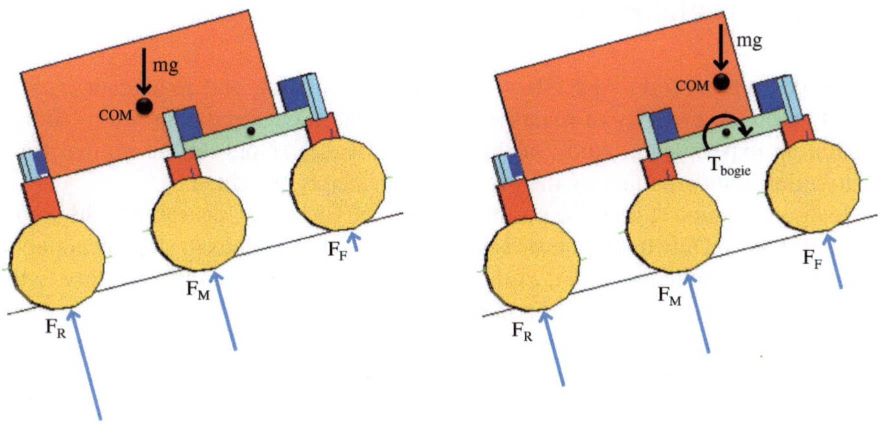

Fig. 5 Uneven load distribution during climbing manoeuvre with the original configuration of the RCP (*left*); improved load distribution after displacing the CoM and introducing a torque between the chassis and the side bogies to reduce NFD (*right*)

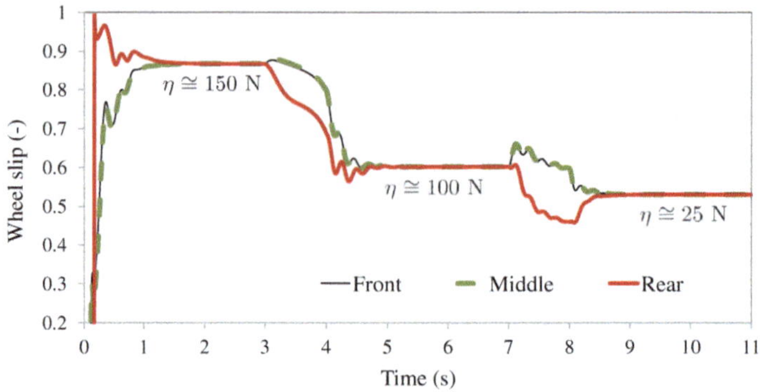

Fig. 6 Slip of the front, middle and rear wheels of the RCP during the climbing of a 12° slope

in which the slip developed to climb the slope in absence of redundant internal actuation exceeds 80 %. Figure 6 displays the results of the simulation. At the beginning of the simulation the rover was placed on a 12° slope and the wheels where commanded to move with $\omega = 0.4$ rad/s. Initially, in the absence of redundant internal actuation, the normal force dispersion was around $\eta = 150$ N. The rover reached a steady-state motion after $t = 2$ s, requiring 87 % slip to move forward. At $t = 3$ s, the torque on the bogie joint was increased gradually up to $T = 20$ Nm. A new steady-state ensued after $t = 5$ s. The new normal force distribution, $\eta = 100$ N, brought the slip down to 60 %. An additional increase in T to 50 Nm further improved the load distribution, enabling the rover to climb the same slope with 53 % slip.

4 Experimental Results

In the previous sections the effect of CoM repositioning and redundant internal actuation on the mobility of rovers was studied based on simulation results. In this section an experimental study of the effect of these factors on the normal force distribution and consequently the rover performance is presented.

In the simulation studies the performance of the rover was measured by its ability to climb slopes. Drawbar pull tests can be considered analogous to slope negotiation tests since the application point of the dragging force was chosen to be close to the CoM of the rover, at least in the vertical direction. Drawbar pull experiments are also easier to carry out, because applying a variable external force to the rover requires less resources than building a soft soil slope with variable inclination.

A set of experiments, including drawbar-pull tests with variable load distribution and wheel slip was carried out with the RCP on soft, sandy soil. These experiments took place in the Mars Dome which is a testing facility located in the UTIAS (University of Toronto Institute for Aerospace studies) campus. All the tests used for this study were carried out with 60 % slip and the load distribution was modified via CoM

Fig. 7 Electric winch used for controlling the wheel slip in drawbar pull experiments (*left*) and design modification of RCP to add the redundant internal actuation option (*right*)

repositioning and redundant internal actuation. The objective of these experiments was to study the effect of load dispersion on the ability of the rover in developing drawbar pull. The slip ratio was controlled by connecting the RCP to the winch shown in Fig. 7. By specifying the winch rotary speed the translational velocity of the rover and consequently its slip ratio were controlled.

Redundant internal actuation was realized by mounting two pneumatic linear actuators on each side of the chassis [17]. One end of each actuator was connected to the front tip of the bogie and the other end to the main body. The force generated by the linear actuator resulted in a moment about the revolute joint between the body and the bogie. The actuator force was regulated by the input air pressure. Therefore, the load distribution between the front and the middle wheels was directly controlled via the pneumatic actuators. This also made the online modification of the load distribution during each test possible. The pneumatic actuator added to the original design of the RCP is shown in Fig. 7.

The RCP is equipped with six triaxial force-torque sensors mounted on each of its legs. These sensors measured the normal, tangent, and lateral terrain reactions on the wheels. However, for the purpose of online measurement of load distribution only normal force sensing is required. A digital force scale was used to measure the net drawbar pull developed by the rover. One end on the force scale was connected to the rope of the winch and the other end to the rear bogie of the rover.

In drawbar pull experiments the RCP travelled on a straight line on soft soil. The motion input was the angular velocity of the wheels, which was set to $\omega = 0.4$ rad/s. The rover was connected to the winch and its translational velocity was set to 0.027 m/s which resulted in about 60 % wheel slip. The normal force readings from the sensors of the front, middle, and rear wheels of the port side of the rover are plotted in Fig. 8. The rover started its motion with the additional mass elements attached to the front of the rover and no actuation was applied on the bogies. The plot shows a very uneven distribution of the load among the wheels, with the middle wheel carrying most of the load. The second part of the motion started at $t = 120$ s,

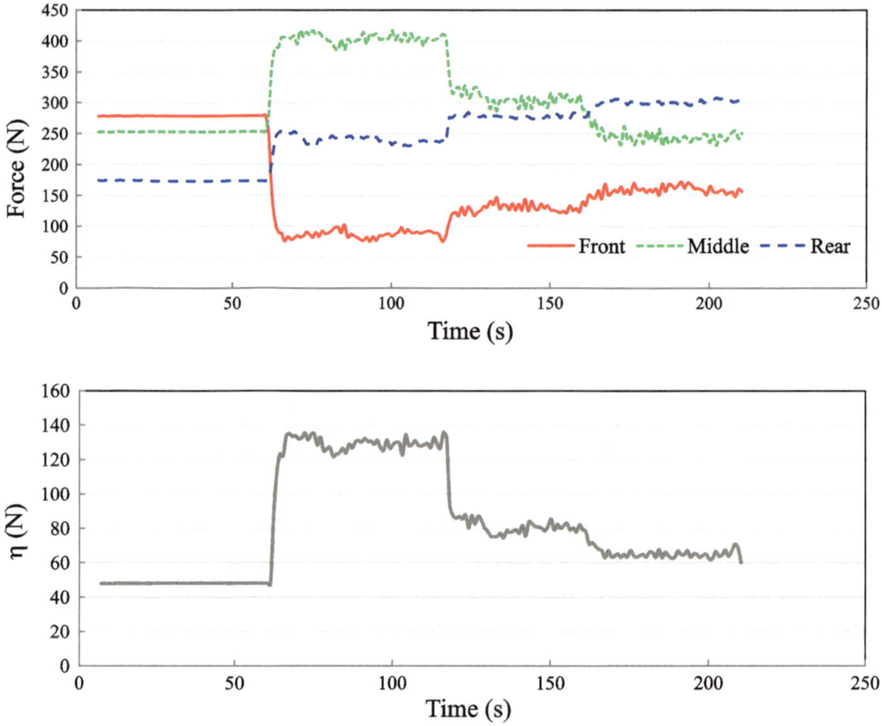

Fig. 8 Effect of redundant internal actuation on normal forces (*upper plot*) and on normal force distribution (*lower plot*)

where a 16-Nm moment about the bogie joints was introduced via the pneumatic actuators. The effect of the actuation on the normal forces can be clearly seen in the plot. The load on the middle wheel was significantly reduced and transferred to the front and rear wheels.

To magnify this effect the actuation was increased to 32 Nm at $t = 160$ s. As expected, this modification further balanced the load distribution among the wheels.

The normal force dispersion was computed for the duration of this test and is plotted in Fig. 8. The results show that only with the aid of the bogie actuation and without CoM repositioning the NFD was reduced significantly in this experiment. Online adjustment of the redundant internal actuation is specially useful for rover manoeuvres on terrain with various slopes. Data from force sensors can be used internally during the rover operation to calculate the required actuation for online tuning of the load distribution.

The presented results shows that redundant internal actuation has a significant effect on the normal force distribution. The final objective in this study, however, is the mobility improvement of rovers, in which the ability of the rover to develop a higher drawbar pull plays a key role. To this end, a similar set of experiments were conducted to study the way drawbar pull changes with variation of the NFD. In

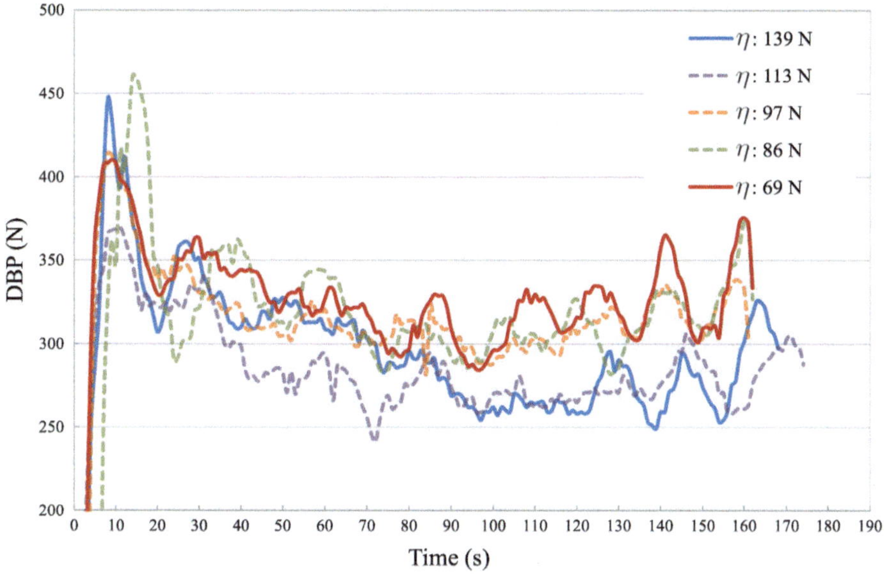

Fig. 9 Effect of normal force dispersion on development of drawbar pull

these experiments NFD was modified by a combination of CoM repositioning and redundant internal actuation. The time history of drawbar pull during these tests is illustrated in Fig. 9.

The comparison of the experimental results shows that for the same slip ratio, the rover configuration with lower NFD provides more drawbar pull compared to the configurations with higher NFD. It was shown in [11] that the relation between the normal and tangential force generated at the wheel-terrain interface follows a non-linear curve. The shape of the curve is a function of the wheel slip and soil and wheel properties. Consequently, the relation between NFD and drawbar pull is also non-linear. The average value of the drawbar pull for each test along with the value of NFD corresponding to the rover configuration in that test are tabulated in Table 2.

Among the reported experiments four cases were selected to be simulated with the generic multibody dynamics library with the soil properties summarized in Table 1. The same angular and linear velocity specifications for the rover in the experiments were used for the simulation tests. Table 3 includes details of the configuration and

Table 2 Experimental results of drawbar pull for different values of normal force dispersion (averaged for each test)

Case	1	2	3	4	5	6	7
η (N)	69.0	82.3	85.9	97.3	111.3	113.2	139.4
DBP (N)	322.9	313.9	307.9	306.8	281.8	272.6	268.2

Table 3 Operation conditions of experimental tests

Experiment	Mass element position	Bogie actuation (N.m)	η (N)
1	Rear	0	141
2	Front	0	125
3	Front	9	70
4	Front	32	63

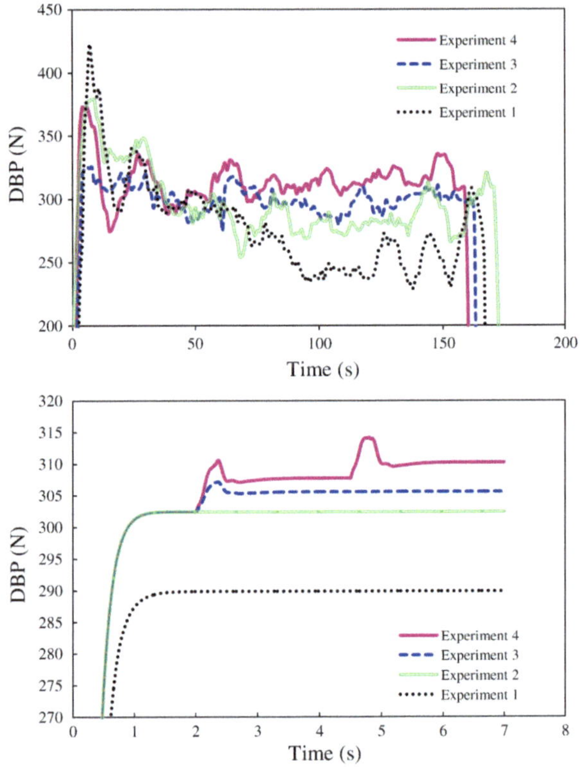

Fig. 10 Experimental (*upper plot*) and simulation (*lower plot*) results from the drawbar pull experiments

redundant internal actuation for the selected tests. Figure 10 shows the experimental and simulation results of drawbar pull in these tests.

Experiments 1 and 2 only differed in the position of the mass elements, which resulted in a more uniform load distribution in the latter. Both experiment and simulation results showed that due to the lower value of NFD in experiment 2 more drawbar pull is generated. In experiment 3 the position of the CoM was the same as in experiment 2. However, after the initial period of the manoeuvre the pneumatic actuators exerted a 9-Nm torque on each bogie, reducing NFD for the rest of the motion. During this phase of the motion the drawbar pull increased in both simulation and experiment. In experiment 4 the actuation was changed in two steps during the motion: first, to 16 Nm and then to 32 Nm. After each increase in the value of the redundant internal actuation the rover reached a more uniform load distribution among the wheels, leading to its improved ability in developing drawbar pull. Both simulation and experiments confirm that lower values of NFD have a positive effect on the developed drawbar pull. The simulation results capture the same trends that can be appreciated in the experiments. The differences between the two plots are explained by simplifications introduced in the multibody modelling, such as not considering the wheel grousers and chassis flexibility, and also to the uncertainty and variability of the terrain parameters.

5 Conclusions

The ability of a wheeled robot to generate traction on soft terrain can be quantified by means of the normal force dispersion. This performance indicator allows one to compare different vehicle configurations and actuation strategies in terms of their suitability to improve the mobility for a given manoeuvre. In the reported research, the performance of a planetary exploration rover prototype was studied with simulation and experiments. Results consistently showed that reducing the normal force dispersion resulted in a better vehicle mobility. A low value of NFD allows the vehicle to develop less slip when climbing a given slope. Two strategies to reduce NFD were designed and tested on the rover prototype: changing the vehicle configuration by displacing its centre of mass, and introducing redundant internal actuation between suspension components. Both strategies proved effective in the reduction of the NFD and therefore, enhancing the vehicle mobility on soft terrains.

Acknowledgments The research work reported here was supported by the Natural Sciences and Engineering Research Council of Canada, and MDA Space Missions. This support is gratefully acknowledged. The second author would like to acknowledge the support of the Spanish Ministry of Economy through its post-doctoral research program Juan de la Cierva, contract No. JCI-2012-12376.

References

1. Apostolopoulos, D.: Analytical configuration of wheeled robotic locomotion. Ph.D. thesis, Carnegie Mellon University (2001)
2. Lamon, P., Krebs, A., Lauria, M., Siegwart, R., Shooter, S.: Wheel torque control for a rough terrain rover. In: Proceedings of the 2004 IEEE International Conference on Robotics and Automation, ICRA 2004. New Orleans, LA, USA (2004)
3. Thueer, T., Krebs, A., Siegwart, R., Lamon, P.: Performance comparison of rough-terrain robots—simulation and hardware. J. Field Robot. **24**(3), 251–271 (2007). doi:10.1002/rob.20185
4. Grand, C., BenAmar, F., Plumet, F., Bidaud, P.: Stability and traction optimization of reconfigurable vehicles. Application to a hybrid wheel-legged robot. Int. J. Robot. Res. **23**(10–11), 1041–1058 (2003)

5. Freitas, G., Gleizer, G., Lizarralde, F., Hsu, L., dos Reis, N.R.S.: Kinematic reconfigurability control for an environmental mobile robot operating in the Amazon rain forest. J. Field Rob. **27**(2), 197–216 (2010)
6. Michaud, S., Richter, L., Patel, N., Thüer, T., Huelsing, T., Joudrier, L., Siegwart, R., Ellery, A.: RCET: rover Chassis Evaluation Tools. In: Proceedings of the 8th ESA Workshop on Advanced Space Technology for Robotics and Automation (ASTRA), paper O-01. Noordwijk, The Netherlands (2004)
7. Thueer, T., Siegwart, R.: Mobility evaluation of wheeled all-terrain robots. Robot. Auton. Syst. **58**(5), 508–519 (2010). doi:10.1016/j.robot.2010.01.007
8. Iagnemma, K., Dubowsky, S.: Traction control of wheeled robotic vehicles in rough terrain with application to planetary rovers. Int. J. Robot. Res. **23**(10–11), 1029–1040 (2004). doi:10.1177/0278364904047392
9. Ghotbi, B., González, F., Kövecses, J., Angeles, J.: Vehicle-terrain interaction models for analysis and performance evaluation of wheeled rovers. In: Proceedings of the 2012 IEEE/RSJ International Conference on Intelligent Robots and Systems (IROS), pp. 3138–3143. Vilamoura, Portugal (2012). doi:10.1109/IROS.2012.6386208
10. Ghotbi, B., González, F., Kövecses, J., Angeles, J.: A novel concept for analysis and performance evaluation of wheeled rovers. Mech. Mach. Theor. **83**, 137–151 (2015). doi:10.1016/j.mechmachtheory.2014.08.017
11. Ghotbi, B., González, F., Kövecses, J., Angeles, J.: Effect of normal force dispersion on the mobility of wheeled robots operating on soft soil. In: Proceedings of the 2014 IEEE International Conference on Robotics and Automation (ICRA). Hong Kong, China (2014). doi:10.1109/ICRA.2014.6907835
12. Wong, J.Y.: Theory of Ground Vehicles, 4th edn. Wiley, New Jersey (2008)
13. Ghotbi, B., González, F., Azimi, A., Bird, W., Kövecses, J., Angeles, J., Mukherji, R.: Analysis, optimization, and testing of planetary exploration rovers: challenges in multibody system modelling. In: Proceedings of Multibody Dynamics 2013—ECCOMAS Thematic Conference. Zagreb, Croatia (2013)
14. Azimi, A., Hirschkorn, M., Ghotbi, B., Kövecses, J., Angeles, J., Radziszewski, P., Teichmann, M., Courchesne, M., Gonthier, Y.: Terrain modelling in simulation-based performance evaluation of rovers. Can. Aeronaut. Space J. **57**(1), 24–33 (2011). doi:10.5589/q11-005
15. Bauer, R., Barfoot, T., Leung, W., Ravindran, G.: Dynamic simulation tool development for planetary rovers. Int. J. Adv. Rob. Syst. **5**(3), 311–314 (2008)
16. Lindemann, R.A., Voorhees, C.J.: Mars exploration rover mobility assembly design, test and performance. In: Proceedings of the IEEE International Conference on Systems. Man and Cybernetics, pp. 450–455. Waikoloa, HI, USA (2005)
17. MacMahon, S.: Modelling and contact analysis of planetary exploration rovers. Master's thesis, McGill University (2016)

Taming the North: Multi-camera Parallel Tracking and Mapping in Snow-Laden Environments

Arun Das, Devinder Kumar, Abdelhamid El Bably and Steven L. Waslander

Abstract Robot deployment in open snow-covered environments poses challenges to existing vision-based localization and mapping methods. Limited field of view and over-exposure in regions where snow is present leads to difficulty identifying and tracking features in the environment. The wide variation in scene depth and relative visual saliency of points on the horizon results in clustered features with poor depth estimates, as well as the failure of typical keyframe selection metrics to produce reliable bundle adjustment results. In this work, we propose the use of and two extensions to Multi-Camera Parallel Tracking and Mapping (MCPTAM) to improve localization performance in snow-laden environments. First, we define a snow segmentation method and snow-specific image filtering to enhance detectability of local features on the snow surface. Then, we define a feature entropy reduction metric for keyframe selection that leads to reduced map sizes while maintaining localization accuracy. Both refinements are demonstrated on a snow-laden outdoor dataset collected with a wide field-of-view, three camera cluster on a ground rover platform.

1 Introduction

A wide range of challenging and remote tasks have been proposed as possible field robotics applications, from wilderness search and rescue, to pipeline and infrastructure inspection, to environmental monitoring. Particularly in Northern climates, these activities require autonomous navigation in snow-laden environments, which present

A. Das (✉) · D. Kumar · A.E. Bably · S.L. Waslander
University of Waterloo, Waterloo, ON, Canada
e-mail: arun.das@uwaterloo.ca; adas@uwaterloo.ca

D. Kumar
e-mail: devinder.kumar@uwaterloo.ca

A.E. Bably
e-mail: ahelbably@uwaterloo.ca

S.L. Waslander
e-mail: stevenw@uwaterloo.ca

D.S. Wettergreen and T.D. Barfoot (eds.), *Field and Service Robotics*,
Springer Tracts in Advanced Robotics 113, DOI 10.1007/978-3-319-27702-8_23

distinct perception challenges for autonomous vehicles. The possibility of tree cover precludes reliance on GPS alone for positioning, and both obstacle detection and accuracy requirements further drive the need for alternate localization methods.

Both visual and laser based simultaneous localization and mapping methods can provide such improved localization. Although laser scanners are not significantly affected by snow, their relatively large costs can be prohibitive for many applications. In this work, we consider the problem of deploying a feature based visual SLAM system known as Multi-Camera Parallel Tracking and Mapping (MCPTAM) in a snowy, outdoor environment. MCPTAM employs an arbitrary cluster of cameras with wide field of view, with or without overlap, to track point features in the environment, and has been demonstrated to provide accuracy better than 1 % of distance traveled in both indoor and outdoor environments [1–3].

The primary challenge with outdoor and snowy environments is that large areas of the image are relatively feature poor due to limited geometric structure, overcast skies and large regions of uniform snow cover. Without employing expensive high dynamic range cameras, this leads to difficulties tracking features near the robot and clusters the points used for map generation along the horizon. The result is poor translational tracking and a susceptibility to map optimization failures if features are incorrectly corresponded.

To address these limitations, we introduce two extensions to our previous work. First we investigate changes to MCPTAM's front-end, by pre-processing the camera frames to extract more robust features. We use as motivation some of the works of [4, 5] which use region based contrast equalization and horizon detection [6] to fulfill this goal. Second, we propose core changes to MCPTAM's backend which allow for more informed keyframe selection based on the expected entropy reduction of uncertainty in the map points. These modifications directly impact the quality of the localization solution by creating a more robust set of features to track and optimize against for mapping.

2 Related Works

To date, there have been comparatively few instances of autonomous robotic deployments in snowy conditions. The CoolRobot is a mobile sensor station deployed both in Greenland and on the Antarctic plateau, and relies on solar power and GPS waypoint navigation to move through primarily flat terrain [7]. Similarly, both the Nomad [8] and MARVIN [9] rely on GPS guided navigation with a laser scanner and vision for local collision avoidance in polar environments. The SnoMote platform seeks to augment GPS with visual localization and terrain drivability estimation for detailed ice sheet mapping [5].

Closely related to visual navigation in snow-covered terrain is use of computer vision for planetary exploration. The visual localization challenges are similar in both environments, with limited local features, large variations in scene depth, and unreliable features in the sky portion of images. For example, stereo localization has been used on lengthy datasets collected in Devon Island, Canada [10], where repetitive ground terrain and a lack of rotation invariant features led the authors to note the concentration of features on the horizon. Similarly, stereo and/or laser scan data was employed in a large range of planetary analog terrains for localization and drivability analysis [11]. In both cases, the image quality both near the robot and at a distance was not often an issue for feature extraction.

The MCPTAM method builds on the foundation of Parallel Tracking and Mapping [12], which splits the localization and mapping problem into separate pose tracking and keyframe based feature mapping processes. This divide prevents pose estimation from being delayed by the batch optimization required as a part of the mapping bundle adjustment. Features are tracked between images and localization is performed relative to the known map, while map updates are performed when new keyframes are selected to be inserted into the global map.

Many visual mapping techniques use keyframes in order to reduce the computational burden of the mapping process. Existing approaches generally insert keyframes based on point triangulation baseline [12], or other heuristics such as the co-visibility of features [13], or the overlap in the number of tracked points [14]. These heuristics attempt to insert keyframes in order to maintain the map integrity, yet do not directly attempt to minimize the uncertainty in the map. The work most related to ours generates image features off-line, creates a buffer of the image frames, and selects keyframes based on saliency in order to reduce content redundancy [15]. In contrast, our approach is a real-time, online system, and attempts to reduce feature uncertainty while the camera is in motion.

In addition to keyframe selection, the identification of strong and stable visual features is both important and challenging in snowy environments. The Snomote [5] integrates a pre-processing technique of contrast limited adaptive histogram equalization (CLAHE) to enhance the contrast of the captured images. A slope finding method is applied to mask out the mountain peaks or other structures from the background and SIFT features are detected mainly from the foreground.

Applying feature detection methods to the entire image is problematic, however, as environments with trees and foliage result in self similar image features which are difficult to match. Instead, horizon detection can be used to apply specific feature detection criteria in the snow-laden region of the image. Existing methods (e.g. [16]) do not explicitly consider the snow-laden case, with the exception of the SnoMote [6], which uses a weighted sum of weak and strong visual cues to identify fairly precise horizon lines. The method is overly computationally expensive for our application, and so we present a simplified method based on the Hough transform in this work.

3 Multiple Camera Parallel Tracking and Mapping

MCPTAM is a real-time, feature-based, visual slam algorithm which extends Klein and Murray's Parallel Tracking and Mapping (PTAM) [12] in five ways. First it allows multiple, non-overlapping field-of-view (FOV), heterogeneous cameras in any fixed configuration to be successfully combined. MCPTAM's novel initialization mechanism allows for scale to be recovered, even with non-overlapping cameras. Second it extends the PTAM's pinhole camera model to work with fish-eye and omnidirectional lenses through the use of the Taylor camera model [17]. The ultra-wide FOV coupled with the multi-camera cluster prevents feature starvation due to occlusions and textureless frames in any single camera. Third, PTAM's backend has been replaced with the g2o optimizer allowing for faster and more flexible optimization structures [18]. Finally, MCPTAM introduces both an improved update process based on box-plus manifolds and a novel feature parameterization using spherical co-ordinates anchored in a base-frame [3].

A brief overview of the MCPTAM formulation proceeds as follows. Denote a point in the global frame, $p \in \mathbb{R}^3$ as $p = [p_x\ p_y\ p_z]^T$ where p_x, p_y, p_z represent the x, y, and z components of the point, respectively. Let the map, P, be a set of points, defined as $P = \{p_1, p_2, \ldots, p_n\}$. Denote the re-projection function as $\Pi : \mathbb{R}^3 \mapsto \mathbb{R}^2$, which maps a point in the global 3D frame to a pixel location on the image plane.

In the standard pinhole camera model, light rays are represented as lines which converge at the center of projection and intersect with the image plane. In order to accommodate the large radial distortion caused by fisheye lenses, the Taylor model uses a spherical mapping where the elevation and azimuth angles to a 3D point, $s = [\theta, \phi]^T$, are modeled as half lines which pass through the sphere's center, which are then mapped to the image plane through a polynomial mapping function.

In order to track the camera cluster pose, $\omega^c \in \mathbb{SE}(3)$, the map points, P, are reprojected into the image frames of the cameras. Given a set of feature correspondences, the camera cluster pose parameters are found through a weighted nonlinear least squares optimization which seeks to determine the pose parameters such that the re-projection error between corresponding points is minimized. By re-observing features, the point locations in the map can be refined using additional measurements, and new map points can be inserted into the map. To perform these tasks, MCPTAM uses *keyframes*, which are a snapshot of the images and point measurements taken from a point along the camera cluster's trajectory. Since MCPTAM performs tracking using multiple cameras, it extends the idea of key-frames to *multi-keyframes*, which are simply a collection of the key-frames from the individual cameras at a particular instant in time.

We shall define a multi-keyframe, M, as collection of keyframes, $M = \{K_1, \ldots, K_m\}$, corresponding to the m individual cameras which are part of the multi-camera cluster. Each multi-keyframe is associated with its pose in $\mathbb{SE}(3)$. In order to insert a new multi-keyframe into the map, the point measurements from each observing keyframe are collected, and the parameters of the point locations, as well as the keyframe poses are optimized using a bundle adjustment procedure.

Entropy Computation for a Gaussian PDF: The Shannon entropy is a measure of the unpredictability or uncertainty of information content. Suppose $X = \{x_1, x_2, \ldots, x_n\}$ is a discrete random variable. The Shannon entropy for X, $H(X)$ is given as $H(X) = -\sum_{x_i \in X} P(x_i) \log P(x_i)$, where $P(x_i)$ denotes the probability of event x_i occurring. The Shannon entropy provides a scalar value that quantifies the average variance of the discrete random variable X. The base of the logarithm denotes the units of the entropy. In the case where the base of the logarithm is 2, the units are referred to as *bits*, and when performed using the natural logarithm, the units are referred to as *nats*. It is also possible to compute the Shannon entropy for a continuous random variable. In the case where the continuous random variable is modeled as a Gaussian distribution, the entropy can be computed as

$$h_e(Y) = \frac{1}{2}\ln((2\pi e)^n \, |\Sigma|), \tag{1}$$

where Σ is the covariance matrix of the multivariate Gaussian distribution, $|\cdot|$ denotes the determinant operator, and $h_e(Y)$ is used to denote that the logarithm was taken with base e. Note that unlike the entropy for discrete random variables, it is possible for the entropy of continuous random variables to be less than zero.

4 Proposed Approach

Our approach involves both pre-processing of images to improve feature tracking despite the limitations of images acquired in snow-covered environments, and improvements to the keyframe selection process that help maintain map quality throughout the test datasets.

4.1 Pre-processing Pipeline

The pre-processing pipeline that is used to enhance the captured image for detecting good features for localization of our mobile robot consists of snow segmentation, histogram equalization, and feature selection phases.

Snow Segmentation: We first apply a Canny edge detector [19] to remove the undesired information from the image while still retaining the structural information. This is applied prior to a Hough Line transform, which is used to detect the line that segments out the snow from the rest of the regions in image.

Consider a line represented in the polar form $\rho = x\cos\theta + y\sin\theta$ where ρ is the radial distance from the origin and θ is the angle formed by this radial line and the horizontal axis measured in the counter-clockwise direction. The Hough Line transform uses a 2D accumulator array to detect the existence of lines in the edge based image from the Canny edge detector using a voting based method to output ρ and θ. Each element, (ρ, θ), in the output represents a line. For our task, we select the element with the highest value as the horizon, which indicates the straight line that is the most strongly represented in the input image. It is important to note that for our concerned task, we only detect horizontal lines in the image.

Histogram Equalization: Before feeding the input image to MCPTAM we use histogram equalization to enhance the global contrast of the image. Since snow laden environments lead to low contrast images, enhancing the contrast can significantly improve the detection of stable features. The global histogram equalization (GHE) transform, $T(r)$, can be represented as

$$T(r) = (L-1)\sum_{j=0}^{L-1} p_r(r_j), \quad (2)$$

where L represents the number of gray level intensities present in the image, j is the intensity level varying from 0 to $L-1$, and $p_r(r_j)$ is the probability distribution function (pdf) of intensity level j.

The pdf is defined by:

$$p_r(r_j) = \frac{N_j}{N_t}, \quad (3)$$

where N_j is the number of pixels with intensity level j and N_t is the total number of pixels present in the image. We also implemented contrast limited adaptive histogram equalization (CLAHE) [20] for comparison. Instead of accounting for global illumination changes and coming up with single histogram, CLAHE computes several histograms each belonging to a different part of the image and uses this information for changing the *local* contrast of the image. CLAHE also contains a contrast limiting function that limits the amplification of noise.

Feature Selection: We take this enhanced image obtained after histogram equalization and input it into MCPTAM system where we detect coarse, mid level and fine FAST features in the images for each camera. FAST features are used because of their computational efficiency and ability to detect stable corner features [21]. Using the (ρ, θ) obtained from the Hough Line transform, we select fine features from the segmented snow region below the horizon, and coarse features from the rest of the image. The large structural features in the snow laden environments are generally trees or far away buildings, and generating fine features from these image regions are not helpful as the features generated are not sufficiently distinguishable to produce correct correspondences. The nearby features in snow on the ground can be better localized, however, and therefore become very important to the mapping process. Hence we detect and track fine features in snow and coarse features from far away structures for localization and mapping.

4.2 Entropy Based Keyframe Selection

The quality of the map point parameter estimation is heavily dependent on the triangulation baseline between the measurement viewpoints. Many visual SLAM techniques use heuristics based on the point triangulation baseline to perform keyframe insertion, however no existing approaches attempt to perform keyframe selection through direct minimization of the point estimate covariance.

We propose a covariance update on the point with the assumption that the keyframe candidate's location is known and fixed. Although the keyframe's pose parameters are in fact updated through bundle adjustment once inserted into the map, the fixed keyframe parameter assumption allows for rapid evaluation of the point covariance update, and is reasonable so long as the tracker pose estimate is sufficiently accurate.

In order to determine when a multi-keyframe should be inserted into the map, we inspect the uncertainty of the current camera cluster provided by the tracking process. The covariance of the tracking pose parameters is given by $\Sigma^c = (G^T W G)^{-1}$, where $G = \frac{\partial \Pi}{\partial \omega^c}$ is the Jacobian of the map re-projection error with respect to the cluster state, and W is the matrix of weights associated with the measurements. To assess the current tracking performance, we extract the x, y, and z diagonal components of covariance matrix Σ^c, denoted as $\sigma_x, \sigma_y, \sigma_z$, respectively. The rotational covariances are ignored at this stage, as generally the rotations of the camera cluster can be tracked accurately using points of varying depth, whereas accurate positional tracking requires relatively close points in order to resolve the scale of the motion. Finally, a multi-keyframe is added when any element of the positional entropy is above a user defined threshold, ε, or

$$\max(h_e(\sigma_x), h_e(\sigma_y), h_e(\sigma_z)) > \varepsilon, \tag{4}$$

where $h_e(\cdot)$ is computed using Eq. (1). When a multi-keyframe addition is triggered, the next step is to determine which multi-keyframe should be added. For this, multi-keyframe candidates are maintained in a buffer and scored based on the expected reduction in point depth entropy if added to the map through a bundle adjustment process.

As the tracking thread operates, each successfully tracked frame, along with its corresponding set of point feature measurements and global pose estimate, are added as multi-keyframe candidates in a buffer. Suppose the tracking thread is currently operating at time t, and the last multi-keyframe insertion occurred at time k. Denote the set of multi-keyframe candidates which are buffered between times t and k as

$$\Phi = \{M_t, M_{t-1}, M_{t-2}, \ldots, M_{t-k}\}. \tag{5}$$

Since each of the multi-keyframe candidates are saved from the tracking thread, an estimate of the global pose of each candidate is available from the tracking solution. Therefore, it is possible to determine the subset of map points observed in the individual keyframes within each multi-keyframe candidate. Denote the set of map points from P, visible in $K_l \in M_i$, as $\tilde{P}_{K_{il}} \subset P$.

Since each map point position is estimated through a standard bundle adjustment approach, the map point parameters are modeled as a Gaussian distribution with an associated mean and covariance. We denote the estimate for point p_j as $\hat{p}_j$, and the associated covariance matrix $\Sigma_j \in \mathbb{R}^{3\times3}$.

Suppose point $p_j \in \tilde{P}_{K_{il}}$ is observed in keyframe $K_l \in M_i$. Our method seeks to determine the updated covariance of point p_j, if triangulated using an additional measurement from keyframe K_l. This is accomplished using a covariance update step as per the Extended Kalman Filter.

Denote the Jacobian of the re-projection function with respect to the point parameters, p, evaluated at point $\hat{p}_j$, as

$$J_j = \frac{\partial \Pi}{\partial p}|_{\hat{p}_j}. \tag{6}$$

The Jacobian, J_j, describes how perturbations in the point parameters for $\hat{p}_j$ map to perturbations in the image re-projections. Using the Jacobian, J_j, and the prior point covariance Σ_j, the predicted point covariance is given as

$$\bar{\Sigma}_j = (I - \Sigma_j J_j^T (J_j \Sigma_j J_j^T + R)^{-1} J_j)\Sigma_j. \tag{7}$$

The predicted covariance $\bar{\Sigma}_j$ provides an estimate of the covariance for point p_j, if the observing keyframe was inserted into the bundle adjustment process. Note that Eq. (7) can be evaluated rapidly for each point, as the computational bottleneck is the inversion of a 3 by 3 matrix.

Although comparison of the predicted covariance to the prior covariance provides information on reduction of point parameter uncertainty for one point, the covariance representation does not allow for a convenient way to asses the uncertainty reduction across all of the points observed in the multi-keyframe. To that end, we propose evaluation of the uncertainty reduction using the point *entropy*.

Denote the entropy corresponding to the point's prior and predicted covariance as $h_e(\hat{p}_j)$ and $\bar{h}_e(\hat{p}_j)$, respectively. The reduction in entropy for point p_j is given as $\Lambda(p_j) = h_e(\hat{p}_j) - \bar{h}_e(\hat{p}_j)$. Using the expected entropy reduction for a single point, the expected entropy reduction for all of the points observed in multi-keyframe M_i is given as $\Psi(M_i) = \sum_{K_l \in M_i} \sum_{p \in \tilde{P}_{K_{il}}} \Lambda(p)$. Finally, when a multi-keyframe needs to be inserted into the map, all of the multi-keyframes within the buffer, Φ, are evaluated for total point entropy reduction. The multi-keyframe selected for insertion, M_i^*, is the one from the buffer which maximizes the point entropy reduction:

$$M_i^* = \underset{M_i \in \Phi}{\operatorname{argmax}}\ \Psi(M_i). \tag{8}$$

Once the optimal keyframe from the buffer is selected, it is inserted into the map through bundle adjustment, and the multi-keyframe buffer, Φ, is cleared.

Although it is possible perform keyframe selection using heuristics which rely on the geometric relationships between point observation baselines, such approaches do not account for possible degradation of point re-projection sensitivity that is also dependent on the camera model. For example, an image taken from a wide field of view fisheye lens camera will generally have significant distortion and spatial compression near the image edges. To illustrate this point, consider a uniform, 2D, planar grid of points, positioned at unit depth from a camera. Figure 1a, b show the projection of the grid onto the image plane using the pinhole and Taylor models,

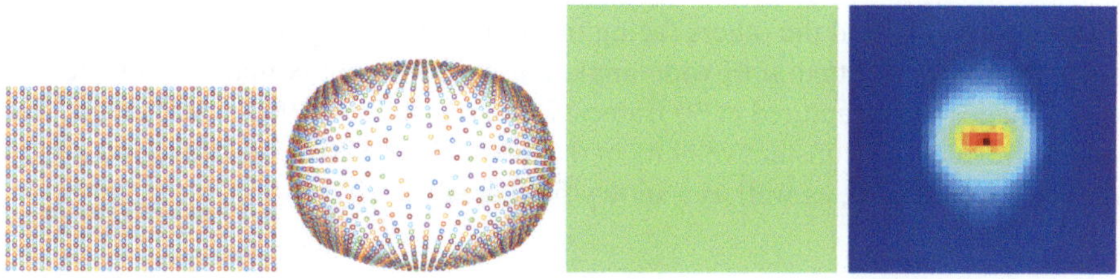

(a) Pinhole **(b)** Taylor **(c)** Jacobian (pinhole) **(d)** Jacobian (Taylor)

Fig. 1 Comparison of image re-projection sensitivity between pinhole and Taylor camera models. **a** and **b** illustrate the projection of 3D points onto the image plane, using the pinhole and Taylor camera models, respectively. The image compression around the edges results in reduced sensitivity of image projection Jacobian in the outer edge areas, as seen in (**d**), where as the pinhole camera model displays uniform strength in the image re-projection Jacobian, as seen in (**c**)

respectively. The pinhole projection preserves the uniform spatial distribution of the 3D grid on the image plane, while the Taylor model spatially compresses the points near the boundaries of the image plane. Such compression suggests that with a large FOV lens described using the Taylor camera model, the point projections which fall near the boundaries of the image are less sensitive to perturbations of the 3D point location. This insight is illustrated in Fig. 1c, d, which show the norm of the projection Jacobian with respect to perturbations in the x direction of the 3D point grid. It is evident that the pinhole camera model maintains uniform sensitivity to point perturbations across the image plane, while the Taylor camera model has reduced sensitivity as the points are projected farther from the image center.

Our proposed keyframe selection method is able to account for the properties of the lens model being used, as the point projection Jacobian, given by Eq. (6), is dependent on the underlying camera model. For example, using the Taylor model, Eq. (6) can be expanded as

$$\frac{\partial \Pi}{\partial p} = \frac{\partial \Pi}{\partial s}\frac{\partial s}{\partial r}\frac{\partial r}{\partial p} \tag{9}$$

where $r \in \mathbb{R}^3$ is the position of point p with respect to the observing frame, $\frac{\partial \Pi}{\partial s}$ relates the image re-projection to the point's projection on the unit sphere, $\frac{\partial s}{\partial r}$ relates the perturbations of a point projection on the unit sphere to perturbations of the point position in the observing keyframe, and $\frac{\partial r}{\partial p}$ relates the changes of the point in the observing keyframe to changes of the point parameters.

5 Experimental Results

To verify our proposed methods, experiments were conducted using field data collected in a snow laden environment. A Clearpath Robotics Husky platform was equipped with three Ximia xiQ cameras, arranged in a rigid cluster, with one camera looking forwards, and the others facing off to the left and right sides of the vehicle. The cameras were fitted with wide angle lenses, with approximately 160° field of view. Images were captured at 30 frames/s, at a resolution of 900 × 600 pixels. The vehicle traveled at a constant velocity of 0.5 m/s for over 120 m, and traversed a snow and ice covered path, as well as a snowy field area.

5.1 Image Pre-processing

We compare GHE and CLAHE in terms of the features that result after pre-processing. The FAST features detected on snow in the enhanced images are shown in Fig. 2. It is evident that the largest number of features detected in snow were found with GHE. To quantitatively compare the two histogram equalization techniques, we calculated the number of features detected below the horizon. The total number of features obtained for a video sequence of 1497 frames from our dataset were 407,665 for GHE, 83,650 for CLAHE and 4,919 without any histogram equalization, demonstrating the advantage of GHE in terms of FAST feature detection in snow.

For segmenting the snow from the rest of the image, representative results are shown in Fig. 3, which includes the output of our snow segmentation algorithm (the red line) for the single frontal view (camera 1) and features detected on snow in Canny edge images for the three camera cluster. Our approach produces a rough segmentation of each image in 0.015 s, on average, over the entire dataset, which has an image resolution of 900 × 600. To compare our approach with the state of the art result [6], we decrease the resolution of our captured dataset to 640 × 480. A naive implementation of our approach took on average 0.0098 s/frame, whereas the method proposed in [6] requires 0.0296 s/frame.

5.2 MCPTAM Using Histogram Equalization

We next compare MCPTAM mapping performance with different equalization methods. As evident in Fig. 4, GHE provides the most consistent feature map, compared

Fig. 2 Comparison of FAST features detection on **a** a normal image, **b** image ehanced by global histogram equalization, **c** image ehanced by CLAHE

Fig. 3 The result of snow segmentation from camera 1 (*lower left*) and FAST features detected on snow in Canny edge images for the three cameras

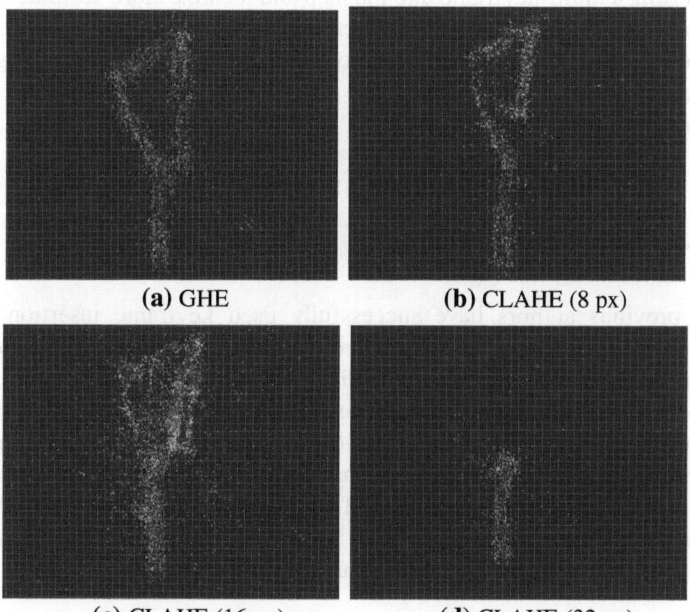

(a) GHE **(b)** CLAHE (8 px)

(c) CLAHE (16 px) **(d)** CLAHE (32 px)

Fig. 4 Comparison of feature maps with different histogram equalization techniques. *Red* points denote fine features, while *blue* and *green* points denote coarse features. **a** shows the resulting map when the images are processed using GHE. **b–d** present maps generated using CLAHE with different patch sizes. Note that large patch sizes cause instability in the feature tracking due to mismatched points

to the CLAHE methods. As the patch size for the CLAHE methods increase, the resulting map exhibits signs of scale drift, as well as poor feature matches. It is also worth noting that GHE results in the recovery of a greater number of fine fea-

Table 1 Summary of results for histogram equalization experiments

	GHE	CLAHE (8)	CLAHE (16)	CLAHE (32)
Max. tracking entropy (nats)	–2.4851	–2.2738	–1.5941	–1.6170
No. map points	2777	2960	6115	–
No. MKFs	168	175	240	–

tures, compared to the adaptive method. This is likely because GHE maintains more consistent illumination between the inserted keyframes, resulting in better feature matches over local methods.

Table 1 presents a summary of the results. It is evident that GHE resulted in a feature map with the fewest number of inserted multi-keyframes and the fewest number of points. This suggests the robot was able to travel longer distances on average before inserting a multi-keyframe into the map and localize more accurately with the features that were included, which is further verified by the reported maximum tracking entropy over the trail. The GHE method resulted in the lowest tracking entropy (as calculated by Eq. (4)), suggesting the generated map provided stable points to track against throughout the test run.

5.3 *Multi-keyframe Selection*

Although previous authors have successfully used keyframe insertion methods related to feature overlap and the number of features tracked, such approaches were completely unsuccessful for our application due to intermittent feature tracking experienced in snowy environments. Instead, we compare our entropy based (EB) approach to a movement threshold on the vehicle, where a multi-keyframe is inserted once the camera cluster moves a user defined threshold distance from the previously inserted multi-keyframe. Only a threshold on the position is used; the rotation need not be considered due to the nearly 360° view of the multi-camera cluster, which tends to maintain consistent orientation based on stable, persistent horizon features.

Table 2 Summary of Results for multi-keyframe selection experiments

	EB-MKF	1 m threshold	2 m threshold
Max. tracking entropy (nats)	–2.76771	–2.0314	–2.4113
No. map points	2316	4001	2897
No. MKFs	150	175	162

Figure 6 presents a comparison of the multi-keyframe selection methods tested. The EB approach provides consistent mapping results, while the 2 m threshold approach fails midway through the path. This is likely because the non-entropy based approaches do not consider any improvements in the map points, and merely assume that the multi-keyframe insertion will improve the map and provide stable

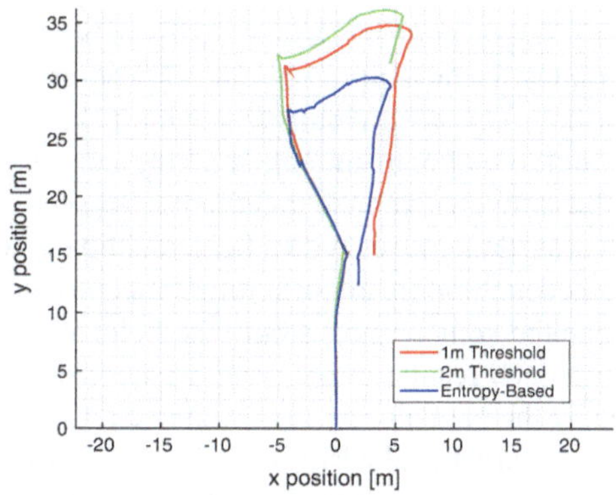

Fig. 5 Comparison of the recovered vehicle motion using different multi-keyframe selection methods. Note that the EB approach demonstrates the lowest scale drift in the trajectory

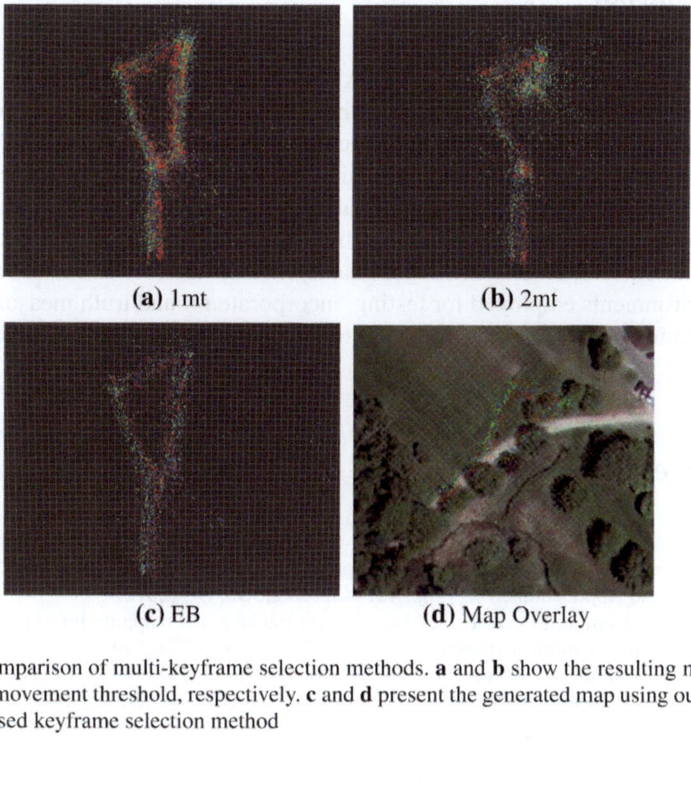

(a) 1mt **(b)** 2mt

(c) EB **(d)** Map Overlay

Fig. 6 Comparison of multi-keyframe selection methods. **a** and **b** show the resulting map using a 1 and 2 m movement threshold, respectively. **c** and **d** present the generated map using our proposed entropy based keyframe selection method points to track against. Our approach, on the other hand, actively seeks to insert multi-keyframes such that the map integrity is maintained, providing the camera cluster with stable and well estimated point features for localization. Although the map generated by the 1 m threshold (1 mt) policy (Fig. 6a) is qualitatively similar to the one generated by the EB approach, the 1 mt map contains approximately 42 % more points compared to our proposed method, as summarized in Table 2. From Table 2, it is also clear that the EB method results in the lowest tracking entropy, along with the fewest inserted multi-keyframes. This is because our approach only adds new multi-keyframes when required by the tracker, and seeks to improve the points which exist in the map. As a result, fewer multi-keyframes are added, and fewer points are required to maintain suitable tracking integrity.

Figure 5 presents the recovered vehicle trajectories. As seen in Fig. 6d, the vehicle traverses along a path area, then moves onto a field, and finally joins up with the path again. All of the evaluated methods result in similar trajectories over the path area, but exhibit differences once the vehicle moves onto the field. We see that the EB multi-keyframe selection approach results in the smallest scale drift while traversing the field, as demonstrated by the path closely rejoining itself. Conversely, the 1 and 2 mt approaches both exhibit a larger scale drift in the trajectory solution, since the static threshold policies do not account for map integrity when inserting multi-keyframes.

6 Conclusion

In this work, two extensions to the MCPTAM visual localization method are shown to significantly improve the performance of the system in snow laden environments. We demonstrate that a pre-processing pipeline that uses GHE to improve FAST feature detection in snow, as well as horizon detection and a tailored feature selection process, results in improved feature tracking. We also show that point entropy reduction can be used as a keyframe selection metric, which leads to fewer keyframes and reduced map drift when compared to existing methods. In the future, we intend to expand the set of environments employed for testing, incorporate ground truth measurement of vehicle motion, and investigate the persistence and accurate localization of features in the map.

References

1. Harmat, A., Sharf, I., Trentini, M.: Parallel tracking and mapping with multiple cameras on an unmanned aerial vehicle. In: Proceedings of the International Conference on Intelligent Robotics and Applications, vol. 1, pp. 421–432. Montreal, QC (2012)
2. Harmat, A., Trentini, M., Sharf, I.: Multi-camera tracking and mapping for unmanned aerial vehicles in unstructured environments. J. Intell. Rob. Syst. 1–27 (2014)
3. Tribou M.J., Harmat, A., Wang, D., Sharf, I., Waslander, S.L.: Multi-camera parallel tracking and mapping with non-overlapping fields of view. Int. J. Robot. Res. **34**(12), 1480–1500 (2015). doi:10.1177/0278364915571429

4. Kim, A., Eustice, R.M.: Real-time visual slam for autonomous underwater hull inspection using visual saliency. IEEE Trans. Robot. **29**(3), 719–733 (2013)
5. Williams, S., Howard, A.M.: Developing monocular visual pose estimation for arctic environments. J. Field Robot. **27**(2), 145–157 (2010)
6. Williams, S., Howard, A.M.: Horizon line estimation in glacial environments using multiple visual cues. In: IEEE International Conference on Robotics and Automation (ICRA), pp. 5887–5892. IEEE (2011)
7. Ray, L.E., Lever, J.H., Streeter, A.D., Price, A.D.: Design and power management of a solar-powered cool robot for polar instrument networks. J. Field Robot. **24**(7), 581–599 (2007). doi:10.1002/rob.20163. http://dx.doi.org/10.1002/rob.20163
8. Apostolopoulos, D.S., Wagner, M.D., Shamah, B.N., Pedersen, L., Shillcutt, K., Whittaker, W.L.: Technology and field demonstration of robotic search for antarctic meteorites. Int. J. Robot. Res. **19**(11), 1015–1032 (2000)
9. Gifford, C.M., Akers, E.L., Stansbury, R.S., Agah, A.: Mobile robots for polar remote sensing. In: The Path to Autonomous Robots, pp. 1–22. Springer (2009)
10. Furgale, P., Barfoot, T.D.: Visual teach and repeat for long-range rover autonomy. J. Field Robot. **27**, 534560 (2010)
11. Wettergreen, D., Wagner, M.: Developing a framework for reliable autonomous surface mobility. In: International Symposium on Artificial Intelligence, Robotics, and Automation in Space (iSAIRAS) (2012)
12. Klein, G., Murray, D.: Parallel tracking and mapping for small AR workspaces. In: Proceedings of the IEEE and ACM International Symposium on Mixed and Augmented Reality (ISMAR), pp. 225–234 (2007)
13. Stalbaum, J., Song, J.B.: Keyframe and inlier selection for visual slam. In: 2013 10th International Conference on Ubiquitous Robots and Ambient Intelligence (URAI), pp. 391–396 (2013)
14. Leutenegger, S., Furgale, P.T., Rabaud, V., Chli, M., Konolige, K., Siegwart, R.: Keyframe-based visual-inertial slam using nonlinear optimization. In: Robotics: Science and Systems (2013)
15. Dong, Z., Zhang, G., Jia, J., Bao, H.: Keyframe-based real-time camera tracking. In: 2009 IEEE 12th International Conference on Computer Vision, pp. 1538–1545. IEEE (2009)
16. Ettinger, S.M., Nechyba, M.C., Ifju, P.G., Waszak, M.: Towards flight autonomy: Vision-based horizon detection for micro air vehicles. In: Florida Conference on Recent Advances in Robotics, vol. 2002 (2002)
17. Scaramuzza, D., Martinelli, A., Siegwart, R.: A flexible technique for accurate omnidirectional camera calibration and structure from motion. In: IEEE International Conference on Computer Vision Systems (ICVS), pp. 45–45. IEEE (2006)
18. Kümmerle, R., Grisetti, G., Strasdat, H., Konolige, K., Burgard, W.: g2o: a general framework for graph optimization. In: Proceedings of the IEEE International Conference on Robotics and Automation (ICRA) (2011)
19. Canny, J.: A computational approach to edge detection. IEEE Trans. Pattern Anal. Mach. Intell. **8**(6), 679–698 (1986)
20. Zuiderveld, K.: Contrast limited adaptive histogram equalization. In: Heckbert, P.S. (ed.) Graphics Gems IV, pp. 474–485. Academic Press Professional, San Diego (1994)
21. Rosten, E., Drummond, T.: Fusing points and lines for high performance tracking. In: Tenth IEEE International Conference on Computer Vision. ICCV 2005, vol. 2, pp. 1508–1515. IEEE (2005)

Four-Wheel Rover Performance Analysis at Lunar Analog Test

Nathan Britton, John Walker, Kazuya Yoshida, Toshiro Shimizu, Tommaso Paniccia and Kei Nakata

Abstract A high fidelity field test of a four-wheeled lunar micro-rover, code-named Moonraker, was conducted by the Space Robotics Lab at a lunar analog site in Hamamatsu Japan, in cooperation with Google Lunar XPRIZE Team Hakuto. For the target mission to a lunar maria region with a steep slope, slippage in loose soil is a key risk; a prediction method of the slip ratio of the system based on the angle of the slope being traversed using only on-board telemetry is highly desirable. A ground truth of Moonraker's location was measured and compared with the motor telemetry to obtain a profile of slippage during the entire four hour 500 m mission. A linear relationship between the slope angle and slip ratio was determined which can be used to predict the slip ratio when ground truth data is not available.

1 Introduction

The focus of this paper is on the soft soil traveling performance of a four-wheel skid-steer rover, code-named Moonraker—one of a dual-rover system intended for a mission to explore lunar caves by tethered-descent. Moonraker was designed specifically for travel over soft loose soil, where slippage is a primary mobility and localization concern. This section introduces Moonraker and the development background of its intended mission.

N. Britton (✉) · J. Walker · K. Yoshida · T. Shimizu · T. Paniccia · K. Nakata
Tohoku University, Sendai, Miyagi, Japan
e-mail: nathan@astro.mech.tohouk.ac.jp

J. Walker
e-mail: john@astro.mech.tohoku.ac.jp

D.S. Wettergreen and T.D. Barfoot (eds.), *Field and Service Robotics*,
Springer Tracts in Advanced Robotics 113, DOI 10.1007/978-3-319-27702-8_24

Fig. 1 The Space Robotics Lab's dual-rover lunar rover system; the parent rover, Moonraker (*left*) and tethered child rover, Tetris (*right*)

1.1 Moonraker

Moonraker (Fig. 1) is 8 kg, and was designed to be as light as possible while not sacrificing mobility performance over lunar regolith, especially of steep slopes of 20° or more that may be encountered around cave entrances [1]. Actuation points were kept minimal, reducing mass and failure modes. The key design features are large relative wheel size and a single non-actuated catadioptric camera (implemented with a hyperbolic mirror) on the top of the rover. There is an additional TOF laser scanner in the front of the rover for detecting obstacles that the camera might fail to. These sensors can also be used to track the location of the rover and assess slippage [2].

Using four wheels, as opposed to six, allows for twice the wheel diameter, assuming the same volume constraints [1]; Larger wheel diameter reduces slip on loose soil. The use of grousers also dramatically reduces slip on loose soil, up to the point at which the grousers no longer penetrate the soil completely [4]. The wheels were therefore designed to be 20 cm in diameter with 2.25 cm grousers. Laboratory experiments indicate that a slip ratio of under 0.1 should be expected with these wheels on slopes of up to 10°.

With a single actuation axis per wheel, turning maneuvers are performed by skid steering, where the wheels on one side turn at different speeds than the other. This can take the form of a spot turn, where both sides spin in opposide directions at the same speeds, or as various degrees of course corrections where one side simply spins forward at a slower rate. Because of this maneuvering dynamic, for the purpose of calculating the total travel distance through wheel odometry, an average of each of the wheels' rotations (as measured by motor encoders) is used.

1.2 Dual Micro Rover System

The Space Robotics Lab has also developed a small, 2 kg child rover, codenamed Tetris, that together with Moonraker composes a dual rover system. Tetris will be tethered to Moonraker, which will serve as an anchor for exploration into pits and down steep cliffs. Cliff traversal experiments and Tetris mobility tests were also conducted, but are not the focus of this paper.

1.3 Team Hakuto

The Space Robotics Lab has partnered with Team Hakuto, a competitor in the Google Lunar XPRIZE (GLXP). These rovers have passed space qualification tests and are intended to be launched on a lunar surface exploration mission in 2017. The field tests reported here were conducted as part of the demonstration round of the GLXP Milestone Mobility Prize, which was awarded to Team Hakuto in January 2015.

2 Field Test

This section introduces the field test conducted at the Nakatajima Sand Dunes in Hamamatsu, Japan on December 19th, 2014. The requirements and conditions of the test as well as the equipment used are presented.

The mission was conducted over the span of 5 h, from 11:30 am until sunset at 16:30. A total travel distance of 550 m (570 m as estimated by wheel encoder odometry) was traversed. Results and performance analysis are presented in Sect. 3.

2.1 High Fidelity Requirements

In preparation for Team Hakuto's planned lunar surface mission (Sect. 1.3), the field test was conducted in "high-fidelity", or as close to the conditions of an actual lunar mission as possible. The test was set up to begin with blind deployment; the rover was placed inside the stowage envelope that will be attached to the lunar lander. After opening the envelope remotely, deploying to the surface, the test was conducted with the following requirements:

- Total travel distance of at least **500 m**
- Travel distance must be proven solely by telemetry

Fig. 2 Moonraker climbs a 10° slope at the Nakatajima Sand Dunes

- No manual reset or human intervention with the hardware after deployment
- Operators forbidden to view the test site, and forbidden contact with anyone who did view the test site during the test
- A time lag of 1.3 s introduced on both the operator laptop and on the rover itself, in order to emulate the communication lag due to the distance between the Earth and Moon.
- Transmission data rate limited to 100 kbps in order to ensure similar bandwidth restrictions to a lunar mission.

Due to legal restrictions in Japan, a 920 MHz radio that is planned for the flight model could not be used for this field test. Instead, a 2.4 GHz 802.11 wireless module was used, dramatically reducing the range of travel from the emulated lander. This had the consequence of limiting the operation range to a 30 m radius from the emulated lander's relay radios.

2.2 *Test Site*

The Nakatajima Sand Dunes is a seaside region of Hamamatsu City and a protected natural environment. The key features are sprawling hills and valleys of loose sand between sections of sparse vegetation. Erosion at the borders of grassy areas create natural steep cliffs. The test site was selected to provide access to sandy slopes, rocky areas, obstacles, and a cliff (Fig. 2).

These macro-features are considered representative of the potential hazards and features of the intended lunar mission. The environment around the target skylight in lacus mortis is in a maria region, with dune-like rolling slopes, and exposed rocks in

Fig. 3 Moonraker maneuvering in a rocky area. A Leica 360° prism, used for ground truth measurements, is visible on top of the mirror

Fig. 4 The Leica TDRA6000 Total Station tracking system used as a ground truth to the wheel odometry

shallow craters and at the edge of the skylight. The average slope down the ramp of the skylight is expected to be an average of 15°, with unpredictable local maxima [1].

The sand itself at Nakatajima, common Earth beach sand, is not as good a match of the target environment. It is well sorted (near-homogenous) granule sizes of 0.1 mm to 1 mm, which is distinct from lunar regolith with very poorly sorted (heterogenous) granule sizes down to the nanometer scale. The sand is quite susceptible to slippage, however, which is sufficient for the goal of assessing relative slip performance.

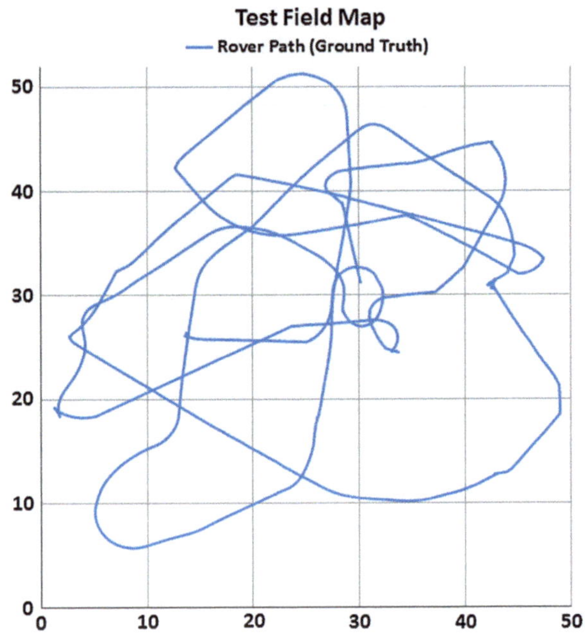

Fig. 5 Map of the rover's path around the field test site. The envelope from which the rover is deployed is at the center of the figure, coordinates (30,30)

2.3 Ground Truth Equipment

In order to evaluate the accuracy of the telemetry gathered by the rover, an external measurement of the rover's location was conducted to serve as a ground truth. A Leica Total Station surveying tool was used with a 360° prism (Fig. 3) attached to the top of the rover. The Total Station unit is equipped with a time of flight laser range finder and a pan/tilt mechanism; when used in conjunction with a reflecting prism, the target can be tracked at 7 points/s, with 3 mm accuracy (Fig. 4). The ground truth data was used to create the map in Fig. 5 tracing the rover's movement.

3 Data Analysis

The data presented in this section corresponds to the first 500 m of the field test, over the course of 4 h. 98 % of the test was captured properly with the Total Station (Sect. 2.3), with the exception of a 5 min gap due to a tracking error at 1 h into the test. Fortunately no significant turns or maneuvers were made during this period of time.

3.1 Travel Distance

The total distance traveled over time is shown in Fig. 6, with the altitude of the rover (above sea level) overlaid to visualize where in the test the major slopes were encountered. The slope angle profile is also presented in Fig. 7, normalized to the travel distance rather than time. The maximum average slope over 2 s periods never exceeded 10° for this segment of the experiment.

In Fig. 6, both the distance as estimated by the motor odometry, and the ground truth data are displayed together; their divergence over time is small but readily apparent. By the end of the ground truth data collection at the 3:50 mark, the wheel

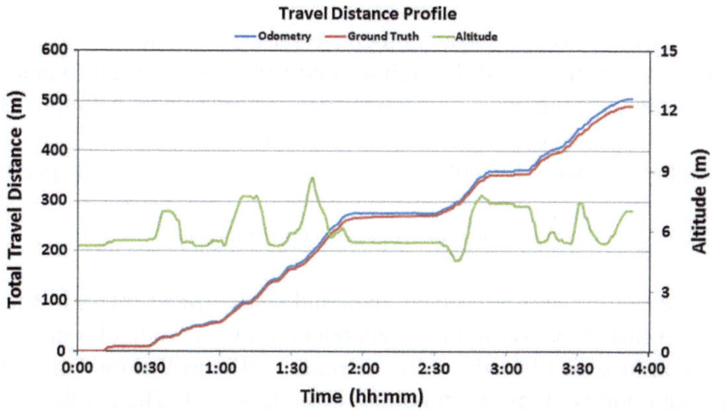

Fig. 6 The total cumulative travel distance over time, as measured by Moonraker's on-board odometry vs externally measured ground truth. Altitude is displayed to indicate the location of slopes

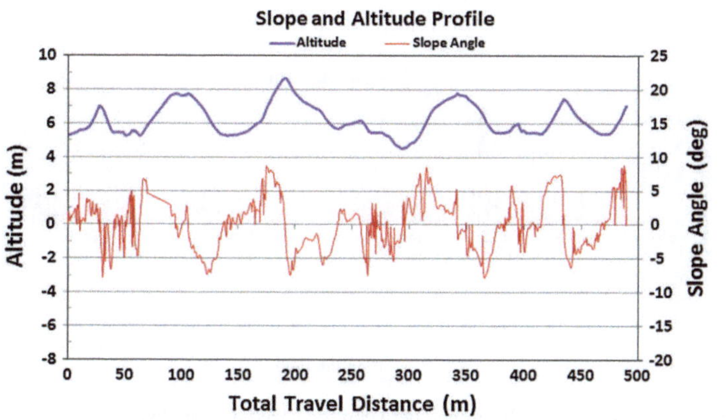

Fig. 7 The altitude of Moonraker's trajectory throughout the test is presented (above), relative to the travel distance. The slope angle at each point is also presented as the derivative of the altitude

odometry-based distance estimation indicated a total 505 m distance traveled, while the ground truth measured 489.6 m.

This represents a total average slip ratio of 0.03, which is consistent with lab-based sandbox tests for the wheel configuration used (Sect. 1.1) for up to 10° slope environments. However due to the uneven and randomly undulating terrain, a total average value does not provide enough data to accurately determine the rover's total travel distance at any given time. It is useful to know what the slip ratio profile is throughout the mission, ideally in real time using only rover telemetry [3].

3.2 Slip Ratio

The speed of Moonraker at each moment, both as estimated by averaging each wheels' encoder odometry and through ground truth, can be used to determine the slip. Figure 8 displays the error of the odometry estimation as a simple difference. These speed data were used to calculate the slip ratio of the rover as a whole every 2 s according to the following formula (ignoring 0 and near-zero wheel speed values):

$$1 - (rover\ speed/wheel\ speed) \tag{1}$$

where *rover speed* is the ground truth data and *wheel speed* refers to the speed of the rover as estimated by averaging the encoder odometry from all four wheels. This result was then median filtered to remove outliers. Figure 9 shows the slip ratio as clusters of data points along the timeline of the field test. The resulting slip ratios occasionally vary widely, but also cleanly cluster together. The slope angles from Fig. 7 are included for comparison; as expected, the slip ratios have a clear relationship with the corresponding slope angle.

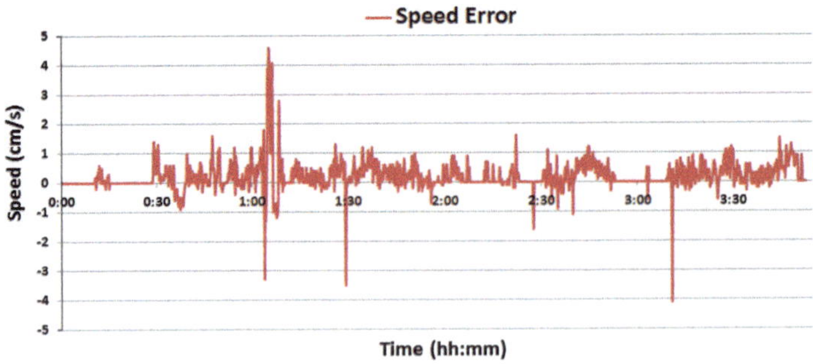

Fig. 8 This graph displays the difference between the speed as calculated by the wheel odometry and by ground truth, which indicates slippage

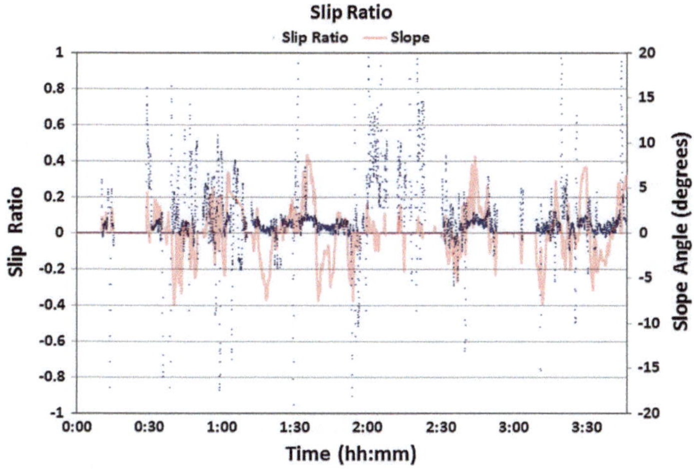

Fig. 9 The slip ratio is calculated every 2 s throughout the field test, and is displayed here in conjunction with the corresponding slope angle. The data points cluster around strong clear slippage events

3.3 *Slope Angle-Slip Ratio Relationship*

The relationship between the slip ratio and the slope angel is presented in Fig. 10, where each of the 6800 slip ratio data points is plotted according to the angle of the slope at the time the measurement was taken. The majority of points are clustered neatly between 0 and 0.2 slip ratio. There are many data points outside of any pattern and artifacts that follow unpredictable trajectories across the graph. Some of these

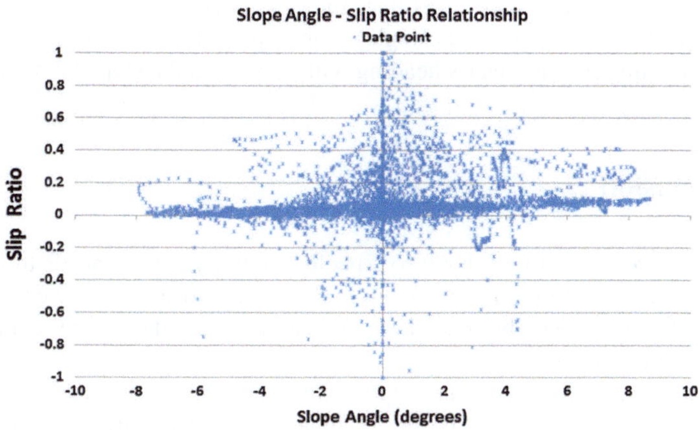

Fig. 10 A cloud of slip ratio data points relative to the angle of the slope they were measured at. Each point represents a 2 s period of time. Figure 11 shows a magnified view

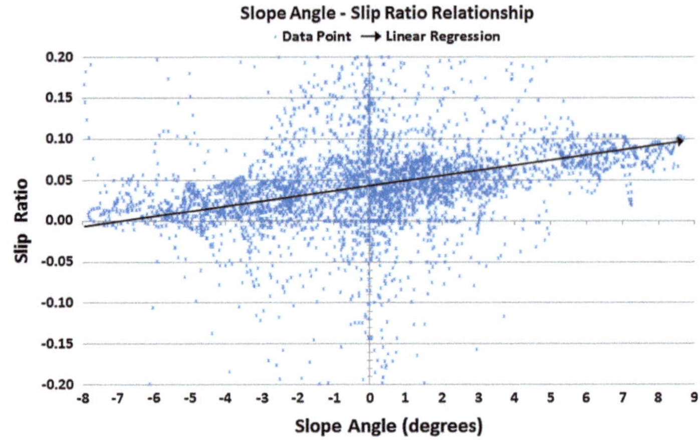

Fig. 11 A magnified view of Fig. 10; a linear trendline from 0 to 0.1 slip ratio is indicated

appear to be due to lateral slip (which is unaccounted for in this study), while others are due to transient-state slip ratios during turning maneuvers.

Figure 11 shows a linear regression calculated after data from turning maneuvers with negligible forward movement (e.g. 2:00–2:30 in Fig. 6) are removed. The linear regression equation is as follows:

$$y = 0.0066x + 0.0426 \tag{2}$$

The correlation coefficient is 0.1122, with a standard deviation of 0.03. This is a loose, but significant linear trend from near-zero to slight negative slip (slipping forward) on downward slopes of 9° to 0.1 slip ratio at 9° upward slope. This linear trendline can therefore be used to estimate the slip ratio of the rover at any given time using only an IMU to determine the angle of the slope that the rover is traversing, even before accounting for the rover's heading with respect to the slope being traversed.

4 Conclusion

Slippage is a very important threat to wheeled mobility, which needs to be understood and accounted for in rover missions to the lunar mare and similar environments on Mars [3]. Controlled laboratory tests are useful for validating the relative effectiveness of different mobility configurations, but field validation is necessary for determining the actual performance in a real environment. At our field test in a lunar analog environment, we traveled over 500 m, and measured a high precision ground truth in order to perform a moment-by-moment slip analysis.

Our results indicate a linear relationship between the angle of the slope being traversed at any given time and the slippage occurring. This linear relationship gives valuable insight into the extent of slippage that can be expected based on a simple easily measurable characteristic of the rover's environment—slope angle, without concern to the heading of the rover with respect to the slope. The data used in this investigation, having come from a high fidelity field test at a lunar analogue environment, gives us high confidence that this linear relationship can be a useful component of navigation systems implemented for lunar and martian wheeled rover systems.

4.1 Future Work

This information can be used in navigation systems to correct rovers wheel odometry in real time. By extracting the heading of the rover from the camera data, a system to account for lateral slopes/slip would further improve the accuracy of wheel odometry for navigation systems. There is also room to investigate refining or defining this linear relationship for different soils without the use of ground truth equipment.

References

1. Britton, N., Yoshida, K., Walker, J., Nagatani, K., Taylor, G., and Dauphin, L. Lunar micro rover design for exploration through virtual reality tele-operation. In: Field and Service Robotics (2013)
2. Maimone, M., Cheng, Y., Matthies, L.: Two years of visual odometry on the mars exploration rovers. J. Field Robot. **24**(3), 169–186 (2007). special issue on Space Robotics
3. Reina, G., Ojeda, L., Milella, A., Borenstein, J.: Wheel slippage and sinkage detection for planetary rovers. IEEE Trans. Mechatron. **11**(2), 185–195 (2006)
4. Sutoh, M.: Traveling Performance Analysis of Lunar/Planetary Robots on Loose Soil. PhD thesis, Tohoku University, 2013

Energy-Aware Terrain Analysis for Mobile Robot Exploration

Kyohei Otsu and Takashi Kubota

Abstract This paper presents an approach to predict energy consumption in mobility systems for wheeled ground robots. The energy autonomy is a critical problem for various battery-powered systems. Specifically, the consumption prediction in mobility systems, which is difficult to obtain due to its complex interactivity, can be used to improve energy efficiency. To address this problem, a self-supervised approach is presented which considers terrain geometry and soil types. Especially, this paper analyzes soil types which affect energy usage models, then proposes a prediction scheme based on terrain type recognition and simple consumption modeling. The developed vibration-based terrain classifier is validated with a field test in diverse volcanic terrain.

1 Introduction

As robotics technology develops rapidly, a number of applications are deployed into real fields. These real-world robotic applications are typically subject to the interaction with a challenging environment, which is characterized by its dynamic and unknown properties. The robots deployed in these fields should have the capability to percept, model, and interact with surrounding situations, in order to enable safe and efficient operations under several hardware restrictions. Such autonomy is to some extent required for any independent systems, especially for robots in extreme environments including planetary surfaces and active volcanoes. Significant examples for extreme terrain operations include the Mars rovers developed by NASA/JPL.

K. Otsu (✉)
Department of Electrical Engineering and Information Systems,
The University of Tokyo, 7-3-1 Hongo, Bunkyo, Tokyo, Japan
e-mail: kyon@ac.jaxa.jp

T. Kubota
Institute of Space and Astronautical Science, Japan Aerospace Exploration Agency,
3-1-1 Yoshinodai, Chuo, Sagamihara, Kanagawa, Japan
e-mail: kubota@isas.jaxa.jp

D.S. Wettergreen and T.D. Barfoot (eds.), *Field and Service Robotics*,
Springer Tracts in Advanced Robotics 113, DOI 10.1007/978-3-319-27702-8_25

The autonomous navigation system have shown successful results on the remote planetary surface without intensive human intervention [2, 11].

Besides the interaction with surroundings, the energy autonomy is an essential technique for battery-powered embedded systems. To enable long-term operations, the robots should be capable to obtain power either from mounted generators or external energy hotspots, and use it properly to perform all assigned tasks. Since the energy budget is severely limited, system designers will face difficult challenges for efficient energy utilization. A battery-powered system is known to survive longer by appropriate scheduling of energy-consuming tasks. For example, an decreasing load profile improves the battery behavior and makes the lifetime longer than an inverse profile [14]. This battery characteristic leads to the following idea: if the energy consumption can be predicted prior to the execution, and the task scheduling is appropriately performed, the exploration period and range might be extended.

The aim of this research is a priori estimation of energy consumption in mobility systems. The energy consumption is associated in some way with the robot mechanical properties and terrain characteristics. One of the challenging problems is to estimate the interaction between a robot and terrain since the soil behavior cannot be modeled uniformly. In the proposed method, a self-supervised scheme is adopted to make a simple model for energy prediction. Firstly, a vibration-based classifier provides the estimation of terrain class which characterize the interaction model. Then, given the class as teacher data, a vision-based classifier gives a priori estimation of the class through machine learning techniques. Finally, the energy consumption is predicted using the terrain class and geometry data.

This paper presents the concept of the proposed scheme and detailed description of energy-aware terrain classification based on vibration signals. The algorithm is tested by real-world data obtained with a four-wheeled vehicle in diverse volcanic terrain.

2 Related Work

The core part of this research belongs to the terrain classification problem. Specifically, vibration-based terrain classification has been conducted by several research groups after it is initially suggested by Iagnemma and Dubowsky [9]. Sadhukhan et al. and DuPont et al. developed a neural network approach using FFT-based vibration analysis, which distinguishes different terrain types [8, 15, 16]. Brooks et al. proposed a classification framework using contact microphones, which estimates terrain components such as sand and gravel [3, 4]. Weiss et al. proposed a feature-based compact representation, which is fairly relevant to this research, classifying different terrain types [20]. Ojeda et al. developed a neural network method applicable to other sensors [12]. Similarly, road roughness estimation was performed for high-speed vehicles by Stavens et al. [17]. These works extract descriptive vectors from raw vibration signals, and utilize machine learning to compute terrain labels. Many of the works are conducted in the frequency domain. Comparisons of different classification methods are given by Weiss et al. [19] and Coyle et al. [7], where they mention high accuracy of the SVM (Support Vector Machine) classifier when paired with proper kernel functions.

Recently, the self-supervised scheme is actively studied and applied to robotics applications. The self-supervised classification is an automatic training of a classifier using estimated labels from other classifiers. The classifier to be trained is usually using remote sensors such as cameras and LIDARs. Angelova et al. performed vision-based unsupervised clustering to obtain terrain labels, then the labels are used to train their slip estimator [1]. Krebs et al. enabled an on-line learning of mobility attributes by combining vision and inertial/mechanical measurements using a Bayesian framework [10]. Brooks et al. proposed a framework to predict mechanical properties of distant terrain by empirical learning of wheel-terrain interaction [5]. Those works successfully predicted terrain attributes of distant terrain.

The research presented in this article also employs self-supervised learning in order to predict energy consumption before the robot actually drives over the terrain. The attribute to be estimated is apparently important for energy-aware behavior planning. However, it is difficult to make accurate estimation since required power is determined by a complex function of the robot and terrain interaction. This paper analyzes the relation of energy consumption and robot-terrain interaction and develops a simple inference model using a vibration sensor and cameras. Based on the model, the energy consumption is predicted for typical wheeled vehicles.

3 Technical Approaches

This section describes the conceptual overview of the system. Then, the detailed technical description is given for the energy-aware terrain analysis using vibration measurements.

3.1 *Self-supervised Scheme for Inferencing Energy Consumption*

The energy consumption in mobility systems depends on both terrain types and geometry. Let E be the energy consumption, A, G be the appearance and geometry information obtained from cameras, and V be the vibration measurement from an IMU. Assuming the consumption model is specific for finite terrain types, the regression function of energy from inputs can be expressed as

$$f(E|A, G, V) = \sum_T P(T|A, G, V) f(E|T, A, G, V) \tag{1}$$

$$= \sum_T P(T|A, V) f(E|T, G) \tag{2}$$

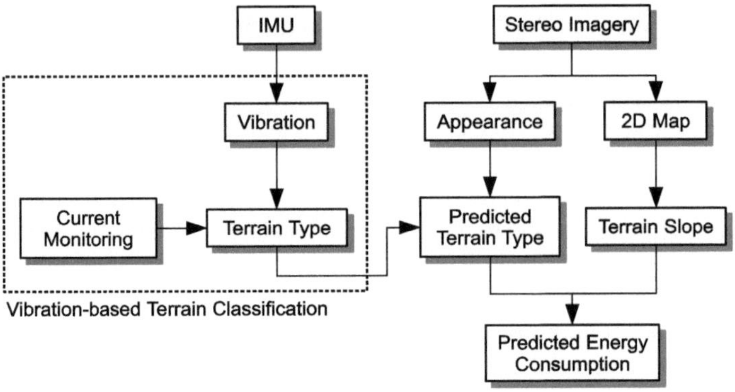

Fig. 1 Overview of self-supervised scheme

for terrain type T and $\sum_T P(T|A, V) = 1$. From this equation, the problem can be split into the terrain type recognition problem ($P(T|A, V)$) and the energy consumption inference problem ($f(E|T, G)$). For recognizing terrain types, a robot classifies terrain using appearance and vibration measurements. The self-supervised scheme is used in this part, i.e., the terrain labels from the vibration-based classifier are used as teacher data for the vision-based predictive classifier. On the other hand, the regression function is determined empirically from experimental data. The function is developed based on the physical model of typical wheeled robots.

The illustration of the proposed self-supervised scheme is given in Fig. 1. The remainder of this paper focuses on the method to estimate energy consumption based on vibration analysis. Firstly, the function $f(E|T, G)$ is formulated from a physical model. It shows the energy consumption is a linear function that depends on the robot-terrain interaction. Next, the self-supervised classification based on vibration analysis is explained. The classified result is processed in a winner-take-all manner, and combined with the formulated energy equation to provide accurate energy prediction.

3.2 Energy Consumption Model for Wheeled Vehicles

The amount of energy consumption depends on the soil type and the terrain geometry. In this section, the model is explained based on a physical model of wheeled vehicles.

Consider a robot driving in a velocity v over a slanted pseudo plane with angle θ_p (Fig. 2). The vehicle dynamics is expressed by

$$\sum_j f_{dj} - \sum_j f_{rj} - Mg \sin \theta_p = M\dot{v} \tag{3}$$

where F_{dj} is the driving force of each wheel, f_{rj} is the driving resistance of each wheel, M is the total mass of the robot, and g is the gravity constant. The driving

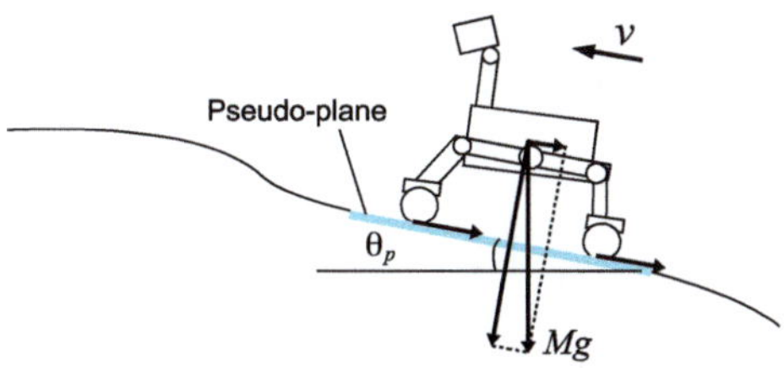

Fig. 2 Wheeled robot model on slanted pseudo plane

resistance f_{rj} is expressed as the sum of rolling resistance between the wheel and terrain $f_{wj}(v)$ and friction loss in bearing and gear $f_{gj}(v)$.

$$F_{rj}(v) = f_{wj}(v) + f_{gj}(v) \tag{4}$$

The resistance depends on the robot velocity. To simplify the problem, let us put an affordable assumption that the robot drives at an arbitrary constant speed v_0 within a small distance. Then, the Eq. (3) becomes

$$F_d = \sum_j \left[f_{wj}(v_0) + f_{gj}(v_0) \right] + Mg \sin\theta_p \tag{5}$$

where F_d is the sum of all driving forces.

On the other hand, the driving force can be computed from the motor torque

$$f_{dj} = \frac{\eta\gamma T_j}{R} \tag{6}$$

where η is the transmission efficiency, γ is the gear reduction ratio, T_j is the generated torque, and R is the wheel radius. Since the torque is proportional to current

$$T_j = k_t I_j \tag{7}$$

the electrical energy consumption is expressed by

$$E_e = \frac{VR\left(\sum_j \left[f_{wj}(v_0) + f_{gj}(v_0) \right] + Mg \sin\theta_p\right)}{\eta\,\gamma\,k_t} \tag{8}$$

where V represents the source voltage. Under the assumption that the traversable slope for wheeled robots is small, the equation can be simplified to

$$E_e \simeq \alpha_{r,t} + \beta_r \theta_p \tag{9}$$

where $\alpha_{r,t}$ and β_r is constant values. Note that $\alpha_{r,t}$ depends on both robots and terrain, while β_r depends on only robot systems. However, in the real natural environments, the slope angle observation θ is not consistent with the pseudo plane angle θ_p due to the terrain deformation. In this paper, the deformation effect is modeled by a linear equation as $\theta_p = \gamma_{r,t}\theta$. Hence, the final inference model is expressed as

$$E_e \simeq \alpha_{r,t} + \beta_r \gamma_{r,t}\theta \tag{10}$$

$$= \alpha_{r,t} + \delta_{r,t}\theta \tag{11}$$

The above model suggests that we can infer the energy consumption using two constants and a slope angle measurement. The constants are estimated empirically from experiments. In the preliminary study, they depends on soil types, which can be classified by vibration-based machine learning. On the other hand, the slope angle is computed geometrically from stereo vision. There are several efficient methods to recover terrain geometry from images [13].

3.3 Vibration-Based Terrain Classification

In order to know the terrain class and the associated constants which affect energy consumption, a vibration-based terrain classifier is proposed. The reason to choose vibration is that it well represents the wheel-terrain interaction as presented in the previous studies [3, 20], whereas the direct measurement of motor currents does not work due to its high dependency on the terrain geometry (which can also be seen in (11)).

The proposed classifier employes the feature-based SVM similar to [20]. However, the feature representation described here is computed in the frequency domain, and designed to work for a real outdoor robot.

At first, vibration data is collected from an accelerometer rigidly attached to the robot body. Using 3-axis acceleration data the signal power is computed and then subtracted by the short-time averages. The processed time-series acceleration vector $\boldsymbol{a} = [a_1, \ldots, a_t, \ldots]$ is converted to the time-frequency domain by continuous wavelet transform [18].

$$\boldsymbol{A} = \begin{bmatrix} A_{f_1,1} & \cdots & A_{f_1,t} & \cdots \\ \vdots & \ddots & \vdots & \\ A_{f_m,1} & \cdots & A_{f_m,t} & \cdots \end{bmatrix} \tag{12}$$

In this representation, each column corresponds to the signal spectrum for each time, and each row corresponds to the time-series of a single frequency.

The raw matrix can be used to train the classifiers. However, in this paper, the raw matrix is subsampled to $2 \times N$ matrix for sake of efficiency. The rows and columns are selected so that the characteristic elements are preserved. For the frequency domain, the natural frequency f_n and its octave $2f_n$ are preserved. The signal power for the natural frequency is dominant in vibration analysis of the robot locomotion. For the time domain, samples on the grouser-to-grouser interval t_g are selected. Typically, all-terrain robots have grousers to obtain traction. The symmetric arrangement

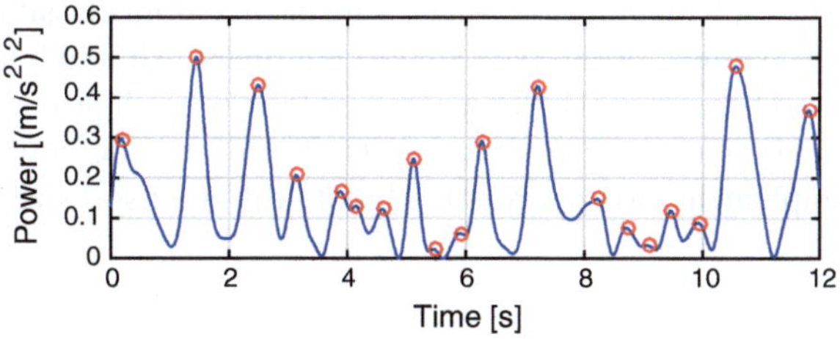

Fig. 3 Time-series signal power corresponding to the natural frequency. Detected positive peaks caused by grouser-to-grouser intervals are marked with *red* circles

of grousers causes periodical characteristics to signal spectra as shown in Fig. 3. The local peak positions are utilized to describe the time-domain characteristics. N samples around designated time t are extracted.

After the subsample process, the following $2 \times N$ matrix is obtained.

$$\begin{bmatrix} \boldsymbol{x}_{t,f_n} \\ \boldsymbol{x}_{t,2f_n} \end{bmatrix} = \begin{bmatrix} A_{f_n,1} & \cdots & A_{f_n,N} \\ A_{2f_n,1} & \cdots & A_{2f_n,N} \end{bmatrix} \tag{13}$$

For each row vector $\boldsymbol{x}_t$, the following features are extracted.

- The mean μ_t of the vector. The mean is roughly 0 for smooth surfaces, while it becomes grater for rough surfaces.

$$\mu_t = \frac{1}{N}\sum_{i=1}^{N} x_i \tag{14}$$

- The standard deviation σ_t. The larger deviation represents the terrain is not uniformly composed.

$$\sigma_t = \sqrt{\frac{1}{N}\sum_{i=1}^{N}(x_i - \mu_t)^2} \tag{15}$$

- The maximum value m_t of the vector. It corresponds to the strength of the shock from the terrain.

$$m_t = \max(\boldsymbol{x}_t) \tag{16}$$

- The coefficient of variation c_t. It is the relative variance to the signal strength.

$$c_t = \frac{\sigma_t}{\mu_t} \tag{17}$$

Using these four types of features, the feature vector for each time is acquired as follows.

$$\boldsymbol{X}_t = \left[\mu_{t,f_n}\ \sigma_{t,f_n}\ m_{t,f_n}\ c_{t,f_n}\ \mu_{t,2f_n}\ \sigma_{t,2f_n}\ m_{t,2f_n}\ c_{t,2f_n}\right]^\top \tag{18}$$

The 8-element feature vectors are used to train classifiers. Each classifier detects one pre-defined terrain type against all others. Although unsupervised clustering can be used here as in [1], the supervised learning still provides accurate enough estimation of the energy-related constants. Therefore, the supervised SVM is employed for implementation in order to classify different soil sizes.

4 Experiment

In the previous section, the energy inference method based on vibration measurements is presented. The field experiment described in this section shows the validity of the approach and evaluates the performance.

4.1 Setup

The rover used in the field experiment is shown in Fig. 4. It is a four-wheeled unmanned vehicle with a customized suspension system. The dimensions are $0.88 \times 0.83 \times 1.50$ m and it weighs 50 kg. Four aluminum wheels with silicon grousers are driven by DC motors at a rate 7.6 rpm. The wheel radius is 0.10 m and the grouser-to-grouser distance is 0.05 m. Attached to the body, a 3-axis accelerometer Crossbow CXL17LF3 measures vibration data at 100 Hz. The consumption energy is computed from motor currents.

Izu-Oshima island in Japan is selected as the experimental field. The formation of the place is based on an active volcano Mt. Mihara. The geological features have been created by volcanic eruptions and water penetrations; therefore, diverse soil types are mixed in local regions. Three terrain types that can be seen in Fig. 5 are defined as follows.

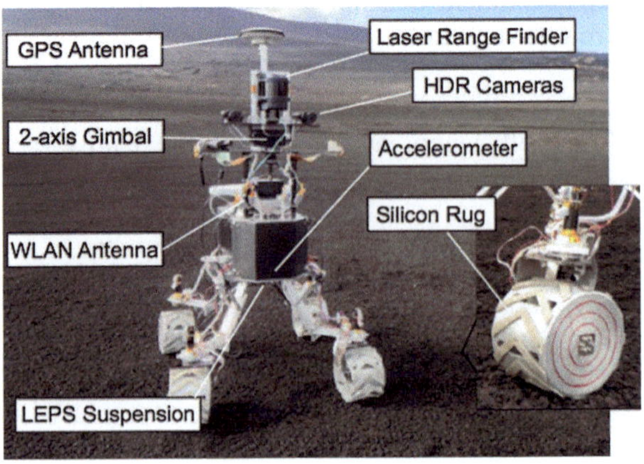

Fig. 4 The AKI rover. Four wheeled all-terrain robots with custom suspension mechanism. Silicon rugs are attached to the aluminum wheels

Fig. 5 Experimental fields with various soil types. Terrain types are labeled manually. (*Green* Dense Sand, *Blue* Fine Gravel, *Red* Coarse Gravel)

1. **Dense Sand**: very small particles are packed and form hard terrain.
2. **Fine Gravel**: gravels of a few centimeters are loosely packed.
3. **Coarse Gravel**: larger gravels are piled and form deformable terrain.

The detailed appearance and sample vibration data are shown in Fig. 6. Each terrain types have distinct signal properties in terms of strength, periodicity, and so on.

The algorithm is implemented in MATLAB. For the wavelet transform to extract features, the software provided in [18] is used. The Morlet wavelet is selected as the mother wavelet. For the terrain classification, LIBSVM [6] is used. It employs the radial basis function kernel with optimal parameters tuned by 5-fold cross validation.

4.2 *Classifying Terrain Based on Vibration Signals*

The wavelet transform results for various terrain (i.e., $\boldsymbol{A}$) are shown in Fig. 7. The natural frequency $f_n = 6.8\,\text{Hz}$ and its octave show significant properties. Time-domain periodicity can be observed in correspondence with the grouser-to-grouser interval $t_g = 0.63\,\text{s}$. In the algorithm, 2×20 matrices are extracted from these results to generate 8-element feature vectors.

The classification result by the vibration-based classifier is shown in Table 1. The dataset size is 191, 300, and 225, respectively. In the experiment, 10-fold cross validation is used to compute the average accuracy and variance. The 64-point FFT features similar to [15, 19] are used as reference. The accuracy was 76.80 % for 3-class classification which is slightly inferior to the FFT features. However, the difference is small considering the number of elements is eight times smaller. Moreover,

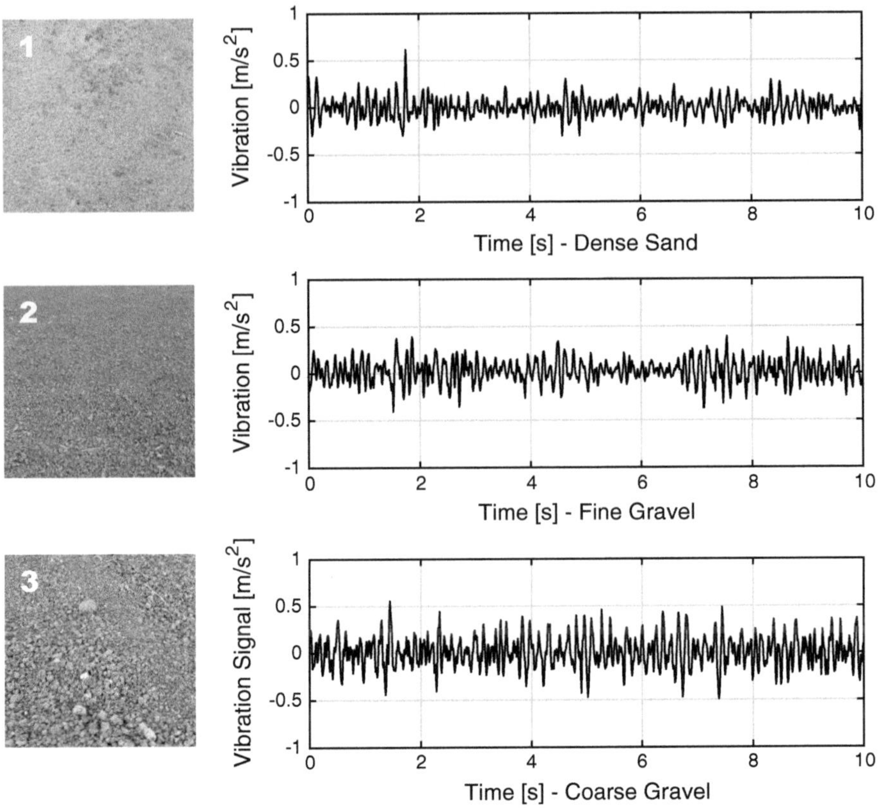

Fig. 6 Vibration signal example for 10 s traversal. Three terrain types are (*1*) Dense Sand, (*2*) Fine Gravel, and (*3*) Coarse Gravel. Each terrain presents distinct properties in signal strength, periodicity, etc

higher classification accuracy is achieved for some classes. In fact, the error rate for dense sand terrain is less than 3 %.

The confusion matrix for 3-class test data is shown in Table 2. There is confusion in fine and coarse gravels. This is because the separability in the feature space was relatively small. One reason will be the ambiguity of human hand-labeling. Introducing pre-training and new data might improve the classification.

4.3 Modeling Energy Usage

Two parameters $\alpha_{r,t}$ and $\delta_{r,t}$ in the energy consumption model in (11) is empirically estimated. From average consumption of all 1 m segments in a 773 m trajectory, the linear regression model is estimated. Obtained data points and parameters are

Fig. 7 Wavelet analysis of vibration signals for 50 s. The signal power corresponding to the natural frequency $f_n = 6.8$ Hz and its octave $2f_n$ shows significant characteristics

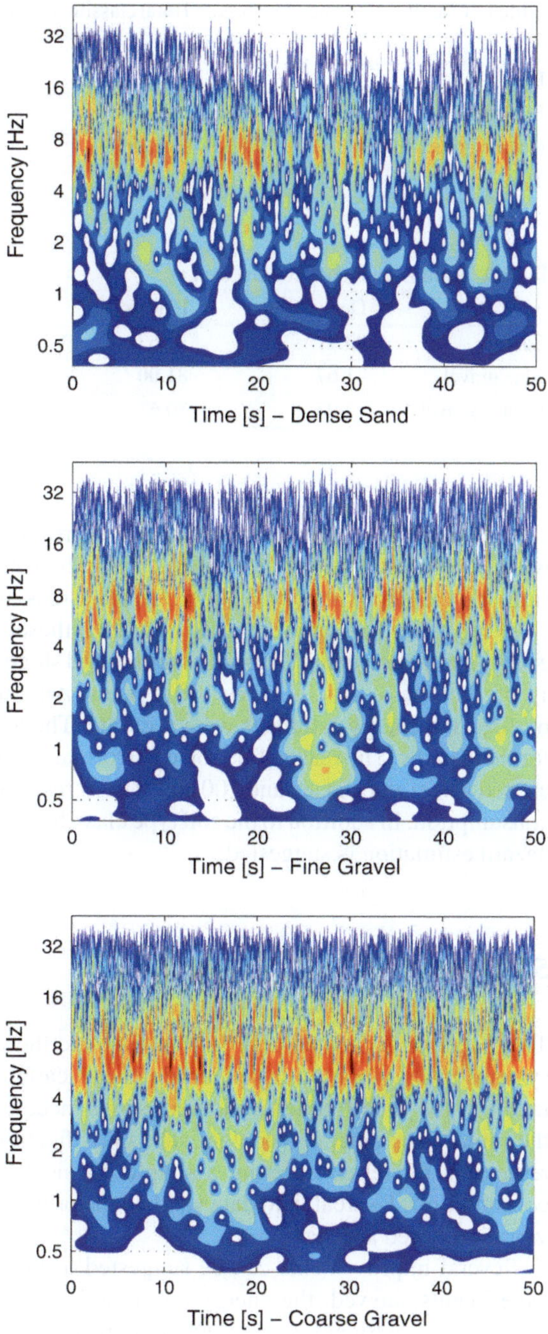

Table 1 Classification rates per class and total classification accuracy (%) in 10-fold cross validation

	Proposed	64pt-FFT
Dense sand	97.21 ± 1.85	95.45 ± 3.71
Fine gravel	79.59 ± 5.42	86.81 ± 8.42
Coarse gravel	83.80 ± 3.38	80.90 ± 7.67
Total	76.80 ± 4.59	78.18 ± 7.67

Table 2 Confusion matrix for test data

	Dense sand	Fine gravel	Coarse gravel	Unclassified
Sand	**93.71**	2.63	0.53	3.13
Fine gravel	1.67	**81.00**	8.33	9.00
Coarse gravel	2.23	30.61	**58.74**	8.42

presented in Fig. 8. The result shows that the terrain in the largest consumption (coarse gravel) requires more than 15 % times grater than the smallest (dense sand). This fact supports the importance of distinguishing classes in the energy-aware context.

Along with the vibration-based classifier, these regression functions produce the energy estimation using a vibration sensor and slope measurement. Figures 9, 10, and 11 present the results for three 100 m paths. Although terrain has various elevation profiles, the energy estimates were accurate. The RMS errors are 3.42, 3.06, 5.56 W for three paths. The reason for worse performance in Fig. 11 is that geometrical steps caused wheel stuck at around 700 and 1000 s, resulting in the rapid increase of energy consumption. In addition to the soil type classification, the importance of geometrical hazard estimation is suggested.

5 Conclusion

This paper presented an approach to estimate the energy consumption of mobility systems using vibration-based terrain classification. The compact feature representation in the time-frequency domain shows accurate classification performance in the multi-class labeling problem. The classification results are combined with the regression model considering a simple physical model to estimate actual energy consumption. The real field data validate the promising performance of the proposed vibration-based approach.

Several improvements can be suggested to the current inference model. As the experiments showed, the energy consumption drastically changes in the presence of (non-)geometrical hazards such as steps or slip-inducing terrain. The regression model should consider those hazards in order to improve robustness. Moreover, the

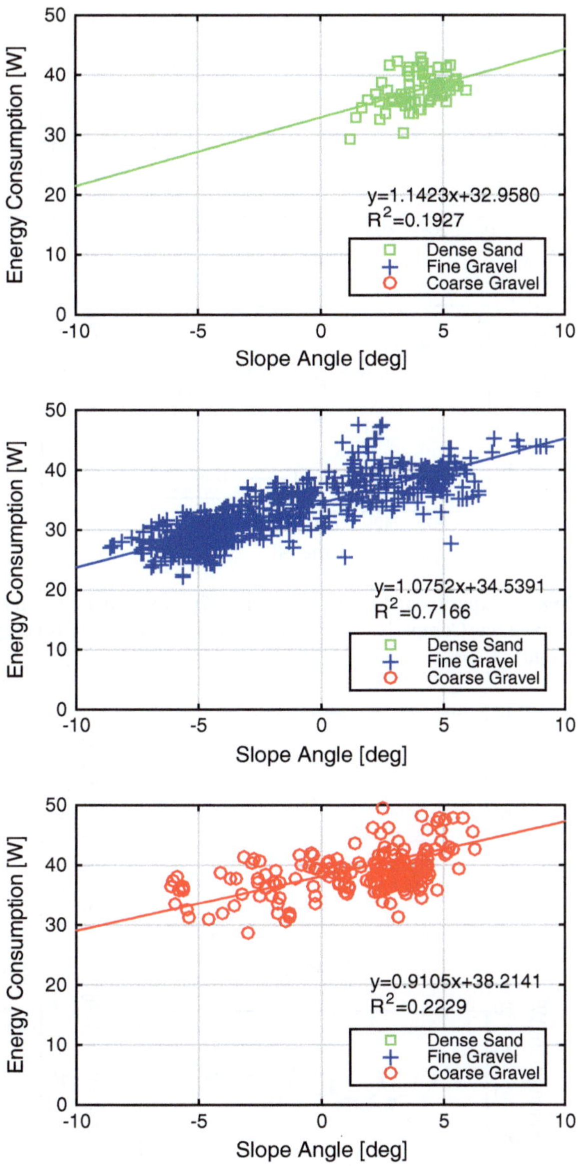

Fig. 8 Relationship between slope angle and energy consumption for 1 m traversal on natural terrain. The linearity can be observed in every terrain types. The estimated constants are shown in the figure

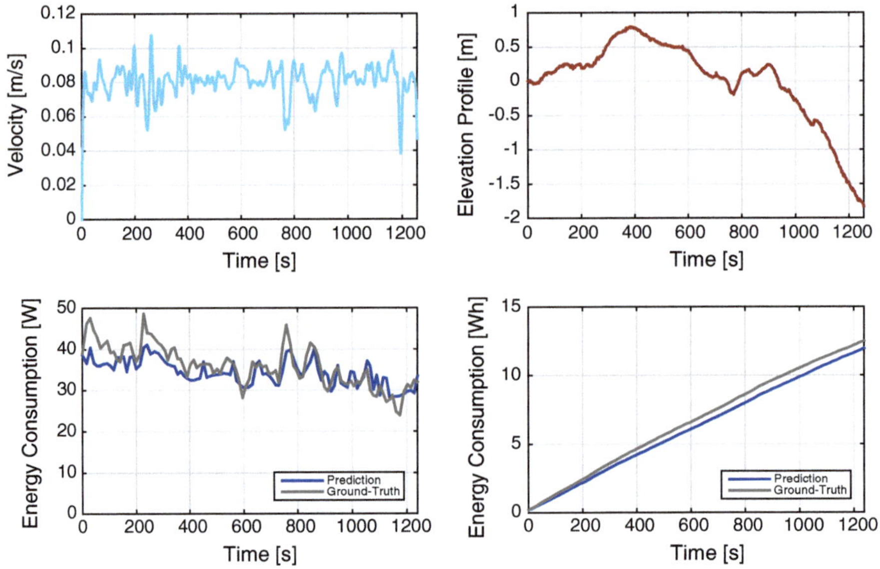

Fig. 9 Experimental result for path 1. *Top row* actual velocity (*left*) and elevation profile (*right*). *Bottom row* Comparison between predicted and measured energy consumption (*left*) and its integral (*right*)

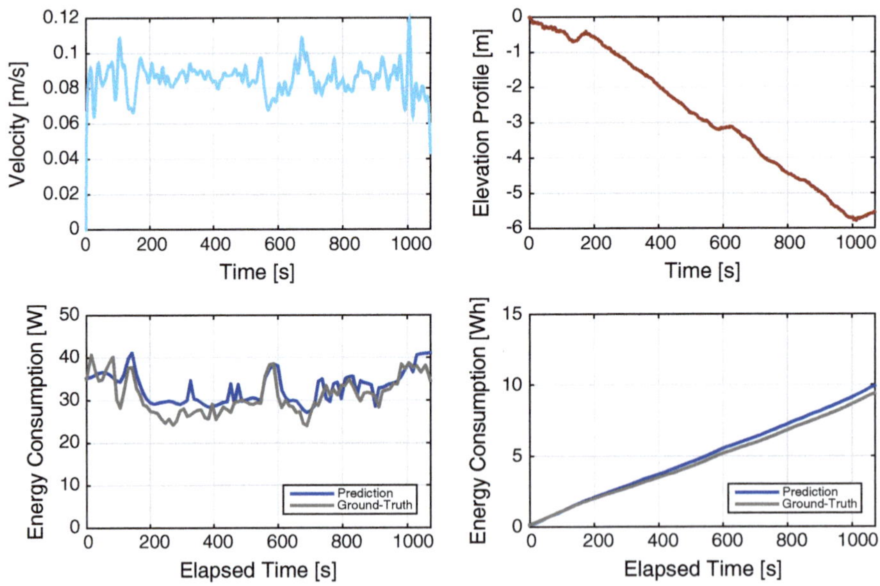

Fig. 10 Experimental result for path 2

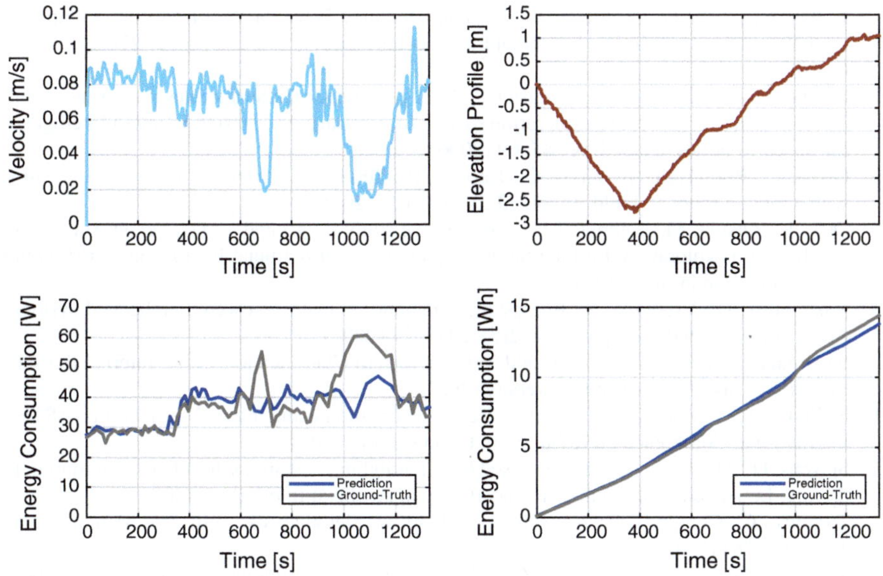

Fig. 11 Experimental result for path 3. Note that the error grows at 700 and 1000 s due to the geometrical step hazards

confusion in similar terrain types may be improved by introducing pre-training, or handling visual information at the same time.

References

1. Angelova, A., Matthies, L., Helmick, D., Perona, P.: Learning and prediction of slip from visual information. J. Field Robot. **24**(3), 205–231 (2007)
2. Biesiadecki, J., Maimone, M.: The Mars exploration rover surface mobility flight software: driving ambition. In: IEEE Aerospace Conference (2006)
3. Brooks, C., Iagnemma, K.: Vibration-based terrain classification for planetary exploration rovers. IEEE Trans. Robot. **21**(6), 1185–1191 (2005)
4. Brooks, C., Iagnemma, K., Dubowsky, S.: Vibration-based terrain analysis for mobile robots. In: IEEE International Conference on Robotics and Automation, pp. 3415–3420 (2005)
5. Brooks, C.A., Iagnemma, K.: Self-supervised terrain classification for planetary surface exploration rovers. J. Field Robot. **29**(3), 445–468 (2012)
6. Chang, C., Lin, C.: LIBSVM: a library for support vector machines. http://www.csie.ntu.edu.tw/cjlin/libsvm (2001)
7. Coyle, E., Collins, E.: A comparison of classifier performance for vibration-based terrain classification. In: 26th Army Science Conference (2008)
8. DuPont, E.M., Moore, C.A., Collins, E.G., Coyle, E.: Frequency response method for terrain classification in autonomous ground vehicles. Auton. Robots **24**(4), 337–347 (2008)
9. Iagnemma, K.D., Dubowsky, S.: Terrain estimation for high-speed rough-terrain autonomous vehicle navigation. In: SPIE Conference on Unmanned Ground Vehicle Technology IV,vol. 4715, pp. 256–266 (2002)

10. Krebs, A., Pradalier, C., Siegwart, R.: Adaptive rover behavior based on online empirical evaluation: rover-terrain interaction and near-to-far learning. J. Field Robot. **27**(2), 158–180 (2010)
11. Maimone, M., Biesiadecki, J.J., Tunstel, E., Cheng, Y., Leger, C.: Surface navigation and mobility intelligence on the Mars exploration rovers. In: Howard, A., Tunstel, E. (eds.) Intelligence for Space Robotics, chap. 3, pp. 45–69 (2006)
12. Ojeda, L., Borenstein, J., Witus, G., Karlsen, R.: Terrain characterization and classification with a mobile robot. J. Field Robot. **23**(2), 103–122 (2006)
13. Otsu, K., Otsuki, M., Kubota, T.: A comparative study on ground surface reconstruction for rough terrain exploration. In: International Symposium on Artificial Intelligence for Robotics and Automation in Space (2014)
14. Rakhmatov, D., Vrudhula, S.: Energy management for battery-powered embedded systems. ACM Trans. Embed. Comput. Syst. **2**(3), 277–324 (2003)
15. Sadhukhan, D.: Autonomous Ground Vehicle Terrain Classification Using Internal Sensors. Master's thesis, Florida State University (2004)
16. Sadhukhan, D., Moore, C., Collins, E.: Terrain estimation using internal sensors. In: IASTED International Conference on Robotics and Applications (2004)
17. Stavens, D., Thrun, S.: A self-supervised terrain roughness estimator for off-road autonomous driving. In: Annual Conference on Uncertainty in Artificial Intelligence (2006)
18. Torrence, C., Compo, G.P.: A practical guide to wavelet analysis. Bull. Am. Meteorol. Soc. **79**(1), 61–78 (1998)
19. Weiss, C., Fechner, N., Stark, M., Zell, A.: Comparison of different approaches to vibration-based terrain classification. In: European Conference on Mobile Robots, pp. 7–12 (2007)
20. Weiss, C., Frohlich, H., Zell, A.: Vibration-based terrain classification using support vector machines. In: IEEE/RSJ International Conference on Intelligent Robots and Systems, pp. 4429–4434 (2006)

Part IV
Aerial

Vision and Learning for Deliberative Monocular Cluttered Flight

Debadeepta Dey, Kumar Shaurya Shankar, Sam Zeng, Rupesh Mehta, M. Talha Agcayazi, Christopher Eriksen, Shreyansh Daftry, Martial Hebert and J. Andrew Bagnell

Abstract Cameras provide a rich source of information while being passive, cheap and lightweight for small Unmanned Aerial Vehicles (UAVs). In this work we present the first implementation of receding horizon control, which is widely used in ground vehicles, with monocular vision as the only sensing mode for autonomous UAV flight in dense clutter. Two key contributions make this possible: novel coupling of perception and control via relevant and diverse, multiple interpretations of the scene around the robot, leveraging recent advances in machine learning to showcase anytime budgeted cost-sensitive feature selection, and fast non-linear regression for monocular depth prediction. We empirically demonstrate the efficacy of our novel

D. Dey (✉) · K.S. Shankar · S. Zeng · S. Daftry · M. Hebert · J.A. Bagnell
The Robotics Institute, Carnegie Mellon University, Pittsburgh, PA, USA
e-mail: debadeep@ri.cmu.edu

K.S. Shankar
e-mail: kumarsha@ri.cmu.edu

S. Zeng
e-mail: samlzeng@ri.cmu.edu

S. Daftry
e-mail: daftry@ri.cmu.edu

M. Hebert
e-mail: hebert@ri.cmu.edu

J.A. Bagnell
e-mail: dbagnell@ri.cmu.edu

R. Mehta
NVIDIA Corporation, Santa Clara, CA, USA
e-mail: rupeshm@nvidia.com

M.T. Agcayazi
George Mason University, Fairfax, VA, USA
e-mail: magcayaz@gmu.edu

C. Eriksen
Harvey Mudd College, Claremont, CA, USA
e-mail: ceriksen@hmc.edu

D.S. Wettergreen and T.D. Barfoot (eds.), *Field and Service Robotics*,
Springer Tracts in Advanced Robotics 113, DOI 10.1007/978-3-319-27702-8_26

pipeline via real world experiments of more than 2 kms through dense trees with an off-the-shelf quadrotor. Moreover our pipeline is designed to combine information from other modalities like stereo and lidar.

1 Introduction

Unmanned Aerial Vehicles (UAVs) have recently received a lot of attention by the robotics community. While autonomous flight with active sensors like lidars has been well studied [2, 28], flight using passive sensors like cameras has relatively lagged behind. This is especially important given that small UAVs do not have the payload and power capabilities for carrying such sensors. Additionally, most of the modern research on UAVs has focussed on flying at altitudes with mostly open space [9]. Flying UAVs close to the ground through dense clutter [26, 28] has been less explored (Fig. 1).

Receding horizon control [18] is a classical deliberative scheme commonly used in autonomous ground vehicles including five out of the six finalists of the DARPA Urban Challenge [5]. Figure 2 illustrates receding horizon control on our UAV in motion capture. In receding horizon control, a pre-selected set of dynamically feasible trajectories of fixed length (the horizon), are evaluated on a cost map of the environment around the vehicle and the trajectory that avoids collision while making most progress towards a goal location is chosen. This trajectory is traversed for a bit and the process repeated again.

We demonstrate the *first* receding horizon control with monocular vision implementation on a UAV. Figure 1 shows our quadrotor evaluating a set of trajectories on the projected depth image obtained from monocular depth prediction and traversing the chosen one.

This is motivated by our previous work [26], where we used imitation learning to learn a purely reactive controller for flying a UAV using only monocular vision through dense clutter. While good obstacle avoidance behavior was obtained, there are certain limitations of a purely reactive layer that a more deliberative approach like receding horizon control can ameliorate. Reactive control is by definition myopic, i.e., it concerns itself with avoiding the obstacles closest to the vehicle. This can lead to it being easily stuck in cul-de-sacs. Since receding horizon control plans for longer horizons it achieves better plans and minimizes the chances of getting stuck [20]. Another limitation of pure reactive control is the difficulty to reach a goal location or direction. In a receding horizon control scheme, trajectories are selected based on a score which is the sum of two terms: first, the collision score of traversing it and second, the heuristic cost of reaching the goal from the end of the trajectory. By weighting both these terms suitably, goal-directed behavior is realized while maintaining obstacle-avoidance capability. Though it is to be noted that reactive control can be integrated with receding horizon for obtaining the best of both worlds in terms of collision avoidance behavior.

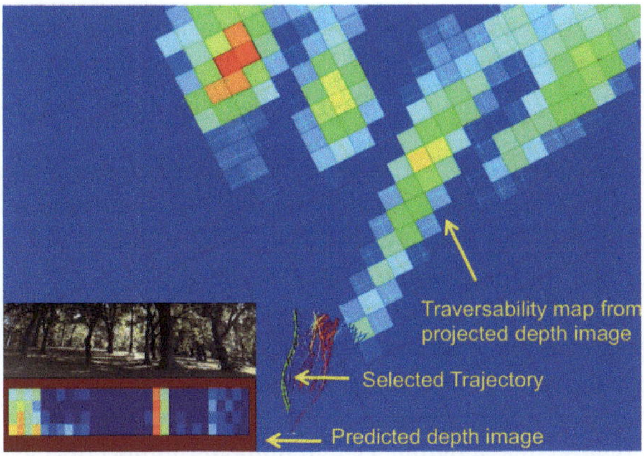

Fig. 1 Example of receding horizon with a quadrotor using monocular vision. The *lower left* images show the view from the front camera and the corresponding depth images from the monocular depth perception layer. The rest of the figure shows the overhead view of the quadrotor and the traversability map (built by projecting out the depth image) where *red* indicates higher obstacle density. The grid is $1 \times 1\,\text{m}^2$. The trajectories are evaluated on the projected depth image and the one with the least collision score (*thick green*) trajectory followed

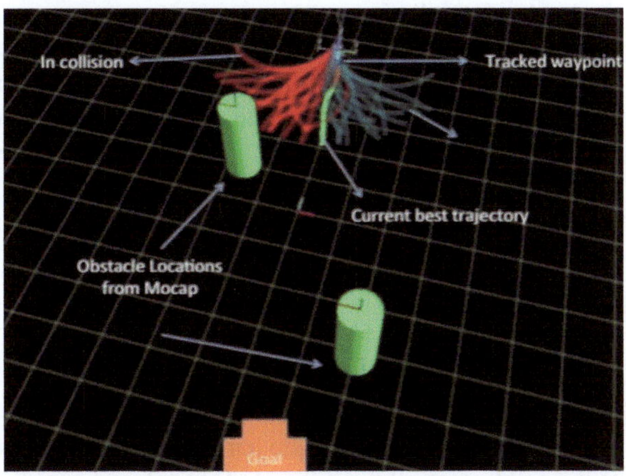

Fig. 2 Receding horizon control on UAV in motion capture. A library of 78 trajectories of length 5 m are evaluated to find the best collision-free trajectory. This is followed for 1 m at 1 m/s and the process repeated

Receding horizon control needs three working components

1. *A method to estimate depth*: This can be obtained from stereo vision [23, 29] or dense structure-from-motion (SfM) [32]. But these are not amenable for achieving higher speeds due to high computational expense. We note that in the presence of enough computation power, information from these techniques can be combined with monocular vision to improve overall perception.
   Biologists have found strong evidence that birds and insects use optical flow to navigate through dense clutter [30]. Optical flow has been used for autonomous flight of UAVs [4]. However, it is difficult to directly derive a robust control prin-

ciple from flow. Instead we follow the same data driven principle as our previous work [26] and use local statistics of optical flow *as features* in the monocular depth prediction module. This allows the learning algorithm to derive complex behaviors in a data driven fashion.
2. *A method for relative pose estimation*: To track the trajectory chosen at every cycle, the pose of the vehicle must be tracked. We demonstrate a relative pose estimation system using a downward facing camera and a sonar, which is utilized by the controller for tracking the trajectory (Sect. 2.5).
3. *A method to deal with perception uncertainty*: Most planning schemes either assume that perception is perfect or make simplistic assumptions of uncertainty. We introduce the concept of making multiple, relevant yet diverse predictions for incorporating perception uncertainty into planning. The intuition is predicated on the observation that avoiding a small number of ghost obstacles is acceptable as long as true obstacles are not missed (high recall, low precision). The details are presented in Sect. 2.4. We demonstrate in experiments the efficacy of this approach as compared to making only a single best prediction.

In summary our list of contributions are:

- Budgeted near-optimal feature selection and fast non-linear regression for monocular depth prediction.
- Real time relative vision-based pose estimation.
- Multiple predictions to efficiently incorporate uncertainty in the planning stage.
- First complete receding horizon control implementation on a UAV with monocular vision.

2 Approach

2.1 Hardware and Software Overview

In this section we describe the hardware platforms used in our experiments. Developing and testing all the integrated modules of receding horizon is very challenging. Therefore we assembled a rover (Fig. 3a) in addition to a UAV (Fig. 3b) to be able to test various modules separately. The rover also facilitated parallel development and testing of modules. Here we describe the hardware platforms and overall software architecture.

2.1.1 Rover

The skid-steered rover (Fig. 3a) uses an Ardupilot microcontroller board which takes in high level control commands from the planner and controls the four motors to achieve the desired motion.

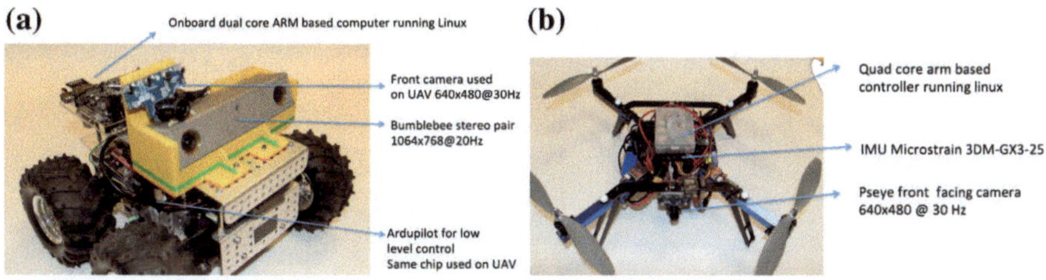

Fig. 3 Rover and Quadrotor platforms used in experiments. **a** Rover assembled with the same control chips and perception software as UAV for rapid tandem development and validation of modules. **b** Quadrotor used as our development platform

Other than the low-level controllers, all other aspects of the rover are kept exactly the same as the UAV to allow seamless transfer of software. For example, the rover has a front facing PlayStation Eye camera which is also used as the front facing camera on the UAV.

A Bumblebee color stereo camera pair (1024×768 at 20 fps) is rigidly mounted with respect to the front camera using a custom 3D printed fiber plastic encasing. This is used for collecting data with groundtruth depth values (Sect. 2.2) and validation of planning (Sect. 2.6). We calibrate the rigid body transform between the front camera and the left camera of the stereo pair. Stereo depth images and front camera images of the environment are recorded simultaneously while driving the rover around using a joystick. The depth images are then transformed to the front camera's coordinate system to provide groundtruth depth values for every pixel. The training depth images are from a slightly different perspective than encountered by the UAV during flight, but we found in practice that depth prediction performance generalized well. Details in Sect. 2.2.

2.1.2 UAV

Figure 3b shows the quadrotor we use for our experiments. Figure 4 shows the schematic of the various modules that run onboard and offboard. The base chassis, motors and autopilot are assembled using the Arducopter kit. Due to drift and noise of the IMU integrated in the Ardupilot unit, we added a Microstrain 3DM GX3 25 IMU which is used to aid real time pose estimation. There are two PlayStation Eye cameras: one facing downwards for real time pose estimation, one facing forward. The onboard processor is a quad-core ARM based computer which runs Ubuntu and ROS [25]. This unit runs the pose tracking and trajectory following modules. A sonar is used to estimate altitude. The image stream from the front facing camera is streamed to the base station where the depth prediction module processes it; the trajectory evaluation module then finds the best trajectory to follow to minimize probability of collision and transmits it to the onboard computer where the trajectory following module runs a pure pursuit controller to do trajectory tracking [6]. The

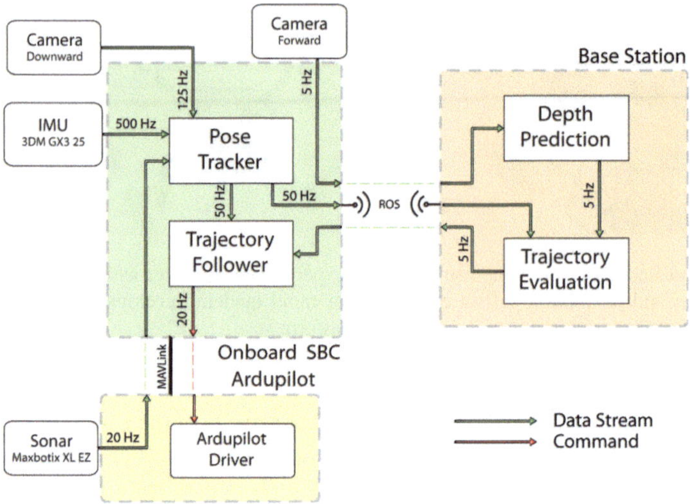

Fig. 4 Schematic diagram of hardware and software modules

resulting desired velocity control commands are sent to the Ardupilot which sends low level control commands to the motor controllers to achieve the desired motion.

2.2 *Monocular Depth Prediction*

In this section we describe the depth prediction approach from monocular images, and the fast non-linear regression method used for regression.

An image is first gridded up into non-overlapping patches. We predict the depth in meters at every patch of the image (Fig. 5 yellow box). For each patch we extract separate features which describe the patch, the full column containing the patch (Fig. 5 green box) and the column of three times the patch width (Fig. 5 red box), centered around the patch. The final feature vector for a patch is the concatenation of the feature vectors of all three regions. When a patch is seen by itself it is very hard to tell the relative depth with respect to the rest of the scene. But by adding the features of the surrounding area of the patch, more context is available to aid the predictor.

2.2.1 Description of Features

In this part we describe in brief the features used to represent the patch. We mainly borrow the features as used in previous work on monocular imitation learning [26] for UAVs, which are partly inspired by the work of Hoiem et al. [15] and

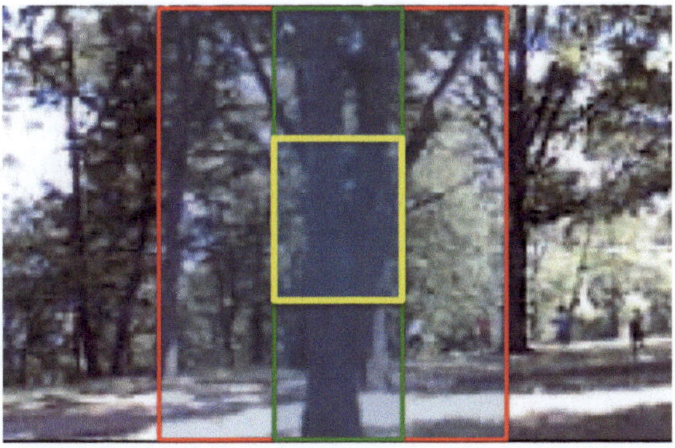

Fig. 5 The *yellow box* is an example patch, the *green box* is the column of the same width surrounding it, and the *red box* is the column of 3 times the patch width surrounding it. Features are extracted individually at the patch, and the columns are concatenated together to form the total feature representation of the patch

Saxena et al. [27]. We predict the depth at every patch which is then used by the planning module.

- *Optical flow*: We use the Farneback dense optical flow [11] implementation in OpenCV to compute for every patch the average, minimum and maximum optical flow values.
- *Radon Transform*: The radon transform captures strong edges in a patch [14].
- *Structure Tensor*: The structure tensor describes the local texture of a patch [13].
- *Laws' Masks*: These describe the texture intensities [8]. For details on radon transform, structure tensor and Laws' masks usage see [26].
- *Histogram of Oriented Gradients (HoG)*: This feature has been used widely in the computer vision community for capturing texture information for object detection [7]. For each patch we compute the HoG feature in 9 orientation bins.
- *Tree feature*: We use the per pixel fast classifier by Li et al. [22] to train a supervised tree detector. Li et al. originally used this for real time hand detection in ego-centric videos. For a given image patch we use this predictor to output the probability of each pixel being a tree. This information is then used as a feature for that patch.

2.2.2 Data Collection

RGB-D sensors like the Kinect, currently do not work outdoors. Since camera and calibrated nodding lidar setup is expensive and complicated we used a rigidly mounted Bumblebee stereo color camera and the PlayStation Eye camera for our outdoor data collection. This setup was mounted on the rover (Fig. 3a). We collected data at 4 different locations with tree density varying from low to high, under varying illumination conditions and in both summer and winter conditions. Our corpus of imagery with stereo depth information is around 16000 images and growing. We will make this dataset publicly available in the near future.

2.2.3 Fast Non-linear Prediction

Due to harsh real-time constraints an accurate but fast predictor is needed. Recent linear regression implementations are very fast and can operate on millions of features in real time [21] but are limited in predictive performance by the inherent linearity assumption. In very recent work Agarwal et al. [1] develop fast iterative methods which use linear regression in the inner loop to obtain overall non-linear behavior. This leads to fast prediction times while obtaining much better accuracy. We implemented Algorithm 2 in [1] and found that it lowered the error by 10 % compared to just linear regression, while still allowing real time prediction.

2.3 Budgeted Feature Selection

While many different visual features can be extracted on images, they need to be computed in real time. The faster the desired speed of the vehicle, the faster the perception and planning modules have to work to maintain safety. Additionally the limited computational power onboard a small UAV imposes a budget within which to make a prediction. Each kind of feature requires different time periods to extract, while contributing different amounts to the prediction accuracy. For example, radon transforms might take relatively less time to compute but contribute a lot to the prediction accuracy, while another feature might take more time but also contribute relatively less or vice versa. This problem is further complicated by the "grouping" effects where a particular feature's performance is affected by the presence or absence of other features.

Given a time budget, the naive but obvious solution is to enumerate all possible combinations of features within the budget and find the group of features which achieve minimum loss. This is exponential in the number of available features. Instead we use the efficient approach developed by Hu et al. [17] to select the near-optimal set of features which meet the imposed budget constraints. Their approach uses a simple greedy algorithm that first whitens feature groups and then recursively chooses groups by the reduction in explained variance divided by the time to achieve that reduction. A more efficient variant of this with equivalent guarantees, chooses features by computing gradients to approximate the reduction in explained variance, eliminating the need to "try" all feature groups sequentially. For each specified time budget, the features selected by this procedure are within a constant factor of the optimal set of features which respect that budget. Since this holds across all time budgets, this procedure provides a recursive way to generate feature sets across time steps.

Figure 6 shows the sequence of features that was selected by Hu et al. [17] feature selection procedure. For any given budget only the features on the left up to the specified time budget need to be computed (Fig. 7).

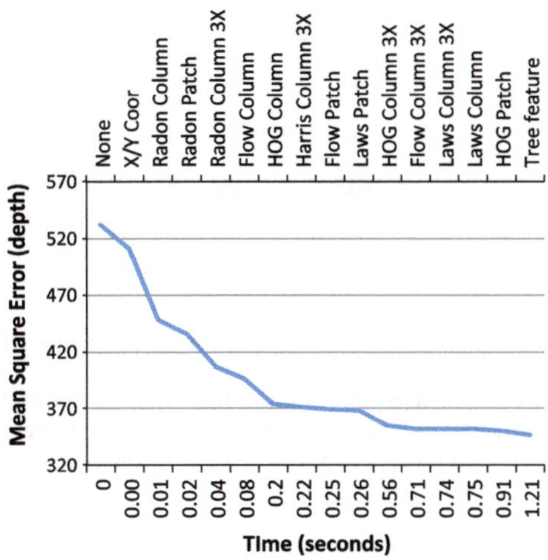

Fig. 6 On the upper x-axis the sequence of features selected by Hu et al.'s method [17] and the lower x-axis shows the cumulative time taken for all features up to that point. The near-optimal sequence of features rapidly decrease the depth prediction error. For a given time budget, the sequence of features to the left of that time should be used

Fig. 7 Depth prediction examples on real outdoor scenes. Closer obstacles are indicated by *red*

2.4 Multiple Predictions

The monocular depth estimates are often noisy and inaccurate due to the challenging nature of the problem. A planning system must incorporate this uncertainty to achieve safe flight. Figure 8 illustrates the difficulty of trying to train a predictive method for

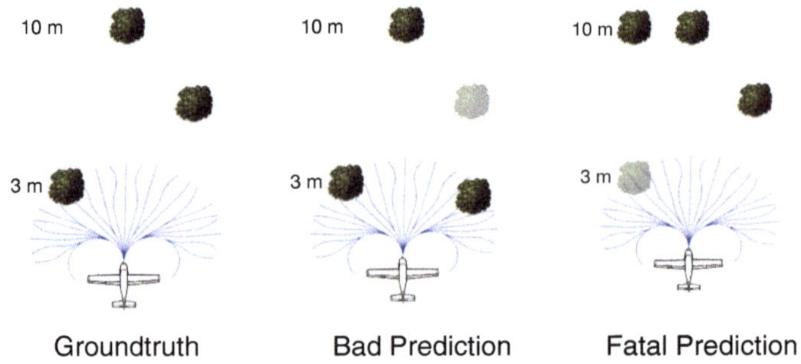

Fig. 8 Illustration of the complicated nature of the loss function for collision avoidance. (*Left*) Groundtruth tree locations. (*Middle*) Bad prediction where a tree is predicted closer than it actually is located. (*Right*) Fatal prediction where a tree close by is mispredicted further away

building a perception system for collision avoidance. Figure 8 (left) shows a ground truth location of trees in the vicinity of an autonomous UAV. Figure 8 (middle) shows the location of the trees as predicted by the perception system. In this prediction the trees on the left and far away in front are predicted correctly but the tree on the right is predicted close to the UAV. This will cause the UAV to dodge a ghost obstacle. While this is bad, it is not fatal because the UAV will not crash but make some extraneous motions. But the prediction of trees in Fig. 8 (right) is potentially fatal. Here the trees far away in front and on the right are correctly predicted whereas the tree on the left originally close to the UAV, is mispredicted to be far away. This type of mistake will cause the UAV to crash into an obstacle it does not know is there.

Ideally, a vision-based perception system should be trained to minimize loss functions which will penalize such fatal predictions more than other kind of predictions. But even writing down such a loss function is difficult. Therefore most monocular depth perception systems try to minimize easy to optimize surrogate loss functions like regularized L_1 or L_2 loss [27]. We try to reduce the probability of collision by generating multiple interpretations of the scene to hedge against the risk of committing to a single potentially fatal interpretation as illustrated in Fig. 8. Specifically we generate 3 interpretations of the scene and evaluate the trajectories in all of them. The trajectory which is least likely to collide on average in all interpretations is then chosen as the one to traverse.

One way of making multiple predictions is to just sample the posterior distribution of a learnt predictor. In order to truly capture the uncertainty of the predictor, a lot of interpretations have to be sampled and trajectories evaluated on each of them. A large number of samples will be from around the peaks of this distribution leading to wasted samples. This is not feasible given the real time constraints of the problem.

In previous work [10], we have developed techniques for predicting a budgeted number of interpretations of an environment with applications to manipulation, planning and control. Batra et al. [3] have also applied similar ideas to structured prediction problems in computer vision. These approaches try to come up with a small number of relevant but diverse interpretations of the scene so that at least one of them

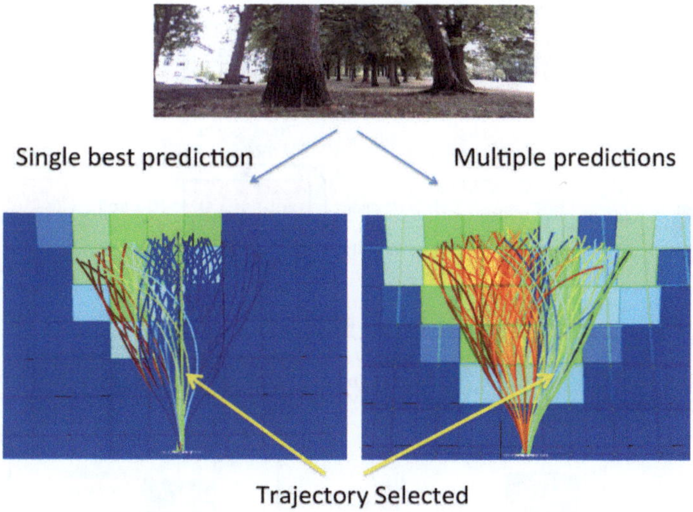

Fig. 9 The scene at *top* is an example from the front camera of the UAV. On the *left* is shown the predicted traversability map (*red* is high cost, *blue* is low cost) resulting from a single interpretation of the scene. Here the UAV has selected the straight path (thick, *green*) which will make it collide with the tree right in front. While on the right the traversability map is constructed from multiple interpretations of the image, leading to the trajectory in the right being selected which will make the UAV avoid collision

is correct. In this work, we adopt a similar philosophy and use the error profile of the fast non-linear regressor described in Sect. 2.2 to make two additional predictions: The non-linear regressor is first trained on a dataset of 14500 images and it's performance on a held-out dataset of 1500 images is evaluated. For each depth value predicted by it, the average over-prediction and under-prediction error is recorded. For example the predictor may say that an image patch is at 3 m while it is actually either, on average, at 4 m or at 2.5 m. We round each prediction depth to the nearest integer, and record the average over and under-predictions as in the above example in a look-up table (LUT). At test time the predictor produces a depth map and the LUT is applied to this depth map, producing two additional depth maps: one for over-prediction error, and one for the under-prediction error.

Figure 9 shows an example in which making multiple predictions is clearly beneficial compared to the single best interpretation. We provide more experimental details and statistics in Sect. 3.

2.5 Pose Estimation

As discussed before, a relative pose-estimation system is needed to follow the trajectories chosen by the planning layer. We use a downward looking camera in

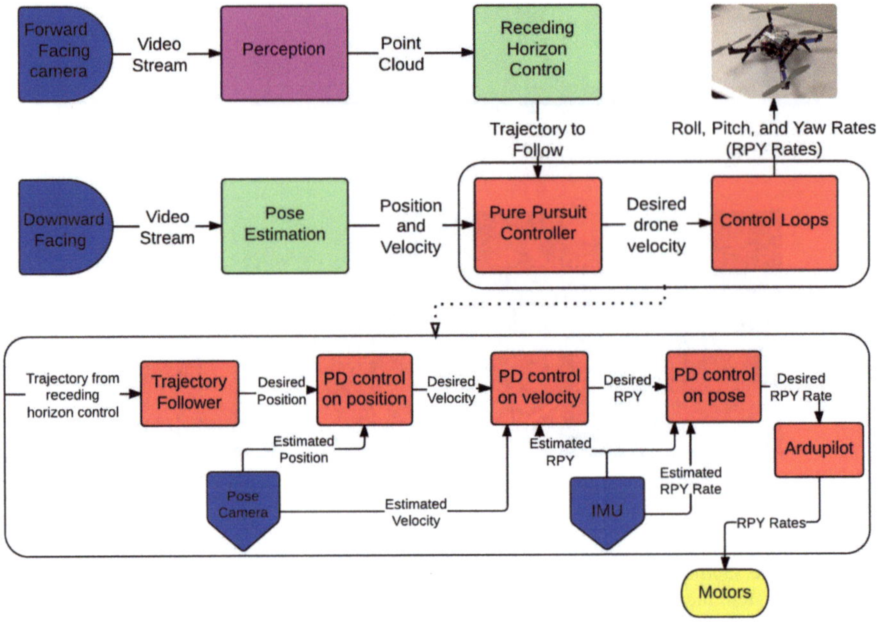

Fig. 10 The overall flow of data and control commands between various modules. The pure pursuit trajectory follower and low level control loops (*red boxes*) are shown in greater detail at the *bottom*

conjunction with a sonar for determining relative pose. Looking forward to determine pose is ill-conditioned due to a lack of parallax as the camera faces the direction of motion. There are still significant challenges involved when looking down. Texture is often very self similar making it challenging for traditional feature based methods [19, 24] to be employed (Figs. 10 and 11).

In receding horizon, absolute pose with respect to some fixed world coordinate system is not needed, as one needs to follow trajectories for short durations only. So as long as one has a relative, consistent pose estimation system for this duration (3 s in our implementation), one can successfully follow trajectories.

We used a variant of a simple algorithm that has been presented quite often, most recently in [16]. This approach uses a Kanade-Lucas-Tomasi (KLT) tracker [31] to detect where each pixel in a grid of pixels moves over consecutive frames, and estimating the mean flow from these after rejecting outliers. We do the outlier detection step by comparing the variation of the flow vectors obtained for every pixel on the grid to a specific threshold. Whenever the variance of the flow is high, we do not calculate the mean flow velocity, and instead decay the previous velocity estimate by a constant factor (Fig. 12).

This estimate of flow however tries to find the best planar displacement between the two patches, and does not take into account out-of-plane rotations, due to motion of the camera. Camera ego-motion is compensated using motion information from the IMU. Finally the metric scale is estimated from sonar. We compute instantaneous

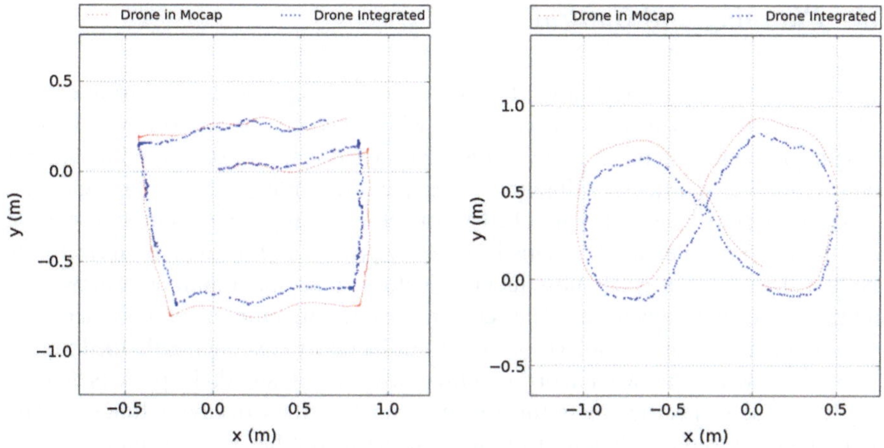

Fig. 11 Comparison of the differential flow tracker performance versus ground truth in MOCAP. *Red* tracks are the trajectories in MOCAP, *blue* are those determined by the algorithm. Note that the formulation of the receding horizon setup is such that mistakes made in following a specific trajectory are forgiven up to an extent since we replan every few seconds

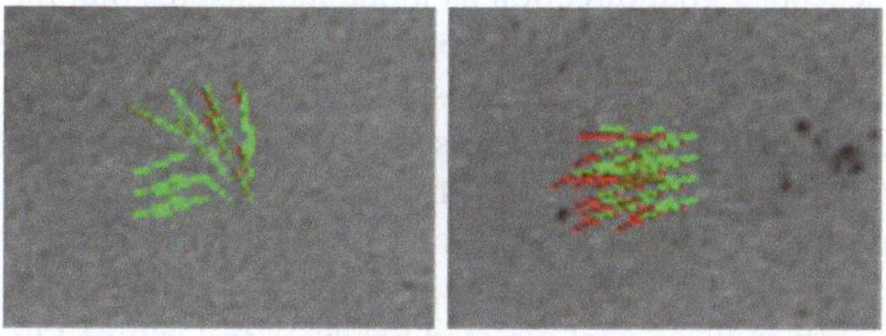

Fig. 12 Instances of failure of the pose tracking system over challenging surfaces. Note the absence of texture in these 320 × 240 images. The figure shows the flow tracks corresponding to the points on the grid. *Red* tracks show the uncorrected optical flow, while the *green* tracks (superimposed) show the flow vectors 'unrotated' using the IMU

relative velocity between the camera and ground which is integrated over time to get position.

This process is computationally inexpensive, and can be run at very high frame rates. Higher frame rates lead to smaller displacements between pairs of images, which in turn makes tracking easier.

We evaluated the performance of the flow based tracker in motion capture and compared the true motion capture tracks to the tracks returned by flow based tracker. The resulting tracks are shown in Fig. 11.

2.6 Planning and Control

Figure 10 shows the overall flow of data and control commands. The front camera video stream is fed to the perception module which predicts the depth of every pixel in a frame, projects it to a point cloud representation and sends it to the receding horizon control module. A trajectory library of 78 trajectories of length 5 m is budgeted and picked from a much larger library of 2401 trajectories using the maximum dispersion algorithm by Green et al. [12]. This is a greedy procedure for selecting trajectories, one at a time, so that each subsequent trajectory spans maximum area between it and the rest of the trajectories. The receding horizon module maintains a score for every point in the point cloud. The score of a point decays exponentially the longer it exists. After some time when it drops below a user set threshold, the point is deleted. The decay rate is specified by setting the time constant of the decaying function. This fading memory representation of the local scene layout has two advantages: (1) It prevents collisions caused by narrow field-of-view issues where the quadrotor forgets that it has just avoided a tree, sees the next tree and dodges sideways, crashing into the just avoided tree. (2) It allows emergency backtracking maneuvers to be safely executed if required, since there is some local memory of the obstacles it has just passed.

Our system accepts a goal direction as input and ensures that the vehicle makes progress towards the goal while avoiding obstacles along the way. The score for each trajectory is the sum of three terms: (1) A sphere of the same radius as the quadrotor is convolved along a trajectory and the score of each point in collision is added up. The higher this term is relative to other trajectories, the higher the likelihood of this trajectory being in collision. (2) A term which penalizes a trajectory whose end *direction* deviates from goal direction. This is weighted by a user specified parameter. This term induces goal directed behavior and is tuned to ensure that the planner always avoids obstacles as a first priority. (3) A term which penalizes a trajectory for deviating in *translation* from the goal direction.

The pure pursuit controller module (Fig. 10) takes in the coordinates of the trajectory to follow and the current pose of the vehicle from the optical flow based pose estimation system (Sect. 2.5). We use a pure pursuit strategy [6] to track it. Specifically, this involves finding the closest point on the trajectory from the robot's current estimated position and setting the target waypoint to be a certain fixed lookahead distance further along the trajectory. The lookahead distance can be tuned to obtain the desired smoothness while following the trajectory; a larger lookahead distance leads to smoother motions, at the cost of not following the trajectory exactly. Using the pose updates provided by the pose estimation module, we head towards this moving waypoint using a generic PD controller. Since the receding horizon control module continuously replans (at 5 Hz) based on the image data provided by the front facing camera, we can choose to follow arbitrary lengths along a particular trajectory before switching over to the latest chosen one.

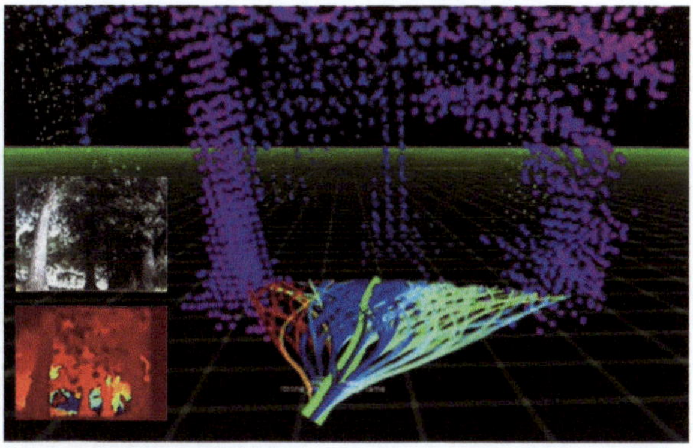

Fig. 13 Receding horizon control validation with rover using depth images from stereo. The *bright green* trajectory is the currently selected trajectory to follow. *Red* trajectories indicate that they are more likely to be in collision

2.6.1 Validation of Modules

We validated each module separately as well as in tandem with other modules where each validation was progressively integrated with other modules. This helped reveal bugs and instabilities in the system.

- *Trajectory Evaluation and Pure Pursuit Validation with Stereo Data on Rover*: We tested the trajectory evaluation and pure pursuit control module by running the entire pipeline (other than monocular depth prediction) with stereo depth images on the rover (Fig. 13).
- *Trajectory Evaluation and Pure Pursuit Validation with Monocular Depth on Rover*: This test is the same as above but instead of using depth images from stereo we used the monocular depth prediction. This allowed us to tune the parameters for scoring trajectories in the receding horizon module to head towards goal without colliding with obstacles.
- *Trajectory Evaluation and Pure Pursuit Validation with Known Obstacles in Motion Capture on UAV*: While testing of modules progressed on the rover we assembled and developed the pose estimation module (Sect. 2.5) for the UAV. We tested this module in a motion capture lab where the position of the UAV as well of the obstacles was known and updated at 120 Hz. (See Fig. 2).
- *Trajectory Evaluation and Pure Pursuit Validation with Hardware-in-the-Loop (HWIL)*: In this test we ran the UAV in an open field, fooled the receding horizon module to think it was in the midst of a point cloud and ran the whole system (except perception) to validate planning and control modules. Figure 14 shows an example from this setup.
- *Whole System*: After validating each module following the evaluation protocol described above, we ran the whole system end-to-end. Figure 1 shows an example scene of the quadrotor in full autonomous mode avoiding trees outdoors. We detail the results of collision avoidance in Sect. 3.

Fig. 14 Hardware-in-the-loop testing with UAV in open field. The receding horizon module was fooled into thinking that it was in the midst of a real world point cloud while it planned and executed its way through it. This allowed us to validate planning and control without endangering the UAV

3 Experiments

We analyze the performance of our proposed deliberative approach in this section. All the experiments were conducted in a densely cluttered forest area, while restraining the drone through a light-weight tether.

Quantitatively, we evaluate performance by recording the average distance flown autonomously by the UAV over several runs (at 1 m/s), before an intervention. An intervention, in this context, is defined as the pilot overriding the autonomous system to prevent the drone from an inevitable crash. Experiments were performed using the multiple predictions approach and single best prediction. The comparison has been shown in Fig. 15. Tests were performed in regions of high and low clutter density (approx. 1 tree per $6 \times 6\,\text{m}^2$ and $12 \times 12\,\text{m}^2$, respectively). Multiple predictions results in significantly better performance. In particular, the drone was able to fly without intervention over a 137 m distance for low density regions. The difference is even higher in case of high-density regions where committing to a single prediction can be even more fatal.

Further, we evaluate the success rate for avoiding large and small trees using our proposed approach (Table 1). We are able to avoid 96 % of all trees over a total covered distance of more than 1 km. Failures are broken down by the type of obstacle the UAV failed to avoid, or whether the obstacle was not in the field-of-view (FOV). Overall, 39 % of the failures were due to large trees and 33 % on hard to perceive obstacles like branches and leaves. As expected, the narrow FOV is now the least contributor to failure cases as compared to a more reactive control strategy [26]. This is intuitive, since the reactive control is myopic in nature and our deliberate approach helps overcome this problem as described in the previous sections. Figure 16 shows some typical intervention examples.

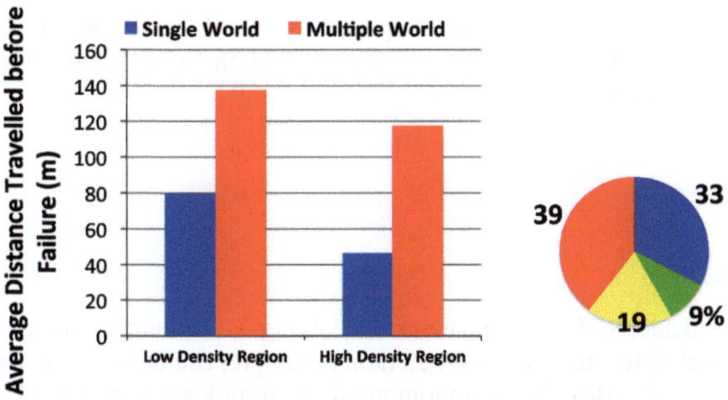

Fig. 15 **a** Average distance flown by the drone before a failure. **b** Percentage of failure for each type. *Red* Large Trees, *Yellow* Thin Trees, *Blue* Foliage, *Green* Narrow FOV

Table 1 Success rate of avoiding trees

	Multiple predictions	Single prediction
Total distance	1020 m	1010 m
Large trees avoided	93.1 %	84.8 %
Small trees avoided	98.6 %	95.9 %
Overall accuracy	**96.6 %**	**92.5 %**

Fig. 16 Examples of interventions: (*Left*) Bright trees saturated by sunlight from behind (Second from *left*) Thick foliage (Third from *left*) Thin trees (*Right*) Flare from direct sunlight. Camera/lens with higher dynamic range and more data of rare classes should improve performance

4 Conclusion

In ongoing work we are moving towards complete onboard computing of all modules to reduce latency. We can leverage other sensing modes like sparse, but more accurate depth estimation from stereo, which can be used as "anchor" points to improve dense monocular depth estimation. Similarly low power, light weight lidars can be actively foveated to high probability obstacle regions to reduce false positives and get exact depth. Another central future effort is to integrate the purely reactive [26] approach with the deliberative scheme detailed here, for better performance.

Acknowledgments This work has been supported by ONR MURI grant N00014-09-1-1051 through the "Provably Stable Vision Based Control of High Speed Flight through Forests and Urban Environments" project.

References

1. Agarwal, A., Kakade, S.M., Karampatziakis, N., Song, L., Valiant, G.: Least squares revisited: scalable approaches for multi-class prediction. arXiv preprint arXiv:1310.1949 (2013)
2. Bachrach, A., He, R., Roy, N.: Autonomous flight in unknown indoor environments. Int. J. Micro Air Veh. (2009)
3. Batra, D., Yadollahpour, P., Guzman-Rivera, A., Shakhnarovich, G.: Diverse m-best solutions in markov random fields. In: Computer Vision-ECCV 2012, pp. 1–16. Springer (2012)
4. Beyeler, A., Zufferey, J.C., Floreano, D.: Vision-based control of near-obstacle flight. Auton. Robots (2009)
5. Buehler, M., Iagnemma, K., Singh, S.: Special issue on the 2007 darpa urban challenge, part i, ii, iii. JFR (2008)
6. Coulter, R.C.: Implementation of the pure pursuit path tracking algorithm. Tech. rep, DTIC Document (1992)
7. Dalal, N., Triggs, B., Schmid, C.: Human detection using oriented histograms of flow and appearance. In: ECCV (2006)
8. Davies, E.R.: Machine vision: theory, algorithms, practicalities (2004)
9. Dey, D., Geyer, C., Singh, S., Digioia, M.: A cascaded method to detect aircraft in video imagery. IJRR (2011)
10. Dey, D., Liu, T.Y., Hebert, M., Bagnell, J.A.D.: Contextual sequence optimization with application to control library optimization. In: RSS (2012)
11. Farnebäck, G.: Two-frame motion estimation based on polynomial expansion. In: Image Analysis (2003)
12. Green, C., Kelly, A.: Optimal sampling in the space of paths: preliminary results. Tech. Rep. CMU-RI-TR-06-51, Robotics Institute, Pittsburgh, PA (2006)
13. Harris, C., Stephens, M.: A combined corner and edge detector. In: Alvey Vision Conference (1988)
14. Helgason, S.: Support of radon transforms. Adv. Math. (1980)
15. Hoiem, D., Efros, A.A., Hebert, M.: Geometric context from a single image. In: ICCV (2005)
16. Honegger, D., Meier, L., Tanskanen, P., Pollefeys, M.: An open source and open hardware embedded metric optical flow cmos camera for indoor and outdoor applications. In: ICRA (2013)
17. Hu, H., Grubb, A., Bagnell, J.A., Hebert, M.: Efficient feature group sequencing for anytime linear prediction. arXiv:1409.5495 (2014)
18. Kelly, A., et al.: Toward reliable off road autonomous vehicles operating in challenging environments. IJRR (2006)
19. Klein, G., Murray, D.: Parallel tracking and mapping for small ar workspaces. In: ISMAR (2007)
20. Knepper, R., Mason, M.: Path diversity is only part of the problem. In: ICRA (2009)
21. Langford, J., Li, L., Strehl, A.: Vowpal Wabbit (2007)
22. Li, C., Kitani, K.M.: Pixel-level hand detection in ego-centric videos. In: CVPR (2013)
23. Matthies, L., Brockers, R., Kuwata, Y., Weiss, S.: Stereo vision-based obstacle avoidance for micro air vehicles using disparity space. In: ICRA (2014)
24. Newcombe, R.A., Lovegrove, S.J., Davison, A.J.: Dtam: dense tracking and mapping in real-time. In: ICCV (2011)
25. Quigley, M., Conley, K., Gerkey, B.P., Faust, J., Foote, T., Leibs, J., Wheeler, R., Ng, A.Y.: Ros: an open-source robot operating system. In: ICRA Workshop on Open Source Software (2009)

26. Ross, S., Melik-Barkhudarov, N., Shankar, K.S., Wendel, A., Dey, D., Bagnell, J.A., Hebert, M.: Learning monocular reactive uav control in cluttered natural environments. In: ICRA (2013)
27. Saxena, A., Chung, S.H., Ng, A.Y.: Learning depth from single monocular images. In: NIPS (2005)
28. Scherer, S., Singh, S., Chamberlain, L.J., Elgersma, M.: Flying fast and low among obstacles: methodology and experiments. IJRR (2008)
29. Schmid, K., Lutz, P., Tomić, T., Mair, E., Hirschmüller, H.: Autonomous vision-based micro air vehicle for indoor and outdoor navigation. JFR (2014)
30. Srinivasan, M.V.: Visual control of navigation in insects and its relevance for robotics. Curr. Opin. Neurobiol. (2011)
31. Tomasi, C., Kanade, T.: Detection and tracking of point features. School of Computer Science, Carnegie Mellon University (1991)
32. Wendel, A., Maurer, M., Graber, G., Pock, T., Bischof, H.: Dense reconstruction on-the-fly. In: CVPR (2012)

Robust Autonomous Flight in Constrained and Visually Degraded Environments

Zheng Fang, Shichao Yang, Sezal Jain, Geetesh Dubey, Silvio Maeta, Stephan Roth, Sebastian Scherer, Yu Zhang and Stephen Nuske

Abstract This paper addresses the problem of autonomous navigation of a micro aerial vehicle (MAV) inside a constrained shipboard environment for inspection and damage assessment, which might be perilous or inaccessible for humans especially in emergency scenarios. The environment is GPS-denied and visually degraded, containing narrow passageways, doorways and small objects protruding from the wall. This makes existing 2D LIDAR, vision or mechanical bumper-based autonomous navigation solutions fail. To realize autonomous navigation in such challenging environments, we propose a fast and robust state estimation algorithm that fuses estimates from a direct depth odometry method and a Monte Carlo localization algorithm with other sensor information in an EKF framework. Then, an online motion planning algorithm that combines trajectory optimization with receding horizon control framework is proposed for fast obstacle avoidance. All the computations are done in real-time onboard our customized MAV platform. We validate the system by running experiments in different environmental conditions. The results of over 10 runs

Z. Fang (✉) · S. Yang · S. Jain · G. Dubey · S. Maeta · S. Roth · S. Scherer
Y. Zhang · S. Nuske
Robotics Institute, Carnegie Mellon University, 5000 Forbes Ave,
Pittsburgh, PA 15213, USA
e-mail: zhengf@andrew.cmu.edu; fangzheng81@gmail.com

S. Yang
e-mail: shichaoy@andrew.cmu.edu

S. Jain
e-mail: sezal@andrew.cmu.edu

G. Dubey
e-mail: gdubey@andrew.cmu.edu

S. Scherer
e-mail: basti@andrew.cmu.edu

Z. Fang
Northeastern University, Shenyang 110819, Liaoning, China

Y. Zhang
Zhejiang University, Hangzhou 310027, Zhejiang, China

D.S. Wettergreen and T.D. Barfoot (eds.), *Field and Service Robotics*,
Springer Tracts in Advanced Robotics 113, DOI 10.1007/978-3-319-27702-8_27

show that our vehicle robustly navigates 20 m long corridors only 1 m wide and goes through a very narrow doorway (66 cm width, only 4 cm clearance on each side) completely autonomously even when it is completely dark or full of light smoke.

1 Introduction

Over the past few years, micro aerial vehicles (MAVs) have gained a wide popularity in both military and civil domains. Surveillance and reconnaissance is one area where they have made a huge impact. In this paper, we aim to develop a MAV that is capable of autonomously navigating through a ship to aid in fire control, damage assessment and inspection, which might be dangerous or inaccessible for humans. Such a constrained and GPS-denied environment poses various challenges for navigating though narrow corridors and doorways, especially because it might be visually degraded: potentially dark and smoke-filled. An illustrative picture is shown in Fig. 1.

For successful operation in such environments, we need to address several challenging problems. *First*, the MAV should be small enough to travel in the narrow corridors with narrower doorways (66 cm width). Therefore, only lightweight sensors can be used, which provide limited measurement range and noisy data. *Second*, the onboard computational resources are very limited while every module should run in real-time, posing great challenges for pose estimation and motion planning. *Third*, since the practical environment is potentially a dark and smoke-filled environment, it prevents us from using state-of-the-art visual navigation methods. Though putting LED lights can give better illumination, it might not output a usable RGB image under smoky conditions. Besides, clear corridors with few geometric features or corridors with many small objects on the wall pose great difficulty for accurate pose estimation and obstacle avoidance. In addition, air turbulence from the MAV in confined spaces poses difficulty for precise control.

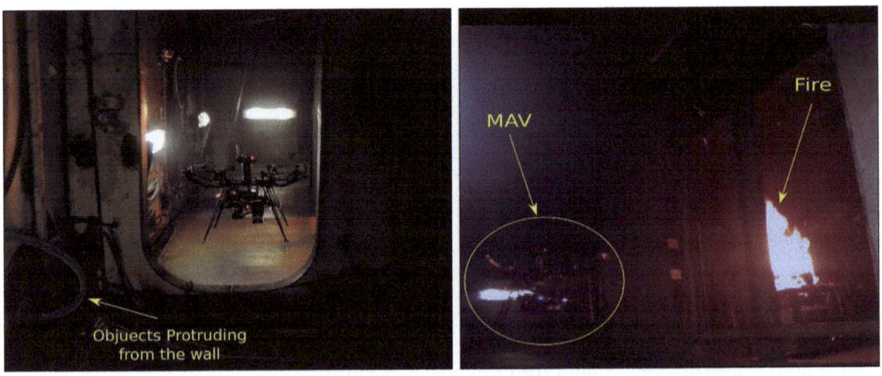

Fig. 1 Autonomous MAV for fire-detection inside a ship: The *left picture* shows MAV's autonomous flight through doorways. The *right picture* shows a testing scenario with fire

To address the above challenges, we build a robust and efficient autonomous navigation system with the following contributions.

- A real-time 6DoF pose estimation system that can directly recover the relative pose from a series of depth images and estimate the absolute pose of the MAV in a given 3D map.
- A data fusion framework of odometry and absolute pose with other sensors to provide fast and robust state estimation.
- An online motion planning algorithm using a modified trajectory optimization method under receding horizon control framework.

We demonstrate the effectiveness of our system through both simulation and field experiments. The field experiment is performed in a constrained shipboard environment containing a 20 m long, 1 m wide corridor and a 66 cm wide doorway. The width of the vehicle is 58 cm leaving only 4 cm clearance on both sides. We conducted more than 10 runs in various environment conditions, from normal to complete dark and smoke-filled environments to demonstrate autonomous navigation capabilities of the MAV.

2 Related Work

In recent years, a number of autonomous navigation solutions have been proposed for MAVs. Those solutions mainly differ in the sensors used for perception in the autonomous navigation problem, the amount of processing that is performed onboard/offboard and the assumptions made about the environment.

2D LIDAR has been extensively and successfully used for autonomous navigation for its accuracy and low latency [1–3]. However, those systems are usually only suitable for structured or 2.5D environments. Recently, there are also many vision-based navigation systems since cameras can provide rich information and have low weight, etc. For example, a stereo camera is used in [4, 5] and a monocular camera with IMU is used in [6–8], but vision is sensitive to illumination changes and could not work in dark or smoky environments. More recently, RGB-D cameras have become very popular for autonomous navigation of indoor MAVs [9–11] because they can provide both image and depth. For example, in [10] a RGB-D visual odometry method is proposed for real-time pose estimation of a MAV and a 3D map is created offline. In [11], a fast visual odometry method is used for pose estimation and 3D visual SLAM is used for constructing a 3D octomap in real-time.

Unfortunately, the existing autonomous navigation methods can not work in our case since our application environment is a confined, complex visually degraded 3D environment that may be very dark or filled with smoke. For example, for state estimation, vision-based methods [8, 10] could not work in our case due to that it is a potentially dark and smoky environment. Besides, for obstacle avoidance, 2D LIDAR-based methods are also unqualified for this complex environment since it only perceives planar information while there are many small objects (e.g. slim cables and pipes) protruding from the wall in our environment. In addition, many above papers' motion planning methods either compute paths offline [2, 11] or heavily rely on prior maps [1]. Some papers online generate steering angles to avoid obstacles by vector field histogram [5] or waypoints by sampling based planners (e.g. RRT*) [3]. However, steering angle is not suitable for precise control and RRT* path is usually not smooth and not fast enough.

In this paper, we present a robust autonomous navigation system that can work in challenging practical environments, which is based on our previous work [12]. However, our previous work only deals with the pose estimation problem while this paper presents all the details of the whole system. In our system, we mainly use depth images for odometry estimation, localization and motion planning, which can work in completely dark or even light smoke-filled environments. Besides, all the components of the system run onboard on an ARM based embedded computer.

3 Approach

3.1 Real-Time Pose Estimation

Pose estimation is required to allow the robot to be self aware of its placement in the surroundings and hence allows it to plan appropriate paths to maneuver around obstacles in the corridor.

3.1.1 Low-Frequency Pose Estimation

Low frequency pose estimates are primarily based on the RGB-D sensor. This includes relative ego-motion of the robot calculated from depth images as well as the absolute pose of robot calculated from the point cloud and a given 3D map.

Relative Pose Estimation A direct method based on [12, 13] is used to calculate the relative pose estimation, which is much faster than state of the art ICP method [14]. Let a 3D point $R = (X, Y, Z)^T$ (measured in the depth camera's coordinate system) be captured at pixel position $r = (x, y)^T$ in the depth image Z_t. This point undergoes a 3D motion $\Delta R = (\Delta X, \Delta Y, \Delta Z)^T$, which results in an image motion Δr between frames t_0 and t_1. Given that the depth of the 3D point will have moved by ΔZ, the depth value captured at this new image location $r + \Delta r$ will have consequently changed by this amount:

$$Z_1(r + \Delta r) = Z_0(r) + \Delta Z \tag{1}$$

This equation is called *range change constraint equation*.

For a pin hole camera model, any small 2D displacement Δr in the image can be related directly to the 3D displacement ΔR which gave rise to it by differentiating the perspective projection equation with respect to the components of the 3D position:

$$\frac{\partial r}{\partial R} = \frac{\Delta r}{\Delta R} = \begin{bmatrix} \frac{f_x}{Z} & 0 & -X\frac{f_x}{Z^2} \\ 0 & \frac{f_y}{Z} & -Y\frac{f_y}{Z^2} \end{bmatrix} \tag{2}$$

where f_x and f_y are the normalised focal lengths.

Under small rotation assumption, if the camera moves with instantaneous translational velocity v and instantaneous rotational velocity ω with respect to the environment, then the point R appears to move with a velocity

$$\frac{dR}{dt} = -v - \omega \times R \tag{3}$$

with respect to the sensor.

Taking the first-order Taylor expansion of the term $Z_1(r + \Delta r)$ in Eq. 1 and substituting Eqs. 3 and 2 into it gives us Eq. 4 where $\nabla Z_1(r) = (Z_x, Z_y)$ are the spatial derivatives of $Z_1(r)$. This equation generates a pixel-based constraint relating the gradient of the depth image ∇Z_1 and the temporal depth difference to the unknown pixel motion and the change of depth. In practice, in order to improve the computation speed, the depth image is downsampled to 80×60 which is sufficient to get an accurate estimation. Using Eq. 4, fast odometry can be calculated from depth images.

$$\begin{bmatrix} -Y - Z_y f_y - Z_x XY\frac{f_x}{Z^2} - Z_y Y^2\frac{f_y}{Z^2} \\ X + Z_x f_x + Z_x X^2\frac{f_x}{Z^2} + Z_y XY\frac{f_y}{Z^2} \\ -Z_x Y\frac{f_x}{Z} + Z_y X\frac{f_y}{Z} \\ Z_x\frac{f_x}{Z} \\ Z_y\frac{f_y}{Z} \\ -1 - Z_x X\frac{f_x}{Z^2} - Z_y Y\frac{f_y}{Z^2} \end{bmatrix}^T \begin{bmatrix} \omega_x \\ \omega_y \\ \omega_z \\ v_x \\ v_y \\ v_z \end{bmatrix} = Z_0(r) - Z_1(r) \tag{4}$$

where ω_x, ω_y, ω_z and v_x, v_y, v_z are components of the rotation and translation vectors.

However, in environments with few geometric features, this method will suffer from the degeneration problem, for example when the camera can only see a ground plane or parallel walls. In these "ill-conditioned"cases which are really common in indoor environments, the proposed method will produce inaccurate estimates. We use the "condition number" [15] to measure the degeneration degree of Eq. 4. When severe degeneration happens, the estimation outputs a failure signal.

Absolute Pose Estimation To obtain the vehicle's absolute pose in a given 3D map, a Monte Carlo Localization (MCL) [16] algorithm is used. Though MCL has been successfully used on ground robots [16], 6DoF pose state $S = (x, y, z, \phi, \theta, \psi)$ necessary for MAVs increases the complexity of the problem. We show that by carefully designing the motion and observation model, MCL can work very well on an embedded computer. More details can be found in our previous work [12].

(1) Motion Model For each subsequent frame, we propagate the previous state estimate according to the motion model $p(S_t|S_{t-1}, u_t)$. The motion command u_t is the visual odometry computed from Eq. 4. To account for unexpected motion, the prediction step adds a small amount of Gaussian noise to the motion command for each particle. The propagation equation is of the form:

$$S_t = S_{t-1} + u_t + e_t \qquad e_t \sim N(0, \sigma^2) \tag{5}$$

where u_t is the odometry and e_t is the Gaussian noise. When odometry estimation fails, we propagate the particle set using a noise-driven dynamical model

$$S_t = S_{t-1} + e'_t \qquad e'_t \sim N(0, \sigma'^2) \tag{6}$$

where σ' is much bigger than σ.

(2) Observation Model The belief of vehicle's 6DoF state is updated according to three different sources of sensor information in one observation O_t, namely depth measurements d_t from depth camera, roll $\tilde{\theta}_t$ and pitch $\tilde{\phi}_t$ measurements from IMU and height measurement $\tilde{z}_t$ from ground plane detection or the point laser. The final observation model is:

$$p(O_t|S_t) = p(d_t, \tilde{z}_t, \tilde{\phi}_t, \tilde{\theta}_t|S_t) = p(d_t|S_t) \cdot p(\tilde{z}_t|S_t) \cdot p(\tilde{\phi}_t|S_t) \cdot p(\tilde{\theta}_t|S_t) \tag{7}$$

The likelihood formulation is given by a Gaussian distribution. To improve the computation efficiency, an endpoint observation model [16] is used for calculating $p(d_t|S_t)$.

3.1.2 High-Frequency Pose Estimation

For real-time control, low latency, accurate, fast and robust estimate of the position and orientation is required. We fuse data from all of the sources providing motion information to output high frequency state estimate as shown in the Fig. 2. We run a high rate attitude estimator at 250 Hz on the flight controller unit (FCU) to stabilize the robot's angular motion. We also designed a robust full state with 9DoF position estimator capable of fusing data from optical flow (downward facing camera), odometry and localization (running onboard computer), height measurement and inertial sensors using an EKF running on the FCU at 100 Hz. Such a setup allows to maintain a reliable estimate of full current pose even when data from a sensor/estimator degrades due to change in environment e.g. if optical flow fails to find enough features on the

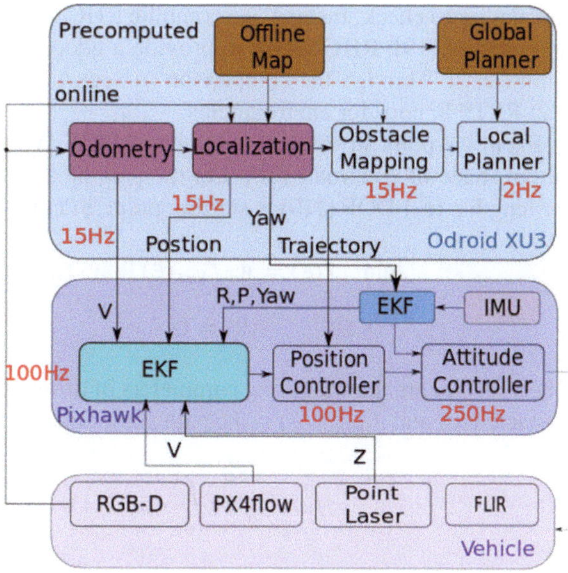

Fig. 2 Software architecture showing main modules with update rates

floor to generate odometry, other sensors/estimators provide enough information to estimate the current pose, therefore maintaining system redundancy and allowing smooth operation of the motion controller. Also, the input signals to the position estimator is pre-processed to produce a smooth input and reject any outliers eg. a moving average with outlier rejection is used for sonar. All these techniques together ensure the filter running on pixhawk FCU doesn't diverge due to outliers.

3.2 Online Motion Planning

Online motion planning is needed to keep MAV safe by quickly avoiding the obstacles which are represented by an online updated 3D occupancy grid [17]. Global mission points for motion planning are specified by a human or a high level mission planner. Here, we focus on local motion planning to generate collision-free trajectories, which is divided into two steps: *path planning* to generate optimal waypoints and *spline fitting* to generate optimal polynomial trajectories through waypoints.

3.2.1 Path Planning

We first search an optimal path, containing a series of safe waypoints to avoid the obstacles. We adopt the receding horizon control (RHC) framework, which searches the best path among an offline library [18]. In order to get a good path for different environments, the library is usually dense with large amounts of paths which is time consuming to check. Instead, we combine RHC with a modified CHOMP optimization method [19]. RHC serves to provide a good initial guess and CHOMP further optimizes it. Through the comparison in Sect. 4.3, this method is faster and better than RRT* in corridor environments.

Each waypoint in the path contains 4 DOF $\{x, y, z, \psi(yaw)\}$, namely the flat output space of quadrotor [20]. Let the path be $\xi(s) : [0, 1] \mapsto R^4$ mapping from arc length s to 4 DOF ($\xi(0)$ is starting point, $\xi(1)$ is ending point) such that:

$$\begin{aligned} \min_{\xi} \quad & J(\xi) = w_1 f_{obst}(\xi) + w_2 f_{smooth}(\xi) + f_{goal}(\xi) \\ \text{s.t.} \quad & \xi(0) = \xi_0 \end{aligned} \tag{8}$$

where w_1, w_2 are the weighting parameters of different cost functions. $f_{obst}(\xi)$ is the obstacle cost and $f_{smooth}(\xi)$ is the path smoothness cost as defined in CHOMP [19]:

$$f_{obst}(\xi) = \int_0^1 c_{obs}(\xi(s)) \| \frac{d}{dt} \xi(s) \| ds \tag{9}$$

$$f_{smooth}(\xi) = \frac{1}{2} \int_0^1 \| \frac{d}{ds} \xi(s) \|^2 ds \tag{10}$$

$f_{goal}(\xi)$ is the cost-to-go heuristic measuring distance between path endpoint $\xi(1)$ to global mission point ξ_g. We add this heuristic to free the endpoint for optimization, while CHOMP doesn't.

$$f_{goal}(\xi) = \| \xi(1) - \xi_g \|^2 \tag{11}$$

As mentioned before, we create an offline path library L containing 27 specifically designed paths shown in Fig. 3a. It is based on the structure property of corridor, where obstacles usually lie on two sides of walls. So it is easy and fast to find a safe path from the library.

We align the library with current pose then select $\xi^* = \arg\min_{\xi \in L} J(\xi)$ as the initial guess and optimize it through modified CHOMP. We keep discrete waypoint parametrization $\xi_0, \ldots, \xi_n$ of the path as in [19] instead of a continuous path to speed up the optimization. An optimization example during turning is shown in Fig. 3b, where the gradient pushes the path away from obstacles. Note that the end point is freed for optimization, different from standard CHOMP algorithm because our method is planning within a horizon and doesn't directly search a path from start to goal. A short horizon makes the optimization faster and more reactive.

3.2.2 Spline Fitting

After getting path waypoints $\xi_0, \ldots, \xi_n$, we fit a continuous spline $\xi(t)$ through them. It specifies the pose MAV should be at each time. The polynomial spline allows us to analytically compute feedforward control input for quadrotor [20], which guarantees

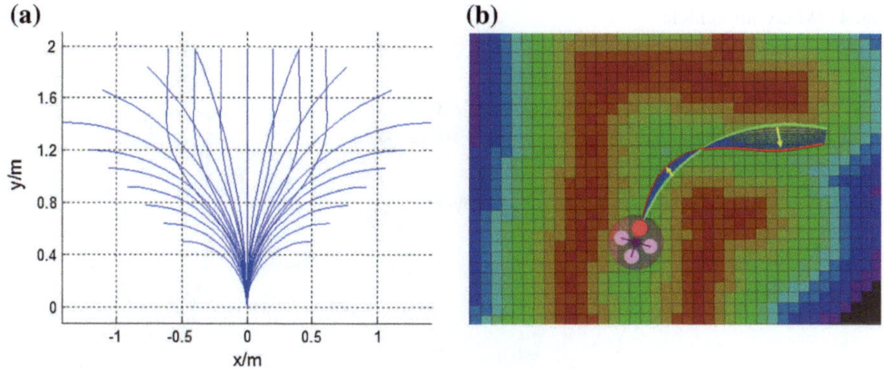

Fig. 3 **a** Initial path library. It is manually designed for the corridor environment where obstacles usually lie on two sides. It includes *straight line*, *turning arcs* with different curvatures and lane changing curves with parallel ending direction, corresponding to the three common flight patterns in the corridor. **b** Path optimization in turning. The *color grid* represents the distance map, computed from online 3D occupancy map [17]. The *green curve* represents the initial best path, *blue curves* are the paths during optimization based on the gradient (*yellow*). The final optimized path is in *red*

exponential tracking stability of the controller while waypoint following or steering angle methods cannot.

We represent the spline as 5 segments of 6th order polynomials. To find the optimal polynomial coefficients, we formulate it as a quadratic programming (QP) problem similar to [20, 21]. The cost function is to minimize the integration of L_2 norm of snap, namely the 4th order derivative (wrt. time). The constraints are passing through waypoints and keeping derivative $c^1, \ldots, c^4$ continuous. A closed form solution of QP with equality constraint could be found using Lagrange multipliers. Tikhonov regularization [22] is used in case of QP matrix ill-condition problem.

4 Experiments

4.1 System Setup

The platform we use for our experiment is a customized MAV as shown in Fig. 4. It's mainly composed of two computation units. One unit is an ARM-based Quadcore embedded computer (Odroid XU3), responsible for high-level task processing, such as odometry estimation, localization and motion planning, etc. The other one is the Pixhawk FCU which is used for multi-sensor data fusion and real-time control. A forward-looking RGB-D camera is used for pose estimation and motion planning. A downward-looking optical flow camera is used for velocity estimation and a point laser is used for height estimation. Besides, a FLIR camera is used for fire detection.

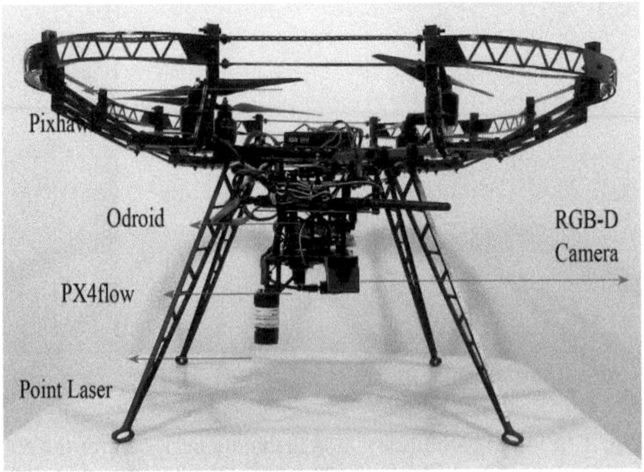

Fig. 4 Micro air vehicle platform

We first conduct some experiments to validate the performance of our state estimation and motion planning algorithms using the datasets recorded by carrying the robot in the ship. Then, field experiments were performed on the ex-USS shadwell to test the performance of the whole system. In the experiments, the RGB-D images are streamed at frame rate of 15 Hz with QVGA resolution. We create the offline 3D maps by using LOAM system [23] and the map resolution is set to 4 cm.

4.2 *Pose Estimation Experiments*

We test the odometry and localization algorithms by manually carrying our customized MAV system. The experiment is conducted in a constrained and visually degraded shipboard environment, which has a size of 16 m × 25.6 m × 4.04 m. In this environments, most of the time the RGB images are very dark as shown in Fig. 5, while the depth images are still very good. There are also some challenging locations where the depth camera can only see the ground plane, one wall or two parallel walls, or even nothing when it is very close to the wall (minimum range >0.5 m). In such situations, the depth-based odometry will suffer from the degeneration problem. In our algorithm, we monitor the degeneration status. If the degeneration is too severe, the odometry estimation method will not output motion estimation results, but a failure indicator. Then, our localization algorithm will use the noise-driven motion model to propagate the MCL particle set. In our experiment, we find that if the odometry failure is relatively short in duration (less than 3 s), it is possible for the localization algorithm to overcome this failure entirely. The localization result is shown in Fig. 5. From the experimental result, we can see that our robot can robustly localize its self even the odometry is not good.

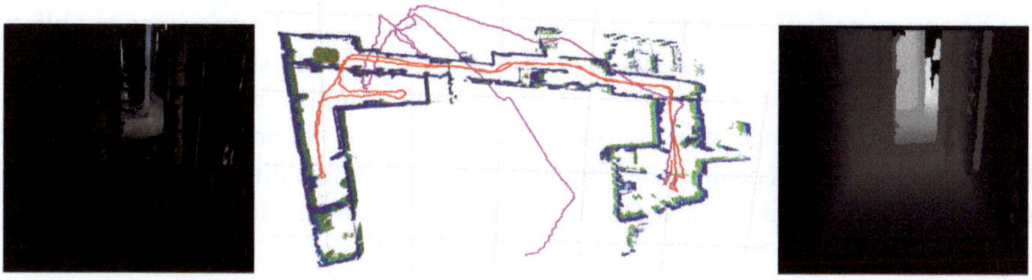

Fig. 5 Localization in degraded visual environment: *Pink* Odometry, *Red* Localization. The *center plot* shows the odometry, localization results with the 3D octomap. Pictures on both sides show the RGB and depth images from onboard RGB-D camera

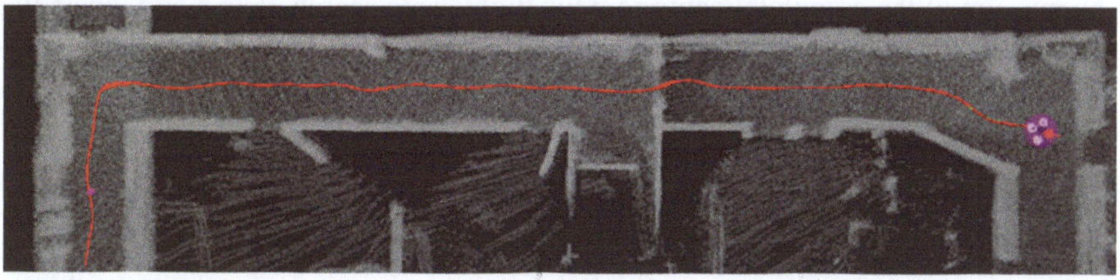

Fig. 6 An example trajectory calculated using path optimization with receding horizon control through a simulated shipboard environment

4.3 Planning Experiments

A simulated depth camera based on 3D point cloud map is used to create an occupancy grid. The mission planner then provides some goal points based on the prior map, about 5 m away from each other and local planner keeps replanning to reach them. The pose history during simulation is shown as red curve in Fig. 6.

To demonstrate the quality of our method, we compare our path planning method with RRT* and keep spline fitting part (minimizing snap) as the same. To bias RRT*, the local goal points are set closer to each other ($\sim$2 m) to greatly decrease the search space. The comparison is implemented on the embedded computer and the result is shown in Table 1. With bias, RRT* still needs more time than our method to generate a valid path and the quality in terms of obstacle cost and snap cost is higher than ours. This is mostly due to the fact that the corridor is a structured environment where obstacles usually lie on two sides. So our path set method could quickly find a smooth and safe path while RRT* needs many random samples in order to get a valid and smooth path.

Table 1 Path planning comparison with RRT*. Dist stands for vehicle distance to the obstacle

Methods	Time (ms)	Mean dist (m)	Min dist	Mean snap (m/s^4)	Max snap
RRT*	70	0.46	0.16	1.46	14.02
Our	30	0.47	0.18	0.58	2.50

An end-to-end offline path could also be computed from the prior map but blindingly following it tends to cause a collision if there is big state estimation error. Instead, the proposed online obstacle mapping and motion planning can guarantee the safety. The goal points in our planner should be set properly so as to avoid being trapped in local dead-ends. Though offline path with online modification could relieve the problem, it is not applicable in other unknown environments.

4.4 Autonomous Flight Test Results

The mission of the completely autonomous flights is to search, detect and locate fire using only onboard sensors and computation resources. In our tests, the MAV needs to operate in a variety of environments:

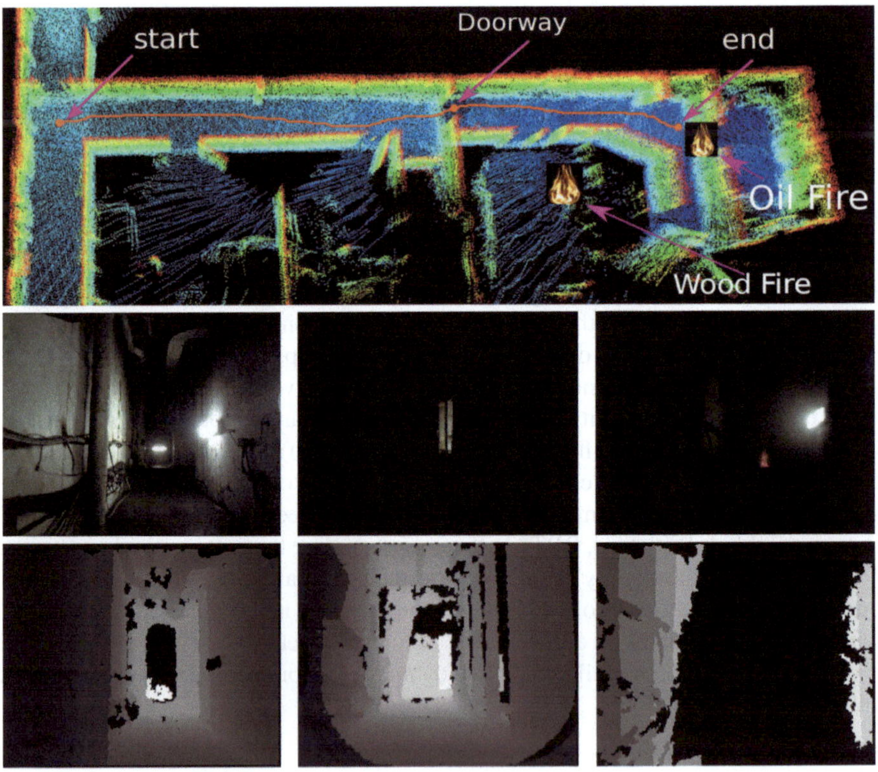

Fig. 7 Map and RGB and depth images of each environment condition from onboard RGB-D cameras. From *left* to *right* with lights on, with lights off, with fire and dense smoke

1. Narrow passageways and doorways: The most common shipboard environment. The space constraints limit the vehicles size.
2. Areas with poor or no lighting: Become visually degraded. Performance of optical flow sensor decreases.
3. Areas filled with smoke and fire: Smoke density varies with fuel source. It strongly affects the depth image and optical flow sensor.

Figure 7 shows the created offline point cloud map of the testing area and typical sensor images in each environment. MAV is launched around the 'start point' and flies autonomously in the 1 m wide, 20 m long corridor, with a tight doorway (66 cm wide, 8 cm clearance) and reaches the 'end point', while detecting fires.

We performed 20 experiments in this testing area under the three environment conditions. The vehicle pose of one experimental run is shown in Fig. 8. The success ratio of 20 runs is shown in Table 2. Failure cases are usually due to quadrotors being slightly rotated and stuck in the tight doorway. It is difficult to cross the door in dense smoke because the depth image is corrupted by smoke making it difficult for state estimation and obstacle detection. Results show that our robot can work very well in all the conditions except very dense smoke.

Runtime performance is also very important for MAVs since the onboard computation abilities are limited. We record the performance including CPU usages of some key algorithms on the Odroid system shown in Table 3. We use 300 particles for MCL localization. When all the system modules are running, the total CPU usage is between 60 and 65 %. The experiment result shows our navigation system can run in real-time by only using the onboard computation resources.

For fire detection, we use a lightweight FLIR-tau thermal camera to measure the temperature of the environment. We segment the appropriate range of temperature for fire, people etc. based on the thermal images. Anything over 100 °C is considered to have a high probability of being fire or close to fire. Similarly, segmented blobs

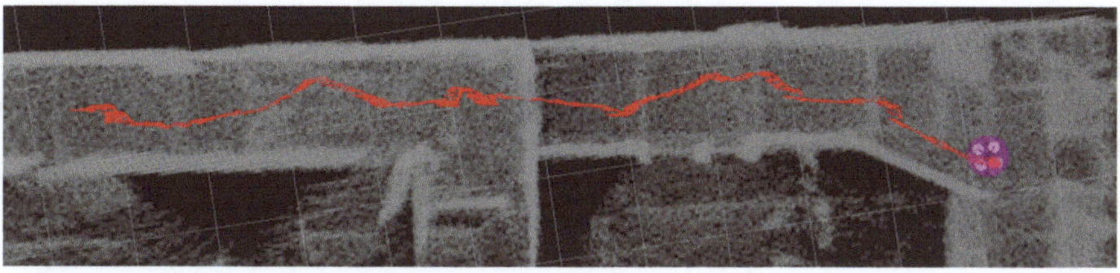

Fig. 8 Localization result from one autonomous flight

Table 2 Autonomous flight results

Environment	Total run	Succeeded	Rate (%)
Normal	4	4	100
Dark	7	5	71.4
Smoky	9	5	55.5

Table 3 Per-frame runtime performance on the embedded computer

Name	Algorithm runtime			
	Mean (ms)	Min (ms)	Max (ms)	StdDev (ms)
Odometry	18.3	8	25.8	5.2
Localization	65.8	45.8	97	16.5
Local planning	29.2	15.2	37.8	6.7

with temperature close to 30 °C is considered to belong to a human being. The video of a field experiment at Shadwell in Nov 2014 can be found at https://www.youtube.com/watch?v=g3dWQCECwlY.

5 Conclusion

In this paper we have shown the feasibility of an autonomous fire detection MAV system in a GPS denied environment with tough visibility conditions. This was achieved without the need of any additional infrastructure on the ship. We achieved autonomous flight with fully online and onboard state estimation and planning through 1 m wide passages while crossing doorways with only 8 cm clearance. We demonstrated 10 consecutive runs where the vehicle crossed completely dark, light smoky passageway respectively and ended by detecting wood and diesel fires.

The next challenges are to increase the robustness and safety of the vehicle while increasing flight time. This will involve improvements in both hardware and software. The current size of vehicle is a little large, resulting in a very tight fit through the ship doorways. In future, we intend to move from a quadrotor design to a single/coaxial ducted rotor design to decrease size but increase flight time efficiency. Currently, our sensor suite loses reliability in dense smoke conditions. We plan on adding sensors which extend the range of environments our robot can successfully navigate and inspect. On the software side, one important goal is to decrease the dependency on a prior map for state estimation to make the system more adaptable to changing or damaged environments. Pursuing exploration and mapping in a damaged environment poses many interesting research challenges.

References

1. Grzonka, S., Grisetti, G., Burgard, W.: A fully autonomous indoor quadrotor. IEEE Trans. Robot. **28**(1), 90–100 (2012)
2. Dryanovski, I., Valenti, R.G., Xiao, J.: An open-source navigation system for micro aerial vehicles. Auton. Robots **34**(3), 177–188 (2013)
3. Shen, S., Michael, N., Kumar, V.: Autonomous multi-floor indoor navigation with a computationally constrained MAV. In: 2011 IEEE International Conference on Robotics and Automation (ICRA), pp. 20–25. IEEE (2011)

4. Schauwecker, K., Zell, A.: On-board dual-stereo-vision for the navigation of an autonomous MAV. J. Intell. Robot. Syst. Theory Appl. **74**, 1–16 (2014)
5. Fraundorfer, F., Heng, L., Honegger, D.: Vision-based autonomous mapping and exploration using a quadrotor MAV. In: IEEE International Conference on Intelligent Robots and Systems, pp. 4557–4564 (2012)
6. Wu, A.D., Johnson, E.N., Kaess, M., et al.: Autonomous flight in GPS-denied environments using monocular vision and inertial sensors. J. Aerosp. Inf. Syst. **10**, 172–186 (2013)
7. Scaramuzza, D., Achtelik, M., Doitsidis, L., et al.: Vision-controlled micro flying robots: from system design to autonomous navigation and mapping in GPS-denied environments, pp. 26–40 (2014)
8. Weiss, S., Scaramuzza, D., Siegwart, R.: Monocular-slam-based navigation for autonomous micro helicopters in GPS-denied environments. J. Field Robot. **28**(6), 854–874 (2011)
9. Flores, G., Zhou, S., Lozano, R., Castillo, P.: A vision and GPS-based real-time trajectory planning for a MAV in unknown and low-sunlight environments. J. Intell. Robot. Syst. **74**(1–2), 59–67 (2014)
10. Huang, A.S., Bachrach, A.: Visual odometry and mapping for autonomous flight using an RGB-D camera. Int. Symp. Robot. Res. 1–16 (2011)
11. Valenti, R.G., Dryanovski, I., Jaramillo, C.: Autonomous quadrotor flight using onboard RGB-D visual odometry. In: 2014 IEEE International Conference on Robotics and Automation, pp. 5233–5238. IEEE (2014)
12. Fang, Z., Scherer, S.: Real-time onboard 6DoF localization of an indoor MAV in degraded visual environments using a RGB-D camera. In: 2015 IEEE International Conference on Robotics and Automation, May 2015
13. Horn, B.K.P., Harris, J.G.: Rigid body motion from range image sequences. CVGIP Image Underst. **53**(1), 1–13 (1991)
14. Pomerleau, F., Colas, F., Siegwart, R., Magnenat, S.: Comparing ICP variants on real-world data sets. Auton. Robots **34**(3), 133–148 (2013)
15. Callaghan, K., Chen, J.: Revisiting the collinear data problem: an assessment of estimator Ill-conditioning in linear regression. Pract. Assess. Res. Eval. **13**(5), 5 (2008)
16. Thrun, S., Burgard, W., Fox, D.: Probabilistic Robotics (Intelligent Robotics and Autonomous Agents). The MIT Press (2005)
17. Scherer, S., Rehder, J., Achar, S., et al.: River mapping from a flying robot: state estimation, river detection, and obstacle mapping. Auton. Robots **33**(1–2), 189–214 (2012)
18. Green, C.J., Kelly, A.: Optimal sampling in the space of paths: Preliminary results (2006)
19. Ratliff, N., Zucker, M., Bagnell, J.A., et al.: Chomp: gradient optimization techniques for efficient motion planning. In: 2009 IEEE International Conference on Robotics and Automation, pp. 489–494 (2009)
20. Mellinger, D., Kumar, V.: Minimum snap trajectory generation and control for quadrotors. In: 2011 IEEE International Conference on Robotics and Automation, pp. 2520–2525 (2011)
21. Richter, C., Bry, A., Roy, N.: Polynomial trajectory planning for quadrotor flight. In: International Conference on Robotics and Automation (2013)
22. Golub, G.H., Hansen, P.C., O'Leary, D.P.: Tikhonov regularization and total least squares. SIAM J. Matrix Anal. Appl. **21**(1), 185–194 (1999)
23. Zhang, J., Singh, S.: LOAM : Lidar Odometry and Mapping in Real-time. In: Robotics: Science and Systems Conference (RSS) (2014)

Autonomous Exploration for Infrastructure Modeling with a Micro Aerial Vehicle

Luke Yoder and Sebastian Scherer

Abstract Micro aerial vehicles (MAVs) are an exciting technology for mobile sensing of infrastructure as they can easily position sensors in to hard to reach positions. Although MAVs equipped with 3D sensing are starting to be used in industry, they currently must be remotely controlled by a skilled pilot. In this paper we present an exploration path planning approach for MAVs equipped with 3D range sensors like lidar. The only user input that our approach requires is a 3D bounding box around the structure. Our method incrementally plans a path for a MAV to scan all surfaces of the structure up to a resolution and detects when exploration is finished. We demonstrate our method by modeling a train bridge and show that our method builds 3D models with the efficiency of a skilled pilot.

1 Introduction

The goal of this work is to use a MAV to rapidly model large outdoor structures with arbitrary geometry. As-built 3D models are increasingly used in a number of industries to detect structural problems, assess damage, design renovations, and to organize other types of data. Although the industry standard ground-based lidar are accurate, building models with them is slow and they suffer from occlusion problems. Modeling large structures found in outdoor, open air environments is an excellent application for MAV-based lidar which can reach almost any vantage point. Compared to stationary ground-based lidar, MAV-based lidar output a model with lower resolution and higher uncertainty as the vehicle must use lightweight sensors and must estimate its position. The benefit of a MAV-based lidar is coverage

This research is funded by the National Science Foundation under grant IIS-1328930.

L. Yoder (✉) · S. Scherer (✉)
Carnegie Mellon University, Pittsburgh, PA, USA
e-mail: lyoder@cmu.edu

S. Scherer
e-mail: basti@cmu.edu

D.S. Wettergreen and T.D. Barfoot (eds.), *Field and Service Robotics*,
Springer Tracts in Advanced Robotics 113, DOI 10.1007/978-3-319-27702-8_28

and rapid deployment. In many environments flying scanners can reach the vantage points required to achieve complete coverage, and do so very quickly. Rather than competing with existing terrestrial lidar, we imagine such a system supporting new applications that require low resolution and complete coverage point clouds.

A number of MAV systems are commercially available for building 3D models of outdoor environments. Acquiring data with these systems, however, is still a manual process requiring a skilled pilot. Not only is safely piloting a MAV near a structure difficult, scanning complex structures under manual control is prone to error as it is difficult for a human pilot to remember what has and what has not been scanned. We propose a practical solution where a user draws a 3D bounding box around a structure, then a small flying vehicle autonomously scans all surfaces of the structure, producing a 3D point cloud.

One challenge in 3D modeling infrastructure with an autonomous MAV is developing path planning algorithms. If a prior model of the infrastructure is available, "inspection" or "coverage" planning could lead the MAV to efficiently scan all surfaces [2, 7]. In many applications, however, obtaining a prior model is impractical. Requiring a prior map will limit the adoption of a robotic planning algorithm for infrastructure modeling. Without a priori models to use for path planning, the remaining problem is one of choosing an exploration path through a partially observed environment with the goal of maximizing exploration efficiency.

The contribution of this paper is a simple yet effective 3D path planning algorithm for completely scanning complex 3D environments with a range sensor attached to a MAV. Specifically, we present the surface frontier, a fundamental geometric aspect of 3D surface exploration, and we present an incremental path planning algorithm using surface frontiers to guide the observation of unknown surface until the surface is completely observed. Finally, we present real world results showing that our system performs as well as or better than a skilled pilot (Fig. 1).

Fig. 1 A MAV autonomously modeling a structure (*left*) and the resulting point cloud (*right*)

2 Related Work

Work related to ours does not rely on a prior model of the environment but iteratively plans exploration paths through a partially observed environment. Common to almost all of the following approaches is discretizing the continuous world into an occupancy grid where cells in the occupancy grid are either occupied, free, or unknown to the robot. The difference between most of the following methods is in how the information in the occupancy grid is used to guide exploration.

Frontier exploration [16] is a simple 2D exploration algorithm used extensively on ground robots. In frontier exploration, information is gained by traveling to frontiers, a heuristic that has proven successful at guiding a robot to a vantage point where unobserved cells can be observed. Extensions of frontier exploration to 3D without reducing occupancy grid resolution are too computationally costly [5] to run in real time on a computationally constrained MAV. Shade et al. [12] propose a 3D frontier exploration and integrated path planning algorithm that runs in real time but is designed for exploration of free space, not modeling outdoor surfaces. Several groups achieve outdoor exploration and 3D mapping using MAVs by limiting exploration to 2D. Heng et al. [6] implement frontier based exploration and wall following exploration, which they validate in outdoor urban environments. Jain et al. [8] propose a frontier shoreline exploration algorithm that they validate on a MAV exploring rivers.

Some methods estimate entropy reduction over a set of possible paths to determine the best exploration path. Stachniss et al. [15] propose generating paths that lead to frontiers and paths that lead to previously visited locations to improve localization. The path that minimizes the sum of map entropy and pose entropy is chosen.

"Next best view" planning algorithms have been developed [9, 11] to incrementally plan sensor views for modeling objects. Next best view algorithms generally constrain the search space of views around the object and then find a view that maximizes some utility function. The utility function for evaluating views might include unknown cells visible from the view and overlap with previously acquired data for data registration.

To reduce computational cost, a number of techniques attempt to find exploration goals without managing an occupancy grid. One method designed for indoor 3D exploration [13] simulates particles expanding from free space according to Newtonian dynamics. Exploration goals are set in regions of high expansion. This algorithm is designed as a simplification of frontier exploration allowing the authors to successfully explore indoor environments with a computationally constrained MAV. Another simulated particle-based method [1] simulates liquid falling on a 3D outdoor scene and detects exploration goals by finding areas where the simulated liquid leaks through the point cloud. This algorithm is limited to exploring 2.5D terrain because the algorithm is not designed for detecting holes in vertical walls and under overhangs.

Our method is closely related to next best view techniques although it employs a simpler utility function. Our utility function is based on the visibility of 2D frontiers on the 2D surface of a 3D object, which are related to the frontiers used in frontier exploration.

3 Problem Formulation

This section provides a representation for the state of a MAV and a partially observed environment. In this section we also describe the problem that we solve in Sect. 5, as well as some assumptions.

Since we are exploring outdoors with the ability to navigate in 3D, we start by constraining exploration to a region of interest $R \subseteq W$ where $W \subseteq \mathbb{R}^3$ is the world. The MAV's position in W is defined by its state $X = [x, y, z, \varphi, \theta, \psi, c]$ where (x, y, z) and (φ, θ, ψ) are position and orientation in world coordinates, respectively. The variable c is the configuration of the sensors on the vehicle which can be multidimensional. We will refer to proposed trajectories T which symbolize time parametrized state such that $X_i = T(t_i)$ for $t_i \in \left[t_o, t_f\right]$.

We represent W as a 3D occupancy grid where cells C^j can be unknown C_u, free C_f, or occupied C_o. The goal of our exploration planning algorithm is to classify all observable C_o in R. Supporting this goal, we define surface information as

$$I_s = \sum_{j=1}^{m} \begin{cases} 1, & \text{if } C^j = C_o \text{ and } C^j \subseteq R \\ 0, & \text{otherwise} \end{cases} \tag{1}$$

where m is the number of cells in R. We define I_s^* as the surface information when all surfaces in the environment are observed. The problem that we would like to solve is to find an optimal trajectory T^* from which the sensor observes all surfaces (i.e. $I_s = I_s^*$) in minimum time. Since an optimal coverage path in an environment that is only partially observed may not be possible to compute, we are left with the goal of achieving high exploration efficiency in terms of surface information gain per unit time.

We need to make a few assumptions. First, we assume that free space in R is one connected set reachable without leaving R. Second, we assume the vehicle's state estimation is accurate enough to not require active localization. Some exploration methods [15] estimate the MAV's ability to localize over proposed trajectories to help maintain low uncertainty in the map. We find that localization accuracy around large structures using our laser odometry approach [17] is accurate enough for exploration using large occupancy grid cells. For this work, we assume that active localization is not necessary.

We use the occupancy grid representation to describe our method in Sect. 5 but first we describe the surface frontiers that guide our exploration approach.

4 Surface Frontiers

Our goal is to sense a structure's surface, not the free space around the structure. Traditional frontier exploration would lead a MAV to explore free space in addition to surface. Because of this, frontier exploration in an outdoor environment may be inefficient. Our method, proposed in Sect. 5, prioritizes surface modeling by using the boundary between known and unknown surface to guide exploration. In this section we describe topological properties of known and unknown space that motivate our method. This section starts by introducing the concept of a surface frontier, then goes on to discuss its benefit for exploration.

4.1 Definition

We consider W as topological space where all points $p \in W$ are either occupied p_o or free p_f. All p are static and all p_o belong to one connected set $P_o = \{p_o\}$. Since W is either unobserved or partially observed a priori, points unknown to the MAV are designated p_u and known points p_k. In this section we assume the MAV is equipped with an ideal range sensor capable of classifying unoccluded volume in its field of view as occupied or free. We define p_k^* as points that can be observed from the free space connected set that the MAV navigates through. As described in Sect. 3, we assume $p_k^* \cap R$ is one connected set. Since a range sensor can only observe the surface of occupied space,

$$S_k^* = P_o \cap \{p_k^*\}$$

is a surface, which does not change during exploration. S_k^* is the object surface observable from the free space connected set that the MAV navigates through, though not necessarily observable by the MAV if some free space is not reachable. During exploration a subset of S_k^* are the known surfaces

$$S_k = P_o \cap \{p_k\}$$

Since S_k is a subset of the surface S_k^* during exploration, we know that the boundary of known surface ∂S_k is also a boundary of unknown surface, where ∂ is the boundary operator. In other words, an unobserved surface must be present just beyond the known surface boundary. We call ∂S_k surface frontiers. During exploration, ∂S_k is a set of one dimensional manifolds as shown in Fig. 2. If we can guide a robot to observe surface frontiers, it is almost certain that we will observe unknown surface.

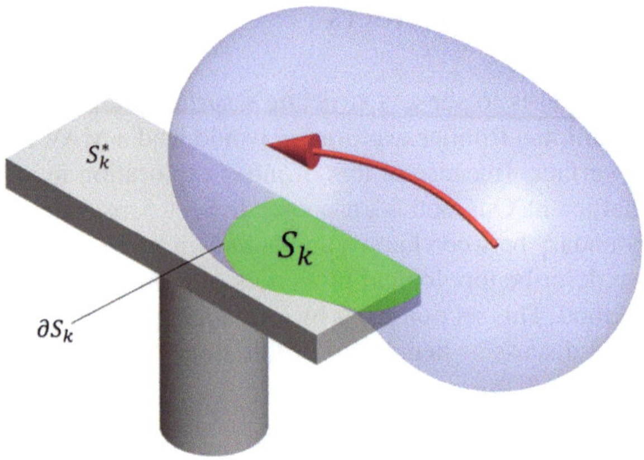

Fig. 2 Surface frontiers at the beginning of exploration

4.2 *Surface Frontier Exploration*

For practical purposes we would like a simple way to define what volume in the world should be explored. Then, given a volume to be explored, we would like to detect when exploration is complete. This subsection introduces a region of interest bounding the volume to be explored, then shows how we can terminate exploration when all surfaces in the volume are observed without exhaustively exploring unknown space.

We set a region of interest grouping points into the connected set R which is the volume containing the structure to be modeled. As an example, R could be the volume inside a cuboid defined by someone using the system. The goal is to observe all surfaces inside the bounding box $S_k^* \cap R$.

We define the region of interest boundary ∂R intersecting the connected free space $\{p_k^*\} \cap \{p_f\}$ as the observable region of interest boundary

$$B = \partial R \cap \{p_k^*\} \cap \{p_f\}$$

Combining surface B with all object surfaces observable by the MAV gives us the surface to be explored

$$E = B \cup S_k \cap R$$

If $\{p_k \in R\} = \{p_k^* \in R\}$ then E becomes a closed connected surface denoted E^*. We can use this property to determine when exploration is complete. If, during exploration, E is a set of connected surfaces instead of a closed connected surface, exploration is not complete as there still may be unobserved surfaces in the bounding box.

Alternatively, if we assume that $S_k^* \cap R$ is a single connected surface then exploration is complete when $\partial S_k^* \cap R$ is an empty set. This assumption relieves the MAV of having to exhaustively search B.

Only using surface frontiers to guide exploration means we are not making any assumptions about a structure's geometry. Including a priori information about the structure (e.g. max curvature) or other heuristics (e.g. number of C_u visible from a view) could improve performance but would also increase algorithm complexity and possibly reduce generality.

5 Method

In the following we formulate an exploration planning strategy for a MAV tasked with the exploration problem introduced in Sect. 3. Unlike the frontier exploration algorithm [16], we cannot directly navigate to surface frontiers as doing so might lead to a collision. We also may not be able to observe frontiers if the MAV cannot reach a state where the frontier is visible. To use surface frontiers for exploration, we compute $T(t_i)$ from which we can observe frontiers, and terminate exploration when surface frontiers are not observable by the MAV. In this section we start by describing how to detect surface frontiers in the occupancy grid. We then introduce a utility function for estimating the utility of views for sensing surface frontiers. Finally, we describe how we can plan exploration paths using an objective function that trades off between a view's utility and the path cost of navigating to the view.

5.1 Occupancy Grid Surface Frontiers

Given an occupancy grid with occupied, free, and unknown cells, surface frontiers can be detected by finding connections between all three cell classes. An occupancy grid surface frontier can be defined as a free cell C_f with a known occupied neighbor C_o and an unknown neighbor C_u where C_o and C_u are also neighbors. An example of surface frontiers in the occupancy grid representation is shown as the blue cells in Fig. 6.

5.2 View Utility

To guide the observation of surface frontiers we create an exploration utility function that estimates the number of surface frontier observations that can be made at a given view.

First we make a simplification based on our vehicle's sensor configuration. Our vehicle's lidar is nearly omni-directional as shown in Fig. 3. Only a small blind spot is present behind the vehicle. We assume that the lidar scans in a spherical pattern sampling in a circular uniform distribution, which allows us to reduce the degrees of

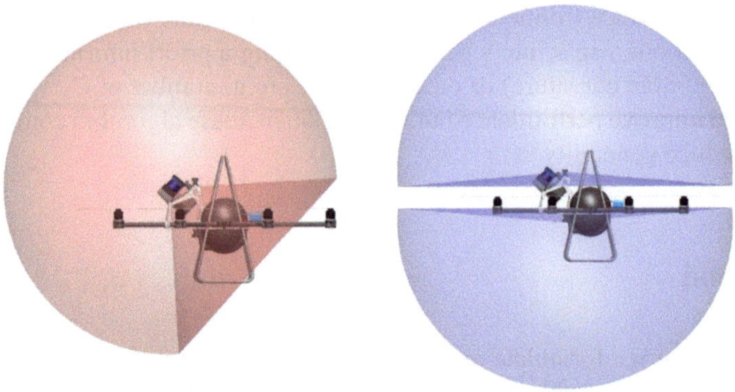

Fig. 3 lidar field of view (*left*) and camera field of view (*right*)

freedom of the sensor field of view. Assuming omni-directional sensing allows us to simplify our MAV's state to $X = [x, y, z]$.

Assuming one complete scan is collected we can approximate the number of lidar rays that will hit one unoccluded surface frontier cell. We also consider safe flying distance from the structure d_s and a maximum desired measurements per cell t_m. Given the cell height h, the distance from the sensor to the cell r, and the number of points per scan N, the utility of a view for observing a single surface frontier is

$$f(r) = \begin{cases} 0, & \text{if } r < d_s \\ t_m, & \text{if } d_s \leq r < d_m \\ \dfrac{A_c N}{A_s}, & \text{if } d_m \leq r \end{cases} \tag{2}$$

where $d_m = \sqrt{\frac{h^2 N}{4\pi t_m}}$, $A_c = h^2$ is the area of one face of the cell and $A_s = 4\pi r^2$ is the area of a sphere. Equation 2 is plotted in Fig. 4 using our vehicle's sensor parameters and the thresholds.

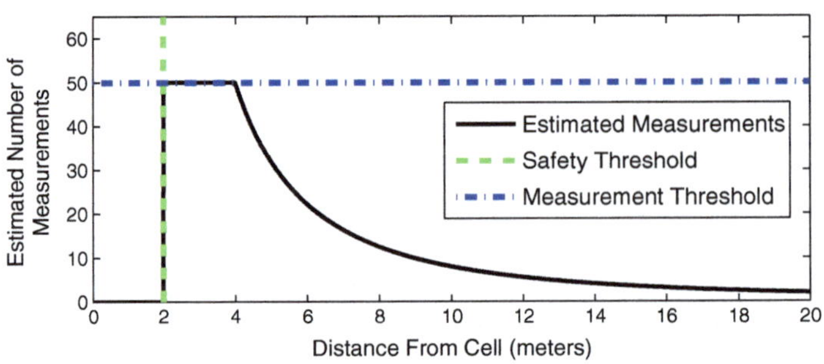

Fig. 4 Equation 2 evaluated with $h = 0.5$ m, $N = 40000$, $d_s = 2$ m, and $t_m = 50$

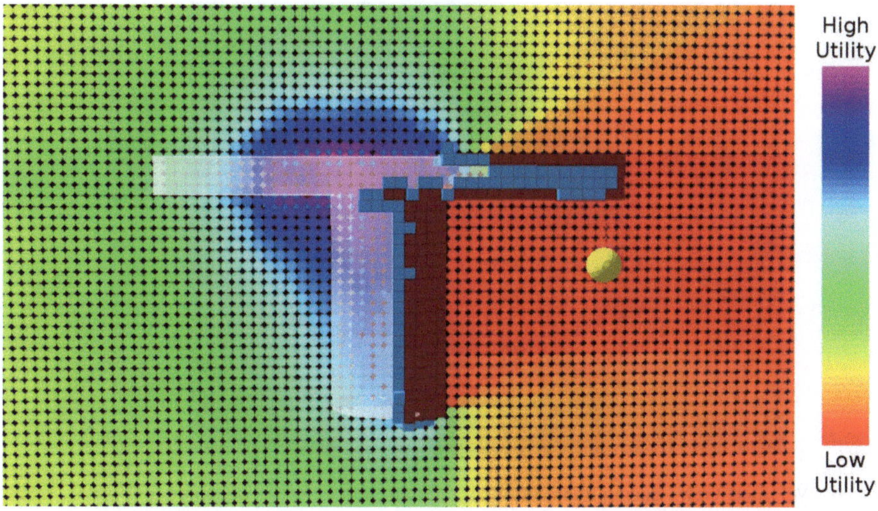

Fig. 5 Equation 2 evaluated during a simulated exploration over the *center line* of a simple bridge-like object. *Red cells* are occupied and *blue cells* are surface frontiers. The *yellow sphere* is a goal that the robot has almost reached

For a given view we can evaluate Eq. 2 for each unocculded surface frontier and sum the results to give us a view utility. We can repeat this over a set of views to create a 3D utility function. To demonstrate this utility function we evaluate Eq. 2 densely in Fig. 5 without thresholds (i.e. $t_m = \infty$ and $d_s = 0$).

Unfortunately, it is expensive to evaluate the utility function densely over the map due to the ray tracing required. From Fig. 5, we observe that the surface frontier observation utility decreases as distance from the surface increases. If our goal is to maximize a view's utility, the view can be offset from the surface between the safety distance d_s and d_m.

5.3 View Planning

Our view planning approach offsets the occupied cells using a distance transform then uniformly samples potential views on the offset surface. The interested reader is invited to read more about the SPARTAN path planner [4] which details our distance transform and view sampling approach. For each view found using SPARTAN, we determine which surface frontiers are visible by ray tracing. For a given view we sum Eq. 2 evaluated for each visible surface frontier cell. This gives us a set of views weighted by their utility as shown in Fig. 6. Given a robot position, partial map, and exploration views, we plan to a views using SPARTAN.

If we want to consider the utility of a proposed path we could incrementally sum the utility of views along a path, updating the map after summing each view by simulating frontier observations. In particular, an approach such as [14] could be

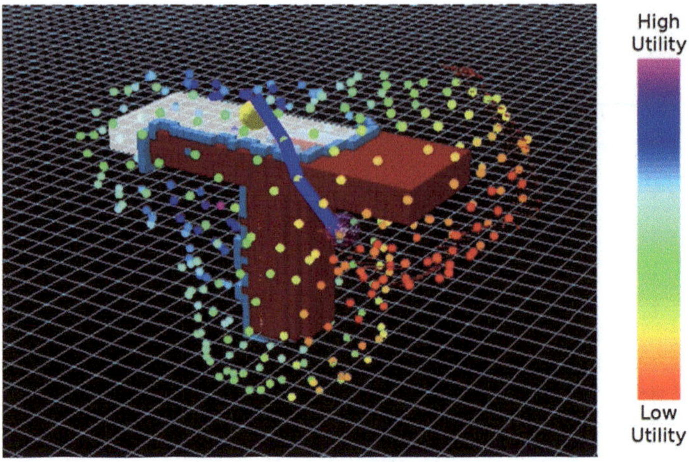

Fig. 6 Weighted views during exploration simulation. *Red cells* are occupied and *blue cells* are surface frontiers. The *yellow sphere* is a goal that the robot is navigating towards

implemented. Unfortunately, such solutions are computationally expensive and may quickly change due to new (real) observations. Instead of explicitly computing the utility of paths, we focus on using path cost to trade off between nearby exploration goals and distant exploration goals. We can then use Eq. 3 to determine the highest value exploration view. For a given map, the exploration value V of a view X_f is

$$V(X_f, X_o) = \frac{U(X_f)}{U_{max}}\alpha - \frac{C(X_f, X_o)}{C_{max}}(1-\alpha) \tag{3}$$

where α is a tuning parameter, $U(X_f)$ is a view utility, and $C(X_f, X_o)$ is the path cost for navigating from the current state X_o to view X_f. We can trade off between path cost and view utility by tuning $0 \leq \alpha \leq 1$. In large environments setting $\alpha = 1$ leads the MAV to inefficiently travel back and fourth across free space as it chooses maximum utility views. Reducing α increases the value of local view utility leading the robot to explore a region before travelling long distances to high utility views.

Once a view is chosen, the MAV replans paths at a fixed frequency until (a) the view is reached, (b) the view utility is reduced to zero, (c) the view is deleted when the distance transform is updated with new observations, or (d) the MAV cannot reach the view after a reasonable amount of time. When one of these conditions is met, a new view is chosen. If all samples are close to zero, the exploration planner terminates. There may be surface frontiers left when the algorithm terminates, but they will not be observable from reachable views.

To begin exploration the MAV could search the bounding box until a surface is detected to begin exploration. In our current implementation we assume that the MAV starts with the structures surface in the sensor field of view and we assume the structures surface in the bounding box $S_k^* \cap R$ is one connected set. This speeds up data collection by making searching the boundary unnecessary.

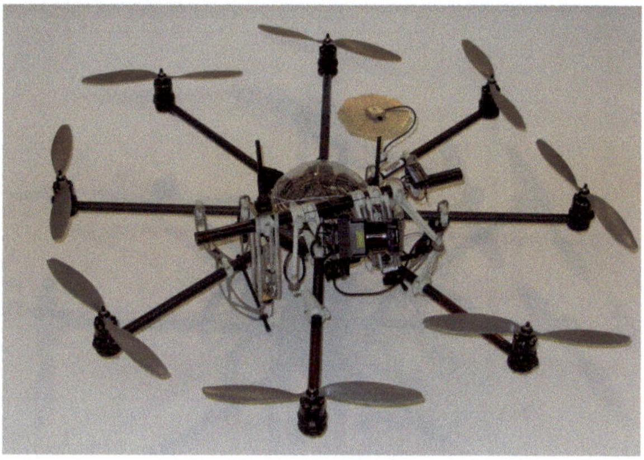

Fig. 7 MAV platform

6 MAV Platform

The vehicle is built on top of an oct-rotor platform shown in Fig. 7. All processes are run on an onboard flight computer using a Intel i7 dual core 2.5 GHz processor. Our mapping, planning, and controls processes communicate using the Robot operating system [10]. The flight computer sends yaw, pitch, roll, and thrust commands to a Mikrokopter flight controller. A cascaded PID controller is used to follow paths at a fixed 0.5 m/s. The vehicle weighs 5 kg and has a flight time of approximately 15 min.

The range sensor used for mapping is a Hokuyo UTM-30LX-EW scanning lidar with a custom gimbal that sweeps the laser in a spherical pattern. All lidar data is stored for point cloud generation, but only measurements within 15 m are added to the occupancy grid. Upward and downward facing IDS imaging UI-1241LE-C-HQ cameras using Sunex DSL215B 185° fisheye lenses give the vehicle a spherical field of view. The cameras are downsampled to 480×480 pixels. The vehicle has a barometric pressure sensor and Microstrain 3DM-GX3-35 IMU reporting readings at 20 and 50 Hz respectively. All sensors are time synchronized using a time server microcontroller. Depth enhanced visual odometry [17] is run online at 30 Hz. An unscented kalman filter [3] is used to fuse IMU, Visual odometry, barometric pressure, and GPS.

7 Results

We validate our algorithm by autonomously modeling a train bridge in Pittsburgh, Pennsylvania. The bridge, shown in Fig. 9, is a 50 m long steel and concrete structure. The environment is vegetated in some places and confined by bridge structural members in others. Although there was little wind during the tests, the environment was challenging for a MAV with intermittent GPS signal. In some areas the vehicle only had a 2 m margin between its position and the obstacles.

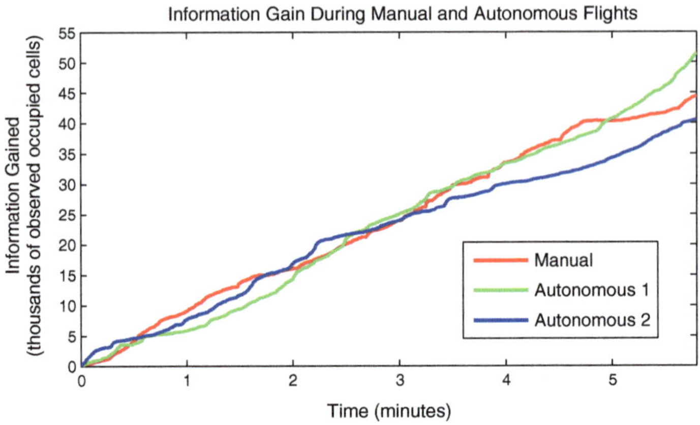

Fig. 8 Autonomous and manual flight surface information gain

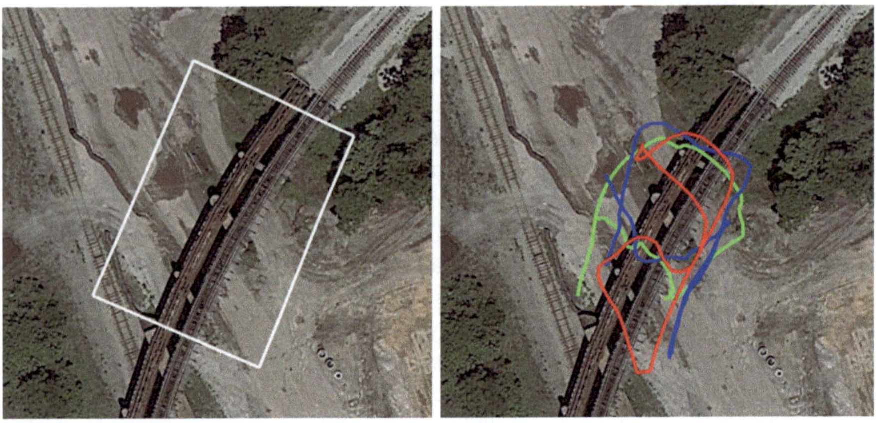

Fig. 9 MAV path and point cloud built during autonomous flights (*blue*, *green*) and manual flight (*red*)

At the start of the trial, the MAV was initialized with the bridge in sensor range. The bounding box parameters were loosely defined around the bridge and sent to the MAV. For this trial, the MAV flew to next best views without considering path cost (i.e. $\alpha = 1$). After manual takeoff the MAV was switched into autonomous mode and began exploration. Since all processes run on board, communication between the MAV and a ground station is not used. The autonomous run lasted 6 min and resulted in a five million point model of the bridge, shown in Fig. 1.

We compare autonomous exploration against a manual flight by skilled pilot. The pilot was instructed to scan all surfaces of the bridge using only a remote control to guide the vehicle. Figure 8 shows surface information versus time for the autonomous and manual trials. The results show that the autonomous exploration planner varies in performance by ±8000 observed occupied cells when compared to the performance of the single flight by a skilled pilot.

Qualitatively, the MAV's behavior during the autonomous trials was similar to manually guided trial. During the autonomous trials the MAV maintained safe distance from bridge surface and obstacles like tree branches and tall grass. Figure 9 shows that the MAV chose a exploration path that varied from the human operator's path, spending most of the mission flying along the sides of the bridge instead of systematically weaving under the bridge like the human operator.

8 Conclusion and Future Work

In this work we demonstrate an exploration planning algorithm that requires simple operator input to generate complex exploration paths for building a 3D model of an arbitrary structure outdoors. We show that a prior map is not necessary for planning paths for such a system. Finally, we demonstrate that autonomous vehicles using our exploration algorithm can perform as well as a skilled pilot.

There are a number of limitations in our method that present challenges for future work. The large 0.5 m occupancy grid cell size used in this work limit the MAV's ability to detect small surface frontiers. If we can significantly decrease cell size using a map representation like an octree we could ensure coverage up to a resolution defined by the user. This would allow the user to trade off between point cloud resolution and flight time. In larger scale environments the travel cost between view points will become significantly higher making the consideration of path cost more important. Our system already supports trading off between view value versus path cost, but thorough analysis is needed to justify this approach in larger scale environments. Finally, this work assumes that the vehicle's position estimate is accurate. This is a reasonable assumption considering the environment used in this paper and the required map accuracy. A higher mapping accuracy would make the system useful in more applications. To do this we would like to employ active localization techniques as well as improved sensors in future work.

References

1. Adler, B., Xiao, J., Zhang, J.: Autonomous exploration of urban environments using unmanned aerial vehicles. J. Field Robot. **31**(6), 912–939 (2014)
2. Blaer, P.S., Allen, P.K.: Data acquisition and view planning for 3-d modeling tasks. In: IEEE/RSJ International Conference on Intelligent Robots and Systems, 2007. IROS 2007, pp. 417–422 (2007)
3. Chambers, A.D., Scherer, S., Yoder, L., Jain, S., Nuske, S.T., Singh, S.: Robust multi-sensor fusion for micro aerial vehicle navigation in gps-degraded/denied environments. In: Proceedings of American Control Conference, Portland, OR (2014)

4. Cover, H., Choudhury, S., Scherer, S., Singh, S.: Sparse tangential network (spartan): motion planning for micro aerial vehicles. In: International Conference on Robotics and Automation (2013)
5. Dornhege, C., Kleiner, A.: A frontier-void-based approach for autonomous exploration in 3d. In: IEEE International Symposium on Safety, Security, and Rescue Robotics (SSRR), pp. 351–356 (2011)
6. Heng, L., Honegger, D., Lee, G.H., Meier, L., Tanskanen, P., Fraundorfer, F., Pollefeys, M.: Autonomous visual mapping and exploration with a micro aerial vehicle. J. Field Robot. **31**(4), 654–675 (2014)
7. Hollinger, G.A., Englot, B., Hover, F.S., Mitra, U., Sukhatme, G.S.: Active planning for underwater inspection and the benefit of adaptivity. Int. J. Robot. Res. **32**(1), 3–18 (2013)
8. Jain, S., Nuske, S.T., Chambers, A.D., Yoder, L., Cover, H., Chamberlain, L.J., Scherer, S., Singh, S.: Autonomous river exploration. In: Field and Service Robotics, Brisbane (2013)
9. Null, B.D., Sinzinger, E.D.: Next best view algorithms for interior and exterior model acquisition. In: Proceedings of the Second International Conference on Advances in Visual Computing-Volume Part II, ISVC'06, pp. 668–677. Springer, Berlin (2006)
10. Quigley, M., Conley, K., Gerkey, B.P., Faust, J., Foote, T., Leibs, J., Wheeler, R., Ng, A.Y.: Ros: an open-source robot operating system. In: ICRA Workshop on Open Source Software (2009)
11. Scott, W.R., Roth, G., Rivest, J.-F.: View planning for automated three-dimensional object reconstruction and inspection. ACM Comput. Surv. **35**(1), 64–96 (2003)
12. Shade, R., Newman, P.: Choosing where to go: complete 3d exploration with stereo. In: 2011 IEEE International Conference on Robotics and Automation (ICRA), pp. 2806–2811 (2011)
13. Shen, S., Michael, N., Kumar, V.: Autonomous indoor 3d exploration with a micro-aerial vehicle. In: 2012 IEEE International Conference on Robotics and Automation (ICRA), pp. 9–15 (2012)
14. Singh, A., Krause, A., Kaiser, W.J.: Nonmyopic adaptive informative path planning for multiple robots. In: Proceedings of the 21st International Jont Conference on Artifical Intelligence, IJCAI'09, pp. 1843–1850. Morgan Kaufmann Publishers Inc., San Francisco (2009)
15. Stachniss, C., Grisetti, G., Burgard, W.: Information gain-based exploration using rao-blackwellized particle filters. In: Proceedings of Robotics: Science and Systems (RSS), Cambridge, MA, USA (2005)
16. Yamauchi, B.: A frontier-based approach for autonomous exploration. In: 1997 IEEE International Symposium on Computational Intelligence in Robotics and Automation, 1997. CIRA'97, Proceedings, pp. 146–151 (1997)
17. Zhang, J., Kaess, M., Singh, S.: Real-time depth enhanced monocular odometry. In: Intelligent Robots and Systems (IROS), Chicago, IL, USA, (2014)

Long-Endurance Sensing and Mapping Using a Hand-Launchable Solar-Powered UAV

Philipp Oettershagen, Thomas Stastny, Thomas Mantel, Amir Melzer, Konrad Rudin, Pascal Gohl, Gabriel Agamennoni, Kostas Alexis and Roland Siegwart

Abstract This paper investigates and demonstrates the potential for very long endurance autonomous aerial sensing and mapping applications with AtlantikSolar, a small-sized, hand-launchable, solar-powered fixed-wing unmanned aerial vehicle. The platform design as well as the on-board state estimation, control and path-planning algorithms are overviewed. A versatile sensor payload integrating a multi-camera sensing system, extended on-board processing and high-bandwidth communication with the ground is developed. Extensive field experiments are provided including publicly demonstrated field-trials for search-and-rescue applications and long-term mapping applications. An endurance analysis shows that AtlantikSolar can provide full-daylight operation and a minimum flight endurance of 8 h throughout the whole year with its full multi-camera mapping payload. An open dataset with both raw and processed data is released and accompanies this paper contribution.

1 Introduction

The field of aerial robotics has seen rapid growth in the last decade. Prerequisite technologies have developed to the point that we are not far from the day when utilization of aerial robots is prevalent in our society. With an application range that includes infrastructure inspection [13], surveillance for security tasks [6], disaster relief [8, 25], crop monitoring [7], mapping [1], and more, Unmanned Aerial Vehicles (UAVs) already provide added value to several critical and financially significant applications, and are widely acknowledged for their potential to achieve a large impact in terms of development and growth. Examples of compelling existing use-cases include the mapping of the Colorado flood area in 2003 [4], the 3D reconstruction of the "Christ the Redeemer" statue in Brazil and the Matterhorn mountain reconstruction [20], and the live offshore flare inspection that took place in the North Sea [3].

P. Oettershagen (✉) · T. Stastny · T. Mantel · A. Melzer · K. Rudin · P. Gohl · G. Agamennoni · K. Alexis · R. Siegwart
Autonomous Systems Lab, ETH Zurich, Zurich, Switzerland
e-mail: philipp.oettershagen@mavt.ethz.ch

D.S. Wettergreen and T.D. Barfoot (eds.), *Field and Service Robotics*,
Springer Tracts in Advanced Robotics 113, DOI 10.1007/978-3-319-27702-8_29

While these are impressive achievements, there are still major factors that limit the applicability of UAVs. One such factor is their relatively low endurance. Indeed, long-endurance flight capabilities are crucial for applications such as large-scale Search-and-Rescue support, industrial pipeline monitoring, atmospheric research, offshore inspection, precision agriculture and wildlife monitoring. This new class of problems exposes a practical limitation in the majority of currently available aerial robot configurations.

Solar-powered flight is a key enabling technology for long-endurance operations. By harnessing the sun's energy and storing solar power during the day, flight times can be significantly prolonged. In cases of extreme designs, sustained flight can even be achieved through night time and/or cloudy conditions. An existing example of extreme endurance is the QinetiQ Zephyr UAV (22 m wingspan), which broke records, sustaining flight for 2 weeks [24]. However, scaling down from the high-altitude "pseudo satellite" class to more manageable, rapidly deployable and low-altitude designs is not trivial.

Motivated by the increasing industrial, scientific and societal demand for persistent automatic aerial sensing and surveillance, long-endurance, solar-powered fixed-wing aircrafts have been a research priority in the Autonomous Systems Lab (ASL) at ETH Zurich. With the most recent development being the AtlantikSolar UAV, our aim is to extend the current technological state of the art with a robust and versatile platform capable of significantly longer term sensing and mapping on the order of days or even weeks. Figure 1 depicts the AtlantikSolar UAV and its sensing capabilities. The detailed design of this UAV platform has been described in [18]. This paper extends our previous design-oriented work by investigating and characterizing possible application scenarios for our platform. More specifically, we present a set

Fig. 1 The AtlantikSolar UAV is capable of very long-endurance operation in missions including mapping, surveillance, victim detection and infrastructure inspection

of field trials that are enabled by a diverse sensor payload recently integrated into the UAV. This on-board sensor payload includes RGB and grayscale camera systems and a thermal vision sensor in combination with a complete suite of sensors that enable the vehicle to navigate autonomously.

The remainder of this paper is organized as follows: We present a description of the AtlantikSolar vehicle in Sect. 2, its sensing and mapping capabilities in Sect. 3, field experiment results in Sect. 4, and derived conclusions in Sect. 5. We also provide a detailed discussion of our experiences from both search-and-rescue as well as mapping missions, and release a dataset containing raw as well as post-processed data.

2 AtlantikSolar Unmanned Aerial Vehicle

2.1 Platform Overview

The AtlantikSolar UAV (Fig. 2, Table 1) is a small-sized, hand-launchable, low-altitude long-endurance (LALE), solar-powered UAV optimized for large-scale aerial mapping and inspection applications. A detailed overview of the conceptual design of AtlantikSolar is given in [18]. The design methodology is based on the work in [10, 16] with extensions on optimizing solar-powered UAVs for a range of deteriorated meteorological conditions (e.g. cloud obstruction of sun radiation) as given in [18]. The platform owes much of its configuration to the optimization of power consumption. Lightweight composite materials are used in the fabrication of a torsionally resistant cylindrical carbon fibre spar, tapered carbon fibre tail boom, and fibreglass fuselage body. The AtlantikSolar prototype UAV used for the flight tests in this paper features 88 SunPower $E60$ cells with an efficiency of $\eta_{sm} = 0.23$. Energy is stored in 2.9 kg of cylindrical high energy density Li-Ion batteries (Panasonic NCR18650b,

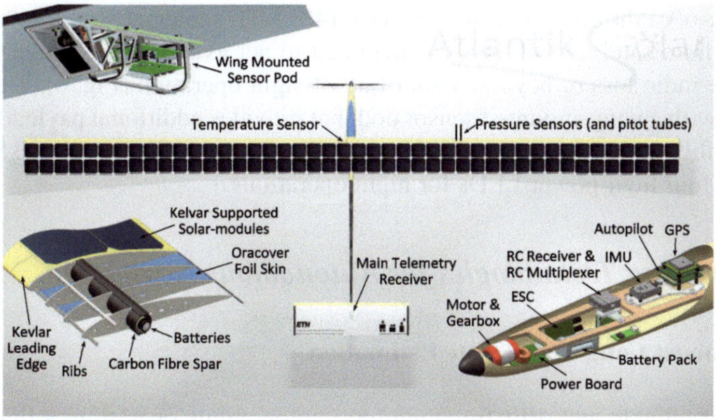

Fig. 2 AtlantikSolar system overview

Table 1 Summary of AtlantikSolar design and performance characteristics

Specification	Value/unit
Wing span	5.65 m
Mass	7.5 kg
Nominal cruise speed	9.7 m s^{-1}
Max. flight speed	20 m s^{-1}
Min. endurance (no payload)[a]	13 h
Design endurance (no payload)	10 days

[a]On battery-power only

243 Wh kg^{-1}, 700 Wh total) that are integrated into the wing spar for optimal weight distribution. The two ailerons, the elevator and the rudder are driven by brushless Volz DA-15N actuators with contactless position feedback. The propulsion system consists of a foldable custom-built carbon-fibre propeller, a 5:1 reduction gearbox and a 450 W brushless DC motor.

A Pixhawk PX4 Autopilot, an open source/open hardware project started at ETH Zurich [21], is the centerpiece of the avionics system. It employs a Cortex M4F microprocessor running at 168 MHz with 192 kB RAM to perform autonomous flight control and state estimation. Major hardware modifications include the integration of the ADIS16448 IMU and the Sensirion SDP600 differential pressure as well as re-writing of the estimation and control algorithms.

2.2 Operational Concept

AtlantikSolar is hand-launched to enable rapid deployment and operation in remote or uneven terrain. It is operated by a two-person team consisting of the safety-pilot and an operator for high-level mission management through the ground control station (GCS) interface (QGroundControl [23]). The GCS allows automatic loitering and autonomous waypoint following of user-defined or pre-computed paths. For visual-line-of-sight operation, the primary (434 MHz) telemetry link is sufficient, but an Iridium satellite link is also integrated to act as a backup link in the event of primary radio loss or beyond-visual-line-of-sight operation (Fig. 3). The UAV is equipped with a wing-mounted sensor pod, but provides additional payload capacity and versatility within its total payload budget of $m_{pld,max} \approx 800$ g. AtlantikSolar also integrates four high-power LEDs for night operations.

2.3 Enabling Technologies for Autonomous Navigation

2.3.1 Robust Long-Term State Estimation

To provide reliable and drift-free long-term autonomous operation, a light-weight EKF-based state estimator, as presented in [11], is implemented on the autopilot.

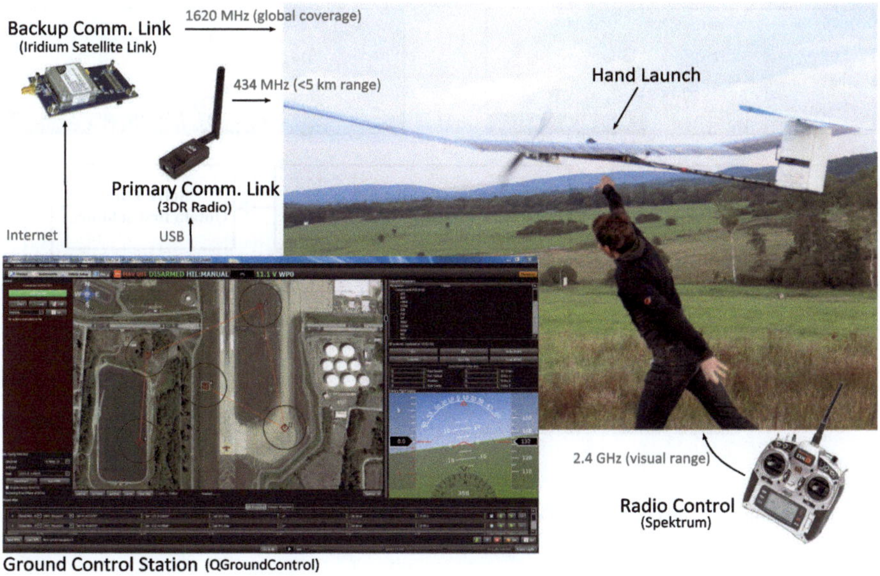

Fig. 3 Communications and ground control

It fuses data from a 10-DoF Inertial Measurement Unit (IMU) with GPS-Position, GPS-velocity and airspeed measurements in order to successively estimate position, velocity, orientation (attitude and heading), QFF as well as accelerometer and gyro biases. Robustness against temporal GPS losses is enhanced through the inclusion of airspeed measurements from a differential barometer. To increase flight safety, the algorithm estimates the local three-dimensional wind vector and employs an internal aircraft aerodynamics model to estimate the current sideslip angle and Angle of Attack (AoA), which can in turn be used by the flight controller to apply implicit flight regime limits, as in the case of the authors' previous work [17].

2.3.2 Flight Control

AtlantikSolar's flight control system features automatic tracking of waypoints along pre-defined paths, allows extended loitering around areas of interest and implements safety-mechanisms such as automatic Return-To-Launch (RTL) in case of prolonged remote control or telemetry signal losses. The baseline control is a set of cascaded PID-controllers for inner-loop attitude control [2]. Output limiters are applied to respect the aircraft flight envelope, dynamic pressure scaling of the control outputs is used to adapt to the changing moment generation as a function of airspeed and a coordinated-turn controller allows precise turning. Altitude control is based on a Total Energy Control System that also allows potential energy gains in thermal updraft while it implements safety mechanisms such as automatic spoiler deploy-

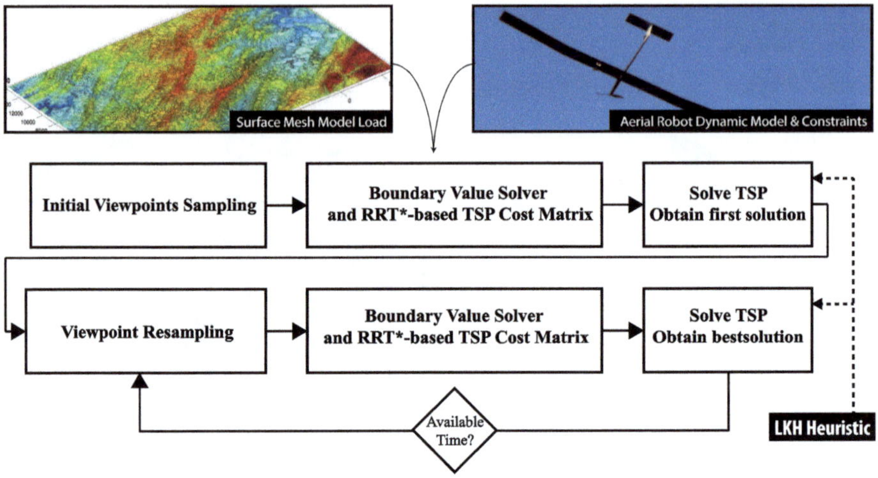

Fig. 4 Summary of the employed 3D inspection path-planning algorithm

ment during violation of maximum altitude limits. Waypoint-following is performed using an extended version of the $\mathcal{L}_1$-nonlinear guidance logic [19]. The detailed implementation and verification of our autopilot is described in [18].

2.3.3 Inspection Path-Planner

An inspection path-planning algorithm is integrated into the system in order to enable automated inspection and mapping of large scale 3D environments. The algorithm is inherently tailored for structural inspection and computes full coverage and collision-free paths subject to a model of the nonholonomic constraints of the vehicle. The overall approach is illustrated in Fig. 4, while a detailed description is available in the authors' previous work [1]. It essentially corresponds to an explicit algorithm that computes an inspection path based on a mesh-model representation of the desired world. It iteratively tries to compute viewpoint configurations that provide full coverage while at the same time employing the Lin-Kernighan heuristic [12] in the search for the best route that visits all of them subject to the motion constraints of the vehicle. Via a viewpoint resampling technique that employs randomized sampling, the designed algorithm allows for an iterative improvement of the path cost while always retaining complete coverage. Fast collision-free navigation is achieved via a combination of a Boundary Value Solver for the considered vehicle model with the RRT* [9] motion planner.

Fig. 5 The sensor pod as it is currently used on the AtlantikSolar UAV. The pod's fairing has been omitted for better visibility of the components

3 Sensor Pod

The sensor pod (see Fig. 5) features a grayscale (Aptina MT9V034) camera with a high dynamic range and a long-wavelength infrared (LWIR) camera (FLIR Tau 2) for thermal imaging, both mounted with an oblique field of view (FOV), as well as a nadir facing RGB camera (uEye XS 2). An IMU (Analog Devices ADIS16448) is also included, measuring linear accelerations, angular velocities, and the magnetic field in all three axes. All sensors are integrated with a Skybotix VI-sensor [27], allowing tight hardware synchronization and timestamping of the acquired data [15]. Furthermore, a state of the art embedded computer (Kontron COMe-mBT10), with an Intel Atom CPU (4 cores, 1.91 GHz) and a thermal design power (TDP) of 10 W, is interfaced with the VI-sensor and the PX4 autopilot board of the UAV. The on-board Atom computer further communicates with the PX4 in order to receive all global pose estimates and raw sensor data and transmit waypoints. The acquired data is processed on-board and communication with the ground control station is achieved over Wi-Fi. As shown in Fig. 5, all components are mounted on a lightweight aluminum construction ensuring a rigid connection between the cameras and the IMU, thus guaranteeing high quality extrinsic calibration of the sensors, a key element for accurate visual-inertial localization.

The on-board computer runs a standard Ubuntu Linux operating system, allowing quick adaptation to different kinds of missions. Furthermore, it enables rapid testing of new algorithms, e.g. for localization and mapping. It has been utilized to evaluate monocular localization [10] while the original stereo version of the VI-sensor is actively used for localization of rotary-wing UAVs in possibly GPS-denied environments [14]. Within the framework of the research projects ICARUS and SHERPA [8, 26], the described sensor pod is used for area mapping, victim detection, and situational awareness tasks. The data of the visible light cameras is combined with the pose estimates and fed to post-processing software [20] to derive accurate 3D reconstructions of the environment. Active research is ongoing for aerial victim detection at altitudes on the order of 100 m.

4 Flight Experiments

AtlantikSolar is a key component of several research projects and has actively participated in multiple large-scale demonstration events. Within this paper, indicative results from the ICARUS project [8] public field-trials event at Marche-en-Famenne, Belgium and a long-endurance mapping mission in Rothenthurm, Switzerland are presented along with flight endurance related tests and evaluations. A dataset is also released and documented to accompany this paper. It contains the vehicle state estimates, IMU and GPS raw data, the camera frames from all the on-board modules as well as post-processed reconstructions of the environment for the field-trials described in Sect. 4.2. This rich dataset is publicly available at [5].

4.1 Search-and-Rescue Application Demonstration

During the ICARUS project field-trials in Marche-en-Famenne [8], the AtlantikSolar UAV was commanded to autonomously execute inspection paths that ensured the complete coverage of a predefined area in order to assist the area monitoring, mapping, victim detection and situational awareness necessities of Search-and-Rescue rapid response teams. Employing the path-planner overviewed in Sect. 2.3.3 and based on the long-endurance capabilities of the UAV, the area was scanned repeatedly over multiple hours. An example inspection path is depicted in Fig. 6 and corresponds to an optimized solution for the case of the oblique-view mounted thermal camera, FOV (56°, 60°) for the horizontal and vertical axes, respectively. The mounting orientations of the grayscale and the thermal camera are identical, but the FOV of the grayscale camera is larger in all directions (70°, 100°), thus the planned path provides full coverage for both vision sensors.

During the execution of these inspection paths, the two camera-system and the pose estimates of the aircraft were uniformly timestamped and recorded in a ROS bag. Subsequently, post-processing of the grayscale images was conducted in order to derive a dense point-cloud of the area using the Pix4D software [20]. An image of the derived result is shown in Fig. 7, while additional results of autonomously executed inspection paths may be found in our previous work [1].

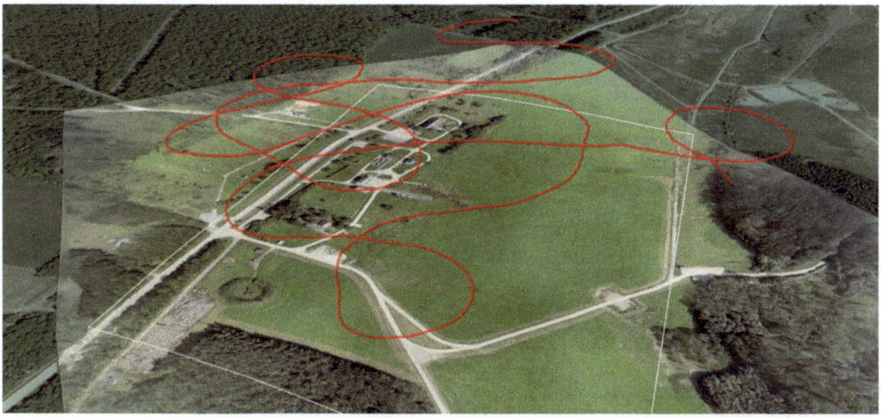

Fig. 6 Inspection path full area coverage using the oblique-view mounted thermal vision and grayscale cameras of the AtlantikSolar sensor pod. The *colored* mosaic was derived using an additional very large field of view nadir-facing camera (HDR-AS100VW)

Fig. 7 Reconstructed dense point cloud based on the combination of the oblique-view grayscale camera images with the vehicle position estimates. The reconstruction was achieved using the Pix4D mapping software

4.2 Area Coverage Application Demonstration

In this specific field experiment, the AtlantikSolar UAV's capabilities for long-term area coverage, inspection and mapping were evaluated. Within 6 h of flight, the system performed multiple lawn-mowing and other paths like those presented in Fig. 8. With a camera frame recording rate set at $F_c = 1$ Hz, synchronization with the vehicle pose estimates and properly designed waypoint distances to ensure coverage and sufficient overlap for all cameras, a solid reconstruction result was achieved. Within this flight,

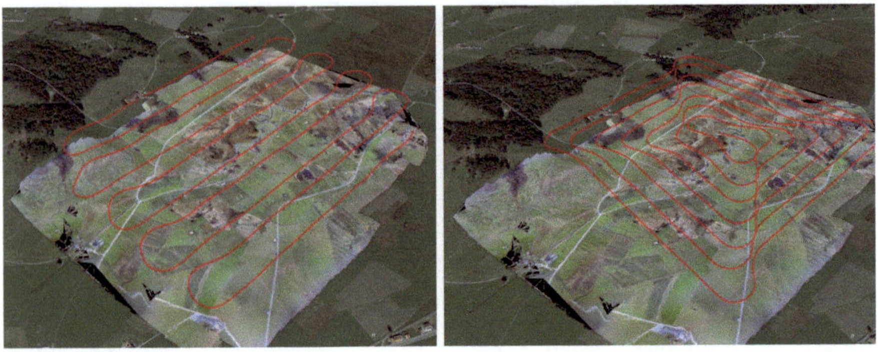

Fig. 8 The lawn-mowing path executed by the AtlantikSolar UAV overlayed on the reconstructed mosaic of the environment, incorporated in Google maps

Fig. 9 The reconstructed point-cloud of the Rothenthurm area based on the combination of the RGB and grayscale camera data as well as the UAV pose estimates collected during the lawn-mowing path and subsequently processed using the Pix4D software

all three cameras were employed and Fig. 9 depicts the reconstructed point cloud using a combination of the geo-tagged nadir-facing RGB camera of the sensor pod with the, likewise, geo-tagged oblique-view grayscale images, while Fig. 10 shows false-colored thermal images that our team is currently aiming to employ for victim detection, extending previous work [22] at ASL. An open dataset containing 1 h of raw data and post-processed results is released to accompany this paper and may be found at [5].

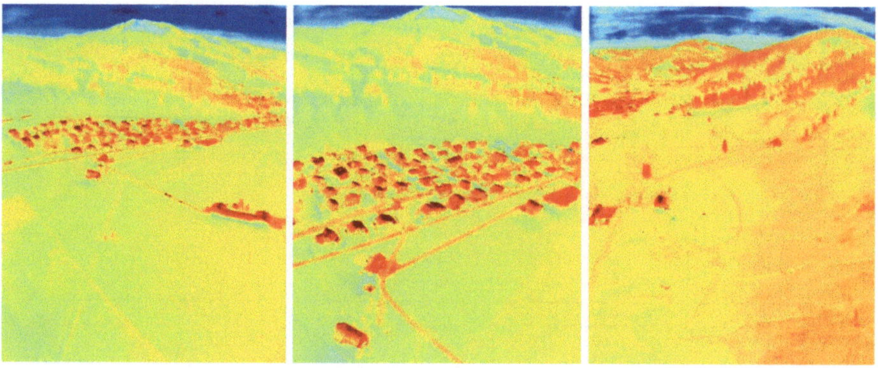

Fig. 10 False-colored thermal camera images recorded using the on-board sensor pod of the AtlantikSolar UAV

4.3 Full-Payload Flight Endurance and Range

After having shown a flight endurance of more than 12 h without payload in Summer 2014 [18], the area coverage demonstration in Rothenthurm on November 21st was used to determine AtlantikSolar's maximum flight endurance with the full sensor pod payload of $m_{payload} = 610$ g during winter conditions. Figure 11 shows the corresponding power income, output, and battery state. The average power consumption during the flight is $P_{aircraft} = 69.7$ W plus $P_{payload} = 15$ W for the sensor pod. After take-off at 10:25 a.m. local time at 94 % battery state-of-charge (SoC), the heavily attitude-dependent solar power income increases but reaches only a maximum of 80 W at noon due to the limited insolation in winter. Nevertheless, as indicated by the SoC, the system power is mostly drawn from the solar panels for more than 3 h of the flight. Power income decreases towards the afternoon: The solar panel maximum power point trackers (MPPTs) are still operating, but the panel voltage has decreased significantly and the MPPTs deliver currents below the measurement threshold. However, the remaining SoC during landing shortly before sunset (4:28 p.m. local time) is still 52 %. Extrapolating using the total power consumption of $P_{tot} = P_{aircraft} + P_{payload} = 69.7\,\text{W} + 15\,\text{W} = 84.7\,\text{W}$ yields an additional 4.32 h of remaining flight endurance assuming zero-radiation conditions and thus a total flight endurance of ca. 10 h with full payload for the installed 700 Wh battery during these winter conditions.

The recorded power consumption of $P_{tot} = 84.7$ W was taken as the input for the flight endurance simulation in Fig. 12. Assuming launch of the airplane exactly at sunrise, full-daylight flight endurance is provided throughout the full year including winter under most atmospheric conditions. More specifically, full-daylight flight capability is only lost when $CLR = P_{Solar}/P_{Solar,ClearSky}$ is smaller than ca. 0.3 in summer and ca. 0.15 in winter, which corresponds to severe cloud coverage or fog that may hinder flight operations independently of energy considerations. The maximum endurance of AtlantikSolar with the full payload is 22.4 h on June 21st, which

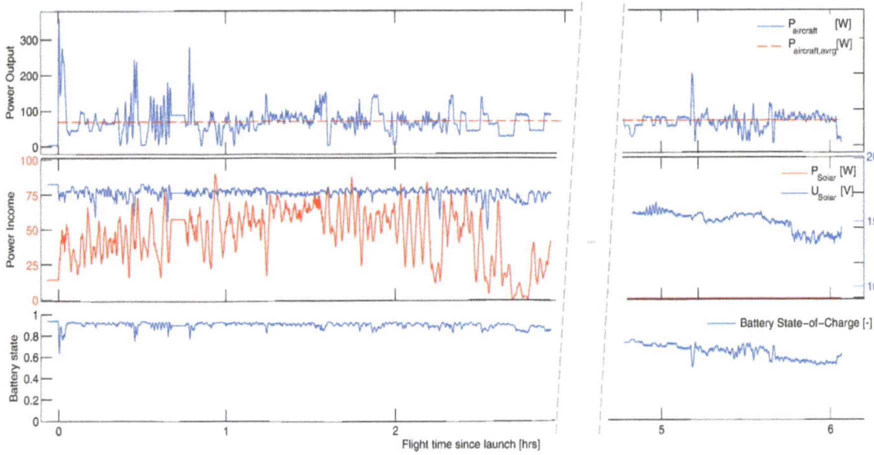

Fig. 11 Power income, output and battery state-of-charge for the Rothenthurm mapping flight. AtlantikSolar covered 243 km ground-distance, was airborne for 6 h 3 min and landed shortly before sunset (4:28 p.m. local time) with 52 % battery capacity remaining

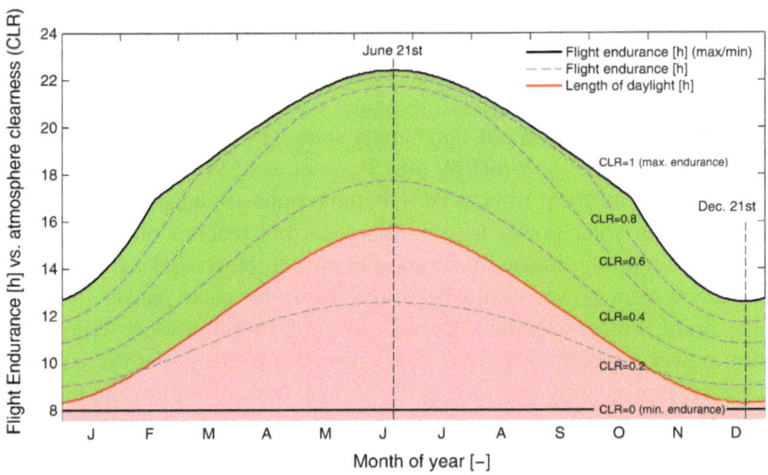

Fig. 12 AtlantikSolar's flight endurance with 610 g/15 W payload versus atmospheric clearness (CLR) for $\phi = 47°$ N (Rothenthurm, CH) when assuming launch at sunrise with SoC = 100 %. Full daylight flight is possible throughout the whole year (*green area*), and only severe cloud coverage can reduce endurance below the daylight duration (*red area*). In all cases 8 h of minimum endurance are achieved

means that perpetual flight is not possible. Note that in all atmospheric conditions, a minimum endurance of 8 h can be guaranteed through battery-powered flight alone. At the chosen airspeed of $v_{air} = 11.02$ m/s, AtlantikSolar can thus cover a ground distance of 317 km (min. endurance) to 888 km (max. endurance). Note that this airspeed provides the maximum range (optimal glide ratio), but is not the power-optimal airspeed (lowest rate of sink). Flying strictly at the power-optimal airspeed found in [18] would e.g.increase the endurance to 23.9 h on June 21st, with battery energy depleting shortly before sunrise. This means that perpetual flight with the full sensor payload can theoretically be achieved through minor aircraft optimizations, e.g. through a slight increase of the available battery capacity. However, note increasing endurance through power-optimal airspeed selection in the non-perpetual flight endurance case comes at a cost of range, and should be considered per application.

5 Conclusions

In this work, we have demonstrated a significant leap in long-endurance, low-altitude aerial sensing and mapping. Utilizing optimized solar aircraft design methodologies, low power consumption electronics, a robust autonomous navigation framework, and a versatile, modular, and self-contained sensor payload, the AtlantikSolar system, as a whole, provides a baseline to address quickly approaching societal needs related to long-term aerial robotic operations. Extensive field-trial experience indicates that solar power is a promising solution towards providing long endurance to small-sized, low-altitude UAVs, and integrated sensor suites, when used in tandem with autonomous navigation and planning methods, can provide wealth of valuable information to end users in an efficient manner. Still, there is great room for improvement, especially in the directions of autonomous navigation close to terrain, where a combination of advanced perception and planning algorithms have to be employed. Also in terms of superior robustness, as required for multi-hour or even multi-day flight.

Acknowledgments This work was supported by the European Commission projects ICARUS (#285417) and SHERPA (#600958) under the 7th Framework Programme. Further information at http://www.atlantiksolar.ethz.ch/.

References

1. Bircher, A., Alexis, K., Burri, M., Oettershagen, P., Omari, S., Mantel, T., Siegwart, R.: Structural inspection path planning via iterative viewpoint resampling with application to aerial robotics. In: 2014 IEEE International Conference on Robotics and Automation (ICRA) (2015) (accepted)
2. Brian, L.S., Frank, L.L.: Aircraft Control and Simulation. Wiley Interscience (1992)
3. Cyberhawk—Remote aerial inspection and land surveying specialists (2015). http://www.thecyberhawk.com/
4. Falcon UAV (2015). http://www.falconunmanned.com/
5. FSR 2015—Solar-powered UAV Sensing and Mapping Dataset (2015). http://projects.asl.ethz.ch/datasets/doku.php?id=fsr2015
6. Girard, A., Howell, A., Hedrick, J.: Border patrol and surveillance missions using multiple unmanned air vehicles. In: 43rd IEEE Conference on Decision and Control, 2004. CDC (2004)

7. Hunt, E.R., Hively, W.D., Fujikawa, S.J., Linden, D.S., Daughtry, C.S.T., McCarty, G.W.: Acquisition of nir-green-blue digital photographs from unmanned aircraft for crop monitoring. Remote Sens. **2**(1), 290–305 (2010)
8. ICARUS: Unmanned Search and Rescue (2015). http://www.fp7-icarus.eu/
9. Karaman, S., Frazzoli, E.: Incremental sampling-based algorithms for optimal motion planning. CoRR (2010). http://www.abs/1005.0416
10. Leutenegger, S.: Unmanned solar airplanes: Design and algorithms for efficient and robust autonomous operation. PhD thesis, ETH Zurich (2014)
11. Leutenegger, S., Melzer, A., Alexis, K., Siegwart, R.: Robust state estimation for small unmanned airplanes. In: IEEE Multi-conference on Systems and Control (2014)
12. Lin, S., Kernighan, B.W.: An effective heuristic algorithm for the traveling-salesman problem. Oper. Res. **21**(2), 498–516 (1973)
13. Metni, N., Hamel, T.: A UAV for bridge inspection: visual servoing control law with orientation limits. Autom. Constr. **17**(1), 3–10 (2007)
14. Nikolic, J., Burri, M., Rehder, J., Leutenegger, S., Huerzeler, C., Siegwart, R.: A UAV System for Inspection of Industrial Facilities. In: IEEE Aerospace Conference (2013)
15. Nikolic, J., Rehder, J., Burri, M., Gohl, P., Leutenegger, S., Furgale, P.T., Siegwart, R.Y.: A Synchronized Visual-Inertial Sensor System with FPGA Pre-Processing for Accurate Real-Time SLAM. In: IEEE International Conference on Robotics and Automation (ICRA) (2014)
16. Noth, A.: Design of solar powered airplanes for continuous flight. PhD thesis, ETH Zurich (2008)
17. Oettershagen, P., Melzer, A., Leutenegger, S., Alexis, K., Siegwart, R.: Explicit Model Predictive Control and $\mathcal{L}_1$-Navigation Strategies for Fixed-Wing UAV Path Tracking. In: 22nd Mediterranean Conference on Control & Automation (MED) (2014)
18. Oettershagen, P., Melzer, A., Mantel, T., Rudin, K., Lotz, R., Siebenmann, D., Leutenegger, S., Alexis, K., Siegwart, R.: A Solar-Powered Hand-Launchable UAV for Low-Altitude Multi-Day Continuous Flight. In: IEEE International Conference on Robotics and Automation (ICRA) (2015)
19. Park, S., Deyst, J., How, J.P.: A new nonlinear guidance logic for trajectory tracking. In: AIAA Guidance, Navigation, and Control Conference and Exhibit, pp. 16–19 (2004)
20. Pix4D (2015). http://pix4d.com/
21. Pixhawk Autopilot (2015). http://pixhawk.org/
22. Portmann, J., Lynen, S., Chli, M., Siegwart, R.: People detection and tracking from aerial thermal views. In: IEEE International Conference on Robotics and Automation (ICRA), pp. 1794–1800 (2014)
23. QGroundControl (2015). http://www.qgroundcontrol.org/
24. QinetiQ: QinetiQ files for three world records for its Zephyr Solar powered UAV. QinetiQ Press Release (2010). http://www.qinetiq.com/media/news/releases/Pages/three-world-records.aspx
25. Rudol, P., Doherty, P.: Human body detection and geolocalization for uav search and rescue missions using color and thermal imagery. In: Aerospace Conference, 2008 IEEE, pp 1–8 (2008)
26. SHERPA Project (2015). http://www.sherpa-project.eu/
27. Skybotix AG (2015). http://www.skybotix.com/

Aerial Vehicle Path Planning for Monitoring Wildfire Frontiers

Ryan C. Skeele and Geoffrey A. Hollinger

Abstract This paper explores the use of unmanned aerial vehicles (UAVs) in wildfire monitoring. To begin establishing effective methods for autonomous monitoring, a simulation (FLAME) is developed for algorithm testing. To simulate a wildfire, the well established FARSITE fire simulator is used to generate realistic fire behavior models. FARSITE is a wildfire simulator that is used in the field by Incident Commanders (IC's) to predict the spread of the fire using topography, weather, wind, moisture, and fuel data. The data obtained from FARSITE is imported into FLAME and parsed into a dynamic frontier used for testing hotspot monitoring algorithms. In this paper, points of interest along the frontier are established as points with a fireline intensity (British-Thermal-Unit/feet/second) above a set threshold. These interest points are refined into hotspots using the Mini-Batch K-means Clustering technique. A distance threshold differentiates moving hotspot centers and newly developed hotspots. The proposed algorithm is compared to a baseline for minimizing the sum of the max time untracked $J(t)$. The results show that simply circling the fire performs poorly (baseline), while a weighted-greedy metric (proposed) performs significantly better. The algorithm was then run on a UAV to demonstrate the feasibility of real world implementation.

1 Introduction

Recent developments in sensing technology have made possible low cost, reliable unmanned aerial vehicles (UAVs). These field robots are being implemented in various application domains, but specifically show promise in applications hazardous

This work was supported in part by NASA grant NNX14AI10G and ONR grant N00014-14-1-0509.

R.C. Skeele (✉) · G.A. Hollinger
Robot Decision Making Laboratory, School of Mechanical, Industrial & Manufacturing Engineering, Oregon State University, Corvallis, OR 97331, USA
e-mail: skeeler@onid.oregonstate.edu

G.A. Hollinger
e-mail: geoff.hollinger@oregonstate.edu

D.S. Wettergreen and T.D. Barfoot (eds.), *Field and Service Robotics*,
Springer Tracts in Advanced Robotics 113, DOI 10.1007/978-3-319-27702-8_30

for humans. Studying wildfires has an obvious benefit when considering the human cost spent combating them. One of the main issues in combating wildfires is monitoring the progression of the fire over time [13]. Live fire frontier monitoring can help produce quicker decisions and result in better resource allocation and fire management [12]. During wildfires, the information available to the Incident Commander (IC) is critical. Current methods of tracking a fire involve a human pilot flying several miles away from the fire and verbally reporting to the IC what trends they see in the fire. Satellite imaging is also available but is often rendered useless by smoke. In 2012, there were a total of 67,774 fires, destroying 9.3 million acres, and costing over $1.9 billion to suppress in the U.S. alone [6]. Large aircraft can negatively affect the fireline, for example if flown too low (below 1,000 ft), wake vortices from the windtips produce wind gusts which can cause torching and spotting [9].

This paper presents tests of different hotspot monitoring algorithms to gather important information for the Incident Commander (IC) managing the wildfire. This research aims to help improve a UAV's effectiveness in gathering valuable information for the IC. To simulate wildfires, a program developed by the Department of Agriculture and Forest Service is used. FARSITE is a free program used by the U.S. Forest Service, National Park Service, and more specifically ICs, to predict the fire's behavior using data on the topography, weather, wind, moisture, and fuel [8]. FARSITE exports various characteristics of the fire. While our simulation (FLAME) uses fireline intensity data (BTU/ft/s), other fire metrics like flame length and rate of spread also be incorporated. See Fig. 1 for a fireline intensity map of a simulated fire.

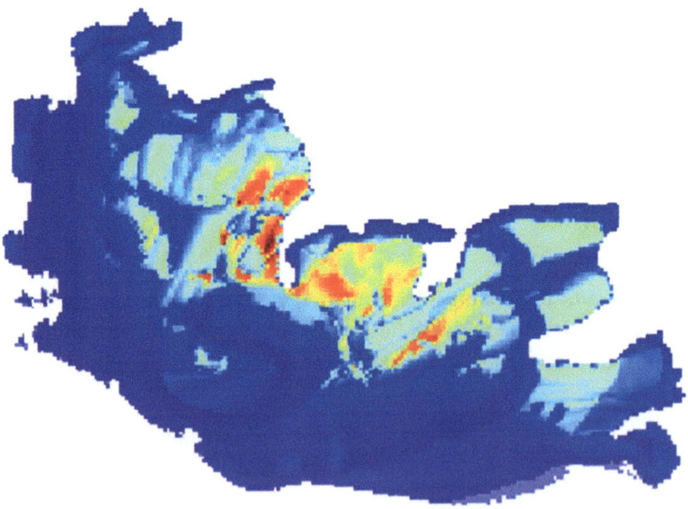

Fig. 1 Wildfire simulation example (*red areas* correspond to hotter areas of the fire). We propose a weighted-greedy algorithm for optimizing the monitoring trajectories of aerial vehicles in wildfire scenarios

Robotic monitoring has become a hot research topic in recent years, due to robots playing a more integral role in collecting environmental data. This has led to a variety of monitoring algorithms [4, 14, 18–20]. Wildfires are highly unpredictable, acting as a unique dynamic frontier. Dynamic monitoring has been explored [2, 18], but fire frontier monitoring is a largely unexplored domain. Our simulator (FLAME) models a dynamic fire frontier and uses techniques like Mini-Batch K-Means Clustering to achieve a similar problem formulation as related monitoring research.

Tracking the most volatile locations on the fireline will give valuable information for the IC. These hotspots will be intelligently monitored by the UAV, using algorithms for minimizing the time hotspots are left unmonitored.

While wildfires were the chosen domain, this research is not limited to wildfires. Similar application domains with dynamic frontiers include: algae blooms, pollution spills, and military battles [16, 22]. These similar domains can also be analyzed using the techniques developed in this paper.

The main novelties of this paper include: (1) a simulation (FLAME) which uses realistic fire modeling software for accurate fire characteristics, (2) a novel fire tracking algorithm which outperforms existing methods, (3) the first investigation into adaptive monitoring of hotspots along a dynamic frontier, and (4) hardware experiments demonstrating the ability to implement our work with existing technology. Taken together, these contributions provide a new approach to the general problem of monitoring dynamic frontiers.

The remainder of this paper is organized as follows. First, we will establish the current state of similar research (Sect. 2). Following that, we will overview the problem and assumptions we made during our investigation (Sect. 3). Next, we describe the simulation and a novel approach to frontier monitoring (Sect. 4). Finally, the algorithm is described in detail in (Sect. 5), and the simulation results are presented (Sect. 6), and the hardware experiments are discussed (Sect. 7).

2 Related Work

Work on autonomous information gathering began with early work in sequential hypothesis testing [23], which focused on determining which experiments could efficiently classify the characteristics of an unknown. This line of research developed into more general approaches and evolved into the field of active perception [1]. Similar insights led to using optimization techniques on robotic information gathering problems, and researchers later developed algorithms for minimizing long-term information uncertainty [4, 5].

Robotic systems are becoming more commonly used as mass data-gathering tools by scientists [7, 10, 18]. Robots are already collecting large datasets on environmental change. Algae blooms, pollution, and other climate variables are application domains for persistent monitoring techniques. Persistent/adaptive monitoring in robotics is currently a growing research topic. Prior work has explored different approaches to monitoring stationary and dynamic feature points. While these adaptive sampling techniques focus on optimizing uncertainty levels in static [4, 14, 19] and in dynamic environments [18, 20], prior work often focuses on systems in obstacle-free environments. Some research has examined collision avoidance [11, 21], but adaptive sampling along dynamic frontiers remains an ongoing research problem.

In [3], fire frontier tracking was integrated into a simulation for determining UAV tracking accuracy of the fire perimeter. Their UAVs follow a circular path around the fire similar to our baseline. However, our metric is to track the most active parts of the fire. We compare the baseline against our weighted-distance algorithm. Our research presents the first investigation into adaptive monitoring of hotspots along a dynamic frontier. We span the domains of hotspot monitoring and dynamic frontier tracking to evaluate path planning techniques in our FLAME simulator. This line of work allows us to test new algorithms in real-world scenarios.

3 Problem Formulation

We will now formally introduce the problem domain and the assumptions we made. We will also introduce the metric we use to evaluate our algorithm against the baseline.

We assume that GPS and communication between the IC and the vehicle are always available. This means the UAV can always localize itself and never needs to return to the starting location to transfer collected data. We assume the UAV always has the simulated fire frontier in order to find the hotspot locations. Additionally, the UAV is assumed, for comparison purposes, to have unlimited endurance.

Each hotspot has a corresponding time since last tracked by the UAV and the maximum time its been left untracked (ϕ) in the past. The sum of ϕ of all hotspots was chosen as the metric to evaluate the effectiveness of an algorithm. In this paper, fireline intensity is used as the crucial information needed by the IC. The intensity is monitored through the clustering into hotspots, directly relating to the goal of providing the IC with up-to-date information about the fire progression.

$$J(t) = \sum_{i=0}^{hotspots} \phi_i \,, \tag{1}$$

where ϕ is max time untracked.

The goal is to minimize the metric $J(t)$, which corresponds to timely hotspot monitoring, through an optimized trajectory for the UAV.

4 FLAME Simulation

We will now explain our simulation and how we developed each of the different components. Figure 2 should be used as a reference of the state transitions in the simulation. There are two aspects to the weighted-greedy algorithm, picking which hotspot to go to, and how to get there.

Fire data is generated using FARSITE, the wildfire simulator currently used by ICs during wildfire management [8]. The data is exported in the form of time of arrival, and a measurable characteristic of the fire. In this work we use the fireline intensity at each location. As stated above, the task is to minimize the sum of max time untracked (ϕ) over all hotspots. At mission start, the UAV must first find the fire and begin identifying the hotspot regions.

Tracking a hotspot is done by calculating the distance between a previous set of hotspots relative to a new set. To determine when a hotspot moved as the fire progressed, a threshold is implemented. If a hotspot is not within the distance threshold of any previous hotspots, it is then classified as a new hotspot. Even after careful tuning, this approach can still lead to some untracked hotspots where the hotspot existence is too short for any response by the UAV.

To identify hotspots, all points along the frontier with a fireline intensity above a normalized threshold are parsed using a clustering technique called Mini-Batch K-means [15]. K-means clustering was chosen because it directly relates the number of interest points (how active the fire is) to the number of cluster centers (hotspots). The desirable amount of clusters (K) changes as the fire evolves. We actively determine the K value for adaptive hotspot extraction with the following formula. With K as

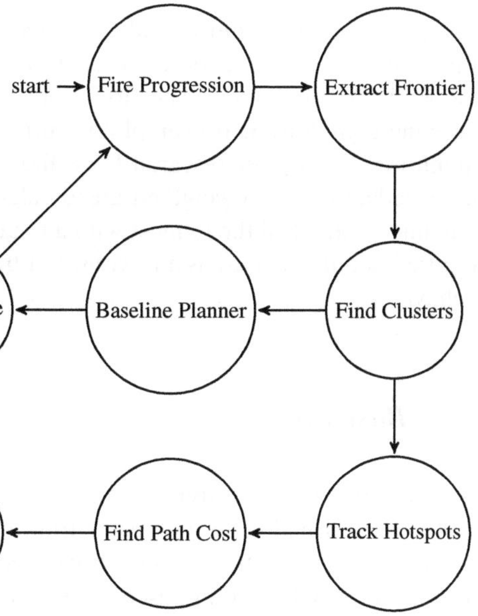

Fig. 2 State diagram of FLAME

the number of centers, and N as the number of interest points,

$$K = \sqrt{N/2} \tag{2}$$

FLAME uses A* path planning for generating paths from the UAV to hotspot locations around the fire. This method works better for estimating path cost over a simple Euclidean distance estimate due to the spherical tendency of the fire spread. Other similar methods where explored to increase efficiency, such as Jump Point Search. Jump Point Search gives respectable speed gains in environments with large open spaces, while the UAVs path remained mostly along the fire frontier. Methods like wall-following could provide faster simulation times, but lack expandability to more complex frontiers, and provide less accurate path costs. Due to the shape of the fire, any benefits of these alternatives were determined to be inconsequential. It was therefore determined to use the A* search algorithm as the UAVs path planner.

A cost map is passed to the A* algorithm, and is generated by applying a blur to the map of the fire up to that point in time and assigning a high cost to areas within the fire. This helps ensure the path generated for the UAV is not within dangerous proximity of the fire, but can still be navigated close enough to monitor the hotspots. The algorithms were tested over seven different fires generated in FARSITE. The baseline and proposed algorithm are described in pseudo code in Algorithms 1 and 2. A state diagram of FLAME is provided in Fig. 2. The algorithm state is weighted, but may be replaced with any tracking algorithm for testing.

5 Algorithms

The proposed algorithm is evaluated against a baseline in the following tests. The following sections will describe each algorithm and how it was implemented in the FLAME simulation. The first monitoring technique described is used as the baseline comparison. It exemplifies current tactics utilized in real world wild fire monitoring, and prior research UAV fire monitoring [3]. This is compared to our proposed approach, a weighted-greedy algorithm that moves to the hotspot that has remained untracked the longest with a tunable parameter of distance from the UAV. Figure 3 should be used as a reference of the difference between the two algorithms behaviors.

5.1 Baseline

Traveling parallel to the dynamic fire frontier is used as a baseline model. Calculating a 90° transformation of the vector from the UAVs current location to the nearest point on the fire frontier gives the travel vector of the UAV. Maximum and minimum distance thresholds are imposed on the UAV so it can then move along the frontier

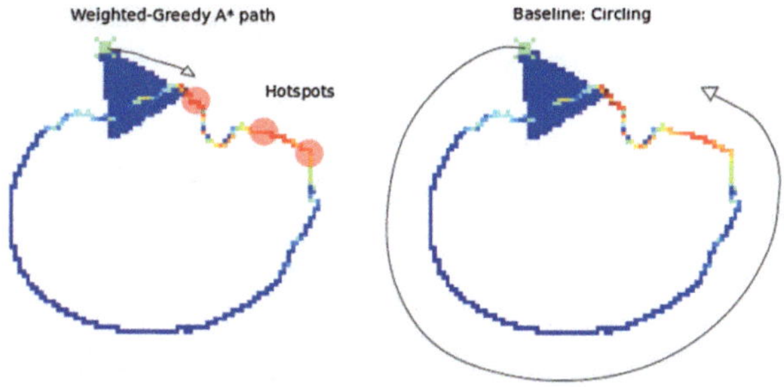

(a) (b)

Fig. 3 **a** UAV monitoring the fire using proposed algorithm identifies and tracks the most important part of the fire. **b** UAV monitoring the fire by constant circling will continue regardless of the state of the fire

monitoring hotspots while maintaining a safe distance from the fire. We use this as a baseline comparison based on the work of [3].

Algorithm 1 Baseline Algorithm

Inputs: UAV_Location, frontier
for all points in frontier **do**
    points.distance = $\sqrt{(\text{points.x - UAV\_location.x})^2 + (\text{points.y - UAV\_location.y})^2}$
end for
closest_point = min(points.distance)
vector_to_nearest = ([UAV_location.x - closest.point.x], [UAV_location.y - closest.point.y])
normalized_vector = vector_to_nearest / distance_to_nearest
if dist_to_nearest > max_distance_to_fire **then**
    travel_vector = vector_to_nearest
else if dist_to_nearest < min_distance_to_fire **then**
    travel_vector = -vector_to_nearest
else
    travel_vector = (-vector_to_nearest.x, vector_to_nearest.y)
end if
path = travel_vector

Algorithm 2 Weighted Algorithm

Inputs: hotspots{location, time_untracked}, α, UAV_location
for all h in hotspots **do**
    h.path, h.path_cost = ASTAR(h.location, UAV_location)
    h.score = h.time_untracked $-\ \alpha *$ path_cost(h)
    if hotspot.score > target_hotspot.score **then**
        target_hotspot = h
    end if
end for
path = target_hotspot.path

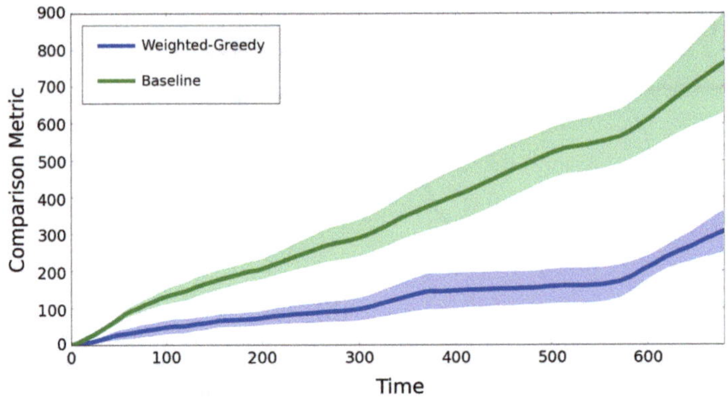

(a) Weighting parameter α = .5 Hotspot Threshold β = .35

Fig. 4 Wildfire simulation, where the comparison metric is J(t) or the sum of max time untracked of all hotspots. Lower is better. The weighting parameter α is set at 0.5. The normalized threshold β, for a spot along the fire to be considered an interest point, is set to 0.35. A lower β corresponds to more hotspot locations. Error bars are one SEM

5.2 Weighted-Greedy

The weighted-greedy algorithm checks the time untracked of every live hotspot, calculates the distance to it, and targets the one with the highest score. Unlike the baseline, the weighted-greedy algorithm makes target decisions based on the current state of the hotspots.

This is done using the following formula where $\mathcal{H}$ is the target hotspot, $\mathcal{T}$ is the time untracked of each hotspot and $\mathcal{C}$ is the path cost to each hotspot:

$$\mathcal{H} = \underset{h}{\operatorname{argmin}}\ \mathcal{T}_h - \alpha * \mathcal{C}_h \tag{3}$$

The proposed algorithm accounts for the distance to each hotspot when choosing the targeted hotspot. The weighting factor α is a parameter evaluated in Figs. 4, 5, and 6. The use of a weighting factor addresses some sub-optimality of using just a greedy algorithm. The weighting parameter helps intelligently pick a hotspot that may not be the longest untracked but is closer to the vehicle. A greedy algorithm will immediately move towards the hotspot with the longest time left untracked, disregarding any nearby hotspots that may not have been untracked for nearly quite as long.

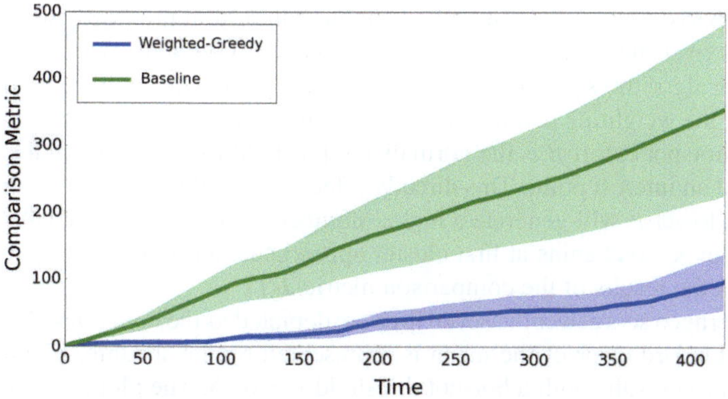

(a) Weighting parameter $\alpha = .5$ Hotspot Threshold $\beta = .25$

Fig. 5 Wildfire simulation, where the comparison metric is J(t) or the sum of max time untracked of all hotspots. Lower is better. The weighting parameter α is set at 0.5. The normalized threshold β, for a spot along the fire to be considered an interest point, is set to 0.25. A lower β corresponds to more hotspot locations. Error bars are one SEM

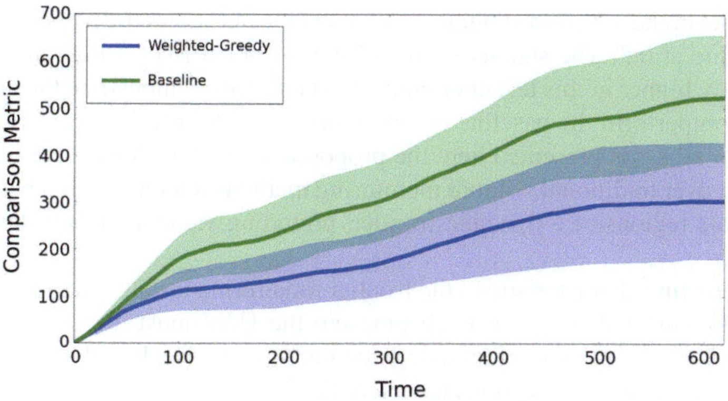

(a) Weighting parameter $\alpha = .5$ Hotspot Threshold $\beta = .45$

Fig. 6 Wildfire simulation, where the comparison metric is J(t) or the sum of max time untracked of all hotspots. Lower is better. The weighting parameter α is set at 0.5. The normalized threshold β, for a spot along the fire to be considered an interest point, is set to 0.45. A lower β corresponds to more hotspot locations. Error bars are one SEM

6 Results

Using our FLAME simulator, we can compare our proposed weighted-greedy approach with traditional methods of monitoring of wildfires. The simulation was run on an Intel i7-4702HQ processor with 8 gigabytes of RAM. The UAV's decision and planning methods took an average of 0.74 s to complete. This is fast enough for a UAV to implement in the field (see Sect. 7).

In comparison to the baseline, the weighted algorithm provided substantial improvement over the course of the trials. The plots in Figs. 4, 5, and 6 show the two monitoring algorithms performance with different parameter settings. As previously discussed, the weighting parameter (α) is multiplied by path cost to the hotspot location. The hotspot cutoff β is the normalized threshold for a spot along the fire to be considered an interest point. This directly affects the total number of hotspots. Tests ran with a lower β will generate a higher number of hotspots for the UAV to track. Time on the X axis begins at first cluster appearance during the simulation. The Y axis shows the results of the comparison metric $J(t)$.

The averaged score over the seven fires are depicted as the bold lines. Around each line, the standard error of the mean is represented by the shading. Figure 4 shows the simulation results with a hotspot threshold $\beta = 0.35$. The plot shows the results with a corresponding α value of 0.5. Our proposed algorithm performs significantly better than currently used approach.

In Fig. 5 the simulation is run with a β equal to 0.25. The β value (0.25) is the lowest used and Fig. 5 shows the performance of both algorithms in an environment with the corresponding large set of hotspots.

Figure 6 depicts the simulation results with a hotspot threshold β at 0.45. This trial uses the highest β (fewest number of hotspots), and shows the plots performance with α value at 0.5. The standard error of the mean (SEM) for both algorithms is significantly higher in this test environment. The results demonstrate the algorithms ability to outperform the baseline in environments with only few clusters, or many clusters. In all cases presented here the proposed algorithm showed significant improvement over traditional wildfire monitoring methods. Our algorithm better tracks the dynamic regions of a dynamic frontier, providing valuable data to better track the frontier.

An interesting characteristic of the frontier monitoring is that it may be simplified into a 1 dimensional problem. Each timestep the UAV must decide between two options, if it wishes to move clockwise or counter-clockwise. It will be worth further investigation into leveraging this characteristic.

7 Hardware Experiments

To demonstrate the feasibility of the proposed algorithm, we implemented the algorithm on a live test. To test on hardware we set up the FLAME simulation as a ground station that acted as live satellite data would for a real fire. The algorithm then sent a live stream of coordinates to a UAV to monitor the fire. While a real fire was not used for purpose of this test (for safety reasons), we are able to demonstrate that a UAV can effectively perform these tasks.

We converted FLAME into a ROS package to use the MAVROS plugins [17]. MAVROS acted as a communication bridge between FLAME and the flight controller on the UAV. This allowed us to update the UAVs path in time with the simulation. We used a tethered IRIS+ quadcopter as the platform for these experiments Fig. 7.

(a) A 3d view of our flight log in Google Earth showing the flight distance and trajectories.

(b) Experiment set up, computer running live simulation and IRIS+ flying autonomously (tethered).

Fig. 7 Experimental setup and flight log results

(a) The UAV first starts off a safe distance away from the fire and must travel to the frontier.

(b) Upon reaching the frontier and identifying a hotspot the UAV stays outside the burn area as it grows.

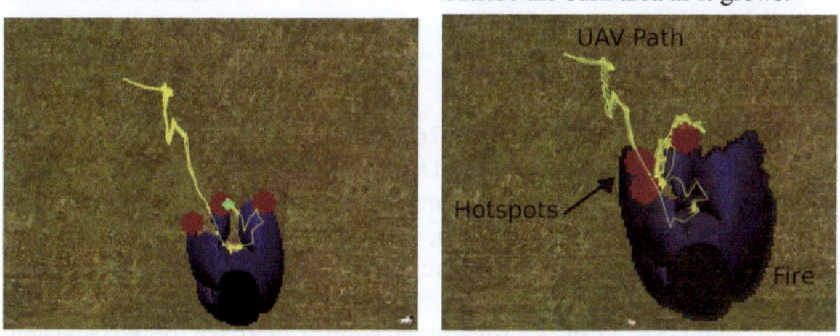

(c) The UAV moves from one hotspot to another to reduce the time untracked.

(d) Final fire size and flight log of our field experiments.

Fig. 8 Four images demonstrating the algorithm path planning during field tests

We ran the experiment for over 10 min, about half the max flight time of the vehicle. The experiment was performed outdoors in about a 60 ft × 60 ft area. The simulation coordinates were scaled and transformed to GPS degrees to support sending waypoints. We present the path of the vehicle around the fire in Fig. 8.

The UAV successfully followed the trajectories generated in the simulation to the best locations along the fire to monitor it as it spread. This illustrates our ability to begin introducing robotic monitoring into these dynamic monitoring situations and gather valuable data from it.

8 Conclusion

In this paper, we have introduced FLAME, a simulation developed for testing monitoring techniques on a dynamic frontier, or more specifically a wildfire. The two algorithms tested in the simulation have demonstrated that there is significant benefit in a weighted-greedy over the baseline method of flying around the fire frontier. Using Mini-Batch K-Means Clustering for identifying hotspots, our proposed weighted-greedy algorithm optimized for $J(t)$, the sum of max time untracked of all hotspots. Three different normalized hotspot thresholds (β) (0.25, 0.35, 0.45) were used. Data results showed the weighted-greedy algorithm with significant improvements over the baseline.

These algorithms depend on global knowledge of the fire, or more specifically where the hotspots are. Future work will include implementing a probabilistic model of hotspot locations and studying the exploration/exploitation trade-off for tracking and updating the model. In this paper we assume the UAVs have unlimited flight time. However, the cost of flight with limited endurance is an important factor. Additionally, hotspots are not all equal, and things such as risk to critical areas will need to be considered. Continuation of the project will also focus on implementation of multiple UAVs and the introduction of common fire monitoring challenges, including smoke and adverse weather conditions.

References

1. Bajcsy, R.: Active perception. Proc. IEEE **76**(8), 966–1005 (1988)
2. Bertozzi, A.L., Kemp, M., Marthaler, D.: Determining environmental boundaries: asynchronous communication and physical scales. In: Cooperative Control, pp. 25–42. Springer (2005)
3. Casbeer, D.W., Beard, R., McLain, T., Li, S.M., Mehra, R.K.: Forest fire monitoring with multiple small uavs. In: Proceedings of the American Control Conference 2005, pp. 3530–3535. IEEE (2005)
4. Cassandras, C., Ding, X.C., Lin, X.: An optimal control approach for the persistent monitoring problem. In: 2011 50th IEEE Conference on Decision and Control and European Control Conference (CDC-ECC), pp. 2907–2912 (2011)
5. Cassandras, C., Lin, X., Ding, X.: An optimal control approach to the multi-agent persistent monitoring problem. IEEE Trans. Autom. Control **58**(4), 947–961 (2013)
6. Center, N.I.F.: Federal fire fighting costs (2015). http://www.nifc.gov/fireInfo/fireInfo_documents/SuppCosts.pdf. Accessed 26 Sep 2014
7. Dunbabin, M., Marques, L.: Robots for environmental monitoring: significant advancements and applications. IEEE Robot. Autom. Mag. **19**(1), 24–39 (2012)
8. Farsite: Fire, fuel and smoke (2014). http://www.firelab.org/project/farsite. Accessed 27 Sep 2014
9. Group, N.W.C.: Wildland fire suppression tactics reference guide (1996). http://www.coloradofirecamp.com/suppression-tactics/suppression-tactics-guide.pdf. Accessed 26 Sep 2014
10. Hollinger, G., Choudhary, S., Qarabaqi, P., Murphy, C., Mitra, U., Sukhatme, G., Stojanovic, M., Singh, H., Hover, F.: Underwater data collection using robotic sensor networks. IEEE J. Sel. Areas Commun. **30**(5), 899–911 (2012)

11. Hollinger, G.A., Sukhatme, G.: Sampling-based motion planning for robotic information gathering. In: Robotics: Science and Systems (2013)
12. Koulas, C.E.: Extracting wildfire characteristics using hyperspectral, lidar, and thermal ir remote sensing systems. In: SPIE Defense, Security, and Sensing, pp. 72,983Q–72,983Q (2009)
13. Kremens, R., Seema, A., Fordham, A., Luisi, D., Nordgren, B., VanGorden, S., Vodacek, A.: Networked, autonomous field-deployable fire sensors. In: Proceedings of the International Wildland Fire Safety Summit (2001)
14. Lan, X., Schwager, M.: Planning periodic persistent monitoring trajectories for sensing robots in gaussian random fields. In: 2013 IEEE International Conference on Robotics and Automation (ICRA), pp. 2415–2420 (2013)
15. Lloyd, S.: Least squares quantization in pcm. IEEE Trans. Inf. Theory **28**(2), 129–137 (1982)
16. Marthaler, D., Bertozzi, A.L.: Tracking environmental level sets with autonomous vehicles. In: Recent developments in cooperative control and optimization, pp. 317–332. Springer (2004)
17. Quigley, M., Conley, K., Gerkey, B., Faust, J., Foote, T., Leibs, J., Wheeler, R., Ng, A.Y.: Ros: an open-source robot operating system. In: ICRA Workshop on Open Source Software, vol. 3, p. 5 (2009)
18. Smith, R.N., Schwager, M., Smith, S.L., Jones, B.H., Rus, D., Sukhatme, G.S.: Persistent ocean monitoring with underwater gliders: adapting sampling resolution. J. Field Robot. **28**(5), 714–741 (2011)
19. Smith, S.L., Rus, D.: Multi-robot monitoring in dynamic environments with guaranteed currency of observations. In: 2010 49th IEEE Conference on Decision and Control (CDC), IEEE, pp. 514–521 (2010)
20. Smith, S.L., Schwager, M., Rus, D.: Persistent robotic tasks: monitoring and sweeping in changing environments. IEEE Trans. Robot. **28**(2) (2012)
21. Soltero, D.E., Smith, S., Rus, D.: Collision avoidance for persistent monitoring in multi-robot systems with intersecting trajectories. In: 2011 IEEE/RSJ International Conference on Intelligent Robots and Systems (IROS), pp. 3645–3652 (2011)
22. Susca, S., Bullo, F., Martínez, S.: Monitoring environmental boundaries with a robotic sensor network. IEEE Trans. Control Syst. Technol. **16**(2), 288–296 (2008)
23. Wald, A.: Sequential tests of statistical hypotheses. Ann. Math. Stat. **16**(2), 117–186 (1945)

Part V
Underground

Multi-robot Mapping of Lava Tubes

X. Huang, J. Yang, M. Storrie-Lombardi, G. Lyzenga and C.M. Clark

Abstract Terrestrial planetary bodies such as Mars and the Moon are known to harbor volcanic terrain with enclosed lava tube conduits and caves. The shielding from cosmic radiation that they provide makes them a potentially hospitable habitat for life. This motivates the need to explore such lava tubes and assess their potential as locations for future human outposts. Such exploration will likely be conducted by autonomous mobile robots before humans, and this paper proposes a novel mechanism for constructing maps of lava tubes using a multi-robot platform. A key issue in mapping lava tubes is the presence of fine sand that can be found at the bottom of most tubes, as observed on earth. This fine sand makes robot odometry measurements highly prone to errors. To address this issue, this work leverages the ability of a multi-robot system to measure the relative motion of robots using laser range finders. Mounted on each robot is a 2D laser range finder attached to a servo to enable 3D scanning. The lead robot has an easily recognized target panel that allows the follower robot to measure both the relative distance and orientation between robots. First, these measurements are used to enable 2D (SLAM) of a lava tube. Second, the 3D range measurements are fused with the 2D maps via ICP algorithms to construct full 3D representations. This method of 3D mapping does not require odometry measurements or fine-scale environment features. It was validated in a building hallway system, demonstrating successful loop closure and mapping errors on the order of

X. Huang (✉) · J. Yang · M. Storrie-Lombardi · G. Lyzenga · C.M. Clark
Harvey Mudd College, Claremont, California, CA 91711, USA
e-mail: xhuang@hmc.edu

J. Yang
e-mail: jyang@hmc.edu

M. Storrie-Lombardi
e-mail: mstorrielombardi@hmc.edu

G. Lyzenga
e-mail: lyzenga@hmc.edu

C.M. Clark
e-mail: clark@hmc.edu

D.S. Wettergreen and T.D. Barfoot (eds.), *Field and Service Robotics*,
Springer Tracts in Advanced Robotics 113, DOI 10.1007/978-3-319-27702-8_31

0.63 m over a 79.64 m long loop. Error growth models were determined experimentally that indicate the robot localization errors grow at a rate of 20 mm per meter travelled, although this is also dependent on the relative orientation of robots localizing each other. Finally, the system was deployed in a lava tube located at Pisgah Crater in the Mojave Desert, CA. Data was collected to generate a full 3D map of the lava tube. Comparison with known measurements taken between two ends of the lava tube indicates the mapping errors were on the order of 1.03 m after the robot travelled 32 m.

1 Introduction

It is understood that within our solar system, Mars shares an environment similar in many respects to that of Earth, and it is possible that there might exist traces of life. The surface of Mars is relatively inhospitable and is constantly bombarded by cosmic radiation due to the thin atmosphere and lack of planetary magnetic field. Furthermore, the surface temperature ranges from 215 to 160 K from the equator to the poles. The temperature also fluctuates greatly within a day. Despite these harsh conditions, many scientists predict the existence of a saline groundwater system in the shallow subsurface of the planet, and therefore the subsurface may provide or may have provided a suitable environment for life. NASA's Astrobiology Roadmap objectives include investigating biosignatures in subsurface rocks, modeling subsurface habitable environments, and developing robotic drilling systems to access subsurface environments on Mars [11].

Lava tubes on Mars have gained considerable interest in the astrobiological community because they offer protection from the harsh conditions experienced on the planet's surface. There have been many attempts to characterize these lava tubes to determine the best sites for future exploration and to study the geomicrobiology in lava tubes. To achieve these goals remote-sensing techniques are required [11]. The lava tubes often have many openings, uneven terrain and variation in floor texture. Therefore, while radar instruments have already been used to drill to the subsurface to detect such characteristics, existing sensing methods often lack the resolution necessary to detect exact positions of interest in each individual lava tube.

These challenges motivate the goal of developing autonomous robots that can explore lava tubes and conduct in-situ scientific measurements. Such robots would need to construct 3D maps of the tubes to not only allow the robot to localize in-situ sample measurements with respect to a coordinate frame fixed to the tube, but also to enable the robot to localize itself with respect to the tube and carry out autonomous robot navigation.

Constructing 3D maps with robots has been well studied in the Simultaneous Localization and Mapping (SLAM) community. Many SLAM strategies have used a single robot that fuses odometry and range measurements via filtering algorithms to localize the robot and map the environment [1, 14]. While these methods are reliable, they are limited by the conditions of the exploration environment. The susceptibility

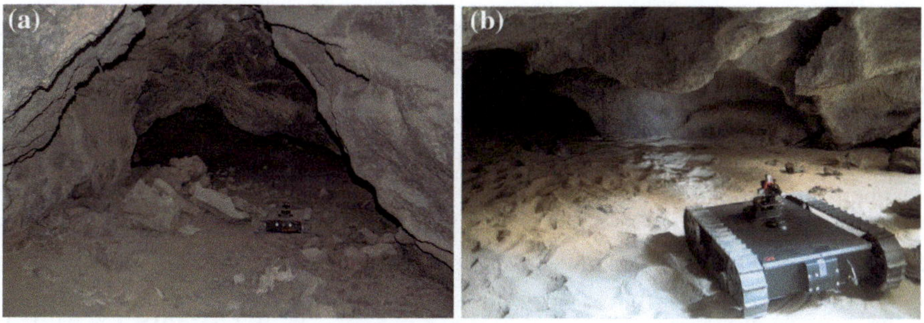

Fig. 1 Image of the Jaguar robot **a** at the entrance of a lava tube **b** on the sandy ground

of the encoder odometry measurements to error resulting from the fine sand found on the lava tube floor further challenges the SLAM problem Fig. 1.

Proposed here is a multi-robot mapping framework that allows robots to cooperatively map lava tubes which (a) have poor odometry measurements due to the fine sand of the tube floor, and (b) lack fine-scale features that reduce dead reckoning errors. Section 2 of this paper presents related work. A three-step solution called *Platoon SLAM* is proposed in Sect. 3, where in the first two steps range finder measurements of the relative distance and bearing-angle orientations between robots are used to update their positions, and in the last step these updated positions are used to seed ICP algorithm queries, that both localize the robot in 2D and construct maps in 3D. Implementation of these techniques are documented in Sect. 4, where results from hallway and lava tube mapping scenarios are presented. Finally, conclusions from these results are drawn in Sect. 5 and possible future work is proposed in Sect. 6.

2 Background

The problem of Simultaneous Localization and Mapping (SLAM) involves constructing a map of an unknown environment while localizing the position of the robot. SLAM is a maturing research area, with work most related to this project including advancements made in the sub-disciplines of 3D SLAM, ICP, 3D mapping in tube like structures, and multi-robot 3D SLAM.

A variety of approaches to 3D mapping in SLAM have been implemented that combine different localization and mapping techniques. Initially, 3D maps were built using multiple 2D scanners with different orientations to construct the 3D map. Thrun et al. [19] used measurements from two laser scanners, oriented perpendicular with respect to each other to form 3D point clouds. However other methods mentioned below give higher resolution of the generated 3D map, including visual SLAM using cameras or 3D range sensing methods are used in autonomous mapping [4, 9, 10, 14, 23, 24]. One popular 3D scanning method uses a pair of cameras with RGB-D cameras in 3D sensing [9]. This method is not well suited for the low lighting environments and low power requirements encountered during the exploration. The more common sensing method is to use 3D laser range finders. These laser range finders are commonly made by spinning the 2D laser scanner to obtain 3D data in the form of 3D point clouds [4, 14, 23, 24]. There are also several attempts to combine a 3D sensor with a 6D localization method. Nuchter et al. used a 3D scanner in combination with 6DOF IMU data to produce an error-minimized map [14]. Borrmann et al. [3] provides a detailed summary of current advancements in SLAM using 2D and 3D scanning mechanisms and explores 6D SLAM with scan matching.

There exist different techniques to register the point clouds into a 3D map, including 3D-FFT methods [12]. The registration currently used in this work, Iterative Closest Point (ICP), is one of the most common ways to register point clouds to represent maps in 3D space. Developed by Besyl et al. [2] and Chen et al. [6], it has been used in many occasions to register 3D maps [14, 16, 19]. There have also been findings on improvements for ICP in terms of processing, such as the 2D-NDT and 3D-NDT method [13], where the data is stored after computing in normal distributions. In addition, there are alternatives for ICP as described by Fischer et al. [8] and Pathak et al. [15] for pose registration which are not as commonly used.

Single robot 3D SLAM demonstrating successful loop closure in underground mine mapping started with Schedling [18]. These mines are similar to lava tubes in that they are long, winding, and without line-of-sight to GPS satellites. Huber et al. [10] used a high resolution 3D scanner on a cart to create an 3D map of an underground mine without additional sensors. Nuchter et al. also used multiple 3D SICK scanners in a stop-and-go method on robots to localize the robot and create a map of the environment through scan matching with ICP with point clouds [14]. Zlot et al. used an iterative matching algorithm to first construct an open-loop map of the mine tunnel, and then a closed loop model [24]. The method relies on pose measurement data and uses a global registration algorithm instead of landmark detection for localization.

Multi-robot systems offer increased spatio-temporal coverage which can be leveraged when exploring and mapping unknown environments [5, 7]. For example, Burgard's group had individual robots simultaneously explore different regions of an unknown environment. The work employed a probabilistic approach for the coordination of multiple robots to reduce the overall exploration time. An algorithm for multi-robot SLAM with sparse extended information filters was presented in Thrun's work [20]. The alignment of local maps into a single global maps was achieved by a tree-based algorithm that searches for similar-looking local landmark configurations. More relevant to this project is the work done by Rekleitis, where a pair of robots observe each other, and act in concert to reduce odometry errors [17]. However, this method relies on video camera observations, which is not suitable for underground lava tubes mapping.

3 Platoon SLAM

The goal of this work is to map the 3D environment of a lava tube using two robots equipped with 2D laser range finders. The lava tubes of interest are greater than 20 m in length, and range in height between 0.30 and 3.0 m. The tube walls are unpredictable, lacking sharp distinct corners. The tube floor consists of fine sand that causes encoder measurements to be highly unreliable due to slipping. Low light conditions in the mapping environment cause image processing techniques to require structured lighting that may increase payload weight and power consumption. Due to the shielding property of the lava tubes, no radiation communication such as GPS can be established between the robots in the tubes and the outside world. Therefore, a local-based SLAM solution is required.

Our core approach to this problem, called *Platoon SLAM*, uses two robots to navigate through the lava tube in a lead-follower formation. Each robot is equipped with a 2D laser range finder mounted on a servo to enable 3D range scanning. The lead robot will also have an easily observed target panel that can be detected by the follower robot's laser range finder. The primary role of the lead robot is to take 3D scans of the environment. The role of the follower robot is to measure the relative position and orientation changes of the robots as they traverse the length of the tube.

3.1 Platoon Actions

The two robots are tightly synchronized to repeat a sequence of 3 actions depicted in Fig. 2. In step 1, the lead robot moves forward a set distance and then the follower robot takes a stationary laser scan to detect the target panel on the lead robot. This scan measures the relative position of the lead robot. In step 2, the follower robot moves forward to a location just behind the lead robot. The follower robot again scans and detects the target panel to measure the relative position of the lead robot. In step 3, the lead robot takes a stationary 3D scan of the environment Fig. 3.

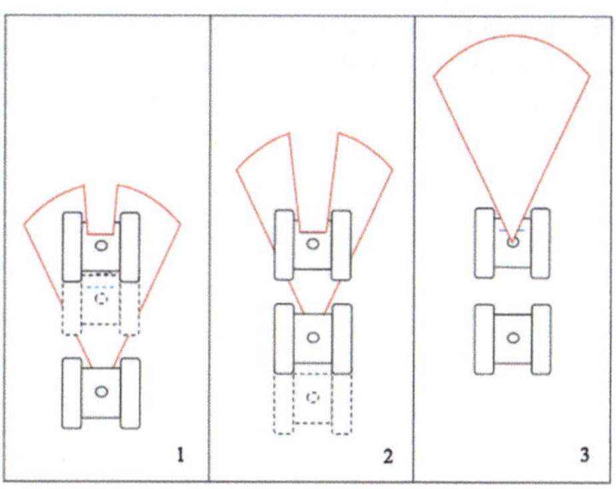

Fig. 2 Three step sequence: **1** lead robot moves and its state change is measured, **2** follower robot moves and its state change is measured, and **3** lead robot takes a 3D scan

Fig. 3 Image of a two-robot system. The robots are Dr. Robot Jaguar Lite platforms. The lead robot is equipped with a target panel

3.2 Robot Position Measurements

Steps 1 and 2 of the action sequence are used to obtain accurate measurements of the robots as they move forward to explore the lava tube. The follower robot obtains laser scans similar to that depicted in Fig. 4. The target panel is easily recognized in the center of this scan and is detected by an algorithm that searches for similar consecutive range measurements. The output of this algorithm is a series of range and bearing tuples $[\rho_i, \alpha_i]$ associated with reflections from the lead's target panel. Here ρ_i represents the relative distance between the two robots and α_i represents the relative bearing angle of the lead robot with respect to the follower robot, as shown in Fig. 5. Each $[\rho_i, \alpha_i]$ tuple is taken with respect to the follower robot's coordinate frame and can be converted to the relative position $[\Delta x_i, \Delta y_i]$ within this local frame.

Fig. 4 A robot-obtained laser scan taken from the lava tube

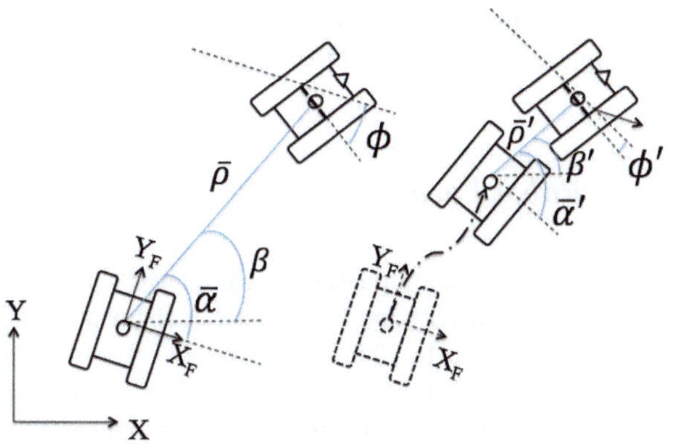

Fig. 5 Geometric representations for steps 1 and 2 of one sequence

The mean relative position $[\bar{\Delta x}, \bar{\Delta y}]$ can be calculated and used to determine a mean relative range and bearing $[\bar{\rho}, \bar{\alpha}]$ from the follower to the lead robot.

To calculate the yaw angle θ_L of the lead robot in the global frame, the difference in bearing angles between the two robots ϕ must first be extracted as the arctangent of the slope of the line fit to the $[\Delta x_i, \Delta y_i]$ tuples. Then, for the first step of the tth action sequence, the lead robot's state $[x_L\ y_L\ \theta_L]_t^T$ can be updated from the follower robot's previous state $[x_F\ y_F\ \theta_F]_{t-1}^T$:

$$\beta = \bar{\alpha} + \theta_F - \frac{\pi}{2} \tag{1}$$

$$\begin{bmatrix} x_L \\ y_L \\ \theta_L \end{bmatrix}_t = \begin{bmatrix} x_F \\ y_F \\ \theta_F \end{bmatrix}_{t-1} + \begin{bmatrix} \bar{\rho}\cos\beta \\ \bar{\rho}\sin\beta \\ \phi \end{bmatrix}_t \tag{2}$$

In Eqs. (1) and (2), β is the angle of the ray connecting the follower to the lead robot, as calculated with respect to the global coordinate frame. Figure 5 depicts the geometry of these calculations.

For the second step of the tth action sequence, the follower robot's state $[x_F\ y_F\ \theta_F]_t^T$ can be updated after its forward movement using its detection of the lead robot's target. In this case, the target data produces similar measurements to the first step, but we denote the second step measurements with $'$, i.e. $\bar{\rho}', \bar{\alpha}', \beta', \phi'$.

$$\beta' = \bar{\alpha}' + \theta_L - \frac{\pi}{2} \tag{3}$$

$$\begin{bmatrix} x_F \\ y_F \\ t_F \end{bmatrix}_t = \begin{bmatrix} x_L \\ y_L \\ t_L \end{bmatrix}_t - \begin{bmatrix} \bar{\rho}'\cos\beta' \\ \bar{\rho}'\sin\beta' \\ \phi' \end{bmatrix} \tag{4}$$

The proposed solution assumes the lead robot's target can always be detected by the follower robot. This can be achieved by ensuring the lead robot takes relatively small steps forward and by subsequently modifying the pitch angle of the follower's 2D laser range finder until the target is detected within a 2D scan.

3.3 Robot Localization

Once the robot state updates are calculated using inter-robot range and bearings as described in Eqs. (2) and (4), the robot states are further refined using environment range measurements. This refinement, or correction, is accomplished using a method called Iterative Closest Point (ICP). ICP attempts to find the relative transformation between two data sets. In this case, each data set corresponds to a single 3D scan taken by the lead robot during step 3. The scan consists of 3D points indicating the position of the lava tube contour with respect to the lead robot. Hence if the ICP algorithm is applied to two consecutive 3D scans taken by the lead robot, the algorithm will output a transformation that represents the lead robot's movement between the consecutive scans.

To initialize the ICP algorithm, an estimate of the transformation between lead robot scans is required. In this case, the relative movements calculated in Sect. 3.2, e.g. $x_{L,t} - x_{L,t-1}$, are used to initialize the ICP algorithm. To reduce the run time complexity, ICP is conducted only on the range data points that lie within some threshold of the horizontal plane that intersects with the robot sensor, as the elevation change between two consecutive scan positions is relatively small. To determine the horizontal plane, IMU data is used to calculate the roll and pitch angles of the robot relative to the initial pose of the robot to which the origin of the global coordinate frame is anchored.

The effect of running the 2D ICP implementation is illustrated in Fig. 6a, b, where the points clouds (blue) from two scans are plotted. The red and pink dots indicate the points determined to be within the 2D horizontal plane of two consecutive 3D scans. It is clear that running ICP to refine the position of the two 3D scans in Fig. 6a improves the alignment of the two subsequent scans in Fig. 6b, with pink dots and red dots overlapping.

3.4 Lava Tube Mapping

As described in the previous two sections, the first two steps in the 3 step sequence are used to estimate the lead robot's state at every 3D scan location with respect to a global coordinate frame. In last step, where the lead robot obtains a 3D scan of the environment, data is collected for constructing the 3D map of the lava tube. Each 3D scan produces a 3D point cloud that is added to the map to create a single global point cloud map representing the entire lava tube. After each scan, the positions of two robots are updated according to point registration results by ICP.

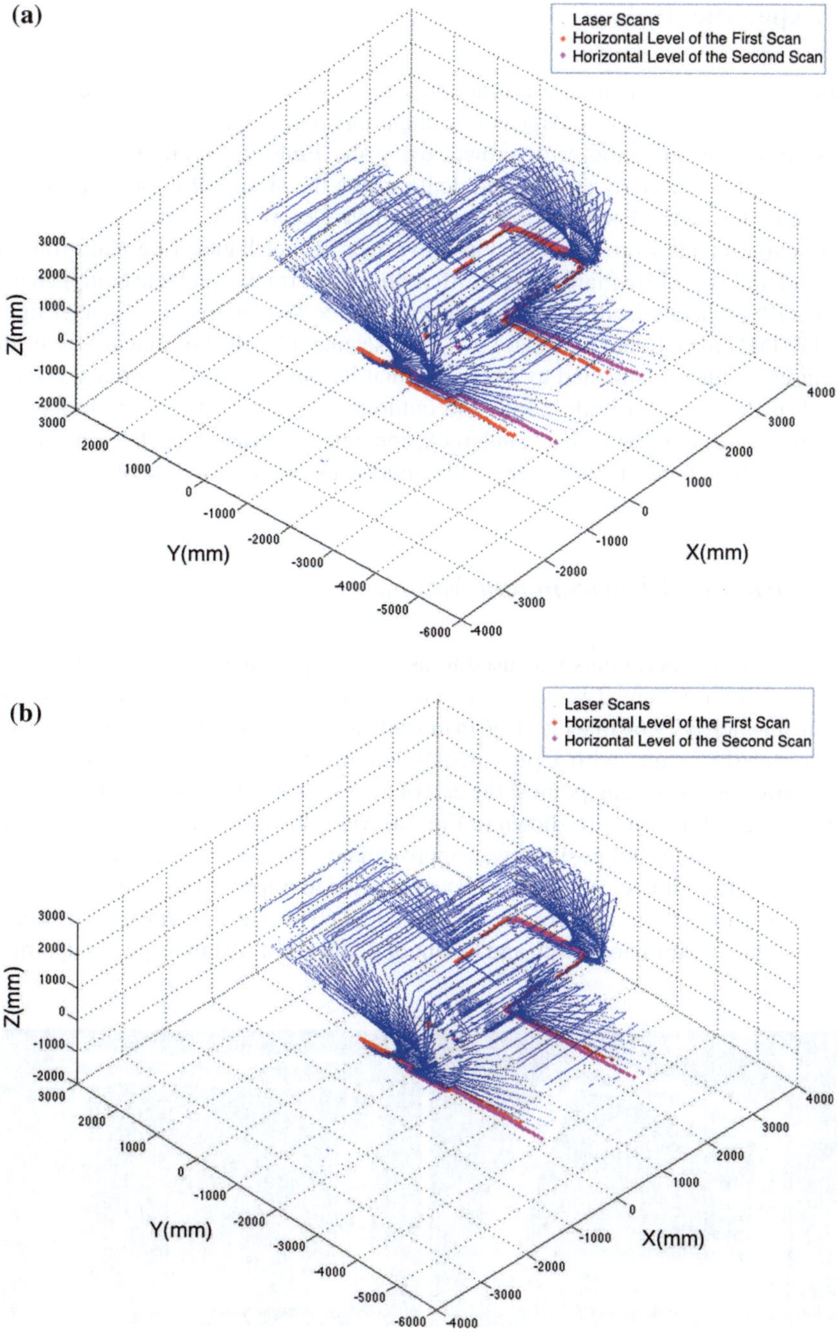

Fig. 6 Two consecutive point clouds **a** before registration **b** after registration

4 Experiments

In this section, experimental results are presented that validate the ability of *Platoon SLAM* to demonstrate loop closure while mapping a hallway system of known dimensions, allow for modeling error growth using the *Platoon SLAM* methodology in environments with sandy terrain, and demonstrate the ability of a robot pair to map a lava tube located at Pisgah Crater in the Mojave Desert, CA.

All experiments were conducted using two Dr. Robot Jaguar Lite platforms (see Fig. 3). The Jaguar Lite Platform is a differential drive tracked vehicle equipped with a 5 Hz GPS, wheel encoders, a color camera (640 × 480, 30 fps), two header lights, a 9DOF IMU from Razor and a Hokuyo laser scanner (20–4000 mm with 3 % error). The laser scanner is attached to a servo so that it could be tilted to obtain 3D laser data. It is designed for both indoor and outdoor navigation and is able to navigate through various terrains such as sand, rock, concrete, grass and gravel. Each platform is powered by a 6-cell LiPo battery with a maximum operating time of 4 h.

4.1 *Structured Environment Mapping*

The first set of experiments was used to assess mapping ability in a controlled and structured environment. Two robots travelled around a rectangular hallway, the total length of which is 79.64 m with 21.96 m in width and 17.86 m in height. The lead robot took a total of 85 scans, with approximately 1 m travelled between consecutive scans, and returned to its starting point at the end of the experiment. Sample maps produced with the logged data set are shown in Fig. 7a. After 80 m travel, the error associated with the final lead robot position was approximately 5 m when ICP was not used to refine the state estimate. When the ICP was applied to improve the localization error, the end position estimation error was reduced to 0.63 m. The hallway map created by ICP has a mean estimated width and height of 22.59 and 17.91 m respectively. Image

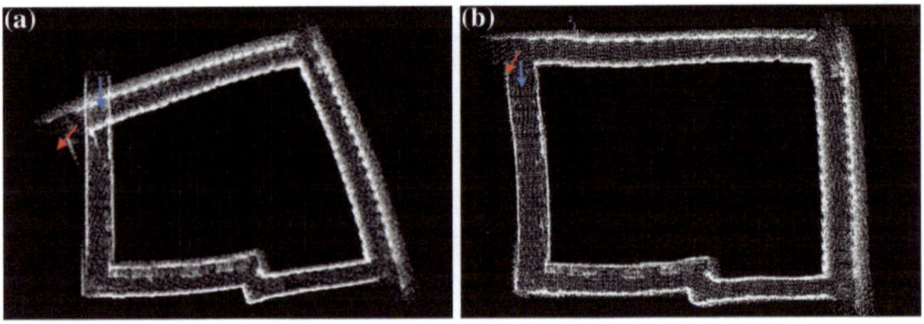

Fig. 7 Hallway map **a** created with multi-robot SLAM **b** after corrected by ICP. The robots started at *blue arrow* and stopped at *red arrow*

of the 2D localization conducted with ICP is shown in Fig. 7b. It can be observed that using ICP allows for loop closure. The loop closure occurs when a new point cloud, after being registered to its previous scan, finds a second matched point cloud among the earlier recorded point clouds.

4.2 Error Model and Lava Tube Mapping

To model the error growth as a function of distance travelled by the platoon, the actual and measured relative positions between two robots were logged. Two robots were placed in a sand pit located near Harvey Mudd College. The lead robot was fixed at a stationary location, and the follower robot was placed (and replaced 4 times) at 49 different positions in the sand pit. The measurement error, calculated by taking the difference between estimated and real distances for each position, is shown in Fig. 8, where a 4th order function has been used to model the estimation error as a function of the follower robot's relative position and angle. It can be seen that the error remains low (on the order of 0.02 m) when the relative distance is less than 2.5 m and the relative angle is less than 30° between two robots.

This model can be propagated over a series of scans to determine error growth as a function of distance travelled. In the same sand pit, the lead and follower robots were driven to follow a rectangular path. The real error growth and model predicted error growth have been plotted in Fig. 9. It can be seen that the actual error growth modeled by a linear fit is predicted by the error propagation function.

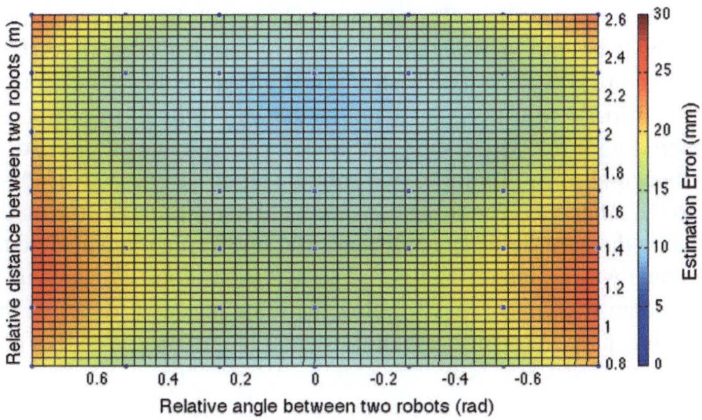

Fig. 8 Estimation error as a function of relative position and relative angle between two robots on the sand pit

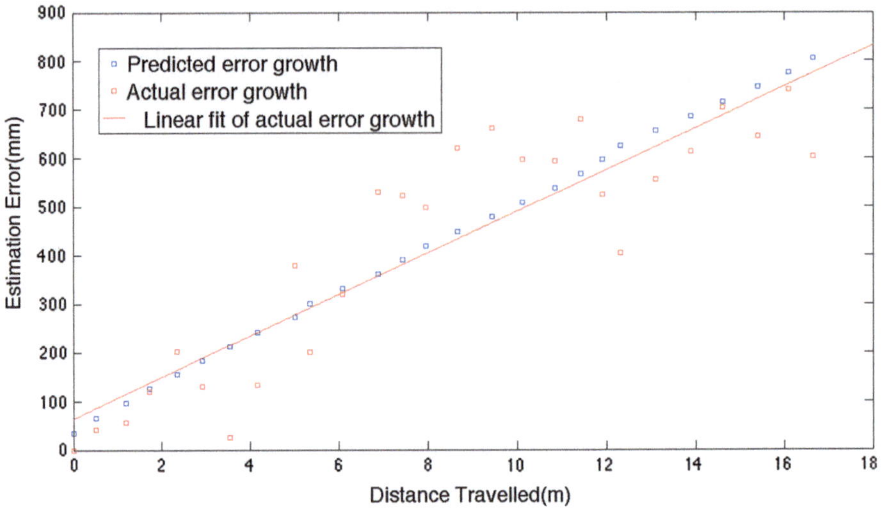

Fig. 9 Predicted error growth and actual error growth versus distance travelled. The actual error growth is modeled by a linear fit (*red line*)

4.3 Lava Tube Mapping

Final experiments were conducted in lava tubes located at Pisgah Center in the Mojave Dessert, CA. The tubes are shielded from external radiation by thick walls of lava rock. The main tube explored is 0.30–3 m high, 2–4 m wide, 32 m long, and 6 m down to the dessert surface. The elevation change of the tube ground is no more than 0.5 m. The temperature inside the tube during summer is about 25 °C while the surface temperature is 40 °C. There is almost no light in the tube. The ceiling consists of near vertical rocks with irregular features that are difficult to characterize. The floor is covered with fine silica sand and rocks, which makes it easy for the tread wheeled robot to slip. In this tube, two robots started at one end of the tube and navigated to the other end. The robot camera could not see anything with the header lights on due to the poor lighting conditions. The maximum pitch change relative to horizontal plane was no larger than 20°. The lead robot took 37 scans in 40 min to construct the map shown in Fig. 10. Using the map, the total length of the tube is 30.97 m which is just over 1 m less than the actual length measured by GPS data.

4.4 Lessons Learned

Several lessons were learned from the lava tube deployment. First, it is important to protect the robot platform against sand. During the experiment, it was found that the fine sand penetrated the robot parts as well as accumulated on the tracks.

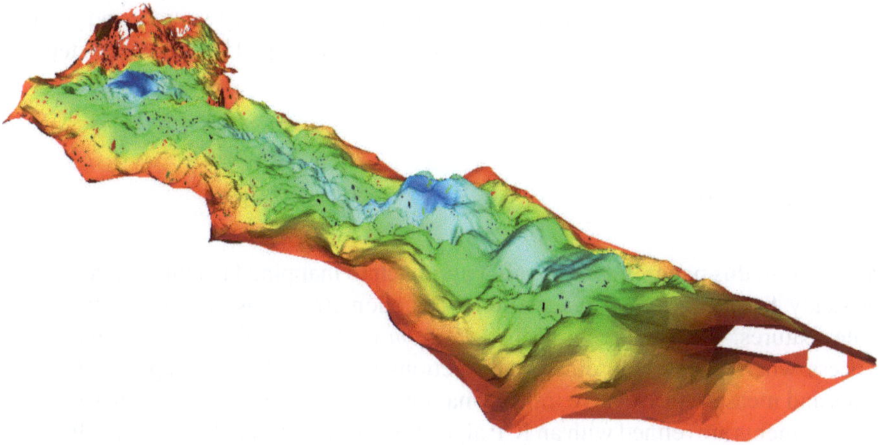

Fig. 10 Top view of the lava tube ceiling model in a top isometric view

Fig. 11 Odometry estimation error as a function of distance travelled on the sand pit

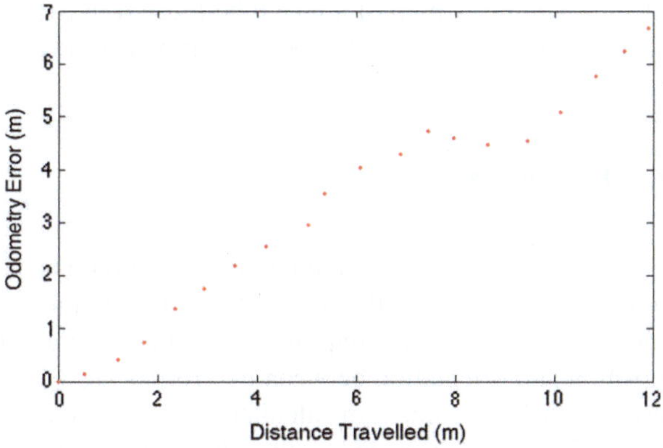

In consequence, the robot track had increased slipping, slower movement and fast odometry error growth, as shown in Fig. 11. Thus it is suggested that all holes on the robots and laptops should be covered, all screws should be tightened, and the track should be cleaned up before each experiment. As well, in order to reduce slip and increase travel speed, a track spoke with larger diameter is recommended, since it will add more contacting area between the track surface and ground.

The battery life is a crucial resource during the experiment, as it is hard to charge the battery in the middle of the dessert or on Mars. Therefore, in order to save battery life, efficient and faster algorithms are recommended for the robot control system. The team used 2D ICP instead of 3D ICP to register point clouds for this reason. Also, it was mentioned earlier that the team tilted the 2D laser scanner to obtaining 3D point clouds. The tilt step size was determined to be 5°. A smaller step size will take longer time to obtain data and thus require more battery source, and a larger step size will lose information when constructing the map. Therefore, the step size needs to be carefully selected.

5 Conclusions

Presented in this paper is a multi-robot approach to mapping lava tube environments on sandy floors that yield inaccurate robot odometry measurements without fine scale features. The approach, termed *Platoon SLAM*, involves an iterative 3 step process where robots coordinate their actions to allow them to capture 3D range scans and measure the relative transformations between scans. These transformation measurements are refined with an ICP algorithm. To construct 3D maps, the 3D scans are translated to point clouds that are added to a global map. The maps created with this system demonstrate error growth on the order of 3 % per meter travelled. Mapping loop closure was successfully demonstrated in a hallway system of approximately 80 m in length. A map of a lava tube located in Mojave Desert was created and the tube length was estimated to be 30.97 m when the actual length was 32 m.

6 Future Work

Future work involves implementing autonomous path planning. One important assumption in our solution is that the lead robot's target can always be detected by the follower robot. This requires a path planning algorithm that ensures the relative position and orientation between two robots are within some threshold to minimize error growth. The function calculated in Sect. 4 suggests using movements with less than 2.5 m in distance and less than 30 in degrees relative orientation between robots. The height of the lava tube along the planned path should also be considered in the algorithm so that both robots can pass through the tube. This can be achieved by analyzing the 3D map generated by the lead robot.

Additional work includes occupancy grid map generation. Currently a mesh file is created as the 3D map. This can be helpful for determining the shape and size of the lava tube. However, with an occupancy grid map, control parameters such as resolution, memory, as well as complexity can be controlled so maps can be generated according to different circumstances and restrictions. Additionally, as many off-the-shelf algorithms use an occupancy grid map representation, it will give future researchers more leverage after they map the environment.

The current work can be easily extended to more than two robots. The follower robots in the platoon will be able to provide more 3D scans and thus produce a more accurate map by advancing through the lava tube in the platoon manner. Specifically, point clouds generated from each robot can be matched and then merged together to increase map accuracy.

The ultimate goal for this project will be moving towards autonomous multi-robot 6DOF SLAM in lava tubes. For the robot system to be able to navigate on steep slopes, the follower robot should have 3D scanning capabilities to detect the target panel on the lead robot on terrains with significant changes in slope. To be able to localize with a 6DOF state, IMU data will likely be needed to further integrated to the state estimation of the robot system.

Acknowledgments The authors would like to thank all the people who contributed to the project: Samuel Yim, Shreyasha Paudel, Phuong Nguyen, Sean Messenger and Kevin Smith.

References

1. Aulinas, J., et al.: The SLAM problem: a survey. In: Proceedings of the 2008 Conference on Artificial Intelligence Research and Development: Proceedings of the 11th International Conference of the Catalan Association for Artificial Intelligence, pp. 363–371 (2008)
2. Besl, P.J., et al.: Method for registration of 3-D shapes. In: Robotics-DL Tentative. International Society for Optics and Photonics, pp. 586–606 (1992)
3. Borrmann, D., et al.: Globally consistent 3D mapping with scan matching. Robot. Auton. Syst. **56** (2), 130–142 (2008)
4. Bosse, M., et al.: Continuous 3D scan-matching with a spinning 2D laser. In: ICRA'09. IEEE International Conference on Robotics and Automation, pp. 4312–4319 (2009)
5. Burgard, W., et al.: Collaborative multi-robot exploration. In: Proceedings. ICRA'00. IEEE International Conference on Robotics and Automation, vol. 1, pp. 476–481 (2000)
6. Chen, Y., et al.: Object modeling by registration of multiple range images. In: Proceedings of the IEEE International Conference on Robotics and Automation, pp. 2724–2729 (1991)
7. Fenwick, J., et al.: Cooperative concurrent mapping and localization. In: Proceedings. ICRA'02. IEEE International Conference on Robotics and Automation, 2002, vol. 2, pp. 1810–1817 (2002)
8. Fischer, D., Kohlhepp, P.: 3D geometry reconstruction from multiple segmented surface descriptions using neuro-fuzzy similarity measures. J. Intell. Rob. Syst. **29**(4), 389–431 (2000)
9. Henry, P., et al.: RGB-D mapping: using depth cameras for dense 3d modeling of indoor environments. Exp. Robot. Springer Tracts Adv. Robot. **79**, 477–491 (2014)
10. Huber, D.F., Vandapel. N.: Automatic three-dimensional underground mine mapping. IJRR **25**(1), 7–17 (2006)
11. Leveille, R., et al.: Lava tubes and basaltic caves as astrobiological targets on earth and mars: a review. Planet. Space Sci. **58**, 592 (2012)
12. Lucchese, L., et al.: A frequency domain technique for range data registration. IEEE Trans. Pattern Anal. Mach. Intell. **24**, 1468–1484 (2002)
13. Magnusson, M., et al.: Scan registration for autonomous mining vehicles using 3D-NDT. J. Field Robot. **10**(24), 803–827 (2007)
14. Nutcher, H., et al.: 6D SLAM-3D mapping outdoor environments. J. Field Robot. **24**, 699–722 (2007)
15. Pathak, K., et al.: Fast registration based on noisy planes with unknown correspondences for 3D mapping. IEEE Trans. Robt. **26**, 424–441 (2010)
16. Pomerleau, F., et al.: Comparing ICP variants on real-world data sets. Auton. Robot. **34**(3), 133–148 (2013)
17. Rekleitis, I., Dudek, G., Milios, E.: Multi-robot collaboration for robust exploration. Ann. Math. Artif. Intell. **31**, 7–40 (2001)
18. Scheding, S.: Experiments in autonomous underground guidance. In: Proceedings, 1997 IEEE International Conference on Robotics and Automation, 1997, vol.3, pp. 1898–1903 (1997)

19. Thrun, S., et al.: A real-time algorithm for mobile robot mapping with applications to multi-robot and 3d mapping. In: IEEE International Conference on Robotics and Automation, 2000. Proceedings. ICRA'00, vol. 1, pp. 321–328 (2000)
20. Thrun, S., et al.: Simultaneous localization and mapping with sparse extended information filters. Algorithmic Found. Robot. V **7**, 363–380 (2004)
21. Tong, C., et al.: Three-dimensional SLAM for mapping planetary work site environments. J. Field Robot. **29**, 381–412 (2012)
22. Vaskevicius, N., et al.: Efficient representation in 3D environment modeling for planetary robotic exploration. Adv. Robot. **24**, 1169–1197 (2010)
23. Weingarten, J., et al.: EKF-based 3D SLAM for structured environment reconstruction. In: IEEE 2005 IEEE/RSJ International Conference on Intelligent Robots and Systems, 2005. (IROS 2005), pp. 3834–3839 (2005)
24. Zlot, R., Bosse, M.: Efficient large-scale 3D mobile mapping and surface reconstruction of an underground mine. Field Serv. Robot.: Springer Tracts Adv. Robot. **92**, 479–493 (2012)

Admittance Control for Robotic Loading: Underground Field Trials with an LHD

Andrew A. Dobson, Joshua A. Marshall and Johan Larsson

Abstract In this paper we describe field trials of an admittance-based Autonomous Loading Controller (ALC) applied to a robotic Load-Haul-Dump (LHD) machine at an underground mine near Örebro, Sweden. The ALC was tuned and field tested by using a 14-tonne capacity Atlas Copco ST14 LHD mining machine in piles of fragmented rock, similar to those found in operational mines. Several relationships between the ALC parameters and our performance metrics were discovered through the described field tests. During these tests, the tuned ALC took 61 % less time to load 39 % more payload when compared to a manual operator. The results presented in this paper suggest that the ALC is more consistent than manual operators, and is also robust to uncertainties in the unstructured mine environment.

1 Introduction

In this paper we document the tuning and evaluation of an admittance-based Autonomous Loading Controller (ALC) by using the Atlas Copco ST14 Load-Haul-Dump (LHD) machine in the underground mine shown in Fig. 1a. A smaller 1-tonne robotic loader was initially used for ALC development prior to the work reported in this paper. Diesel-hydraulic LHDs are used in underground mines to move fragmented rock (in mining *muck*) from draw points to ore passes or trucks, so the rock can be removed from the mine. Current robotic LHDs can haul and dump autonomously [1], but require an operator to load rock manually (usually by tele-remote). The ALC test results presented in Sect. 4 show a 39 % increase in payload mass and a 61 %

A.A. Dobson (✉)
Clearpath Robotics, Kitchener, ON N2R 1H2, Canada
e-mail: adobson@clearpathrobotics.com

J.A. Marshall
Mining Systems Laboratory, Queen's University, Kingston, ON K7L 3N6, Canada
e-mail: joshua.marshall@queensu.ca

J. Larsson
Division Rocktec Automation, Atlas Copco Rock Drills AB, SE-701 Örebro, Sweden
e-mail: johan.larsson@se.atlascopco.com

D.S. Wettergreen and T.D. Barfoot (eds.), *Field and Service Robotics*,
Springer Tracts in Advanced Robotics 113, DOI 10.1007/978-3-319-27702-8_32

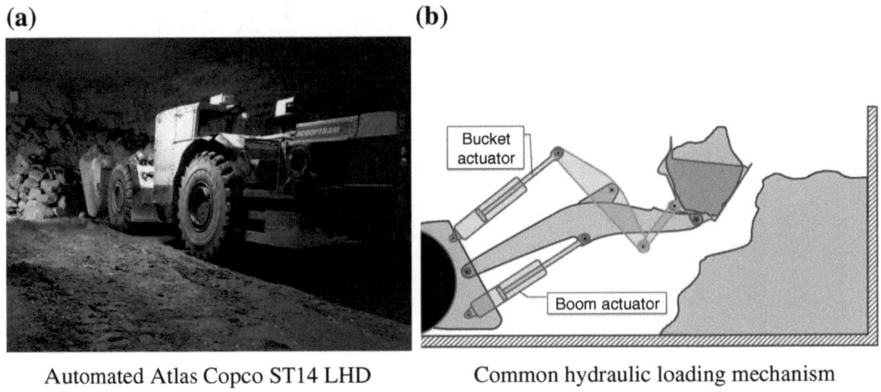

Fig. 1 The ALC has been tested on a 1-tonne wheel loader (not shown), and a 14-tonne Atlas Copco ST14 LHD (**a**). Both vehicles use a boom and a bucket actuator to hoist and curl the bucket respectively (**b**). The ST14 field experiments described in this paper were carried out in an underground mine on a roadway consisting of a gravel and clay over a limestone subsurface

reduction in dig time compared to an expert operator loading from the ST14 cab. The greater efficiency of the ALC over manual loading has implications for increasing mine productivity, and for decreasing costs by moving operators farther from potentially hazardous and remote mines [2].

Both the 1-tonne loader and the ST14 have similar hoisting and curling mechanisms as shown in Fig. 1b. Hoisting (vertical bucket motion) is controlled by altering the extension of the *boom* actuator, while curling (rocking the bucket forward and back) is controlled by altering the extension of the *bucket* actuator. The ALC admittance controller uses the forces sensed in the *boom* actuator to control the extension of the *bucket* actuator, and consequently, the curl of the bucket.

Others have proposed using scripted dig paths, lookup tables, Artificial Intelligence (AI), and impedance control to automate the digging process. Many of these methods were tested in homogeneous materials (e.g., soil, sand, and gravel, but not fragmented rock) by using sub-scale excavators. The scripted and lookup table methods [3–6] require pre-defined dig paths or bucket velocity targets, and did not perform well when sub-surface obstacles were encountered. The AI methods [7–11] attempt to overcome this deficiency by using heuristically-derived digging rules, but these rules are generally difficult to develop and reproduce. These methods were also less efficient and consistent than human operators.

Impedance control [12–15] is well-suited to tasks like trenching and landscaping, where the final target shape is more important than filling the bucket efficiently. This realization led Marshall in [16] to propose adapting Seraji's general admittance controller [17] for loading by controlling the *admittance* between the robot and the muck pile. Marshall's proposed admittance controller for loading was never tested, but was ultimately used as the starting point for the ALC presented in this paper. It is worth noting that despite a long history of research and development in robotic excavation, at the time of writing, there exists no widely-available commercial technology for autonomous digging in mining applications.

2 Admittance-Based Autonomous Loading Controller (ALC)

The proposed admittance controller modulates the error e_f between a preselected target force f_T and the sensed forces f_S by altering the velocity v_A of the bucket actuator, and consequently the bucket curl rate. In this way, the controller seeks to control the mechanical admittance Y between the bucket and the muck pile, where

$$Y = \frac{v_A}{f_T - f_S} = \frac{v_A}{e_f}. \tag{1}$$

Intuitively, this approach is believed suitable for robotic loading in fragmented rock because a typical muck pile contains irregular rocks, having a range of sizes, with varying cohesion due to moisture content and other factors, which cause force variations as the bucket is moved through the pile. These conditions are not as suitable, for example, for path-tracking controllers where these disturbance cause large deviations from the desired path. Also, the muck pile itself is expected to comply during the excavation process unlike in impedance control where the robot complies to the target. For example, a window washing robot must comply to its target to prevent breaking the uncompliant glass. Hence the widow washing problem is better solved by using an impendence controller. When loading rock the opposite situation occurs since the target rock must comply to the motion of the bucket. This inverse compliance relationship makes the loading problem better suited to admittance control.

The admittance controller is implemented in one of the four states of the ALC finite state machine. Each state in the ALC is executed in order, as follows:

State 0—Go to entry pose
State 1—Drive into pile until entry forces are above entry force target
State 2—Activate admittance controller until bucket has curled to breakout
State 3—Go to the weighing pose and terminate

Note that *breakout* occurs when the bucket curls past the point where additional material can easily enter the bucket, and is accompanied by a drop in digging forces [18]. In State 0, the ALC moves the boom and bucket to an appropriate entry pose before switching to State 1. In State 1, the LHD is commanded to drive forward until the bucket encounters enough resistance (as measured by the boom hydraulic cylinder pressures) to activate the admittance controller, at which point it switches to State 2. In State 2, the admittance controller controls admittance by referencing a target actuator force f_T. A high-level block diagram for this admittance control scheme is shown in Fig. 2. While any controller C could be used to map the force error e_f to the actuator velocities, a simple proportional-type (P) admittance relationship was evaluated in the field experiments presented in this paper. In this instance, the constant admittance relationship is given by

$$v_A = k_A \cdot e_f, \tag{2}$$

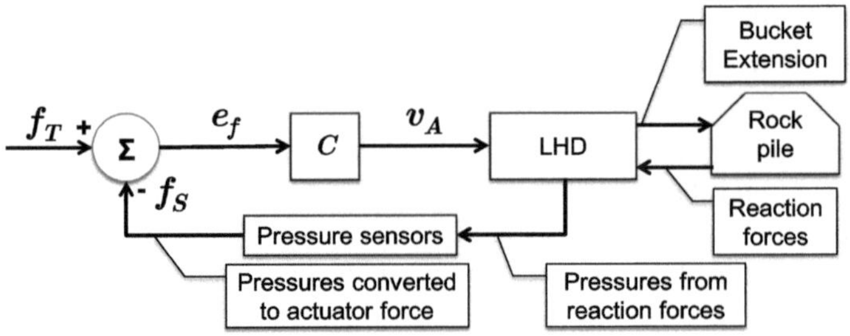

Fig. 2 The admittance controller uses any suitable controller C to map the error between the desired and sensed actuator forces to the range of possible actuator velocities

where v_A is the actuator velocity, $k_A > 0$ is the (admittance) gain, and the force error e_f is given by $e_f = f_T - f_S$.

The bucket motion direction depends on both the reaction forces f_S, and the element used to sense f_S. The ALC admittance controller alters the *bucket actuator* velocity by using the *boom* actuator to measure f_S. In a conventional admittance controller the actuator velocity is controlled by using the forces sensed in the same actuator. We use the forces sensed in the boom actuator because (1) the actuator loading in Fig. 3 shows that the boom actuator will tend to sense increasing forces as the bucket is curled up, (2) the boom stops on the ST14 tend to unload the boom actuator, which biases the ALC toward the breakout condition, and (3) the boom forces were generally cleaner than the bucket forces (see Figs. 5, 6, and 8).

In Fig. 3a, curling up tends to decrease the forces sensed in the bucket actuator, while increasing the forces sensed in the boom actuator. When f_S decreases e_f increases, which causes the admittance controller in Eq. (2) to respond by increasing v_A until the bucket actuator velocity limit is reached. Reaching the velocity limit saturates the ALC, which means the ALC can no longer control the admittance between the bucket and muck pile. Hence it is better to sense f_S in the boom actuator

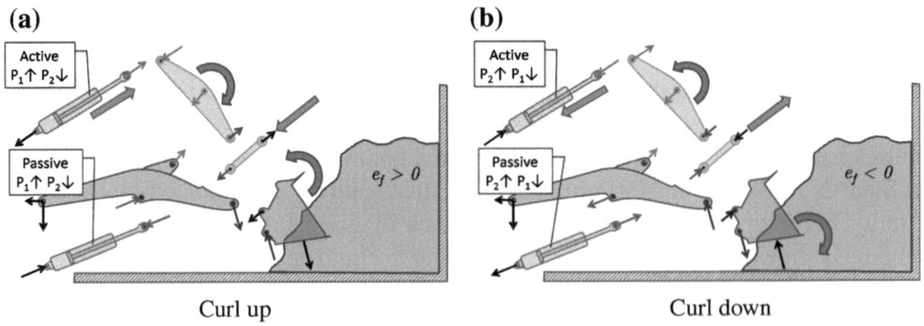

Fig. 3 When the bucket actuator extends in (**a**) P_1 goes up and the bucket curls back. The pile resists by putting the boom actuator in compression, which increases P_1 and the sensed reaction force. When the bucket curls down in (**b**) the inverse load case occurs and the boom actuator experiences tension. This tension manifests as a force drop because P_2 increases relative to P_1

since curling up increases boom loading, which decreases e_f and hence v_A. However, reducing e_f to zero is also not desirable since this condition will result in no bucket velocity and no breakout. This situation is prevented in part by selecting an f_T above the highest f_S, which ensures that $e_f > 0$ as discussed in Sect. 4.1. Stalling is also prevented because the boom arms tend to be driven downwards as the bucket fills. This downward motion is eventually arrested by two boom stops. Once these stops are encountered part of the load flowing through the boom actuator is redirected through the boom stops, which tends to decrease f_S, and increase e_f and v_A right at the end of the dig when it is most beneficial for ensuring breakout.

The digging forces are generated by both the bucket motion and the forward thrust of the LHD. When the forces sensed in the boom actuator are below f_T the admittance controller will increase f_S by curling up. Curling up increases the sensed forces in the boom because the boom actuator experiences compression in addition to the compression caused by the load in the bucket. Curling down tends to relieve this compression, which reduces f_S.

State 2 terminates when the bucket has curled passed the point where rock can easily enter the bucket (i.e., the breakout condition). Once breakout has occurred, the controller switches to State 3 where the LHD stops thrusting into the pile, raises the boom to the weighing pose, and finishes curling the bucket to settle the load. Once the dig cycle is complete, the dig time, total actuator work, and final payload are computed to determine dig efficiency.

2.1 Dig Efficiency

We define overall dig efficiency ε_d as

$$\varepsilon_d \left(t_d, W_d, M_d\right), \tag{3}$$

which is a combination of three parameters: (1) the dig time t_d; (2) the actuator work expended while digging W_d; and (3) the mass of rock in the bucket at the end of the dig attempt M_d. Together these three parameters define a point in 3D-space with time, work and mass axes (e.g., as shown in two 2D-space plots in Fig. 7).

The payload mass M_d was calculated by using a proprietary load weighing system described by Grahn [19]. This load weighing system calculates M_d by

$$M_d = k \cdot \left(P_C - P_R\right), \tag{4}$$

where P_C and P_R are the boom actuator cylinder and rod pressures, and k is a calibration constant for a specific weighing pose. According to Grahn, the load weighing system is calibrated to a precision of ±0.5 t. The average ALC payload was 14.47 ± 1.09 t, and the rated payload limit for the ST14 is 14 t.

Work and dig time are calculated between entry (after the entry force target is reached), and breakout. Let n be the total number of sensor readings and let the subscript i denote the time index associated with each sensor reading. Thus, the total work W_d was estimated by

$$W_d = \frac{1}{2}\sum_{i=1}^{n-1}\left[\left(F_{h,i} + F_{h,i+1}\right)\cdot|d_{h,i} - d_{h,i+1}| + \left(F_{c,i} + F_{c,i+1}\right)\cdot|d_{c,i} - d_{c,i+1}|\right], \tag{5}$$

where F_h and F_c are the hoist and curl forces in the boom and bucket actuators respectively, and d_h and d_c are the displacements for each actuator. Note that this work estimate includes only the work done by the actuators, and not the drive train, which thrusts the loader into the pile.

3 Apparatus and Methodology

This section introduces the operating environment and test equipment used at the Kvarntorp Mine near Örebro, Sweden. Kvarntorp is an underground limestone room-and-pillar mine that is no longer in production. The test area is located approximately 30 m below surface, where the tunnels (called mine *drifts*) are approximately 10–12 m wide and 6 m tall. Over 200 t of fragmented granite was added to the end of Drift 165 while the pile in Drift 159 consisted of several hundred tonnes of limestone from previous blasts in the mine. Drift 159 was primarily used for controller development and preliminary tuning, while Drift 165 was used for all manual digs and all final ALC digs. Figure 4a, b show the muck piles along the wall of Drift 159 and at the end of Drift 165 respectively. The largest visible dimension of the muck in Drift 159 ($\pm 1\sigma$) was 0.20 ± 0.09 m. The muck in Drift 165 was over twice as large, with double the standard deviation (0.48 ± 0.19 m).

The Atlas Copco Scooptram ST14 is a 38 t vehicle with a 14 t, 6.4 m^3 bucket. The nominal dimensions of the vehicle are 10.8 m long, 2.6 m tall, and 2.8 m wide [20]. The ST14 used for these tests was equipped for teleoperation [21]. However, the ALC only uses the actuator extension and pressure measurement sensors that are available on the stock ST14. The pressure measurements are taken on the rod and cylinder sides of the boom actuator. These pressures combined with the rod and cylinder areas (A_R and A_C respectively) can be used to calculate f_S by

$$f_S = P_C \cdot A_C - P_R \cdot A_R. \tag{6}$$

Both the manual and final ALC dig trials were conducted at the end of Drift 165 in the granite muck pile. The actuator pressure and extension measurements were logged for both manual and autonomous operating modes, and were used to generate the digging histograms in Sect. 4. The vehicle was warmed up for 10 to 20 min at the beginning of each test day. Each dig began by positioning the vehicle in front of the muck pile as shown in Fig. 4c. For the manual dig trials, our expert operator "Frank" was instructed to dig (1) normally by using both boom and bucket actuators;

Fig. 4 The limestone muck pile along the wall of Drift 159 (**a**) was used for all preliminary logic tests and tuning, while the manual tests, final tuning, and ALC evaluation tests were conducted in the granite muck pile at the end of Drift 165 (**b**). The mean $\pm$ one standard deviation rock size distribution estimates were 0.20 ± 0.09 m in Drift 159 and 0.48 ± 0.19 m in Drift 165. (**c**) The ST14 began each dig in the start position, which was approximately 11 m from the toe of the pile. (**d**) The ST14 was moved into position by using the operator station within the automation office. Following automated loading, the operator weighed and dumped the material manually

(2) in a manor similar to the ALC using only the bucket actuator; and (3) by using 50 % throttle. The 50 % throttle setting was selected to determine if there were any advantages to digging at lower throttle. The manual dig efficiency results shown in Fig. 7 indicate that digging at lower throttle should be avoided and hence the ALC throttle setting was set to 100 % to better match the bucket only and both actuator manual digging methods. Similarly, the entry velocity was also selected to match the manual dig attempts and averaged 5.0 km/hr. This velocity corresponds to 100 % throttle, first gear, and 0 % brake.

In all tests, Frank controlled the vehicle from inside the ST14. The ALC digs began by switching the ST14 to "automation mode". The operator then left the vehicle, and entered the automation office shown in Fig. 4d. After uploading the desired tuning parameters to the ST14, the ALC was initiated. When the ALC reached its final state, the ST14 was switched to teleremote mode so that the bucket could be lifted, weighed, and dumped. The same weighing and dumping procedure was also performed by Frank following his dig attempts. After dumping, the ST14 was driven back to the approximate start position.

4 Field Experiments Results

The Autonomous Loading Controller (ALC) tuning tests were used to find final values for the ALC parameters, which were then held constant for all performance tests. These performance tests were conducted to compare the ALC to manual digging. The ALC parameters that were tuned were f_T, k_A, the breakout condition, as well as the entry and weighing poses. Additionally, field tuning revealed key information about controller saturation, ground detection, and ALC performance.

4.1 Force Target f_T

Figure 5 shows the ALC digging response as f_T was reduced from 11 MN to 9 MN. An initial guess for k_A was 0.001, which was selected by using

$$k_A \approx r \cdot \frac{v_{A\max}}{f_{S\max}} \tag{7}$$

where $r = \frac{1}{8}$, $v_{A\max}$ is the maximum bucket actuator velocity (0.08 m/s), and $f_{S\max}$ is the maximum force sensed in the boom actuator (10, MN). r is an arbitrary scalar that sets the minimum increment between no gain and a gain that results in complete actuator saturation. Initial tuning results (in Sect. 4.2) indicated that the controller was unacceptably saturated when k_A was increased to 0.002. Saturation should be avoided because it means the admittance controller is no longer maintaining the desired admittance dictated by Eq. (2). The manual results (in Sect. 4.4 and specifically Fig. 8) show that digging without compensating for the digging forces tends to result in less overall payload and more payload variability.

Decreasing the dig target increased dig time, decreased bucket velocity, and decreased bucket actuator control valve saturation. When f_T dropped to 9.5 MN, the dig time increased from 8 to 30 s, the bucket velocity was much slower, and the sensed forces were barely high enough to bias the admittance controller toward the breakout condition. At $f_T =$ 9.0 MN these effects became so severe that the dig failed because the ALC could then reduce the force error e_f close to zero. Figure 5a, b also illustrate that more controller saturation leads to higher, more irregular forces. The 11-MN and 10-MN test results indicate that these irregular forces generated higher payloads, but also more payload variability. It should also be noted that for the 11-MN and 10-MN tests the bucket curls down (see between 10 and 15 s) when the boom forces exceed their respective f_T values. While curling down may seem counter productive, it allows the bucket to circumvent force concentrations and dig deeper into the pile. We believe that this results in increased payload and less payload variability because the admittance between the bucket and muck pile is maintained, and hence each dig trajectory is tailored to the unique force environment encountered within the pile.

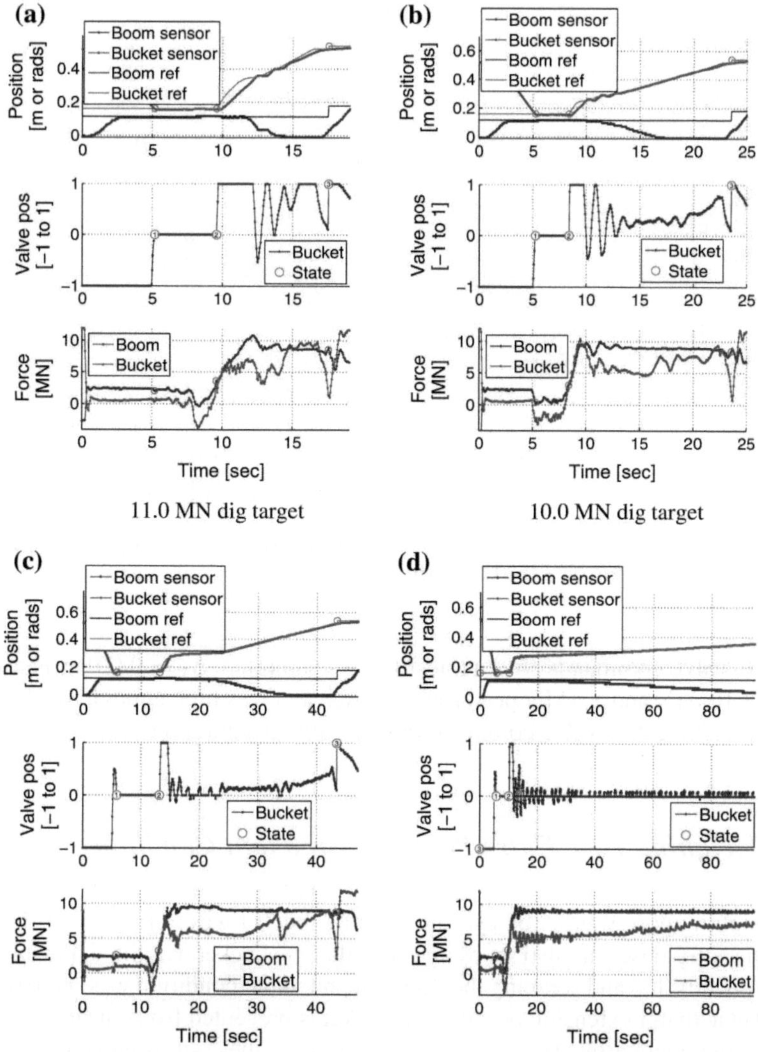

11.0 MN dig target

10.0 MN dig target

9.5 MN

9.0 MN dig target

Fig. 5 Finding a dig force target—At 11 MN (**a**) the ALC was more saturated than in the 10 MN digs (**b**), but both completed successfully. The 10 MN digs took twice as long as the 11 MN digs, and the 9.5 MN digs (**c**) took three times longer than the 11 MN digs. At 9 MN (**d**), the dig failed because the ALC was able to reduce the error to 0.0 and the curl rate dropped too low for the ALC to finish in a reasonable time

4.2 Admittance Gain k_A

Figure 6 shows the ALC responses when the admittance gain k_A was raised from 0.001 to 0.002 while f_T was maintained at 10 MN. $k_A = 0.002$ was too high since

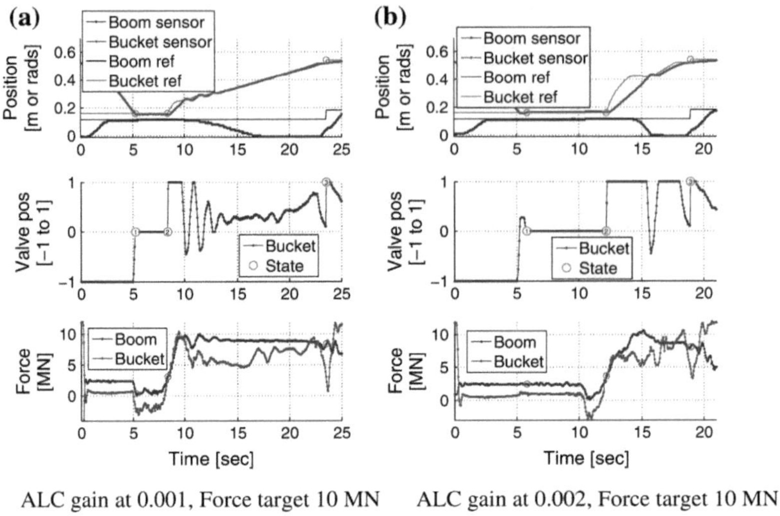

Fig. 6 ALC gain selection—The ALC gain at 0.001 (**a**) issues excellent valve commands with little saturation compared to the 0.002 gain (**b**), which was almost always saturated

the ALC valve commands were almost always saturated. $k_A = 0.001$ was used for both the 10-MN and 11-MN performance tests, and was high enough to cover both positive and negative valve command ranges without saturation.

4.3 Weighing Pose Entry Pose and Breakout Condition

The weighing pose was set by eye such that the bucket was in free space above the pile. The entry pose was also set by eye such that the bucket was tilted downwards at approximately 15° and scraping the floor. The breakout condition was set to 0.500 m of bucket actuator extension because the bucket is prevented from curling further by stops on the boom arms. However, as the boom rises these stops move further back. Midway through the tuning process, the bucket breakout extension was increased from 0.500 to 0.520 m, which increased payload to 12.50 t from 10.13 t. This increase occurred because the bucket curled back farther as soon as the boom started to lift, which kicked more material into the bucket. Only a few tests were performed at each breakout setting in the muck pile in Drift 159, before moving to the ALC performance tests. These performance tests were conducted in Drift 165, which contained the larger, higher density rock fragments. Several runs were made at both 10-MN and 11-MN dig targets and all other ALC parameters were kept constant so that the ALC could be compared to a manual operator.

4.4 ALC Performance

The dig efficiency results from the 26 autonomous and 28 manual digs are shown in Fig. 7. The number of tests was dictated by the availability of the apparatus, operator, and test site. The manual digs with the highest dig efficiencies were Frank's bucket-only, and low-throttle digs. The autonomous digs with the highest efficiencies were the 11-MN digs. While the 10-MN autonomous digs were also excellent, six of these digs failed. The likely cause of these failures was low entry force due to striking the ground or spillage before entry.

Figure 7a shows the payload and dig times for the 54 dig attempts. While the autonomous dig attempts were tightly clustered, there was much more variability in the manual dig times and payloads. Figure 7b shows the payload and work expended for the same 54 dig attempts. The autonomous dig attempts were again tightly clustered, while there was much more variability in the manual digs. Work also increased as payload increased. All dig efficiency results are summarized in Table 1.

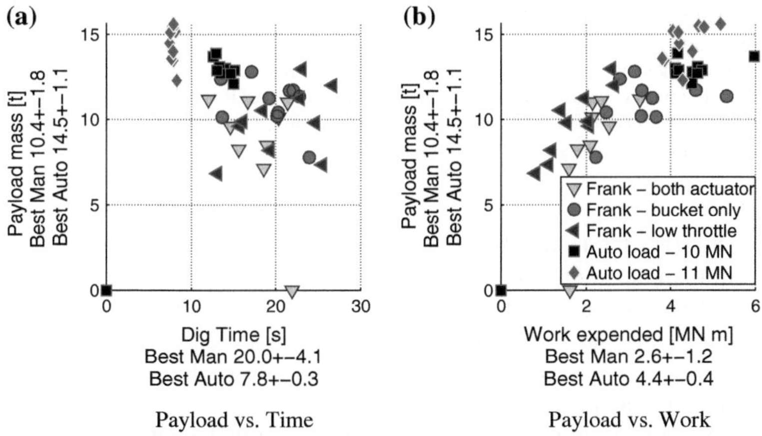

Fig. 7 The payload versus dig time (**a**) and the payload versus work (**b**) dig efficiency *plots* show that the only autonomous dig attempts that were less than 12 t were the six 10 MN dig target digs that failed due to insufficient entry force to trigger the admittance controller. The manual dig attempts had much greater variability in payload mass, dig time, and actuator work than the tightly clustered autonomous dig attempts. There is also a clear trend towards increasing work as payload increases (**b**)

Table 1 The ALC loaded 39 % more payload in 61 % less time, but required 68 % more work than the best expert operator digs

ε_d	Manual	Autonomous	Difference (%)
t_d [s]	20.03 ± 4.10	7.82 ± 0.26	−61
W_d [MN m]	2.59 ± 1.17	4.36 ± 0.43	+68
M_d [t]	10.41 ± 1.77	14.47 ± 1.09	+39

The ALC was also much more consistent

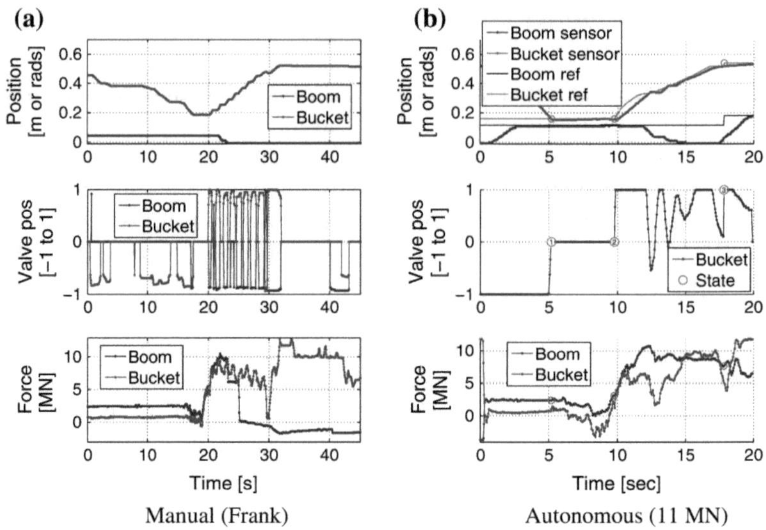

Fig. 8 In (**a**) Frank gave the bucket regular oscillating command signals that resulted in a jagged force profile, and severe valve position oscillations between 1 and 0. The ALC in (**b**) sent much smoother commands that used partial valve positions to regulate the speed of the bucket. As a result, the forces were much smoother than the manual dig attempts. Additionally the bucket curled down at 12 s to reduce the forces below the dig target. This behaviour caused the bucket to dig deeper into the pile, and ultimately increased the final payload

Figure 8 shows the results for an excellent manual, and typical autonomous dig attempt. In both digs only the bucket was actuated either by Frank or by the ALC. Frank oscillated the bucket rhythmically while the ALC only oscillated when the forces were below the 11 MN target force. This reduced oscillation resulted in smoother force and valve command profiles, and ultimately greater bucket velocity control, and more payload in less time.

Tests were conducted in both a settled and an unsettled muck pile, as well as the two muck piles with different rock types and size distributions. The average payload dropped from 14.47 ± 1.09 t in the unsettled pile (11-MN autonomous tests), to 12.50 t in the settled pile. Only one test was performed in the settled pile since due to the time it takes for the pile to settle. The rock type and size distribution had little effect on the ALC because the force profiles resulting from digging in the two piles were nearly identical. The resulting payload change was slight, going from 11.40 t in the lower density 0.20 ± 0.09 m limestone rock in Drift 159, to an average of 12.93 ± 0.55 t in the higher density 0.48 ± 0.19 m granite rock in Drift 165.

5 Conclusion

An Autonomous Loading Controller (ALC) based on constant admittance control was tuned and compared to manual dig trials at the Kvarntorp underground mine by using an Atlas Copco ST14 LHD, and various limestone and granite muck piles. In this paper, the admittance controller within the ALC prescribed a constant admittance relationship that used the forces sensed in the boom to alter the bucket velocity. Preliminary ALC tuning tests revealed that the dig target and admittance gain must be set such that the admittance controller can never fully reduce the force error to zero, which ensures that the ALC is biased toward breaking out of the muck pile. Biasing the ALC toward breakout made the ALC surprisingly robust to disturbances caused by changing much pile conditions. The performance comparisons between admittance-based and manual (expert operator) digs are the most important outcomes from these field experiments. However, vital insight was also gained into the digging process, as well as how to tune the ALC to match the machine to the test environment. The ALC had 61 % better dig time and 39 % greater payload, but required 68 % more actuator work. The ALC dig efficiency variability was greatly reduced compared to the manual digs, which should make mass flow rates out of the mine easier to predict. Some 10-MN digs failed due to the uneven roadway. Ideally this variability in the roadway should be compensated for by maintaining a bucket hight relative to the ground.

Acknowledgments The authors would like to thank Atlas Copco Rock Drills AB for facilitating this research collaboration. This research was supported in part by the Natural Sciences and Engineering Research Council of Canada (NSERC) under grant 371452-2009.

References

1. Marshall, J.A., Barfoot, T.D., Larsson, J.: Autonomous underground tramming for center-articulated vehicles. J. Field Robot. **25**(6–7), 400–421 (2008)
2. Zlotnikov, D.: Mining in the extreme. CIM **7**(5), 50–56 (2012)
3. Sarata, S., Koyachi, N., Sugawara, K.: Field test of autonomous loading operation by wheel loader. In: Proceedings of the IEEE/RSJ International Conference on Intelligent Robots and Systems, pp. 2661–2666. IEEE, Acropolis convention center, Nice, France (2008)
4. Almqvist, H.: Automatic bucket fill. Masters thesis, Linköping University, Linköping, Sweden (2009)
5. Shull, A.G.: Digging control system. US Patent No. 8160783 (2012)
6. Rocke, D.J.: Control system for automatically controlling a work implement of an earthworking machine to capture material. US Patent No. 5528843 (1996)
7. Lever, P.: An automated digging control for a wheel loader. Robotica **19**(5), 497–511 (2001)
8. Shi, X., Lever, P., Wang, F.: Autonomous Rock Excavation: Intelligent Control Techniques and Experimentation. Series in intelligent control and intelligent automation, World Scientific (1998)
9. Dasys, A., Geoffroy, L., Drouin, A.: Sensor feedback control for automated bucket loading. US Patent No. 5941921 (1999)

10. Schmidt, D., Proetzsch, M., Berns, K.: Simulation and control of an autonomous bucket excavator for landscaping tasks. In: IEEE International Conference on Robotics and Automation, pp. 5108–5113 (2010)
11. Bradley, D.A., Seward, D.W.: The development, control and operation of an autonomous robotic excavator. J. Intell. Robot. Syst. **21**(1), 73–97 (1998)
12. Bernold, L.: Motion and path control for robotic excavation. J. Aerosp. Eng. **6**(1), 1–18 (1993)
13. Ha, Q., Santos, M., Nguyen, Q., Rye, D., Durrant-Whyte, H.: Robotic excavation in construction automation. IEEE Robot. Autom. Mag. **9**(1), 20–28 (2002)
14. Salcudean, S., Tafazoli, S., Hashtrudi-Zaad, K., Lawrence, P.: Evaluation of impedance and teleoperation control of a hydraulic mini-excavator. Lecture Notes in Control and Information Sciences 232, 229–240 (1998)
15. Maeda, G.J., Rye, D.C., Singh, S.P.N.: Iterative autonomous excavation. In: Yoshida, K., Tadokoro, S. (eds.) Field and Service Robotics. Springer Tracts in Advanced Robotics, vol. 92, pp. 369–382. Springer, Berlin (2013)
16. Marshall, J.A.: Toward autonomous excavation of fragmented rock: Experiments, modelling, identification and control. Master of science: Engineering, Queen's University, Kingston, On. Canada (2001)
17. Seraji, H.: Adaptive Admittance Control: an approach to explicit force control in compliant motion. In: Proceedings of theIEEE International Conference on Robotics and Automation, vol. 4, pp. 2705–2712 (1994)
18. Marshall, J.A., Murphy, P.F., Daneshmend, L.K.: Toward autonomous excavation of fragmented rock: full-scale experiments. IEEE Trans. Autom. Sci. Eng. **5**(3), 562–566 (2008)
19. Grahn, F.: Specification for load weighing. Internal T2 0330, Atlas Copco Rock Drills AB, Örebro Sweden (2005)
20. Atlas Copco.: Atlas Copco Underground loaders: Scooptram ST14 Technical specification. Atlas Copco (2012). http://www.atlascopco.ca/images/technical_specification_scooptram_st14_9851_2350_01_tcm836-1532858.pdf
21. Larsson, J., Broxvall, M., Saffiotti, A.: An evaluation of local autonomy applied to teleoperated vehicles in underground mines. In: Proceeding of the IEEE International Conference on Robotics and Automation, pp. 1745–1752 (2010)

From ImageNet to Mining: Adapting Visual Object Detection with Minimal Supervision

Alex Bewley and Ben Upcroft

Abstract This paper presents visual detection and classification of light vehicles and personnel on a mine site. We capitalise on the rapid advances of ConvNet based object recognition but highlight that a naive black box approach results in a significant number of false positives. In particular, the lack of domain specific training data and the unique landscape in a mine site causes a high rate of errors. We exploit the abundance of background-only images to train a k-means classifier to complement the ConvNet. Furthermore, localisation of objects of interest and a reduction in computation is enabled through region proposals. Our system is tested on over 10 km of real mine site data and we were able to detect both light vehicles and personnel. We show that the introduction of our background model can reduce the false positive rate by an order of magnitude.

1 Introduction

While the mining industry pushes for greater autonomy, there still remains a need for human presence on many existing mine sites. This places significant importance on the safe interaction between human occupied and remotely operated or autonomous vehicles. In this work, we investigate a vision based technique for detecting other vehicles and personnel in the workspace of heavy vehicles such as haul trucks.

Traditionally, methods for detecting light vehicles and personnel from heavy mining equipment have relied on radio transponder based technologies. Despite transponder based sensors being mature and reliable for ideal conditions, in practice their

A. Bewley (✉)
School of Electrical Engineering and Computer Science,
Queensland University of Technology, Brisbane, Australia
e-mail: aj.bewley@qut.edu.au

B. Upcroft
ARC Centre of Excellence for Robotic Vision, School of Electrical Engineering
and Computer Science, Queensland University of Technology, Brisbane, Australia
e-mail: ben.upcroft@qut.edu.au
URL: http://www.roboticvision.org/

D.S. Wettergreen and T.D. Barfoot (eds.), *Field and Service Robotics*,
Springer Tracts in Advanced Robotics 113, DOI 10.1007/978-3-319-27702-8_33

reliability is circumvented by practical issues around their two way active nature, portable power requirements, limited spatial resolution and human error. Using computer vision offers a unique alternative that is passive and readily available on existing remotely operated vehicles.

Vision based object recognition has made tremendous progress as measured by standard benchmarks [4, 16]. The major advancements in this area can be attributed to both the availability of huge annotated datasets [4, 7, 16, 26] and developments in data driven models such as deep convolutional networks (ConvNets) [13, 24]. In this work we utilise the ConvNet of [13] which has shown astonishing performance on the ImageNet recognition benchmark [4] and extend it to data collected from mine sites with minimal training.

Using ConvNets in different domains requires a large training set relevant to the target task [29]. When the amount of training data is small, data driven approaches tend to over-fit the training samples and not generalise to unseen images. In this work we utilise a pre-trained ConvNet using millions of images from ImageNet and address how to map the original ImageNet classes to mining classes with minimal training effort.

Another consideration regarding this application is that cameras are rigidly coupled to the vehicles orientation and configured with a fixed focal length. This distinguishes it from the ImageNet recognition problem where typical images collected were implicitly pointed at regions of interest and appropriately zoomed. Additionally, due to the wide field of view the majority of the images are background with zero to potentially multiple objects of interest visible in any given frame. To locate the objects, we follow a similar strategy to [10] and apply an initial step for finding likely object locations through a region proposal process before performing object recognition with the ConvNet.

Given that the majority of the images collected in a mine site dataset have zero objects of interest in them, we can provide a standard classifier with a huge amount of labelled background data. Using this newly trained classifier in conjunction with the ConvNet ensures robustness and drastically reduces spurious detections. This classifier is based on k-means clustering offering a convenient way to partition the background data into different categories. This approach accurately captures the characteristics of the background, enabling the discovery of novel non-background objects.

The contributions of this paper are:

- adapting ConvNets to new scenes in a mining context,
- complementing the powerful classification provided by ConvNets with a simple classifier trained on background mine data for increased robustness,
- a novelty detector using ConvNet feature clustering.

This paper is organised with a short review of related literature before describing the proposed method in greater detail. We then analyse the performance of the proposed method on a challenging set of mining videos and conclude with a discussion of the learnt outcomes and avenues for future improvement.

2 Related Work

Here we briefly review object detection methods that are not reliant on two way communication before covering some related work using ConvNets for generic object detection. Early work has focused on range based techniques such as LiDAR [17, 22] commonly used for mapping fixed obstacles such as buildings or underground tunnel walls. Applying these sensors to detecting personnel and vehicles fitted with retro-reflectors, is found to be sensitive to the dynamics of the sensor platform [20]. In this work we focus specifically on detecting potentially dynamic obstacles including vehicles and particularly people from vision based data. To this end, the more relevant prior work is that of [18] which exploits the standardised requirement for personnel on mine sites to wear high-visibility clothing equipped with retro-reflector strips. This enables a single IR camera with active flash to highlight personnel in view which can then be used for tracking [19].

Recent popularity of big data and deep learning have dominated the object recognition problem. Among these data driven approaches, deep convolutional neural networks (ConvNets) with recognition performance quickly approaching human levels [5, 13, 21, 23] are selected for use in this work. ConvNets themselves have been used for over 20 years [14] for tasks such as character recognition. Over recent years ConvNets have made an astonishing impact on the computer vision community [5, 6, 10, 13, 21] thanks to the availability of huge labelled image sets such as ImageNet [3].

Recognising what objects are in an image is only half of the object detection problem. The other half is locating the objects within the image. Sermanet et al. [23] sample over multiple scales and exploit the inherently spatially dense nature of the convolutions within ConvNets to identify regions with high responses. Similarly, [6] also perform convolutions over multiple scales and combine the responses over superpixel segmentation [9]. Another popular approach and the one that we base this work off is the region convolutional neural network (RCNN) of [10]. The RCNN framework efficiently combines the ConvNet of [13] with an object proposal method: selective search [27]. Generic object proposal methods aim to efficiently scan the entire image at different scales and aspect ratios to reduce potentially millions of search windows down to hundreds [11] of the most likely candidates. In this work we use edge box object proposals [30] as the accuracy is higher while also running at an order of magnitude faster [11].

3 Methodology

In this section we outline our detection pipeline and how it differs from [10]. Our method consists of three key phases: (1) Region proposals with non-maximum suppression (NMS), (2) ConvNet recognition and finally, (3) Detections are validated by checking for novelty against the background model. See Fig. 1 for a high-level overview of this pipeline. We bypass the problem of over-fitting on a small dataset

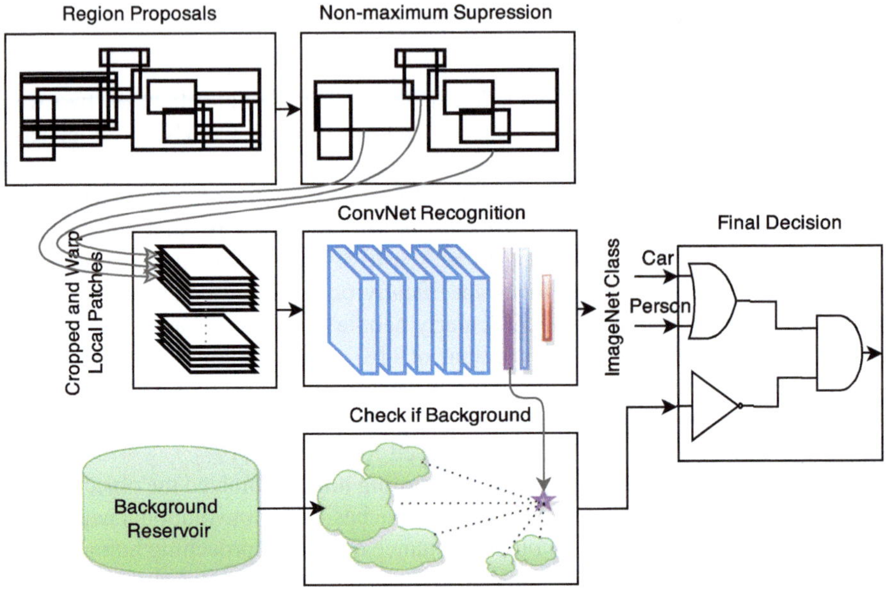

Fig. 1 An illustration of the detection pipeline used in this work. The system parameters are highlighted in *blue* and *green* which are learnt offline from an off-the-shelf network and background only images respectively. Note the *red output layer* of the ConvNet outputs are ImageNet classes (200 different). Any car or person is suppressed if it also matches the background model to minimise the number of false positives

by using a pre-training ConvNet and map its output to mining relevant classes. This method is then extended with our proposed background modelling technique to significantly reduce the number of false positives generated by the system.

3.1 Region Proposals

The aim of region proposals is to efficiently scan the image to eliminate millions of potential windows, keeping only the regions that are likely to contain an object of interest. We use the `EdgeBoxes` region proposal method [30] over the `selective search` [27] used in the original RCNN work as this method is orders of magnitude faster with comparable accuracy. For a detailed comparison of region proposal methods we refer the reader to [11].

The default parameters for `EdgeBoxes` were adjusted to return a fixed 1000 proposals. These region proposals are then further reduced to approximately 100 regions through a process of non-maximum suppression (NMS). The NMS process considers the score produced by the `EdgeBoxes` method and the overlap with other bounding boxes. As the name suggests it then greedily suppresses all but the maximum scoring proposal for all adjacent regions overlapping by 30 % or more. In contrast to applying NMS after the ConvNet [10], this way we can speed up the detection pipeline by reducing the number of proposals going into the ConvNet while maintaining comparable coverage over the image.

3.2 Region Classification

Having selected regions of the image that have the general characteristics of an object, we now perform object recognition to distinguish the object category. For this we apply the ConvNet from RCNN [10] which is based on the winning architecture [13] for the ImageNet Large Scale Recognition challenge in 2012. For this work, we used the RCNN implementation provided with the Convolutional Architecture for Fast Feature Embedded (`caffe`) [12] framework out-of-the-box.

The original detection task for RCNN was to predict one of 200 classes that represent common objects found in images taken from the internet. For this application we are only interested in distinguishing between three high level categories, namely: `background`, `person` and `light vehicles(LV)`. Using this model in a mining context raises several issues that need addressing:

1. Most of the 200 classes are irrelevant, e.g. jellyfish, miniskirt, unicycle etc.
2. How to associate mining classes with ImageNet classes?
3. Semantically the `background` is significantly different from many of the existing object specific classes.

To gain some insight, we use a small validation set of 200 images to investigate the output of the ConvNet out-of-the-box. This set is made up of cropped mine-site images containing the classes `person` and `LV` along with 90 interesting region proposals extracted from background only images. We also included a few `heavy vehicles (HV)` images in this set but keep them as a separate class to identify any correlations. In Fig. 2 we show the results of naively applying the pre-trained RCNN model to this image set. To better visualise the output we applied a soft-max transform to approximate the output class prediction as a probabilistic estimate.[1]

Not surprisingly, the `person` and `LV` classes are well represented and can be directly mapped from the `person` and `car` ImageNet classes used to train the original ConvNet. On the other hand, the `background` closely resembles uniform random sampling of classes as there are no relevant classes in the existing model such as trees, buildings, or road signs etc. Similarly, the `HV` class prediction also mostly resembles a uniformly random distribution with a slight bias towards the ImageNet classes `snowplow`, `cart` and `bus`. As for this application, we are only concerned with distinguishing `person` and `LV` from the background, we simply assign all 198 non `person` or `car` outputs as background.

[1]It is important to note that this is for visualisation purposes only and that the y-axis does not represent the true probability since the final SVM layer of RCNN was not calibrated for probabilistic outputs.

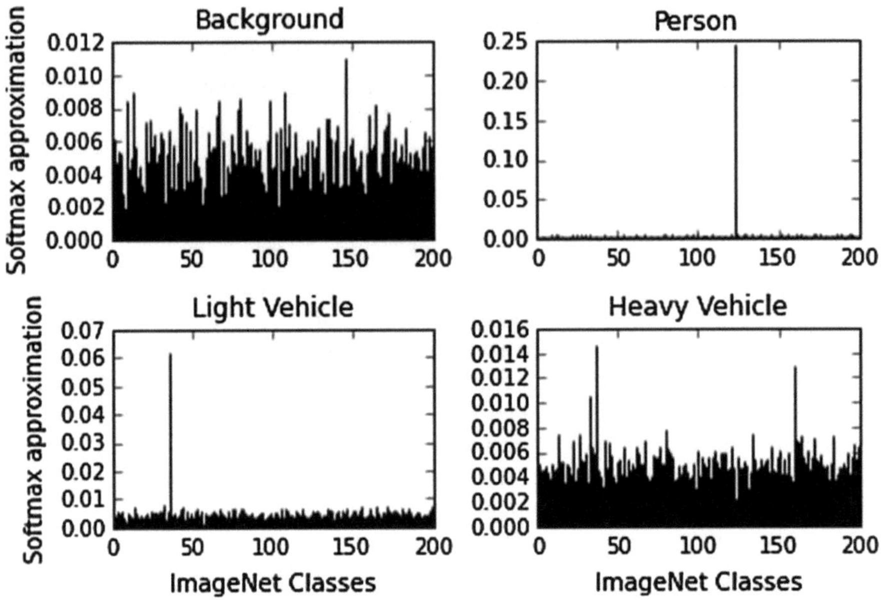

Fig. 2 The average class estimate for a set of mining related images. Notice that person (class 123) and light vehicle/car (class 36) are existing classes for the pre-trained network and can be used directly. The background and the heavy vehicle classes are novel and show a wider spread as they are not modelled with the pre-train ConvNet

With this simple class mapping approach and assuming that falsely picking one of the positive classes is in fact uniformly random, we expect to eliminate 99 % of all the proposed background regions. However, when processing around 100 proposals per frame, the expected false positive rate is once per frame. Next we propose a simple background model that reuses the ConvNet computation to provide a background likelihood estimate for reducing this false positive rate.

3.3 Background Modelling

While on a mine-site the landscape is constantly changing from a geometric perspective, the bleak visual appearance of the background is generally constant. For this, we model the background regions as belonging to one of an arbitrary set of categories, such as the semantic categories of rock, sky, tree etc. If a sample differs significantly from any of these background classes then we can assume it is an object of interest.

Rather than using supervised techniques that require a set of manually annotated images, we instead partition the background data without explicit semantic labels. To do this, we exploit the assumption that intra-category samples generally appear visually similar to each other, yet may be distinctively different to other background categories. Put another way, the background regions form natural clusters enabling us to employ unsupervised techniques to model their visual appearance. See Fig. 3 for an illustration of the natural background clusters found by applying this method to a mining dataset.

To describe the visual appearance of each region, the intermediate layers of the ConvNet provide a free and compact representation suitable for this task. Additionally, these features have been shown to be robust against lighting and viewpoint

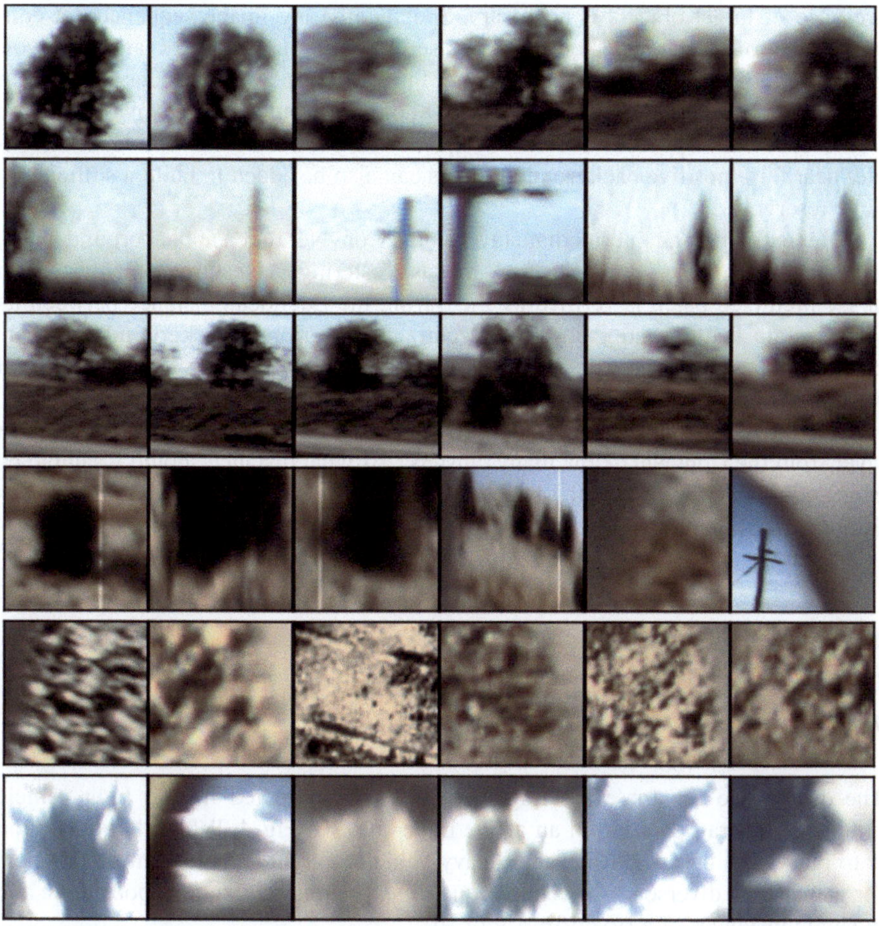

Fig. 3 An illustration showing six of the most common types of background region proposals. The rows represent different clusters while the columns show a random background region which is a member of the associated cluster. Each cluster gathers samples with similar visual appearance such as centred on a tree (*top row*) or centred on sky with an adjacent vertical structure (*second row*)

changes without any re-training [25]. We refer the interested reader to [13] for an illustration of the ConvNet's inner workings. In general, the first layer of a ConvNet extracts simple colour and texture features in the first layer, and through subsequent layers, these features eventually transition to the learnt specific task [29] such as classifying the 200 ImageNet classes. Along the way irrelevant visual information for the original task (e.g. features describing sky) are lost once it reaches the final layer. With this intuition we reuse the transformed data from one of the ConvNet's intermediate layers as an input to our background model.

To learn this cluster based model, a reservoir of negative samples is required. Gathering background data is a relatively simple task since only inspection for the presence of target objects is necessary. Specifically any image sequence not containing any of the target objects can be used to build an extremely large reservoir by extracting proposals from each frame. Furthermore, we only focus on difficult regions by perform hard-negative-mining [8] of background samples by running the ConvNet detection pipeline over these sequences. By lowering the confidence threshold, near false positive background regions can also be added to build a sufficiently large reservoir.

After extracting an intermediate layer of the ConvNet for each background patch, we then cluster these samples using k-means clustering. At test time, each `person` or `LV` predicted patch is verified by measuring the Euclidean distance between its intermediate feature and each cluster centre. If the nearest background cluster is close in this feature space, i.e. is visually similar, then we suppress the detection and regard it as background.

In building this background model the following choices are to be made: Which layer from the ConvNet? How many clusters? At what distance should a sample be considered background? In the following section we address these design choices through experimental validation.

4 Experiments

4.1 Mining Dataset

The dataset we use for evaluating this work was collected from a light vehicle mounted camera operating in an active mine-site, see Fig. 4. While the motivation is to put vision based sensing on a heavy vehicle, a light vehicle is more practical for gathering a diverse set of visual sequences. The dataset contains both static and dynamic instances of a `person`, `LV` or `HV`.

Continuous video was gathered with and without the camera in motion and on various haul roads and a few light vehicle only zones to capture variation in the environment. This video data was captured at 10 fps and partitioned into various sequences. In this work we use 5 sequences where no people or vehicles are visible

Fig. 4 The experimental dataset gathering vehicle with cameras mounted to the bullbar. *Note* all images used in this paper were captured from the camera on the left hand side of the vehicle

to build our background model. Collectively these background sequences make up 8952 frames in total (approximately 14 km).

To evaluate the performance we use another 5 sequences with several instances of `person`, `LV` or `HV`, that we personally annotated using the tool developed by Vondrick et al. [28]. These annotated sequences contain 9405 frames in total (approximately 10 km). In addition to these sequences we made a small validation set of 200 using other images collected on a mine site from various sources including a few captured at night. This set was used to generate Fig. 2.

4.2 Background Model Validation

Here we describe the experiments performed to design our background modelling system explained in the previous section. From the 5 background sequences, we applied the region proposal and ConvNet detection framework to find challenging region proposals from every tenth frame. While some of the false objects may be observed in multiple frames, the time difference is sufficient to capture a variety of view points for these distracting objects. We lowered the detection threshold to collect region proposals if the ConvNet predicted either a `person` or `car` in the top 5 out of 200 class responses. With this configuration we collect around 8000 hard negatives for our background reservoir. We held out 90 of the most interesting background regions and added them to the validation set.

To address the design decisions for this model, we perform an empirical study using the reservoir containing only negatives and the validation set with both negative and positives. We jointly test different combinations of ConvNet layer features and number of clusters by evaluating their performance on the validation set. For the distance threshold we set this to the distance corresponding to a 95 % recall on the positive set. With the recall fixed, the overall performance of the background model is measured by the precision at which it can identify a true negative.

Figure 5 shows the relative performance of sweeping the number of clusters for different ConvNet layers. While `fc6` layer with 2048 clusters achieved the highest precision of 90 % we instead opted to use only 128 clusters with a precision of 89 % which is significantly faster to compute. A detailed view of the distances between the validation samples and the cluster centres can be seen in Fig. 6.

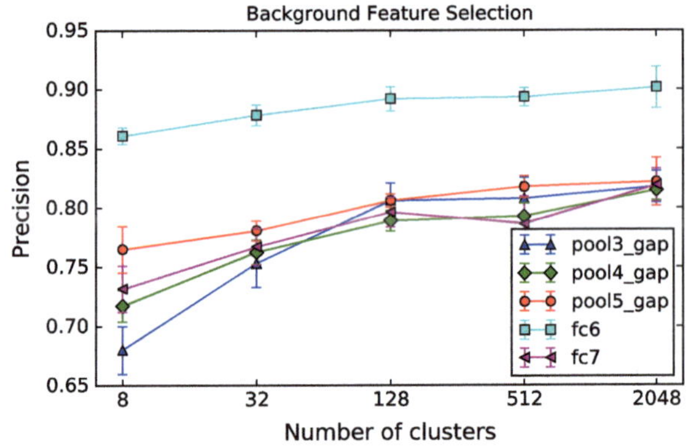

Fig. 5 Cross-validation precision at 95 % recall for different ConvNet layers and the number of clusters used to represent the background. Each point shows average of 5 trials

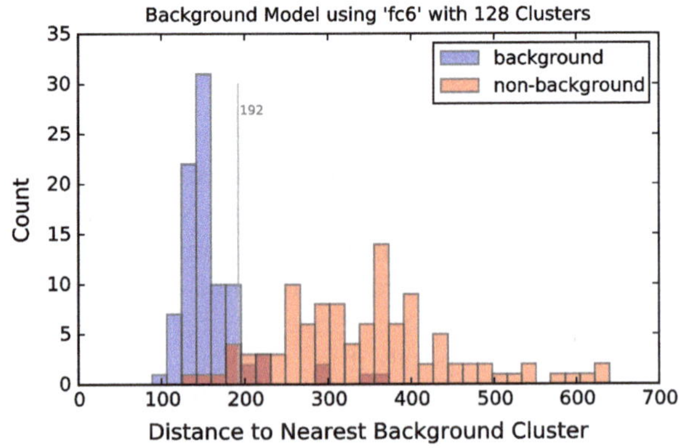

Fig. 6 Detailed view of the distribution of the validation images distance to their nearest background cluster centre. The *grey line* marks the 95 % recall distance threshold

Implementation Detail

The first 5 ConvNet layers produce dense tensor representations which gradually reduce in size. Then there are two `fully connected` layers `fc6` and `fc7` before the final prediction layer. Again we refer the interested reader to [13] for details of the ConvNet structure. Due to the density of data and the computational complexity of computing distances in such high dimensional feature spaces we only evaluate the ConvNet layers 3–7 and compress convolutional layers 3–5 by pooling all filter responses across the feature map for each tensor in [15] this is referred to global average pooling. In Fig. 5 these are marked as pool{3–5}_gap.

The false negatives and some of the false positives are also shown in Fig. 7. The false negatives are mostly night images which can be put down to the fact that similar images are rare if not non-existent in the ImageNet samples used to train the ConvNet. For the false positives, these are mostly signs which make up a minority of the scene. From these samples we can describe our background model as a form of novelty detection where interesting parts of the scene such as signs are distinguished from the general background. This finding along with the unsupervised clustering shown in Fig. 3 are a testament to the ConvNet's expressive capabilities in representing visual similarity.

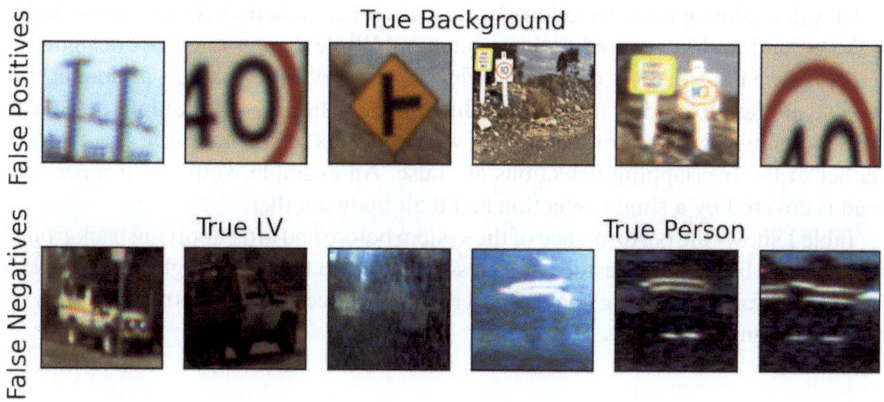

Fig. 7 Validation samples where the background model failed. Images are shown in their warped form, representing the ConvNet input. The four *right* false negatives were collected at night

Table 1 System comparison before and after background suppression (BGS) on mining sequences

Sequence (frames)	F1 score[a] (Precision, Recall)		Mostly hit[b]		Mostly missed[b]		False positives	
	Baseline	With BGS[c]	–	BGS	–	BGS	–	BGS
1 (1462)	0.38 (0.57, 0.29)	**0.40** (0.77, 0.27)	2	2	16	16	242	**87**
2 (2950)	**0.94** (0.96,0.91)	0.93 (0.97, 0.89)	3	3	6	6	73	**47**
3 (599)	0.02 (0.01, 0.09)	**0.06** (1.00, 0.03)	0	0	2	2	349	**0**
4 (2826)	0.64 (0.56, 0.74)	**0.80** (0.95, 0.69)	**2**	1	**4**	5	186	**9**
5 (1568)	**0.68** (0.78, 0.61)	0.43 (0.92, 0.28)	**4**	1	**3**	6	177	**24**
Total			11	7	31	35	1027	**167**

[a]F1, Precision and Recall is computed treating each frame as independent
[b]Mostly indicates where a single object instance was detected or missed 50 % of the time
[c]The proposed background suppression (BGS) is applied to the baseline `EdgeBox` and ConvNet detector

4.3 Detection Evaluation

We now evaluate the system on the set of 5 sequences with `person` or `LV` where the task is to locate objects of interest. In this evaluation we consider a true detection if at least 50 % of the detection region is covered by a single ground truth object. This differs from the intersection-over-union (IOU) definition of overlap, as we accept detecting a `person`'s head and shoulders without their whole body while IOU would count this as both a miss detection and a false positive. It should be also noted that any detection or miss detection of a `person` or `LV` labelled as partially occluded in the ground truth is ignored in this evaluation. While the system is not designed to detect `HV` we consider any detections which overlap with `HV` objects as neither true or false and are excluded from the evaluation. Additionally, if multiple detections overlap a single ground truth instance, we count this as a single true positive and neither of the overlapping detections are false. An example would be if a person's head is covered by a single detection and their body another.

Table 1 shows the performance of the system before and after applying background suppression. From these results we can see that while there is a slight drop in recall our method for suppressing background regions reduces the false positive rate by an order of magnitude.

5 Conclusions and Future Work

In this paper we presented a vision only system that takes advantage of recent developments in computer vision and machine learning to detect both personnel and light vehicles. We circumvented the problem of ConvNet over-fitting on small datasets by reusing a pretrained model directly and mapping its output to mining classes. We further presented a method for exploiting the abundance of background only images to learn a background cluster model leading to a significant reduction in false positives. This sensing approach was evaluated in an active open-pit mine site environment. The experiments show that the in-pit environment is suitable for object proposals along with background modelling techniques such as the one presented here.

While this work is only concerned with single camera based sensor data we see many opportunities to combine techniques incorporating stereo [2] or range-based sensors [20] for improved robustness. As an initial investigation of vision as a possible sensor on a mine we see many opportunities to further improve on the results. As more labelled mining image data becomes available we expect to be able to design and fine-tune a ConvNet that performs better in this domain than the existing network. We also plan to extend this work to fuse information from multiple frames by combining the ConvNet appearance model with recent motion segmentation techniques [1].

Acknowledgments This research was funded by the Australian Coal Association Research Program (ACARP). The authors would also like to acknowledge AngloAmerican for allowing data collection at the Dawson operation. Acknowledgement also goes to the high performance computing group at Queensland University of Technology for both support and use of their services when conducting some of the experiments in this paper.

References

1. Bewley, A., Guizilini, V., Ramos, F., Upcroft, B.: Online self-supervised multi-instance segmentation of dynamic objects. In: International Conference on Robotics and Automation, Hong Kong, China, IEEE (2014)
2. Bewley, A., Upcroft, B.: Advantages of exploiting projection structure for segmenting dense 3D point clouds. In: Australian Conference on Robotics and Automation (2013)
3. Deng, H., Clausi, D.A.: Unsupervised image segmentation using a simple MRF model with a new implementation scheme. Pattern Recognit. **37**(12), 2323–2335 (2004)
4. Deng, J., Dong, W., Socher, R., Li, L.-J., Li, K., Fei-Fei, L.: ImageNet: a large-scale hierarchical image database. In: IEEE Conference on Computer Vision and Pattern Recognition, pp. 248–255, June 2009
5. Donahue, J., Hoffman, J., Rodner, E., Saenko, K., Darrell, T.: Semi-supervised domain adaptation with instance constraints. In: IEEE Conference on Computer Vision and Pattern Recognition, pp. 668–675, June 2013
6. Farabet, C., Couprie, C., Najman, L., LeCun, Y.: Learning hierarchical features for scene labeling. Pattern Anal. Mach. Intell. **35**(8), 1915–1929 (2013)
7. Fei-Fei, L., Fergus, R., Perona, P.: Learning generative visual models from few training examples: an incremental Bayesian approach tested on 101 object categories. Comput. Vis. Image Underst. **106**(1), 59–70 (2007)
8. Felzenszwalb, P.F., Girshick, R.B., McAllester, D., Ramanan, D.: Object detection with discriminatively trained part-based models. IEEE Trans. Pattern Anal. Mach. Intell. **32**(9), 1627–1645 (2010)
9. Felzenszwalb, P.F., Huttenlocher, D.P.: Efficient graph-based image segmentation. Int. J. Comput. Vis. **59**(2), 167–181 (2004)
10. Girshick, R.B., Donahue, J., Darrell, T., Malik, J.: Rich feature hierarchies for accurate object detection and semantic segmentation. In: Computer Vision and Pattern Recognition (CVPR) (2014)

11. Hosang, J., Benenson, R., Schiele, B.: How good are detection proposals, really?. In: British Machine Vision Conference (BMVC) (2014)
12. Jia, Y.: Caffe: an open source convolutional architecture for fast feature embedding (2013). http://caffe.berkeleyvision.org/
13. Krizhevsky, A., Sutskever, I., Hinton, G.E.: ImageNet classification with deep convolutional neural networks. Adv. Neural Inf. Process. Syst. (NIPS) **1**(2), 4 (2012)
14. LeCun, Y., Boser, B., Denker, J.S., Henderson, D., Howard, R.E., Hubbard, W., Jackel, L.D.: Backpropagation applied to handwritten zip code recognition. Neural Comput. **1**, 541–551 (1989)
15. Lin, M., Chen, Q., Yan, S.: Network In Network (2013). arXiv preprint
16. Lin, T.-Y., Maire, M., Belongie, S., Hays, J., Perona, P., Ramanan, D., Dollár, P., Zitnick, C.L.: Microsoft COCO: common objects in context. In: European Conference on Computer Vision (ECCV), vol. 8693, pp. 740–755. Springer International Publishing (2014)
17. Marshall, J.A., Barfoot, T.D.: Design and field testing of an autonomous underground tramming system. Springer Tracts Adv. Robot. **42**, 521–530 (2008)
18. Mosberger, R., Andreasson, H.: Estimating the 3d position of humans wearing a reflective vest using a single camera system. In: International Conference on Field and Service Robotics (FSR) (2012)
19. Mosberger, R., Andreasson, H., Lilienthal, A.J.: Multi-human tracking using high-visibility clothing for industrial safety. In: IEEE/RSJ International Conference on Intelligent Robots and Systems (IROS), pp. 638–644 (2013)
20. Phillips, T., Hahn, M., McAree, R.: An evaluation of ranging sensor performance for mining automation applications. In: IEEE/ASME International Conference on Advanced Intelligent Mechatronics: Mechatronics for Human Wellbeing, AIM 2013, pp. 1284–1289 (2013)
21. Razavian, A.S., Azizpour, H., Sullivan, J., Carlsson, S.: CNN features off-the-shelf: an astounding baseline for recognition. In: The IEEE Conference on Computer Vision and Pattern Recognition (CVPR) Workshops (2014)
22. Roberts, J.M., Corke, P.I.: Obstacle detection for a mining vehicle using a 2D laser. In: Proceedings of the Australian Conference on Robotics and Automation, pp. 185–190 (2000)
23. Sermanet, P., Eigen, D., Zhang, X., Mathieu, M., Fergus, R., LeCun, Y.: Overfeat: integrated recognition, localization and detection using convolutional networks. In: International Conference on Learning Representations (ICLR 2014), December 2014
24. Sermanet, P., Kavukcuoglu, K., Chintala, S., LeCun, Y.: Pedestrian detection with unsupervised multi-stage feature learning. In: IEEE Conference on Computer Vision and Pattern Recognition, pp. 3626–3633. IEEE, June 2013
25. Sunderhauf, N., Dayoub, F., Shirazi, S., Upcroft B., Milford, M.: On the performance of ConvNet features for place recognition. In: arXiv (2015)
26. Torralba, A., Fergus, R., Freeman, W.: 80 Millions tiny images: a large dataset for non-parametric object and scene recognition. IEEE Trans. Pattern Anal. Mach. Intell. **30**, 1958–1970 (2008)
27. Uijlings, J.R.R., Sande, K.E.A., Gevers, T., Smeulders, A.W.M.: Selective search for object recognition. Int. J. Comput. Vis. **104**(2), 154–171 (2013)
28. Vondrick, C., Patterson, D., Ramanan, D.: Efficiently scaling up crowdsourced video annotation. Int. J. Comput. Vis. **101**(1), 184–204 (2012)
29. Yosinski, J., Clune, J., Bengio, Y., Lipson, H.: How transferable are features in deep neural networks?. In: Advances in Neural Information Processing Systems (NIPS) (2014)
30. Zitnick, C.L., Dollár, P.: Edge boxes: locating object proposals from edges. In: European Conference on Computer Vision (ECCV) (2014)

Part VI
Systems

Building, Curating, and Querying Large-Scale Data Repositories for Field Robotics Applications

Peter Nelson, Chris Linegar and Paul Newman

Abstract Field robotics applications have some unique and unusual data requirements—the curating, organisation and management of which are often overlooked. An emerging theme is the use of large corpora of spatiotemporally indexed sensor data which must be searched and leveraged both offline and online. Increasingly we build systems that must never stop learning. Every sortie requires swift, intelligent read-access to gigabytes of memories and the ability to augment the totality of stored experiences by writing new memories. This however leads to vast quantities of data which quickly become unmanageable, especially when we want to find what is relevant to our needs. The current paradigm of collecting data for specific purposes and storing them in ad-hoc ways will not scale to meet this challenge. In this paper we present the design and implementation of a data management framework that is capable of dealing with large datasets and provides functionality required by many offline and online robotics applications. We systematically identify the data requirements of these applications and design a relational database that is capable of meeting their demands. We describe and demonstrate how we use the system to manage over 50TB of data collected over a period of 4 years.

1 Introduction

Lifelong learning for robotic systems requires large quantities of data to be collected and stored over long periods of time. As these data accumulate, they become increasingly difficult to manage and query. Without a scalable system in place, finding

P. Nelson (✉) · C. Linegar · P. Newman
Mobile Robotics Group, Department of Engineering Science,
University of Oxford, Oxford, UK
e-mail: peterdn@robots.ox.ac.uk

C. Linegar
e-mail: chrisl@robots.ox.ac.uk

P. Newman
e-mail: pnewman@robots.ox.ac.uk

D.S. Wettergreen and T.D. Barfoot (eds.), *Field and Service Robotics*,
Springer Tracts in Advanced Robotics 113, DOI 10.1007/978-3-319-27702-8_34

useful data becomes ever more complex and this undermines our goal of achieving long-term autonomy.

Many publications focus on the mechanics of lifelong autonomy but very few explicitly deal with the problem of storing and accessing the required data in a way designed to aid long-term, large-scale mobile autonomy. Given the need for field roboticists to build coherent systems, it is time for this subject to be addressed.

Mobile robotics applications have some unusual data needs that cannot always be anticipated in advance. An example is illustrative. We often want to evaluate the efficacy of a new feature detector for visual odometry and thus testing under differing lighting conditions is vital. What we actually want to do is automatically collate image sequences that satisfy complicated compound queries such as 'find sequences >50 m of stereo images, captured while driving into the sun over wet ground in the absense of dynamic obstacles'. This should run over all data ever collected and return a pristine dataset as if this had been the sole purpose of our experimentation over the past 4 years.

As another example, we need images of traffic lights to train a new state-of-the-art traffic light detector. Instead of wasting time collecting a whole new set of data for this specific purpose, we should first look to our existing data. It is probably the case that we inadvertently have images of traffic lights from previous data collection missions, and therefore we would like the ability to search for them.

To aid in solving this problem, we have designed and implemented a relational database framework that is applicable to a wide range of robotics applications. A data and query model is presented which cleanly distinguishes between sensor data and user-defined metadata. This makes it trivial for a user to decorate the database with their own contributions, and makes those contributions accessible to other users in a consistent way. For example, if someone builds the aforementioned traffic light detector and runs it over 100,000 images, they are then able (and encouraged) to add those results to the database for others to use in the future.

Our framework not only makes offline batch processing tasks easier, but also supports the needs of online tasks, for example the storage and retrieval of maps used by a robot's navigation system at runtime. This massively reduces the overhead of implementing and testing new navigation systems as data back-ends do not need to be written from scratch. A motivating use case that demonstrates this is our Experience Based Navigation (EBN) system [1], a visual navigation framework designed to deal with vast maps that grow continuously over a robot's entire lifetime (see Fig. 1). EBN utilises our database framework in order to fulfil these challenging data storage and retrieval demands.

In the following sections we present the design and implementation of this framework. We analyse the performance of the system and demonstrate its real-world use by EBN.

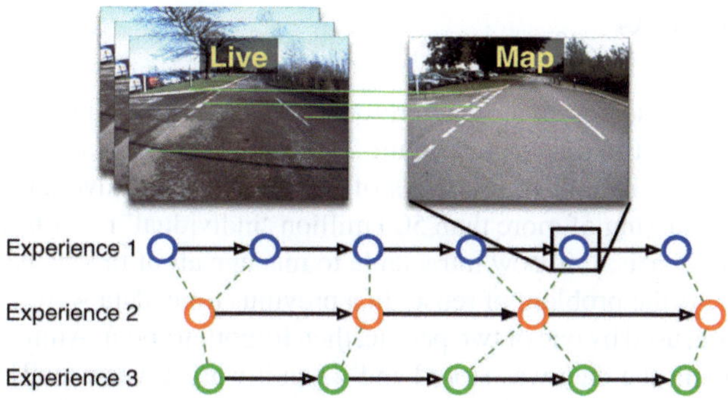

Fig. 1 Experience Based Navigation (EBN) [1] is a state-of-the art visual navigation framework for autonomous robots. Maps, in the form of 'experiences' with associated graph structure, camera images, and landmarks accumulate over many years. This paper aims to answer the question: 'how do we store and retrieve this kind of data in a flexible and efficient way?'

2 Related Work

In the context of lifelong learning for autonomous robots, very little work has so far addressed the problem of organising, maintaining, and utilising the vast amounts of data that will accumulate over long periods of time. Traditionally, (relatively) small datasets are collected for specific purposes and are often discarded or forgotten about once they have served their purpose.

Various datasets have been released to the community along with associated tools and documentation. Some of these have been widely referenced in other publications and serve as convenient benchmarks for comparing performances of related techniques. Examples include the New College Vision and Laser dataset [2] and the DARPA Urban Challenge dataset [3]. Although these resources are invaluable to the robotics and computer vision communities, they are also utterly static—the data are a snapshot of a single period in time and adding to them in a way that is accessible for others to use is difficult. Additionally, as these datasets are often formatted differently and require non-standard tools to use, it is cumbersome to mine data from several at once. To be useful in the context of lifelong learning, we wish to move away from the idea of disjoint, immutable datasets and towards a living, growing repository of everything we record and have ever recorded.

RoboEarth [4] is an ambitious project that proposes to solve some of these problems by building a 'World Wide Web' for robots. A distributed cloud database is used to store machine-readable semantic data about environments, objects, and actions. Generic robots can access this prior knowledge to help complete a task and can upload their own knowledge once they succeed. A modular software architecture enables generic actions (e.g. moving, grasping) to be realised on specific hardware. A subset of RoboEarth's vision is close in spirit to what we want to achieve, however it places more of an emphasis on the storage, retrieval, and reconciliation of knowledge required for high-level planning and reasoning tasks.

3 Requirements

Our own condition serves as motivation for what is to follow, and we suspect these requirements are not unique to us. Figure 2 shows how our data have accumulated exponentially over the past 4 years. As of March 2015 we have amassed over 50 TB of data, comprising of more than 500 million 'individual' records, and more are added on a daily basis. It is now intractable to manage all of this by hand.

Firstly there is the problem of reuse. In a previous time, data were collected for a specific purpose, used by one or two people, then forgotten about. Almost no semantic information about the data was stored and when it was, it was usually done so in a non-standard ad-hoc way (such as in notes on a wiki, or in a readme). Information and metadata (annotations) extracted from processed datasets suffered from the same problems. When gigabytes of data are collected and processed like this it becomes increasingly difficult to reuse what is considered useful and instead is easier to collect and process new data for each new purpose. To address this problem, we require that existing useful data can be found effectively and we therefore need a way to index it, as well as efficient ways to add new data when necessary. We also require a standard method for data annotation and require that these annotations can be easily traced to their underlying data.

Next, we have the problem of retrieval. Once we know a dataset contains relevant data, we must somehow retrieve them. For batch tasks, this would have been done by manually inspecting the dataset, chopping out the required parts and discarding the rest—a time-consuming and boring process. For online tasks, data would typically be accessed through a custom-written ad-hoc back-end, leading to bugs and time wasted on re-implementing common components.

Lastly, we have the problem of physical storage and organisation of data. Previously, datasets were organised only by directory structure and filename. This reliance on physical location presents many problems when datasets must be moved. Users should not need to care which machines their data physically reside on and so aim

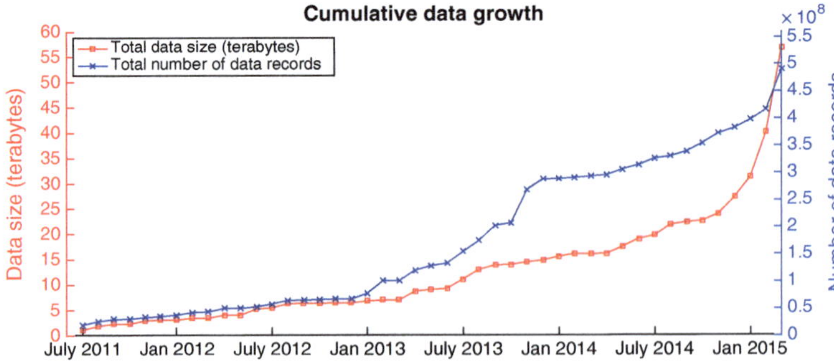

Fig. 2 Cumulative amount of sensor data we have amassed each month from July 2011 to March 2015. Shown in red is the total size of these data in terabytes and shown in blue is the approximate number of 'individual' data records (i.e. single camera images, GPS readings, laser scans)

to build a modular system that is platform and tool agnostic. Users should be able to issue queries and access data using any programming language on any operating system. Ideally, queries should run online on robotic platforms with little or no modification.

4 Data and Query Models

Here we describe the main data and query models that underpin our framework. At the lowest level is *raw sensor data*: typically recorded directly from physical hardware and subjected to minimal, if any, processing. *Annotations* exist as a layer above this and are related directly to the data they annotate. Data that do not fall neatly into either of these two categories follow the standard relational model.

In many ways these categories are arbitrary and some data can fall into multiple categories depending on context. However, they help us identify common types of queries which we take advantage of to reduce complexity. Returning to our traffic light example: raw data include images captured by a camera during a data collection mission. Annotations include the output of a traffic light detector on these images (perhaps bounding boxes and labels). Other data include everything else that isn't directly causally related, for example an OpenStreetMap [5] road network map.

The following subsections make some use of relational algebra, a comprehensive introduction to which is available in [6].

4.1 Raw Sensor Data

Raw sensor data form the bedrock of our data model and many of our subsequent processing needs. As the collection of sensor data is such a fundamental part of our workflow, it seems appropriate that they be given special consideration.

Raw data are characterised by the fact that they represent part of the state of the world, as observed by some robotic platform, at a particular moment in time. A fundamental query we wish to support is to find the *entire* state of the world, as observed by a platform, at a particular moment in time. For example, our detector finds a traffic light in some camera image that was captured at time t. We therefore may want to find, from our collection of raw data, the corresponding GPS reading giving the location of the robot at time $t' \approx t$. This will be the first and most coarse step in many applications' processing pipelines.

Consider a robotic platform $r \in R$ representing an entity equipped with a set of sensors $S(r)$. The output datum of sensor $s^i \in S(r)$ at time t is denoted s^i_t. We assume that a single sensor can be mounted on more than one platform, though not at the same time. We also assume that, by definition and without loss of generality, sensors cannot produce multiple data at the same time (i.e. in cases such as stereo cameras, we treat both images as a single datum). Given these constraints, it can be seen that ⟨platform, sensor, timestamp⟩ tuples map to individual data via $\bar{\Omega}$:

$$s^i_{t'} \leftarrow \bar{\Omega}(r, s^i, t) \tag{1}$$

where t' is the nearest[1] discretised timestamp for which a sensor reading exists. It follows that ⟨platform, timestamp⟩ tuples map to the state of the world as observed by the entire platform via Ω:

$$\{s^i_{t^i}\} \leftarrow \Omega(r, t) \tag{2}$$

where t^i is the nearest discretized timestamp for which an output from s^i exists.

These observations give rise to three relations (tables) whose rows respectively represent individual robotic platforms, individual sensors, and individual data:

$$\text{Platforms}(\underline{\textit{platform\_id}},\ \ldots)$$

$$\text{Sensors}(\underline{\textit{sensor\_id}},\ \ldots)$$

$$\text{Data}(\underline{\textit{data\_id}},\ \textit{platform\_id},\ \textit{sensor\_id},\ \textit{timestamp},\ \textit{data},\ \ldots)$$

Expressed in relational algebra, Eq. (1) becomes a simple selection over rows in the Data relation:

$$\sigma_{\text{platform\_id} = r \,\wedge\, \text{sensor\_id} = s^i \,\wedge\, \text{timestamp} \approx t}(\text{Data})$$

Equation (2) is given by:

$$\sigma_{\text{timestamp} \approx t}(D_{r,s^1} \bowtie_T \ldots \bowtie_T D_{r,s^N})$$

where $\{s^1, \ldots, s^N\} = S(r)$, $D_{r,s^i} = \sigma_{\text{platform\_id} = r \,\wedge\, \text{sensor\_id} = s^i}(\text{Data})$ and $\bowtie_T$ is a *temporal join* operator that associates records whose timestamps are close together.

4.2 Annotations

Annotations are characterised by the fact that they mandatorily relate to other data—whether raw data, other annotations, or otherwise. In other words, they are 'meaningless' without context. To expand on our traffic light example: one detector may create annotations consisting of bounding boxes that identify the locations of traffic

[1] The definition of 'nearest' may differ between queries but note that it is beyond the scope of the system to, for example, interpolate between records or verify annotation correctness—how that is handled is up to individual client applications.

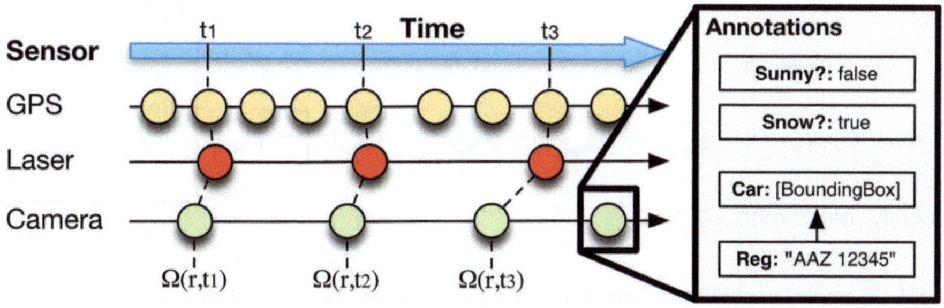

Fig. 3 Visualisation of some raw GPS, laser, and camera data and the relative times they were captured. An example annotation hierarchy for a single image datum is shown—note that the 'Car' annotation has a child that contains the detected registration number. Links between data records that are temporally joined at times t_1, t_2, and t_3 are shown as *dashed lines*

lights in camera images. Another detector might annotate *these* annotations, indicating the state of the traffic lights, or other higher-level properties. Metadata, such as the version of the algorithm used to generate the labels, might also be included.

Annotations for raw data are stored in their own recursive relations:

$$\text{Annotations}(\underline{annotation\_id},\ data\_id,\ parent\_id,\ annotation\_data,\ \ldots)$$

The *data_id* field references a row in the Data relation, and the *parent_id* field references a lower-level annotation. One and only one of these fields must be non-empty. Queries such as 'find all raw data with a particular annotation θ' are simply a selection over the join of the Data and Annotations relations (more complex compound queries can have any number of conditions chained together):

$$\sigma_{\text{annotation\_data}=\theta}(\text{Data} \bowtie \text{Annotations})$$

Figure 3 shows graphically how annotations and raw data are related.

4.3 *Other Data*

This class encompasses any other arbitrary data that do not fall neatly into the two aformentioned categories. For example, it includes completely standalone data, such as OpenStreetMap maps and sensor calibration information. Harnessing the power of the relational model, these data can still link to raw sensor data and annotations if the need arises. For example, one of our traffic light annotations could link to the OpenStreetMap intersection ID representing where the light is located.

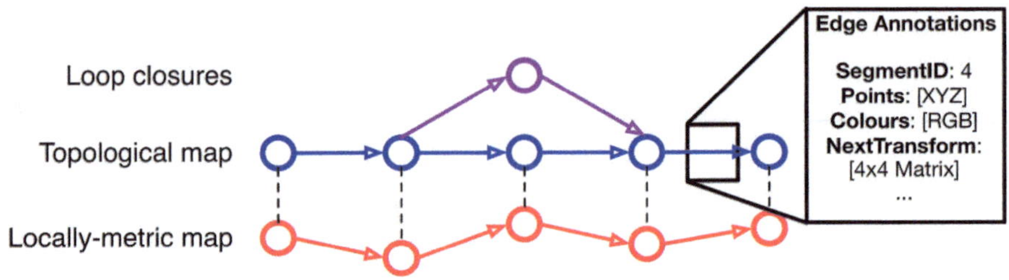

Fig. 4 Example GraphDB structure representing a metric map, topological map, and loop closures. Example annotations linked to a single edge are shown

4.3.1 Multigraph Maps

One ubiquitous pattern that we have explicitly considered is the use of a directed multigraph structure (*GraphDB*) to represent maps for use by localisation and navigation systems, for example EBN. The graph is defined by two relations and is encoded in an edge-list representation. Two additional relations store links between nodes, edges, and annotations:

$$\text{Nodes}(\underline{\mathit{node\_id}})$$

$$\text{Edges}(\underline{\mathit{edge\_id}}, \mathit{node\_from\_id}, \mathit{node\_to\_id})$$

$$\text{NodeAnnotations}(\underline{\mathit{node\_id}, \mathit{annotation\_id}})$$

$$\text{EdgeAnnotations}(\underline{\mathit{edge\_id}, \mathit{annotation\_id}})$$

An example of how a map might be represented using this structure is shown in Fig. 4.

5 Implementation

5.1 How Do We Store Data?

We have standardized the way raw sensor data are stored in an attempt to eliminate fragmentation of our tools and methods. A *monolithic file* contains a sequence of atomic data records. Each record consists of an XML header and a binary blob containing a serialised Google Protocol Buffer[2] message. The XML header contains sensor-agnostic metadata about the message such as a timestamp indicating when it

[2]https://developers.google.com/protocol-buffers/.

was collected, its length in bytes, and the name of the corresponding message format. As sensor data are being collected, monolithic files are constructed on the fly. Raw sensor data are streamed through a driver which converts them into messages of the appropriate format and both the message and header are appended to the output monolithic. As they are simply an unlinked sequence of distinct and atomic records, monolithics can be constructed, split, and concatenated easily, and their length is limited only by the underlying file system.

Raw sensor data are stored in monolithic files in a well-defined directory structure that is organised by the platform and sensors used to collect them, and time of collection. The top-level directory is located on a network drive that allows users to mount it to anywhere in their own filesystem.

5.2 *Relational Data and Query Framework*

We use a relational database management system (DBMS)—either PostgreSQL[3] or SQLite,[4] depending on our particular use case. PostgreSQL is designed for a multi-user environment and has full transactional and locking abilities, allowing many concurrent connections to the database without the risk of data corruption. These guarantees come at the cost of performance—queries must be sent over pipe or network to the PostgreSQL server process which executes them and returns results over the same medium. For real-time, online applications, we therefore prefer SQLite which stores all necessary data in a single file. Applications access this file via a shared library so no costly inter-process or network communication is required.

As SQL is a concrete superset of relational algebra, the schemas and queries described in Sect. 4 can be translated trivially (see [6] for an overview) into appropriate `SELECT` statements. In particular, any ‘linking’ of data is done using foreign key constraints. Compound queries are implemented using joins or subqueries.

As raw data are kept in monolithic files, we only store references to them in the DBMS. They do not contain a built-in index, meaning searching them for specific data requires a slow linear scan. Therefore, we store an index in an SQL table which holds offsets into monolithic files for every record. This table corresponds to the Data relation described in Sect. 4.1.

Annotations are stored in one global table which corresponds to the Annotation relation described in Sect. 4.2. This encodes the annotation hierarchy. Annotation-specific data (for example, the bounding boxes from our traffic light detector) are then stored in their own tables. Records in these tables link to their corresponding records in the global table. Annotations can then be created, updated, and deleted using SQL `INSERT`, `UPDATE`, and `DELETE` statements, respectively.

The GraphDB structure is implemented as tables which correspond to the relations described in Sect. 4.3.1. A dedicated API is provided that handles graph-based

---

[3]http://www.postgresql.org/.

[4]http://www.sqlite.org/.

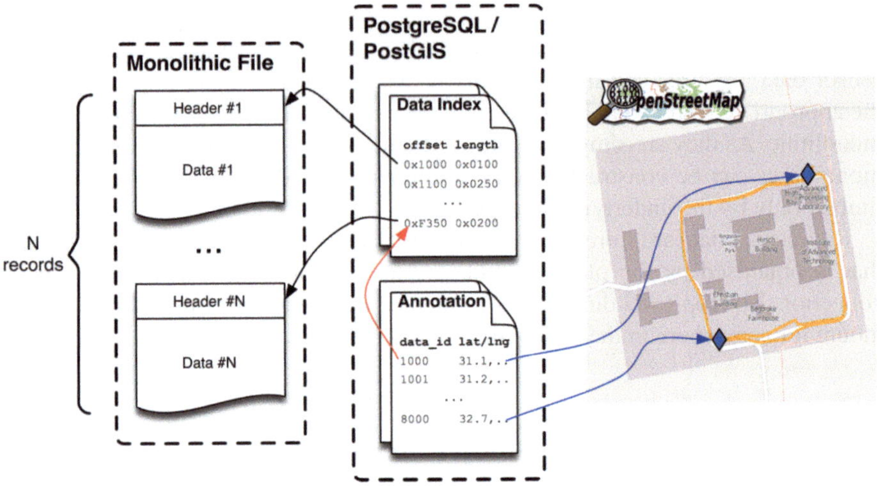

Fig. 5 Overview of the concrete data model structure. At the lowest level, raw data are stored in sequential records in flat 'monolithic' files. An index table in the DBMS holds offsets that point to these records (represented by *black arrows*). Annotations exist in separate tables with each individual annotation pointing to some record in the index (*red arrow*) and optionally to any other data—in this example, to locations in OpenStreetMap (*blue arrows*)

operations such as creating, querying, and deleting nodes, edges, and their respective annotations.

In addition, we keep a subset of OpenStreetMap [5] consisting of all roads and regions in England. This is stored locally in our centralised PostgreSQL instance using the PostGIS[5] extension, allowing us to perform geospatial queries very efficiently.

Figure 5 shows an example of how the aforementioned components of the database interact.

6 Performance

The importance we ascribe to run-time performance of the database system depends on the specific use case in question. For example, batch tasks that process our entire 50TB of data have very different performance expectations to those of navigation systems running on live robots.

Both PostgreSQL and SQLite support indexes on data tables which we use to improve the performance of certain queries. These are implemented using well known B-Tree [7], R-Tree [8], and hash table data structures. For example, we create a B-Tree index on the *data_id* field in the Data table. This reduces the search for a record

[5]http://postgis.refractions.net/.

Table 1 Measured execution times for some example queries

Query	Execution time
Insert approx. 250,000,000 index records	12 h approx.
Select single datum by ID	0.06 ms
Select single datum by platform, sensor, and timestamp	0.06 ms
Temporal join to select data by platform and timestamp	7.5 ms
Temporal join for every record in an approx. 60 GB dataset	14.7 s

Performed using a test PostgreSQL instance containing a 250,000,000-record subset of our data (approximately 20TB)

identified by a specific *data_id* from $O(n)$ to $O(\log n)$. For most long-running batch tasks, the speedups afforded by these indexes are sufficient. Table 1 shows measured execution times for some example queries of this nature.

In the rest of this section we analyse some of the performance characteristics of PostgreSQL and SQLite under particular configurations.

6.1 PostgreSQL Versus SQLite

PostgreSQL is a server-based DBMS, meaning queries are issued via relatively costly interprocess communication or network calls. In return it provides full transactional and fault-tolerant behaviour and therefore allows many users to read and write data simultaneously without interfering with each other.

For real-time use cases, we instead prefer SQLite. It is linked statically and stores all of its data, metadata, and indexes in a single file which vastly reduces the overhead of its API calls.

Here we give performance results for the creation and traversal of a 'worst case' linked list graph structure. Although somewhat contrived, this example allows us to identify certain bottlenecks and observe how different configurations perform relative to one another. In the real world, access patterns and disk speed should also be major considerations.

Experiments were performed on an early 2011-era MacBook pro with 8 GB RAM and a 500 GB hard drive, with a PostgreSQL server running locally.

6.1.1 Experiments and Analysis

To test write performance, we create linear chains (successive nodes connected by single edges) of varying lengths n in the GraphDB. PostgreSQL and SQLite are tested in both *synchronous* (every write is fully flushed to disk) and *buffered* (writes are buffered in memory by the host operating system) modes. In the synchronous case, a write-ahead log is also used. In addition, we compare the creation of all nodes and edges in one operation (*batch*) versus creating them individually. The former case is more likely to be the behaviour of an offline batch task that has all data already available to it, whereas the latter behaviour is more likely to be exhibited by an online task that is constantly processing new data from its environment.

Timing results for write operations are shown in Fig. 6. Unsurprisingly, buffering individual writes is faster for both PostgreSQL and SQLite, although much more dramatically for the latter. Buffering batch writes makes virtually no difference. Batch writes of more than about 10–100 nodes are significantly faster than writing them individually. This is likely due to the overhead of API calls which begins to dominate in longer chains, particularly for PostgreSQL.

To test read performance, we load the previously created linear chains. Again we test how reading the whole graph in one operation (batch) compares with traversing the chain one node at a time.

Timing results for read operations are shown in Fig. 7. Similarly we see that reading in batch is far quicker than traversing nodes individually. Additionally, SQLite is faster than PostgreSQL by almost 1–2 orders of magnitude in all cases.

Overall SQLite consistently outperforms PostgreSQL, although the difference is less marked in batch than individual cases. We believe this is because of the overhead associated with interprocess communication return trips, of which there are $O(1)$ in the batch case compared with $O(n)$ individually. We use this as justification for our use of PostgreSQL as our centralised DBMS, suitable for use with long-running offline tasks—where the multi-user and fault tolerance properties outweigh slight performance gains—and SQLite for real-time tasks.

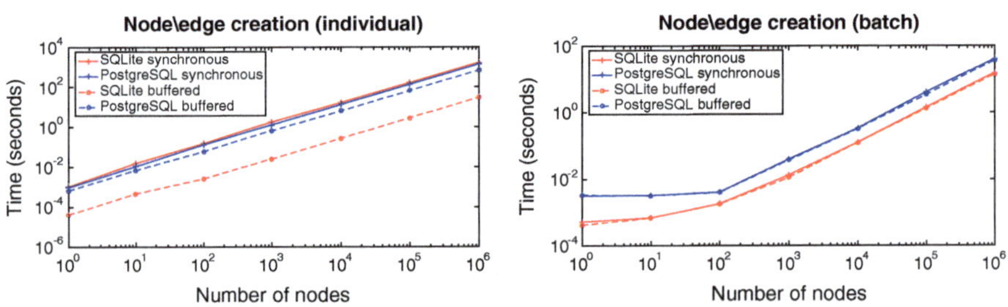

Fig. 6 Log-log plots showing time taken to create linear chains of various lengths in PostgreSQL and SQLite, in both synchronous and buffered modes. *On the left* nodes were all created individually—i.e. *n* calls to a `CreateNode` API function. *On the right* nodes were created in batch using a single API call

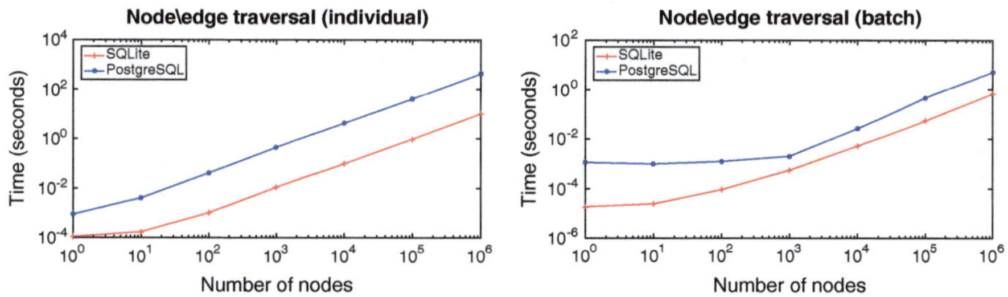

Fig. 7 Log-log plots showing time taken to read linear chains of various lengths in both PostgreSQL and SQLite. *On the left* nodes were all read individually—i.e. n calls to a `GetNextNode` API function. *On the right* the entire chain was read in batch using a single API call

7 Use Case: Experience Based Navigation

Experience-Based Navigation (EBN) [1] is a general-purpose framework for vast-scale, lifelong visual navigation. Outdoor environments exhibit challenging appearance change as the result of variation in lighting, weather, and season. EBN models the world using a series of overlapping 'experiences', where each experience models the world under particular conditions. New experiences are added in real-time when the robot is unable to sufficiently localise live camera images in the map of experiences. Additionally, the robot learns from its previous use of the experience map in order to more accurately retrieve relevant experiences from memory at runtime. The system has been tested on datasets totalling 206 Km of travel, successfully running in real-time at 20 Hz [9].

Since EBN must operate over vast distances and throughout the lifetime of a robot, it is essential that the map of experiences is persisted in long-term storage. In order to maintain reasonable memory requirements, the system needs to be able to selectively load relevant portions of the map into memory, while leaving others on disk. To meet real-time constraints, these data must also be efficiently accessible.

The map of experiences is represented as a graph. Nodes describe the appearance of the environment at particular times and poses, while edges specify topological and metric links between these places (see Fig. 1).

The GraphDB meets these requirements and therefore provides the mechanism for storing and retrieving data. Raw sensor data (e.g. images) and processed data (e.g. visual landmarks) are linked to corresponding nodes as annotations. Edges are annotated with 6DOF relative transformations, giving the graph a local metric structure.

The pipeline for localisation is shown in Fig. 8. The breadth-first search enables EBN to load relevant portions of the map (nodes nearby the robot) into memory at runtime. Since this search may require a large number of read requests, the SQLite implementation of the GraphDB is used in buffered mode to maintain real-time performance.

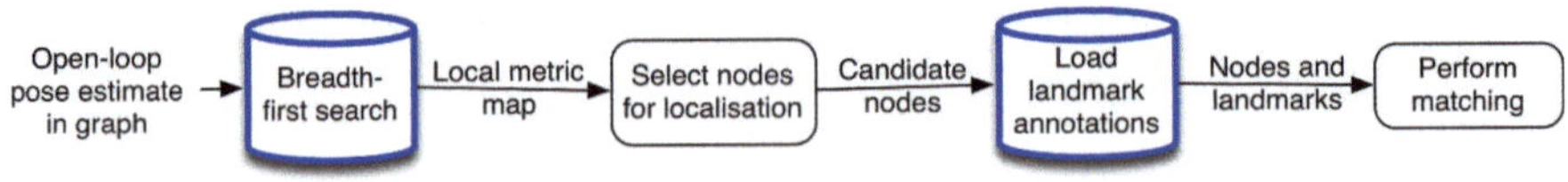

Fig. 8 Overview of the EBN localisation pipeline. This process is strictly required to run at 20 Hz—the system has 50 ms to complete, including data access tasks (highlighted in *blue*)

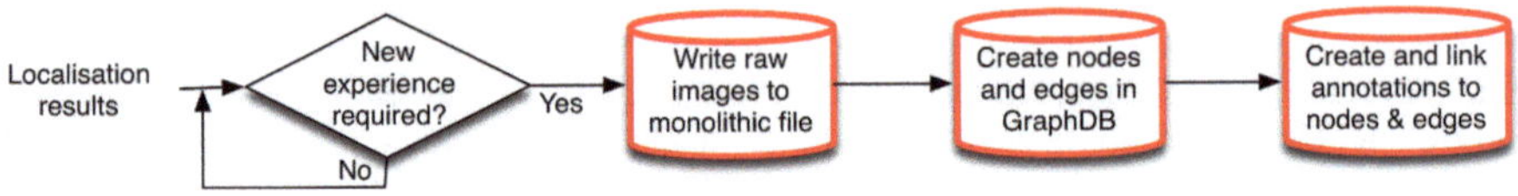

Fig. 9 Overview of the EBN experience creation pipeline. This process has less strict runtime requirements and runs in a background thread. Writes to the database are highlighted in *red*

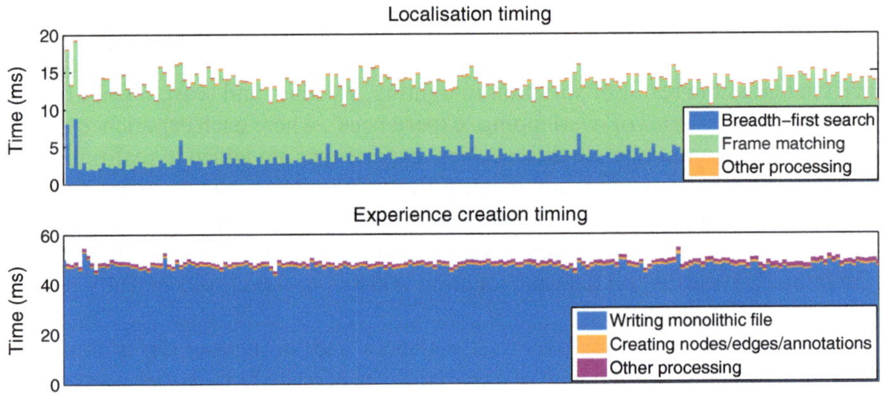

Fig. 10 Sampled timing measurements for the localisation and experience creation pipelines. Dominant database access periods are shown in *blue*. It can be seen that during localisation, these accesses comfortably meet the 50 ms target. Experience creation takes in general longer however still runs under 50 ms for majority of the time. For this task, occasional spikes in disk access latency do not impact on real-time performance as it is handled in an independent thread

A core feature of EBN is the ability to add new experiences to the map in real-time (pipeline shown in Fig. 9). These experience creation tasks are processed in an independent thread so that slow disk write speeds do not impact on real-time performance. Figure 10 shows timing measurements for both the localisation and experience creation pipelines.

8 Conclusion

In this work we have motivated the need to move away from privileged datasets and ad-hoc data storage and annotation. These habits do not allow us to fully utilise our resources. Data take valuable time to collect and annotate and we make the case that it is impossible to predict which will be useful as they are collected. It is often only after the fact—weeks or months later—that we realise the full potential of some data. Without a consistent framework like ours in place, we would likely forget these useful data, not to mention waste time reimplementing separate storage and retrieval back-ends. We firmly believe that a system like this is the way forward for robotics applications—text files, readmes and wikis are no longer sufficient for many of our data management needs.

Acknowledgments Peter Nelson is supported by an EPSRC Doctoral Training Account. Chris Linegar is supported by the Rhodes Trust. Paul Newman is supported by EPSRC Leadership Fellowship EP/I005021/1.

References

1. Churchill, W., Newman, P.: Experience-based navigation for long-term localisation. Int. J. Robot. Res. **32**(14), 1645–1661 (2013)
2. Smith, M., Baldwin, I., Churchill, W., Paul, R., Newman, P.: The New College vision and laser data set. Int. J. Robot. Res. **28**, 595–599 (2009)
3. Huang, A.S., Antone, M., Olson, E., Fletcher, L., Moore, D., Teller, S., Leonard, J.: A high-rate, heterogeneous data set from the DARPA urban challenge. Int. J. Robot. Res. **29**, 1595–1601 (2010)
4. Waibel, M., Beetz, M., Civera, J., D'Andrea, R., Elfring, J., Galvez-Lopez, D., Haussermann, K., Janssen, R., Montiel, J., Perzylo, A., et al.: Roboearth. IEEE Robot. Autom. Mag. **18**(2), 69–82 (2011)
5. Haklay, M., Weber, P.: OpenStreetMap: user-generated street maps. IEEE Pervasive Comput. **7**(4), 12–18 (2008)
6. Ramakrishnan, R., Gehrke, J.: Database Management Systems. Osborne/McGraw-Hill (2000)
7. Bayer, R.: Organization and maintenance of large ordered indexes. Acta Informatica **1**(3), 173–189 (1972)
8. Guttman, A.: R-trees: A Dynamic Index Structure for Spatial Searching, vol. 14. ACM (1984)
9. Linegar, C., Churchill, W., Newman, P.: Work smart, not hard: recalling relevant experiences for vast-scale but time-constrained localisation. In: Proceedings of the IEEE International Conference on Robotics and Automation (ICRA2015) (2015)

Search and Retrieval of Human Casualties in Outdoor Environments with Unmanned Ground Systems—System Overview and Lessons Learned from ELROB 2014

Bernd Brüggemann, Dennis Wildermuth and Frank E. Schneider

Abstract The European Land Robot Trail (ELROB) is a robot competition running for nearly 10 years now. Its focus changes between military and civilian applications every other year. Although the ELROB is now one of the most established competition events in Europe, there have been changes in the tasks over the years. In 2014, for the first time, a search and rescue scenario was provided. This paper addresses this Medical Evacuation (MedEvac) scenario and describes our system design to approach the challenge, especially our innovative control mechanism for the manipulator. Comparing our solution with the other teams' approaches we will show advantages which, finally, enabled us to achieve the first place in this trial.

1 Introduction

Rescuing of a wounded person is an important but also dangerous task not only in military scenarios but also in civil disasters. In any case the rescue of a victim results in high risks for the rescuers themselves or, if these risks are reduced, in an unacceptably long duration until the wounded person can be brought to emergency treatment. Here robots can help not only to locate wounded persons in the first place but also to bring them into safety. Exactly this evacuation task was addressed in ELROB 2014 for the first time. Localization of the wounded person was only a minor part of the scenario because in the organizers' view transporting a wounded person was already novel and a hard enough task to be tackled in a trial.

Since new things often have a strong attraction, there were nine teams altogether trying to accomplish the so-called MedEvac scenario. As, on the one hand, Fraunhofer FKIE acts as scientific advisor for the ELROB competition and, on the other hand,

B. Brüggemann (✉) · D. Wildermuth · F.E. Schneider
Fraunhofer FKIE, Fraunhoferstr. 20, Wachtberg 53343, Germany
e-mail: bernd.brueggemann@fkie.fraunhofer.de

D. Wildermuth
e-mail: dennis.wildermuth@fkie.fraunhofer.de

F.E. Schneider
e-mail: frank.schneider@fkie.fraunhofer.de

D.S. Wettergreen and T.D. Barfoot (eds.), *Field and Service Robotics*,
Springer Tracts in Advanced Robotics 113, DOI 10.1007/978-3-319-27702-8_35

sent a team inside the competition, this paper will present the design of the scenario as well as a system to solve the task. Whereas FKIE's organizing team and the team participating in ELROB were strictly separated before and during the competition the authors can now combine both insights to present results and some lessons learned.

The remainder of the paper is organised as follows: In Sect. 2 we present current system designs to address medical evacuation tasks in general as well as competitions of particular interest for the Search & Rescue (SAR) community. Section 3 will present the MedEvac scenario in detail, describing the scenario design and its realisation during ELROB. Our approach to the MedEvac scenario, the combination of hardware and software, is described in Sect. 4. The performance of our system, also in comparison to other participants, is the topic of Sect. 5. Finally, we close the paper with lessons learned and some conclusions.

2 Related Work

It is generally a problematic task to compare approaches and methods in the field of outdoor robotics [5]. In the majority of cases results are reported only for a specific robotic system. All tasks are carried out in a static and often specially defined environment, making it hard to compare the outcome with results from other research groups, other approaches, and other robots. As one possible solution, robot competitions have been proposed for benchmarking real robot systems [2]. Of course, the difficulties of repeatability and controlled experimentation remain. In outdoor trials, for instance, weather and lighting conditions can dramatically change even for consecutive runs. Starting positions differ and obstacles are not always accurately placed, as exemplarily mentioned in [1]. The authors also notice that new kinds of problems arise. Participants often tend to exploit rules or create special-purpose solutions related only to a specific trial instead of developing adaptive and flexible approaches.

When looking at the Search & Rescue (SAR) domain the very large field of robotic competitions dramatically decreases. Regarding Urban Search and Rescue (USAR) aspects one of the more sophisticated events is the RoboCup Rescue competition, which is part of the annually organized worldwide RoboCup. However, although very well established this competition is far from working in realistic environments. More real-world related is the ongoing DARPA Robotic Challenge (DRC) which is currently in progress. Looking at Europe, one can find the newly founded EURATHLON and, of course, the European Land Robot Trial (ELROB) with its user-centred tasks and real world scenarios. These four competitions will be described in more detail in the following paragraphs.

The RoboCup Rescue is a special part of the worldwide RoboCup competition. The intention of RoboCup Rescue is to promote research and development in interdisciplinary research themes around robot aided search and rescue. The majority of the teams are built by students. The environment used in the competition is constructed based on standard test methods for emergency response robots developed by the U.S.

National Institute of Standards and Technology (NIST). The greatest advantage of these so-called arenas is that they allow repeatable tests in an environment anybody can build [11]. There are color-coded arenas with different levels of difficulty available. In all arenas, the robots have to find simulated victims and generate a map, which helps rescuing personnel to locate and rescue the victims.

The DARPA challenges started with the Grand Challenge in 2004. Initially, the goal was to travel autonomously, first in a desert-like area, later in an urban environment. Especially in the context of USAR the new DARPA Robotics Challenge (DRC) is of relevance. The DRC looks for robots capable of assisting humans in response to natural and man-made disasters. After some preliminary decisions, 16 teams have been elected to participate in the semi-finals in December 2013. Details and results can be found at [17]. The finals will take place in June 2015.

Funded by the European Commission, EURATHLON is an international competition that welcomes university, industry or independent teams from any EU country. EURATHLON provides real-world robotics challenges for outdoor robots in demanding scenarios. The focus of the first EURATHLON competition in 2013 was land robots, and had five scenarios covering a number of the key competencies needed in outdoor disaster response, including mapping the disaster site, searching for objects of potential interest (e.g. survivors), turning off valves (i.e. a gas leak), finding hazardous materials and securing them, and navigating autonomously from one place to another [18]. The focus of EURATHLON 2014 was underwater robots, and EURATHLON 2015 will finally add flying robots. Inspired by the Fukushima accident of 2011, this grand challenge will require cooperating groups of land, sea and flying robots to investigate the scene, collect environmental data, then identify and stabilise critical hazards.

The ELROB trials have been started in 2006 as an annual competition, which alternates its key aspect between military and civilian tasks [16]. In contrast to the DARPA challenges, the teams can choose different scenarios. Among these scenarios are different kinds of reconnaissance and surveillance missions combined with the detection of special objects, or transportation, which can be carried out with a single vehicle or in form of a convoy with at least two vehicles. In the recent years several scenarios from the Search & Rescue domain have been added, e.g. the inspection of partially wrecked urban and semi-urban structures or the search for injured persons [15]. The ELROB 2014 competition and especially the Medical Evacuation (MedEvac) trial are subject of this work and are described in more detail in Sect. 3.

Robotic systems for medical support have been discussed in literature for a couple of years now. Apart from victim transportation, other applications include search and localization of injured persons, direct medical support (e.g. providing water or establishing an audio connection) or even life sign detection [13] (e.g. through infrared cameras or pulse measurement). In [8] a Cognitive Task Analysis (CTA) is used to identify requirements and preconditions for using robots in such medical applications. Although in [13] Robin Murphy describes a payload for medical assessment and very limited support for the victim, for most authors the idea of using a robot for helping injured persons is more or less a long-term vision. Only in the recent years a couple of large research projects, e.g. the European ICARUS project [4], address victim search and support from a more practical side.

In the context of medical evacuation and victim transport only very few robot systems have been actually built. In [12] a small platform for use in fire-fighting situations has been developed. It can be thrown into a fire site to gather environmental information, search displaced people, and show them the best way out. Of course, this approach requires that the persons can still move on their own. For several years the US Army has sponsored research in the military aspects of robotic casualty extraction and evacuation but this research mainly produced concepts [6] and did not lead to a working system. Among others the problem of safely picking up an injured person was not even conceptually solved.

Other authors addressed partial movement and manipulation of the body of injured persons [10, 19], e.g. to bring their head into a better position for breathing. This task allows using smaller robots and, thus, lowers the risk of further injuring a victim. Since this task only solves a partial problem in rescuing the person, Iwano et al. also discussed using a group of such smaller robots for victim transportation [10]. In [9] the same group developed a completely different approach. Instead of using an intelligent robot, they addressed the vehicle design first and improved a normal rescue support stretcher system, allowing a single rescuer to pick up and transport a victim even on difficult terrain like stairs.

3 Task Description

Before describing our approach to the ELROB 2014 MedEvac scenario we will briefly introduce the general idea of ELROB and the ELROB 2014 competition from the organizers' point of view. Afterwards, the newly created MedEvac scenario in which unmanned ground systems (UGV) had to rescue a wounded person out of a hazardous environment is described in detail.

3.1 The European Land Robot Trial and the 2014 Event

The organizers see the European Land Robot Trial (ELROB) as an opportunity to provide an overview of the current state of the art in European unmanned systems technology. ELROB enables participants and visitors to get a glance at the latest research and development in the area of outdoor unmanned ground vehicles (UGV). For participants from industry ELROB allows to evaluate their commercial products in realistic scenarios dealing with dangerous and hazardous environments. Additionally, participants from universities and research institutes guarantee that also cutting edge methods in robotics can be seen. This mixed field of participants results in a community creating process bringing together developers and users.

ELROB 2014 was hosted by the Warsaw Military University of Technology and co-organized by Fraunhofer FKIE. The tasks have been developed in close co-operation with the potential end users and reflect the up-to-date requirements of military forces as well as civil first-responders. Altogether, participating teams could choose from five scenarios:

- Reconnaissance and surveillance in non-urban environments: A specified target area had to be searched for particular markers passing a number of given way-points.
- Mule: A vehicle had to shuttle between the two camps carrying as much payload as possible. The vehicle had to learn the position of the second camp and the route how to get there by following a human guide (teach-in).
- Reconnoitring of structures: An area of interest with a number of small buildings had to be inspected. The robot had to enter the building, partially using stairs, and search for particular markers.
- Medical evacuation: Two wounded persons were lying at two roughly known positions. A vehicle had to approach these positions, locate the dummy and transport it back to the starting point.
- Reconnaissance and disposal of bombs and explosive devices: An area of interest, indoor and outdoor, had to be explored and searched for suspicious objects.

3.2 The MedEvac Scenario

The rescue of wounded persons is an important yet often difficult task in civil catastrophes as well as in military scenarios. During military operations the retrieval of casualties usually takes place in hostile environments, thus leading to severe dangers for the involved soldiers. The use of robotic vehicles, first, to find injured persons and, second, to autonomously pick them up and transport them back to safe areas obviously is a great improvement (see Fig. 1).

In the MedEvac scenario, as well as in all other ELROB scenarios, one operator and on technician are allowed during the run. While the operator has only the information he or she gets from the control station (and e.g. no direct line-of-sight) the technician is allowed to follow the robot. Thus the technician is able to perform an emergency stop to prevent the robot from damage or free the robot if it gets stuck. All interventions by the technician were measured and resulted in penalties.

During the scenario the wounded persons were represented by dummies. Depending on what the robot was capable to transport, participants could choose between 10, 35 or 74 kg dummies. While the 10 kg dummy was only a black bag, both other dummies were in a human-like shape. Additionally, the dummy had a pull strap or loop for easier transportation. In the scenario two wounded persons were hidden at two roughly known positions (named with P_1 and P_2). The participant had to first approach P_1, search and locate the dummy, and then transport it back to the starting

Fig. 1 The MedEvac scenario in an overview: Starting from the marked position on the *bottom left corner*, the participants have to go to each of the marked way-points and search the area for the dummy. After locating the dummy and acquiring a GPS coordinate of it, the robot is supposed to bring the dummy back to the starting position. The whole scenario takes place in a 150 m × 150 m area with a distance from the dummies to the controller's tent of about 75 m

point e.g. by dragging it at the special strap, by pushing it, or by completely lifting it. Afterwards, the same had to be done for the area around P_2.

The environment was characterized as a typical non-urban terrain with obstacles like high grass, ditches, trees and bushes. In the actual scenario the environment appeared as a large grassy area. Most of the grass was waist-high, thus, the organizers decided to cut down some parts to enable participants to use autonomous functions and smaller robots. Nevertheless, one of the two dummies could only be found by entering the high grass area.

In addition to the main task, the rescuing of the wounded persons, participants could gain extra points for additional tasks:

- acquired imagery and exact GPS positions of both dummies,
- transmission of all data to the control station, online or offline after having returned to the starting point,
- transmission of live position and video imagery.

The scenario ended with manoeuvring both imitated wounded persons back to the starting point or with reaching the time limit of 45 min. Transferring any result data had to be done within the scenario time.

4 System Description

In this section we describe the idea how to solve the MedEvac task as it is described in Sect. 3. This includes the question 'How to transport the dummy?' as well as technical decisions and the control method for the robot and especially the manipulator. All decisions were made not only having the task in mind, but also with a focus to perform best in that scenario. This includes to respect the score sheet in a way that bonus points should be achieved and aspects which are not relevant for the points system can be postponed.

4.1 Our Scenario Approach

To optimize the scoring three different aspects had to be considered. Firstly, as ELROB always wants to foster autonomy, more points can be achieved with semi-autonomous and autonomous robots than with simple tele-operation. Secondly, the time needed to complete the task is important, and, thirdly, the weight of the dummy that is handled. Additionally there are no penalties for being rude to the dummy. In fact, as this was the first time MedEvac was offered as an ELROB scenario, the possible solutions should not be narrowed by too much restrictions.

Dealing autonomously or semi-autonomously with the scenario was not possible for us because the preparation time between announcement of the scenario and the actual competition was too short. Thus, we had to focus on speed and power of the resulting system. We agreed that the scenario was not solvable without some kind of manipulation. As we have no manipulator able to handle the 74 kg of the heaviest dummy but a robot which is capable of moving such weights, we realized that the manipulator should be best used to link the wounded person with the robot, and afterwards the robot itself should actually move the dummy. This resulted in a towing approach. The manipulator was used to attach a hook to the gear of the wounded person. This hook was attached with a steel rope to the robot. Thus, after hooking the dummy, the robot was able to tow the dummy back to the starting position.

4.2 The Mobile Platform

Our vehicle is the prototype GARM built by RUAG in Switzerland in collaboration with FKIE's engineers. It is a robot in the 500 kg class with a long-lasting lithium-ion-battery and a tracked drive. In this class it is one of only few robots that have a closed-loop controller for the engines, which allows sending velocities from the computer to the robot and makes autonomous navigation a lot easier. This is quite unique because most other robots of this size are built solely for tele-operated EOD missions and just let the operators control the power of the engines directly. Usually they are not equipped with any odometry sensors at all. The top speed of our robot is roughly 20 km/h and the possible payload is about 200 kg. The chassis is water-resistant, but should not be submerged completely.

We use a payload box developed by FKIE that is equipped with a 7 degrees-of-freedom (DOF) manipulator taken from a telerob telemax EOD robot. It has a parallel gripper that can be opened and closed. The third joint from the base is a prismatic joint that enables the manipulator to extend the upper arm for about 30 cm. Thus, the manipulator has a range of around 1.7 m. For communication freely available

WiFi components where used which are able to cope with distances of up to several hundred metres, so fully sufficient for the described MedEvac scenario. We used standard IEEE 802.11n with flexible channel planning at 2.4 and 5 GHz frequencies.

4.3 Robot Control

The robot control was designed mainly to deal with the task as fast as possible. It consists of three different aspects: fast set-up of the system, easy manipulator control, and robustness against connection failures.

4.3.1 Driving and GUI

As most other research groups we are using the Robot Operating System (ROS) framework. In our solution the robot and the control station are two physically divided systems. This causes problems in ROS if the connection between robot and control station is unreliable. As a solution we use the FKIE Multi-Master extension for ROS, giving us an improved robustness against temporary connection failures. Within the multi-master the existing ROS master is unchanged and executed independently on each robot. To enable the ROS nodes which are registered at different ROS masters to communicate with each other, each node has to be registered at each ROS master. Therefore, the ROS master provides a XML-RPC-interface, so we do not have to change the source code of the ROS master. A so-called sync-node is responsible to register all discovered remote nodes at local ROS masters. Since only the local ROS master is changed by the sync-nodes losses of connection do not result in inconsistent states. To reduce the configuration overhead, a discovery node discovers other discovery nodes by steadily broadcast and received heartbeat messages. The discovery node also monitors the local ROS master and announces the timestamp of last change using heartbeat-messages. So the remote sync-node can detect the changes and update its synchronization. Additionally, the Multi-Master comes with a graphical user interface for managing launch files, greatly helping us to build a quick set-up system. The code of the ROS Multi-Master is published with BSD license at github and the documentary can be found at [14].

The robot GUI is built of rqt widgets. Beside pictures of the three cameras (manipulator hand and turret; overview camera) we display a map of the area, which displays, for example, the given way-points for the scenario. As we expected an environment very difficult for autonomous driving, we included two kinds of driving control: autonomous driving via way-points set in the map, and a simple joystick control.

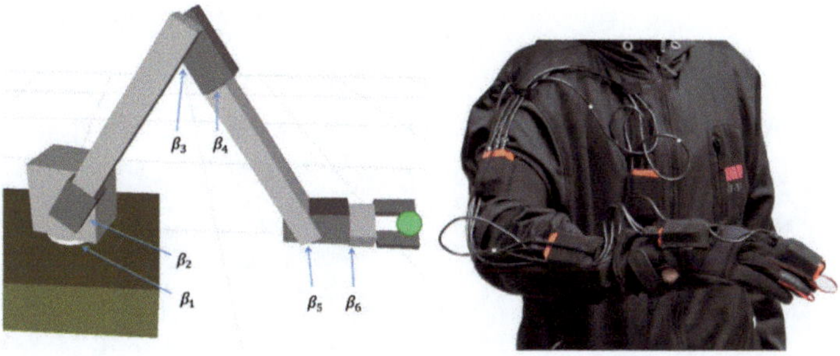

Fig. 2 Directly coupled man-manipulator control. Using several IMUs (*right*) the operator's movement is measured and transferred to the manipulator (*left*)

4.3.2 Manipulator Control

Although the chosen method to pull the wounded person out of the dangerous area looks simple, it yet results in a difficult manipulation task. The hook has to be safely placed at the gear of the dummy but it is not known in advance where a suitable strap will be located. Additionally, the exact position of the dummy is unknown. Thus, we decided to solve the manipulation task purely tele-operated. Whereas typical solutions to manipulator tele-operation include at least a joystick and some combination of direct joint control and tool-center-point control, we introduce a novel system for controlling the manipulator directly by the movement of the operator's arm.

The operator is equipped with a jacket in which an inertial measurement unit (IMU) is placed at each part of the arm (see Fig. 2). By measuring the current orientation of each of those sensors the actual arm position can be calculated. Using also the velocity readings an automatic calibration can be done (see [7]). This enables the operator to wear the jacket during the competition run, access the manipulator control if necessary and switch to other control mechanism without time delay. Additionally, this manipulator control method enables the operator to conduct even complex manipulation tasks in a very intuitive manner, as described in detail in [3].

5 MedEvac at ELROB 2014—The Competition

5.1 Solutions of Other Competitors

As stated before, the MedEvac scenario was part of the ELROB competition and new things are appealing to people for the first time. Thus, nine out of the twelve teams participating in ELROB 2014 took part in this scenario. Two types of solutions were presented: towing/pulling—as FKIE did—and lifting.

Two of the industry teams, Cobham and ELP, also chose to tow the dummy back to the starting point (see Fig. 3). As both robots originally are designed for bomb

Fig. 3 Two other competitors using a similar strategy to our approach: towing the dummy back to the starting point. Due to the size of the robots only the 10 kg dummy (*black bag*) could be moved

disposal, they are small and not able to move high weights. Although they both managed to pull the dummies back to the starting position in time, they were only able to move the small 10 kg dummy.

Lifting the dummy obviously has the advantage that it is much more convenient for the wounded person. The University of Oulu and the team Marek from the Warsaw Military University of Technology (WAT) tried this solution. While Oulu built a pick-up mechanism (see Fig. 4, left) team Marek performed the task with pure power. They tried to use a fork lifter originally designed for moving around heavy loads (see Fig. 4, right). Unfortunately, as they had no GPS localization and visualisation they were not able to locate the dummy. Also Oulu could not evaluate their lifting mechanism because the robot was not able to pull the lifting mechanism over the dummy.

Altogether only three teams were capable of locating the dummies and moving them both back to the starting position within the time limit. All three teams had a tele-operated robot. While two teams used small bomb disposal robots and could only move the small 10 kg dummy, our team successfully moved the heavy (74 kg) one.

Fig. 4 Two teams presented lifting strategies without using a robot arm. While the University of Oulu constructed a lifting mechanism, team Marek used sheer force in form of a large fork lifter

5.2 Our Own Run

Our actual run was preponed due to the withdraw of other teams. Thus, preparations had to be done in a hurry, but within less than ten minutes the control station was set up and the robot was ready to enter the scenario (see Fig. 5, left). First, a dummy in approximately 75 m distance had to be retrieved. Due to the high grass, we decided to operate fully tele-operated and drive the robot directly to the given way-point. Although the GARM is capable of driving with up to 20 km/h, we could only go with a maximum speed of 10 km/h as the vibration heavily disturbed the camera image.

The imitated victim was placed in high grass (see Fig. 5, right), but due to the high viewpoint of the camera (approx. 1.4 m from ground) the dummy could be located already during the approach and no time was needed to search for it. To gain all extra points a camera picture had to be stored at which the dummy could be clearly seen and also the exact GPS coordinates had to be recorded. This could be done manually because the manipulator control jacket still allowed using keyboard and mouse. Nevertheless, an automatic function would have saved another minute. After acquiring the picture we manoeuvred the robot to the left side of the dummy and started the manipulation task. Standing beside the dummy seemed not to be the best position and the hook was released from the manipulator without being tightly secured. To make sure that the hook held during towing the operator picked up the hook once again and moved it to a better position. This was done without any manual intervention from the technician. The dummy was towed back to the starting position with a speed of approximately 3.6 km/h.

When arriving back at the starting position the technician removed the hook from the dummy and attached it back to the manipulator. Although this was done at the starting position and was thought to be in accordance to the rules, the judges counted this action as manual intervention. The second dummy was also immediately seen in the video stream but, as it was surrounded by ditches on three sides, the robot could not easily access it. After acquiring the picture and GPS coordinate, the robot moved to the opening in the ditches and was now located directly at the head of the dummy.

Fig. 5 *Left* The FKIE robot at the starting position. Here the dummy had to be brought back to. *Right* The robot arriving at the first dummy. From here the manipulation task was to hook up the gear

This position was more beneficial and, thus, the hook could be placed securely at the dummy within less then one minute. Towing the dummy back past the ditches took some time but the total run could be finished within 21 min.

5.3 *Results*

The final scoring sheet ranked our team first with team ELP and Cobham as second respectively third. These teams were the only teams able to finish the task in time. Also all of these three teams presented a tele-operated solution. Our team was the only team with penalty for manual intervention, as the judges counted the removal of the hook from the gear of the dummy as manual intervention even though this happened in the save area, were in a real task medical assistance will wait for the wounded person.

Comparing to the second and third place we reached more points due to the fact that we were able to complete the mission in less than half of the maximum time. ELP as runner-up was able to solve the mission in 28 min while Cobham needed more than 34 min to transport both dummies to the starting point. Using a robot which was able to tow the 74 kg dummy equalled out the given penalty for manual intervention. Additionally, it turned out to be important to get the extra points for pictures and GPS positions as this was done by all competitors.

6 Lessons Learned

Competitions are great opportunities to benchmark different systems against each other but they measure always a complete system including hardware, software and the operator. Therefore, some aspects like the robustness of the hardware have a big influence on the overall performance while others, like cutting-edge algorithms, only have an effect if everything else works well. Nevertheless, taking part in a competition is always valuable for the participant to learn interesting lessons about the own system.

One of the main aspects is in our opinion the robustness of the whole system. This includes hardware, software but also an operator who is familiar with the whole system and also the scenario which has to be solved. In the ELROB 2014 MedEvac scenario two participants were not able to present their approaches because of hardware failures. From the retrospective of the last ELROB events this seems to be especially a problem of universities, which are not able to afford expensive hardware platforms. FKIE's cooperation with RUAG resulted in a very robust and sophisticated platform in a robot class (up to 500 kg) which is not really supported by the industry at the moment. Additionally, we use ROS together with the FKIE multi-master extension, a technically mature solution which comes with a graphical user interface for easier system launch management. Especially this graphical user interface results in a robust and fast way to start a complex system with many different software components (ROS nodes).

Our scenario solution, to tow the dummy out of the dangerous area, was a good decision regarding the used scoring system. Nevertheless, in real operation a method has to be found to move a wounded person much gentler. Even if some of the attending relief unit members told us, that there is nothing worse than leaving wounded persons where they are, we expect serious additional injuries by towing the wounded persons over other surfaces than the grass in this scenario.

In our view the novel direct control method for the manipulation task made the real difference to the other teams. Placing the hook at the gear of the dummies was not an easy task, which took a considerable amount of time even for the trained operators of the commercial teams. Having gained a seven minute margin over the other competitors indicates that our control method is feasible for complex tele-operated manipulation with only camera pictures available. It also showed how valuable assistance functions are for the operator in stressful and complex missions. While having such assistance functions for the main tasks (steering the system, controlling the robot and the manipulator), the lack of such automatisms for the bonus tasks (acquiring pictures and GPS coordinates of the victims) was a burden for the operator. The bonus tasks had to be done manually using a lot of different tools and outside the main control architecture. This required a lot of additional concentration and therefore was quite error-prone.

In summary, the authors believe that a successful robot for a competition has to be designed in an easy-to-use way, including the robustness of the hardware, a fast set-up of the system and intelligent assistance functions to reduce the operator's workload. Altogether such a design reduces the error-proneness of the system and increases the chance to present what is unique in your system during the one-shot chance in such a competition.

7 Conclusion

Search and retrieval of human casualties in outdoor environments with unmanned ground systems or, in short, MedEvac was a new and successful scenario in ELROB 2014. Nine teams tried to compete and presented different approaches. Of those nine teams three were able to solve the task. All of those teams used a towing technique to move the simulated wounded person back to a medical care point. Here the fact that there were no penalties for a rough handling of the dummies influenced the solutions. More realistic requirements regarding the victim care will make the scenario more demanding, maybe already in the next ELROB 2016.

Our focus on a robust system together with an intuitive control for the demanding manipulator task not only resulted in winning the scenario but also gave us the special jury award for the "best scientific solution".

References

1. Anderson, J., Baltes, J., Tu, K.Y.: Improving robotics competitions for real-world evaluation of AI. In: AAAI Spring Symposium: Experimental Design for Real-World Systems, pp. 1–8 (2009)
2. Behnke, S.: Robot competitions-ideal benchmarks for robotics research. In: Proceedings of IROS-2006 Workshop on Benchmarks in Robotics Research (2006)
3. Brüggemann, B., Gaspers, B., Ciossek, A., Pellenz, J., Kroll, N.: Comparison of different control methods for mobile manipulation using standardized tests. In: 2013 IEEE International Symposium on Safety, Security, and Rescue Robotics (SSRR), pp. 1–2. IEEE (2013)
4. De Cubber, G., Doroftei, D., Serrano, D., Chintamani, K., Sabino, R., Ourevitch, S.: The EU-ICARUS project: developing assistive robotic tools for search and rescue operations. In: Safety, Security, and Rescue Robotics, pp. 1–4. IEEE (2013)
5. del Pobil A.P.: Why do we need benchmarks in robotics research. In: International Conference on Intelligent Robot and Systems, Beijing, China (2006)
6. Gilbert, G., Turner, T., Marchessault, R.: Army Medical Robotics Research, Army Telemedicine and Advanced Technology Research Center (TATRC) project report, Fort Detrick, MD (2007)
7. Hoffmann, J., Brüggemann, B., Krüger, B.: Automatic calibration of a motion capture system based on inertial sensors for tele-manipulation. In: ICINCO, vol. 2, pp. 121–128 (2010)
8. Humphrey, C.M., Adams, J.A.: Robotic tasks for chemical, biological, radiological, nuclear and explosive incident response. Adv. Robot. **23**(9), 1217–1232 (2009)
9. Iwano, Y., Osuka, K., Amano, H.: Development of rescue support stretcher system with stair-climbing. In: 2011 IEEE International Symposium on Safety, Security, and Rescue Robotics (SSRR), pp. 245–250. IEEE (2011)
10. Iwano, Y., Osuka, K., Amano, H.: Posture manipulation for rescue activity via small traction robots. In: Safety, Security and Rescue Robotics, Workshop, 2005 IEEE International, pp. 87–92. IEEE (2005)
11. Jacoff, A., Messina, E., Evans, J.: A standard test course for urban search and rescue robots. NIST SPECIAL PUBLICATION SP, 253–259 (2001)
12. Kim, Y.D., Kim, Y.G., Lee, S.H., Kang, J.H., An, J.: Portable fire evacuation guide robot system. In: IEEE/RSJ International Conference on Intelligent Robots and Systems, 2009. IROS 2009, pp. 2789–2794. IEEE (2009)
13. Murphy, R.R., Riddle, D., Rasmussen, E.: Robot-assisted medical reachback: a survey of how medical personnel expect to interact with rescue robots. In: 13th IEEE International Workshop on Robot and Human Interactive Communication, 2004. ROMAN 2004, pp. 301–306. IEEE (2004)
14. ROS packages for multimaster support (discovering, synchronizing and management GUI). http://fkie.github.io/multimaster_fkie/
15. Schneider, F.E., Wildermuth, D.: Aims and Outcome of Professional Ground Robotic Competitions: A Systematic Comparison. In: Proceedings of the 3rd NATO EOD Demonstrations and Trials Workshop "New technologies assistance and limitations of the EOD in post-ISAF era", Trencin (2014)
16. Schneider, F.E., Wildermuth, D., Brüggemann, B., Röhling, T.: European Land Robot Trial (ELROB) Towards a Realistic Benchmark for Outdoor Robotics (2010)
17. The DARPA Robotics Challenge (DRC), official website. http://www.theroboticschallenge.org
18. Winfield, A.F., Franco, M.P., Brüggemann, B., Castro, A., Djapic, V., Ferri, G., Viguria, A.: euRathlon Outdoor Robotics Challenge: Year 1 Report. In: Advances in Autonomous Robotics Systems: 15th Annual Conference, TAROS 2014, Birmingham, UK, September 1–3, 2014. Proceedings, vol. 8717, p. 267. Springer (2014)
19. Yim, M., Laucharoen, J., Yim, M., Laucharoen, J.: Towards small robot aided victim manipulation. J. Intell. Robot. Syst. **64**(1), 119–139 (2011)

Monocular Visual Teach and Repeat Aided by Local Ground Planarity

Lee Clement, Jonathan Kelly and Timothy D. Barfoot

Abstract Visual Teach and Repeat (VT&R) allows an autonomous vehicle to repeat a previously traversed route without a global positioning system. Existing implementations of VT&R typically rely on 3D sensors such as stereo cameras for mapping and localization, but many mobile robots are equipped with only 2D monocular vision for tasks such as teleoperated bomb disposal. While simultaneous localization and mapping (SLAM) algorithms exist that can recover 3D structure and motion from monocular images, the scale ambiguity inherent in these methods complicates the estimation and control of lateral path-tracking error, which is essential for achieving high-accuracy path following. In this paper, we propose a monocular vision pipeline that enables kilometre-scale route repetition with centimetre-level accuracy by approximating the ground surface near the vehicle as planar (with some uncertainty) and recovering absolute scale from the known position and orientation of the camera relative to the vehicle. This system provides added value to many existing robots by allowing for high-accuracy autonomous route repetition with a simple software upgrade and no additional sensors. We validate our system over 4.3 km of autonomous navigation and demonstrate accuracy on par with the conventional stereo pipeline, even in highly non-planar terrain.

1 Introduction

Visual Teach and Repeat (VT&R) is an effective tool for autonomously navigating previously traversed paths using only on-board visual sensors. In an initial *teach pass*, a human operator manually drives an autonomous vehicle along a desired route while

L. Clement (✉) · J. Kelly · T.D. Barfoot
Institute for Aerospace Studies, University of Toronto, Toronto, Canada
e-mail: lee.clement@mail.utoronto.ca

J. Kelly
e-mail: jkelly@utias.utoronto.ca

T.D. Barfoot
e-mail: tim.barfoot@utoronto.ca

D.S. Wettergreen and T.D. Barfoot (eds.), *Field and Service Robotics*,
Springer Tracts in Advanced Robotics 113, DOI 10.1007/978-3-319-27702-8_36

the VT&R system uses imagery from a camera to build a map of the route. In the subsequent *repeat pass*, the system localizes against the stored map to autonomously repeat the route, sometimes combining map-based localization with visual odometry (VO) to estimate relative motion in cases where map-based localization is temporarily unavailable (Fig. 1). VT&R is well-suited to repetitive navigation tasks where GPS is unavailable or insufficiently accurate, and has found applications in autonomous tramming for mining operations [14] and sample return missions [8].

The map representation in a VT&R system may be purely topological, purely metric, or a mixture of the two (sometimes called topometric). Purely topological VT&R [9, 15, 20] uses a network of reference images (keyframes) where the navigation goal is to match the current image to the nearest keyframe using a visual homing procedure. These systems are restricted to heading-based control, which only loosely bounds lateral path-tracking error. Purely metric maps are uncommon in VT&R systems due to the high computational cost of creating globally consistent maps for long routes, but successful applications do exist [11, 21]. Topometric systems [8, 14, 22, 23] reap the benefits of both mapping strategies by decoupling map size from path length while still retaining metric information.

Furgale and Barfoot [8] developed the first VT&R system capable of autonomously repeating multi-kilometre routes in unstructured outdoor terrain using only a stereo camera. Their system creates a topometric map of metric keyframes connected by 6DOF VO estimates, which are combined via local bundle adjustment into locally consistent metric submaps for localization in the repeat pass.

Furgale and Barfoot's system has been extended to other 3D sensors such as lidar [16] and RGB-D cameras, but a monocular implementation has not been forthcoming. While monocular cameras are appealing in terms of size, cost, and simplicity, perhaps the most compelling motivation for using monocular vision for VT&R is the plethora of existing mobile robots that would benefit from it. Indeed, vehicles equipped with monocular vision, typically for teleoperation, run the gamut of robotics applications,

Fig. 1 Our field robot during a 140 m autonomous traverse in the UTIAS MarsDome indoor rover testing environment, with the path overlaid for illustration. In order to compare the performance of stereo and monocular VT&R with the same hardware, we equipped our rover with a stereo camera and used only the left image stream for our monocular traverses

and in many cases—search and rescue, mining, construction, and personal assistive robotics, to name a few—would benefit from accurate autonomous route-repetition, especially if it were achievable with existing sensors.

Several techniques exist for accomplishing online 3D simultaneous localization and mapping (SLAM) with monocular vision, ranging from filter-based approaches [4, 5] to online batch techniques that make use of local bundle adjustment [10, 12, 25]. Such algorithms are capable of producing accurate 3D maps, but only up to an unknown scale factor. This scale ambiguity complicates threshold-based outlier rejection, as well as the estimation and control of lateral path-tracking error during the repeat pass, which are essential for achieving high-accuracy route-following.

In this paper, we extend Furgale and Barfoot's VT&R system to monocular vision by using the approximately known position and orientation of a camera mounted on a rover to estimate the 3D positions of keypoints near the ground with absolute scale. Similar techniques have succeeded in computing VO with a monocular camera using both sparse feature tracking [3, 6, 24] and dense image alignment [13], but have not considered the problem of map construction. We show that by treating the ground surface near the vehicle as approximately planar and applying an appropriate uncertainty model, we can generate local metric maps that are accurate enough to achieve centimetre-level accuracy during the repeat pass, even on highly non-planar terrain. Although the flat-ground assumption is not globally valid, it is sufficient for our purposes since VT&R uses metric information only locally.

The main contribution of this paper is an extensive comparison of the performance of monocular and stereo VT&R in a variety of conditions, including an evaluation of their robustness to common failure cases. To this end, we present experimental results comparing the two systems over 4.3 km of autonomous navigation. While our results show that both systems achieve similar path-tracking accuracy when functioning normally, the monocular system suffers a reduction in robustness compared to its stereo counterpart in certain conditions. We argue that, for many applications, the benefit of deploying VT&R without a potentially costly sensor upgrade far outweighs the associated reduction in robustness.

2 Monocular Depth Estimation

We estimate the 3D coordinates of features observed by a camera pointed downward, but not directly at the ground surface, by approximating the local ground surface near the vehicle as planar and recovering absolute scale from the known position and orientation of the camera relative to the vehicle. We account for variations in terrain shape by applying an appropriate uncertainty model. In what follows, $\mathbf{z}_j^i$ denotes the 3D coordinates of feature i expressed in coordinate frame $\mathscr{F}_j$.

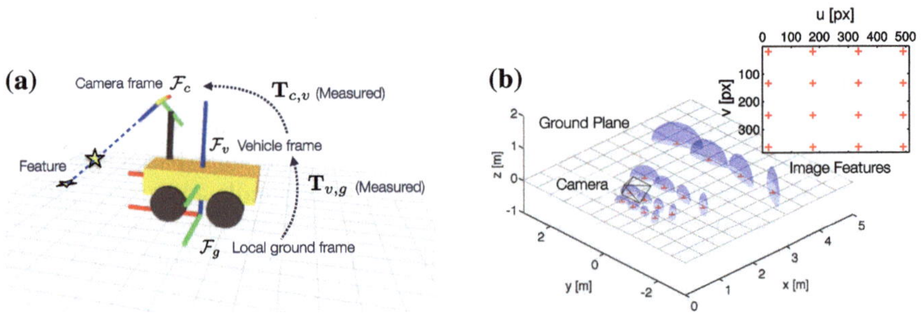

Fig. 2 Geometry and uncertainty model of our monocular depth estimation scheme. **a** Coordinate frames in our monocular depth estimation scheme. The local ground frame $\mathscr{F}_g$ is defined relative to the vehicle frame $\mathscr{F}_v$ and travels with the vehicle. **b** Evenly-spaced synthetic image features (*top right*) and estimated D coordinates with 1σ uncertainity ellipses for the experimental configuration described in Sect. 4

2.1 Locally Planar Ground Surfaces

For a monocular camera observing the ground, we can estimate the 3D coordinates of features near the ground by making the following assumptions (see Fig. 2a):

1. all features of interest lie in the xy-plane of a local ground frame $\mathscr{F}_g$ defined such that its z-axis is normal to the ground and always intersects the origin of the vehicle coordinate frame $\mathscr{F}_v$ (for a ground vehicle, this is the vehicle's footprint);
2. the transformation $\mathbf{T}_{c,v} \in SE(3)$ from $\mathscr{F}_v$ to the camera-centric coordinate frame $\mathscr{F}_c$ is known; and
3. the transformation $\mathbf{T}_{v,g} \in SE(3)$ from $\mathscr{F}_g$ to $\mathscr{F}_v$ is known.

Assuming that incoming images have been de-warped and rectified in a pre-processing step, we can model the camera as an ideal pinhole camera with calibrated camera matrix $\mathbf{K}$ such that the image coordinates $\mathbf{y}^i$ of $\mathbf{z}_c^i$ are given by

$$\mathbf{y}^i := \left[u^i \ v^i \ 1\right]^T = \mathbf{K}\mathbf{p}^i \ , \tag{1}$$

where

$$\mathbf{p}^i := \left[p_x^i \ p_y^i \ 1\right]^T = \frac{1}{z_c^i}\left[x_c^i \ y_c^i \ z_c^i\right]^T \tag{2}$$

represents the (unitless) normalized coordinates of $\mathbf{z}_c^i$ on the image plane. Note that although u^i, v^i represent pixel coordinates, they are not necessarily integer-valued.

By assumption 1, $z_g^i = 0, \forall i$, so we can write

$$\mathbf{z}_c^i := \left[x_c^i \ y_c^i \ z_c^i \ 1\right]^T = \mathbf{T}_{c,g}\left[x_g^i \ y_g^i \ 0 \ 1\right]^T \ , \tag{3}$$

where $\mathbf{T}_{c,g} = \mathbf{T}_{c,v}\mathbf{T}_{v,g}$. We can therefore obtain the feature depth z_c^i as a function of $\mathbf{p}^i$ by substituting $x_c^i = z_c^i p_x^i$ and $y_c^i = z_c^i p_y^i$ according to Eq. (2), and solving the third component of Eq. (3) for z_c^i, yielding

$$z_c^i = \frac{k_1}{k_2 + k_3 p_x^i + k_4 p_y^i}, \tag{4}$$

where, using T_{mn} as shorthand for the mth row and nth column of $\mathbf{T}_{c,g}$,

$$\begin{aligned} k_1 &= T_{11}(T_{22}T_{34} - T_{24}T_{32}) & k_2 &= T_{11}T_{22} - T_{12}T_{21} \\ &\quad + T_{12}(T_{24}T_{31} - T_{21}T_{34}) & k_3 &= T_{21}T_{32} - T_{22}T_{31} \\ &\quad + T_{14}(T_{21}T_{32} - T_{22}T_{31}) & k_4 &= T_{12}T_{31} - T_{11}T_{32}\,. \end{aligned}$$

Finally, using Eqs. (1) and (2) with z_c^i as in Eq. (4), we can express the Cartesian coordinates of $\mathbf{z}_c^i$ in terms of $\mathbf{y}^i$ as

$$\mathbf{z}_c^i = z_c^i \mathbf{K}^{-1}\mathbf{y}^i\,. \tag{5}$$

2.2 Uncertainty Considerations

A crucial component of enabling monocular VT&R using this depth estimation scheme is an appropriate model of the uncertainty in each observation $\mathbf{z}_c^i$. We consider two important factors: uncertainty in image coordinates $\mathbf{y}^i$, and uncertainty in ground shape far from the vehicle. In early experiments, we found that image coordinate uncertainty alone did not permit reliable feature tracking since there was little overlap in 3D feature coordinate estimates across multiple frames.

We model feature coordinates in image space as Gaussian distributions centred on $\mathbf{y}^i$ with covariance $\mathbf{R}_{\mathbf{y}^i} := \text{diag}\{(\sigma_u^i)^2, (\sigma_v^i)^2\}$. We use SURF features [2] in our system and determine σ_u^i, σ_v^i from the image pyramid level at which each feature is detected. To incorporate uncertainty in ground shape far from the vehicle, we represent the ground-to-vehicle transformation as a Gaussian distribution on SE(3) with mean $\mathbf{T}_{v,g}$ and covariance $\mathbf{R}_{\mathbf{T}_{v,g}} := \text{diag}\{\sigma_1^2, \sigma_2^2, \sigma_3^2, \sigma_4^2, \sigma_5^2, \sigma_6^2\}$, where $\sigma_1 \ldots \sigma_6$ are tunable parameters corresponding to the six generators of SE(3). Together these factors form an 8-dimensional Gaussian distribution with covariance $\mathbf{R}_i := \text{diag}\{\mathbf{R}_{\mathbf{y}^i}, \mathbf{R}_{\mathbf{T}_{v,g}}\}$, which we propagate via the combined Jacobian

$$\mathbf{G}_i := \begin{bmatrix} \frac{\partial \mathbf{z}_c^i}{\partial \mathbf{y}^i} & \frac{\partial \mathbf{z}_c^i}{\partial \mathbf{T}_{v,g}} \end{bmatrix}$$

to approximate $\mathbf{z}_c^i$ as a Gaussian in 3D space with covariance $\mathbf{Q}_i = \mathbf{G}_i\mathbf{R}_i\mathbf{G}_i^T$.

Using the Cartesian coordinates of $\mathbf{z}_c^i$ and $\mathbf{y}^i$ to compute the Jacobian, we have

$$\frac{\partial \mathbf{z}_c^i}{\partial \mathbf{y}^i} = \frac{z_c^i}{k_1} \begin{bmatrix} \left(k_1 + k_3 x_c^i\right)/f_u & k_4 x_c^i / f_v \\ k_3 y_c^i / f_u & \left(k_1 + k_4 y_c^i\right)/f_v \\ k_3 z_c^i / f_u & k_4 z_c^i / f_v \end{bmatrix} \tag{6}$$

and

$$\frac{\partial \mathbf{z}_c^i}{\partial \mathbf{T}_{v,g}} = \frac{\partial \mathbf{z}_c^i}{\partial \mathbf{T}_{c,g}} \frac{\partial \mathbf{T}_{c,g}}{\partial \mathbf{T}_{v,g}} = \left[\mathbf{1}\ (-\mathbf{z}_c^i)^\times\right] \mathrm{Ad}(\mathbf{T}_{c,v}) \,. \tag{7}$$

In the above, we adopt the notation of [1]: $\mathbf{1}$ denotes the (3×3) identity matrix, $\mathrm{Ad}(\cdot)$ the adjoint in SE(3), and $(\cdot)^\times$ the skew-symmetric cross-product matrix.

Figure 2b shows 1σ uncertainty ellipses for a number of evenly spaced synthetic image features resulting from a camera configuration similar to that used in the experiments described in Sect. 4.

3 System Overview

This section provides an overview of the VT&R system as it pertains to the methods of the previous section. In particular, we discuss the generic localization pipeline used for both online mapping in the teach pass and local map construction in the repeat pass. Figure 3 shows the stereo and monocular versions of the pipeline, which differ mainly in the front-end image processing used to generate 3D keypoints.

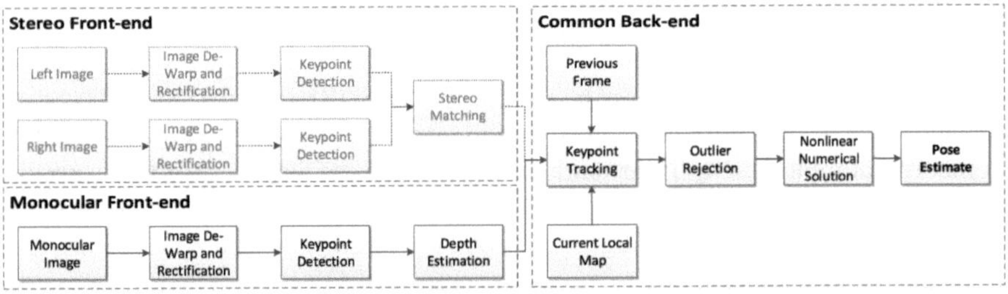

Fig. 3 The major processing blocks of the stereo and monocular localization pipelines. The monocular pipeline shares most of the same processing blocks as its stereo counterpart, differing mainly in the front-end image processing used to generate 3D keypoints. The "Current Local Map" block is only used for keypoint tracking during the repeat pass

3.1 Keypoint Generation

Raw images entering the pipeline first pass through a pre-processing step that uses a calibrated camera model to make them appear as though they were produced by an ideal pinhole camera. A GPU implementation of the SURF detector [2] then identifies keypoints in the de-warped and rectified images. The pipeline estimates the 3D coordinates of each keypoint in the camera frame using a matching procedure in the stereo case or the technique of Sect. 2 in the monocular case. The subsequent behavior of the pipeline differs slightly between the teach pass and the repeat pass.

3.2 Teach Pass

In the teach pass, the system constructs a pose graph whose vertices store lists of 3D keypoints with associated uncertainty and SURF descriptors, and whose edges store lists of matched keypoints and 6DOF pose change estimates. The system first tracks 3D keypoints in the current image against those in the most recent keyframe to generate a list of keypoint matches. These matches form the input to a 3-point RANSAC algorithm [7] that generates hypotheses for the 6DOF interframe pose change and rejects outlying feature tracks. In the context of monocular VT&R, this procedure typically rejects features far from the local ground surface (e.g., walls) since their motion is not adequately captured by the uncertainty model described in Sect. 2.2. The resulting pose change estimate serves as the initial guess in an iterative Gauss-Newton that refines the estimate based on inlying tracks.

3.3 Repeat Pass

The repeat pass begins with a manual initialization at some vertex in the pose graph, and the specification of a destination vertex. The system then reconstructs the vehicle's path from the appropriate chain of relative transformations.

At every timestep, the system identifies the nearest keyframe in the path and performs a local bundle adjustment over a user-specified number of topologically adjacent keyframes, generating a local metric map in the reference frame of the nearest keyframe. The system then forms an augmented keyframe from the adjusted map keypoints against which freshly detected features may be matched. As in the teach pass, the system performs frame-to-frame VO to obtain an initial 6DOF pose estimate at each time step, which it uses as an initial guess to localize against the current local map and refine its pose estimate.

If the system fails to localize against the map, it may rely purely on VO until either a successful localization occurs or the vehicle exceeds some preset distance threshold since the last successful localization. In the latter case, the system will halt the traverse and enter a search mode until it relocalizes or the operator intervenes.

4 Experiments

We conducted two sets of experiments at the University of Toronto Institute for Aerospace Studies (UTIAS), the first outdoors on relatively flat terrain, and the second on the highly non-planar terrain of the UTIAS MarsDome indoor rover testing environment. We compare the performance of our monocular VT&R system to that of the established stereo system [8] over 4.3 km of autonomous navigation. Table 1 reports path lengths, repeat speeds, start times, and autonomy rates for each experiment. We repeated each route using the monocular pipeline first, and conducted each experiment between roughly 10:00 and 14:00 when the sun was highest in the sky to minimize the effects of lighting changes and shadows.

4.1 Hardware

We used a four-wheeled skid-steered Clearpath Husky A200 rover equipped with a PointGrey Bumblebee XB3 stereo camera, which outputs 512 × 384 pixel greyscale images at 15 frames per second. The camera is mounted 1.0 m above the ground and is angled downwards at 47° to the horizontal (Fig. 4). These values were measured by hand since our system functions well even without an especially accurate estimate of $\mathbf{T}_{c,v}$. Small errors in $\mathbf{T}_{c,v}$ are simply absorbed by the uncertainty in $\mathbf{T}_{v,g}$.

Table 1 Summary of experimental results

Trial	Route	Path length (m)	Repeat speed (m/s)	Local start time (UTC-4)			Autonomy rate	
				Teach	Mono	Stereo	Mono (%)	Stereo (%)
1	Outdoor	1370	0.6	09:56:46	10:35:10	12:08:30	99.71[a]	100.00
2	Outdoor	1360	0.6	11:45:40	12:22:26	13:43:49	99.88	100.00
3	Outdoor	1361	0.6	13:26:41	14:00:12	15:20:12	99.74	100.00
4	Indoor	126	0.3	13:32:23	13:40:53	14:02:46	96.28	100.00
5	Indoor	140	0.3	12:18:57	12:32:20	12:59:11	91.60	100.00
				Mono	**Stereo**			
Total distance driven				4298 m[a]	4357 m			
Total distance autonomously traversed				99.41 %	100.00 %			

[a]During the monocular repeat pass of Trial 1, a parked vehicle on the path forced manual driving for 59 m before successful relocalization. We exclude this segment in our analysis and report the monocular autonomy rate for Trial 1 based on a reduced path length of 1311 m

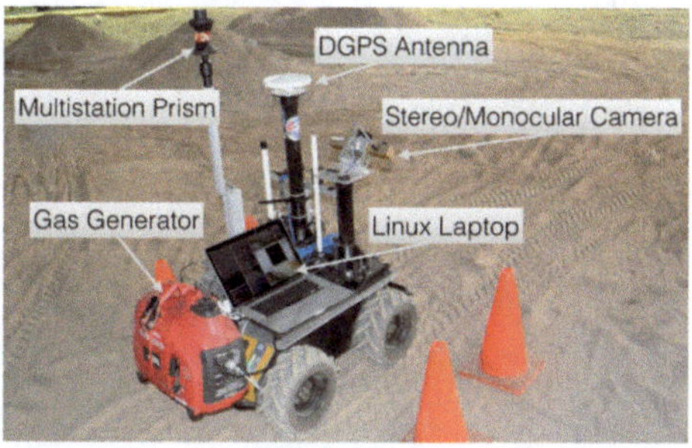

Fig. 4 Clearpath Husky A200 rover equipped with PointGrey Bumblebee XB3 stereo camera, DGPS receiver, Leica Nova MS50 MultiStation prism, 1 kW gas generator, and Linux laptop running ROS [19]

During the teach pass, we recorded stereo images and used them to teach identical paths using both the monocular and stereo pipelines. For the monocular pipeline, we used imagery from the left camera of the stereo pair only. The system detects 600 SURF keypoints in each incoming image and creates new keyframes every 25 cm in translation or 2.5° in rotation. For the monocular pipeline, we assigned standard deviations of 10 cm to the translational components of $\mathbf{T}_{v,g}$ and 10° to its rotational components as these values generally worked well in practice.

4.2 *Outdoor Experiments*

To evaluate the performance of the monocular VT&R system over long distances, we taught three 1.4 km paths through the parking lots and driveways of UTIAS. While these paths consisted mostly of flat pavement, they included many non-planar features such as speed bumps, side slopes, deep puddles, and rough shoulders, as well as other terrain types including gravel, sand, and grass.

We equipped the rover with an Ashtech DG14 Differential GPS unit used in tandem with a second stationary DG14 unit to obtain centimetre-accuracy RTK-corrected GPS data during the outdoor experiments. We used these data purely for ground-truthing purposes; they had no bearing on the behaviour of either pipeline. Figure 5 shows GPS tracks of the teach and repeat passes of one outdoor route.

Figure 6 shows estimated and measured lateral path-tracking errors during the monocular and stereo repeat passes. Both pipelines achieved centimetre-level accuracy in their respective repeat passes and produced similar estimates of lateral path-tracking error. In cases of map localization failure (i.e., when the system relied on pure VO), the monocular pipeline's estimated lateral path-tracking error diverged from ground truth more quickly than that of the stereo pipeline since keypoint position uncertainties are poorly constrained by only two measurements. Note, however, that the vehicle remained within about 20 cm of the taught path at all times.

Figure 7 compares the number of successful feature matches for frame-to-frame VO and map-based localization for both pipelines. Both pipelines track similar numbers of features from frame to frame, but the monocular pipeline generally tracks

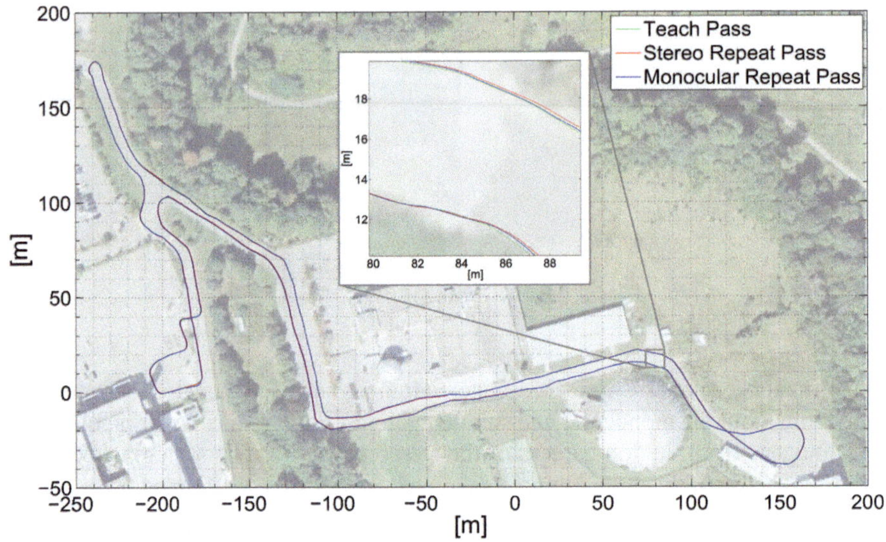

Fig. 5 Comparison of RTK-corrected GPS tracks of the teach pass, stereo repeat pass, and monocular repeat pass of a 1.4 km outdoor route (Trial 3 in Table 1). The zoomed-in section highlights the centimetre-level accuracy of both pipelines (Map data: Google, DigitalGlobe.)

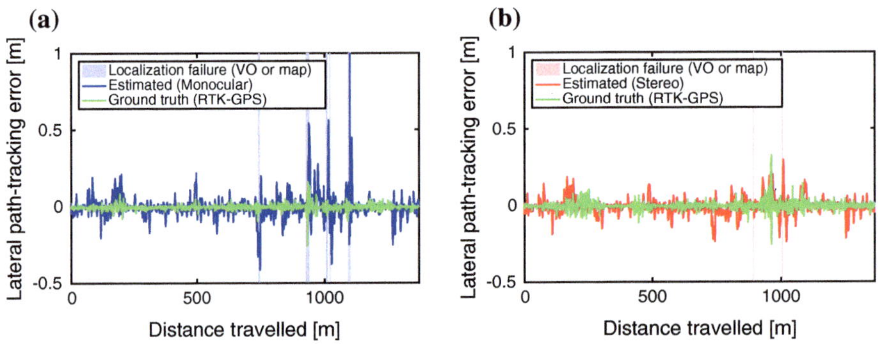

Fig. 6 Estimated and measured lateral path-tracking error during the monocular and stereo repeat passes of the 1.4 km outdoor route shown in Fig. 5 (Trial 3 in Table 1). GPS tracking shows that both monocular and stereo VT&R achieve centimetre-level accuracy, although estimated lateral path-tracking error tends to diverge from the true value in cases of localization failure. **a** Monocular repeat pass. **b** Stereo repeat pass

twice as many map features as its stereo counterpart. This result is most likely due to bad data association during local map construction in the monocular pipeline, which stems from the comparatively large positional uncertainties of distant keypoints.

Bad data association is especially problematic in regions of highly self-similar terrain (e.g., Fig. 11a) since large positional uncertainties exacerbate ambiguity in feature matches. With fewer correctly associated measurements, the bundle adjustment procedure will not maximally constrain the positions of map keypoints, which

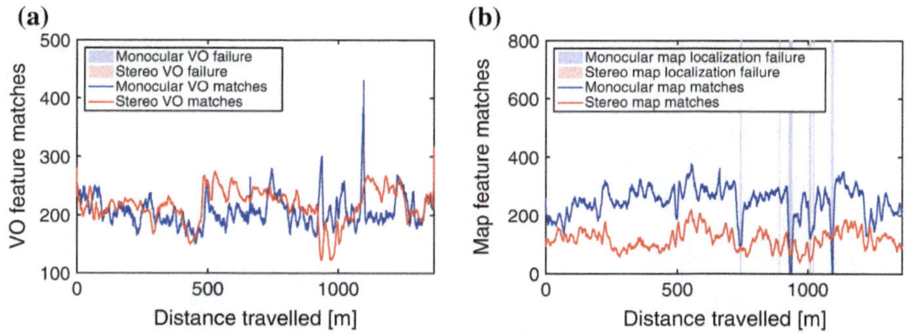

Fig. 7 Keypoint matches during the monocular and stereo repeat passes of the 1.4 km outdoor route shown in Fig. 5 (Trial 3 in Table 1), with localization failures highlighted. A localization failure is defined as less than 10 feature matches. There were no VO failures during either repeat pass. For clarity, we have applied a 20-point sliding-window mean filter to the raw data. **a** VO feature matches. **b** Map feature matches

we would expect to increase the risk of localization failures. Indeed, Fig. 7b shows that the monocular pipeline suffered more serious map localization failures than the stereo pipeline, although these forced manual intervention only once.

4.3 Indoor Experiments

The second set of experiments took place in the more challenging terrain of the UTIAS MarsDome. These routes included a number of highly non-planar features such as hills, large bumps, valleys, and slopes of a similar scale to the vehicle.

Since the MarsDome is an enclosed facility, GPS tracking was not available, and we instead made use of a Leica Nova MS50 MultiStation to track the position of the rover with millimetre-level accuracy. Similarly to the outdoor experiments, we used these data for ground-truthing purposes only. Figure 8 shows MultiStation data of the teach and repeat passes of a 140 m route through the MarsDome, along with images of some of the more challenging terrain features on the route.

Figure 9 shows estimated and measured lateral path-tracking errors for the monocular and stereo repeat passes. As in the outdoor case, both pipelines achieved centimetre-level accuracy, even in difficult terrain. Again, note that although the monocular pipeline's estimated lateral path-tracking error diverged significantly from ground-truth during localization failures, the MultiStation tracks show that the vehicle remained within a few centimetres of the path throughout the traverse.

Figure 10 shows VO and map feature matches for both repeat passes. The monocular pipeline suffered map localization failures more often than the stereo pipeline, the worst failure occurring in the valley and hill regions (see Fig. 8) where the lighting

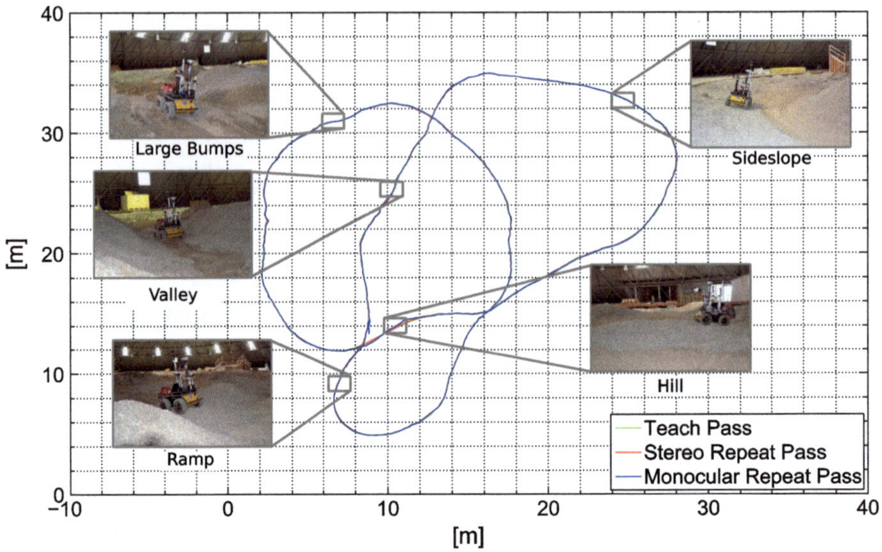

Fig. 8 Comparison of MultiStation tracks of the teach pass, stereo repeat pass, and monocular repeat pass of a 140 m indoor route (Trial 5 in Table 1), with some interesting segments highlighted

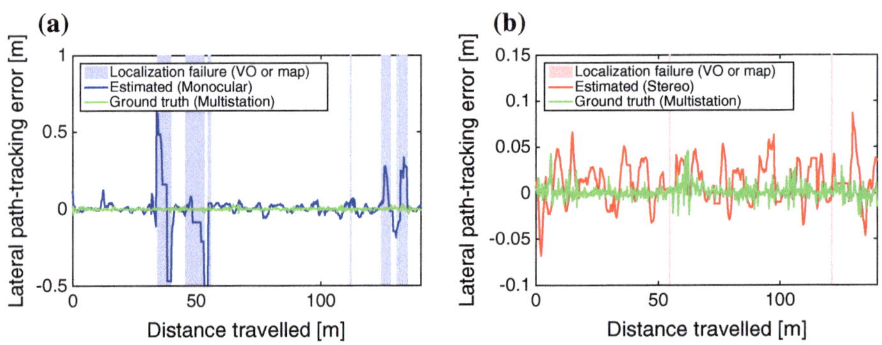

Fig. 9 Estimated and measured lateral path-tracking error during the monocular and stereo repeat passes of the 140 m indoor route shown in Fig. 8 (Trial 5 in Table 1). MultiStation tracking shows that both monocular and stereo VT&R achieve centimetre-level accuracy in highly non-planar terrain, although estimated lateral path-tracking error tends to diverge from the true value in cases of localization failure. Note the difference in scale between the two plots. **a** Monocular repeat pass. **b** Stereo repeat pass

was especially poor. This led to increased motion blur (see Fig. 11b) and poor feature matching due to greater uncertainty in keypoint positions. Both failures necessitated manual intervention over a few metres, however, the system successfully relocalized once the lighting improved.

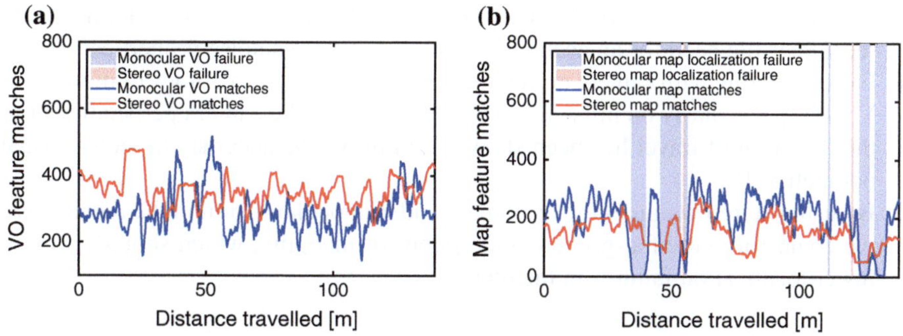

Fig. 10 Keypoint matches during the monocular and stereo repeat passes of the 140 m indoor route shown in Fig. 8 (Trial 5 in Table 1), with localization failures highlighted. A localization failure is defined as less than 10 feature matches. There were no VO failures during either repeat pass. For clarity, we have applied a 5-point sliding-window mean filter to the raw data. **a** VO feature matches. **b** Map feature matches

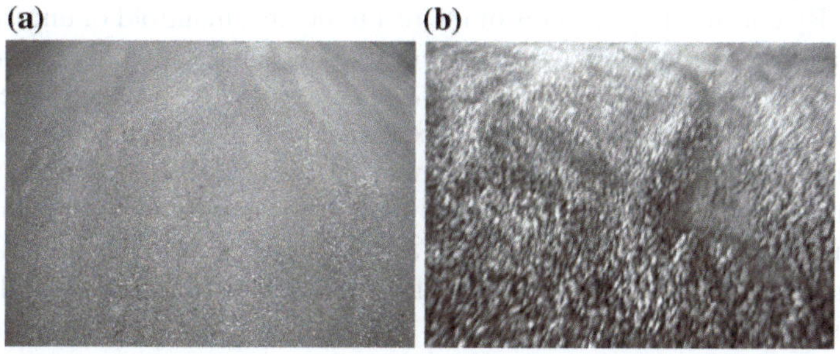

Fig. 11 The most common causes of localization failure were highly self-similar terrain and motion blur. Neither stereo nor monocular VT&R is immune to these conditions, but their effects were exacerbated by high spatial uncertainty in the monocular case. **a** Self-similar terrain. **b** Motion blur

5 Lessons Learned and Future Work

Experiments with our systems led to several useful lessons and possible extensions:

1. With sufficient spatial uncertainty, the flat-ground assumption seems to be usable even in rough driving conditions, provided the scene is well-lit and reasonably textured. Steep hills were problematic for monocular VT&R since the camera would observe features mainly on the horizon or on walls during the ascent.
2. The performance our systems depends on a search (often manual) through a high-dimensional space of tuning parameters, and it is difficult to be certain that an optimal configuration has been found. Iterative learning algorithms such as [17] may present a solution by learning optimal parameters from experience.
3. Data association quality is not a monotonic function of observation uncertainty. Too little uncertainty and good feature matches get rejected; too much and all

matches are equally good (or bad). Both cases result in tracking failure. This reinforces the need for an accurate model of a system's noise properties.
4. Experimenting with camera orientation could improve the accuracy of monocular VT&R, particularly on hills. For example, orienting the camera perpendicular to the direction of travel has been shown to improve the accuracy of stereo visual odometry [18].
5. By using stereo vision in the teach pass and monocular vision in the repeat pass, we could forgo the flat-ground assumption for mapping, which should result in fewer localization failures in the repeat pass.

6 Conclusions

This paper has described a Visual Teach and Repeat (VT&R) system capable of autonomously repeating kilometre-scale routes in rough terrain using only monocular vision. By constraining features of interest to lie on a manifold of uncertain local ground planes, we relax the requirement for true 3D sensing that had prevented the deployment of Furgale and Barfoot's VT&R system [8] on a wide range of vehicles equipped with monocular cameras. Extensive field tests have demonstrated that this system is capable of achieving centimetre-level accuracy on par with its stereo counterpart, but that there is an associated trade-off in robustness. Nevertheless, we believe that the benefit of deploying VT&R on existing vehicles without requiring the installation of additional sensors far outweighs the associated reduction in robustness.

Acknowledgments The authors would like to thank Matthew Giamou and Valentin Peretroukhin of the Space and Terrestrial Autonomous Robotic Systems (STARS) lab for their assistance with field testing, the Autonomous Space Robotics Lab (ASRL) for their guidance in interacting with the VT&R code base, Leica Geosystems for providing the MultiStation, and Clearpath Robotics for providing the Husky rover. This work was supported by the Natural Sciences and Engineering Research Council (NSERC) through the NSERC Canadian Field Robotics Network (NCFRN).

References

1. Barfoot, T., Furgale, P.: Associating uncertainty with three-dimensional poses for use in estimation problems. IEEE Trans. Robot. (T-RO) **30**(3), 679–693 (2014)
2. Bay, H., Ess, A., Tuytelaars, T., Gool, L.V.: Speeded-up robust features (SURF). Comput. Vis. Image Underst. (CVIU) **110**, 346–359 (2008)
3. Choi, S., Joung, J., Yu, W., Cho, J.: Monocular visual odometry under planar motion constraint. In: Proceedings of the International Conference on Control, Automation and Systems (ICCAS), pp. 1480–1485 (2011)
4. Davison, A.J., Reid, I.D., Molton, N.D., Stasse, O.: MonoSLAM: real-time single camera SLAM. IEEE Trans. Pattern Anal. Mach. Intell. (TPAMI) **29**(6), 1052–1067 (2007)
5. Eade, E., Drummond, T.: Scalable monocular SLAM. In: Proceedings of IEEE Conference on Computer Vision and Pattern Recognition (CVPR) (2006)

6. Farraj, F., Asmar, D.: Non-iterative planar visual odometry using a monocular camera. In: Proceedings of the International Conference on Advanced Robotics (ICAR), pp. 1–6 (2013)
7. Fischler, M.A., Bolles, R.C.: Random sample consensus: a paradigm for model fitting with applications to image analysis and automated cartography. Commun. ACM **24**(6) (1981)
8. Furgale, P., Barfoot, T.D.: Visual teach and repeat for long-range rover autonomy. J. Field Robot. (JFR) **27**(5), 534–560 (2010)
9. Goedemé, T., Nuttin, M., Tuytelaars, T., Gool, L.V.: Omnidirectional vision based topological navigation. Int. J. Comput. Vision (IJCV) **74**(3), 219–236 (2007)
10. Holmes, S.A., Murray, D.W.: Monocular SLAM with conditionally independent split mapping. IEEE Trans. Pattern Anal. Mach. Intell. (TPAMI) **35**(6), 1451–1463 (2013)
11. Kidono, K., Miura, J., Shirai, Y.: Autonomous visual navigation of a mobile robot using a human-guided experience. Robot. Auton. Syst. (RAS) **40**(2–3), 121–130 (2002)
12. Klein, G., Murray, D.: Parallel tracking and mapping for small AR workspaces. In: Proceedings of IEEE/ACM International Symposium on Mixed and Augmented Reality (ISMAR) (2007)
13. Lovegrove, S., Davison, A.J., Ibanez-Guzman, J.: Accurate visual odometry from a rear parking camera. In: Proceedings of the Intelligent Vehicles Symposium (IV) (2011)
14. Marshall, J., Barfoot, T.D., Larsson, J.: Autonomous underground tramming for center-articulated vehicles. J. Field Robot. (JFR) **25**, 400–421 (2008)
15. Matsumoto, Y., Inaba, M., Inoue, H.: Visual navigation using view-sequenced route representation. In: Proceedings of the IEEE International Conference on Robotics and Automation (ICRA), pp. 83–88 (1996)
16. McManus, C., Furgale, P., Stenning, B., Barfoot, T.D.: Lighting-invariant visual teach and repeat using appearance-based Lidar. J. Field Robot. (JFR) **30**(2), 254–287 (2013)
17. Ostafew, C., Schoellig, A., Barfoot, T.: Iterative learning control to improve mobile robot path tracking in challenging outdoor environments. In: Proceedings of the IEEE/RSJ International Conference on Intelligent Robots and Systems (IROS), pp. 176–181 (2013)
18. Peretroukhin, V., Kelly, J., Barfoot, T.: Optimizing camera perspective for stereo visual odometry. In: Proceedings of the Conference on Computer and Robot Vision (CRV), pp. 1–7 (2014)
19. Quigley, M., Conley, K., Gerkey, B.P., Faust, J., Foote, T., Leibs, J., Wheeler, R., Ng, A.Y.: ROS: an open-source robot operating system. In: Proceedings of the ICRA Workshop Open Source Software (2009)
20. Remazeilles, A., Chaumette, F., Gros, P.: 3D navigation based on a visual memory. In: Proceedings of the IEEE International Conference on Robotics and Automation (ICRA), pp. 2719–2725 (2006)
21. Royer, E., Lhuillier, M., Dhome, M., Lavest, J.M.: Monocular vision for mobile robot localization and autonomous navigation. Int. J. Comput. Vision (IJCV) **74**(3), 237–260 (2007)
22. Simhon, S., Dudek, G.: A global topological map formed by local metric maps. In: Proceedings of the IEEE/RSJ Intrnational Conference on Intelligent Robots and Systems (IROS), pp. 1708–1714 (1998)
23. Zhang, A.M., Kleeman, L.: Robust appearance based visual route following for navigation in large-scale outdoor environments. Int. J. Robot. Research (IJRR) **28**(3), 331–356 (2009)
24. Zhang, J., Singh, S., Kantor, G.: Robust monocular visual odometry for a ground vehicle in undulating Terrain. In: Proceedings of the Field and Service Robotics (FSR), pp. 311–326 (2012)
25. Zhao, L., Huang, S., Yan, L., Jianguo, J., Hu, G., Dissanayake, G.: Large-scale monocular SLAM by local bundle adjustment and map joining. In: Proceedings of the IEEE International Conference on Control, Automation Robotics and Vision (ICARCV), pp. 431–436 (2010)

In the Dead of Winter: Challenging Vision-Based Path Following in Extreme Conditions

Michael Paton, François Pomerleau and Timothy D. Barfoot

Abstract In order for vision-based navigation algorithms to extend to long-term autonomy applications, they must have the ability to reliably associate images across time. This ability is challenged in unstructured and outdoor environments, where appearance is highly variable. This is especially true in temperate winter climates, where snowfall and low sun elevation rapidly change the appearance of the scene. While there have been proposed techniques to perform localization across extreme appearance changes, they are not suitable for many navigation algorithms such as autonomous path following, which requires constant, accurate, metric localization during the robot traverse. Furthermore, recent methods that mitigate the effects of lighting change for vision algorithms do not perform well in the contrast-limited environments associated with winter. In this paper, we highlight the successes and failures of two state-of-the-art path-following algorithms in this challenging environment. From harsh lighting conditions to deep snow, we show through a series of field trials that there remain serious issues with navigation in these environments, which must be addressed in order for long-term, vision-based navigation to succeed.

1 Introduction

Appearance-based localization and mapping algorithms have enabled mobile robots to navigate autonomously through their environments using inexpensive, commercial sensors. This is appealing in that it opens the door for many exciting applications such as autonomous motor vehicles, search-and-rescue robots, and hazardous exploration robots. However, in order for these applications to succeed, robots must have the

M. Paton (✉) · F. Pomerleau · T.D. Barfoot
University of Toronto Institute for Aerospace Studies, Toronto, Ontario M3H 5T6, Canada
e-mail: mpaton@robotics.utias.utoronto.ca

F. Pomerleau
e-mail: francois.pomerleau@robotics.utias.utoronto.ca

T.D. Barfoot
e-mail: tim.barfoot@utoronto.ca

D.S. Wettergreen and T.D. Barfoot (eds.), *Field and Service Robotics*,
Springer Tracts in Advanced Robotics 113, DOI 10.1007/978-3-319-27702-8_37

ability to navigate reliably through their environments over long periods of time. This poses a serious problem in outdoor environments where lighting, weather, and seasonal changes quickly alter the appearance of the scene.

This problem is exacerbated in winter and polar environments where the appearance of the scene has the potential to change on a daily basis. The low elevation of the sun and short time between sunrise and sunset cause drastic changes in lighting. Light snow forms small patches of texture that melt on sunny days, while heavy snow blankets the environment in a featureless landscape as well as causing issues for path-tracking controllers. Some of these difficulties were recently observed during a field trial in the Canadian High Arctic. In August 2014, our autonomous path-following code was deployed to Alert (Nunavut, Canada) in collaboration with Defence Research and Development Canada (DRDC) (Fig. 1). Results were unsatisfactory due in part to the difficult environment.

Environments with highly variable appearance are especially difficult for applications that require vision-in-the-loop navigation. This specific task requires the vision system to provide constant, accurate, metric localization to the control loop to keep the robot driving. An example of a such a system is Stereo Visual Teach & Repeat (VT&R) [4], an autonomous path-following algorithm that navigates using vision. Proposed solutions for localization across appearance change either provide only topological localization [10, 11], require offline collection of the scene in multiple appearances [2, 9], or have under-performed in winter environments [14]. In this paper, we classify some of the difficulties associated with autonomous path following in winter environments and test the limits of two of our VT&R algorithms [14, 15] in two challenging winter field trials. We also discuss issues that need to be overcome to provide reliable, long-term, outdoor navigation using vision.

The remainder of this paper is outlined as follows. Work related to vision in feature-limited environments and localization across appearance changes is presented in Sect. 2. Brief details of the two tested VT&R systems are discussed in Sect. 3. Field trials, environmental information, and evaluation metrics are described in Sect. 4. Results are presented in Sect. 5. Lessons learned and challenges related to winter field deployments are discussed in Sect. 6 before concluding the paper.

Fig. 1 Multi-Agent Tactical Sentry (MATS) vehicle performing autonomous path following in Alert (Nunavut, Canada). Polar environments cause issues for vision-based systems such as ice, snow, and 24-h sunlight with a peak elevation of 12°. This leads to unsatisfactory results for current vision-based systems

2 Related Work

This paper presents the performance of autonomous path-following techniques in winter environments that are especially difficult for vision algorithms. These environments are difficult for a variety of reasons: snow accumulates and melts at a rapid pace, visual feature detectors do not fire on contrast-free snow, and low sun-elevation accelerates the effects of lighting change, to name a few.

Motion estimation through Visual Odometry (VO) is typically not affected by appearance change, but can suffer in these feature-limited environments. Williams and Howard [17] apply Contrast Limited Adaptive Histogram Equalization (CLAHE) to increase feature matches in images with snowy foregrounds. They show an increase in feature match count by an order of magnitude. Operating in feature-limited, volcanic fields, Otsu et al. [13] extract and track different features depending on the terrain, they show an improvement in feature count and computation speed.

Lighting change is typically the first issue seen by vision-based localization system with regard to appearance change. Color-constant images, which are partially invariant to lighting conditions [16], have recently been used to great success in vision algorithms. Corke et al. [3] show an improvement in place recognition across lighting changes using these images. McManus et al. [7] localize by switching between greyscale and color-constant images. They show improved results on a challenging dataset with significantly different lighting conditions.

While these techniques help overcome issues with lighting, general appearance change over time remains an issue. Naseer et al. [11] align sequences of images through a probabilistic network flow problem. Churchill and Newman [2] treat localization failures as new experiences and build a system of parallel localizers. Neubert et al. [12] build a dictionary that encodes the transformation of a scene between winter and summer. McManus et al. [9] train custom features that describe a specific element of the scene. While these methods are capable of localizing across appearance changes, they are not suitable for applications that require vision-in-the-loop navigation, such as autonomous path following. Some methods only provide topological localization [11], while others require that examples of the scene in multiple appearances are manually collected prior to reliable operation [2, 9, 12].

The autonomous path-following algorithms presented in this paper are built upon the Stereo VT&R work presented by Furgale and Barfoot [4]. Because this system navigates by comparing visual features from greyscale images, it is highly susceptible to lighting change. This can be overcome by using an active sensor. McManus et al. [8] perform VT&R using keypoints formed from lidar-generated intensity images and range data. While it is invariant to lighting conditions, it suffers from motion distortion issues. Krüsi et al. [6] perform autonomous path following through dense, point-cloud registration at the cost of potential failure cases in open spaces that lack geometric information. Vision-based path-following algorithms do not share these limitations, but are less stable in terms of appearance change. This paper examines the performance of the legacy system [4] as well as two improvements to the VT&R

algorithm [14, 15] that attempt to mitigate the effects of appearance change. These are presented in further detail in the following section.

3 VT&R Solutions

As an application context for visual navigation, we selected three previously published variants of VT&R solutions labeled here: *Legacy* [4], *Color-Constant* [14], and *Multi-Stereo* [15]. The details of these solutions are fully described and evaluated in their respective publications. Therefore, we only introduce them at a high level and point out the main differences. Figure 2 overviews the processing pipeline for each solution. A key element to compare is the number of images required for each pipeline, which gives an idea of the computation power required to track the robot position. The color-constant solution is the most expensive with three inputs, but remains within the range of real-time computation [14].

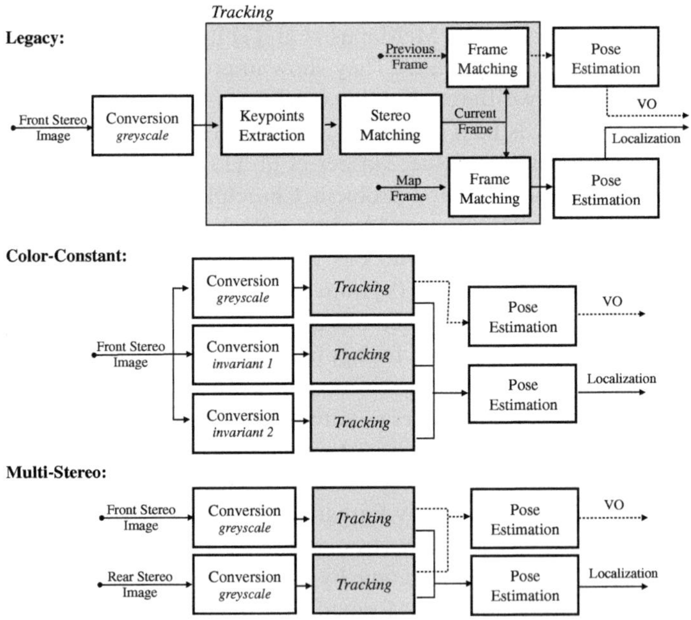

Fig. 2 Localization pipelines for the different stereo VT&R systems under investigation. The input to the system is a *left*/*right* RGB stereo image pair. The output is a pose estimate relative to a small subsection of the map (localization) and a pose estimate relative to the last image (VO). Incoming stereo images are first converted to different sources (i.e., greyscale, Invariant 1, and/or Invariant 2). Keypoints are extracted from each image source independently. Those keypoints are matched *left*-to-*right* for each respective image source to obtain depth for each feature. The 3D keypoints are then matched to a small subsection of the map to obtain feature correspondences between the live keyframe and the map keyframe. The *grey box* named *Tracking* is the same for all three solutions

(1) Legacy VT&R: This appearance-based path-following system is built upon the generation and tracking of keypoints, SURF [1] features with 3D position and uncertainty. Keypoints calculated from a stereo pair are organized into a keyframe. In the teaching phase, a robot is manually driven along a path while building a pose graph of keyframes connected by relative transformations. To repeat the path, the live keyframe, the collection of keypoints observed from the live stereo pair is matched to a map keyframe, a small subset of keyframes from the teach map relaxed into a single privileged coordinate frame. Data associations found between the live and map keyframe are used to obtain an estimate of the pose relative to the path, which is used to control the robot. The localization pipeline for this algorithm is illustrated in the upper section of Fig. 2.

(2) Color-Constant VT&R: Inspired by recent developments in the research area of color constancy, this stereo VT&R algorithm aims at increasing robustness against changes in lighting conditions. Color constancy is the ability to perceive the color of objects as constant under varying illuminations. Changes in the lighting of a scene is a major problem for appearance-based, localization algorithms that use passive sensors. This stereo VT&R pipeline is an autonomous path-following algorithm that is capable of handling significant lighting changes in a variety of outdoor environments. By expanding on the idea introduced by McManus et al. [7], the algorithm combines the accuracy of greyscale images with the robustness of color-constant images to achieve superior localization. This algorithm is identical to the *Legacy* system, with the exception of the generation of a set of two color-constant images that are partially invariant to lighting conditions. The localization pipeline is depicted in the middle section of Fig. 2. Note that tracked keypoints from each image source are fused to a single pose estimate.

(3) Multi-Stereo VT&R: Through multiple field deployments of Color-Constant VT&R, it was observed that failure situations were primarily due to a lack of successfully matched visual features in the environment. In the Alert field trial, we observed the camera pointing directly at the sun, causing glare. The probability of sun glares augments during the winter as the sun stays low on the horizon. The Multi-Stereo solution uses a second camera pointing behind the robot in order to augment the general number of matches and reduce the impact of glare. This pipeline is very similar to the Color-Constant solution, with the exception that image sources are coming from different cameras instead of multiple versions of the same image. Point clouds from all cameras are transformed into one common coordinate frame to obtain a single pose estimate. The localization pipeline is presented in the lower section of Fig. 2.

4 Methodology

As the goal of this paper is to quantify difficulties in harsh conditions, and not to introduce new algorithms, we describe here the datasets and evaluation metrics we explored to quantify the impact of extreme environments on visual navigation.

Throughout all the experiments, two components were kept stable: (1) the hardware and (2) the sky condition. The robot used is the Grizzly RUV from Clearpath Robotics, displayed in different environments in Fig. 3. The Grizzly is equipped with a payload that includes a suite of interoceptive and exteroceptive sensors. For the purpose of this evaluation, only the stereo cameras were used. More precisely, localization and mapping relied solely on forward and/or rear facing PGR Bumblebee XB3 stereo cameras. All experiments were executed outdoors under clear sky conditions (i.e., few or no clouds with the sun casting hash shadows on the ground).

4.1 Datasets

Three datasets demonstrate the impact of winter on visual navigation systems. As a nominal scenario, we included a summer experiment recorded at the *Canadian Space Agency (CSA)* on the Mars Emulation Terrain. We also conducted a set of trials in a *Meadow* and a field covered by *Snow* surrounding the campus of the University of Toronto Institute for Aerospace Studies (UTIAS) with the purpose of testing the limits of vision-based navigation algorithms in challenging winter environments. Displayed in Fig. 3, the two winter environments consisted of open fields with trees and buildings on the horizon, with and without the presence of heavy snowfall.

(1) Canadian Space Agency (CSA): This kilometer-long dataset was recorded during the summer of 2014 in the CSA Mars Emulation Terrain and its surrounding forest. Key components of the environment include a balance of desert, marsh and forest. A continuous trajectory was recorded through those different biomes and autonomously repeated 26 times over the period of four days between sunrise and sunset in late May. We consider this dataset as our nominal scenario in terms of environment complexity and use it for comparison against winter scenarios. More details about this dataset can be found in the work of Paton et al. [14].

Fig. 3 Examples of winter environments that are challenging for vision-based navigation systems. *Left* Winter meadow consisting of dead vegetation, sparse snow patches, and trees at the horizon. *Right* Open field with deep snow cover

(2) Winter Meadow: This dataset was designed to test our system's robustness against lighting change and sun-stare in a challenging environment. The recording occurred in the early winter, before large snow storms covered the entire landscape. Displayed in Fig. 3-*Left*, this environment consists of a large field containing dead vegetation and sparse snow patches surrounded by trees and buildings in the background. This environment is difficult for vision systems for a number of reasons: (i) the dead vegetation is uniform in color and often matted to the ground, producing little contrast, (ii) tall grass moves with the wind, resulting in feature matches that are inconsistent to the movement of the robot, (iii) small patches of snow shrink and change shape as they melt, (iv) the low elevation of the sun accelerates lighting change between traverses and is often directly in the camera's field of view, which significantly changes the exposure of the image. This field trial proceeded by teaching an approximately 100 m loop through this environment. The path was taught when the sun was at its highest elevation point. The robot autonomously repeated the path six times between 15:20 and 16:50 when the sun was setting (i.e., sunset happens much earlier during winter).

(3) Snowy Landscape: This dataset was designed to test our system's robustness against autonomous navigation through snowy environments. Snow is an especially difficult environment for vision-based systems as it is practically contrast free, causing a lack of visual features in most of the scene. Snow cover changes shape quickly as well. It accumulates, melts, turns to ice and can be blown by the wind changing the shape of the ground within minutes. Snow is also highly reflective; on sunny days this can lead a camera's autoexposure to generate images that are overexposed. An example of this environment can be seen in Fig. 3-*Right*, where the Grizzly is traversing through a snow covered field. A 250 m path was manually driven through a large field with fresh snow cover as a teaching pass. During the teach, the sun was at its highest point in the sky, causing significant overexposure of the camera. The path was autonomously repeated approximately 3 h later, when the elevation of the sun was significantly different. The complexity of the deployment and hardware limitations during this cold and windy day lead to a smaller number of repeats compared to the other dataset. Nonetheless, it is enough to draw a comparison with other environments and initial conclusions about winter deployments.

4.2 Evaluation Metrics

To evaluate the impact of extreme conditions on visual navigation, we selected three quantitative metrics: *Feature Quantity*, *Feature Uncertainty*, and *Feature Sparsity*. In this section, we describe these metrics and analyze examples from a nominal scenario (i.e., CSA dataset), which will be used as foundations for the discussion of results in Sect. 5.

(1) Feature Quantity: This is a notion of the amount of total inlier matches observed at any point in time between the live keyframe and the map keyframe during an autonomous traverse. Over the course of a day, this number is guaranteed to decrease

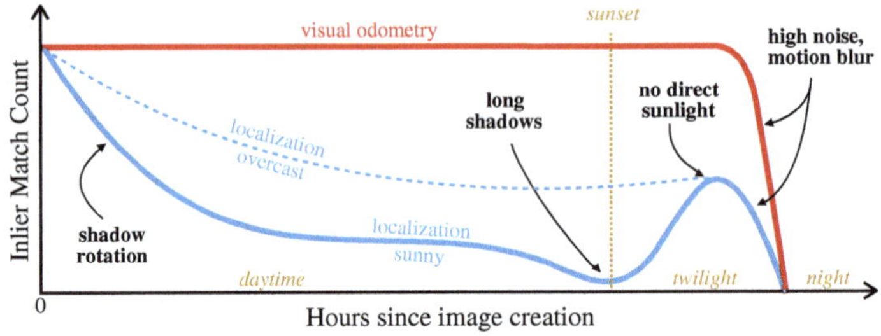

Fig. 4 Illustration of the evolution of the number of inlier feature matches through a nominal day. Time zero corresponds to when the reference images are collected (teaching phase) and the *blue line* represent the typical slow degradation of the number of matches when matching current images to the teaching phase. The difference between a sunny day (*solid line*) and an overcast day (*dashed line*) is also included. The *red line* represents the number of features used during VO, which stays constant up to the limit of the sensor. *Yellow annotations* refer to time events and *black annotations* refer to the main causes of inlier decreases or increases

with time. If the number of inlier matches drops too low, the system will be forced to rely on VO, and eventually fail at following the taught trajectory. Figure 4 shows an illustration of the trend associated with the number of inlier matches typically observed over the course of a day. This figure sums up the experience collected during prior field tests, as reported in [14]. On overcast days, there is a gradual decline in feature matches, because the appearance of the scene is generally constant. This is not true on sunny days, where an early drop is caused by the sun changing position and creating sharp, moving shadows on the ground. Feature quantity begins to rise again at the beginning of twilight, when the light from the sun is not directly observable, generating a shadowless environment similar to an overcast day.

(2) Feature Uncertainty: Only considering the number of features is insufficient to ensure precise trajectory following. 3D landmarks measured with a stereo camera have an uncertainty in their depth associated with the disparity between the left and right feature matches. As this disparity decreases, the uncertainty associated with the depth reconstruction increases. High uncertainty is correlated to features observed far from the camera (i.e., in the background of the image). A reliance on background features leads to a pose estimation that is inaccurate in translation.

An example of the typical distribution of inlier matches observed between the live keyframe and map keyframe during an autonomous traverse with respect to depth uncertainty and measurement location is displayed in Fig. 5. This feature distribution is typical for a forward-looking camera on a moving robot. When moving forward, features close to the lower image border are typically not in the field of view of both the live keyframe and the map keyframe, leading to a skewed distribution of points on the vertical axis. In addition, the platform moves through the environment, generating changes in the re-observed images. On soft ground, a heavy vehicle will generate ruts that modify the deployment area over time.

(3) Feature Sparsity: Lastly, features can be distributed unevenly through a given trajectory. The previously mentioned metrics (i.e., feature quantity and feature

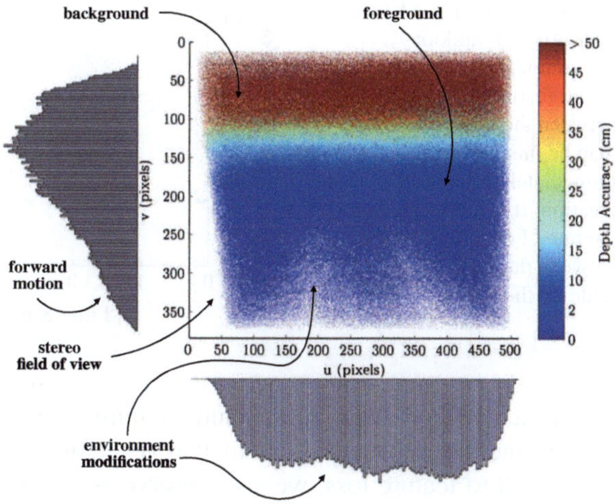

Fig. 5 Matched features with respect to the pixel coordinates aggregated through a full trajectory during the CSA field deployment. Side histograms represent the distributions of matches projected on the vertical (v-axis) and horizontal (u-axis) fields of views. All matches are colored by their depth accuracy with *dark red* being poor (>50 cm) and *dark blue* being optimal. Key elements are labeled in *black*

uncertainty) aggregate the data through a full repeat trajectory, limiting the analysis on consecutive successful localizations. We can indirectly observe this metric using the distance the robot relied on VO before being able to localize within its taught images. A short distance relying on VO is sign of a robust solution against the environment traversed. A system relying entirely on VO for a long period of time will increase its position uncertainty and will drift away from its reference trajectory leading to a mission failure.

5 Results

This section provides an overview of the results of our field trials with respect to the metrics defined in Sect. 4. We first perform a dataset comparison, where we look at the quantity and quality of inlier visual feature matches observed during autonomous traversals of each dataset. Results from the CSA dataset were obtained with the Color-Constant VT&R algorithm, and results from the winter trials were obtained with the Multi-Stereo VT&R algorithm. We note that color-constant images are underperforming in winter environments, and multi-stereo produces better results. We then analyze the performance of our VT&R algorithms with respect to the sparsity of successful localization matches to the map.

Figure 6 shows the rate of feature loss with respect to time since map creation. For each data set, we show the rate of feature degradation from map creation to sunset. The black horizontal line denotes our threshold match count where we can safely localize. It can be seen that when compared to the baseline dataset, the winter datasets have an accelerated decay rate. This can be primarily contributed to lighting having a much higher effect on localization, due to the low elevation of the sun and the poor performance of the color-constant images in these environments. Further reasons

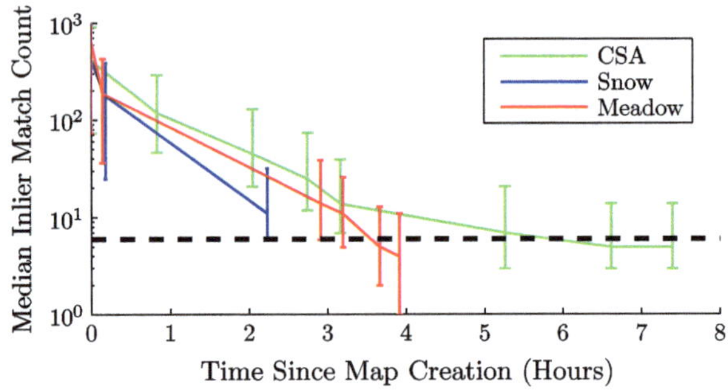

Fig. 6 Feature degradation over time. In outdoor environments, the quantity of visual feature matches between the live view and map begins declining immediately after map creation. It can be seen that the rate of decline varies between data sets. Note log scale on the y-axis

for the accelerated decay rate include featureless snowy foregrounds, overexposed images, melting snow, and dead matted vegetation.

Related to feature loss, we also observe an accelerated migration of the distribution of observed feature matches towards the horizon as time passes in the winter environments. This is displayed in Fig. 7, where the distribution of inlier matches with respect to their vertical pixel coordinates over three repeats is shown for all three datasets. The green line shows this distribution when the map is compared to images collected during map creation. This is the upper limit on feature quantity as well as quality. For each data set, this distribution is nearly uniform. The blue line shows the distribution when the map is compared to the autonomous traversal taken as soon to map creation as possible, and the red line shows when the map is compared to an autonomous traversal several hours after map creation.

The distribution of our baseline comparison, the CSA data set, shows a slight migration towards the horizon after 5.2 h, yet retains a fair amount of foreground matches. In contrast, the winter data sets both show a fast shift to horizon matches only. Looking at the red lines of Fig. 7b,c, there is a significant positive skew of the distribution of matches. This means that after only a few hours in this environment, the majority of matches were obtained from the background of the image. The ramification of this is an increase in uncertainty in our localization estimate.

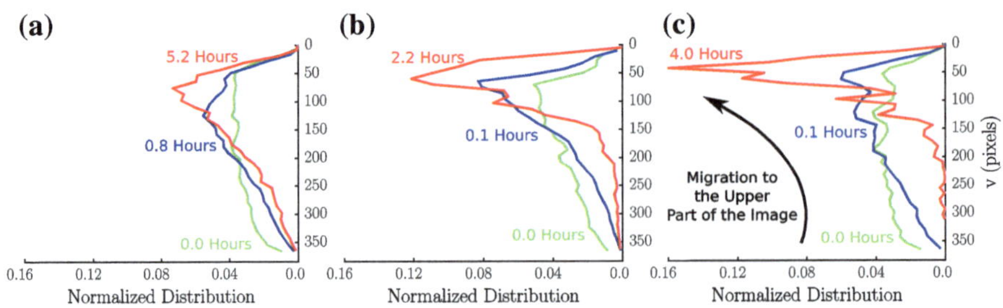

Fig. 7 Vertical distribution of the matched inlier features in the image coordinate frame. On the v-axis, 0 corresponds to a feature at *top* of the image and 360 at the *bottom* of the image. All distributions are normalized and represented over a time period of several hours for different datasets. **a** CSA, **b** Snow, **c** Meadow

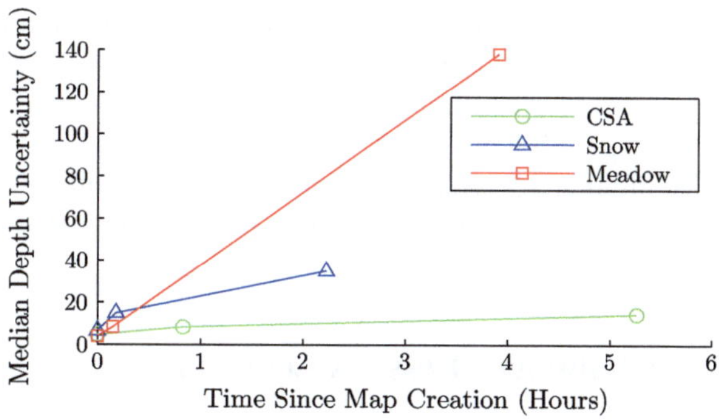

Fig. 8 Median uncertainty of inlier matches over a period of several hours for all data sets. This demonstrates that the migration of inlier matches to the *upper* part of the image leads to an increased uncertainty in the estimation of the feature's depth, which can lead to an inaccurate state estimation

This is confirmed in Fig. 8, where we plot the median uncertainty in our depth estimates for all of the inlier features observed during autonomous traversals of each data set. As expected, the CSA data set maintains a low uncertainty, while the uncertainty seen in matches during the winter data sets quickly rise. The CSA data set maintains a median uncertainty less than 20 cm after 5 h, while in a fraction of the time, the Snow and Meadow data sets reach a median uncertainty level of 40 cm and 1.4 m, respectively.

If the count of inlier feature matches at a specific time step is below our threshold of six features, we discard the localization results and rely on VO for navigation. If navigation relies on dead reckoning for too long, the drift in error will cause the robot to stray from its path. We analyze the distance the robot would have driven on VO using the various VT&R methods detailed in Sect. 3. For results on the baseline CSA dataset, we refer the reader to [14]. Results with respect to sparsity are displayed in Fig. 9. These figures show the Cumulative Distribution Function (CDF) of the distance the robot would have driven on VO during the most difficult traverse of each trial. For the Snowy Landscape, this was the repeat that occurred 2.2 h after map creation, for the Winter Meadow the repeat at 4.0 h was chosen. The figure reads as: "for Y% of the traverse, the robot drove less than X m on VO". The black dashed vertical line denotes the mission failure point of 20 m.

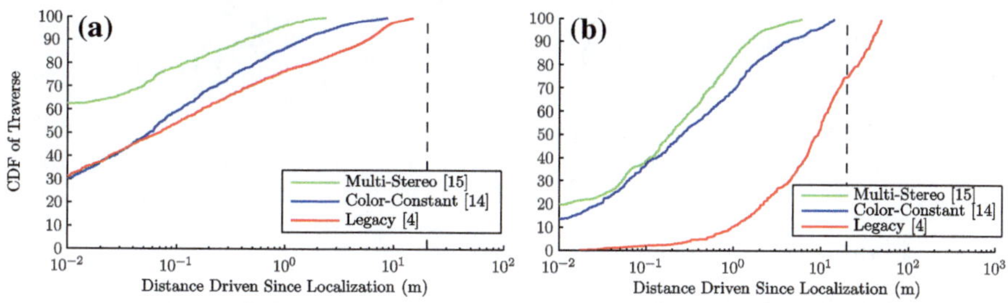

Fig. 9 Cumulative distribution of the distance the robot would have driven on VO for various algorithms on both winter datasets. *Left* Results from the second repeat of the Snow dataset, which occured 2.2 h after map creation. *Right* Results from the fourth repeat of the Meadow dataset, which occured 4.0 h after map creation. Note log scale on the x-axis. **a** Snow, **b** Meadow

For both environments, we see the trend for multi-stereo to outperform color-constant, and color-constant to outperform the legacy system. This comes as no surprise, as color-constant images were shown to underperform in these environments. This is possibly due to a lack in color information in the snow and dead vegetation. The multi-stereo system is based on greyscale images only, but has a wider field of view, having the ability to acquire more stable visual features.

6 Challenges/Lessons Learned

Snow: During the teaching phase of the Snowy Landscape data set, it was bright and sunny. Due to the high reflectivity of the snow, this caused unforeseen issues for our stereo cameras. The brightness of the scene brought the factory settings of the autoexposure algorithm of the Point Grey Research (PGR) Bumblebee XB3 to the limit. The result was saturated images, which reduced details in the foreground.

The Snowy Landscape data set was collected when there was light snow cover. We also attempted to perform autonomous path following in deep snow conditions with unsatisfactory results. In light snow, small vegetation is often visible in the foreground, providing visual features with high contrast. In deep snow, these features are gone and what remains in the foreground is nearly featureless. The only usable matched features were on the horizon not only for localization, but also for VO. This caused frequent inaccurate pose estimates, which caused issues for the path tracker. Figure 10 shows the vertical distribution of features only 0.1 h between the teach and the repeat phase, for deep snow, light snow, and meadow. The majority of matched features in the Deep Snow trial are concentrated on the upper part of the images explaining the poor performance.

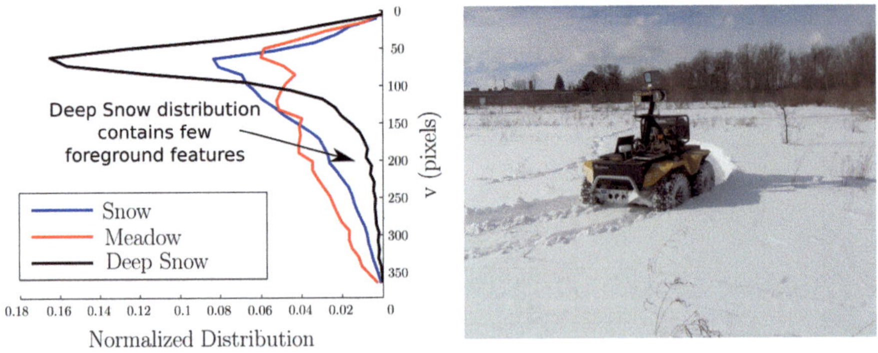

Fig. 10 Figures from the Deep Snow attempt. A lack of visual features in the foreground resulted in poor localization and VO estimates. *Left* Distribution of inlier feature matches with respect to vertical pixel location. The distribution is seen after 0.1 h for all data sets. *Right* Grizzly robot autonomously traversing in the deep snow before the failing point

Glare: An initial hypothesis motivating those field deployments was the assumption that the low elevation of the sun would case glare in the camera, making localization impossible. Due in part to the attitude of the stereo cameras, glare was never an issue. With the cameras tilted to the ground by 20°, the sun was in the worst case only at the top of the image. We even observed cases where sun glare increased the contrast of horizon features, providing a significant boost in feature count. However, glare would be an issue if the cameras were pointed at the horizon.

Color-Constancy: The color-constant images are designed to remove the effects of lighting from the scene. These images were used to great success in the CSA field trials presented in [14]. In these trials, the robot repeated a 1 km route 26 times with an autonomy rate of 99.9 % of distance travelled in nearly every daylight condition. With this prior knowledge, the color transformations were expected to boost performance in the winter field trials presented here, but this was not the case. A hypothesis is that the color-constant images were tuned to perform in green vegetation and red-rocks-and-sand. It is possible that the dead vegetation and snowy landscapes lack the color information to remove the effects of lighting from the images.

Feature-Migration: As explained in Sect. 5, we found that the distribution of features with respect to vertical pixel location migrates to the horizon as time passes. We found that this process is accelerated in winter environments encountered in these trials. This migration results in an increase in the uncertainty of the robot's pose estimate during autonomous navigation. As the depth of observed features increase, the scale estimate becomes only loosely observable, degenerating the problem to localization based on a mono-camera. Further investigation will be required to account for this unforeseen consequence.

7 Conclusion/Future-Work

This paper presented the results of conducting a series of field trials that tested autonomous path-following algorithms in challenging winter environments. When compared to a summer dataset, we show a significant decrease in the quantity and quality of visual features matched over time. Furthermore, color-constant images that increase robustness to changes in lighting conditions have shown to be ineffective in these environments. In order for vision-based navigation to reliably navigate in these environments, we must address some of these difficult issues.

Future avenues of research may involve further classification of appearance-based matching performance in varying environments, variations in camera configurations to mitigate the issue of pose uncertainty as features migrate to the horizon, and the use of image pre-processing [17] and intelligent exposure techniques [5] to increase foreground matching in snowy environments.

Acknowledgments This work was supported by the Natural Sciences and Engineering Research Council (NSERC) through the NSERC Canadian Field Robotics Network (NCFRN). We would also like to thank Jack Collier and the DRDC for conducting a field trial on our behalf in the Canadian High Arctic and Chris Ostafew for assisting in the field trials leading up to this paper.

References

1. Bay, H., Ess, A., Tuytelaars, T., Van Gool, L.: Speeded-up robust features (surf). Comput. Vis. Image Underst. **110**(3), 346–359 (2008)
2. Churchill, W.S., Newman, P.: Experience-based navigation for long-term localisation. Int. J. Robot. Res. **32**(14), 1645–1661 (2013)
3. Corke, P., Paul, R., Churchill, W., Newman, P.: Dealing with shadows: capturing intrinsic scene appearance for image-based outdoor localisation. In: Proceedings of the International Conference on Intelligent Robots and Systems (IROS), Nov 2013
4. Furgale, P., Barfoot, T.D.: Visual teach and repeat for long-range rover autonomy. J. Field Robot. **27**(5), 534–560 (2010)
5. Hrabar, S., Corke, P., Bosse, M.: High dynamic range stereo vision for outdoor mobile robotics. In: Robotics and Automation (ICRA) (2009)
6. Krüsi, P., Bücheler, B., Pomerleau, F., Schwesinger, U., Siegwart, R., Furgale, P.: Lighting-invariant adaptive route following using ICP. J. Field Robot. (2014)
7. McManus, C., Churchill, W., Maddern, W., Stewart, A., Newman, P.: Shady dealings: robust, long-term visual localisation using illumination invariance. In: Robotics and Automation (ICRA) (2014)
8. McManus, C., Furgale, P., Stenning, B., Barfoot, T.D.: Visual teach and repeat using appearance-based lidar. In: Robotics and Automation (ICRA) (2012)
9. McManus, C., Upcroft, B., Newman, P.: Scene signatures: localised and point-less features for localisation. In: Robotics Science and Systems (RSS) (2014)
10. Milford, M.J., Wyeth, G.F.: SeqSLAM: visual route-based navigation for sunny summer days and stormy winter nights. In: Robotics and Automation (ICRA) (2012)
11. Naseer, T., Spinello, L., Burgard, W., Stachniss, C.: Robust visual robot localization across seasons using network flows. In: AAAI (2014)
12. Neubert, P., Sunderhauf, N., Protzel, P.: Appearance change prediction for long-term navigation across seasons. In: Mobile Robots (ECMR) (2013)
13. Otsu, K., Otsuki, M., Kubota, T.: Experiments on stereo visual odometry in feature-less volcanic fields. In: Field and Service Robotics. Springer Tracts in Advanced Robotics, vol. 105, pp. 365–378 (2015)
14. Paton, M., McTavish, K., Ostafew, C., Barfoot, T.D.: It's not easy seeing green: lighting-resistant visual teach & repeat using color-constant images. In: Robotics and Automation (ICRA), May 2015a
15. Paton, M., Pomerlau, F., Barfoot, T.D.: Eyes in the back of your head: robust visual teach & repeat using multiple stereo cameras. In: Computer and Robot Vision (CRV), June 2015b. To Appear
16. Ratnasingam, S., Collins, S.: Study of the photodetector characteristics of a camera for color constancy in natural scenes. J. Opt. Soc. Am. A **27**(2), 286–294 (2010)
17. Williams, S., Howard, A.M.: Developing monocular visual pose estimation for arctic environments. J. Field Robot. **27**(2), 145–157 (2010)

Non-Field-of-View Acoustic Target Estimation in Complex Indoor Environment

Kuya Takami, Tomonari Furukawa, Makoto Kumon and Gamini Dissanayake

Abstract This paper presents a new approach which acoustically localizes a mobile target outside the Field-of-View (FOV), or the Non-Field-of-View (NFOV), of an optical sensor, and its implementation to complex indoor environments. In this approach, microphones are fixed sparsely in the indoor environment of concern. In a prior process, the Interaural Level Difference IID of observations acquired by each set of two microphones is derived for different sound target positions and stored as an acoustic cue. When a new sound is observed in the environment, a joint acoustic observation likelihood is derived by fusing likelihoods computed from the correlation of the IID of the new observation to the stored acoustic cues. The location of the NFOV target is finally estimated within the recursive Bayesian estimation framework. After the experimental parametric studies, the potential of the proposed approach for practical implementation has been demonstrated by the successful tracking of an elderly person needing health care service in a home environment.

1 Introduction

Target localization and tracking, or mobile target estimation, in indoor environments has been a research challenge over several decades due to the existence of a variety of applications in addition to the significance and the difficulty of each application.

K. Takami (✉) · T. Furukawa
Department of Mechanical Engineering, Virginia Tech, Blacksburg, VA, USA
e-mail: kuya@vt.edu

T. Furukawa
e-mail: furukawa@vt.edu

M. Kumon
Department of Mechanical System Engineering, Kumamoto University, Kumamoto, Japan
e-mail: kumon@gpo.kumamoto-u.ac.jp

G. Dissanayake · T. Furukawa
Center for Autonomous Systems, University of Technology, Sydney, NSW, Australia
e-mail: gamini.dissanayake@uts.edu.au

D.S. Wettergreen and T.D. Barfoot (eds.), *Field and Service Robotics*,
Springer Tracts in Advanced Robotics 113, DOI 10.1007/978-3-319-27702-8_38

It is significant in applications such as home security, home health care, and urban search-and-rescue, but its usefulness is limited by the complexity of indoor structures [7, 13]. Complex indoor structures make estimation problems challenging as they can introduce large unobservable regions when an optical sensor such as a camera is deployed. This is because optical sensors' FOV is determined by the Line Of Sight (LOS) and range of the optical sensor, which could be small in highly constrained environments. In addition, there are environments such as personal homes where privacy concerns do not allow for the use of cameras. These limitations on optical sensors give rise to a need for NFOV mobile target estimation.

Recent work for NFOV mobile target estimation has been tackled in three different ways. The first approach deploys target mounted radio-frequency (RF) transmitters and fixed receivers in the environment. In one arrangement, RS receivers form a wireless sensor network (WSN), and numerical techniques are used to localize a NFOV target by processing information of received signals such as signal intensity [3, 6]. An improved arrangement with minimal infrastructure uses "fingerprints" [1, 10]. There is a unique fingerprint at each location in a static environment. A target can thus be localized by feature-matching the fingerprints. Whilst this arrangement can achieve higher accuracy, the critical problem inherent in the RS based approach is its applicability only to near-NFOV target estimation [13, 15].

In the second approach, acoustic sensors are used for target estimation. Since sound signals are reflected by structures, it is possible to localize a NFOV target unlike the RS based approach provided that the sound signals contain information on the target location. The most common approach utilizes the Time-of- Arrival (TOA)/ Time-Difference-of-Arrival (TDOA) information of acoustic signals [2, 11, 18]. The existing acoustic techniques, however, have not achieved true NFOV target estimation to the best of our knowledge. The majority of sound localization challenges have been focused on the direction of sound rather than its position due to the complexity of sound wave propagation [16, 17].

The final approach enhances NFOV target estimation by including a sensor with a limited FOV, such as an optical sensor, by applying a numerical technique. Mauler [12] stated the NFOV estimation problem mathematically, and Furukawa et al. [4, 5] developed a generalized numerical solution. In this technique, the event of "no detection" is converted into an observation likelihood and utilized to positively update probabilistic belief on the target. This belief is dynamically maintained by the recursive Bayesian estimation (RBE). The technique, however, has been found to fail in target estimation unless the target is re-discovered within a short period after being lost. Kumon et al. [9], incorporated an acoustic sensor to maintain belief with no optical detection more reliably. Nevertheless, the technique performed poorly unless the target re-entered the optical FOV since the acoustic sensing is only conducted in an assistive capacity.

This paper presents a new acoustic approach to estimate a NFOV mobile target, and its application and implementation to complex indoor environments. In the approach, microphones are sparsely installed in an indoor environment. In a prior process to the estimation, the IID of observations acquired by a combination of stereo microphone pairs is derived for different target positions and stored as the "fingerprints", or acoustic cues. This a priori data collection process is accelerated by a speaker localization device. With the acquisition of a new sound from the target, an acoustic observation likelihood is computed for dominant pair of microphones by quantifying the correlation of the IID of the new observation to the stored IIDs. The joint likelihood is then created by fusing the acoustic observation likelihoods, and the NFOV target is estimated by recursively updating the belief within the RBE framework using the joint likelihood.

2 Recursive Bayesian Estimation

Consider the motion of a target t, which is discretely given by

$$\mathbf{x}_{k+1}^t = \mathbf{f}^t \left(\mathbf{x}_k^t, \mathbf{u}_k^t, \mathbf{w}_k^t\right) \tag{1}$$

where $\mathbf{x}_k^t \in \mathcal{X}^t$ is the target state at time step k, $\mathbf{u}_k^t \in \mathcal{U}^t$ is the set of control inputs, and $\mathbf{w}_k^t \in \mathcal{W}^t$ is the "system noise". For simplicity, the target state describes the two-dimensional position.

FOV and NFOV are defined by physical properties of a camera s_c where the global state of the optical sensor is assumed to be known as $\tilde{\mathbf{x}}^s \in \mathcal{X}^s$. Note that $\tilde{()}$ is an instance of (). The FOV of the optical sensor can be expressed by the probability of detecting the target $P_d\left(\mathbf{x}_k^t|\tilde{\mathbf{x}}^{s_c}\right)$ as ${}^{s_c}\mathcal{X}_o^t = \left\{\mathbf{x}_k^t|0 < P_d\left(\mathbf{x}_k^t|\tilde{\mathbf{x}}^{s_c}\right) \le 1\right\}$. Accordingly, the target position observed from the optical sensor, ${}^{s_c}\mathbf{z}_k^t \in \mathcal{X}^t$, is given by

$${}^{s_c}\mathbf{z}_k^t = \begin{cases} {}^{s_c}\mathbf{h}^t\left(\mathbf{x}_k^t, \tilde{\mathbf{x}}^s, {}^{s_c}\mathbf{v}_k^t\right), & \text{if } \mathbf{x}_k^t \in {}^{s_c}\mathcal{X}_o^t \\ \varnothing, & \text{otherwise} \end{cases} \tag{2}$$

where ${}^{s_c}\mathbf{h}^t$ is the optical sensor model, ${}^{s_c}\mathbf{v}_k^t$ is the observation noise, and $\varnothing$ represents an "empty element", indicating that the optical observation contains no information on the target or that the target is unobservable when it is not within the observable region. The acoustic sensor can, on the other hand, observe a target on the Non-Line-of-Sight (NLOS) or even in the NFOV with limited accuracy due to the complex behavior of sound signals including reflection, refraction and diffraction. Because of its broad range, the observation region of the acoustic sensor could be considered unlimited when compared to that of the optical sensor. The acoustic sensor model ${}^{s_a}\mathbf{h}^t$ can be then constructed without defining an observable region unlike the optical sensor model:

$${}^{s_a}\mathbf{z}_k^t = {}^{s_a}\mathbf{h}^t\left(\mathbf{x}_k^t, \tilde{\mathbf{x}}^s, {}^{s_a}\mathbf{v}_k^t\right) \tag{3}$$

The RBE updates belief on a dynamical system, given by a probability density, in both time and observation. Let a sequence of observations of a moving target t by a stationary sensor system s from time step 1 to time step k be

${}^{s}\tilde{\mathbf{z}}^{t}_{1:k} \equiv \left\{{}^{s}\tilde{\mathbf{z}}^{t}_{\kappa} | \forall \kappa \in \{1, ..., k\}\right\}$. Given the initial belief $p\left(\mathbf{x}^{t}_{0}\right)$, the sensor platform state $\tilde{\mathbf{x}}^{s}$ and a sequence of observations ${}^{s}\tilde{\mathbf{z}}^{t}_{1:k}$, the belief on the target at any time step k, $p\left(\mathbf{x}^{t}_{k} | {}^{s}\tilde{\mathbf{z}}^{t}_{1:k}, \tilde{\mathbf{x}}^{s}\right)$ can be estimated recursively through the two stage equations. The prediction may be expressed as

$$p\left(\mathbf{x}^{t}_{k} | {}^{s}\tilde{\mathbf{z}}^{t}_{1:k-1}, \tilde{\mathbf{x}}^{s}\right) = \int_{\mathcal{X}^{t}} p\left(\mathbf{x}^{t}_{k} | \mathbf{x}^{t}_{k-1}\right) p\left(\mathbf{x}^{t}_{k-1} | {}^{s}\tilde{\mathbf{z}}^{t}_{1:k-1}, \tilde{\mathbf{x}}^{s}\right) d\mathbf{x}^{t}_{k-1}, \tag{4}$$

whereas the correction takes the form

$$p\left(\mathbf{x}^{t}_{k} | {}^{s}\tilde{\mathbf{z}}^{t}_{1:k}, \tilde{\mathbf{x}}^{s}\right) = \frac{l\left(\mathbf{x}^{t}_{k} | {}^{s}\tilde{\mathbf{z}}^{t}_{k}, \tilde{\mathbf{x}}^{s}\right) p\left(\mathbf{x}^{t}_{k} | {}^{s}\tilde{\mathbf{z}}^{t}_{1:k-1}, \tilde{\mathbf{x}}^{s}\right)}{\int_{\mathcal{X}^{t}} l\left(\mathbf{x}^{t}_{k} | {}^{s}\tilde{\mathbf{z}}^{t}_{k}, \tilde{\mathbf{x}}^{s}\right) p\left(\mathbf{x}^{t}_{k} | {}^{s}\tilde{\mathbf{z}}^{t}_{1:k-1}, \tilde{\mathbf{x}}^{s}\right) d\mathbf{x}^{t}_{k-1}}, \tag{5}$$

where $l\left(\mathbf{x}^{t}_{k} | {}^{s}\tilde{\mathbf{z}}^{t}_{k}, \tilde{\mathbf{x}}^{s}\right)$ represents the likelihood of $\mathbf{x}^{t}_{k}$ given ${}^{s}\tilde{\mathbf{z}}^{t}_{k}$ and $\tilde{\mathbf{x}}^{s}$, which is a probabilistic version of the sensor model; i.e., Eq. (2) if the sensor is optical. It is to be noted that the likelihood does not need to be a probability density since the normalization in Eq. (5) makes the output belief a probability density regardless of the formulation of the likelihood.

3 NFOV Acoustic Target Estimation

3.1 Indoor Installation

Figure 1 shows a schematic for the hardware installation necessary for the proposed acoustic target estimation approach. As shown in the figure, microphones are placed with some distance in the indoor environment. This is a complex environment where optical sensors could not be used effectively as a large number of optical sensors would need to be placed to cover the entire space. Microphones, on the other hand, can collect information on the NFOV. A much lower number of inexpensive sensors need to be installed, for this reason, making the installation efficient in both time and cost.

3.2 Modeling of Acoustic Observation Likelihood

In accordance with the preliminary investigations of the authors [8], the theoretical approach proposed in this paper constructs acoustic cues of the target in the environment of concern a priori to create an acoustic observation likelihood. The assumption of two-dimensional (2D) space and a use of a data collection device in the proposed method reduce the time consumed by a priori data collection. First, the three-dimensional (3D) complex environment can be simplified by assuming the

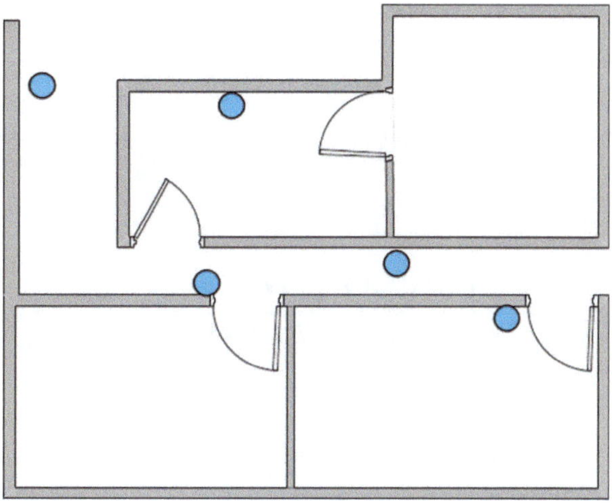

Fig. 1 Schematic of hardware installation for proposed approach where *circles* indicate microphones

omni-directional sound source belongs in the 2D planar domain depicted in Fig. 2. This assumption is realized by placing a sound source at a foot level which generally kept at constant height throughout movement of a human. Second, a priori sound data is collected automatically using a speaker with range finders, which measures the distance to the walls to locate the speaker and emits a white noise when the data collection button is pressed.

Having the data collected into the ILD database in the prior process, Fig. 3 shows a schematic diagram of the main process of the proposed approach. Given the target sound, The acoustic observation likelihood is created for each microphone pair by correlating the observation with IID vectors in the database. The collection of observation likelihood finally yields a joint acoustic observation likelihood. This fusion process only considers a few dominant microphone pairs above the signal-to-noise ratio (SNR) threshold for scalability of the system.

Mathematically, let the estimation of the a priori ith data collection position be $(\tilde{\mathbf{x}}_k^t)_i$. When a target sound is observed by j_{m}-th microphone at $\tilde{\mathbf{x}}_k^s$, the sound is considered "detected" if the SNR of the microphone is greater than the SNR threshold:

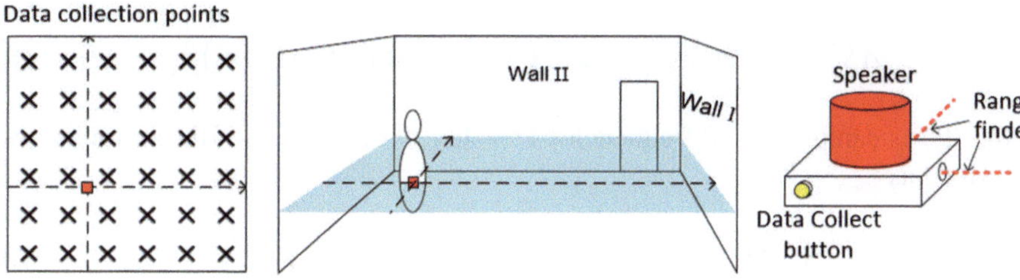

Fig. 2 Data collection and localization

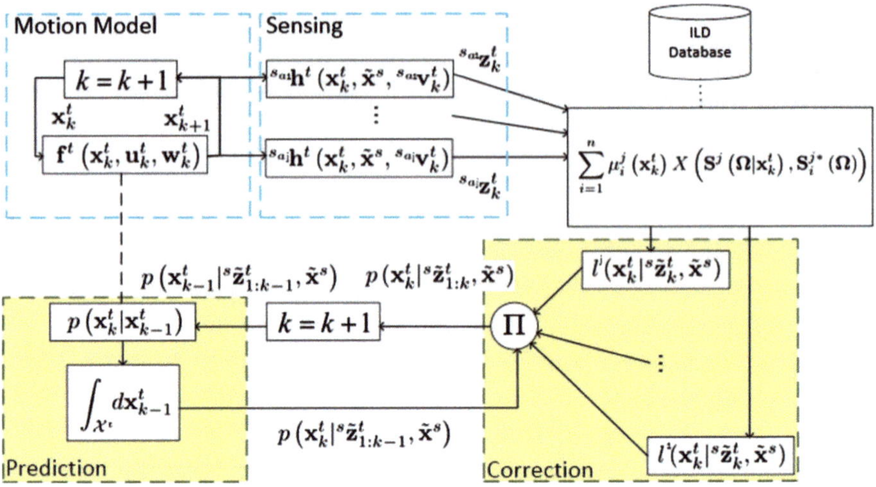

Fig. 3 Schematic diagram of proposed approach within the RBE framework

$$s^{\mathrm{S}}_{j_m} \equiv \frac{s_{j_m}\left(\omega|\left(\tilde{\mathbf{x}}^t_k\right)_i\right)}{s_{j_m}(\omega)_{ambient}} > \delta_{\mathrm{S}} \tag{6}$$

where ω is the sound frequency. Stereo microphone pairs increase with combination of form $\binom{n}{r} = \frac{n!}{r!(n-r)!}$ by choosing stereo pair $r = 2$ from n possible microphones. Figure 4 shows the detectable region of red and yellow microphone as the jth microphone pair $\{j_1, j_2\}$. When the target is located within union of those regions, the ILD of the microphone pair is constructed:

$$\mathbf{x}^t \in {}^{s_a}\mathcal{X}^t_d(\gamma, \delta_{\mathrm{S}}) = {}^{s_{aj_1}}\mathcal{X}^t_d(\gamma_{j_1}, \delta_{\mathrm{S}}) \cap {}^{s_{aj_2}}\mathcal{X}^t_d(\gamma_{j_2}, \delta_{\mathrm{S}}). \tag{7}$$

where γ is acoustic and environmental characteristics. It is reasonable to sort and choose the microphones with largest s^{S} values. The maximum microphone pair is set to be $j_{\max}$. Following the above selection process, the IID of the jth microphone pair $\{j_1, j_2\}$ for the ith position $\left(\tilde{\mathbf{x}}^t_k\right)_i$, $\Delta S^j_i(\omega)$, is then given by

$$\Delta S^j_i(\omega) = 20\log\left|s_{j_1}\left(\omega|\left(\tilde{\mathbf{x}}^t_k\right)_i\right)\right| - 20\log\left|s_{j_2}\left(\omega|\left(\tilde{\mathbf{x}}^t_k\right)_i\right)\right|. \tag{8}$$

If the IID is sampled at N frequencies $\mathbf{\Omega} = [\omega_1, \ldots, \omega_N]^\top$, the IID vector can be described as

$$\mathbf{S}^j_i(\mathbf{\Omega}) = \left[a^j_1 \Delta S^j_i(\omega_1), \ldots, a^j_N \Delta S^j_i(\omega_N)\right]^\top, \tag{9}$$

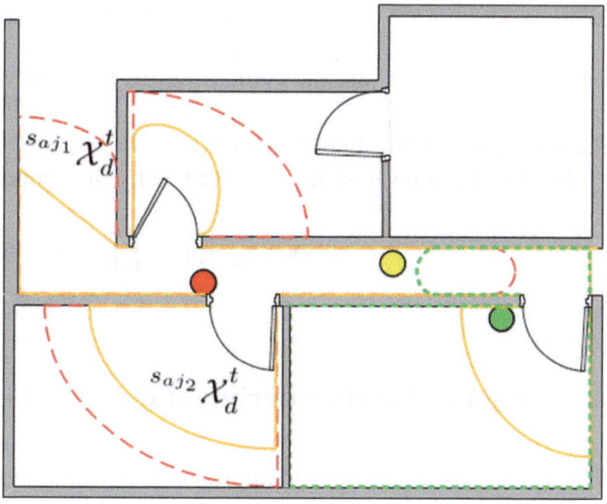

Fig. 4 Detectable region indicated by *lines* for each microphone location

where

$$a_i^j = \left\langle \min\{\left|s_{j1}\left(\omega_N | \left(\tilde{\mathbf{x}}_k^t\right)_i\right)\right|, \left|s_{j2}\left(\omega_N | \left(\tilde{\mathbf{x}}_k^t\right)_i\right)\right|\} - \epsilon \right\rangle. \tag{10}$$

In the equation, $\langle \cdot \rangle$ is Macaulay brackets, and $\min\{\cdot, \cdot\}$ returns the smaller value of the two entities. The acoustic observation likelihood modeling results in the IID vectors for n target positions, i.e., $\boldsymbol{S}_i^{j*}(\boldsymbol{\Omega}), \forall i \in \{1, \ldots, n\}$. They are essentially the acoustic cues to be prepared in advance and used to create the acoustic observation likelihood. The selection of microphone pairs $\boldsymbol{S}_i^{j*}(\boldsymbol{\Omega}) \, \forall j \in \{1, \ldots, j_{\max}\}$ must satisfy the conditions $s_j^{\mathrm{S}} > \delta_{\mathrm{S}}$.

Given the IID vector $\boldsymbol{S}^j\left(\boldsymbol{\Omega}|\mathbf{x}_k^t\right)$ created from ${}^s\tilde{\mathbf{z}}_k^t$ with the unknown target position $\mathbf{x}_k^t$, the proposed technique quantifies its degree of correlation to the ith IID vector as

$$X\left(\boldsymbol{S}^j\left(\boldsymbol{\Omega}|\mathbf{x}_k^t\right), \boldsymbol{S}_i^{j*}(\boldsymbol{\Omega})\right) = \frac{1}{2}\left\{\frac{\boldsymbol{S}^j\left(\boldsymbol{\Omega}|\mathbf{x}_k^t\right)^\top \boldsymbol{S}_i^{j*}(\boldsymbol{\Omega})}{\left|\boldsymbol{S}^j\left(\boldsymbol{\Omega}|\mathbf{x}_k^t\right)\right|\left|\boldsymbol{S}_i^{j*}(\boldsymbol{\Omega})\right|} + 1\right\}. \tag{11}$$

where $0 \le X(\cdot) \le 1$. The acoustic observation likelihood with the particular $\boldsymbol{S}_m\left(\boldsymbol{\Omega}|\mathbf{x}_k^t\right)$ can be finally calculated as

$$l_j^a\left(\mathbf{x}_k^t|{}^s\tilde{\mathbf{z}}_k^t, \tilde{\mathbf{x}}_k^s\right) = \sum_{i=1}^{n} \mu_i^j\left(\mathbf{x}_k^t\right) X\left(\boldsymbol{S}^j\left(\boldsymbol{\Omega}|\mathbf{x}_k^t\right), \boldsymbol{S}_i^{j*}(\boldsymbol{\Omega})\right), \tag{12}$$

where $\mu_i^j\left(\mathbf{x}_k^t\right)$ is a basis function developed by adjacent measurements. One of the suited basis function is a T-spline basis function where $\mu_i^j\left(\mathbf{x}_k^t\right)$ in a T-mesh in parameter space (s, t) can be represented as

$$\mu_{im}(s,t) = g(s)g(t) \tag{13}$$

where $g(s)$, and $g(t)$ are the cubic B-spline basis functions. Further detailed formulations are found in [14]. Similarly to $X(\cdot)$, $l_j^a(\cdot)$ is also bounded as $0 \le l_m^a(\cdot) \le 1$ due to the use of the shape function.

Finally, the joint likelihood is derived by the canonical data fusion formula:

$$l^a\left(\mathbf{x}_k^t|{}^s\tilde{\mathbf{z}}_k^t, \tilde{\mathbf{x}}_k^s\right) = \prod_j l_j^a\left(\mathbf{x}_k^t|{}^s\tilde{\mathbf{z}}_k^t, \tilde{\mathbf{x}}_k^s\right). \tag{14}$$

4 Numerical and Experimental Analysis

The efficacy of the proposed approach was examined experimentally in two steps. The first step was aimed at studying the capabilities and limitations of the proposed acoustic sensing technique by parametrically changing the complexity of the test environment. This was accomplished with an experimental system consisting of a speaker array and a movable/replaceable wall developed specifically for this study. After verifying the feasibility of the acoustic sensing technique for NLOS target localization, the applicability of the proposed approach in a practical indoor scenario was investigated. The investigation looked into not only the performance of the proposed approach but also compared it to a conventional approach.

4.1 Acoustic Observation of NLOS Target

Figure 5a shows the design of the experimental system that changed the complexity of the environment for the evaluation of the proposed approach. The number of microphones was fixed at two to investigate the environmental complexity, and they were located next to an outer wall and faced open space where a speaker array and movable/replaceable wall(s) were placed. The complexity of the environment was changed by varying two parameters of the movable/replaceable wall: the distance of the wall to the edge of speaker array L_d and the length of the wall L_w. The shorter the distance and/or the larger the length, the more complex the environment due to the increased number of reflections of the sound signal.

Speaker locations are shown in Fig. 5a as blue crosses. A microcontroller controlled speakers so that each speaker sequentially emitted white noise for a programmed period. A set of IIDs for a wall setting were thus collected automatically. Once the IIDs were collected, the ability of the proposed approach was evaluated by emitting sound from a speaker at some location within the area of the speaker array and identifying the location in the form of an observation likelihood. This location was different than that of the speakers of the speaker array to demonstrate the ability of the proposed technique to identify the target at an arbitrary position.

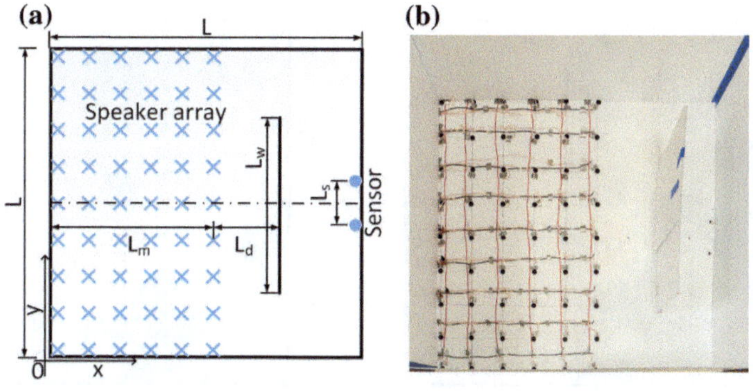

Fig. 5 Experimental system for investigating environmental complexity. **a** Schematic desgin. **b** Developed system

Table 1 Dimensions and other parameters in the experiments

Parameter	Value	Parameter	Value
$\tilde{\mathbf{x}}^t$ single wall	[42, 34] [cm, cm]	L	90 cm
$\tilde{\mathbf{x}}^t$ double wall	[22, 56] [cm, cm]	Height	0 cm
ω_1	0 Hz	L_m	50 cm
ω_N	22 kHz	L_s	10 cm
N	8,192	L_d	{0, 10, 20, 30} cm
ϵ	0.01	L_w	{50, 60, 70} cm
n	54	n_w	{1, 2}

Figure 5b shows the developed experimental system and the dimensions and other parameters used in the experiments are listed in Table 1. The sound was sampled at 8,192 frequency bins within the audible range to capture its behavior accurately. 54 speakers were aligned to cover the open space. The distance and the length of the wall were varied to introduce both lightly NLOS and heavily NLOS environments. The case of two walls ($n_w = 2$) was tested in addition to the single wall case to increase environmental complexity. Only the distance of the wall closer to the acoustic sensor was varied.

Figure 6 shows the resulting acoustic observation likelihoods when the sound target was at position [42, 34] and [22, 56] for the single wall and double wall cases, respectively. The former two cases were with a single wall at different distances. The latter two cases were with two walls with different wall length. The result first indicates that the target location is well estimated when the distance is short or when the length is small. The target is closer to LOS in these conditions since sound reaches the acoustic sensor with a small number of reflections. The identification of the target location in the remaining two cases is difficult due to the number of sound reflections. The identification with two walls is seen to be significantly harder than that with a

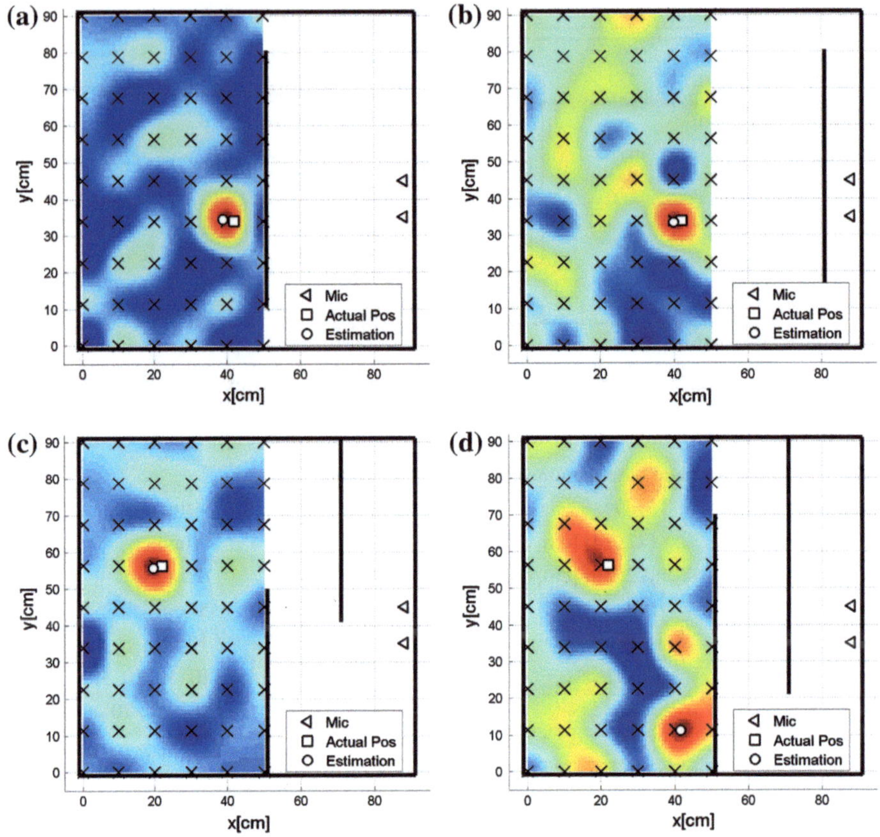

Fig. 6 Acoustic likelihoods for different environmental complexity. **a** $\{L_d, L_w, n_w\} = \{0, 70, 1\}$, **b** $\{L_d, L_w, n_w\} = \{30, 70, 1\}$, **c** $\{L_d, L_w, n_w\} = \{20, 50, 2\}$, **d** $\{L_d, L_w, n_w\} = \{20, 70, 2\}$

single wall for the same reason. While the acoustic observation likelihood is heavily multi-modal with these cases, the target location is still captured by the highest peak or at least by one of the peaks as shown in Fig. 6d. This demonstrates the ability of the proposed approach to identify the location of the NFOV target though with limited accuracy.

Figure 7a, b show the mean error of the acoustic observation likelihood when the distance and the length were varied for single and double wall cases. The mean error is the distance of the nearest peak of the acoustic observation likelihood to the true target location. The result of the mean error shows that the proposed technique could locate the target to within 2 cm error in 11 of the 12 cases for single wall case. The estimation was particularly good when the wall length was small. Figure 7c shows the uncertainty comparison for the two cases, using the differential entropy derived at a point within the normalized likelihood is used as the uncertainty. The mean

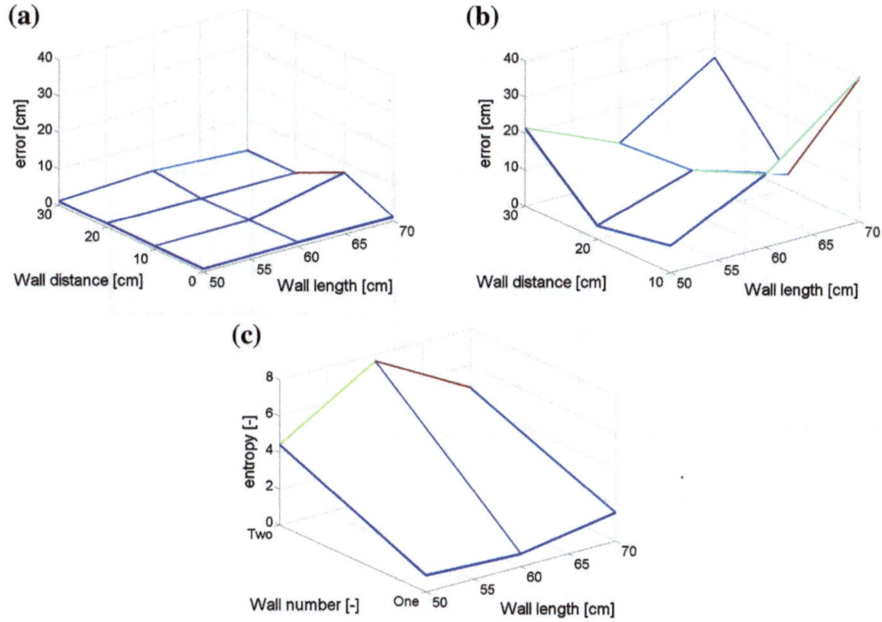

Fig. 7 Mean error and differential entropy of the acoustic observation likelihood with a single and double wall. **a** Single wall mean error. **b** Double wall mean error. **c** Differential entropy for single and double wall

entropies for the two cases show that uncertainty increases with increase in a number of walls for all wall lengths as expected. For the double wall case, the uncertainty is higher with less success in target identification, but the proposed approach could still be used to identify the target location.

4.2 Applicability to Practical Indoor Scenario

4.2.1 Practical Indoor Scenario

Having validated the ability of the proposed acoustic sensing technique, the applicability of the proposed approach in NFOV target estimation to a practical indoor scenario was investigated. Figure 8 shows the actual indoor environment used for the investigation: the apartment of an elderly person who needs home health care service. As shown in the figure, the environment with five separate rooms is so complicated that it is difficult to cover the entire area by cameras. In addition, this is personal home, so cameras are not to be installed. The approximate dimensions of the apartment are 7.1 m in width, 10.4 m in length and 2.5 m in height. Six microphones,

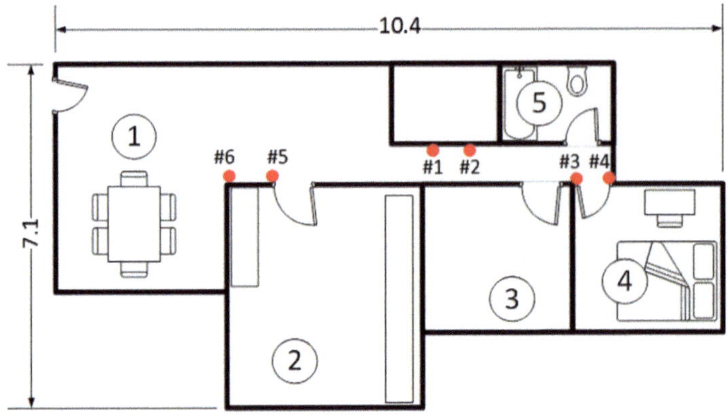

Fig. 8 Map of the test environment dimensions[m] and other details

Table 2 Dimensions and other parameters in the experiments

Parameter	Value	Parameter	Value
ω_1	0 [Hz]	Height	5 [cm]
ω_N	2.7 [kHz] k	n	255
N	2,000	ϵ	0.01
δ_S	2		

shown as red dots, were fixed to cover the entire space. The target person carried a small speaker which emitted sound with white noise. Parameters used for acoustic target estimation are listed in Table 2.

4.2.2 Results

Figure 9 shows the acoustic observation likelihoods created by microphone pairs when the target person walked in Room 3. The square dot indicates the true target position. Only the likelihoods with microphones 1–4 are shown since those with microphones 5 and 6 did not meet the δ_{SNR}. Identified best of the combinations are pairs 2, 3 and 1, 3. Microphones 1–3 have the most direct LOS to Room 3, so the result matched well with the expected observable region. Figure 10 shows the resulting joint likelihood. The target location is accurately identified by filtering uncertainties (Fig. 11).

The result of RBE when the target person walked around is shown in Fig. 12 with the true position again indicated by a square dot. It is seen that the proposed approach accurately tracks the target. The estimated position was less than 15 cm from the true target position in 83 % of the time. Cameras and RF receivers/transmitters cannot be

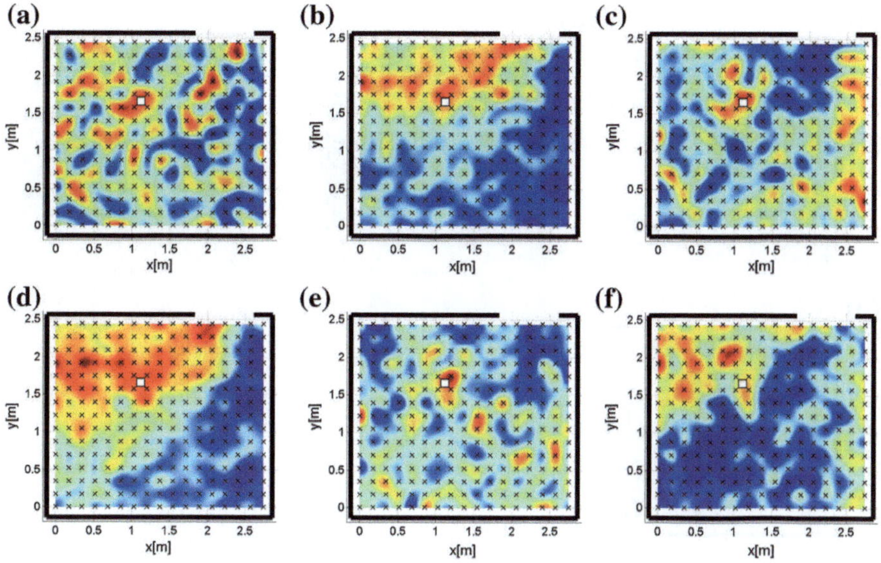

Fig. 9 Acoustic likelihood in room 3 from multiple sensor combinations. **a** microphone pair {#1, #2}, **b** microphone pair {#1, #3}, **c** microphone pair {#1, #4}, **d** microphone pair {#2, #3}, **e** microphone pair {#2, #4}, **f** microphone pair {#3, #4}

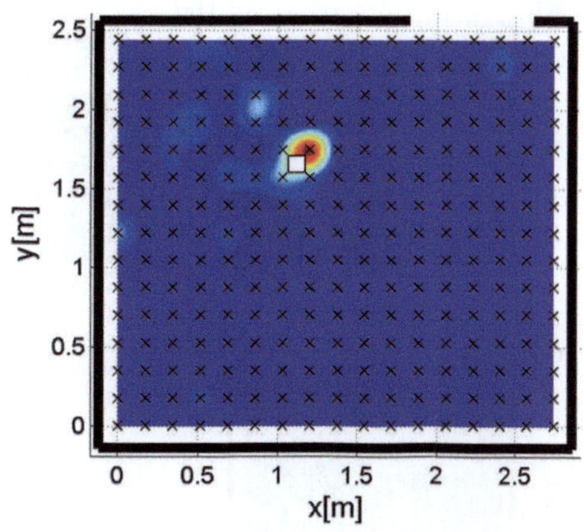

Fig. 10 Joint acoustic observation likelihood

used for such a highly constrained environment, so the conventional acoustic sensing technique based on two microphones was tested as the only comparable approach. As shown in Fig. 11, the conventional approach was not able to identify the target location once it had failed in the localization.

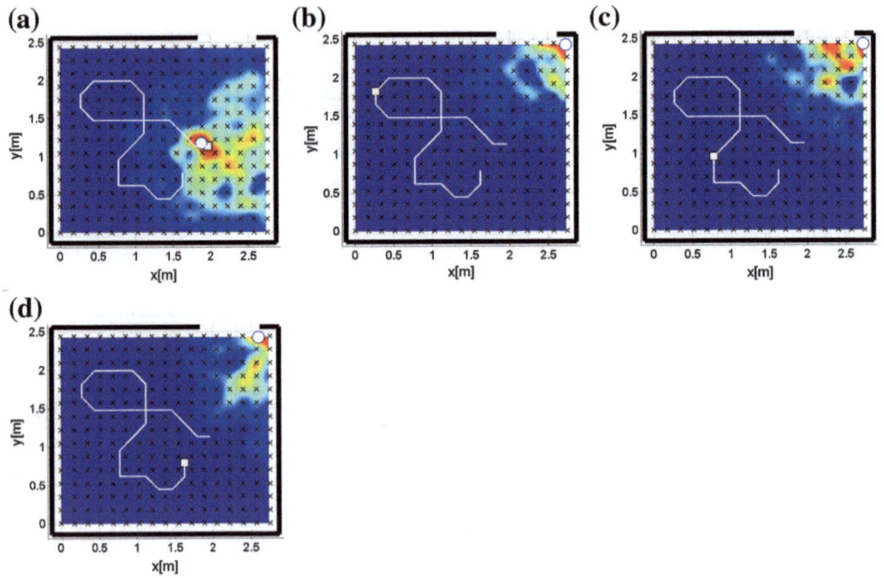

Fig. 11 Acoustic observation likelihood in room 3 with one microphone pair. **a** $k = 1$, [1.96, 1.13], **b** $k = 11$, [0.27, 1.82], **c** $k = 21$, [0.78, 0.96], **d** $k = 29$, [1.63, 0.79]

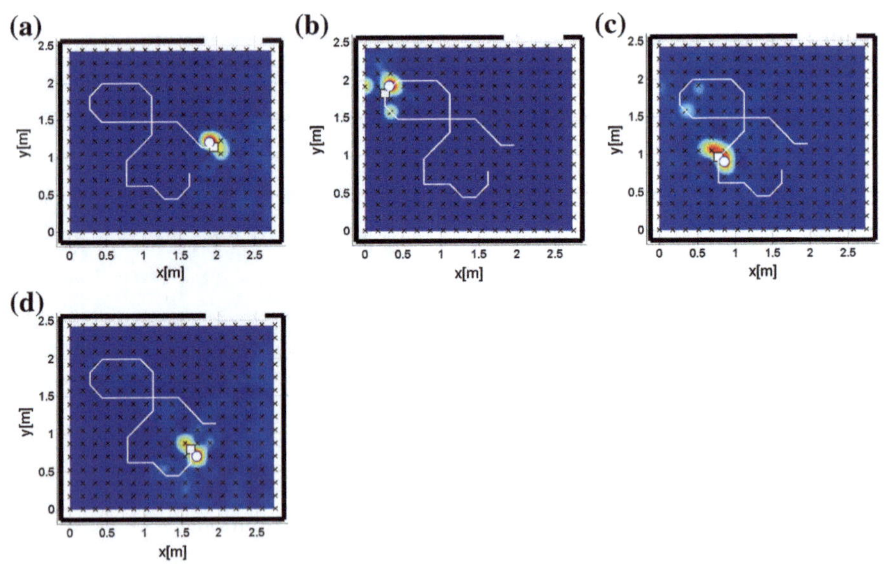

Fig. 12 Proposed Joint acoustic observation likelihood in room 3 with RBE. **a** $k = 1$, [1.96, 1.13], **b** $k = 11$, [0.27, 1.82], **c** $k = 21$, [0.78, 0.96], **d** $k = 29$, [1.63, 0.79]

5 Conclusions

This paper has presented a new approach which uses a set of microphones to localize and track a mobile NFOV target, and its applicability and implementation in complex indoor environments. The proposed approach derives the IID of observations from a selected set of microphones for different target positions and stores the IIDs as acoustic cues. Given a new sound, an acoustic observation likelihood is computed for each pair of microphones by correlating IIDs. The joint likelihood is then created by fusing the acoustic observation likelihoods, and the NFOV mobile target is estimated by the RBE. Following the experimental parametric studies, the proposed approach was applied to track an elderly person needing home health care service, yielding an estimation which was successful to within 15 cm accuracy at 83 % of all the tested positions. These results have conclusively demonstrated the potential of the proposed approach for practical target localization.

The paper has demonstrated the new concept, and many challenges are still open for future study. The issues of immediate interest include the enhancement of acoustic sensing using the Interaural Time Difference (ITD) and the Interaural Phase Difference (IPD) as well as the use of non-white noise sound with sound separation/speech recognition techniques, so that the approach could be used for various applications. For the IID database in a dynamic environment, automated update needs further investigation.

References

1. Bahl, P., Padmanabhan, V.N.: Radar: an in-building rf-based user location and tracking system. In: INFOCOM 2000, Proceedings of the Nineteenth Annual Joint Conference of the IEEE Computer and Communications Societies. vol 2, pp. 775–784. IEEE (2000)
2. Chen, J., Benesty, J., Huang, Y.: Time delay estimation in room acoustic environments: an overview. EURASIP J. Appl. Signal Process. **2006**, 170–170 (2006)
3. Dai, H., Zhu, Z., Gu, X.F.: Multi-target indoor localization and tracking on video monitoring system in a wireless sensor network. J. Netw. Comput. Appl. (2012)
4. Furukawa, T., Bourgault, F., Lavis, B., DurrantWhyte, H.: Recursive bayesian search-and-tracking using coordinated uavs for lost targets. In: Proceedings IEEE International Conference on Robotics and Automation, ICRA 2006, pp. 2521–2526. IEEE (2006)
5. Furukawa, T., Mak, L.C., Durrant-Whyte, H., Madhavan, R.: Autonomous bayesian search and tracking, and its experimental validation. Adv. Robot. **26**(5–6), 461–485 (2012)
6. Guvenc, I., Chong, C.: A survey on toa based wireless localization and nlos mitigation techniques. IEEE Commun. Surv. Tutor. **11**(3), 107–124 (2009)
7. Khoury, H.M., Kamat, V.R.: Evaluation of position tracking technologies for user localization in indoor construction environments. Autom. Constr. **18**(4), 444–457 (2009)
8. Kimoto, D., Kumon, M.: Optimization of the ear canal position for sound localization using interaural level difference. In: 36th Meeting of Special Interest Group on AI Challenges (2012)
9. Kumon, M., Kimoto, D., Takami, K., Furukawa, T.: Bayesian non-field-of-view target estimation incorporating an acoustic sensor. In: IEEE/RSJ International Conference on Intelligent Robots and Systems (IROS), pp. 3425–3432. IEEE (2013)
10. Ladd, A.M., Bekris, K.E., Rudys, A.P., Wallach, D.S., Kavraki, L.E.: On the feasibility of using wireless ethernet for indoor localization. IEEE Trans. Robot. Autom. **20**(3), 555–559 (2004)

11. Mak, L., Furukawa, T.: Non-line-of-sight localization of a controlled sound source. In: IEEE/ASME International Conference on Advanced Intelligent Mechatronics, 2009. AIM 2009, pp. 475–480 (2009)
12. Mauler, R.: Objective functions for Bayesian control-theoretic sensor management, II: MHC-like approximation. Recent Developments in Cooperative Control and Optimizatio, pp. 273–316. Kluwer Academic Publishers, Norwell (2003)
13. Priyantha, N.B., Balakrishnan, H., Demaine, E.D., Teller, S.: Mobile-assisted localization in wireless sensor networks. In: Proceedings of the INFOCOM 2005, 24th Annual Joint Conference of the IEEE Computer and Communications Societies, vol 1, pp. 172–183. IEEE (2005)
14. Sederberg, T.W., Zheng, J., Bakenov, A., Nasri, A.: T-splines and t-nurccs. In: ACM transactions on graphics (TOG), vol 22, pp. 477–484. ACM (2003)
15. Seow, C.K., Tan, S.Y.: Non-line-of-sight localization in multipath environments. IEEE Trans. Mob. Comput. **7**(5), 647–660 (2008)
16. Svaizer, P., Brutti, A., Omologo, M.: Environment aware estimation of the orientation of acoustic sources using a line array. In: Proceedings of the 20th European Signal Processing Conference (EUSIPCO), pp. 1024–1028. IEEE (2012)
17. Tamai, Y., Sasaki, Y., Kagami, S., Mizoguchi, H.: Three ring microphone array for 3d sound localization and separation for mobile robot audition. In: IEEE/RSJ International Conference on Intelligent Robots and Systems, (IROS 2005), pp. 4172–4177. IEEE (2005)
18. Ward, D.B., Lehmann, E.A., Williamson, R.C.: Particle filtering algorithms for tracking an acoustic source in a reverberant environment. IEEE Trans. Speech Audio Process. **11**(6), 826–836 (2003)

Novel Assistive Device for Teaching Crawling Skills to Infants

Mustafa A. Ghazi, Michael D. Nash, Andrew H. Fagg, Lei Ding, Thubi H.A. Kolobe and David P. Miller

Abstract Crawling is a fundamental skill linked to development far beyond simple mobility. Infants who have cerebral palsy and similar conditions learn to crawl late, if at all, pushing back other elements of their development. This paper describes the development of a robot (the Self-Initiated Prone Progression Crawler V3, or SIPPC3) that assists infants in learning to crawl. When an infant is placed onboard, the robot senses contact forces generated by the limbs interacting with the ground. The robot then moves or raises the infant's trunk accordingly. The robot responses are adjustable such that even infants lacking the muscle strength to crawl can initiate movement. The novel idea that this paper presents is the use of a force augmenting motion mechanism to help infants learn how to crawl.

1 Introduction

Cerebral Palsy (CP) is a common physically disabling condition for children in the United States. It is a lifelong physical disability caused by damage of the developing brain and it affects muscle function, postural control, and coordination of skilled movements. According to the Cerebral Palsy International Research Foundation, globally, 17 million people have CP. Among these, 1 in 3 is unable to walk. The US Centers for Disease Control and Prevention has estimated that the cost to care for an individual with CP over their lifetime is nearly $1 million. Children with CP attain developmental milestones, such as independent crawling and walking, late in life, if at all. There is no known cure for CP. Treatments such as physical therapy, medication, and surgery have shown inconsistent improvement in the children's functional status and capabilities. Generally the consensus is that the earlier the treatment is initiated, the better the chances for improvement.

Research shows that the effects of CP are apparent within the first year of life [1, 2]. Common early milestones that are delayed are independent sitting and crawling. Inability to crawl has implications beyond locomotion as it is associated with other

M.A. Ghazi (✉) · M.D. Nash · A.H. Fagg · L. Ding · T.H.A. Kolobe · D.P. Miller
University of Oklahoma, 865 Asp Ave Rm 212, Norman, OK 73072, USA
e-mail: mghazi@ou.edu

D.S. Wettergreen and T.D. Barfoot (eds.), *Field and Service Robotics*,
Springer Tracts in Advanced Robotics 113, DOI 10.1007/978-3-319-27702-8_39

domains of child development that are crucial for learning, such as spatial cognition [3, 4]. Consequently, failure by infants with CP to attain crawling during the first year of life may negatively impact the development in other cognitive and perceptual-motor areas. Crawling also develops during the period of rapid brain growth [5], making it a crucial target for early mobility interventions.

This research aims to develop a device intended to serve two purposes: (1) Facilitate crawling in infants at risk of CP and (2) Measure the learning strategies that these infants use when they learn to crawl. To that end, we have created a robotic system, the SIPPC3 (see Fig. 1), which can move an infant by sensing its intentions, regardless of whether the infant is strong enough to move or not. The system consists of the robot and an operator's laptop for control and datalogging. An infant is placed in the SIPPC3 in a prone position. The robot supports the infant at a pre-set height or can vary the height based on the forces exerted by the infant against the floor. Forces exerted in the horizontal plane are used to generate motions of the robot in the appropriate direction. The effect of these forces can be amplified for weaker infants, if desired.

Our approach capitalizes on the neuronal group selection theory [6] by assisting the infant to crawl early before the crawling-age (experience dependent plasticity). By bypassing some of the constraints experienced by infants with CP, such as decreased muscle strength and incoordination, and rewarding the infant's every effort to move, our device can potentially ameliorate or eliminate the negative consequences caused by the inability to crawl. Other benefits may be improvement in postural control, muscle strength, coordination, and understanding spatial relationships. The use of this device has the potential to improve development, particularly crawling, in a similar way as infants without CP.

Section 2 describes requirements, constraints, and related work. Sections 3 through 5 detail the SIPPC3 robot's mechanical and electrical systems, and control laws. Section 6 describes some preliminary testing of the robot. Finally, conclusions and future work are discussed in Sect. 7.

Fig. 1 SIPPC3 in action. *Photo credit* Sooner Magazine/Hugh Scott

2 Motivating Factors for the SIPPC3 Design

2.1 Previous Approaches

There has been a number of robotic approaches to assist infants with Cerebral Palsy to obtain mobility. Some researchers have created robots that the infant can ride, [7], which, while potentially giving the child some sense of independent mobility, does not develop any motor skills or have any of the other benefits of physical activity. Schoepflin [8], working with somewhat older children (3–4 years) developed an assistive device more similar in action to a robotic pedal cart. Children in a sitting position could activate and control the cart (a seat mounted on a Pioneer robot platform) by using a pedaling-like motion. Kolobe [9], describes some earlier, related work in prone locomotion. This work drew from lessons learned from SIPPC1, which was a passive platform with no assistance in movement. SIPPC2 [9], could amplify some of the movements initiated by the children, but the fixed height put them in an advanced crawling position, regardless of their age or crawling developmental stage.

2.2 Requirements and Constraints

Children learn to crawl in stages. They start in a prone position close to or on the ground. As they develop, they lift more of their body off the ground and eventually move to an alternating pattern on their hands and knees. Orientation of the head is important, especially during the transition in development from lying flat on the ground to the point where the head is lifted above the shoulders [10]. Infants are very interested in their surroundings and will grab at near objects that are within view. Children with or at risk of CP may have reduced muscle strength. If they are interested in objects in their surroundings, they may not be able to generate the force required to mobilize the body during crawling.

Our robotic assistant needs to allow children to be in the prone position, and be as close to the ground as possible, while still providing adequate support for breathing. The robot should be able to assist the child in weight bearing. A crawling infant may use just their arms, legs or coordinated action amongst all four of their limbs when moving, and so the robot should be able to move the child in any direction and rotate around any point. The robot should also be able to constrain those movements and points of rotation in order to encourage more productive crawling behavior. The robot also needs to be able to handle children of different sizes and weights. Finally the robot needs to give an infant a clear view of where he/she is headed, and to give access to objects (e.g., toys) in front of the infant, so that he/she can plan and execute goal-driven movements [11].

3 SIPPC3 Crawler Mechanical Design

The key requirements for the mechanical design of the robot are support for infant movement in any direction along the floor and in the vertical direction. Accordingly, we have developed a system with 4 DOF motion, of which 3DOF along the floor are achieved by using omni-wheels (see Fig. 2). All this needs to be done using a minimum possible number of wheels and supporting structure while giving an infant a wide view of the surroundings.

The mechanical structure of the robot is designed around an infant support platform (see Fig. 3). This platform is mounted to a Y-shaped central frame (see Figs. 3 and 4) with three motion control modules or "legs". We have selected a Y-shape because it allows for 3 legs, which is the smallest number of legs we can use to support the infant. The Y-shape is helpful since having the front two legs spread to the sides gives the infant a reasonably sized, unobstructed view. The infant support platform is a frame with a padded base on which an infant can lie down in a prone position. The padded base is tilted up by 7° to assist infant breathing. A 6 DOF FT sensor with integrated electronics [12, 13] is the mechanical interface between the infant support platform and the central frame (see Fig. 5). This ensures that all forces exerted by the infant below will be transferred to the robot through the FT sensor.

The legs are mounted at the ends of the central frame. Together, the legs provide 4 DOF motion for the infant support platform: one for raising the platform off the floor, and three for moving it in x, y, and yaw around the z-axis (see Fig. 9). Each of the legs contains a linear actuator (see Fig. 5) that can extend to raise the infant support platform. Built-in potentiometers in each actuator provide position feedback. The actuators are not backdrivable so they do not consume power to maintain a given height, nor will they suddenly move if there is an unexpected power loss.

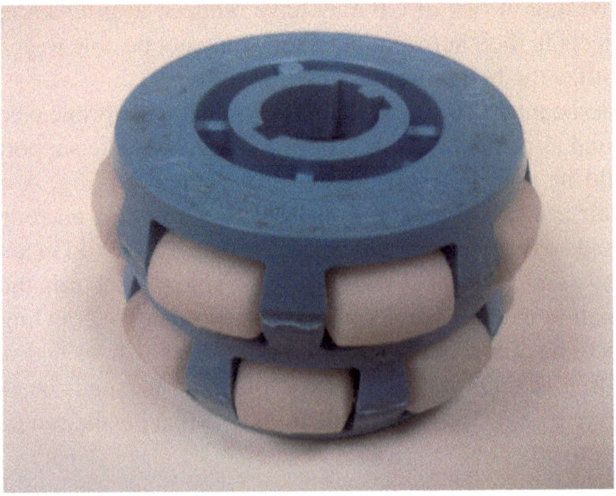

Fig. 2 Omni-wheel drive for holonomic motion

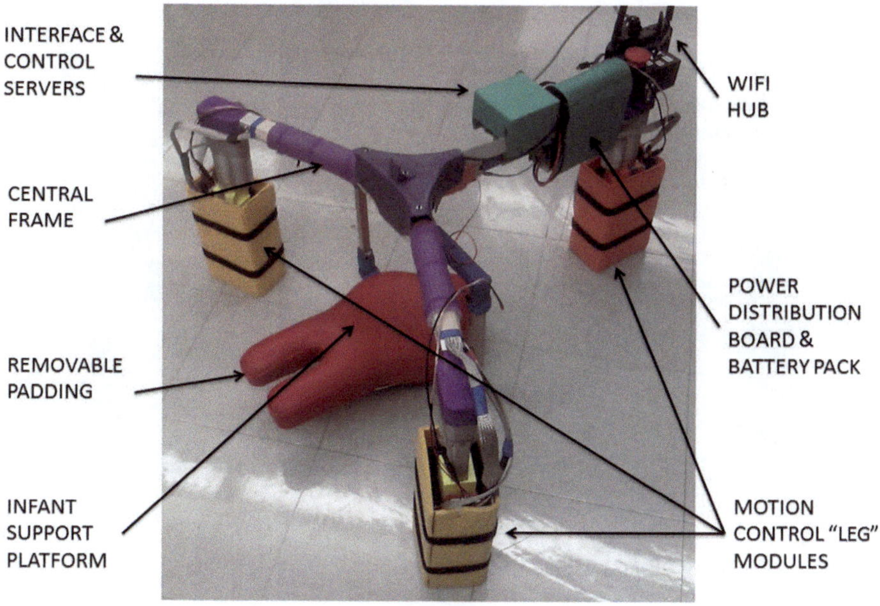

Fig. 3 Overview of the mechanical system

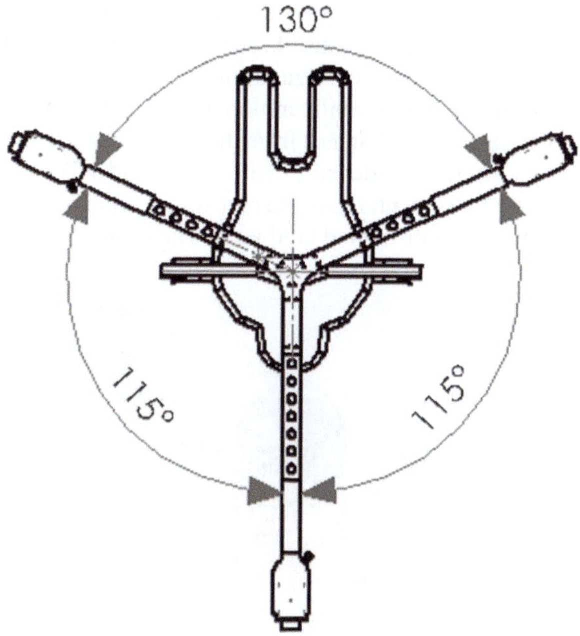

Fig. 4 Relative angles between the Y-frame

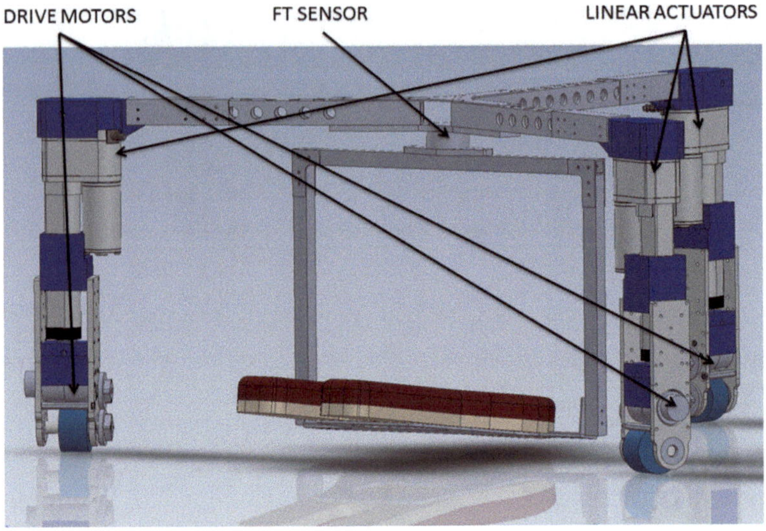

Fig. 5 CAD model of SIPPC3 with leg detail exposed

For motion across the floor, each leg has a 131:1 geared DC motor driving an omni-wheel (see Fig. 2). Omni-wheels were used to allow variability in movement patterns. Built-in quadrature encoders provide rotation feedback of the wheels. Each omni-wheel is oriented such that the axis of rotation passes through the center of the robot. This forms a holonomic drive configuration. Our configuration is different from the typical three-wheel holonomic configuration where all the wheels are positioned 120° apart and at the same radial distance from the center. Instead, the angle between the two front wheels has been widened to 130° (see Fig. 4). The central frame has the front wheels closer to the center than the rear wheel. The wider angle is to allow the infant to have a wider unobstructed field view. The front wheels are closer to the

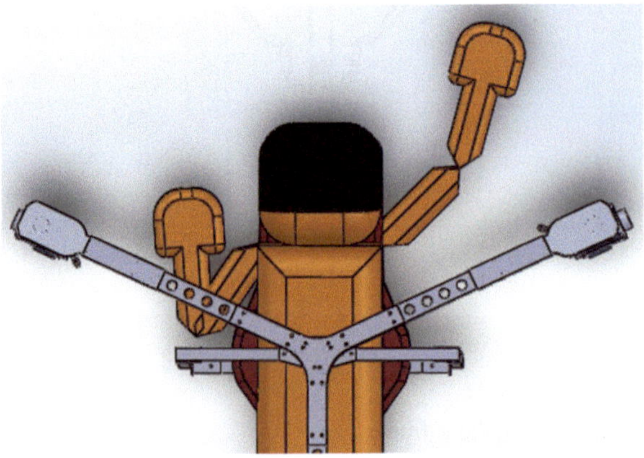

Fig. 6 Overhead view showing typical arm and head positioning

center, making the robot smaller and more maneuverable in homes while maintaining adequate workspace for the subject's arms (Fig. 6).

To protect the baby from the mechanical and electrical parts, the "legs" have been surrounded by aluminum sheet metal enclosures. The sheet metal (as are most of the hard surfaces in the SIPPC3) are covered by soft, brightly colored padding (see Fig. 3).

4 Control Electronics

The electronic subsystems comprise an onboard WiFi hub, an Interface Server, a Control Server, Motion Control "leg" Modules, and the FT sensor (see Fig. 7). These communicate over three different physical layers: ethernet (using TCP-IP), I2C, and Controller Area Network (CAN [14]). Ethernet connects the Interface Server and the Control Server to the WiFi hub. An I2C bus links the three Motion Control Modules and Control Server. The CAN bus connects the Control Server to the FT sensor.

The Interface Server is an ARM® Cortex™-A8 processor (BeagleBone Black [15]) running a stripped-down version of the Ubuntu operating system. The Control Server is an ARM® Cortex™-M3 micro-controller (mbed LPC1768 [16]). Each Motion Control Module is made up of a Cortex™-M4 micro-controller (Teensy 3.1 [17]), a 2-channel motor driver, a linear actuator, and DC motor.

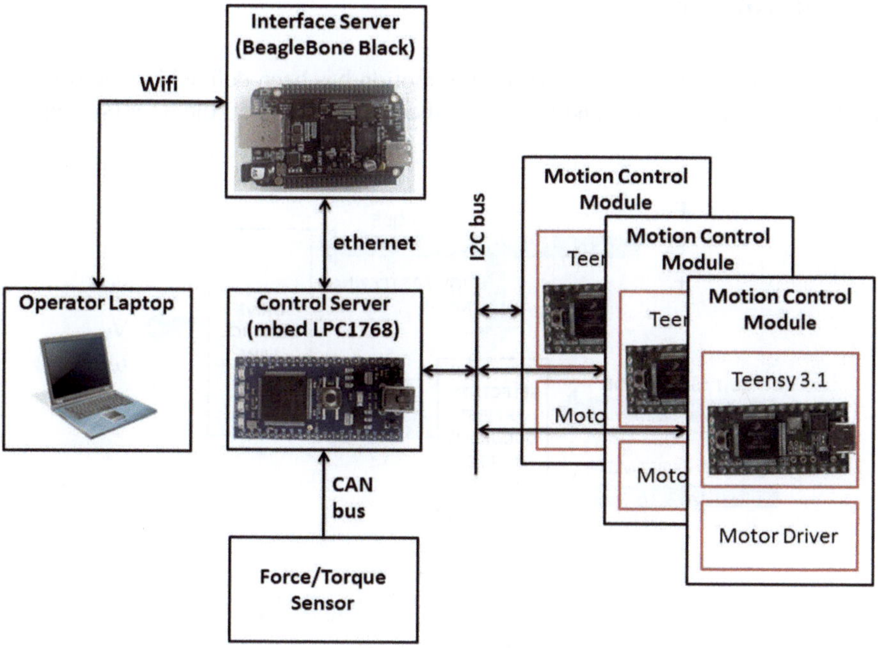

Fig. 7 Overview of the control electronics

The Interface Server receives commands from the operator's laptop over WiFi. It transmits back system health, sensor, and odometry data. The Interface Server generates synchronization signals for external recording devices.

The Control Server is central to the functioning of the robot. It receives commands from the Interface Server and relays back system health and odometry data. It receives the FT data from the 6 DOF FT sensor and computes wheel velocity and actuator height set points for the three legs. These set points are then transmitted to the Motion Control Modules. Each Motion Control Module runs a feedback control loop through the motor driver. Position for the linear actuator is controlled using the potentiometer feedback. Wheel velocity is controlled using quadrature encoders and a movement to omni-wheel speed transformation similar to [18].

The entire system is designed to be portable and fully self-contained. It is powered by a 4-cell LiPo battery pack with a 5000mAh capacity.

Multiple levels of safety features for the infant have been built into the system. At the software level, the operator can issue software emergency E-stop commands over the laptop. An E-stop command issues a stop command for all motors and actuators. If communication with the Control Server is lost, the Motion Control Modules are programmed to stop driving the motors. At the hardware level, a physical E-stop button on the robot cuts power to the motors, causing them to decelerate to an almost immediate stop.

5 Control Laws

The mapping of infant actions onto robot motion has been defined by control laws for driving along the floor, and for raising the infant's trunk off the floor (see Fig. 8).

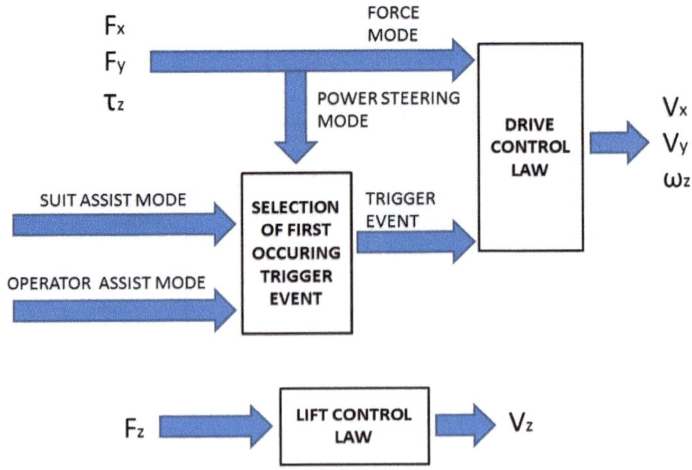

Fig. 8 Control laws defining how the infant interacts with the robot

There are four drive control modes for motion along the floor. These are force mode, operator assist mode, power steering mode and suit assist mode. The force mode and power steering modes allow control through interactions between the infant and the ground. Suit assist is available for gesture-based control for very weak infants using a motion capture system [19] and a novel gesture-recognition system. The operator assist mode allows the operator to intervene in case the infant drives the robot into a spot that is difficult for the infant to extricate themselves from on their own.

These four drive modes can be activated independently, and they work together to generate global robot velocity commands. A generalized equation mapping infant action to global drive velocity commands is provided below; it is used for linear velocities in the x and y directions, and the angular velocity about the z-axis:

$$V_D = K_D F_D + V_A(t) \tag{1}$$

where V_D is the commanded global robot velocity to drive along the floor, K_D is the gain for the force mode, F_D is the driving force or torque induced by the infant, and $V_A(t)$ is the velocity contribution of an "assist event" triggered by the operator assist, power steering, or suit assist modes. $V_A(t)$ is a function of time, and provides a small, short-term motion in a specified direction. Figure 9 illustrates the axes used.

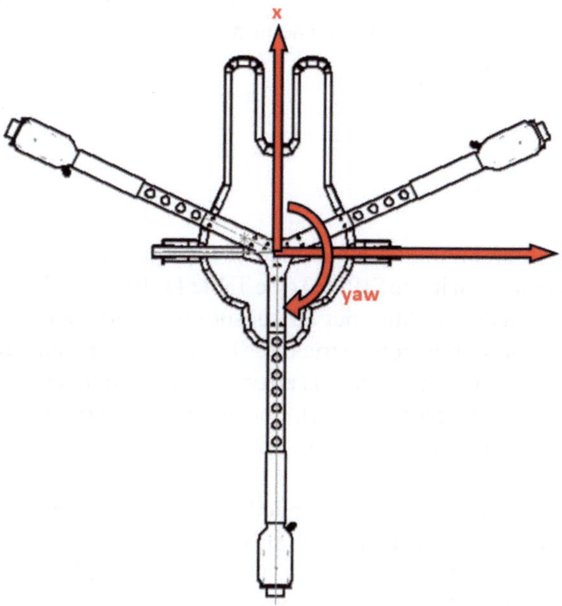

Fig. 9 Robot frame of reference for control kinematics. The x axis points towards the *front* of the robot. The z-axis is into the plane of the page

In the force mode, a force or torque generated by the infant generates a component of the global robot velocity. The other three modes compete with each other to generate the other component of the global velocity. This is done by triggering "assist events." In the operator assist mode, an assist event is triggered by the caregiver through the operator laptop. In power steering mode, a force beyond a certain threshold triggers an assist event. In suit assist mode, a gesture recognized by the wearable motion capture system [19] triggers an assist event.

Once an assist event is triggered, a third order minimum jerk velocity profile is generated for a preset period of time δt_A over a preset distance δs_A. It is followed by a preset refractory period δt_R. During the time $(\delta t_A + \delta t_R)$, other assist events are ignored. For example, if the power steering mode triggers an assist event, a subsequent assist event triggered by the suit assist mode will be ignored.

For lifting the infant off the ground, there is only one mode, which is called the gravity mode. When active, the upward force can trigger an upward movement for the linear actuators. The operator sets a desired lifting force, a force deadband, and a minimum height. If the last δt_L milliseconds have an average lifting force greater than the top end of the band, the linear actuators lift the infant. If the average lifting force is within the deadband, the actuators maintain the current height. If the average lifting force is below the bottom end of the deadband, then the linear actuators settle down towards the preset height.

When gravity mode is deactivated, the preset minimum height is maintained. It can be adjusted at any time through the operator laptop. Currently, the infant's torso can be placed 3–10 cm from the floor. The minimum height is the limit imposed by the padding placed under the infant.

6 Testing

We have measured the minimum magnitude of three different forces that an infant could use to trigger motion in the SIPPC3 (see Table 1). These are based on the thresholds that we have selected to filter out FT sensor noise and undesirable oscillations caused by the dynamics of the robot structure. The uncertainty quoted for each of the forces is based on the measurement bias error and the random error using Student's t-distribution (95 % confidence interval). For the moment arm measurements, only the bias error is quoted, since these were not repeated. The x and y axes are shown on Fig. 9.

The first of these is the force F_x applied in the forward direction through the center of the robot. This is a force that an infant could use to propel himself or herself forward. The load was applied to the frame by pulling on straps used to attach infants onto the padding. The robot was placed on top of a table and a string was attached to the straps. The other end of the string was run over a smooth pivot over the edge of the table and attached to an empty container. Water was gradually poured into the container to apply an increasing steady force. Once the robot started to move, the container was taken off the string and was weighed on a weighing scale with 1 g resolution. The mean threshold force F_x was calculated using acceleration due to gravity as 9.81 m/s^2.

The second force is also a force in x but the line of action is offset from the center. This is similar to a force that an infant could use to propel himself or herself forward using one hand on the floor. From video recordings of infants on the SIPPC3, one line of action of this force is close to the shoulder. For this test, we took the line of action of this force to be one hand breadth away from an infant's shoulder in the y direction. Using mean shoulder breadth and hand breadth data for 6–8 month old infants [20] gave a moment arm of 0.118 ± 0.063 m. Force was applied in a similar manner as above. To provide a rigid offset point for force application, a metal plate with holes was clamped onto the padding such that one of the holes lined up with the desired line of action of the force.

The third force is a force in y with the line of action offset from the center of the robot. This is similar to a force that an infant could use to push against the floor to turn away. From video recordings, one line of action of such a force is slightly above shoulder level. For this test, we took it to be half the length of the upper arm of 6–8 month old infants. Using anthropometric data from [20], the moment arm about the center of the robot is 0.229 ± 0.63 m. Force was applied in a similar manner as the $F_{x,offset}$ force above.

We also performed some tests to evaluate smoothness and response time. The motion from an assist event is smooth because the controller uses a minimum-jerk trajectory to generate smooth motion profiles (see Sect. 5). The motion resulting from the force mode is approximately as smooth as the force applied. For the force mode, we have verified smooth response when applying continuous forces. With infants, however, the motion is not as smooth and continuous. This is because crawling is not a continuous process and infants do not apply continuous, smooth forces. We measured the response time of the robot using video footage. The response time is defined as the time interval between the instant the force is applied, and the instant that the robot starts to move. The response time was 120 ± 18 ms.

In all the above tests, and in all the experimental sessions with infants, there has been no instance of the robot tipping over. The wheels are placed far enough apart and the center of gravity is low enough that an infant cannot tip over the robot.

The robot is currently in use in a study which is planned to test 30 typically developing infants and 20 infants at risk for CP over the next twelve months. Three

Table 1 Force thresholds required to activate the SIPPC3 under force control

	F_x	$F_{x,offset}$	$F_{y,offset}$
Mean force (N)	2.26	2.86	1.78
Error (N)	0.120	0.080	0.061
Standard deviation (N)	0.168	0.126	0.085
Samples	10	12	10

F_x is a simulated forward push, $F_{x,offset}$ is a simulated forward push using one hand, and $F_{y,offset}$ is a simulated turning force applied sideways typically developing infants have completed the study so far using this robot. Subjects start at four to five months of age and have multiple sessions per week with the robot for the subsequent eight weeks. The subjects are able to learn how to engage the robot in order to reach toys that have been placed for them on the ground. The robot has been approved by IRB as safe for testing with typically developing infants and infants with CP (IRB number 3755).

7 Conclusions and Future Work

We have described an assistive crawler robot to supplement the efforts of children who have CP or similar conditions. This robot allows shared and dynamically changing weight bearing and it can adjust the height of the baby from approximately 3 cm (the thickness of the infant support pad) to 10 cm. Together, these features enable the robot to accommodate infants of wide range in height and weight, and let them develop their crawling capabilities in a close-to-natural pose from scooting along the ground to advanced crawling. The holonomic motion capability allows the robot to accommodate turns and motions that are generated by the subjects. These are new capabilities for assistive crawler robots and allow the subjects to learn and develop their prone locomotion skills more naturally.

In addition to traditional methods for monitoring infant development, we are using electroencephalography (EEG)-based neuro-imaging and a wearable motion-capture system (kinematic suit) developed in-house [19]. The kinematic suit can also be used as an interaction interface where an infant's limb motions are used to trigger the robot response. The neuro-imaging with the SIPPC3 is giving additional indications of goal-directed movement [21]. The SIPPC3 body also serves as an advantageous mounting point for cameras to record head, arm and foot movements.

Acknowledgments The authors wish to thank the following individuals for their contributions to this research: Brandon Boos, Rebecca Browder, Steven Grissom, Zac Morris, Johnny O'Keefe, Amanda Porter, Tyler Schmiderberg, Joshua B. Southerland, Jennifer Spillers, Leonard W. Wilson and Ran Xiao. We would also like to thank the infants and their parents who have been volunteering to participate in this research. This work was supported in part by the NSF Division of Information & Intelligent Systems award #1208639 under the National Robotics Initiative.

References

1. Barbosa, V.M., Campbell, S.K., Smith, E., Berbaum, M.: Comparison of test of infant motor performance (TIMP) item responses among children with cerebral palsy, developmental delay, and typical development. Am. J. Occup. Therapy **59**(4), 446–456 (2005)
2. Kolobe, T.H., Bulanda, M., Susman, L.: Predicting motor outcome at preschool age for infants tested at 7, 30, 60, and 90 days after term age using the test of infant motor performance. Phys. Therapy **84**(12), 1144–1156 (2004)

3. Anderson, D.I., Campos, J.J., Witherington, D.C., Dahl, A., Rivera, M., He, M., Uchiyama, I., Barbu-Roth, M.: The role of locomoion in psychological development. Front. Psychol. **4**, 1–17 (2013)
4. Campos, J.J., Anderson, D.I., Barbu-Roth, M.A., Hubard, E.M., Hertenstein, M.J., Witherington, D.: Travel broadens the mind. Infancy **1**(2), 149–219 (2000)
5. Hadders-Algra, M.: The neuronal group selection theory: promising principles for understanding and treating developmental motor disorders. Dev. Med. Child Neurol. **42**(10), 707–715 (2000)
6. Sporns, O., Edelman, G.M.: Solving Bernstein's problem: a proposal for the development of coordinated movement by selection. Child Dev. **64**(4), 960–981 (1993)
7. Smith, M.E., Stansfield, S., Dennis, C.W.: Tots on bots. In: Proceedings of the 6th International Conference on Human-Robot Interaction, HRI'11, New York, NY, USA, pp. 405–406. ACM (2011)
8. Schoepflin, Z.: Pediatric, bio-driven, mobile-assistive devices and their effectiveness in purposeful driving for typically-and atypically-developing toddlers. Master's thesis, University of Delaware, Oct 2010
9. Kolobe, T.H.A., Pidcoe, P.E., McEwen, I., Pollard, V., Truesdell, C.: Self-initiated prone progression in infants at risk for cerebral palsy. Pediatr. Phys. Ther. **18**(1), 93–94 (2007)
10. Goldfield, E.C.: Transition from rocking to crawling: postural constraints on infant movement. Dev. Psychol. **25**(6), 913–919 (1989)
11. McEwan, M.H., Dihoff, R.E., Brosvic, G.M.: Early infant crawling experience is reflected in later motor skill development. Percep. Motor Skills **72**, 75–79 (1991)
12. Parmiggiani, A., Randazzo, M., Natale, L., Metta, G., Sandini, G.: Joint torque sensing for the upper-body of the iCub humanoid robot. In: IEEE International Conference on Humanoid Robots. IEEE (2009)
13. FTSens—6 axis torque and force sensor with CAN Bus communication. http://www.icub.org/images/brochures/iCub_ftSens_flyer_web.pdf
14. Bosch: CAN Specification Version 2.0. http://www.bosch-semiconductors.de/media/ubk_semiconductors/pdf_1/canliteratur/can2spec.pdf. Sept 1991
15. BeagleBone Black. http://beagleboard.org/black
16. mbed LPC1768. https://mbed.org/platforms/mbed-LPC1768/
17. Teensy 3.1—New Features. https://www.pjrc.com/teensy/teensy31.html
18. Siegwart, R., Nourbakhsh, I.: Introduction to Autonomous Mobile Robots. The MIT Press (2004)
19. Southerland, J.B.: Activity recognition and crawling assistance using multiple inexpensive inertial measurement units. Master's thesis, School of Computer Science, University of Oklahoma, May 2012
20. Snyder, R.G., Schneider, L.W., Owings, C.L., Reynolds, H.M., Golomb, D.H., Schork, M.A.: Anthropometry of infants, children and youths to age 18 for product safety design. Technical Report UM-HSRI-77-17, Highway Safety Research Institute, Ann Arbor, Michigan, May 1977
21. Miller, D.P., Fagg, A.H., Ding, L., Kolobe, T.H.A., Ghazi, M.A.: Robotic crawling assistance for infants with cerebral palsy. In: Proceedings of the AAAI'15 Workshop on Assistive Technologies Emerging from Artificial Intelligence Applied to Smart Environments. AAAI Press, Jan 2015

SPENCER: A Socially Aware Service Robot for Passenger Guidance and Help in Busy Airports

Rudolph Triebel, Kai Arras, Rachid Alami, Lucas Beyer, Stefan Breuers, Raja Chatila, Mohamed Chetouani, Daniel Cremers, Vanessa Evers, Michelangelo Fiore, Hayley Hung, Omar A. Islas Ramírez, Michiel Joosse, Harmish Khambhaita, Tomasz Kucner, Bastian Leibe, Achim J. Lilienthal, Timm Linder, Manja Lohse, Martin Magnusson, Billy Okal, Luigi Palmieri, Umer Rafi, Marieke van Rooij and Lu Zhang

R. Triebel (✉) · D. Cremers
Department of Computer Science, TU, Munich, Germany
e-mail: triebel@in.tum.de

D. Cremers
e-mail: cremers@in.tum.de

K. Arras · T. Linder · B. Okal · L. Palmieri
Social Robotics Lab, University of Freiburg, Freiburg im Breisgau, Germany
e-mail: arras@cs.uni-freiburg.de

T. Linder
e-mail: linder@cs.uni-freiburg.de

B. Okal
e-mail: okal@cs.uni-freiburg.de

L. Palmieri
e-mail: palmieri@cs.uni-freiburg.de

R. Alami · M. Fiore · H. Khambaita
LAAS-CNRS: Laboratory for Analysis and Architecture of Systems, Toulouse, France
e-mail: ralami@laas.fr

M. Fiore
e-mail: mfiore@laas.fr

H. Khambaita
e-mail: harmish@laas.fr

L. Beyer · S. Breuers · E. Leibe · U. Rafi
RWTH Aachen, Aachen, Germany
e-mail: beyer@vision.rwth-aachen.de

S. Breuers
e-mail: breuers@vision.rwth-aachen.de

E. Leibe
e-mail: leibe@vision.rwth-aachen.de

U. Rafi
e-mail: urafi@vision.rwth-aachen.de

D.S. Wettergreen and T.D. Barfoot (eds.), *Field and Service Robotics*,
Springer Tracts in Advanced Robotics 113, DOI 10.1007/978-3-319-27702-8_40

Abstract We present an ample description of a socially compliant mobile robotic platform, which is developed in the EU-funded project SPENCER. The purpose of this robot is to assist, inform and guide passengers in large and busy airports. One particular aim is to bring travellers of connecting flights conveniently and efficiently from their arrival gate to the passport control. The uniqueness of the project stems from the strong demand of service robots for this application with a large potential impact for the aviation industry on one side, and on the other side from the scientific advancements in social robotics, brought forward and achieved in SPENCER. The main contributions of SPENCER are novel methods to perceive, learn, and model human social behavior and to use this knowledge to plan appropriate actions in real-time for mobile platforms. In this paper, we describe how the project advances the fields of detection and tracking of individuals and groups, recognition of human social relations and activities, normative human behavior learning, socially-aware task and motion planning, learning socially annotated maps, and conducting empirical experiments to assess socio-psychological effects of normative robot behaviors.

R. Chatila · M. Chetouani · O.A.I. Ramírez
ISIR-CNRS: Institute for Intelligent Systems and Robotics, Paris, France
e-mail: chatila@isir.upmc.fr

M. Chetouani
e-mail: chetouani@isir.upmc.fr

O.A.I. Ramírez
e-mail: islas@isir.upmc.fr

V. Evers · M. Joosse · M. Lohse · L. Zhang
University of Twente, Enschede, The Netherlands
e-mail: v.evers@utwente.nl

M. Joosse
e-mail: m.p.joosse@utwente.nl

M. Lohse
e-mail: m.lohse@utwente.nl

L. Zhang
e-mail: l.zhang@tudelft.nl

H. Hung · L. Zhang
Delft University of Technology, Delft, The Netherlands
e-mail: h.hung@tudelft.nl

T. Kuncer · A.J. Lilienthal · M. Magnusson
Örebro University, Örebro, Sweden
e-mail: tomasz.kucner@oru.se

A.J. Lilienthal
e-mail: achim.lilienthal@oru.se

M. Magnusson
e-mail: martin.magnusson@oru.se

M. van Rooij
University of Amsterdam, Amsterdam, The Netherlands
e-mail: m.m.j.w.vanrooij@uva.nl

1 Introduction

The immensely growing passenger volume in air traffic worldwide poses an enormous challenge for all air carriers and airport operators. With the increasing number of passengers arriving and departing at an airport, the probability of delays and missed connection flights grows accordingly. Furthermore, busy hubs such as the airport of Amsterdam Schiphol are particularly challenging for the growing numbers of first-time air passengers, people with little knowledge of foreign languages or those who need any kind of special attendance. For them and for others, finding a fast and efficient way from an arrival gate to a departure gate for connection can be very difficult, especially if the first, incoming flight was delayed. For air carriers such as the Dutch KLM, missed connecting flights often result in additional cost for rebooking and baggage reloading, while for the passengers it means further delays and the inconveniences associated with them.

This is the main motivation for the launch of the EU-funded project SPENCER, which we present in this paper. In SPENCER, we develop a mobile robotic platform that efficiently guides oversea passengers at Schiphol airport from their arrival gate to the passport control point for further, inner-European connections, the so-called "Schengen barrier". The project is unique in at least two major aspects: First, it addresses a highly relevant business case with a large potential impact for the entire aviation industry, motivated by a growing need for passenger assistance and the decrease of missed connecting flights. And second, in contrast to earlier tour-guide robot systems (e.g. [4, 31]), it addresses topics in *social robotics* by developing new methods to perceive, learn and model human social behavior and to use this knowledge to plan appropriate actions in real-time for a mobile robotic platform. In doing so, SPENCER generates novel scientific contributions in the fields of

- detection, tracking and multi-person analysis of individuals and groups of people,
- recognition of human social relations, social hierarchies and social activities,
- normative human behavior learning and modeling,
- socially-aware task, motion and interaction planning,
- learning socially annotated maps in highly dynamic environments,
- empirically evaluating socio-psychological effects of normative robot behaviors.

In SPENCER, we address these problems jointly and in a multi-disciplinary project team, which enables us to exploit synergies between social science and robot engineering for the implementation of an effective cognitive system that operates robustly and safely among humans. In this paper, we present first encouraging results in all mentioned fields, as well as the insights gained from integrating all relevant system components onto the same common platform.

The paper is organized as follows: First, we present an overall view on the system regarding the platform design and the system architecture. Then, we show results of our socially aware localization and mapping module. In Sect. 4 we describe our people and group tracking component, a major building block for social analysis tools. Section 5 introduces the human-aware task and motion planning module of

SPENCER. Then, we develop important tools to analyse human social behavior and discuss the two main approaches we pursue to implement social behavior on the robot. Finally, Sect. 8 briefly describes the integrated system and concludes the paper.

2 Platform Design and System Architecture

A key element of a socially acting and interacting robot is its physical appearance, because even if the robot's behavior fully complies with socially normative rules, it is of little use if the platform itself appears unfriendly or even threatening. Therefore, a human- or animal-like appearance is often chosen for robots that operate in human environments. However, a completely antropomorphic design has the disadvantage that it implicitly raises expectations regarding certain cognitive capabilities of the platform, which cannot be accomplished with current systems. This can lead to disappointments or to refusal of the system. To avoid this, we decided to use a human-like but abstract appearance, which combines friendliness with believability. The result is a human-size platform (see Fig. 1a, b), where the body resembles the functionality of an information desk, and the head serves as a device for a comprehensible but simplified non-verbal communication (e.g. nodding or orientation towards spokesperson). For physical interaction with the user, the platform has a touchscreen and a boarding pass reader. The sensors consist of two SICK LMS 500 2D laser scanners covering 360° range in total at 0.65 m height, two front and two rear RGB-D cameras, and a stereo camera system at shoulder height. A schematic view of the architecture is given in Fig. 1c. We use the Robot Operating System (ROS, see http://www.ros.org) as a middleware for the software components.

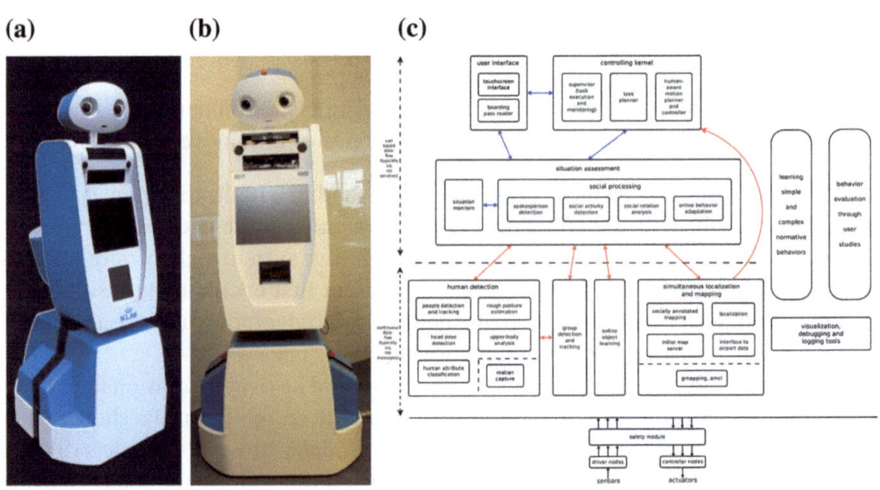

Fig. 1 **a** and **b** Design view and actual appearance of the robot platform. **c** System architecture

3 SLAM and Socially Annotated Mapping

Airports are very dynamic environments, and this poses a big challenge for the localisation and mapping module. Often, large parts of the range sensors' field of view are occluded by people or semi-static objects such as carts or trolleys. When these large semi-static obstacles are placed close to walls they can cause major problems in measuring the true distance to the walls. To build consistent maps in environments with high dynamics, we recently introduced the Normal Distributions Transform Occupancy Map (NDT-OM) [30] and the NDT-OM Fusion algorithm [32]. We have also developed a data structure called the Conditional Transition Map (CTMap) to model typical motion patterns. Here, we present a novel extension of the CTmap, the Temporal CTMap, which can additionally represent motion speeds. CTMaps are very useful for "social" motion planning, as they enable to plan paths that interfere less likely with the flow of passengers.

3.1 Normal Distributions Transform Occupancy Map

NDT-OM [30] combines two established mapping approaches: Normal Distribution Transform (NDT) maps [3, 20] and occupancy grid maps [23]. It has been shown that the NDT-OM Fusion algorithm [32] produces consistent maps in large-scale dynamic environments in real time, and it can handle dynamic changes and provide a set of multi-resolution maps. For map building, the vehicle pose is tracked using a frame-to-model registration, and the sensor data are fused into the NDT-OM, by updating distributions with newly obtained and aligned points. By using submap indexing the system can represent large-scale environments at combined registration and fusion times between 100 ms and 2 s. Evaluations on the public FORD data set [28] yield absolute trajectory errors (ATE) of 1.7 m after 1.5 km (see Fig. 2). Further evaluations on a 10-h data set in a large industrial environment resulted in ATEs of under 0.1 m and update rates of 510 Hz.

3.2 Conditional Transition Maps

NDT-OM can compactly represent dynamic environments, but for social interaction we also need to distinguish directions of motion. For that, we have developed the Conditional Transition Map (CTMap [15]), a grid-based representation that models transitions of dynamic objects in the environment. For each cell **x**, CTMap learns the probability distribution of an object leaving to each neighboring cell, given the cell from which it entered into **x**. Based on these learned patterns, motion directions can then be predicted, which is a very important feature for socially aware navigation. We evaluated the CTMap approach on data from a Velodyne-HDL64 3D laser scanner

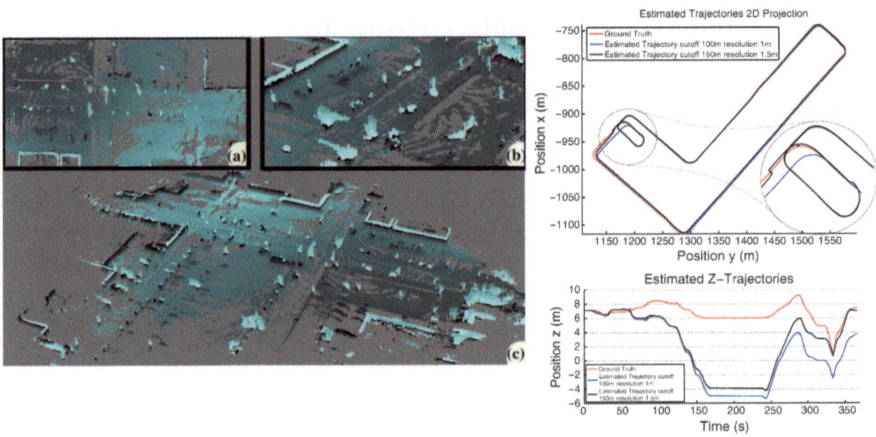

Fig. 2 Mapping and tracking results on the FORD data set. *Left* Maps produced by the system while tracking **a** *top view*, **b** zoomed view of the start in point, **c** overview. The ellipsoids represent height-coded scaled covariance matrices in each map cell from a map at 1 m resolution. *Right* trajectory plots, at the *top* x-y trajectory for the 100 and 150 m cutoff settings, *bottom* estimated z position over time. Note the zoomed-in detail and the re-entry into a previously mapped area

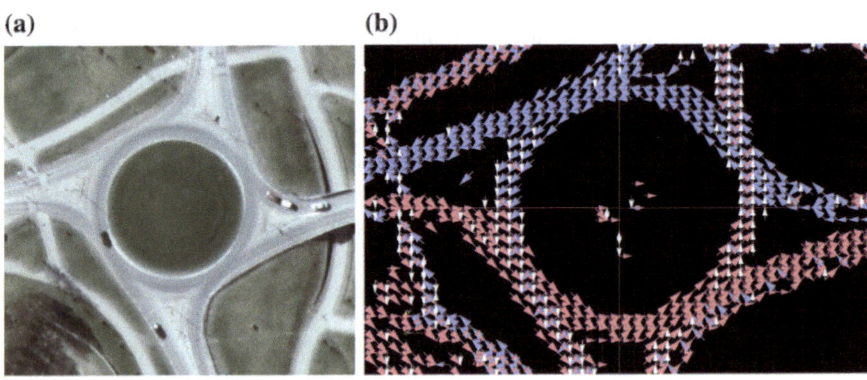

Fig. 3 Visualization of CTMap using data from a roundabout. **a** Overhead view of the environment. **b** Pattern of movement on the roundabout, extracted with CTMap, using a cell size of 2 × 2 m. As a simple denoising step we have removed edges with less than 10 exit events. For clarity, the entry directions are not shown. The colors refer to the orientation of the vectors

that was placed at the center of a roundabout during rush-hour (see Fig. 3a). The obtained CTMap after 1.5 h of observation is shown in Fig. 3b. The arrows show the *most likely* exit directions from each cell. They are distributed along highly dynamic areas and closely correspond to the shape of the roads. We also see that the map is able to capture correct motion patterns of pedestrians on the sidewalks.

As an extension to CTMap, we introduce here the Temporal CTMap. In addition to the set of conditional probabilities of exit directions stored for each entry direction of a cell, the T-CTMap stores a bivariate normal distribution to model the dependencies between entry and exit times. This allows us to not only learn the average motion directions and speeds, but also the *variations* of speed. Thus, in contrast to Pomerleau et al. [29], who average velocities of neighboring points over consecutive frames, the T-CTMap represents a complete distribution of velocities.

4 People and Group Tracking

Another crucial component for a socially compliant robot is a reliable detection and tracking of humans in the environment. As described in Sect. 2, our robot uses 2D laser and RGB-D sensors, and each has benefits and drawbacks. While 2D laser data is more robust against illumination changes and provides a large field of view, it is sparse and has no appearance information. Therefore, we use multiple detection and tracking algorithms that operate on different sensors, as described next.

4.1 2D Range-Based Detection and Tracking

To detect people from 2D laser data, we first segment the data points using agglomerative hierarchical clustering. Then we compute 17 different features for each segment and apply a boosted classifier that was previously trained on 9535 frames of hand-labelled data. The resulting detections are tracked using a multi-hypothesis tracker (MHT), which generates hypotheses by considering all feasible assignments between measurements and tracks, all possible interpretations of measurements as new tracks or errors, and all tracks as being matched, occluded or deleted (see [2]). Each hypothesis represents one possible set of assignments between measurements and track labels. Given a parent hypothesis and new detections, the MHT generates a number of assignment sets, where each produces a new child hypothesis branching off from the parent. To prune the exponentially growing hypothesis tree, a probability is computed recursively for each hypothesis using the measurement likelihood, the assignment set probability and the probability of the parent hypothesis. We use multi-parent k-best branching according to Murty [25] and N-scan back pruning [5]. A Kalman filter with a constant-velocity motion model then predicts the state of tracked people.

We extend this MHT approach in Luber and Arras [19] for the detection and learning of socio-spatial relations and to track social groupings. To do this, layers with group formation hypotheses are interleaved with regular data association hypotheses (see Fig. 4a), each leading to a social network graph (see Fig. 4b). We reason about social groupings recursively to achieve real-time tracking performance. The resulting group information can be fed back into person-level tracking to predict human motion from intra-group constraints and to aid data association with track-specific occlusion probabilities. This leads to an improved occlusion handling and a better trade-off between false negative and false positive tracks. In experiments on large outdoor

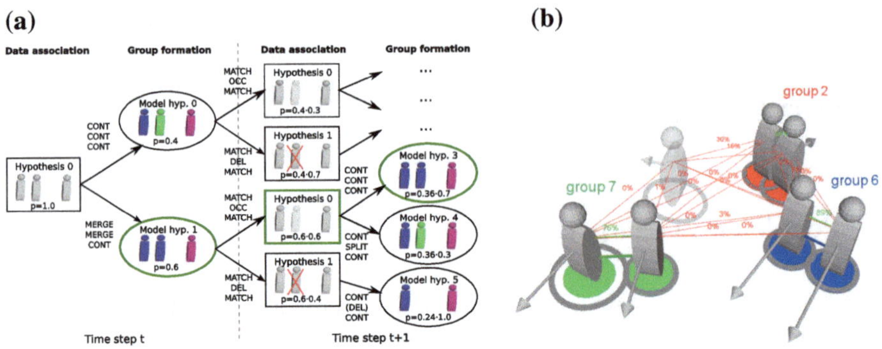

Fig. 4 **a** In our multi-model MHT approach, group formation hypotheses are interleaved between regular data association hypotheses. **b** A social network graph, based on the output of a probabilistic SVM trained on coherent motion indicator features (relative velocity, orientation and distance)

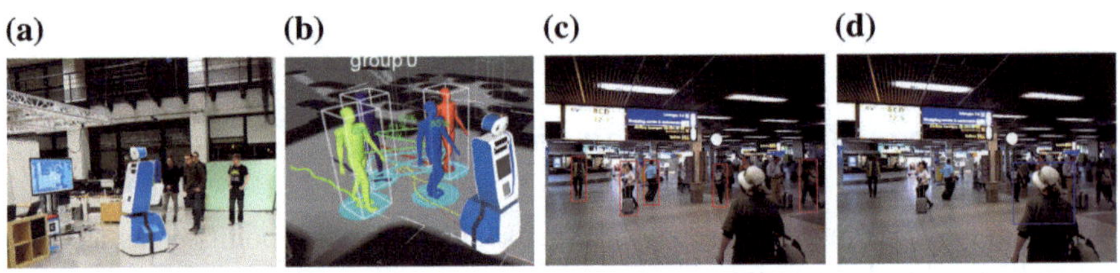

Fig. 5 **a** Person- and group-tracking experiments during a SPENCER integration meeting. The robot tracks and guides a group of people to the other end of a corridor. **b** Group affiliations are displayed as *green lines* connecting the group members. The group is tracked robustly even if individuals are occluded temporarily. **c** The groundHOG detector most likely detects persons in the distance. **d** People near the robot, often partly visible, are detected by the upperbody detector

data sets, we obtain an improved person tracking by a significant reduction of track identifier switches (TIS) and false negative tracks. In Linder and Arras [18], we extend this to RGB-D data, and we show that the approach can track groups with varying sizes over long distances with few TIS. Some results of the combined people and group detection and tracking method are shown in Fig. 5.

4.2 Tracking Based on RGB-D Data

For close-range, appearance-based people detection and tracking we developed a real-time RGB-D based multi-person tracker [10], which aims at making maximal use of the depth information from the RGB-D sensors to speed up computation. It classifies the observed 3D points into *object candidates*, *ground*, and *fixed structures*, e.g. walls. *Ground* points are used to estimate the ground plane, and *object candidates* are passed to an efficient upper-body detector [22], which uses a learned normalized-depth template to find head-shoulder regions. It operates on depth only and is thus limited to the depth range of the RGB-D sensors, i.e. up to 5 m. To obtain also far-range detections for pedestrians, we combine the upper-body detector with a full-body HOG based detector. This second detector runs efficiently on the GPU and uses the estimated ground plane to restrict the search for geometrically valid object regions [33]. Finally, we use the estimated camera motion, the ground plane and the detections from both detectors for tracking based on Leibe et al. [16] (see Fig. 5c, d).

5 Human-Aware Task and Motion Planning

In SPENCER, there are three main components responsible for planning actions, interactions and the motion of the platform: the supervision system, the task and action planner, and the motion planning module. All three operate human-aware, e.g. by aiming for legibility of the paths and collaborative planning, as detailed next.

5.1 *The Supervision System*

The supervision system (SUP) interacts with the user and generates and executes action plans. For interaction, we use the devices 'lights', 'head', 'screen', and 'microphone' and provide three interaction modes: Engaging with potential users before guiding, giving information to guided users, and asking other people to clear the passage. The SUP also receives safety-critical information, e.g. about planning failures or potential dangers for humans, and reacts accordingly. Using the work of Fiore et al. [7], the SUP was built and sucessfully tested in a simplified scenario.

5.2 *Action Planning with Human Collaboration*

Action planning and execution alone is not sufficient for a socially aware robot, because it also needs to consider actions performed by the users. For example, while guiding, the robot has to deal with situations where some members of the guided group purposely don't follow the robot. Therefore, we represent the human's intention as a hidden variable and formulate the problem as a Mixed Observability Markov Decision Process (MOMDP [26]), where in contrast to standard POMDPs some state components are fully observable and others only partially. MOMDPs can be solved much more efficiently than general POMDPs. For cooperation with humans in different tasks we associate to each task a collaboration planner (CP) represented as a MOMDP. To reduce complexity we use a simplified state space, focusing on the intention estimation problem, and let the SUP adapt the MOMDP plans to the current situation. When executing a cooperative action with a human, the SUP gathers observations about the human and updates the corresponding CP, resulting in a high-level action adapted to the situation. In our system, we use a CP for the guiding action and tested it successfully with a single person following the robot. For groups, we currently regard the "most cooperative" behavior, i.e. we consider the group as following as long as a single member follows the robot.

5.3 Socially Compliant Motion Planning

The motion planning module is the system component for which the benefit of complying with social rules is most obvious. Whereas standard planning algorithms mainly aim to find shortest feasible paths, social motion planning trades the shortest path off with the cost of breaking social rules, e.g. when crossing through a group of people instead of deviating it. Therefore, our motion planner extends standard kinodynamic planning in the following ways. First, for global planning we use a human-aware cost map that ensures a path around the detected people, which humans consider as safe. Second, our planning algorithm produces *legible* paths by avoiding abrupt motion changes in presence of dynamic obstacles and by anticipating future collisions and adapting the velocity accordingly. The improved legibility of the produced paths has been experimentally validated in a user study with a robot platform similar to the SPENCER robot (see [14]).

As a further extension to standard motion planning, we investigate RRT*-based planning [13] using low-level vehicle constraints in combination with high-level socially compliant cost maps. Our planner uses a novel extent function for differential-drive robots, which improves the smoothness of the paths and overcomes some limitations of other existing control laws (see [27]). To reduce planning time, we use a learning approach based on a nonlinear parametric model that infers the distance metric for selecting the nearest vertex in RRT*. Results of our improved RRT* planner using a cost map learned with inverse reinforcement learning (IRL, see Sect. 7.2) are shown in Fig. 6a.

(a) **(b)** **(c)**

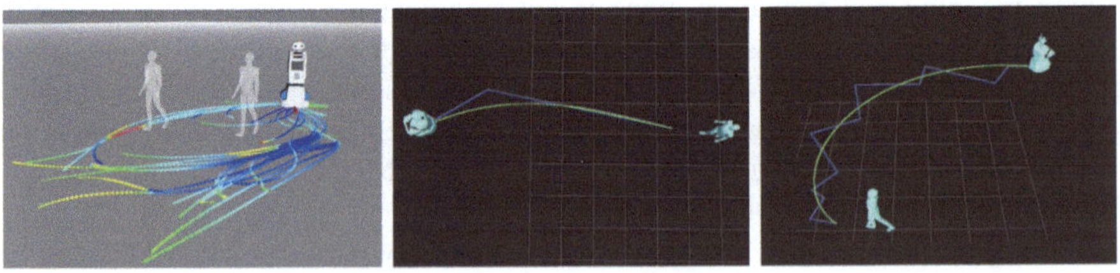

Fig. 6 **a** An example tree generated by the RRT* motion planner on IRL cost maps, when a single relation is in the scene. *Red branches* are high cost actions, low cost actions are displayed in *blue*. **b** and **c** Learning to approach a person using IRL. The *light blue line* is the result of the discretized position and the *green line* is the smoothed path used by the planner

6 Perception of Human Social Attributes

We have shown how information about social human relations is obtained from basic cues such as tracked groups, and how social rules are used to perform human-aware actions and motions. However, for a deeper analysis and recognition of social relations and attributes, more detailed information must be extracted from the sensor data. Therefore, in SPENCER we develop tools for automatic estimation of body postures, classification of human attributes such as gender and age, estimation of head poses, spokesperson detection, and the classification of important objects in the environment. For the latter three, we present details in the following.

6.1 Head Pose Estimation

An important cue for human social interactions is the head orientation. Groups of people can often be recognized as either standing in a circular formation facing towards the centre, or walking next to each other while looking into the same direction. This suggests that the head orientation can be used to support tasks such as group detection and tracking. To estimate the head orientation, we classify a given upper-body detection as looking left, right, front, back or being a false-positive. Our approach computes a feature covariance matrix of the image's Lab colors and applies a Difference of oriented Gaussians (DooG) filter. The result is split into a regular, overlapping grid and a kernel-SVM is trained on a Riemannian approximation to the geodesic distance between covariance matrices in each cell of the grid. We have evaluated various such approximations, which can trade off computational speed for accuracy, with either an accuracy of up to 93.5 % or a two orders of magnitude faster computation than the current state of the art (see [34]).

6.2 Spokesperson Detection

Another key element of analysing social behaviour is the detection of a *spokesperson*, i.e. a group member who is available for interaction and can make decisions on behalf of the group. Examples include parents in a family and teachers in a school class. For the guiding scenario in SPENCER, determining a spokesperson is particularly useful, because other group members will more likely follow the robot when the spokesperson does. Thus, even if some members are not tracked due to occlusions, the robot can still guide the group as long as the spokesperson is following.

To determine a spokesperson, one can use heuristics such as people's height (this excludes children as a spokesperson) or their position relative to the robot. Another approach is to use people's speech patterns to determine dominance in multi-party meetings (see [9]). However, audio-related cues can not be extracted reliably in airports. Cristani et al. [6] use body behavior and gestures to classify a video of four participants having a conversation into intervals of speech or non-speech. The method achieves 72 % accuracy, but the setting is static. However, in an airport people usually move. Also, from our investigations on the same data the movements associated with speech are much shorter-lived than the gesture itself, i.e. different metrics to quantify gesturing are needed. Furthermore, gestures can indicate both speaking and "active listening" behavior. In further experiments with three different implementations of speaker detection using the above data and recordings from speed datings [38], we found that gesturing alone is not a good indication for speech (up to half of the observed speech was not accompanied by strong gesturing), and that the relationship between gesturing and speaking is person-specific. We are therefore investigating the relation of gestures and the length of the subsequent speech period for a more reliable speaker detection. Meanwhile, we use the above mentioned heuristics to determine the spokesperson.

6.3 Efficient Object Classification Using Online Learning

Apart from people and their attributes, the robot must also be aware of relevant objects in the environment. In an airport, these include moving objects such as carts and trolleys, which can be dangerous for the robot. However, instead of employing standard offline learning from previously obtained training data, we develop online learning methods for object classification. Particularly, we focus on *autonomous learning* methods, which have the two major advantages that they are adaptive to new situations, i.e. they can incorporate new information by updating their learned models, and they require less user interaction by selectively choosing the data that is particularly useful for training. Based on the work of Triebel et al. [35, 36], we developed in Mund et al. [24] an efficient online multi-class classifier, that generates less label queries but better classification results than previous methods. This is particularly useful for classifying and learning many different objects online and with only little user interaction, as it is given for the application in SPENCER.

7 Analysis and Learning of Socially Normative Behaviors

So far, we have shown cues to analyse human social behavior, and how social rules can be used to perform a socially compliant robot behavior, particularly during path planning. But how can we obtain these social rules? In principle, there are two different approaches. Either the rules are provided manually by human experts and converted into machine-understandable representations, or they are learned automatically from sensor observations. In SPENCER, we pursue both approaches: High-level, complex rules are established using empirical user studies, and low-level rules are learned automatically from demonstrations. Here, we give two examples.

7.1 User Studies and Contextual Analysis

Airport environments are naturally populated by people from many different cultures. Thus, many different social rules may be required here. One example we investigate is *proxemics* [8], i.e. the distance the robot should keep from a group when interacting. We consider this in the exemplified scenario of a robot approaching a small group of people. The results of an online survey (N = 181), which was distributed to people in China, the U.S.A. and Argentina (see Fig. 7a), show that participants prefer a robot that stays out of their intimate space zone just like a human would be expected to do [11]. However, Chinese participants accepted closer approaches than people from the U.S.A. and Argentinia. This suggests a culturally dependent application of social rules also for SPENCER.

Furthermore, we conducted a contextual analysis at Schiphol Airport to analyze human behavior and to identify observable social rules that the SPENCER robot must be aware of [12]. From video data collected during two consecutive days, we established several typical, highly relevant human behaviors. For example, one such behavior is that groups of people tend to walk in pairs or triads behind each other. Another one is the typical avoidance of areas close to information monitors (see Fig. 7b). These findings have direct implications both for the perception and the planning module of the system, because they potentially lead to a more reliable group tracking and to a more socially appropriate motion of the robot.

7.2 Behavior Learning via Inverse Reinforcement Learning

Inverse Reinforcement Learning (IRL [1]) aims at recovering an objective function that encodes a given behavior from an input reward signal. This is more robust than policy search, because rewards are better generalizable and more succinct (see [37]). We use Bayesian IRL [21] to learn a distribution over the rewards and select the

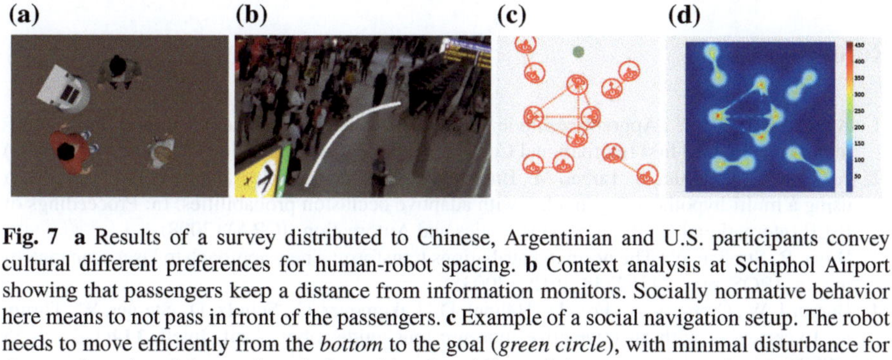

Fig. 7 **a** Results of a survey distributed to Chinese, Argentinian and U.S. participants convey cultural different preferences for human-robot spacing. **b** Context analysis at Schiphol Airport showing that passengers keep a distance from information monitors. Socially normative behavior here means to not pass in front of the passengers. **c** Example of a social navigation setup. The robot needs to move efficiently from the *bottom* to the goal (*green circle*), with minimal disturbance for the people and social groupings indicated by *dotted lines*. **d** A costmap learned with IRL for the setup. Areas around people have high cost, but also the 'social' links between individuals

best reward as the MAP estimate. For experiments we use a custom-made pedestrian simulator based on models from computational social sciences to perform behavior tests with arbitrarily large crowds, because testing on the real robot with large crowds is too costly. Figure 7c shows a typical social navigation setup in a crowded environment. The learned costmap using IRL is shown in Fig. 7d. Such a costmap is then used by the RRT-based motion planner (see Sect. 5.3) to find the desired path for the setup.

Furthermore, we aim at learning relevant social norms when approaching a person. These norms involve a comfortable speed, an appropriate approaching direction and social relations within groups if the person is in a group. Currently, however, we focus on approaching only one person. Again we use IRL, and in particular Gaussian Process IRL [17] to learn a policy from a set of demonstrations given by an expert. In our MDP formulation the states are given by distance and orientation in a human-centered frame, and actions are those performed by the motion planner. Two paths learned from 11 demonstrations are shown in Fig. 6b, c.

8 System Integration and Conclusion

All presented system components are developed independently and simultaneously. However, to also achieve a steady progress of the entire system, all components are integrated and attuned to each other in regular meetings every 6 months. As a result, the platform in its current state already combines the map representation presented in Sect. 3, the laser-based people and group tracker (Sect. 4), and the task and motion planner (Sect. 5). Experiments with the complete system have shown that the robot is able to approach and engage with a person, receive a goal position and guide the person or a group to the goal while keeping track of the following person(s). If a failure of cooperation is detected when the person does not follow any more, it stops and waits for re-engagement. Encouraged by these results, a first deployment of the platform at the Schiphol airport is planned for the near future.

References

1. Abbeel, P., Ng, A.Y.: Apprenticeship learning via inverse reinforcement learning. In: Proceedings of the Twenty-first International Conference on Machine Learning (ICML). ACM (2004)
2. Arras, K.O., Grzonka, S., Luber, M., Burgard, W.: Efficient people tracking in laser range data using a multi-hypothesis leg-tracker with adaptive occlusion probabilities. In: Proceedings of IEEE International Conference on Robotics and Automation (ICRA) (2008)
3. Biber, P., Straßer, W.: The normal distributions transform: a new approach to laser scan matching. In: IROS, pp. 2743–2748. IEEE (2003)
4. Burgard, W., Cremers, A., Fox, D., Hähnel, D., Lakemeyer, G., Schulz, D., Steiner, W., Thrun, S.: Experiences with an interactive museum tour-guide robot. Artif. Intell. **114**(1–2), 3–55 (2000)

5. Cox, I., Hingorani, S.: An efficient implementation of Reid's multiple hypothesis tracking algorithm and its evaluation for the purpose of visual tracking. IEEE Trans. Pattern Anal. Mach. Intell. (PAMI) **18**(2), 138–150 (1996)
6. Cristani, M., Pesarin, A., Vinciarelli, A., Crocco, M., Murino, V.: Look at who's talking: voice activity detection by automated gesture analysis. In: Constructing Ambient Intelligence, pp. 72–80. Springer (2012)
7. Fiore, M., Clodic, A., Alami, R.: On planning and task achievement modalities for human-robot collaboration. In: The International Symposium on Experimental Robotics (2014)
8. Hall, E.T.: The Hidden Dimension. Anchor Books, New York (1966)
9. Hung, H., Huang, Y., Friedland, G., Gatica-Perez, D.: Estimating dominance in multi-party meetings using speaker diarization. Tr. Audio Speech Lang. Process. **19**(4), 847–860 (2011)
10. Jafari, O.H., Mitzel, D., Leibe, B.: Real-time RGB-D based people detection and tracking for mobile robots and head-worn cameras. In: International Conference on Robotics and Automation (ICRA) (2014)
11. Joosse, M., Poppe, R., Lohse, M., Evers, V.: Cultural differences in how an engagement-seeking robot should approach a group of people. In: Proceedings of International Conference on Collaboration Across Boundaries: Culture, Distance and Technology (CABS) (2014)
12. Joosse, M.P., Lohse, M., Evers, V.: How a guide robot should behave at an airport insights based on observing passengers. Technical Report TR-CTIT-15-01, Centre for Telematics and Information Technology, University of Twente, Enschede (2015)
13. Karaman, S., Frazzoli, E.: Incremental sampling-based algorithms for optimal motion planning. In: Proceedings of Robotics: Science and Systems (RSS) (2010)
14. Kruse, T., Khambhaita, H., Alami, R., Kirsch, A.: Evaluating directional cost models in navigation. In: ACM/IEEE International Conference on Human-Robot Interaction (HRI) (2014)
15. Kucner, T., Saarinen, J., Magnusson, M., Lilienthal, A.J.: Conditional transition maps: learning motion patterns in dynamic environments. In: Proceedings of IEEE/RSJ International Conference on Intelligent Robots and Systems (IROS), pp. 1196–1201 (2013)
16. Leibe, B., Schindler, K., Van Gool, L.: Coupled object detection and tracking from static cameras and moving vehicles. IEEE Trans. PAMI **30**(10) (2008)
17. Levine, S, Popovic, Z., Koltun, V.: Nonlinear inverse reinforcement learning with Gaussian processes. In: Shawe-Taylor, J., Zemel, R.S., Bartlett, P.L., Pereira, F., Weinberger, K.Q. (eds.) Advances in Neural Information Processing Systems, vol. 24, pp. 19–27 (2011)
18. Linder, T., Arras, K.: Multi-model hypothesis tracking of groups of people in RGB-D data. In: Proceedings of IEEE International Conference on Information Fusion (FUSION), pp. 1–7 (2014)
19. Luber, M., Arras, K.O.: Multi-hypothesis social grouping and tracking for mobile robots. In: Robotics: Science and Systems (RSS'13), Berlin, Germany (2013)
20. Magnusson, M., Lilienthal, A., Duckett, T.: Scan registration for autonomous mining vehicles using 3D-NDT. J. Field Robot. **24**(10), 803–827 (2007)
21. Michini, B., How, J.P.: Improving the efficiency of Bayesian inverse reinforcement learning. In: Proceedings of IEEE International Conference on Robotics and Automation (ICRA), St. Paul, Minnesota, USA (2012)
22. Mitzel, D., Leibe, B.: Close-range human detection for head-mounted cameras. In: British Machine Vision Conference (2012)
23. Moravec, H., Elfes, A.: High resolution maps from wide angle sonar. In: Proceedings of IEEE International Conference on Robotics and Automation, pp. 116–121 (1985)
24. Mund, D., Triebel, R., Cremers, D.: Active online confidence boosting for efficient object classification. In: Proceedings of IEEE International Conference on Robotics and Automation (ICRA) (2015)
25. Murty, K.G.: An algorithm for ranking all the assignments in order of increasing cost. Oper. Res. **16** (1968)
26. Ong, S.C., Png, S.W., Hsu, D., Lee, W.S.: POMDPs for robotic tasks with mixed observability. In: Proceedings of Robotics: Science and Systems (RSS) (2009)

27. Palmieri, L., Arras, K.: POSQ: a new RRT extend function for efficient and smooth mobile robot motion planning. In: Proceedings of IEEE/RSJ International Conference on Intelligent Robots and Systems (IROS) (2014)
28. Pandey, G., McBride, J.R., Eustice, R.M.: Ford campus vision and lidar data set. Int. J. Robot. Res. **30**(13), 1543–1552 (2011)
29. Pomerleau, F., Krüsi, P., Colas, F., Furgale, P., Siegwart, R.: Long-term 3D map maintenance in dynamic environments. In: IEEE International Conference on Robotics and Automation (ICRA) (2014)
30. Saarinen, J., Andreasson, H., Stoyanov, T., Lilienthal, A.: 3D normal distributions transform occupancy maps: an efficient representation for mapping in dynamic environments. Int. J. Robot. Res. (IJRR) 1627–1644 (2013)
31. Siegwart, R., Arras, K.O., Bouabdallah, S., Burnier, D., Froidevaux, G., Greppin, X., Jensen, B., Lorotte, A., Mayor, L., Meisser, M., Philippsen, R., Piguet, R., Ramel, G., Terrien, G., Tomatis, N.: Robox at Expo. 02: a large-scale installation of personal robots. RAS **42**(3–4), 203–222 (2003)
32. Stoyanov, T., Saarinen, J., Andreasson, H., Lilienthal, A.: Normal distributions transform occupancy map fusion: simultaneous mapping and tracking in large scale dynamic environments. In: Proceedings of IEEE/RSJ International Conference on Intelligent Robots and Systems (IROS), pp. 4702–4708 (2013)
33. Sudowe, P., Leibe, B.: Efficient use of geometric constraints for sliding-window object detection in video. In: International Conference on Computer Vision Systems (ICVS) (2011)
34. Tosato, D., Spera, M., Cristani, M., Vittorio, M.: Characterizing humans on Riemannian manifolds. IEEE Trans. Pattern Anal. Mach. Intell. (2013)
35. Triebel, R., Grimmett, H., Paul, R., Posner, I.: Driven learning for driving: how introspection improves semantic mapping. In: Proceedings of International Symposium on Robotics Research (ISRR) (2013)
36. Triebel, R., Stühmer, J., Souiai, M., Cremers, D.: Active online learning for interactive segmentation using sparse Gaussian processes. In: German Conference on Pattern Recognition (GCPR) (2014)
37. Vasquez, D., Okal, B., Arras, K.O.: Inverse reinforcement learning algorithms and features for robot navigation in crowds: an experimental comparison. In: IROS, Chicago, USA (2014)
38. Veenstra, A., Hung, H.: Do they like me? Using video cues to predict desires during speed-dates. In: Internatioanl Conference on Computer Vision, Workshops, pp. 838–845 (2011)

Easy Estimation of Wheel Lift and Suspension Force for a Novel High-Speed Robot on Rough Terrain

Jayoung Kim, Bongsoo Jeon and Jihong Lee

Abstract In operation of high-speed wheeled robots on rough terrain, it is important to predict or measure the interaction between wheel and ground in order to maintain optimal maneuverability. Therefore, this paper proposes an easy way to estimate wheel lift and suspension force of a high-speed wheeled robot on uneven surfaces. First, a high-speed robot with six wheels with individual steer motors was developed, and with the body of the robot connected to each wheel by semi-active suspensions. In a sensor system, potentiometers, which can measure angle of arms, are mounted at the end of arms and have a critical role in estimating wheel lift and suspension force. A simple dynamic equation of the spring-damper system is used to estimate the suspension force, and the equation is calculated in terms of the suspension displacement by measured angle of arms because the suspension displacement is a function of arm angle in the boundary of the kinematic model of the body–wheel connection. In addition, wheel lift can be estimated using the arm angle. When the robot keeps its initial state without normal force, the arm angle is set as zero point. When the wheels receive the normal force, the link angle is changed to a value higher than zero point. If a wheel does not contact to a ground, then the suspension force goes toward the negative direction as a value. Therefore, if wheel lift happens while driving, the arm angle will follow the zero point or the suspension force will indicate a negative value. The proposed method was validated in ADAM simulations. In addition, the results of the performance were verified through outdoor experiments in an environment with an obstacle using a high-speed robot developed for this purpose.

J. Kim · B. Jeon · J. Lee (✉)
Department of Mechatronics Engineering, Chungnam National University, Daejeon, South Korea
e-mail: jihong@cnu.ac.kr

J. Kim
e-mail: jaya@cnu.ac.kr

D.S. Wettergreen and T.D. Barfoot (eds.), *Field and Service Robotics*,
Springer Tracts in Advanced Robotics 113, DOI 10.1007/978-3-319-27702-8_41

1 Introduction

Research on outdoor robotic vehicles has received significant attention for important tasks involving exploration, reconnaissance, rescue, and so on. In actual applications on outdoor environments, especially rough terrains, it is hard to automatically operate outdoor vehicles or robots because there are many elements that can put them in dangerous situations, such as overturn or stuck wheel. Accordingly, it is a big issue to optimize wheel traction [1, 2] and stability [3, 4] of vehicles on rough terrains and to estimate suspension force of vehicles for achieving the aims, since suspension force is a variable used in order to control traction and to evaluate stability of vehicles [1–10]. Suspension force can be expressed as normal force acting on wheel and body. In previous studies, fully dynamic models of vehicles or robots are applied to estimate the normal force [2–10]. However, it is not easy to derive the dynamic models and it is a laborious task to acquire accurate values of normal force in estimation systems based on the dynamic models since the dynamic models include model uncertainty by complex terrain conditions, and, thus, robot states cannot be correctly estimated in real-time. In addition, when a wheel is taken off the ground (wheel lift) in case of high-speed driving on rough terrains, it is impossible to predict robot states and it may be confronted with a hazardous situation. Therefore, this paper proposes an easy way to estimate wheel lift and suspension force of a high-speed wheeled robot on uneven surfaces. In this paper, only an inexpensive potentiometer was employed to measure angle of arms, which is sufficient to estimate wheel lift and suspension force in this simple method.

2 Estimation of Suspension Force and Wheel Lift

2.1 *Caleb9; Omnidirectional High-Speed Rough Terrain Robot*

In this paper, an outdoor wheeled robot called Caleb9 was developed, as shown in Fig. 1. Caleb9 has six in-wheel motors for driving and six BLDC motors for steering. Semi-active suspensions that can automatically adjust damping force are mounted for connection between wheel and body, independently. Arms of Caleb9 were designed as a structure of four-bar linkage in order to overcome surface obstacles effectively. Brake modules are attached to each wheel for rapid breaking of wheels. Caleb9 controls each driving motor to optimize wheel traction (Terrain-adaptive Slip Control [1]), steering motor to keep the desired steering angle (Position Control), semi- active suspension to adjust damping force (Position Control), and brake module to maintain safety driving (Force Control). Caleb9 can move omnidirectionaly on rough terrains by six driving motors, six steering motors, and six semi-active suspensions. Detailed specification is depicted in Table 1.

Fig. 1 Design of Caleb9 and mounted potentiometer at the end of each arm

Table 1 Specification of Caleb9

Max velocity	10 m/s(40 km/h)	Total weight	800 kg
Max slope	20°	Operating time	1 h 30 min
Steering angle	–90°–90°	Battery	Li-ion 48 V, 24 V
Arm displacement	25 cm	Main board O/S	Linux
Robot size (mm)	1460 × 2180 × 990	Communication	CAN

In a sensor system of caleb9, rotational velocity, torque, and steered position of wheel are acquired from feedback data of motor controllers. Three-dimensional position, velocity, acceleration, and angle of the robot can be estimated by commercial INS/GPS system on the top of the robot. Arm angles can be measured by potentiometers mounted at the end of each arm, as shown in Fig. 1. The potentiometer has a critical role in estimating suspension force and wheel lift by observing changed angle of arms.

2.2 *Easy Method for Estimation of Suspension Force and Wheel Lift*

Suspension force and wheel lift can be estimated from the kinematic relation between arm and suspension in Fig. 2. Simply, when the wheel is raised by a force from the ground (L_D), angle of the arm is changed (θ) and at the same time, the suspension is compressed (x) depending on the angle of the arm θ. Once the displacement x of the suspension is known, then suspension force can be easily estimated using (1). In (1), F_s represents suspension force, K is spring coefficient, C is damper coefficient, and $\dot{x}$ denotes derivative term of the displacement x with respect to sampling time Δx.

$$F_s = Kx + C\dot{x} \tag{1}$$

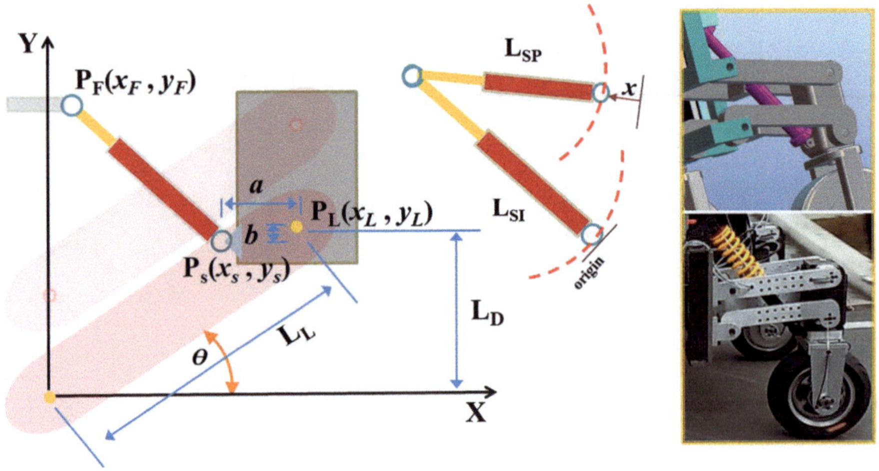

Fig. 2 Kinematic relation between arm and suspension

The displacement x of the suspension can be expressed as a function of angle of the arm, θ. In (1), $\mathrm{L_{SI}}$ denotes initial total length of the suspension without compression, and $\mathrm{L_{SP}}$ represents subsequent total length of the suspension with compression. Accordingly, the displacement of the suspension is calculated by

$$x = \mathrm{L_{SI}} - \mathrm{L_{SP}} \tag{2}$$

Initial total length of the suspension, $\mathrm{L_{SI}}$, is given as a constant. Subsequent total length of the suspension $\mathrm{L_{SP}}$ is changed depending on the starting position $\mathrm{P_S}(x_\mathrm{s}, y_\mathrm{s})$ of the suspension which is a function of angle θ of the arm. In Fig. 2, $\mathrm{P_F}(x_\mathrm{F}, y_\mathrm{F})$ is the end position of the suspension, $\mathrm{P_L}$ (x_L, x_L) represents the end position of arm, $\mathrm{L_L}$ denotes length of arm, and a is the distance in the $x-$direction between $\mathrm{P_L}$ and $\mathrm{P_S}$. b represents the distance in the $y-$direction between $\mathrm{P_L}$ and $\mathrm{P_S}$. $\mathrm{P_F}$ and $\mathrm{L_L}$ are given as constant from design parameters a and b of caleb9, respectively. x-y elements of $\mathrm{P_L}$can be substituted into x-y elements of $\mathrm{P_S}$ by a and b as follows

$$\mathrm{P_S}(x_s, y_s) = \mathrm{P_L}(x_L - a, x_L - b) \tag{3}$$

In addition, x-y elements of $\mathrm{P_L}$ are variables to be calculated according to angle θ of the arm as below

$$\mathrm{P_L}(x_L, y_L);\ x_L = \mathrm{L_L}\cos(\theta), y_L = \mathrm{L_L}\sin(\theta) \tag{4}$$

For the displacement x of the suspension in (2), $\mathrm{L_{SP}}$can be found by calculating the length between $\mathrm{P_F}$ and $\mathrm{P_S}$ as

$$L_{SP} = \sqrt{(x_F - x_s)^2 + (y_F - y_s)^2} \tag{5}$$

Therefore, suspension force can be estimated by (1) based on measurement of angle θ of the arm.

From estimated suspension force, wheel lift can be easily checked. In Fig. 3, the left-side figure describes total forces acting on suspension in the case of contact between wheel and ground (wheel contact). The right-side figure shows total forces acting on suspension in the case of wheel lift. F'_B is the gross force from robot body, F_G expresses the force from ground, F'_G denotes the rotated force of F_G in the direction of suspension, F_W is the force from wheel part, and F'_W denotes the rotated force of F_W in the direction of suspension. In the case of wheel contact, the suspension makes compressed motion and the suspension force can be expressed as the sum of F'_G and F'_B. Suspension force F_s is positive by keeping the compressed motion while driving. In the case of wheel lift, the suspension makes extension movement and the suspension force can be represented as the sum of F'_W and F'_B in the reverse direction to the suspension force; thereby, the suspension force F_s is negative. Additionally, suspension force, F_s, can be zero in case that the displacement x of the suspension becomes zero by kinematical constraints since x cannot be changed toward the negative direction. This situation happens when angle of arm is zero due to wheel lift. Therefore, it is easy to check the wheel lift by observing negative value and zero value of the suspension force as follow, respectively.

$$F_s = F'_G + F'_B, (F_s > 0) \tag{6}$$

$$F_s = -F'_W - F'_B, (F_s \le 0) \tag{7}$$

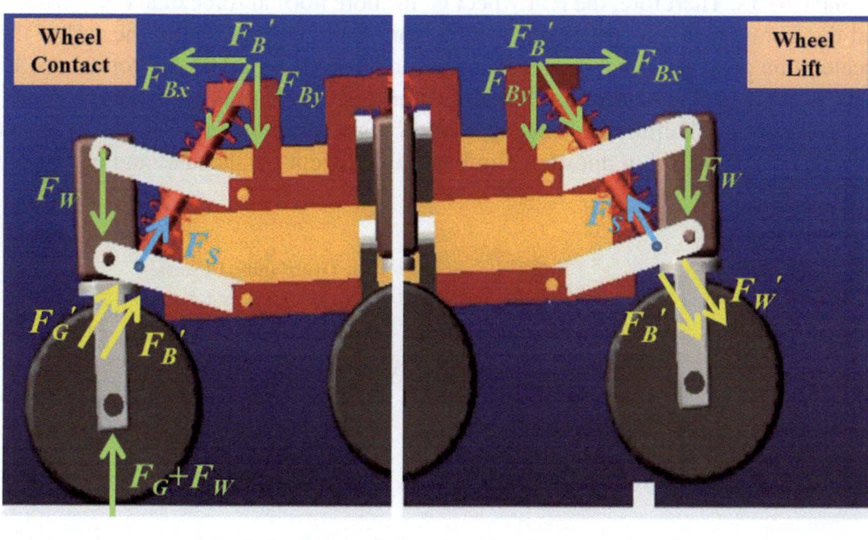

Fig. 3 Total forces acting on suspension of caleb9 in case of wheel contact and wheel lift

3 Validation of Estimation Method on ADAMS Simulations

The purpose of this simulation is to observe the performance of estimating suspension force and wheel lift in comparison between the proposed theory and simulation data on an environment similar to real conditions. The ADAMS simulator was used to validate the proposed method on two types of terrains: (1) Hill climbing (30°) (2) Overcoming obstacles (height 10 and 5 cm, width 5cm), as shown in Fig. 4. Terrain types 1 and 2 were selected to observe estimation performance in case of mild changes and rapid changes of suspension force, respectively. In simulations, the velocity of the robot was controlled at 1, 2, and 3 m/s in the longitudinal direction, and the friction coefficient on the surface was set as 1 to prevent wheel from slippage. The design parameters of virtual robot in the simulation such as size or weight were set as those of the real robot. The spring coefficient and damper coefficient were designated as $K = 8000$ N/m and $C = 2200$ Ns/m, respectively. The needed variables to be acquired on simulations are actual angle, θ, of arm and ideal suspension force while driving on such terrains, and the variables were extracted from simulation data.

3.1 Simulation Results in Case of Hill Climbing

Figure 5 describes actual angles of right-side arms while climbing a hill at 1 m/s. the arm angles are the same as those of of left-side arms because the robot moves in the longitudinal direction and the right-side surface shape the same as the left-side surface shape. At 0 s, the suspension of the robot takes initial posture without compression. After that time, the robot accelerates to meet desired velocity from around 0 to 5 s. Therefore, the rear wheel gains more normal force than other wheels and the front wheel gets the lowest normal force among them. From the end of the acceleration area, the robot moves with uniform velocity until 15 s. From about 15 to 29 s, the robot encounters a hill with 30° and the angle of arms are significantly changed during hill climbing. The angle of the right-middle arm is slightly different from that when the robot does not climb the hill, except for the start and end of the

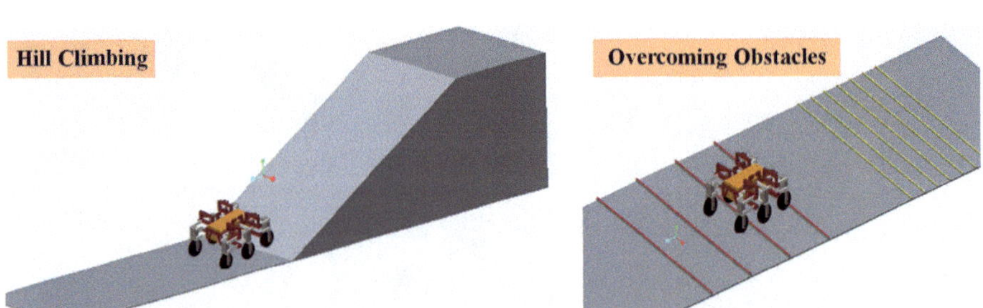

Fig. 4 Simulation environments on ADAMS: (1) Hill climbing and (2) Overcoming obstacles

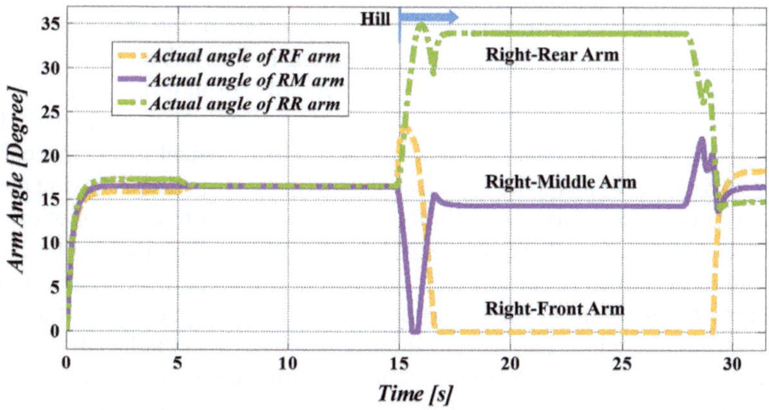

Fig. 5 Measured angles of right-side arms while climbing the hill at 1 m/s

hill. In the vicinity of the start point of the hill, the angle of the right-middle arm reached zero point as the initial state of the suspension. This can be explained by the wheel being taken off from ground, because zero angle of the arm means that normal force was exerted to the wheel. The angle of right-front arm was also reached to zero point during hill climbing. Accordingly, the right-front wheel was lifted off from the surface.

From the angle data in Fig. 5, the suspension force can be estimated by using (1)–(5). Figure 6 shows the estimated data of suspension force in comparison to ideal data of suspension force. Thick lines express the ideal suspension force of right-side arms and thin dot lines represent the estimated suspension force. Figure 6 shows that the estimated suspension forces are well matched with the ideal suspension forces. In case of right-front wheel, the ideal suspension force indicates negative values during

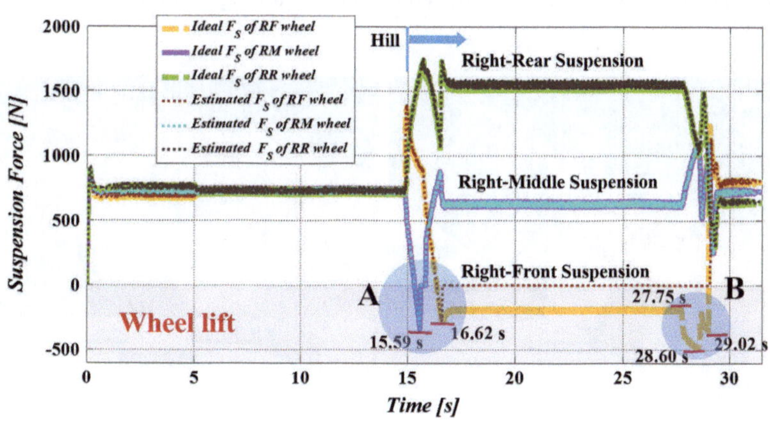

Fig. 6 Estimated suspension forces of right-side wheels while climbing the hill at 1 m/s

hill climbing. However, in the actual situation, the angle of the arm is not changed in the negative direction of the angle, as depicted in Fig. 5. The suspension force of the right-front and middle wheels are momentarily displayed as negative values because of the term related to the damper in (1), especially $\dot{x}$. Nevertheless, the forces returned soon to the zero line, as shown in A of Fig. 6. Wheel lift happened at the right-front and the right-middle wheel, as shown by the suspensions having negative and zero force values during hill climbing. In comparison to actual motion of the right-front wheel, region A expresses the wheel motion in the vicinity of start point of the hill as described in (a) of Fig. 7, and region B indicates the wheel motion in the vicinity of end point of the hill as depicted in (b) of Fig. 7. In A, at around 15 s, the right-middle wheel has wheel lift since the front wheel is faced with the hill and the rear wheel supports the robot against pitch motion of the body. After 1 s, the right-front wheel has wheel lift until around the end of the hill, as shown by the angle of the arm having zero value after about 16.58 s in (a) of Fig. 7. The right-front wheel contacts to the ground at 29.02 s in Fig. 6. The result shows the same performance in (b) of Fig. 7.

3.2 *Simulation Results in Case of Overcoming Obstacles*

Another simulation was performed to validate the proposed method in a flat surface with obstacles at the robot speed 3 m/s. Figure 8 shows the angle of right-side arms while getting over the obstacles. The robot encounters the obstacles with different heights (10 and 5 cm). In Fig. 8, during the initial 7 s, the arm motion is similar to previous motion of arms in Fig. 5 because of the acceleration movement. The RR arm gets the highest angle value among them, and the RF arm has the lowest. After 7 s, the robot meets the high obstacles four times and then, after 11 s, the robot collides with low obstacles seven times. From the front wheel, the arm angle increases in order of position. The change of the RM arm is the smallest among them. In contrast with the

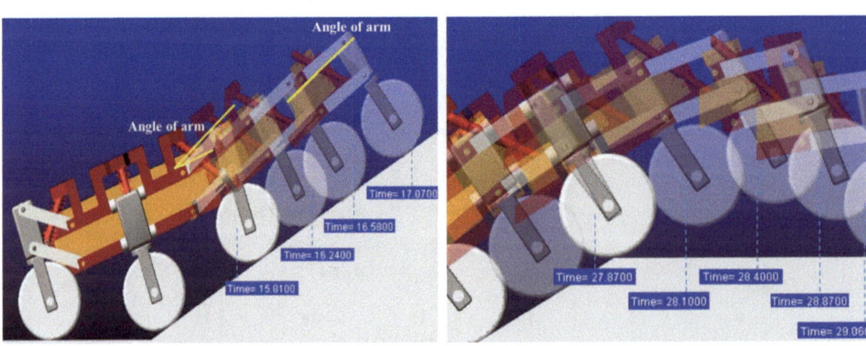

Fig. 7 Motion analysis of wheel lift of the right-front wheel while climbing the hill at 1 m/s

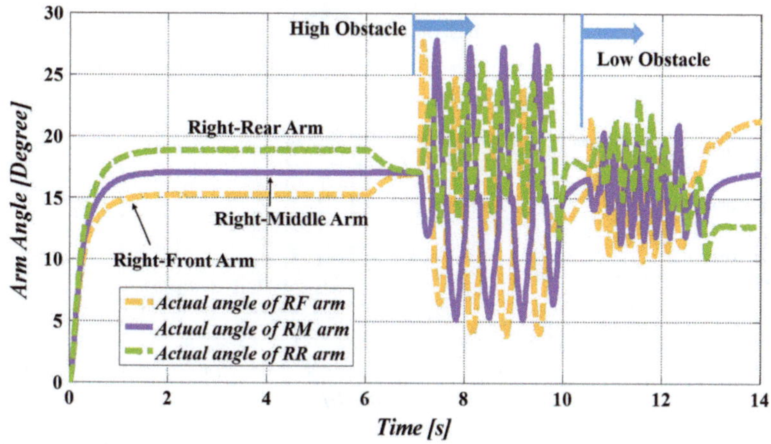

Fig. 8 Measured angles of right-side arms while overcoming obstacle at 3 m/s

angles of RF and RM arms in Fig. 5, all arm angles of the robot were not converged to zero line in this simulation. From the angle data in Fig. 8, the suspension force of all arms can be also calculated by using (1)–(5), as shown in Fig. 9. Although the angle of arms did not reach zero line, the estimated suspension forces of all arms are sometimes changed as negative values in both cases of being faced with high and low obstacles. As a result, wheel lift is shown to occur almost eleven times (i.e., on all obstacles) in order to overcome the obstacles.

For evaluation of validation of measured suspension forces, the wheel motions while the robot leaps and bounds over the obstacle were analyzed. Figure 10 describes the result of comparison between measured suspension force and ideal suspension force of RF wheel. The measured suspension force is well fitted with the ideal one

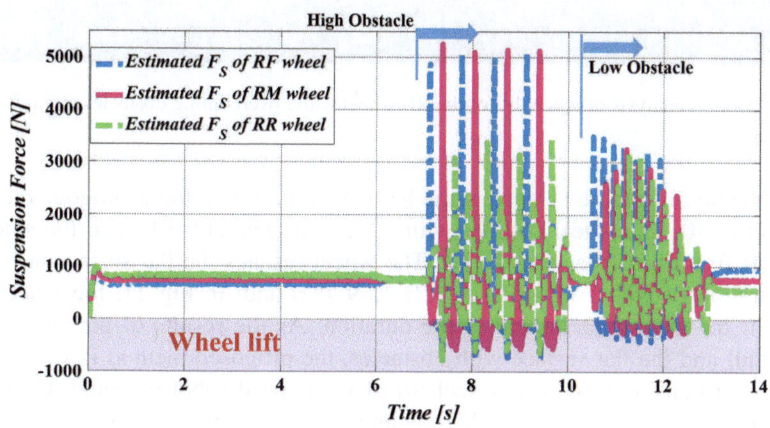

Fig. 9 Estimated suspension forces of right-side wheels while overcoming obstacles at 3 m/s

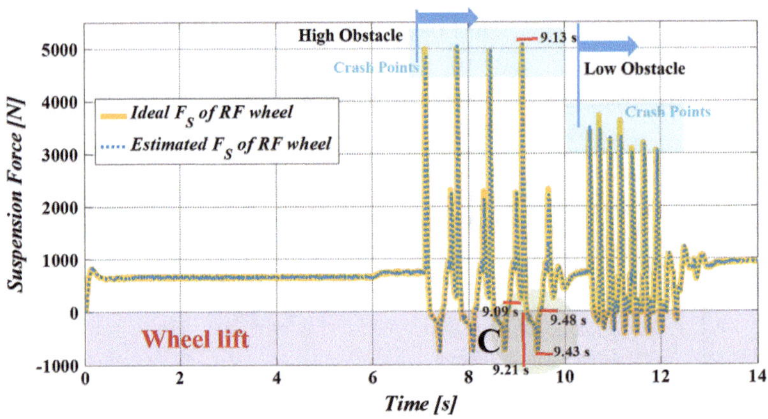

Fig. 10 Estimated suspension force of right-front wheel while overcoming obstacles at 3 m/s

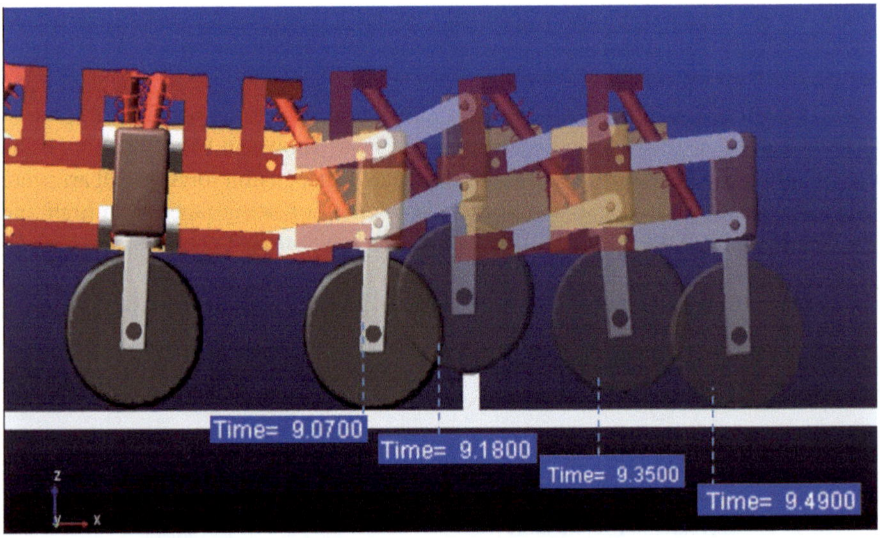

Fig. 11 Motion analysis of wheel lift of the RF wheel while overcoming obstacles at 3 m/s

across the board. Figure 11 depicts the RF wheel motion at the analogous moment to C in Fig. 10. The wheel collides with a high obstacle at 9.13 s and the wheel is taken off from the obstacle at 9.21 s. The wheel reaches the flat surface at 9.48 s. The duration of wheel lift is from 9.21 to 9.48 s and, in Fig. 11, the wheel was lifted off for the similar period to the duration. As the results of the simulations on the hill and the flat surface with obstacles, the proposed method is validated to estimate suspension force and wheel lift of high-speed robot on rough terrain. In the simulations, it was assumed that the angles of all arms, as the key variable for this method, can be accurately measured by a potentiometer mounted at the end of each arm, and it resulted in such performances as mentioned in the sections for the simulation. For an actual verification in outdoor environments, Caleb9 was applied for getting data of suspension force and wheel lift in real time on types of surface similar to the simulated environments.

4 Actual Application of Caleb9 in Outdoor Environments

4.1 Experimental Study for Nonlinear Spring and Damper Coefficients

For an actual application of the proposed method, firstly, spring and damper characteristics should be analyzed to set the coefficients of spring and damper because the suspension system is nonlinear, unlike conditions in the simulation. For the analysis, a force sensor was installed at the end of the suspension on the RR wheel to acquire exact data of suspension force in the same direction to the suspension motion, as shown in Fig. 12. The suspensions mounted on Caleb9 are customized products from a company named "J5 Suspension". Accordingly, the data related to spring and damper characteristics are as obtained from the company. Figures 13 and 14 describe the data of spring and damping force depending on the displacement x and the damping velocity $\dot{x}$, respectively. From the data of Figs. 13 and 14, the spring-damper equations can be derived by a nonlinear regression technique using polynomial equations of (9)–(11). Equation (9) represents the spring force as a function of the displacement x, and Eqs. (10)–(11) indicate the damping force as functions of the damping velocity $\dot{x}$. In (10)–(11), the damping force is divided into two cases of compression (C_{ext}) and extension (C_{com}) of the suspension and it can be determined by observing

Fig. 12 Suspension test to determine spring and damper coefficients on the RR wheel

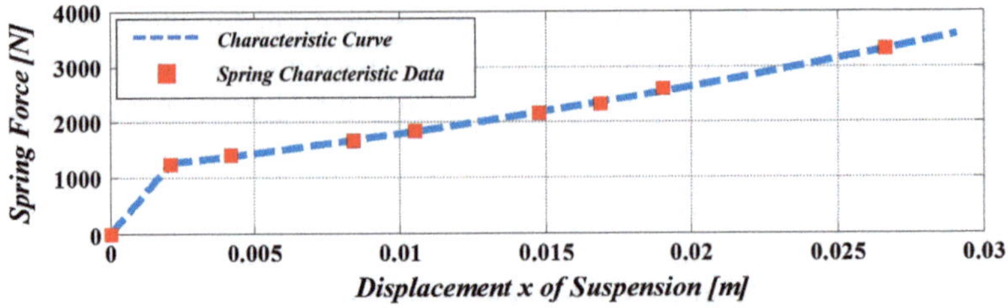

Fig. 13 The spring force depending on the displacement x of the suspension

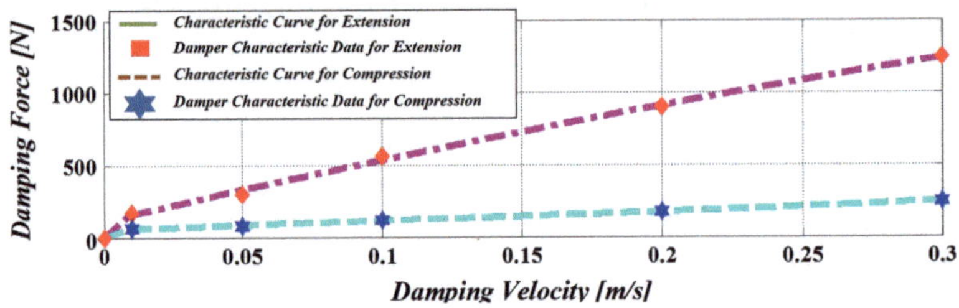

Fig. 14 The damping force depending on damping velocity $\dot{x}$ of the suspension

the positive and negative sign of the damping velocity$\dot{x}$. In (9)–(11), the polynomial constants are $K_1 = 5.9694\text{e}+5$, $K_2 = -3.5295\text{e}+6$, $K_3 = 1.1493\text{e}+6$, $K_4 = 0.0531\text{e}+6$, $K_5 = 0.0011\text{e}+6$, $C_{com1} = 6000$, $C_{com2} = 2.89\text{e}+3$, $C_{com3} = -1.0875\text{e}+3$, $C_{com4} = 0.7289\text{e}+3$, $C_{com5} = 0.0509\text{e}+3$, $C_{ext1} = 17000$, $C_{ext2} = 0.6463\text{e}+3$, $C_{ext3} = -2.5403\text{e}+3$, $C_{ext4} = 4.4513\text{e}+3$, $C_{ext5} = 0.1142\text{e}+3$. Therefore, the suspension force can be estimated by (8).

$$F_s = F_{spring} + F_{damper} \tag{8}$$

$$\begin{array}{ll} F_{spring} = K_1 x, & \text{if } x < 0.0021\,[\text{m}] \\ F_{spring} = K_2 x^3 + K_3 x^2 + K_4 x + K_5, & \text{if } x \geq 0.0021\,[\text{m}] \end{array} \tag{9}$$

$$\text{if } \dot{x} \geq 0 \begin{cases} F_{damping} = C_{com1}\dot{x}, & \text{if } \dot{x} < 0.01\,[\text{m/s}] \\ F_{damping} = C_{com2}\dot{x}^3 + C_{com3}\dot{x}^2 + C_{com4}\dot{x} + C_{com5}, & \text{if } \dot{x} \geq 0.01\,[\text{m/s}] \end{cases} \tag{10}$$

$$\text{if } \dot{x} < 0 \begin{cases} F_{damping} = C_{ext1}\dot{x}, & \text{if } |\dot{x}| < 0.01\,[\text{m/s}] \\ F_{damping} = C_{ext2}\dot{x}^3 + C_{ext3}\dot{x}^2 + C_{ext4}\dot{x} + C_{ext5}, & \text{if } |\dot{x}| \geq 0.01\,[\text{m/s}] \end{cases} \tag{11}$$

For verifying the validity of the equations related to the suspension force, two types of tests were performed; a changing force test (the right-side figure in Fig. 12) and a jump test (the left-side figure in Fig. 12). The changing force test is for reviewing only

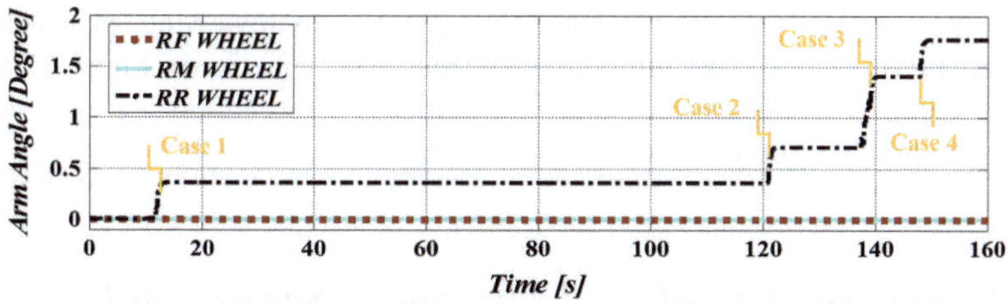

Fig. 15 The arm angle of the right-side wheels under normal force on the RR wheel

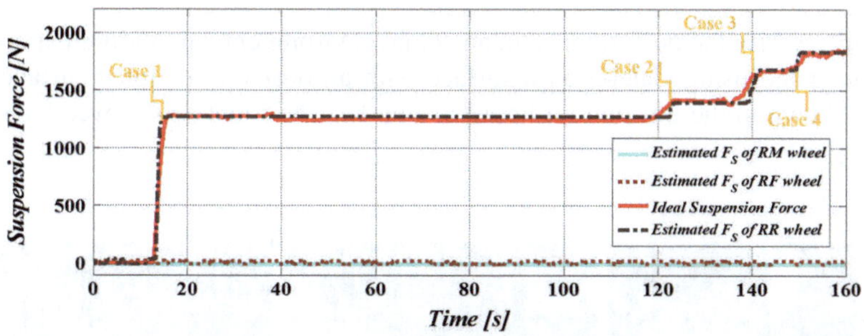

Fig. 16 Comparison between estimated and ideal suspension force using the data in Fig. 13

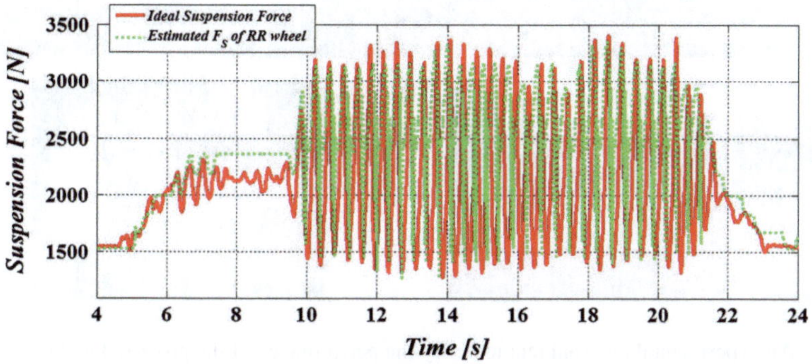

Fig. 17 Verification of estimated suspension force through the jump test

spring characteristics, without a damper effect, by slowly changing the displacement x of the suspension. Figure 15 shows the arm angle of the right-side wheels. The angle of the RM and RF arms are zero since the RM and RF wheel were lifted off from the surface. However, the angle of the RR arm is gradually changed four times (cases 1–4) by concentrated weight of the robot on the rear-side wheels while tilting the body by a crane. Figure 16 depicts the estimated suspension force of the right-side wheels in comparison to the ideal suspension force by the force sensor on the RR wheel. In Fig. 16, the estimated F_S is well matched with the ideal F_S in spite of changing cases from 1 to 4, and it shows that the RM and RF wheels were taken off from the ground. The jump test is for comprehensively reviewing spring-damper characteristics by periodically jumping on the rear of the robot body. Figure 17 shows that the actual F_S is closely estimated to the ideal F_S throughout the test, despite rapidly changing the force by jump.

4.2 Experimental Results of Estimation of Suspension Force and Wheel Lift

In order to verify the performance in an actual environment, an outdoor experiment was conducted using Caleb9 on a surface with an obstacle (inclined surface with 25°) as shown in the left-bottom figure of Fig. 18. The robot was moved backward

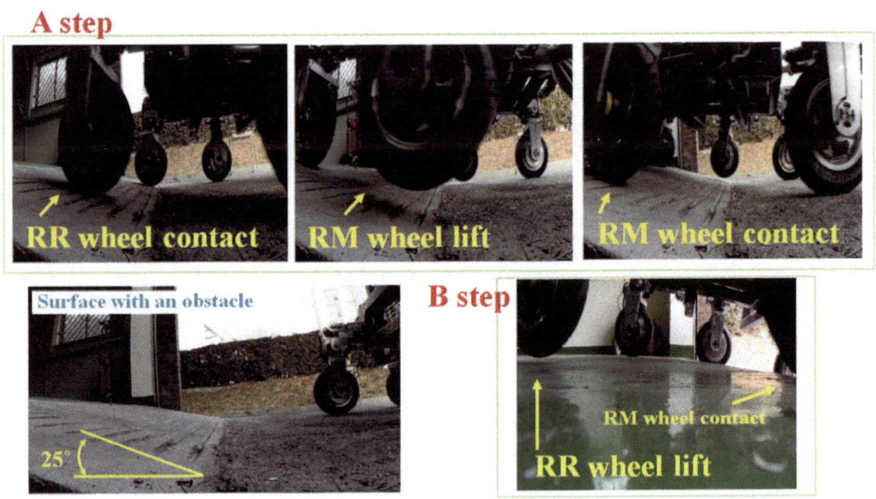

Fig. 18 An experimental environment to verify the performance of the proposed method

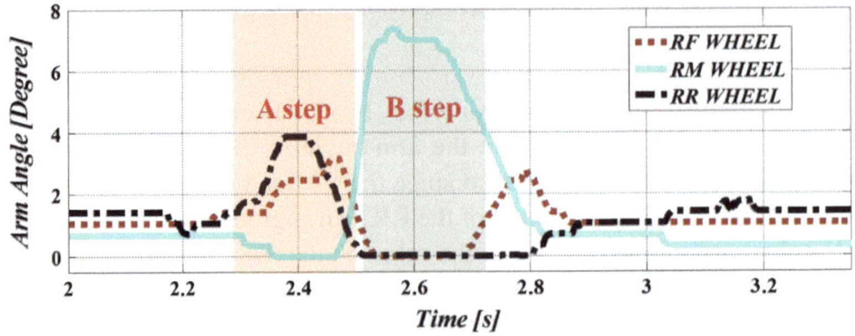

Fig. 19 Measured angles of the right-side arms while overcoming an obstacle at 3 m/s

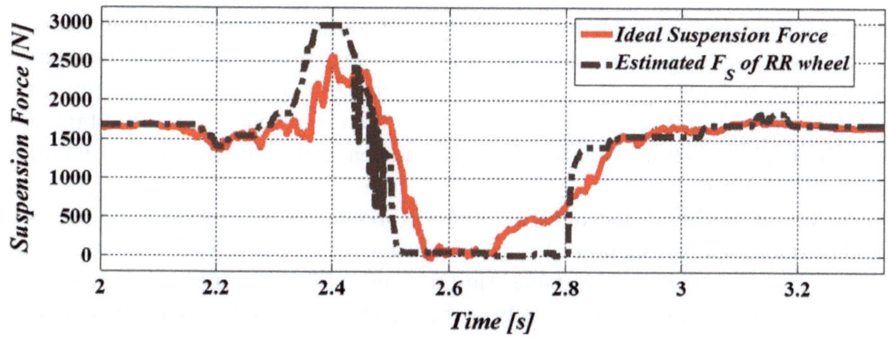

Fig. 20 Estimated suspension force of the right-rear wheel under the experimental conditions

at almost 3 m/s and the wheels move as from A step to B step while overcoming the obstacle. In A step, firstly, the RR wheel encountered the obstacle and, secondly, the RM wheel was lifted off by the effect of the surface shape, although the RR and RF wheels remained in contact with the surface. Finally, the RM wheel was reached on the ground. Then, /in B step, the RR wheel was also taken off by the force on the RM wheel supported by the robot weight at the surface of the obstacle. The process of the motion from A step to B step is described as the measured data of the angle of the right-side arms as depicted in Fig. 19. In A step of Fig. 19, the angle of the RR and RF arms increased by the collision with the obstacle; thereby, the angle of the RM arm was converged to the zero point, which means wheel lift. In B step of Fig. 19, the RM wheel was colliding with the obstacle and the angle of the RM arm increased sharply at that time, during which the angle of the RR and RF arms reached the zero line for about 0.3 s. The validity of the results can be verified in Fig. 20. In Fig. 20, the estimated F_S of the RR wheel increases until 3000 N (collision) and decreases until 0 N (wheel lift). After overcoming the obstacle, the estimated F_S returns to the initial suspension force. Comparison between the estimated and the ideal F_S in Fig. 20 shows that the estimated physical phenomenon is quite analogous to the ideal one. In addition, these motions of the wheel and the arms are considerably similar to the simulations of the hill climbing in Figs. 5, 6 and 7.

5 Conclusion

For actual applications of rough terrain robots, it is important to know the present state of wheels, especially wheel lift and suspension force related to wheel traction and body stability, to maintain optimized maneuverability. For this reason, this paper proposed an easy way to estimate wheel lift and suspension force of a high-speed wheeled robot (Caleb9) on uneven surfaces. For the achievement of this goal, inexpensive potentiometers were applied to estimate wheel lift and suspension force by measuring the angle of each arm in real-time. In addition, the simple spring-damper system was employed, and the equations related the suspension was derived based on the data provided by the manufacturing company. The proposed method was validated through two types of simulations on the environments: hill climbing and overcoming obstacles. It was also verified through actual experiments of overcoming the inclined surface.

As future works, in the outdoor mobile robotics, it is of great importance to predict stabilities for traction of wheel and rollover of body. Such the studies are closely related to the research measuring the normal force or the suspension force. Therefore, the proposed method in this paper can be employed in dynamical outdoor environments in order to evaluate the stability based on the more exact force data from this method than estimating actual force using dynamic models.

Acknowledgments The authors gratefully acknowledge the support from Basic Science Research Program through the National Research Foundation of Korea (NRF) funded by the Ministry of Education, Science and Technology (2015R1D1A3A03020805).

References

1. Kim, J., Lee, J.: Intelligent slip-optimization control with traction-energy trade-off for wheeled robots on rough terrain. In: IEEE/RSJ International Conference on Intelligent Robots and Systems (IROS) (2014)
2. Krebs, A., Risch, F., Thueer, T., Maye, J., Pradalier, C., Siegwart, R.: Rover control based on an optimal torque distribution—application to 6 motorized wheels passive rover. In: IEEE/RSJ International Conference on Intelligent Robots and Systems (IROS) (2010)
3. Bouton, N., Lenain, R., Thuilot, B., Martinet, P.: A new device dedicated to autonomous mobile robot dynamic stability: application to an off-road mobile robot. In: IEEE International Conference on Robotics and Automation (ICRA) (2010)
4. Peters, S.C., Iagnemma, K.: Stability measurement of high-speed vehicles. J. Veh. Syst. Dyn. **47**(6), 701–720 (2009)
5. Mann, M., Shiller, Z.: Dynamic stability of off-road vehicles: quasi-3D analysis. In: IEEE International Conference on Robotics and Automation (ICRA) (2008)
6. Doumiati, M., Charara, A., Victorino, A., Lechner, D.: Vehicle Dynamics Estimation using Kalman Filtering. Automation—Control and Industrial Engineering Series (2013)
7. Ishigami, G., Kewlani, G., Iagnemma, K.: Statistical mobility prediction for planetary surface exploration rovers in uncertain terrain. In: IEEE International Conference on Robotics and Automation (ICRA) (2010)
8. Joo, S.H., Lee, J.H., Park, Y.W., Yoo, W.S., Lee, J.: Real time traversability analysis to enhance rough terrain navigation for an 6 $\times$ 6 autonomous vehicle. J. Mech. Sci. Technol. **4**(27), 1125–1134 (2013)
9. Matthew, S., Yoji, K., Steven, D., Karl, L.: Hazard avoidance for high-speed mobile robots in rough terrain. J. Field Robot. **5**(23), 311–331 (2006)
10. Krid, M., Benamar, F.: Design and control of an active anti-roll system for a fast rover. In: IEEE/RSJ International Conference on Intelligent Robots and Systems (IROS) (2011)

Application of Multi-Robot Systems to Disaster-Relief Scenarios with Limited Communication

Jason Gregory, Jonathan Fink, Ethan Stump, Jeffrey Twigg, John Rogers, David Baran, Nicholas Fung and Stuart Young

Abstract In this systems description paper, we present a multi-robot solution for intelligence-gathering tasks in disaster-relief scenarios where communication quality is uncertain. First, we propose a formal problem statement in the context of operations research. The hardware configuration of two heterogeneous robotic platforms capable of performing experiments in a relevant field environment and a suite of autonomy-enabled behaviors that support operation in a communication-limited setting are described. We also highlight a custom user interface designed specifically for task allocation amongst a group of robots towards completing a central mission. Finally, we provide an experimental design and extensive, preliminary results for studying the effectiveness of our system.

1 Introduction

Humanitarian assistance and disaster relief (HA/DR) has long been appreciated as one of the most compelling applications of robotics technology, giving responders tools to sense and act in dangerous environments [24]. For example, the use of robots in the aftermath of the Fukushima Daiichi nuclear disaster has been well documented [19, 25], and analysis of the response suggests that action at one of several "inflection points" of the crisis would have probably averted further catastrophe [31] if those actions had not been deemed too dangerous at the time. Partly inspired by these implications, the DARPA Robotics Challenge was conceived to catalyze the focused development of solutions for solving the myriad of challenges related to locomotion, manipulation, perception, and human interface that are needed to build a robot that can act as a stand-in for humans at such "inflection points" in the future.

Though this "avatar" concept inspires the imagination, we would argue that robotics has an even more important role to play in the broader HA/DR mission as the backbone for the required information-gathering activities that lie at the heart of any coordinated response. As an illustration, the *Foreign Humanitarian Assistance*

J. Gregory (✉) · J. Fink · E. Stump · J. Twigg · J. Rogers · D. Baran · N. Fung · S. Young
U.S. Army Research Laboratory, Adelphi, MD, USA
e-mail: jason.m.gregory1.civ@mail.mil

D.S. Wettergreen and T.D. Barfoot (eds.), *Field and Service Robotics*,
Springer Tracts in Advanced Robotics 113, DOI 10.1007/978-3-319-27702-8_42

manual published by the U.S. Department of Defense [35] identifies that the military will primarily assist in a few ways to a disaster requiring government response: with the first-responder Crisis Action Team tasked as the immediate responder and *assessor* for the regional commander; and with the Humanitarian Assistance Survey Team whose primary responsibility is assessment, such as dislocated populations, degree of property damage, and remaining communications infrastructure. These are all activities that feed into the planning phase that must happen *before* any larger action can be carried out. Though not quite as exciting as a humanoid robot that wades through a flooded disaster site to extinguish a critical fire, we believe a heterogeneous, multi-robot team that can quickly navigate through an environment to quantify an emerging situation is more important to the timeliness and success of the larger response.

Two important focal points of multi-robot systems deployed in a primarily information-gathering sense have been the Robocup Rescue League [14] and the MAGIC 2010 competition [15, 26]. From these activities, we learn that, although physical platform capabilities play a role, the majority of the system complexity is derived from the overarching operational problems of team management and communication.

Toward this end, this work establishes a preliminary formal problem description that places an HA/DR-inspired, information-gathering mission in an operations research context (Sect. 2). The primary contribution of this work is to provide documentation and analysis of a multi-robot system capable of performing intelligence-gathering tasks in communications-limited, disaster-relief scenarios. We present the design of such a system (Sect. 3), a set of autonomy-enabled behaviors that can be used to address the HA/DR mission in a relevant environment (Sect. 4), and a user interface that allows a human operator to task the system (Sect. 5). Finally, we report extensive experimental results, which address the current capabilities of our system with respect to the implementation of a solution to the HA/DR mission (Sect. 6).

2 Problem Statement

Within the scope of information-gathering activities required for planning a response to a HA/DR scenario, we focus on simultaneously solving two specific problems: the evaluation of damage to infrastructure in the environment, e.g., traversability of roads; and localizing particular targets of interest, e.g., a potentially injured "very important person" (VIP) who we discover through sensing a radio signal, such as a cell phone. This problem statement contains both a priori goals (key assessment sites established from prior maps) and *dynamic* goals (the existence and possible locations of targets), and a solution must focus on effectively balancing between these two types of goals. Moreover, we address the issues of unreliable autonomy and limited communications through incorporation of dynamically uncovered costs, and we cast the entire problem as a dynamic variant of the Capacitated Team Orienteering Problem with details discussed below.

If we considered only the problem of efficiently visiting a set of locations derived from prior maps of the environment, a classical formulation would suffice. Initially it could be as a well-studied Vehicle Routing Problem (VRP): with known travel costs between sites, find paths for multiple vehicles to visit all sites that minimize total travel costs. However, since we may assume that the mission is time-critical and some sites are likely to be more interesting than others, we could instead formulate it as a Team Orienteering Problem (TOP): with known travel costs between sites and known rewards for visitation, find paths that maximize the total gathered reward with a fixed cost bound [36]. The environment limitations suggest one final modification.

Because the environment is communications-limited, we conjecture that as we send robots to visit sites and gather information, we need them to eventually return to communications range in order to offload their information before it becomes too outdated. This is most closely modeled as a Capacitated Team Orienteering Problem (CTOP): as a TOP but with a constraint on the total reward that any individual vehicle may gather on a single trip [13].

A key component of the problem is the dynamic goals that arrive because of detecting unknown targets. We model these as dynamically-updated rewards available at the visitation sites of the CTOP, and we assign the value of these rewards according to the expected information gain about the target location using the available sensing, similar to information-guided exploration strategies [30]. If we assign a distribution to these rewards initially or as the mission progresses, there is prior work on solving TOPs with stochastic rewards [32] that could apply.

The last challenge is to incorporate the effects of unreliable autonomy, which we model as unknown travel costs between visitation sites: we may have some intuition about how likely it is for a given site-to-site navigation to be successful, but ultimately we build a navigation risk model during operation in the environment. It is important to note that failed navigation is not necessarily fatal because we assume we have backup behaviors to return to a known safe location. If we assign a distribution to these costs, there is prior work on solving TOPs with stochastic costs [16] that could apply.

Our preliminary formal problem formulation is thus as a Capacitated Team Orienteering Problem with stochastic (unknown) costs and rewards. We ask: what value is it to have such a formal problem given that we are not developing an online planner to demonstrate through these experiments? The answer is that having the solution for any specific mission instance gives us an upper-bound on how well any autonomy or human could perform at the task and therefore gives us a metric to know when the system is improving. Even for the case of unknown costs and rewards, we can solve the plan as if the costs/rewards were known up front or solving it in a receding-horizon fashion as information is uncovered. Developing these upper-bounds for this experiment remains future work.

3 Experimental Multi-robot System

We present a heterogeneous, multi-robot system with a rich sensor suite, composed of hardware and software components for autonomous operations in relevant environments. In particular, our focus is on moving from small-scale systems operating in controlled laboratory environments to the study of interacting systems and the development of algorithms that can robustly operate in real-world scenarios.

3.1 Hardware

Two robotic platforms are used in this work: an iRobot PackBot [8] and a Clearpath Robotics Husky [3]. The PackBot, seen in Fig. 1a, is a military-grade, tracked platform capable of speeds up to 2 m/s and traversing both indoor and outdoor terrains. To enable autonomous operation, the PackBot is outfitted with a processing payload containing a Quad-Core Intel i7 ICOM express board and a 256 GB solid-state drive (SSD). The PackBot collects 3D point cloud data by nodding a Hokuyo UTM-30LX-EW LiDAR [5] with a Dynamixel servo. This Hokuyo LiDAR has a 270° field of view, 30 m range, and 1 mm resolution. Accurate state information is achieved using a MicroStrain 3DM-GX3-25 inertial measurement unit (IMU) [6] mounted on a custom-made vibration isolator. Additionally, a Garmin 18x PC GPS sensor [4] is elevated on a mast in an effort to receive better GPS measurements. Finally, an ASUS Xtion Pro Live provided RGB data [1].

The second robot used in this work, the Clearpath Husky seen in Fig. 1b, is a larger, wheeled platform that is limited to a maximum velocity of 1 m/s and is best suited for outdoor operations. Similar to the PackBot, the Husky employs a MicroStrain 3DM-GX3-25 IMU and a Garmin 18x PC GPS. The Husky is equipped with two Quad-Core Intel i7 Mini-ITX processing payloads, each with a 256 GB SSD. The Husky has a Velodyne HDL-32E LiDAR [12], which generates a 360° point cloud

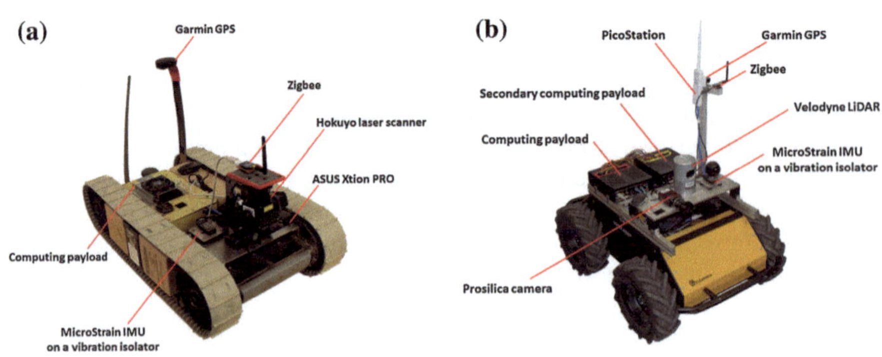

Fig. 1 The hardware configurations of **a** the iRobot PackBot and **b** the Clearpath Husky

of 700,000 points per second at a range of 70 m and an accuracy of up to ±2 cm. Finally, the Husky collects imagery data using a Prosilica GT2750C, 6 megapixel CCD color camera [9].

Both robots use Ubuntu 14.04 (Trusty) and leverage the open-source *Robotics Operating System* (ROS) Indigo [27] to support higher-level algorithms for mapping, navigation, and autonomous capabilities.

To provide the necessary wireless connectivity, we utilize off-the-shelf *IEEE 802.11.g* radios operating in the 2.4 GHz frequency band and capable of 28 dBm transmit power. The PackBot and Husky are equipped with Ubiquiti RouterStation Pro and PicoStation2HP respectively [11]. Each wireless radio operates in *AdHoc* mode and runs of the open-source embedded Linux distribution *OpenWRT* [7] with end-to-end connectivity supported by the *B.A.T.M.A.N.* mesh routing protocol [2]. Since the focus of these experiments was not on teaming or inter-robot communication, we allocated each robotic platform with a unique frequency for communication and placed the "base station" in an advantaged location, i.e., a tower approximately 20 m above the ground [10]. The placement of the "base station," environment complexity, and the fact that each robot's radio was placed very close to the ground induced a communication environment within our experimental facility that clearly exhibited regions of high-bandwidth reliable communication, intermittent unreliable communication, and no communication at all. While the *B.A.T.M.A.N.* routing protocol supports multi-hop communication, we restricted all communication in this experiment to be over a single wireless link in order to simplify the modeling of communication capabilities.

The search for an injured VIP can be represented by localizing a radio frequency beacon, e.g., a cell phone. In fact, a variety of spatial information-gathering tasks, including chemical and radiation analysis, can be emulated with radio signal propagation from one or more beacons. We use a low-power IEEE 802.15.4 XBee radio, shown in Fig. 2, to broadcast a beacon once per second at 2.4 GHz. Each robot also carries a XBee radio and records radio signal-strength when it successfully receives packets from the beacon while traversing the environment in pursuit of the other data-collection tasks.

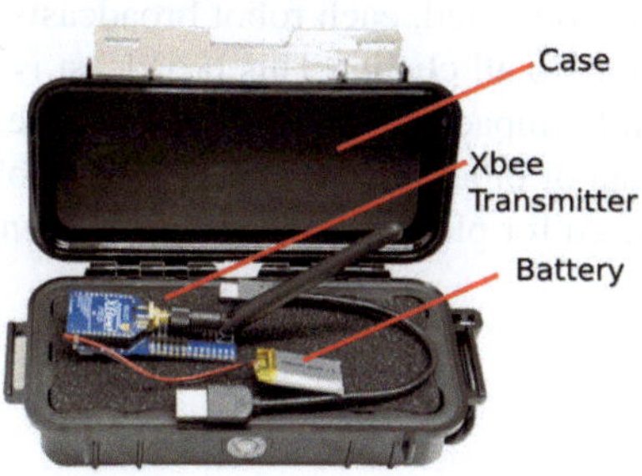

Fig. 2 XBee "beacon signal" transmitter with protective case

3.2 Mapping

The simultaneous localization and mapping (SLAM) problem focuses on the requirement for precise, consistent knowledge of the robot's trajectory as it gathers sensor measurements and has been studied for some time in the robotics literature [22, 33]. We adopt a modern *graph-based* solution to the SLAM problem based on the square-root smoothing and mapping ($\sqrt{\text{SAM}}$) technique [17] and the *GTSAM* software library developed at the Georgia Institute of Technology [18]. Our technique leverages the Generalized Iterative Closet Point (ICP) algorithm [29] for dense interframe matching of point cloud data and loop closure constraints. GPS measurements, when available, are robustly incorporated into our solution based on the techniques described in our previous work [28].

We refer to our SLAM system as *OmniMapper* due to its ability to integrate sensor data from a variety of sensor sources including laser scanners and 3D cameras. We divide the components of this system into a backend, the *OmniGraph*, which is responsible for solving the factor graph representation of the SLAM problem, and a frontend, the *OmniCache*, which is responsible for managing sensor data and performing computations that yield the probabilistic factors connecting nodes in the factor graph. The *OmniGraph* solves for the robot's optimal trajectory using the *GTSAM* library; the frontend tasks of data association and generating relevant measurements is handled by the *OmniCache*. The point-cloud *OmniCache* used in this work receives local point-cloud data aggregated over small time windows based on the odometry of the robot and serves two primary purposes. First, it can respond to queries about the relative pose of two local point-clouds via ICP algorithms in order to generate measurement factors. Second, it acts as a pipeline for generating a series of data products based on the underlying local point-cloud data. This includes a set of intrinsic products, i.e., ones that are invariant to the global pose of a local point-cloud, such as per-cloud terrain classification, occupancy grid rendering, and terrain height estimation. Other products are extrinsic, i.e., ones that must be recomputed after optimization of the factor graph yields a new optimal trajectory for the robot, including an aggregated point cloud and composite occupancy grid map. A block diagram of the relevant components of the *OmniMapper* can be seen in Fig. 3. Once an optimized trajectory is computed, each robot broadcasts its current location in a GPS-based reference frame to all clients. This broadcast is at a low enough rate so that it does not significantly impact the bandwidth available to other services on the network. The position data of other agents are inserted as obstacles into the robot's costmap, which is later used for planning and trajectory generation.

3.3 Navigation

We use a three-stage architecture, consisting of a global motion planner, a local planner, and a local controller, to drive our software design within the ROS framework.

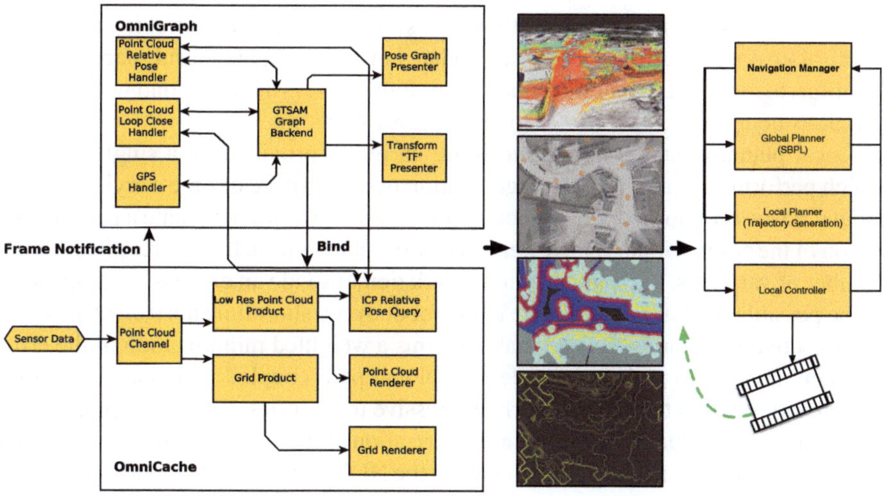

Fig. 3 Architecture for autonomous mapping and navigation

Each stage of the navigation system depicted in Fig. 3 is implemented as a node, or independent software process, which provides an *ActionServer* interface that responds to an abstraction of the navigation problem. *ActionServer* interfaces are a ROS construct used to deal with long-running tasks and include an internal state machine to manage the setting of goals, task feedback, and eventual completion state, i.e., success or failure. For instance, the global planner provides a *ComputePlan* action, which takes as input a starting and goal pose—given the current map, it returns an optimal, kinematically feasible path. The local planner provides a *ComputeLocalPlan* action, which takes a global plan as input and uses the robot's current pose and a local map of dynamic obstacles to find a short-term high-resolution path that follows the global plan. In this formulation, the local planner is capable of generating high-resolution plans over a short time-horizon while the global planner helps prevent the system from being trapped in local minima caused by non-convex environments. Finally, the local controller provides a *ControlToPlan* action, which takes the current local plan and the current state of the robot to compute control inputs, which can be sent to the underlying platform.

Sequencing of the actions is performed by a *NavigationManager* process, which presents an external interface to the user or application. The software architecture presented above is designed to maximize flexibility in implementing different solutions to not only each component of the navigation system, but also provide flexibility in how the external interface to navigation is presented.

For this experiment, we rely on the Search-Based Planning Library (SBPL) [23] to perform global planning actions. We generate a custom set of motion primitives based on our platform's kinematics and use of 0.2 and 0.3 m occupancy grids for the PackBot and Husky, respectively. We use the ARA* planner algorithm and compute reverse plans so that computations can be reused as the robot drives for fast re-planning actions. Re-planning allows the system to quickly correct its path in the event of errors in platform control or updates of the occupancy grid map. Feasible solutions to most initial planning queries are found in less than a second with optimal solutions being found in a few seconds for most scenarios.

Local planning and control actions are currently provided by a single process, which performs optimal trajectory generation over the space of time-varying control inputs. Based on prior work in trajectory generation [20], we formulate a parameterization of the control input for a differential-drive platform such that a relatively small number of variables, 4 in our current instantiation, provide an expressive description of the possible trajectories available to the robot over a short time horizon of $T = 3$ s. An objective function is devised that performs a weighted minimization of the error between the robot's path and the desired global path coupled with some curvature minimization terms to prevent overly aggressive trajectories. The final optimization problem, including bounds on the parameterization of the control input, can be solved with a variety of algorithms implemented in the NLOPT library [21]. We are typically able to solve the trajectory generation optimization for a time horizon of $T = 3$ s in 5–10 ms, allowing for a control frequency of 10 Hz. We are able to directly execute the optimized time-varying control inputs, thus simultaneously addressing the local planning and control problems.

4 Behaviors Supporting Autonomy

In this section, we describe how we build automata to sequence basic capabilities of our multi-robot system in order to provide higher-level autonomous actions and begin to address the data-collection mission described in Sect. 2. While the behaviors described here are fairly simplistic, the underlying architecture allows for complex collections of actions.

For the purposes of this work, all of our navigation behaviors build on the canonical *GotoRegion* action in which the robot plans and drives to an arbitrary pose within a defined region of the environment. The design decision to rely on region-based navigation is based on the observation that navigation to a precise pose in the environment leads to brittle solutions and that many data-collection problems can in fact be satisfied with large degrees of flexibility. Take for example, the image collection problem—there are many viewpoints from which to obtain a suitable image of a target in $\mathbb{R}^3$. While the complexity of solving this viewpoint problem is beyond the scope of this work, we believe many future data-collection problems can be generalized to a desired region in the environment.

At their core, the behaviors generated by sequencing basic capabilities are meant to aid the operator in tasking the robot when it must go outside the area of reliable communication. Thus, we begin by defining the *GuardedNavigation* behavior to be one where a *goal* region and *safe* region are defined. If execution of navigation to the *goal* fails, the robot navigates back to the *safe* region where communication is known to be reliable and the operator can continue to task the robot. Clearly, the

GuardedNavigation behavior can be extended to support sequences of *goal* regions such that a failure at any point in the sequence results in returning to the *safe* region.

With the addition of a simple *Collect* action that causes the robot to capture and store an image, the operator can immediately begin to address the data-collection mission from Sect. 2. By specifying a sequence of *goal* regions with accompanying *Collect* actions, the operator instructs the robot to visit a number of sites at which it will record high-resolution images. When it completes visiting the sequence of *goal* regions or deems a leg of the task to be infeasible, the robot returns to the *safe* region with its known reliable communication and transmits all of the images to the operator. For now, the operator selects *safe* regions based on previous locations from which the robot has successfully transmitted data.

5 Operator Interface

We rely on a simple graphical user interface (GUI) that enables a human operator to task one or more robots. Our GUI is based on the RViz application that is included in ROS for 3D rendering of sensor-data visualizations, tools for on-screen interactions, and an extensible plugin architecture. In addition to software components that allow for visualization of experiment-specific data, we developed tools for creating and interacting with generic graph-embeddings on $\mathbb{R}^2$, which are used to specify autonomous behaviors. It should be noted that our design and implementation of an operator interface is driven by necessity in order to evaluate our system in appropriately relevant scenarios rather than as an example of best practices in terms of human-robot interaction.

For this work, we used RViz to display a top-down orthographic view of satellite imagery of our experimental facility, predefined GPS locations throughout the site, the occupancy grid produced by the 3D mapping techniques described in Sect. 3.2, and the current positions of all the robots during a mission. We rely on a generic graph structure because it presents an intuitive representation for a variety of tasks including patrol, exploration, and data-collection. For the purposes of this work, we focus on the data-collection task and implicitly add edges to create linear topologies along a sequence of nodes, which are defined by a disk with a center position and radius. After the operator has annotated each node as *safe* or *goal*, we can easily map a graph onto the behaviors described in Sect. 4. After defining a graph in RViz, the system runs a verification to ensure that there are one or more *goal* regions and only one *safe* region for each task. The mission definition is then communicated to each robot where the resulting state machine is executed (Fig. 4).

As each robot drives near the radio beacon marking the location of an injured VIP, it will successfully receive transmissions and be able to record the signal strength. Aggregating the signal-strength measurements from multiple robots in many locations across the environment, the operator can infer an estimate of the beacon location from the maximum of the signal-strength field. This task is complicated by the fact that radio-signal propagation is notoriously challenging to model in complex

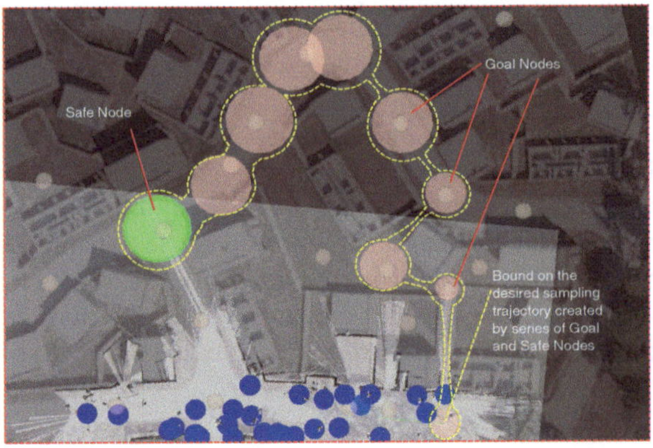

Fig. 4 An example of the user interface for a single data-collection task in a trial. The map is overlaid on *top* of a satellite image with *small pink disks* representing the predefined GPS mission nodes. The *blue disks* indicate that the robot has measured poor received signal strength data thus far. The *large orange* and *green disks* are the goal and safe nodes, respectively, as set by the operator. Note, the *red lines*, *white text*, and *yellow dotted lines* have been manually added for clarity

urban environments due to the phenomena of shadowing and multi-path. Furthermore, a high frequency beacon transmission may make complete reconstruction of the signal-strength measurements at the operating station impractical. We employ a segmentation-based approach for modeling that allows each robot to maintain efficient models of the received signal strength [34]. These compressed models can be transmitted to the operator and visualized to allow adaptive exploration of the environment with the goal of accurately localizing the VIP beacon.

6 Experimental Results

We conducted a series of experimental trials using the 175×175 m environment pictured in Fig. 5 to evaluate the capability of our system to address missions defined according to the problem statement in Sect. 2. Each experiment consisted of one or two robotic platforms and mission operators tasked with the mission of capturing an image at as many of the defined collection sites as possible within the time limit of 20 min. Experiments were designed such that the visitation of some collection sites require traversal over a variety of terrain complexities and that robots must travel outside of communication to motivate the use of autonomy. While collecting images, each robot monitors the received signal strength from a radio beacon carried by a mock VIP that is hidden in a static location for the duration of an experimental trial. Localization of the VIP through received signal strength at the end of each 20 min experiment is an auxiliary intelligence-gathering task that further guides the exploration strategies employed by the mission operator.

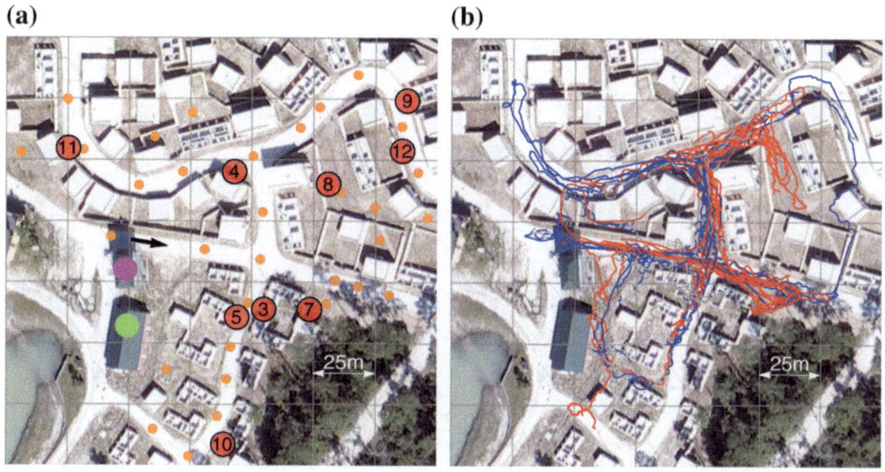

Fig. 5 A satellite overview of the experimental facility overlaid with **a** experiment annotations (*green* operating center, *purple* elevated base station antenna, *orange* mission-specified sites, *red* VIP location for each trial) and **b** the aggregated paths driven by robots over all trials

While we envision a multi-robot system capable of autonomous traversal of the complete mission with high degrees of reliability, i.e., suitable for tasking by an autonomous agent that dynamically optimizes vehicle routes; this is beyond the scope of state-of-the art algorithms when implemented in a realistic field environment. The use of a safety operator not constrained by unreliable communication, i.e., following the robot through the environment, who is able to intermittently intervene and control the robot's actions, drastically improves our ability to collect information on the system performance across an entire mission execution. As such, evaluation of the frequency and duration of these interventions serves as a primary benchmark in terms of rating current autonomous capability.

We report on the results of 9 experimental trials with respect to the number of sites visited and mock VIP localization accuracy in Table 1. The trajectories traversed by both robots across all experiments are overlaid in Fig. 5b to depict the breadth of experiments conducted. In most experimental trials, the robots drove more than 90 % of their total distance while autonomously executing *GuardedNavigation*-based sub-missions designed by the human operators to gather high-resolution images and VIP signal strength data.

Figure 6 depicts the trajectories of both robots, sites visited, and measured VIP signal strength for two specific examples of experimental trials. Note that in both of these trials, in addition to visiting a number of sites and collecting images, signal-strength data were collected that provide good estimates of the VIP beacon location. Indeed, in trial 11 an image of the VIP was captured, providing the system operator with direct evidence as to the VIP's location and well-being.

Figure 6c, d depict the reliability of operator communication with each robot during experiments as measured by analysis of the reception of periodic diagnostic

Table 1 Results from each experimental trial

Trial	Interventions		Intervention distance (m)		Autonomous distance (m)		Percent of mission autonomous		Sites	VIP localization error (m)
	PackBot	Husky	PackBot	Husky	PackBot	Husky	PackBot (%)	Husky (%)		
3	17	4	14.5	2.6	101.2	167.1	87.5	98.5	4	60
4	20	7	51.6	21.9	386.4	336.4	84.1	93.9	13	15
5	22	9	34.2	77.9	175.5	494.4	83.7	86.4	9	3
7	5	1	9.9	0.5	162	169.9	94.2	99.7	7	0
8	10	6	25.6	1.7	334.3	378.4	92.9	99.6	15	2
9	8	16	26.1	15.2	403.1	371.4	93.9	96.1	17	45
10	13	5	61.5	0.1	454.4	426.2	88.1	99.9	15	53
11	24	0	48.1	0.0	446.3	342.7	90.3	100.0	12	0
12	13	11	107.0	125.7	605.9	326.0	85.0	72.2	17	8

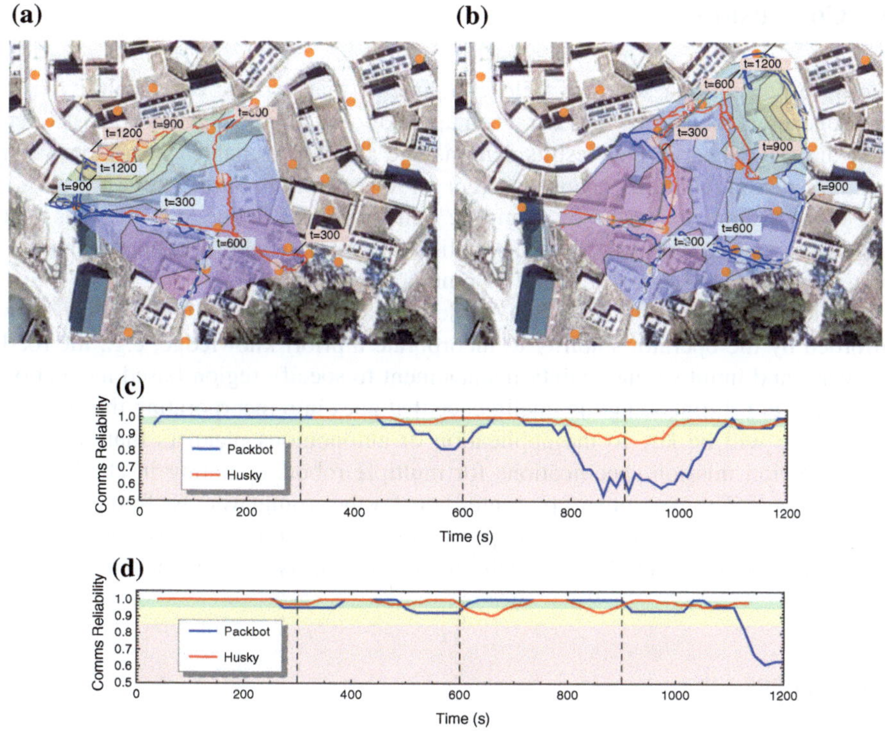

Fig. 6 Experimental trials 11 (**a**) and 12 (**b**). Robot trajectories are shown for the PackBot (*blue*) and Husky (*red*). The colormap indicates interpolated signal strength from the VIP beacon (*red* indicates high signal strength). The communication reliability for trials 11 and 12 are depicted in (**c**) and (**d**), respectively, where background colors indicate teleoperation (*green*), command (*yellow*), and position-only (*red*) communication thresholds

packets sent by each robot to the operating center. For the purposes of these experiments, we define three levels of communication—reliability exceeding 95 % allows for teleoperation, within 85–95 % robot sub-missions can be commanded and map data are updated after some delay, and below 85 % provides no guarantee on useful communication but robot position data may occasionally be available. In all experimental trials, the use of sub-mission specifications using the *GuardedNavigation* capability allowed operators to task robots routinely into regions of the environment with 85–95 % reliable communication and, in several cases, enabled collection of data in the 0–85 % reliability regime.

7 Conclusion

We have presented a series of field experiments that explore the capability of a heterogeneous multi-robot system when applied to intelligence-gathering tasks in a post-disaster scenario. Our results demonstrate autonomy-enabled operation when communication reliability is not sufficient for teleoperation. Furthermore, by allowing the operators to on-the-fly compose behaviors and define sub-missions that respond to new conditions such as navigation failure, we enable safe operation completely outside the range of reliable communication.

It should be noted that there is a subtle increase in the reliability of our system afforded by the operator's ability to incorporate a priori knowledge, e.g., the road network, and intuitive uncertainty management to specify region-based navigation as seen in Fig. 4. Encoding the intelligence that goes into incorporating this a priori knowledge will be key to the application of autonomous planners that schedule the collection mission specifications for multiple robots operating in challenging environments. The experiments presented here lay the ground work for future systems that allow a minimal set of human operators to intelligently task large numbers of robotic platforms for intelligence-gathering tasks in disaster-relief scenarios.

References

1. ASUS Xtion Pro Live. http://www.asus.com/us/Multimedia/Xtion_PRO_LIVE/
2. Better Approach to Mobile Ad-Hoc Networking. http://www.open-mesh.org
3. Clearpath Robotics Husky. http://www.clearpathrobotics.com/husky/
4. Garmin GPS. https://buy.garmin.com/en-US/US/oem/sensors-and-boards/gps-18x-oem/prod27594.html
5. Hokuyo LiDAR. http://www.hokuyo-aut.jp/02sensor/07scanner/download/products/utm-30lx-ew/
6. MicroSrain IMU. http://www.microstrain.com/inertial/3DM-GX3-25
7. OpenWRT. https://openwrt.org/
8. PackBot. http://www.irobot.com/For-Defense-and-Security/Robots/510-PackBot.aspx
9. Prosilica Camera. http://www.alliedvisiontec.com/us/products/cameras/gigabit-ethernet/prosilica-gt/gt2750.html
10. RouterStation. http://wiki.ubnt.com/RouterStation
11. Ubiquiti Networks. http://www.ubnt.com
12. Velodyne LiDAR. http://velodynelidar.com/lidar/hdlproducts/hdl32e.aspx
13. Archetti, C., Feillet, D., Hertz, A., Speranza, M.G.: The capacitated team orienteering and profitable tour problems. J. Oper. Res. Soc. **60**(6):831–842 (2008)
14. Balakirsky, S., Carpin, S., Kleiner, A., Lewis, M., Visser, A., Wang, J., Ziparo, V.A.: Towards heterogeneous robot teams for disaster mitigation: results and performance metrics from robocup rescue. J. Field Robot. **24**(11–12), 943–967 (2007)
15. Butzke, J., Daniilidis, K., Kushleyev, A., Lee, D.D., Likhachev, M., Phillips, C., Phillips, M.: The University of Pennsylvania MAGIC 2010 multi-robot unmanned vehicle system. J. Field Robot. **29**(5), 745–761 (2012)
16. Campbell, A.M., Gendreau, M., Thomas, B.W.: The orienteering problem with stochastic travel and service times. Ann. Oper. Res. **186**(1), 61–81 (2011)

17. Dellaert, F., Kaess, M.: Square Root SAM: simultaneous localization and mapping via square root information smoothing. Int. J. Robot. Res. **25**(12), 1181–1203 (2006)
18. Dellaert, F.: Factor graphs and GTSAM: a hands-on introduction. Technical Report, September, GT RIM (2012)
19. Guizzo, E.: Fukushima robot operator writes tell-all blog. In: IEEE Spectrum. http://spectrum.ieee.org/automaton/robotics/industrial-robots/fukushima-robot-operator-diaries
20. Howard, T.M., Kelly, A.: Optimal rough terrain trajectory generation for wheeled mobile robots. Int. J. Robot. Res. **26**(2), 141–166 (2007)
21. Johnson, S.G.: The NLopt nonlinear-optimization package. http://ab-initio.mit.edu/nlopt
22. Leonard, J.J., Durrant-Whyte, H.F.: Simultaneous map building and localization for an autonomous mobile robot. In: IEEE/RSJ International Workshop on Intelligent Robots and Systems (1991)
23. Likhachev, M.: Search-Based Planning Library. https://github.com/sbpl/sbpl
24. Murphy, R.R.: Disaster Robotics. MIT Press (2014)
25. Nagatani, K., Kiribayashi, S., Okada, Y., Otake, K., Yoshida, K., Tadokoro, S., Nishimura, T., Yoshida, T., Koyanagi, E., Fukushima, M., Kawatsuma, S.: Emergency response to the nuclear accident at the Fukushima Daiichi Nuclear Power Plants using mobile rescue robots. J. Field Robot. **30**(1), 44–63 (2013)
26. Olson, E., Strom, J., Morton, R., Richardson, A., Ranganathan, P., Goeddel, R., Bulic, M., Crossman, J., Marinier, B.: Progress toward multi-robot reconnaissance and the MAGIC 2010 competition. J. Field Robot. **29**(5), 762–792 (2012)
27. Quigley, M., Conley, K., Gerkey, B., Faust, J., Foote, T.B., Leibs, J., Wheeler, R., Ng, A.Y.: ROS: an open-source Robot Operating System. In: International Conference on Robotics and Automation, Open-Source Software workshop (2009)
28. Rogers, J.G., Fink, J.R., Stump, E.A.: Mapping with a ground robot in GPS denied and degraded environments. In: American Control Conference (2014)
29. Segal, A.V., Haehnel, D., Thrun, S.: Generalized-ICP. In: Robotics: Science and Systems (2009)
30. Stachniss, C., Burgard, W.: Exploring unknown environments with mobile robots using coverage maps. In: IJCAI, pp. 1127–1134 (2003)
31. Strickland, E.: 24 hours at Fukushima. IEEE Spectr. **48**(11), 35–42 (2011)
32. Taylan, I., Iravani, S.M.R., Daskin, M.S.: The orienteering problem with stochastic profits. IEE Trans. **40**(4), 406–421 (2008)
33. Thrun, S.: The graph SLAM algorithm with applications to large-scale mapping of urban structures. Int. J. Robot. Res. **25**(5–6), 403–429 (2006)
34. Twigg, J.N., Fink, J., Yu, P.L., Sadler, B.M.: Efficient base station connectivity area discovery. Int. J. Robot. Res. (2013)
35. U.S. Department of Defense: Foreign humanitarian assistance. Joint Publication, 3–29 Jan 2014
36. Vansteenwegen, P., Souffriau, W., Van Oudheusden, D.: The orienteering problem: a survey. Eur. J. Oper. Res. **209**(1), 110 (2011)

MIX
Papier aus verantwortungsvollen Quellen
Paper from responsible sources
FSC® C105338

If you have any concerns about our products,
you can contact us on
ProductSafety@springernature.com

In case Publisher is established outside the EU,
the EU authorized representative is:
Springer Nature Customer Service Center GmbH
Europaplatz 3, 69115 Heidelberg, Germany

Printed by Libri Plureos GmbH
in Hamburg, Germany